FUNCTION THEORY OF SEVERAL COMPLEX VARIABLES

SECOND EDITION

FUNCTION THEORY OF SEVERAL COMPLEX VARIABLES

SECOND EDITION

BY

STEVEN G. KRANTZ

AMS CHELSEA PUBLISHING
American Mathematical Society • Providence, Rhode Island

2000 *Mathematics Subject Classification*. Primary 32-02, 32Bxx, 32Dxx, 32Fxx, 32Hxx, 32Txx, 32Uxx, 32Wxx; Secondary 35N15.

Library of Congress Cataloging-in-Publication Data

Krantz, Steven G. (Steven George), 1951–
    Function theory of several complex variables / Steven G. Krantz.—2nd ed.
    p. cm.
    Includes bibliographical references and index.
    ISBN 0-8218-2724-3 (alk. paper)
    1. Functions of several complex variables. I. Title.

QA331.7 .K74   2000
515′.94—dc21
                                                        00-059363

to Randi

Contents

5   Solution of the Levi Problem and Other Applications of $\bar{\partial}$ Techniques   213

6   Cousin Problems, Cohomology, and Sheaves   243

9   Constructive Methods   381

10   Integral Formulas for Solutions to the $\bar{\partial}$ Problem and Norm Estimates   403

PREFACE TO THE SECOND EDITION

It is a pleasure to take this opportunity to revise, correct, and update the original edition of *Function Theory of Several Complex Variables* that was published in 1982. I find that, taking into account the developments of the last ten years, I have not fundamentally changed my view of what constitutes the foundations of this subject. Accordingly, Chapters 1–8 have been changed only to the extent of updating some references and correcting some errors. In the process, I have been able to clarify and amplify certain passages.

Chapter 11 now briefly describes some of the ground-breaking work of Lempert on the Kobayashi metric. We still do not understand its full significance; but its impact on the subject is undeniable. This chapter now also includes a brief introduction to the concept of finite type and its connections with mapping theory. It concludes with a quick introduction to complex analytic dynamics. I am grateful to J. E. Fornæss for providing me with an outline of this latter material.

The new Chapter 9 treats what I call "constructive methods." By this I mean the solution of the inner functions problem and the attendant techniques (developed by Forstneric, Løw, Rudin, and Ryll-Wojtacek) for constructing holomorphic functions and mappings with specified characteristics. Although the initial surge of activity following the solution of the inner functions problem has subsided, I think that the long-term effects of this methodology will prove to be substantial.

Although there have been exciting developments in the theory of biholomorphic and proper mappings, this area rapidly degenerates into a zoology of specialized results and technical methodologies. The methods of Bell are still the most far-reaching in the subject; they provide a suitable introduction to mappings for a first book on several complex variables. We have, accordingly, retained that treatment in Chapter 11.

We might have said more about the $\bar{\partial}$-Neumann problem and Condition R, but those are treated in detail in the forthcoming volume *Partial Differential Equations and Complex Analysis* by this author. The $\bar{\partial}$-problem is treated in Chapters 4 and 10 of the present book.

Next, a word about Chapter 1. Just as the Cauchy integral formula is the driving force in the function theory of one complex variable, so I see integral formulas as occupying a central position in the theory of several complex variables. Much of the history of several complex variables could be interpreted as an effort to circumvent an inadequate understanding of integral formulas (indeed, we now realize that we could have solved the Levi problem a long time ago if we

xiv Preface to the Second Edition

had had a good grip on the Cauchy-Fantappiè formalism). We now have a firm understanding of both the canonical and the constructible integral formulas on strongly pseudoconvex domains—the most important domains in the subject. Thus I believe that integral formulas deserve a prominent, indeed a motivating, position in my exposition.

But many users, especially students, have found my Chapter 1 discouraging. The heavy use of differential forms, together with the plethora of different integral formulas, can be confusing to the neophyte. The good news is that one *can* skip all of Chapter 1 (except for a glance at Sections 1.1 and 1.2) on a first reading. There are only a few portions of Chapters 2 through 6 that depend directly on the Bochner-Martinelli kernel or the Bergman projection, and the reader may refer back to Chapter 1 for key ideas when they are needed. In the end, the reader will certainly want to become well acquainted with Chapter 1. But Chapters 0, 2, 3, 4, 5, and 6 by themselves constitute a self-contained and logically presented first course in the function theory of several complex variables.

Also a word about the references: We have made every effort to provide an extensive bibliography. This should prove especially useful to those beginning work in the subject. But it would foolish to claim that it is exhaustive. Even allowing for limitations of the author, the unfortunate barriers to communication that have existed between eastern Europe and the western world for so many years (and which, fortunately, are now being removed) have the side effect that the important contributions of the Russian and other schools tend to appear to be minimized. Sincere apologies for any omissions in this wise.

It is a pleasure to thank the many friends and colleagues who have helped me with this second edition. Eric Bedford and John D'Angelo have given me permission to use exercises from courses they have taught. A number of people have brought misprints and corrections in the first edition to my attention. I mention particularly Gerardo Aladro, Andrew Balas, David Barrett, Eric Bedford, Steven Bell, Harold Boas, Dan Burns, David Catlin, Urban Cegrell, Kevin Clancey, John D'Angelo, Guy David, Katy Diaz, Fausto Di Biase, Peter Duren, John E. Fornæss, Siqi Fu, Paul Gauthier, Estela Gavosto, Robert E. Greene, Reese Harvey, Xiaojun Huang, Piotr Jakòbczak, Marek Jarnicki, Morris Kalka, S. Kobayashi, Jacob Korevaar, Mark Kruelle, John M. Lee, Eric Løw, Bao Luong, Daowei Ma, John Millson, Alex Nagel, Marco Peloso, Wiesław Pleśniak, John Polking, Walter Rudin, Stephen Semmes, Nancy Stanton, Lee Stout, Jim Walker, Sid Webster, Jan Wiegerinck, H. H. Wu, and Chen Zhenhua. Jiye Yu contributed several suggestions for the new edition. I am grateful to Richard Laugesen, Marco Peloso, Paul Vojta, and Tomasz Wolniewicz for reading much of the new edition inch by inch and helping to correct both the mathematics and the exposition. Responsibility for all remaining errors resides, of course, entirely with me.

—S. G. K.

PREFACE TO THE
FIRST EDITION

The strong interplay between harmonic analysis and the theory of several complex variables (hereinafter referred to as SCV) that has come about in the last 20 years has stimulated desire for a text that introduces the classically oriented analyst to holomorphic functions on $\mathbb{C}^n$. This is such a text. It is addressed to those with an interest in partial differential equations, Fourier analysis, and integral operators.

The organization of the material is accordingly function-theoretic. The omission of certain topics (e.g., analytic spaces, the *proofs* of Cartan's theorems A and B) also reflects the prejudice toward classical analysis. We have included a detailed chapter and a half on sheaf cohomology, written strictly for the ingénue, but this material may be safely skipped with no loss in continuity.

Novelties in the text include discussions of the Szegö and Bergman kernels, peaking functions for algebras of holomorphic functions, the Henkin integral formulas and estimates for the $\bar{\partial}$ equation, foliation by complex analytic varieties, the Fefferman mapping theorem, invariant metrics, the characterization of the ball by its automorphism group, the notion of admissible convergence of H^p functions, the edge-of-the-wedge theorem, nonisotropic Lipschitz spaces, extension phenomena, and a detailed discussion of convexity. Since this is a *text*, there are many examples and exercises. Analogies and nonanalogies are drawn with the one-variable theory (e.g., there are infinitely many Cauchy integral formulas in SCV, whereas there is only one in the classical one-variable theory). Several-variable phenomena are examined from the one-variable point of view (e.g., why is every open subset of $\mathbb{C}$ pseudoconvex?). Chapter 0 consists of a long exposition of the differences between one and several complex variables. Numerous open problems (e.g., the inner functions problem, the biholomorphic mapping problem, the corona problem) are considered in both the text and the exercises. Much of the material in Chapters 5 and 7 through 10 has never appeared in book form.

The exercises are of two kinds. The ones in the body of the text are fairly routine. Some consist of "details left to the reader," whereas others contain additional ideas. The Miscellaneous Exercises at the end of each chapter are highly nonuniform in character and difficulty. Some are straightforward whereas others are fairly challenging. Some exposit results from published or unpublished papers (references are included). Others are results from the mathematical folklore. Difficult exercises are distinguished by an asterisk (*).

In places this book is computational: We have computed the behavior of a convex domain under a biholomorphic map, computed detailed examples of pseudoconvex domains, constructed smoothly varying peak functions on strong pseudoconvex domains, computed numerous integral formulas, and computed estimates for the $\bar{\partial}$ operator. The purpose of these computations is to force the reader to do some work. We hope this feature will facilitate the reader's rapid and serious involvement in the subject. Also, we wish to provide a repository for several calculations that are not readily available elsewhere.

As do all books, this one makes certain demands on the reader. We assume familiarity with the elements of real variable theory, measure theory, one complex variable, and functional analysis. We adopt the custom of using the same letter to denote constants whose specific values may change from line to line (to which fact the unaccustomed reader must become acclimated). What is perhaps more significant is that this book uses the theory of differential forms consistently and from the very outset. We expect, however, only that the reader should have an acquaintance with differential forms at the level of advanced calculus. The only way to gain facility with the language of forms is to apply it, and this is the ideal subject in which to do so. Based on experience in teaching this subject, we have included appendices (which are both *ad hoc* and extremely terse) on the four prerequisites with which students seem to experience the greatest discomfort: manifolds, surface measure, exterior algebra, and differential forms.

A few comments on the organization of the book are in order. Chapters 0 through 5 constitute a basic course in the analytic parts of several complex variables theory. Chapter 6 and Sections 7.1 through 7.3 contain the rudiments of the algebraic theory. Sections 7.4, 7.5, and Chapters 8 through 10 are dessert. They contain more advanced topics in harmonic analysis, partial differential equations, differential geometry, and holomorphic mapping theory. These topics may be read selectively and in almost any order.

Many of the results in this book have been supplied with names (suggestive of the mathematicians who proved them), and many corresponding references have been given. This bibliographic work is *not* based on an exhaustive search of the literature and is not meant to be complete. It is meant only to help the reader begin to explore the vast number of research papers in existence. Any omissions of references or failure to attribute theorems reflects only my ignorance and not any malicious design.

—Steven G. Krantz

0 An Introduction to the Subject

0.1 Preliminaries

The following standard notation will be used: $\mathbb{R}$ denotes the real numbers; $\mathbb{C}$ denotes the complex numbers; $(x_1, \ldots, x_N)$ denotes an element of $\mathbb{R}^N$; and

$$(z_1, \ldots, z_n) \approx (x_1 + iy_1, \ldots, x_n + iy_n) \approx (x_1, y_1, \ldots, x_n, y_n)$$

is an element of $\mathbb{C}^n$. The partial differential operators on $\mathbb{C}^n$ given by

$$\frac{\partial}{\partial z_j} \equiv \frac{1}{2} \left(\frac{\partial}{\partial x_j} - i \frac{\partial}{\partial y_j} \right), \quad j = 1, \ldots, n,$$

$$\frac{\partial}{\partial \bar{z}_j} \equiv \frac{1}{2} \left(\frac{\partial}{\partial x_j} + i \frac{\partial}{\partial y_j} \right), \quad j = 1, \ldots, n$$

will prove very useful. While at first perhaps puzzling in their form, these definitions are forced upon us by the logical requirements that

$$\frac{\partial}{\partial z_j} z_j = 1, \quad \frac{\partial}{\partial \bar{z}_j} \bar{z}_j = 1,$$

$$\frac{\partial}{\partial z_j} \bar{z}_k = 0, \quad \frac{\partial}{\partial \bar{z}_j} z_k = 0,$$

and, when $j \neq k$,

$$\frac{\partial}{\partial z_j} z_k = 0, \quad \frac{\partial}{\partial \bar{z}_j} \bar{z}_k = 0.$$

Likewise, we have the differentials

$$dz_j \equiv dx_j + idy_j, \quad j = 1, \ldots, n,$$

$$d\bar{z}_j \equiv dx_j - idy_j, \quad j = 1, \ldots, n.$$

Notice that

$$\left\langle dz_j, \frac{\partial}{\partial z_j} \right\rangle = 1, \qquad \left\langle d\bar{z}_j, \frac{\partial}{\partial \bar{z}_j} \right\rangle = 1,$$

$$\left\langle dz_j, \frac{\partial}{\partial \bar{z}_k} \right\rangle = 0, \qquad \left\langle d\bar{z}_j, \frac{\partial}{\partial z_k} \right\rangle = 0,$$

and, when $j \neq k$,

$$\left\langle dz_j, \frac{\partial}{\partial z_k} \right\rangle = 0, \qquad \left\langle d\bar{z}_j, \frac{\partial}{\partial \bar{z}_k} \right\rangle = 0.$$

Next, since we shall deal with a great many integrals on both $\mathbb{R}^N$ and $\mathbb{C}^n$, we introduce the standard Euclidean volume form on $\mathbb{C}^n$ (in perhaps slightly unfamiliar notation):

$$
\begin{aligned}
dV(z) = dV_n(z) \quad &= \quad \left(\frac{1}{2i}\right)^n (d\bar{z}_1 \wedge dz_1) \wedge \ldots \wedge (d\bar{z}_n \wedge dz_n) \\
&= \quad \left(\frac{1}{2i}\right)^n (2i\, dx_1 \wedge dy_1) \wedge \ldots \wedge (2i\, dx_n \wedge dy_n) \\
&= \quad (dx_1 \wedge dy_1) \wedge \ldots \wedge (dx_n \wedge dy_n).
\end{aligned}
$$

The symbol dV is also used to denote the usual Euclidean volume on $\mathbb{R}^N$.

We shall have occasion throughout this book to use many standard concepts from the classical function theory of one complex variable: Cauchy integrals, normal families, power series, and so forth will be used virtually without comment. The classic text Ahlfors [1] is a good reference for these matters. However, all several complex variables concepts will be defined and developed here from first principles.

If $z^0 \in \mathbb{C}^n, r > 0$, then define the *open ball*

$$B(z^0, r) = \{z \in \mathbb{C}^n : |z - z^0| < r\}$$

and the *open polydisc*

$$D^n(z^0, r) = \{z \in \mathbb{C}^n : |z_j - z_j^0| < r, j = 1, \ldots, n\}.$$

If $P \in \mathbb{C}$ and $r > 0$, then $D(P, r)$ denotes the disc $\{z \in \mathbb{C} : |z - P| < r\}$. We reserve the symbol D for the unit disc $D(0, 1)$. The notations $\bar{B}(z^0, r), \bar{D}^n(z^0, r)$ denote, respectively, the closures of $B(z^0, r), D^n(z^0, r)$. We will also will have occasion to consider polydiscs of the form

$$D(z_1, r_1) \times \cdots \times D(z_n, r_n),$$

but shall introduce no special notation for them at this time.

Any theory of functions of several variables requires multi-index notation. Let $\mathbb{Z}^+ = \{0, 1, 2, \ldots\}, \mathbb{N} = \{1, 2, 3, \ldots\}$. A *multi-index* α is an element of $(\mathbb{Z}^+)^n$. If $\alpha = (\alpha_1, \ldots, \alpha_n)$ is a multi-index, $w = (w_1, \ldots, w_n)$, then

$$
\begin{aligned}
w^\alpha &\equiv w_1^{\alpha_1} \cdots w_n^{\alpha_n}, \\
\bar{w}^\alpha &\equiv \bar{w}_1^{\alpha_1} \cdots \bar{w}_n^{\alpha_n}, \\
\left(\frac{\partial}{\partial w} \right)^\alpha &\equiv \left(\frac{\partial}{\partial w_1} \right)^{\alpha_1} \cdots \left(\frac{\partial}{\partial w_n} \right)^{\alpha_n}, \\
\left(\frac{\partial}{\partial \bar{w}} \right)^\alpha &\equiv \left(\frac{\partial}{\partial \bar{w}_1} \right)^{\alpha_1} \cdots \left(\frac{\partial}{\partial \bar{w}_n} \right)^{\alpha_n}.
\end{aligned}
$$

Also, $\alpha! \equiv \alpha_1! \cdots \alpha_n!$ and $|\alpha| \equiv \sum \alpha_j \geq 0$. If $\alpha, \beta \in (\mathbb{Z}^+)^n$, then $\alpha < \beta$ means that $\alpha_j < \beta_j$ for all j (and similarly for $\alpha \leq \beta$).

Finally, we consider differential forms on $\mathbb{C}^n$. For any $j \geq 1$, let $\alpha \in (\mathbb{Z}^+)^j$. Then $dz^\alpha \equiv dz_{\alpha_1} \wedge \cdots \wedge dz_{\alpha_j}$ and $d\bar{z}^\alpha \equiv d\bar{z}_{\alpha_1} \wedge \cdots \wedge d\bar{z}_{\alpha_j}$. There is a similar notation for real forms. If $\alpha = (\alpha_1, \ldots, \alpha_j)$ is a multi-index used to index a form (rather than a derivative), $dz^\alpha = dz_{\alpha_1} \wedge \cdots \wedge dz_{\alpha_j}$, then $|\alpha|$ denotes j, the length of α. Context will make clear which meaning of $|\alpha|$ is intended.

The theory of functions of one complex variable is old and well developed. Many theorems are proved on fairly arbitrary domains. The function theory of several complex variables is a much younger one; correspondingly, we are not at a stage where we can profitably study domains with pathological boundaries. In most (but not all) contexts in this book, we shall restrict attention to bounded open sets with some boundary regularity. Our terminology is as follows. A *domain* is a connected open set. A domain Ω in $\mathbb{R}^N$ (resp. $\mathbb{C}^n$) with boundary $\partial\Omega$ is said to have C^k boundary, $k \geq 1$, if there is a k times continuously differentiable function ρ defined on a neighborhood U of the boundary of Ω such that

   a. $\Omega \cap U = \{z \in U : \rho(z) < 0\}$;
   b. $\nabla\rho \neq 0$ on the boundary of Ω.

We call the function ρ a C^k *defining function* for Ω. It is an exercise with the implicit function theorem to see that, for $k \geq 1, \Omega$ has a C^k defining function if and only if $\partial\Omega$ is a C^k manifold. See Hirsch [1] for more on these matters.

This definition is so important that it bears some discussion. It is intuitively appealing to think of the C^k boundary of a domain $\Omega \subseteq \mathbb{R}^N$ as locally the graph of a C^k function. That is, if $P \in \partial\Omega$, then we select a neighborhood $U_P \ni P$ in $\mathbb{R}^N$ and a Euclidean coordinate system $t_1, \ldots, t_N$ on U_P, so that the positive t_N-axis points into the domain. We say that $\partial\Omega$ is locally the graph of the C^k function $\phi_P(t_1, \ldots, t_{N-1})$ if

$$\partial\Omega \cap U_P = \{(t_1, \ldots, t_{N-1}, \phi_P(t_1, \ldots, t_{N-1})) \in U_P\}.$$

Then, on U_P, the role of the defining function ρ is played by $\rho_P(t) = \phi_P(t_1, \ldots, t_{N-1}) - t_N$. A defining function on a neighborhood of all of $\partial\Omega$ is obtained by using a partition of unity to patch together the local defining functions.

The concept of defining function ρ allows us to avoid (the somewhat artificial) reference to local coordinates when describing a boundary. Along with guaranteeing that the boundary is a C^k manifold, the condition that $\nabla\rho \neq 0$ implies that $\nabla\rho$ will be a nontrivial outward normal to the boundary at each point. The books J. Munkres [1] and M. Hirsch [1] give a thorough treatment of these ideas. Appendix I also treats other methods of thinking about C^k boundaries.

Note in passing that once a defining function is given on a neighborhood of the boundary of a domain, it is then a simple matter, using a partition of unity, to extend the defining function to all of space. It is sometimes convenient for us to assume that our defining function is given on all of space, and we do so without comment.

0.2   What is a Holomorphic Function?

Now we can begin our study of the function theory of several complex variables. How shall we define the concept of holomorphic (i.e., analytic) function? Let us restrict attention to functions $f : \Omega \to \mathbb{C}$, Ω a domain in $\mathbb{C}^n$, that are locally integrable (denoted $f \in L^1_{\text{loc}}$). That is, we assume that $\int_K |f(z)| dV(z) < \infty$ for each compact subset K of Ω. In particular, in order to avoid aberrations, we shall only discuss functions that are distributions (to see what happens to function theory when such a standing hypothesis is not enforced, see the work of E. R. Hedrick [1]). Distribution theory will not, however, play an explicit role in this book.

We now offer four plausible definitions of holomorphic function on a domain $\Omega \subseteq \mathbb{C}^n$.

DEFINITION I   A function $f : \Omega \to \mathbb{C}$ is *holomorphic* if for each $j = 1, \ldots, n$ and each fixed $z_1, \ldots, z_{j-1}, z_{j+1}, \ldots, z_n$ the function

$$\zeta \mapsto f(z_1, \ldots, z_{j-1}, \zeta, z_{j+1}, \ldots, z_n)$$

is holomorphic, in the classical one-variable sense, on the set

$$\Omega(z_1,\ldots,z_{j-1},z_{j+1},\ldots,z_n) \equiv \{\zeta \in \mathbb{C} : (z_1,\ldots,z_{j-1},\zeta,z_{j+1},\ldots,z_n) \in \Omega\}.$$

In other words, we require that f be holomorphic in each variable separately.

DEFINITION II   A function $f : \Omega \to \mathbb{C}$ that is continuously differentiable in each complex variable separately on Ω is said to be *holomorphic* if f satisfies the Cauchy-Riemann equations in each variable separately.

This is just another way of requiring that f be holomorphic in each variable separately.

DEFINITION III   A function $f : \Omega \to \mathbb{C}$ is *holomorphic* (viz. complex analytic) if for each $z^0 \in \Omega$ there is an $r = r(z^0) > 0$ such that $\bar{D}^n(z^0,r) \subseteq \Omega$ and f can be written as an absolutely and uniformly convergent power series

$$f(z) = \sum_\alpha a_\alpha (z - z^0)^\alpha$$

for all $z \in D^n(z^0,r)$.

DEFINITION IV   Let $f : \Omega \to \mathbb{C}$ be continuous in each variable separately and locally bounded. The function f is said to be *holomorphic* if for each $w \in \Omega$ there is an $r = r(w) > 0$ such that $\bar{D}^n(w,r) \subseteq \Omega$ and

$$f(z) = \frac{1}{(2\pi i)^n} \oint_{|\zeta_n - w_n| = r} \cdots \oint_{|\zeta_1 - w_1| = r} \frac{f(\zeta_1,\ldots,\zeta_n)}{(\zeta_1 - z_1)\cdots(\zeta_n - z_n)} \, d\zeta_1 \cdots d\zeta_n$$

for all $z \in D^n(w,r)$.

Fortunately Definitions I–IV, as well as several other plausible definitions, are equivalent. This matter is developed in detail in the next several sections. We ask, for now, that the reader accept this fact in order to appreciate the forthcoming overview of some of the important questions of the subject of several complex variables.

In Section 1.2 we fix one formal definition of holomorphic function of several variables and proceed logically from that definition.

0.3 Comparison of $\mathbb{C}^1$ and $\mathbb{C}^n$

At a quick glance, one might be tempted to think of the analysis of several complex variables (or several real variables, for that matter) as being essentially one variable theory with the additional complication of multi-indices. This perception turns out to be incorrect. Deep new phenomena and profound (and as yet

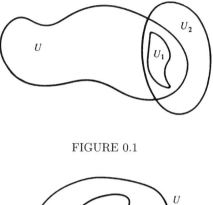

FIGURE 0.1

FIGURE 0.2

unsolved) problems present themselves in the theory of several variables. The purpose of this section is to give the reader an appreciation for some of these new features.

0.3.1   Domains of Holomorphy

Call an open set $U \subseteq \mathbb{C}^n$ a *domain of holomorphy* if the following property holds:

> There *do not* exist nonempty open sets U_1, U_2, with U_2 connected, $U_2 \not\subseteq U$, $U_1 \subseteq U_2 \cap U$, such that for every holomorphic function h on U there is a holomorphic h_2 on U_2 such that $h = h_2$ on U_1. (Refer to Figure 0.1).

The definition of domain of holomorphy is complicated because we must allow for the possibility (when dealing with an arbitrary open set U rather than a smooth domain Ω) that ∂U may intersect itself; see Figure 0.2. Setting aside this technicality for the moment, we see that an open set U is *not* a domain of holomorphy if there is an open set $\hat{U}$ that properly contains U and such that *every* holomorphic function h on U analytically continues to a holomorphic function $\hat{h}$ on $\hat{U}$. *Note that the open set $\hat{U}$ is not supposed to depend on h.*

In the theory of holomorphic functions of one complex variable, *every* open set is a domain of holomorphy. Here is a simple proof that the unit disc $D \subseteq \mathbb{C}$

is a domain of holomorphy. Let

$$f_0(\zeta) = \sum_{j=0}^{\infty} 2^{-j} \zeta^{2^j}.$$

Then the series defining f_0 converges absolutely and uniformly on $\bar{D}$ by the Weierstrass M-test. Therefore, it is continuous on $\bar{D}$ and holomorphic on D. However the boundary function

$$\theta \mapsto \sum_{j=0}^{\infty} 2^{-j} e^{i2^j \theta}$$

is the Weierstrass nowhere differentiable function (see Y. Katznelson [1]), so that f_0 cannot be extended even *differentiably*, much less analytically, to a larger open set. Since there exists one analytic function on D that cannot be continued to a larger open set, it follows that D is a domain of holomorphy. Now if Ω is a simply connected, smoothly bounded domain then, by the Riemann mapping theorem, there is a conformal mapping $\Phi : \Omega \to D$. By an old result of Painlevé (see M. Tsuji [1] or S. Bell and S. G. Krantz [1]), the mapping Φ continues to be continuously differentiable from $\bar{\Omega}$ to $\bar{D}$. Likewise the mapping Φ^{-1} extends to be continuously differentiable from $\bar{D}$ to $\bar{\Omega}$. Then $f_0 \circ \Phi$ is holomorphic on Ω and continuous on $\bar{\Omega}$, yet could not continue to be even differentiable on any strictly larger open set. It follows that Ω is a domain of holomorphy.

For the general case, let U be *any* connected open subset of $\mathbb{C}$. Let $\{z_j\} \subseteq U$ be a sequence of points that has no accumulation point in U but that accumulates at every boundary point of U (the reader may construct such a sequence by exploiting the distance function to the boundary). By Weierstrass's theorem (see subsection 0.3.4) there is a nonconstant analytic function h on U that vanishes at each z_j and nowhere else. If h were to analytically continue to an analytic function $\hat{h}$ on some strictly larger open set $\hat{U}$, then it would necessarily hold that $\hat{U}$ contains a boundary point P of U. But then $\hat{h}$ would have zero set with interior accumulation point P. It follows that $\hat{h} \equiv 0$, which is a contradiction. Thus U is a domain of holomorphy.

The situation in several complex variables is decidedly different. Some open sets are trivially domains of holomorphy; if

$$\Omega = D^2(0, 1) \subseteq \mathbb{C}^2,$$

then let f_0 be the function on the disc that we constructed above and set $u(z_1, z_2) = f_0(z_1) \cdot f_0(z_2)$. This function is clearly analytic in Ω and does not analytically continue to any strictly larger open set. *If this simple product construction were the only way to construct domains of holomorphy in several variables, then there would be little justification for studying this subject. But, in* fact, the characterization of domains of holomorphy in several complex variables is quite subtle.

It turns out that the unit ball $B(0,1) \subseteq \mathbb{C}^n$ *is* a domain of holomorphy, but is not a product domain. The proof of this statement requires considerable ingenuity, and we shall explore it later. The important lesson now is that *not all domains in several complex variables are domains of holomorphy.* We now explain this important discovery of F. Hartogs [1].

Let

$$\Omega \equiv D^2(0,3) \setminus \bar{D}^2(0,1) \subseteq \mathbb{C}^2.$$

We claim that every holomorphic function on Ω analytically continues to the domain $\hat{\Omega} \equiv D^2(0,3)$. This assertion is a special case of a more general phenomenon (the Kugelsatz) that we explore later. For now we give an *ad hoc* proof, using only elementary one-variable facts about power series, of the continuation result.

Now let h be holomorphic on Ω. For z_1 fixed, $|z_1| < 3$, we write

$$h_{z_1}(z_2) = h(z_1, z_2) = \sum_{j=-\infty}^{\infty} a_j(z_1) z_2^j, \qquad (0.3.1.1)$$

where the coefficients of this Laurent expansion are given by

$$a_j(z_1) = \frac{1}{2\pi i} \oint_{|\zeta|=2} \frac{h(z_1, \zeta)}{\zeta^{j+1}} \, d\zeta.$$

In particular, $a_j(z_1)$ depends holomorphically on z_1 (by Morera's theorem, for instance). But $a_j(z_1) = 0$ for $j < 0$ and $1 < |z_1| < 3$. Therefore, by analytic continuation, $a_j(z_1) \equiv 0$ for $j < 0$. But then the series expansion $(0.3.1.1)$ becomes

$$\sum_{j=0}^{\infty} a_j(z_1) z_2^j$$

and this series *defines* a holomorphic function $\hat{h}$ on all of $D^2(0,3)$ that agrees with the original function h on Ω. Thus Ω is *not* a domain of holomorphy— all holomorphic functions on Ω continue to the larger domain $D^2(0,3)$. This completes the proof of the "Hartogs extension phenomenon."

One of the principal jobs for us in a basic course of several complex variables is to solve the Levi problem: to characterize the domains of holomorphy in terms of some geometric properties of the boundary (which should make no reference to holomorphic functions on Ω). This highly nontrivial problem was first solved by K. Oka [1] for $n = 2$ and by H. Bremerman [1], F. Norguet [1], and K. Oka [3] for $n \geq 3$. There are many useful approaches to the problem, including the sheaf-theoretic methods of the authors just mentioned and the functional analysis methods of A. Andreotti and H. Grauert [1], L. Ehrenpreis [1], Grauert [2], and R. Narasimhan [2] (to name only a few). The Levi problem for complex analytic manifolds and for analytic spaces has been solved by H. Grauert [1], R. Narasimhan [2], R. Remmert and K. Stein [1], and others. These results are both of theoretical interest and are extremely useful in applications. Further progress on the Levi problem has been made by H. Skoda [2], J. P. Demailly [1], Y. T. Siu [1], and others. A number of important problems still remain open.

The present text will present the partial differential equations approach to the problem of characterizing domains of holomorphy. This technique has been developed primarily by C. B. Morrey [1], J. J. Kohn [1], and L. Hörmander [1], although modern contributions have been made by J. E. Fornæss [1], R. M. Range [3], and others. For the record, we note that the difficult part of characterizing domains of holomorphy—namely, showing that a certain differential geometric condition on the boundary is sufficient for a domain to be a domain of holomorphy—is known as the *Levi problem,* in honor of E. E. Levi. This problem will be enunciated in great detail in Chapters 3 and 5.

It turns out that all convex domains are domains of holomorphy. This would be a lovely necessary and sufficient condition, for it is purely geometric and makes no reference to holomorphicity. However convexity is not preserved under holomorphic mappings: the domain $D \subseteq \mathbb{C}$ is convex, but its image under the mapping $\zeta \mapsto (4 + \zeta)^4$ is not. Thus some less rigid geometric condition is required to characterize domains of holomorphy. We shall learn that there is a biholomorphically invariant version of convexity, known as *pseudoconvexity,* which characterizes domains of holomorphy. The notion of convexity will command much of our attention in Chapters 3, 4, and 5. That is why we have spent so much time discussing it now. It is the paradigm for a number of positivity conditions that play an important role in the complex analysis of several variables.

We conclude this section by giving rather more brief coverage to several other problems that distinguish several complex variables from one complex variable.

0.3.2   Biholomorphic Mappings

The Riemann mapping theorem says that every proper, simply connected open subset Ω of $\mathbb{C}$ is biholomorphic to the disc: that is, there is a one-to-one, onto, holomorphic mapping $\Phi : \Omega \to D$. (It is redundant to point out, but we do so for the record, that such a mapping automatically has a holomorphic inverse.) There is no analogue for this result in $\mathbb{C}^n, n \geq 2$. Indeed "most" (in the sense of category) domains in $\mathbb{C}^n$ that are close to the ball—in any topology you like, say, the C^∞ topology—are *not* biholomorphic to the ball. In fact the set of biholomorphic equivalence classes, formed among domains close to the ball in any reasonable sense, is uncountable in number (see D. Burns, S. Shnider, and R. Wells [1], and R. E. Greene and S. G. Krantz [1, 2]).

Historically, the first result that suggested the failure of a Riemann mapping theorem in several complex variables was proved by Poincaré; he proved that, in any dimension $n \geq 2$, the ball $B(0,1)$ and the polydisc $D^n(0,1)$ are not biholomorphic. Poincaré's technique was to calculate the group of biholomorphic self-maps of each domain and to demonstrate that these groups are not isomorphic *as groups*. We shall treat that proof and also some more modern, geometric proofs.

The discussion in the preceding two paragraphs implies that there is no canonical topologically trivial domain in $\mathbb{C}^n, n \geq 2$, as there is in $\mathbb{C}^1$ (namely, the disc). This means that the analysis of holomorphic functions of several variables depends on the domain in question. Even for topologically complex domains in $\mathbb{C}^1$, the uniformization theorem (see H. Farkas and I. Kra [1]) often enables one to reduce analytic questions on a planar domain to analytic questions on the disc. This approach is explored in detail in S. Fisher [1]. But there is also no uniformization theorem in several complex variables.

In spite of the complex state of affairs discussed in the last paragraph, there has been some heartening progress made in the search for substitutes in several variables for the Riemann mapping theorem. We mention particularly the work of Fridman [3], L. Lempert [2], R. E. Greene and S. G. Krantz [1, 2], Daowei Ma [1], and S. Semmes [1].

The subject of biholomorphic and, more generally, *proper* holomorphic mappings has been an active area of research for the last 20 years. The most fundamental question is the following: If $\Omega_1, \Omega_2 \subseteq \mathbb{C}^n$ are open sets in $\mathbb{C}^n$ with smooth boundaries and if $\Phi : \Omega_1 \to \Omega_2$ is a biholomorphic (resp. proper holomorphic) mapping, that is, $\Phi(z) = (\Phi_1(z), \ldots, \Phi_n(z))$ where each Φ_j is holomorphic and Φ is one-to-one and onto (resp. each Φ_j is holomorphic and $\Phi^{-1}(K)$ is compact in Ω_1 for each compact K in Ω_2), is it true that Φ extends to be a C^∞ mapping of $\bar{\Omega}_1$ to $\bar{\Omega}_2$?

The answer to this last question is known to be yes in a number of important cases. There are no known counterexamples. We discuss some of these matters in detail in Section 11.4. Once the answer to this basic question is known to be yes, then one might hope to calculate differential boundary invariants for biholomorphic mappings. In turn, this will, in principle, allow one to identify domains that are biholomorphically inequivalent, at least locally. There has been some heartening progress in this program of Poincaré (see S. S. Chern and J. Moser [1], and C. Fefferman [2]), and we shall touch on it in Chapter 11.

0.3.3   Zero Sets of Analytic Functions

If $\Omega \subseteq \mathbb{C}^n$ is a domain, $f : \Omega \to \mathbb{C}$ is holomorphic, and $\mathcal{N}_\Omega(f) = \{z \in \Omega : f(z) = 0\}$, then what properties (geometric, topological, or otherwise) does the set $\mathcal{N}_\Omega(f)$ have? In one complex variable, the Weierstrass theorem says that any (denumerable) set $S \subseteq \Omega$ without interior accumulation point is the zero set of a holomorphic function. The converse is true as well and is rather trivial. A more interesting question in one complex variable is to characterize the zero sets of holomorphic functions satisfying a certain growth condition. For instance, let $H^\infty(D)$ be the collection of bounded, holomorphic functions on the unit disc $D \subseteq \mathbb{C}$. Then a necessary and sufficient condition for $\{z_j\}_{j=1}^\infty \subseteq D$ to be the zero set of an $f \in H^\infty(D)$ is that $\sum(1 - |z_j|) < \infty$.

Matters are vastly more complicated in the theory of several complex variables. The zeros of a holomorphic function $f : \mathbb{C}^n \to \mathbb{C}$, $n \geq 2$, are *never* isolated. [If f had an isolated zero at $z = 0$, then $1/f$ would be holomorphic on $D^n(0,3) \setminus D^n(0,1)$ yet *not* continuable to $D(0,3)$, contradicting the Hartogs phenomenon.] Roughly speaking, the zero set of a holomorphic function of several complex variables is no more complicated than the zero set of a holomorphic polynomial of several variables (the Weierstrass preparation theorem, treated in Chapter 6, makes this statement precise). Yet the study of zero sets of polynomials of several variables is the principal subject of algebraic geometry, one of the deepest branches of mathematics.

Speaking somewhat imprecisely, we may say that the zero set of a holomorphic function on a domain $\Omega \subseteq \mathbb{C}^n$ is locally (but for an exceptional set of lower dimension) a complex manifold of complex dimension $(n-1)$ (think about what this says in dimension one!). To amplify and clarify this true, but rather vague, description of a complex analytic variety is hard work (see R. C. Gunning [2]). We shall say more about this matter in Chapters 6 and 7. Recently some results have been obtained about zero sets for holomorphic functions satisfying a growth condition, but they are hard to state and much more so to prove (see G. M. Henkin [6], H. Skoda [3], N. Th. Varopoulos [1, 2, 3], W. Rudin [3]).

0.3.4   Divisors

Recall the Mittag-Leffler theorem of one complex variable theory.

THEOREM 0.3.1 (Mittag-Leffler)   Let $\Omega \subseteq \mathbb{C}$ be a proper open subset and $\{z_j\}_{j=1}^\infty \subseteq \Omega$ a discrete set (no interior accumulation points). Let $\{f_j\}$ be meromorphic in a neighborhood of z_j, each j. Then there exists a meromorphic function f on Ω such that f is analytic on $\Omega \setminus \{z_j : j = 1, 2, \ldots\}$ and $f - f_j$ is analytic near z_j for each j.

Likewise, the Weierstrass theorem, which we mentioned earlier in this chapter, is as follows:

THEOREM 0.3.2 (Weierstrass)   Let $\Omega, \{z_j\}_{j=1}^\infty \subseteq \Omega$ be as in the preceding theorem. Let $\{k_j\} \subseteq \mathbb{Z}$. There is a meromorphic function f on Ω, analytic and nonzero off $\{z_j : j = 1, 2, \ldots\}$, such that $f(z)/(z - z_j)^{k_j}$ is analytic and nonzero near z_j, each j.

It is of interest to have analogues of these theorems in the function theory of several complex variables. Since we know from subsection 0.3.3 on zero sets of holomorphic functions that the singularities of holomorphic functions are never isolated, it follows that there is difficulty in even *formulating* analogues of the theorems.

Here is another way to phrase these types of problems. Let $S \subseteq \Omega$ have the property that it is locally the zero set of an analytic function; that is, to each $w \in S$ is associated a small open set $\Omega_w \subseteq \Omega$ and a holomorphic function f_w on Ω_w such that

$$S \cap \Omega_w = \mathcal{N}_{\Omega_w}(f_w).$$

Does it follow that there is a holomorphic function F on Ω with $S = \mathcal{N}_{\Omega}(F)$? Problems formulated in this way lend themselves to solution by means of (the known solutions of) the Cousin problems. The Cousin problems were a precursor of sheaf theory.

Both sheaf theory and the Cousin problems will enter into our formulations and our proofs of analogues of the Weierstrass and Mittag-Leffler theorems in Chapters 6 and 7.

0.3.5   Factoring Holomorphic Functions

Let $H^\infty(D)$ be the space of bounded analytic functions on the unit disc. If $f \in H^\infty(D)$, then (Section 8.1) it is possible to write $f = F \cdot B$, where

1. F, B are analytic on D;
2. B is bounded by 1;
3. $\lim_{r \to 1^-} B(re^{i\theta})$ exists and has modulus 1 for almost every $\theta \in [0, 2\pi)$;
4. F is nonvanishing.

Stated crudely, F carries the "size" of f, whereas B carries the zeros. It turns out that B is a Blaschke product, a special case of the concept of "inner function." It is known (see L. A. Rubel and A. Shields [1]) that a factorization like $f = F \cdot B$ is impossible in several complex variables. However there *do* exist nonconstant inner functions: bounded analytic functions on the ball in $\mathbb{C}^n$ with unimodular radial boundary limits almost everywhere. In fact, it was an open problem, studied intensely from 1965 until 1982, whether nonconstant inner functions exist in dimensions greater than one. Most people, including the original formulators of the problem (Rudin and Vitushkin), were convinced that inner functions could not exist. A fairly large body of work was amassed to support that belief. Rather surprisingly, Aleksandrov [1] and Løw [1] constructed nonconstant inner functions in 1982.

While inner functions have proved a useful tool for constructing holomorphic functions with prescribed behavior (see W. Rudin [9] for an exhaustive treatment), they have not provided the rich structure that inner functions and Blaschke products have given us in dimension one (see K. Hoffman [1] for the classical theory). In Chapter 9 we shall construct inner functions of several complex variables and present related techniques in constructive function theory.

A related problem, also in the spirit of factoring holomorphic functions, is ascribed to A. Gleason. For a domain $\Omega \subseteq \mathbb{C}^n$, any n, let $A(\Omega)$ denote the space of functions that are continuous on $\bar{\Omega}$ and holomorphic on Ω. If $f \in A(D)$ and $f(0) = 0$, then of course we can write $f(z) = z \cdot g(z)$, with g also an element of $A(D)$. Gleason proposed that the analogue in $\mathbb{C}^n$ should be that if $f \in A(B)$ and $f(0) = 0$, then we can write

$$f(z) = z_1 g_1(z) + \cdots + z_n g_n(z)$$

with $g_j \in A(B), j = 1, \ldots, n$. This result was proved not only for the ball, but for convex domains with C^2 boundary, in Leibenzon [1]. It was proved for more general domains in N. Kerzman [2], N. Kerzman and A. Nagel [1], N. Sibony and J. Wermer [1], and J. E. Fornæss and N. Øvrelid [1]. We will treat these matters in Chapters 5, 7, and 10.

0.3.6   Domains of Convergence for Power Series

Consider the power series

$$\sum_{j=0}^{\infty} a_j z^j$$

for $z \in \mathbb{C}$. It is elementary to see that there is an $r, 0 \leq r \leq \infty$, such that the series converges absolutely and uniformly for $|z| < r$ and diverges for $|z| > r$. Observe that the domain of convergence is always a disc.

Matters are not so simple, geometrically speaking, in several variables. The series

$$\sum_{j=0}^{\infty} (z_1 + z_2)^j$$

converges absolutely and uniformly for $|z_1 + z_2| < 1$ and diverges for $|z_1 + z_2| > 1$. However, the series

$$\sum_{j=0}^{\infty} (z_1 \cdot z_2)^j$$

converges absolutely and uniformly for $|z_1 \cdot z_2| < 1$ and diverges when $|z_1 \cdot z_2| > 1$. Finally, the series

$$\sum_{j=0}^{\infty} z_1^j$$

converges absolutely for $|z_1| < 1, z_2$ arbitrary and diverges for $|z_1| > 1$, z_2 arbitrary.

Whereas in one complex variable the domain of convergence of a power series is always a disc, in several complex variables the domain of convergence does not have a simple geometric description. Since power series are central to the way that we represent holomorphic functions, it is clearly of interest to understand this matter. We shall address it in Chapter 2.

0.3.7   Extending Holomorphic Functions

Let $B \subseteq \mathbb{C}^2$ be the unit ball and define

$$\tilde{\mathbf{d}} = B \cap \left\{ (z_1, z_2) \in \mathbb{C}^2 : z_2 = \frac{1}{2} \right\}.$$

Of course $\tilde{\mathbf{d}}$ is just a copy of the disc $D^1(0, \sqrt{3/4})$. Suppose that ϕ is a holomorphic function of one complex variable on $D(0, \sqrt{3/4})$. Then, by defining $f(z_1, \frac{1}{2}) = \phi(z_1)$, we can construct an analytic function on $\tilde{\mathbf{d}}$. The question is: Is ϕ necessarily the restriction to $\tilde{\mathbf{d}}$ of a function F on all of B? If ϕ is bounded, can we find an F that is bounded?

In any program to construct holomorphic functions with specified properties, an affirmative answer to these questions would be useful. There is an affirmative answer on the ball, and on any domain of holomorphy, but a great deal of the machinery of several complex variables must be used to provide a construction. [Note, in the example posed in the last paragraph, that the trivial extension $F(z_1, z_2) = f(z_1, \frac{1}{2})$ cannot work since it makes no sense for $|z_1| \geq \frac{\sqrt{3}}{2}$.] We shall explore methods for treating extension problems in Chapter 5.

Observe that the types of extension problems being discussed here are trivial in complex dimension one.

0.3.8   Approximation Problems

Let $\Omega \subseteq \mathbb{C}^n$ be a domain and $f : \Omega \to \mathbb{C}$ a holomorphic function. Let $K \subseteq \Omega$ be compact. Is f the uniform limit, on K, of a sequence of functions holomorphic on a neighborhood of $\overline{\Omega}$?

This is an example of what we call an *approximation problem*. In one complex variable, much is known. Two examples of classical results of this type are as follows.

THEOREM 0.3.3 (Runge)   Let $\Omega \subseteq \mathbb{C}$ be a domain. Let f be holomorphic on $\Omega \subseteq \mathbb{C}$ and let $K \subseteq \Omega$ be compact. Let $\eta > 0$. Then there is a rational function (quotient of polynomials) $r(z)$ with poles in $^c K$ such that

$$\sup_K |f(\zeta) - r(\zeta)| < \eta.$$

THEOREM 0.3.4 (Mergelyan)   Let K be a compact subset of $\mathbb{C}$ whose complement in $\mathbb{C}$ is connected. Let $f : K \to \mathbb{C}$ be continuous on K and holomorphic on the interior of K. Then f is the uniform limit on K of holomorphic polynomials.

Notice that the hypotheses of both these theorems are purely topological in nature. In several complex variables, to the limited extent that we have any results of this type, the hypotheses depend on differential geometric data of Ω and K and also *on how Ω and K are imbedded in space.* Sections 5.4 and 10.4 treat these matters briefly.

0.3.9 The Cauchy-Riemann Equations

Let ϕ be a continuously differentiable function on a domain Ω in $\mathbb{C}$ and taking values in $\mathbb{C}$. Then ϕ is holomorphic if and only if $(\partial/\partial\bar{\zeta})\phi(\zeta) = 0$. For if we write $\phi = \phi_1 + i\phi_2$—the decomposition of ϕ into its real and imaginary parts—then

$$\begin{aligned}
\frac{\partial\phi}{\partial\bar{\zeta}} &= \frac{1}{2}\left(\frac{\partial}{\partial x} + i\frac{\partial}{\partial y}\right)(\phi_1 + i\phi_2) \\
&= \frac{1}{2}\left(\frac{\partial\phi_1}{\partial x} - \frac{\partial\phi_2}{\partial y}\right) + \frac{1}{2}i\left(\frac{\partial\phi_1}{\partial y} + \frac{\partial\phi_2}{\partial x}\right).
\end{aligned}$$

Thus the condition that $(\partial/\partial\bar{\zeta})\phi(\zeta) = 0$ corresponds to the simultaneous vanishing of the real and imaginary parts of the displayed equations. But that is just the classical Cauchy-Riemann equations for ϕ.

It is of interest to be able to solve the inhomogeneous equation

$$\frac{\partial u}{\partial\bar{\zeta}} = \psi$$

when ψ is a given function on $\mathbb{C}$, that is, say, C^1 with compact support. It turns out (as we shall prove in the next chapter) that the formula

$$u(\zeta) = -\frac{1}{\pi}\int \frac{\psi(\xi)}{(\xi - \zeta)}\, dV(\xi)$$

always supplies a solution to this problem. However, *even though ψ is compactly supported,* no solution of the equation $\partial u/\partial\bar{\zeta} = \psi$ can be compactly supported if $\int\psi(\zeta)dV(\zeta) \neq 0$. For if there were a compactly supported u, say $u(\zeta) = 0$

when $|\zeta| \geq R$, that satisfied the equation, then

$$
\begin{aligned}
0 &= \oint_{|\xi|=R} u(\xi)\, d\xi \\
&\overset{\text{(Stokes)}}{=} \int_{|\xi|<R} \frac{\partial u}{\partial \bar{\xi}}(\xi)\, d\bar{\xi} \wedge d\xi \\
&= 2i \int_{|\xi|<R} \psi(\xi)\, dV(\xi) \\
&\neq 0
\end{aligned}
$$

if $D(0, R)$ contains the support of ψ. That is a contradiction.

Now the analogous problem in several complex variables is as follows: Let $\psi_1, \ldots, \psi_n$ be given functions that are in $C_c^1(\mathbb{C}^n)$ (that is, C^1 with compact support in $\mathbb{C}^n$). We seek a C^2 function u such that

$$
\frac{\partial u}{\partial \bar{z}_j} = \psi_j, \quad j = 1, \ldots, n. \tag{0.3.2}
$$

Since

$$
\frac{\partial^2 u}{\partial \bar{z}_j \partial \bar{z}_k} = \frac{\partial^2 u}{\partial \bar{z}_k \partial \bar{z}_j},
$$

it is clearly necessary, for (0.3.2) to have a solution, that

$$
\frac{\partial \psi_j}{\partial \bar{z}_k} = \frac{\partial \psi_k}{\partial \bar{z}_j}, \quad \text{all } j, k = 1, \ldots, n.
$$

Subject to this compatibility condition, we shall prove in Section 1.1 that the system is always solvable and a solution is given by

$$
u(z) = -\frac{1}{\pi} \int \frac{\psi_j(z_1, \ldots, z_{j-1}, \xi, z_{j+1}, \ldots, z_n)}{(\xi - z_j)}\, dV(\xi),
$$

for any choice of $j = 1, \ldots, n$. This solution is *always* compactly supported when the dimension $n > 1$. Suppose for simplicity that $j = 1$. Now $\partial u/\partial \bar{z}_\ell = \psi_\ell = 0$, all ℓ, when z is large. Thus u is holomorphic in each variable separately, hence holomorphic, for z large—say when $|z| > R$. But, looking at the definition of u, we see that $u(z)$ itself must be zero when z is large. By analytic continuation, $u(z) = 0$ on the unbounded component of $^c(\cup_j \text{supp}\, \psi_j)$ In other words, u is compactly supported.

Although it is not evident now, it will become plain later that many of the differences between the the function theories of one and several complex variables can be accounted for by the support behavior of solutions of the inhomogeneous Cauchy-Riemann equations. These equations will be a principal tool for us in constructing holomorphic functions with specified properties. Since the theory of one complex variable has so many other tools for constructing functions—Blaschke products, Weierstrass products, Runge and Mittag-Leffler theorems, and so on—the Cauchy-Riemann equations have played a less prominent role in that subject. But in several complex variables these equations are central.

We now terminate this list of introductory problems, since the intended point has been established—namely, we have seen that there are significant differences between the theories of one and several complex variables, and we have introduced a number of the problems to be considered in succeeding chapters. We will see later that these problems are by no means disjoint; many of the same tools are used to attack all of them. This is a reflection both of the beauty of the subject and of the limited nature of our knowledge.

1 Some Integral Formulas

Our mission in this chapter is a bit eccentric when compared to the mission of other books on the subject. But it is well motivated: Whereas many classical treatments of the function theory of one or several complex variables imply by their presentation that complex analysis is a subject unto itself, with its own language and its own notation, our point of view is rather different. We demonstrate in this chapter that Stokes's theorem is the first step and that all our basic analytic tools follow from it. Thus our presentation will entail the repetition of a bit of classical textbook material; for instance, we derive Green's theorem in N dimensions *directly from* Stokes's theorem. We derive several important integral formulas using first principles directly from Green's theorem. We show that the classical Cauchy integral formula, and some of its generalizations to several complex variables, is an aspect of Stokes's theorem.

Although many of the steps of our treatment may be found in some form in some other text (generally in a much more abstract setting), it is our purpose here to present the entire development in a self-contained fashion, beginning with ideas of advanced calculus. Since the function theory of several complex variables can become rather abstract rather quickly, it is well to begin our journey with concrete material such as this. For another quite concrete development, see W. Rudin [7]. The reader who is impatient may safely skip Sections 1.3–1.6, referring back to this material when necessary. We do encourage the impatient reader to look over the material in Sections 1.1 and 1.2 before proceeding to Chapter 2.

1.1 The Bochner-Martinelli Formula

If ω is a differential form on $\Omega \subseteq \mathbb{R}^N$, then, in coordinates, ω can be written as a finite sum of terms of the form $\omega_\alpha dx^\alpha$, where α is a multi-index and ω_α is a smooth function. Differential forms on $\mathbb{C}^n$ may be written in this fashion also, since $\mathbb{C}^n$ is canonically identified with $\mathbb{R}^{2n}$. However it is much more convenient to use the complex notation introduced in Chapter 0. Thus if $\Omega \subseteq \mathbb{C}^n$ and ω is a differential form on Ω, then ω is a sum of terms of the form $\omega_{\alpha\beta} dz^\alpha \wedge d\bar{z}^\beta$, where α, β are multi-indices with $|\alpha| \le n, |\beta| \le n$. If $0 \le p, q \le n$ and

$$\omega = \sum_{|\alpha|=p, |\beta|=q} \omega_{\alpha\beta} \, dz^{\alpha} \wedge d\bar{z}^{\beta},$$

then ω is said to be a differential form of *type* (or *bidegree*) (p, q).

In classical advanced calculus only a differential form of total degree m may be integrated on a space or surface or manifold of (real) dimension m (see Appendix IV). Likewise, in our new notation, only forms of type (p, q) with $p + q = m$ may be integrated on a space or surface or manifold of (real) dimension m.

If $\omega = \sum_{\alpha,\beta} \omega_{\alpha\beta} \, dz^{\alpha} \, d\bar{z}^{\beta}$, then we define

$$\partial\omega = \sum_{j=1}^{n} \sum_{\alpha,\beta} \frac{\partial\omega_{\alpha\beta}}{\partial z_j} \, dz_j \wedge dz^{\alpha} \wedge d\bar{z}^{\beta},$$

$$\bar{\partial}\omega = \sum_{j=1}^{n} \sum_{\alpha,\beta} \frac{\partial\omega_{\alpha\beta}}{\partial \bar{z}_j} \, d\bar{z}_j \wedge dz^{\alpha} \wedge d\bar{z}^{\beta}.$$

Letting d denote the usual exterior differential operator on forms (see W. Rudin [1], H. Federer [1], and Appendix IV), we see by a straightforward calculation that $d = \partial + \bar{\partial}$.

Notice that when f is a C^1 *function* (or $(0,0)$ form), then

$$\bar{\partial}f = \sum_j \frac{\partial f}{\partial \bar{z}_j} \, d\bar{z}_j.$$

Since the differentials $d\bar{z}_j$ are linearly independent, we conclude that $\bar{\partial}f \equiv 0$ on Ω if and only if $\partial f/\partial \bar{z}_j \equiv 0$ for $j = 1, \dots, n$; that is, $\bar{\partial}f \equiv 0$ on Ω if and only if f is holomorphic in each variable separately.

The language of differential forms is needed in order to formulate Stokes's theorem.

THEOREM 1.1.1 (Stokes)   Let $\Omega \subseteq \mathbb{C}^n$ be a bounded open set with C^1 boundary. Let ω be a differential form of bidegree (p, q) with coefficients in $C^1(\bar{\Omega})$. Then

$$\int_{\partial\Omega} \omega = \int_{\Omega} d\omega = \int_{\Omega} \partial\omega + \bar{\partial}\omega.$$

Standard references for Stokes's theorem are W. Rudin [1] and G. de Rham [1]. The Stokes's theorem that we have recorded here is the standard one simply expressed in complex notation.

The *full Cauchy integral formula* in $\mathbb{C}$ is a formula not just about holomorphic functions but about all continuously differentiable functions. We now derive this more general result for all $\mathbb{C}^n, n \geq 1$, and learn what consequences it has for the function theory of both one and several variables.

DEFINITION 1.1.2   On $\mathbb{C}^n$ we let

$$\omega(z) \equiv dz_1 \wedge dz_2 \wedge \cdots \wedge dz_n,$$

$$\eta(z) \equiv \sum_{j=1}^{n}(-1)^{j+1}z_j dz_1 \wedge \cdots \wedge dz_{j-1} \wedge dz_{j+1} \wedge \cdots \wedge dz_n.$$

The form η is sometimes called the *Leray form*. We will often write $\omega(\bar{z})$ to mean $d\bar{z}_1 \wedge \cdots \wedge d\bar{z}_n$ and, likewise, $\eta(\bar{z})$ to mean $\sum_{j=1}^{n}(-1)^{j+1}\bar{z}_j d\bar{z}_1 \wedge \cdots \wedge d\bar{z}_{j-1} \wedge d\bar{z}_{j+1} \wedge \cdots \wedge d\bar{z}_n$.

The genesis of the Leray form is explained by the following lemma.

LEMMA 1.1.3   For any $z_0 \in \mathbb{C}^n$, any $\epsilon > 0$, we have

$$\int_{\partial B(z^0,\epsilon)} \eta(\bar{z}) \wedge \omega(z) = n \int_{B(z^0,\epsilon)} \omega(\bar{z}) \wedge \omega(z).$$

Proof   Notice that $d\eta(\bar{z}) = \bar{\partial}\eta(\bar{z}) = n\omega(\bar{z})$. By Stokes's theorem,

$$\int_{\partial B(z^0,\epsilon)} \eta(\bar{z}) \wedge \omega(z) = \int_{B(z^0,\epsilon)} d[\eta(\bar{z}) \wedge \omega(z)].$$

Of course the expression in [ ] is saturated in dz's, so, in the decomposition $d = \partial + \bar{\partial}$, only the term $\bar{\partial}$ will not die. Thus the last line equals

$$\int_{B(z^0,\epsilon)} [\bar{\partial}(\eta(\bar{z}))] \wedge \omega(z) = n \int_{B(z^0,\epsilon)} \omega(\bar{z}) \wedge \omega(z). \qquad \square$$

Remark: Notice that, by change of variables,

$$\begin{aligned}
\int_{B(z^0,\epsilon)} \omega(\bar{z}) \wedge \omega(z) &= \int_{B(0,\epsilon)} \omega(\bar{z}) \wedge \omega(z) \\
&= \epsilon^{2n} \int_{B(0,1)} \omega(\bar{z}) \wedge \omega(z).
\end{aligned}$$

A straightforward calculation shows that

$$\int_{B(0,1)} \omega(\bar{z}) \wedge \omega(z) = (-1)^{q(n)} \cdot (2i)^n \cdot (\text{volume of the unit ball in } \mathbb{C}^n \approx \mathbb{R}^{2n}),$$

where $q(n) = [n(n-1)]/2$. We denote the value of this integral by $W(n)$. $\square$

THEOREM 1.1.4 (Bochner-Martinelli)   Let $\Omega \subseteq \mathbb{C}^n$ be a bounded domain with C^1 boundary. Let $f \in C^1(\bar{\Omega})$. Then, for any $z \in \Omega$, we have

$$
\begin{aligned}
f(z) \;=\; & \frac{1}{nW(n)} \int_{\partial\Omega} \frac{f(\zeta)\eta(\bar{\zeta} - \bar{z}) \wedge \omega(\zeta)}{|\zeta - z|^{2n}} \\
& - \frac{1}{nW(n)} \int_{\Omega} \frac{\bar{\partial}f(\zeta)}{|\zeta - z|^{2n}} \wedge \eta(\bar{\zeta} - \bar{z}) \wedge \omega(\zeta).
\end{aligned}
$$

Proof   Fix $z \in \Omega$. We apply Stokes's theorem to the form

$$
L_z(\zeta) \equiv \frac{f(\zeta)\eta(\bar{\zeta} - \bar{z}) \wedge \omega(\zeta)}{|\zeta - z|^{2n}}
$$

on the domain $\Omega_{z,\epsilon} \equiv \Omega \setminus \bar{B}(z, \epsilon)$, where $\epsilon > 0$ is chosen so small that $\bar{B}(z, \epsilon) \subseteq \Omega$. Note that Stokes's theorem does not apply to forms that have a singularity; thus we may not apply the theorem to L_z on any domain that contains the point z in either its interior or its boundary. This observation helps to dictate the form of the domain $\Omega_{z,\epsilon}$. As the proof develops, we shall see that it also helps to determine the outcome of our calculation.

Notice that

$$
\partial(\Omega_{z,\epsilon}) = \partial\Omega \cup \partial B(z, \epsilon)
$$

but that the two pieces are equipped with opposite orientations (Figure 1.1). Thus, by Stokes,

$$
\begin{aligned}
\int_{\partial\Omega} L_z(\zeta) - \int_{\partial B(z,\epsilon)} L_z(\zeta) \;=\; & \int_{\partial\Omega_{z,\epsilon}} L_z(\zeta) \\
\;=\; & \int_{\Omega_{z,\epsilon}} d_\zeta(L_z(\zeta)). \qquad (1.1.4.1)
\end{aligned}
$$

Notice that we consider z to be fixed and ζ to be the variable. Now

$$
\begin{aligned}
d_\zeta L_z(\zeta) \;=\; & \bar{\partial}_\zeta L_z(\zeta) \\
\;=\; & \frac{\bar{\partial}f(\zeta) \wedge \eta(\bar{\zeta} - \bar{z}) \wedge \omega(\zeta)}{|\zeta - z|^{2n}} \\
& + f(\zeta) \cdot \left[\sum_{j=1}^{n} \frac{\partial}{\partial\bar{\zeta}_j} \left(\frac{\bar{\zeta}_j - \bar{z}_j}{|\zeta - z|^{2n}} \right) \right] \omega(\bar{\zeta}) \wedge \omega(\zeta). \qquad (1.1.4.2)
\end{aligned}
$$

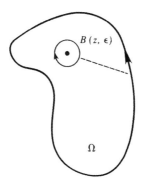

FIGURE 1.1

Observing that

$$\frac{\partial}{\partial \bar{\zeta}_j} \left(\frac{\bar{\zeta}_j - \bar{z}_j}{|\zeta - z|^{2n}} \right) = \frac{1}{|\zeta - z|^{2n}} - n\frac{|\bar{\zeta}_j - \bar{z}_j|^2}{|\zeta - z|^{2n+2}},$$

we find that the second term on the far right of (1.1.4.2) dies and we have

$$d_\zeta L_z(\zeta) = \frac{\bar{\partial} f(\zeta) \wedge \eta(\bar{\zeta} - \bar{z}) \wedge \omega(\zeta)}{|\zeta - z|^{2n}}.$$

Substituting this identity into (1.1.4.1) yields

$$\int_{\partial\Omega} L_z(\zeta) - \int_{\partial B(z,\epsilon)} L_z(\zeta) = \int_{\Omega_{z,\epsilon}} \frac{\bar{\partial} f(\zeta) \wedge \eta(\bar{\zeta} - \bar{z}) \wedge \omega(\zeta)}{|\zeta - z|^{2n}}. \qquad (1.1.4.3)$$

Next we remark that

$$\begin{aligned}
\int_{\partial B(z,\epsilon)} L_z(\zeta) &= f(z) \int_{\partial B(z,\epsilon)} \frac{\eta(\bar{\zeta} - \bar{z}) \wedge \omega(\zeta)}{|\zeta - z|^{2n}} \\
&\quad + \int_{\partial B(z,\epsilon)} \frac{(f(\zeta) - f(z))\,\eta(\bar{\zeta} - \bar{z}) \wedge \omega(\zeta)}{|\zeta - z|^{2n}} \\
&\equiv T_1 + T_2. \qquad (1.1.4.4)
\end{aligned}$$

Since $|f(\zeta) - f(z)| \le C|\zeta - z|$ (and since each term of $\eta(\bar{\zeta} - \bar{z})$ has a factor of some $(\bar{\zeta}_j - \bar{z}_j)$), it follows that the integrand of T_2 is of size $O(|\zeta - z|)^{-2n+2} \approx \epsilon^{-2n+2}$. Since the surface over which the integration is performed has area $\approx \epsilon^{2n-1}$, it follows that $T_2 \to 0$ as $\epsilon \to 0^+$.

By Lemma 1.1.3, we also have

$$
\begin{aligned}
T_1 &= \epsilon^{-2n} f(z) \int_{\partial B(z,\epsilon)} \eta(\bar{\zeta} - \bar{z}) \wedge \omega(\zeta) \\
&= n\epsilon^{-2n} f(z) \int_{B(0,\epsilon)} \omega(\bar{\zeta}) \wedge \omega(\zeta) \\
&= nW(n)f(z).
\end{aligned}
\tag{1.1.4.5}
$$

Finally, (1.1.4.3)–(1.1.4.5) yield that

$$
\left(\int_{\partial\Omega} L_z(\zeta) \right) - nW(n)f(z) + o(1) = \int_{\Omega_{z,\epsilon}} \bar{\partial}f(\zeta) \wedge \left[\frac{\eta(\bar{\zeta} - \bar{z})}{|\zeta - z|^{2n}} \right] \wedge \omega(\zeta).
$$

Since

$$
\left| \frac{\eta(\bar{\zeta} - \bar{z})}{|\zeta - z|^{2n}} \right| = O(|\zeta - z|^{-2n+1}),
$$

the last integral is absolutely convergent as $\epsilon \to 0^+$ (remember that $\bar{\partial}f$ is bounded). Thus we finally have

$$
f(z) = \frac{1}{nW(n)} \int_{\partial\Omega} L_z(\zeta) - \frac{1}{nW(n)} \int_{\Omega} \bar{\partial}f(\zeta) \wedge \frac{\eta(\bar{\zeta} - \bar{z})}{|\zeta - z|^{2n}} \wedge \omega(\zeta).
$$

This is the Bochner-Martinelli formula. □

COROLLARY 1.1.5 If $\Omega \subseteq \mathbb{C}$ is a bounded domain with C^1 boundary and if $f \in C^1(\bar{\Omega})$, then, for any $z \in \Omega$,

$$
f(z) = \frac{1}{2\pi i} \int_{\partial\Omega} \frac{f(\zeta)}{\zeta - z} d\zeta - \frac{1}{2\pi i} \int_{\Omega} \frac{(\partial f(\zeta)/\partial\bar{\zeta})}{\zeta - z} d\bar{\zeta} \wedge d\zeta.
$$

Proof   It is necessary only to note that, when $n = 1$,

$$
\omega(\zeta) = d\zeta, \qquad \eta(\bar{\zeta} - \bar{z}) = \bar{\zeta} - \bar{z}, \quad \text{and} \quad nW(n) = 2\pi i. \qquad □
$$

COROLLARY 1.1.6 With hypotheses as in Corollary 1.1.5 and the additional assumption that $\bar{\partial}f = 0$ on Ω, we have

$$
f(z) = \frac{1}{2\pi i} \int_{\partial\Omega} \frac{f(\zeta)}{\zeta - z} d\zeta.
$$

Remark: Corollary 1.1.6 contains the familiar Cauchy integral formula from analysis of one variable. It is a wonderful fluke of one-variable analysis that when $n = 1$, the expression $(\bar{\zeta}_j - \bar{z}_j)/|\zeta - z|^{2n}$ becomes $(\bar{\zeta} - \bar{z})/|\zeta - z|^2 = 1/(\zeta - z)$. In particular, *the kernel of the classical Cauchy integral formula is holomorphic in z.* This provides an important technique for creating holomorphic functions (namely, integrating this kernel against a measure) in one-variable analysis that at this stage is missing from multi-variable analysis.   □

COROLLARY 1.1.7   If $\Omega \subseteq \mathbb{C}^n$ is bounded and has C^1 boundary and if $f \in C^1(\bar{\Omega})$ and $\bar{\partial} f = 0$ on Ω, then

$$f(z) = \frac{1}{nW(n)} \int_{\partial \Omega} \frac{f(\zeta)\eta(\bar{\zeta} - \bar{z})}{|\zeta - z|^{2n}} \wedge \omega(\zeta). \qquad (1.1.7.1)$$

It is rather surprising that the nonholomorphic kernel in (1.1.7.1) somehow manages to reproduce $\bar{\partial}$-closed, or holomorphic, functions. What makes (1.1.7.1) of limited utility is that it *does not create* holomorphic ($\bar{\partial}$-closed) functions. Although it is true that the *form*

$$\frac{\eta(\bar{\zeta} - \bar{z})}{|\zeta - z|^{2n}} \wedge \omega(\zeta)$$

is $\bar{\partial}$-closed in z for $z \neq \zeta$ (in fact, this kernel is a fundamental solution for the $\bar{\partial}$ operator—see R. Harvey and J. Polking [1]), it is not the case that this expression is *holomorphic* in the free variable z unless $n = 1$. In other words, the form is $\bar{\partial}$-closed away from the singularity, but its coefficients, as functions, are not holomorphic in z.

It is a fairly recent development, primarily due to G. M. Henkin [2], E. Ramirez [1], and H. Grauert and I. Lieb [1], that fairly explicit holomorphic n-dimensional reproducing kernels have been constructed (a nonconstructive way to find such kernels is given in Sections 1.4 and 1.5). We consider these constructions in Chapters 5 and 10.

In spite of its limitations, a few useful facts can be gleaned from Corollary 1.1.7. In particular, one may check by differentiation under the integral sign that a $\bar{\partial}$-closed function is in fact C^∞, indeed real analytic, on its domain of definition because the integral kernel has these properties.

Now we derive the explicit integral solution formula for the inhomogeneous Cauchy-Riemann equations that we discussed in subsection 0.3.9. We will see that the holomorphicity of the one-dimensional Cauchy kernel will play an important role in the derivation of this formula. First we give a careful definition of the support of a form.

DEFINITION 1.1.8   If λ is a differential form on $\mathbb{R}^N$ or $\mathbb{C}^n$, then the *support* of λ, written supp λ, is the complement of the union of all open sets on which the coefficients of λ vanish identically.

THEOREM 1.1.9   Let $\psi \in C_c^1(\mathbb{C})$. The function defined by

$$u(\zeta) = -\frac{1}{2\pi i} \int \frac{\psi(\xi)}{\xi - \zeta} \, d\bar{\xi} \wedge d\xi = -\frac{1}{\pi} \int \frac{\psi(\xi)}{\xi - \zeta} \, dV(\xi)$$

satisfies

$$\bar{\partial} u(\zeta) = \frac{\partial u}{\partial \bar{\zeta}}(\zeta) \, d\bar{\zeta} = \psi(\zeta) \, d\bar{\zeta}.$$

Proof   Let $D(0, R)$ be a large disc that contains the support of ψ. Then

$$\begin{aligned}
\frac{\partial u}{\partial \bar{\zeta}}(\zeta) &= -\frac{1}{2\pi i} \frac{\partial}{\partial \bar{\zeta}} \int_{\mathbb{C}} \frac{\psi(\xi)}{\xi - \zeta} \, d\bar{\xi} \wedge d\xi \\
&= -\frac{1}{2\pi i} \frac{\partial}{\partial \bar{\zeta}} \int_{\mathbb{C}} \frac{\psi(\xi + \zeta)}{\xi} \, d\bar{\xi} \wedge d\xi \\
&= -\frac{1}{2\pi i} \int_{\mathbb{C}} \frac{\frac{\partial \psi}{\partial \bar{\xi}}(\xi + \zeta)}{\xi} \, \delta\bar{\xi} \wedge d\xi \\
&= -\frac{1}{2\pi i} \int_{D(0,R)} \frac{\frac{\partial \psi}{\partial \bar{\xi}}(\xi)}{\xi - \zeta} \, \delta\bar{\xi} \wedge d\xi.
\end{aligned}$$

By Corollary 1.1.5, this last equals

$$\psi(\zeta) - \frac{1}{2\pi i} \int_{\partial D(0,R)} \frac{\psi(\xi)}{\xi - \zeta} \, d\xi = \psi(\zeta).$$

Here we have used the support condition on ψ. This is the result that we wish to prove.   □

Recall now that the *convolution* of two L^1 functions on $\mathbb{R}^N$ is defined by

$$(f * g)(x) = \int_{\mathbb{R}^N} f(x - t)g(t) \, dV(t) = \int_{\mathbb{R}^N} g(x - t)f(t) \, dV(t).$$

The same notion applies in $\mathbb{C}^n \approx \mathbb{R}^{2n}$. Let $K(\xi) = -1/(\pi\xi)$ on $\mathbb{C}$. Then Theorem 1.1.9 asserts that for any $\psi \in C_c^1$, it holds that $u(\zeta) = K * \psi(\zeta)$ satisfies

$$\frac{\partial u}{\partial \bar{\zeta}}(\zeta) = \psi(\zeta).$$

In other words,

$$\left(\frac{\partial K}{\partial \bar{\zeta}} \right) * \psi = \psi$$

for all $\psi \in C_c^\infty$. Elementary distribution theory then implies that

$$\frac{\partial K}{\partial \bar{\zeta}} = \delta,$$

the Dirac δ mass at 0. In particular, $\partial K / \partial \bar{\zeta}$ has no support away from 0, so K must be holomorphic away from 0. Of course we can see by inspection that this last assertion is true, but our argument shows that it is necessary for *any* convolution kernel that solves the Cauchy-Riemann equations to be holomorphic away from zero. (The reader who is adept at distribution theory can give an alternative proof of Theorem 1.1.9 by showing directly that $\partial K / \partial \bar{\zeta} = \delta$ in the sense of distributions.)

It is an important and nontrivial fact that the equation $\partial K / \partial \bar{\zeta} = \delta$ has an analogue in several complex variables, where the role of K is played by the Bochner-Martinelli kernel. That is, the Bochner-Martinelli kernel is a fundamental solution for the $\bar{\partial}$ operator (see R. Harvey and J. Polking [1]). We shall not explore this point of view further.

Now let us set up the machinery for studying the inhomogeneous Cauchy-Riemann equations, or $\bar{\partial}$-problem, in several variables and see how things change. The reader will want to refer back to the introductory material presented in Section 0.3.

Let ψ be a $(0,1)$ form on $\mathbb{C}^n$ with C_c^1 coefficients. We seek solutions u to the equation

$$\bar{\partial} u = \psi$$

where u should be (at least) a C^1 function on $\mathbb{C}^n$. If λ is any form of type (p,q), then

$$0 = d^2 \lambda = (\partial + \bar{\partial})^2 \lambda = \partial^2 \lambda + (\partial\bar{\partial} + \bar{\partial}\partial)\lambda + \bar{\partial}^2 \lambda \equiv I + II + III.$$

Since I is a form of type $(p+2, q)$, II is a form of type $(p+1, q+1)$, and III is a form of type $(p, q+2)$, it follows that each of $I, II,$ and III must be zero. We conclude that

$$\partial^2 = 0, \qquad \bar{\partial}^2 = 0, \qquad \partial\bar{\partial} + \bar{\partial}\partial = 0.$$

Now if we apply the operator $\bar{\partial}$ to the equation $\bar{\partial} u = \psi$ that we wish to solve, then we find that a necessary condition for its solvability is that $\bar{\partial}\psi = 0$.

Exercise for the Reader

1. Compare the compatibility conditions for the $\bar{\partial}$-problem discovered in subsection 0.3.9 with the condition $\bar{\partial}\psi = 0$ that we just determined to see that they say the same thing.

2. Begin with Theorem 1.1.4 and attempt to imitate the proof of Theorem 1.1.9 to find a solution operator for the $\bar{\partial}$-problem in several complex variables. What sort of formula do you get?

3. The compatibility condition $\bar{\partial}\psi = 0$ is always trivially satisfied for $(0,1)$ forms in $\mathbb{C}^1$. Why?

We now present a formula that will solve the inhomogeneous $\bar{\partial}$-equation for compactly supported data in $\mathbb{C}^n$. Although we shall later learn that this is in fact not the most interesting type of inhomogeneous Cauchy-Riemann equation to solve, it still has some short-term utility.

THEOREM 1.1.10   Let $\psi = \sum_{j=1}^n \psi_j(z)\, d\bar{z}_j$ be a $(0,1)$ form on $\mathbb{C}^n$, with C_c^1 coefficients, that is $\bar{\partial}$-closed. Then for any choice of $j, 1 \leq j \leq n$, the function

$$u_j(z) = -\frac{1}{2\pi i} \int_{\mathbb{C}} \frac{\psi_j(z_1, \ldots, z_{j-1}, \zeta, z_{j+1}, \ldots, z_n)}{\zeta - z_j}\, d\bar{\zeta} \wedge d\zeta$$

satisfies $\bar{\partial}u_j = \psi$. For any j and $j', 1 \leq j, j' \leq n$, it holds that $u_j = u_{j'}$.

If $n > 1$, then the functions u_j are compactly supported; indeed, $u_j \equiv 0$ on the connected component of $^c(\operatorname{supp}\psi)$ that contains ∞.

Proof   The assertion about compact support was proved in subsection 0.3.9. Given this, we note that the C^1 function $u_j - u_{j'}$ is annihilated by $\bar{\partial}$; thus it is holomorphic in each variable separately. Hence, because of its compact support, it must be identically zero. It remains to prove the first assertion.

We write

$$
\begin{aligned}
\frac{\partial}{\partial \bar{z}_\ell} u_j(z) &= -\frac{1}{2\pi i}\frac{\partial}{\partial \bar{z}_\ell} \int_{\mathbb{C}} \frac{\psi_j(z_1, \ldots, z_{j-1}, \zeta + z_j, z_{j+1}, \ldots, z_n)}{\zeta}\, d\bar{\zeta} \wedge d\zeta \\
&= -\frac{1}{2\pi i} \int_{\mathbb{C}} \frac{\frac{\partial \psi_j}{\partial \bar{z}_\ell}(z_1, \ldots, z_{j-1}, \zeta + z_j, z_{j+1}, \ldots, z_n)}{\zeta}\, d\bar{\zeta} \wedge d\zeta \\
&= -\frac{1}{2\pi i} \int_{\mathbb{C}} \frac{\frac{\partial \psi_\ell}{\partial \bar{z}_j}(z_1, \ldots, z_{j-1}, \zeta + z_j, z_{j+1}, \ldots, z_n)}{\zeta}\, d\bar{\zeta} \wedge d\zeta.
\end{aligned}
$$

In this last equality we have exploited the compatibility condition $\bar{\partial}\psi = 0$. For fixed z, let $D(0, R)$ be a large disc in $\mathbb{C}$ that contains the support of $\psi(z_1, \ldots, z_{j-1}, \cdot, z_{j+1}, \ldots, z_n)$. Then the last integral can be written as

$$-\frac{1}{2\pi i} \int_{D(0,R)} \frac{\frac{\partial \psi_\ell}{\partial \bar{z}_j}(z_1, \ldots, z_{j-1}, \zeta, z_{j+1}, \ldots, z_n)}{\zeta - z_j}\, d\bar{\zeta} \wedge d\zeta.$$

Now the argument that we used to complete the proof of Theorem 1.1.9, using Corollary 1.1.5, shows that this last expression equals $\psi_\ell(z)$. In other words,

$$\bar\partial u_j = \psi,$$

as required. □

Exercise for the Reader
Use integration by parts to prove that if ψ is a $\bar\partial$-closed $(0,1)$ form in $\mathbb{C}^n$ with C_c^k coefficients, $k \in \mathbb{N}$, then $u_1 = \cdots = u_n$ is a C^k function.

The integral formulas that we have derived will have interesting applications that we will present in the next section. However, it turns out to be much more important to solve the equation $\bar\partial u = \psi$ on a domain Ω when the data ψ are supported in Ω, but *not compactly so*. The existence problem for this differential equation, subject to the usual compatibility condition that $\bar\partial\psi = 0$ on Ω, is a delicate one that is bound up with geometric properties of $\partial\Omega$. In fact the problem is essentially equivalent with the Levi problem—that is, the problem of characterizing domains of holomorphy. A large part of this book is directed to this circle of ideas.

The next exercise for the reader will utilize a piece of notation that will prove useful throughout the remainder of this book. If U is an open set and E is a subset of U, then we will write $E \subset\subset U$ to mean that the closure of E in space is a compact subset of U—in other words, the notation means that E is relatively compact in U.

Exercise for the Reader
Let $\{f_j\}_{j=1}^\infty$ be a sequence of C^1 functions on a domain $\Omega \subset\subset \mathbb{C}^n$ such that $\bar\partial f_j = 0$ for each j and suppose that the sequence converges *uniformly on compact sets* to a limit function f. Prove that the limit function f is C^1 and satisfies $\bar\partial f = 0$ on Ω. (*Hint*: Apply the Bochner-Martinelli formula on a smoothly bounded subdomain $\Omega' \subset\subset \Omega$ to the functions f_j. By differentiating under the integral sign, prove that $f_j \to f$ in the C^1 topology on any subdomain $\Omega'' \subset\subset \Omega'$.)

We conclude this section by giving a brief and informal discussion of the role of the Bochner-Martinelli kernel as a fundamental solution for the Cauchy-Riemann operator. (This discussion is not necessary for further reading of the text, and those unfamiliar with distribution theory may wish to skip it.) Our source for this material is Harvey and Polking [1], which is an excellent reference for those interested in integral formulas (see also R. M. Range [3]).

If P is a linear partial differential operator, then a *fundamental solution* for P is a function (or sometimes a distribution) K that satisfies

$$P(K) = \delta.$$

Here δ is the Dirac mass at the origin. If ϕ is a C^∞ function with compact support, then define

$$u = \phi * K.$$

It follows that

$$P(u) = \phi * (PK) = \phi * \delta = \phi. \qquad (1.1.11)$$

Thus, in some sense, K gives rise to a right inverse for P. This is the most important feature of a fundamental solution. As previously noted, the Cauchy kernel is a fundamental solution for the $\bar{\partial}$ operator in $\mathbb{C}^1$.

Because the Cauchy-Riemann equations are overdetermined in dimensions greater than one, the notion of fundamental solution must then be modified. That is, we can no longer expect to solve $\bar{\partial}u = \phi$ for *any* ϕ; therefore, we replace equation (1.1.11) by the *chain homotopy* formula (1.1.12). First we need a new notion of convolution.

In Harvey and Polking [1], the notion of convolution is extended to forms as follows. If

$$A = \sum_{\alpha,\beta} a_{\alpha\beta}\, dz^\alpha \wedge d\bar{z}^\beta \qquad B = \sum_{\alpha,\beta} b_{\alpha\beta}\, dz^\alpha \wedge d\bar{z}^\beta$$

are differential forms, then we define the convolution of these forms to be

$$A \# B = \int_\Omega A(z - \zeta) \wedge B(\zeta).$$

In order to obtain maximum flexibility from this concept, we do not mandate that $(a_{\alpha\beta}d\zeta^\alpha \wedge d\bar{\zeta}^\beta) \wedge (b_{\alpha'\beta'}d\zeta^{\alpha'} \wedge d\bar{\zeta}^{\beta'})$ integrates to zero if either $|\alpha| + |\alpha'| \neq n$ or $|\beta| + |\beta'| \neq n$. If either of the sums is *greater* than n, then, of course, the term dies. But if the sum is less than n, then we think of the form as living on a lower-dimensional manifold (such as the boundary of the domain). The proper language for these ideas is that of *currents* (see H. Federer [1]), but these are beyond the scope of this book.

A form K on $\mathbb{C}^n$ is called a *fundamental solution* for the $\bar{\partial}$ operator if, for any form ϕ with smooth coefficients of compact support, it holds that

$$\bar{\partial}(K \# \phi) + K \# (\bar{\partial}\phi) = \phi. \qquad (1.1.12)$$

Notice that when ϕ is $\bar{\partial}$-closed, then the second term on the left drops out and (1.1.12) looks more like the classical formula (1.1.11) for a fundamental solution. It can be checked that the Bochner-Martinelli kernel B, considered as a form of type $(n, n-1)$, is a fundamental solution for the operator $\bar{\partial}$. In particular, one can then prove that if ϕ is a $\bar{\partial}$-closed $(0, 1)$ form with smooth coefficients *of compact support*, then $\bar{\partial}(B \# \phi) = \phi$.

If one extends the foregoing concepts to forms with coefficients that are distributions (that is, to currents), then one can apply the chain homotopy formula to $\chi_\Omega \phi$, where χ_Ω is the characteristic function for a domain Ω and ϕ is now smooth on $\bar{\Omega}$ (not necessarily compactly supported inside). From these considerations one can recover the Bochner-Martinelli formula of Theorem 1.1.4.

1.2   Applications of Cauchy Theory and the $\bar{\partial}$ Equation

Now we fix once and for all a definition of holomorphic function in $\mathbb{C}^n$ and develop some elementary properties of these functions. We finesse for now some niceties about minimal hypotheses on these functions. We shall see later, by a theorem of F. Hartogs (much as Goursat proved for analytic functions of one variable), that there is no loss to assume as we do that holomorphic functions are C^1.

DEFINITION 1.2.1   A C^1 function f defined on an open set $U \subseteq \mathbb{C}^n$ is said to be *holomorphic* if $\bar{\partial} f = 0$ on U.

THEOREM 1.2.2 (Cauchy Formula for Polydiscs)   Let $w \in \mathbb{C}^n$ and $r_1, r_2, \ldots, r_n > 0$. Suppose that f is continuous on $\bar{D}^1(w_1, r_1) \times \cdots \times \bar{D}^1(w_n, r_n)$ and holomorphic on $D^1(w_1, r_1) \times \cdots \times D^1(w_n, r_n)$. Then, for any $z \in D^1(w_1, r_1) \times \cdots \times D^1(w_n, r_n)$, it holds that

$$f(z) = \frac{1}{(2\pi i)^n} \int_{|\zeta_n - w_n| = r_n} \cdots \int_{|\zeta_1 - w_1| = r_1} \frac{f(\zeta_1, \ldots, \zeta_n)}{(\zeta_1 - z_1) \cdots (\zeta_n - z_n)} \, d\zeta_1 \cdots d\zeta_n.$$

Proof   By repeated application of the one-variable Cauchy integral formula, we obtain

$$
\begin{aligned}
f(z) \\
&= \frac{1}{2\pi i} \int_{|\zeta_n - z_n| = r_n} \frac{f(z_1, z_2, \ldots, z_{n-1}, \zeta_n)}{(\zeta_n - z_n)} \, d\zeta_n \\
&= \frac{1}{2\pi i} \frac{1}{2\pi i} \int_{|\zeta_n - z_n| = r_n} \int_{|\zeta_{n-1} - z_{n-1}| = r_{n-1}} \frac{f(z_1, z_2, \ldots, z_{n-1}, \zeta_n)}{(\zeta_{n-1} - z_{n-1})(\zeta_n - z_n)} \, d\zeta_{n-1} d\zeta_n \\
&= \cdots \\
&= \frac{1}{(2\pi i)^n} \int_{|\zeta_n - z_n| = r_n} \cdots \int_{|\zeta_1 - z_1| = r_1} \frac{f(\zeta_1, \ldots, \zeta_n)}{(\zeta_1 - z_1) \cdots (\zeta_n - z_n)} \, d\zeta_1 \cdots d\zeta_n. \qquad \square
\end{aligned}
$$

COROLLARY 1.2.3   If f is holomorphic on $\Omega \subseteq \mathbb{C}^n$, then f is C^∞ on Ω.

Proof   Apply the Cauchy integral formula on any closed polydisc contained in Ω. Then differentiate under the integral sign.   □

COROLLARY 1.2.4   If f is holomorphic on $\Omega \subseteq \mathbb{C}^n$, then f has a convergent power series expansion about each element $w \in \Omega$.

Proof   Assume without loss of generality that $w = 0$. Apply the theorem on some $\bar{D}^n(0, r) \subseteq \Omega$. We may assume that $r = 1$. As in the corresponding one-variable proof, write each $(\zeta_j - z_j)^{-1}$ as

$$\zeta_j^{-1}(1 - z_j\zeta_j^{-1})^{-1} = \zeta_j^{-1} \sum_{\ell=0}^{\infty} (z_j\zeta_j^{-1})^\ell,$$

which converges for $|z_j\zeta_j^{-1}| < 1$.   □

Remark: The hypothesis that $f \in C^1(\Omega)$ in the definition of holomorphic function was *not* used in the proof of the Cauchy integral formula for polydiscs. However, to differentiate under the integral sign or to compute the power series in Corollary 1.2.4, Fubini's theorem must be applied. Therefore, one needs to assume, for instance, that f is bounded on compact sets. Our hypothesis that $f \in C^1$ is much stronger than is needed.

Historically, holomorphic functions on $\Omega \subseteq \mathbb{C}^n$ were defined to be functions that are holomorphic in each variable separately and bounded on compact sets; then the Cauchy integral for polydiscs implies that they are infinitely differentiable. It was Hartogs who proved, in 1906, that the hypothesis of boundedness on compacta is superfluous. His proof, using a lemma of Osgood and some subharmonic function theory, is very ingenious and a bit difficult to fathom. To this day there is not an essentially simpler proof.   □

To avoid interrupting the flow of this section, we now state Hartogs's result, but we isolate its proof in Section 2.4. This technical proof may be skipped with no resulting loss in continuity.

THEOREM 1.2.5 (Hartogs)   Let $U \subseteq \mathbb{C}^n$ be an open set and $f : U \to \mathbb{C}$. Suppose that for each $j = 1, \ldots, n$ and each fixed $z_1, \ldots, z_{j-1}, z_{j+1}, \ldots, z_n$ the function

$$\zeta \mapsto f(z_1, \ldots, z_{j-1}, \zeta, z_{j+1}, \ldots, z_n)$$

is holomorphic, in the classical one-variable sense, on the set

$$U(z_1, \ldots, z_{j-1}, z_{j+1}, \ldots, z_n) \equiv \{\zeta \in \mathbb{C} : (z_1, \ldots, z_{j-1}, \zeta, z_{j+1}, \ldots, z_n) \in U\}.$$

Then f is continuous on U.

Proof   Deferred to Section 2.4.   □

Theorem 1.2.5 certainly solves our problem, for continuous functions are bounded on compacta. We now know that all four definitions of holomorphic function posited in Section 0.2 are equivalent. The reader should verify that this is so.

Remark: The "real variable analogue" of Hartogs's theorem is that a function

$$f : \mathbb{R}^N \to \mathbb{R}$$

such that

$$t \mapsto f(x_1, \ldots, x_{j-1}, t, x_{j+1}, \ldots, x_N)$$

is in $C^\infty(\mathbb{R})$ for each j, and for each fixed set of $x_1, \ldots, x_{j-1}, x_{j+1}, \ldots, x_N$, is continuous as a function on $\mathbb{R}^N$. This assertion is false (it is even false if we replace C^∞ by real analytic—see Krantz and Parkes [2]). The reader should construct an example to see that this is the case.

The most one can say is that a function satisfying these hypotheses must be in the $(N-1)^{\text{st}}$ Baire class. In particular, it will be measurable. (See C. Kuratowski [1] for more on these matters.)   □

The reader who has come this far will now be rewarded with our first truly n-dimensional theorem using a typical $\bar{\partial}$-argument. The scheme used here is generally attributed to Ehrenpreis and to Serre.

THEOREM 1.2.6 (Hartogs's Phenomenon Again)   Let $\Omega \subseteq \mathbb{C}^n$ be a bounded domain, $n > 1$. Let K be a compact subset of Ω with the property that $\Omega \setminus K$ is connected. If f is holomorphic on $\Omega \setminus K$, then there is a holomorphic F on Ω such that $F|_{\Omega \setminus K} = f$.

Proof   Let ϕ be a function in $C_c^\infty(\Omega)$ that is identically 1 on a neighborhood of K. Define

$$\tilde{f}(z) = \begin{cases} (1 - \phi(z)) \cdot f(z) & \text{if } z \in \Omega \setminus K, \\ 0 & \text{if } z \in K. \end{cases}$$

Then $\tilde{f} \in C^\infty(\Omega)$. Finally, set

$$\psi(z) = \bar{\partial}\tilde{f}(z).$$

Then ψ satisfies the following crucial properties:

1. ψ has C^∞ coefficients;
2. $\bar{\partial}\psi = \bar{\partial}^2 \tilde{f} \equiv 0$;
3. supp ψ is a *compact* subset K_0 of Ω.

The first two of these properties are obvious, and the last follows because $\tilde{f}$ is holomorphic in $\Omega \cap$ (neighborhood of $\partial\Omega$).

By Theorem 1.1.10, there is a function $u \in C_c^\infty(\mathbb{C}^n)$, with support compact in Ω, so that $\bar{\partial}u = \psi$. In particular, the function u is identically 0 in a neighborhood U of $\partial\Omega$. We define $F = \tilde{f} - u$. Then

$$\bar{\partial}F = \bar{\partial}\tilde{f} - \bar{\partial}u = \psi - \psi = 0,$$

so F is holomorphic on Ω. Also, shrinking U if necessary,

$$F|_U = (\tilde{f} - u)\big|_U = \tilde{f}\big|_U = f|_U .$$

Therefore, F agrees with f near $\partial\Omega$. Since $\Omega \setminus K$ is connected, we may conclude by the uniqueness of analytic continuation that $F = f$ on $\Omega \setminus K$. The proof is complete. $\square$

Remark: Let us examine this proof and see why it works. The idea is that $\tilde{f}$ is a C^∞ extension of f to Ω. It could not, in general, be holomorphic, because it is identically 0 on K (which could have interior). What we hope is that we can subtract a function u from $\tilde{f}$ to make it holomorphic. That is, we want $\bar{\partial}(\tilde{f} - u) = 0$ or $\bar{\partial}u = \psi$.

The equation $\bar{\partial}u = \psi$ has many solutions because $\bar{\partial}$ is a linear operator with a large null space (all holomorphic functions). One solution is the function $\tilde{f}$ itself; but this would be a poor choice, since then $\tilde{f} - u$ would be identically zero. This is why it is so important that we be able to find a solution u with support compact in Ω. For since $\tilde{f}$ has support near $\partial\Omega$, then $\tilde{f} - u$ is not the zero function and agrees with the original function f near $\partial\Omega$. $\square$

Exercises for the Reader

1. Construct a counterexample to Theorem 1.2.6 for the case $n = 1$. At what crucial step does the proof break down?

2. Construct a counterexample to Theorem 1.2.6 for the case when $\Omega \setminus K$ is disconnected.

3. Let f be holomorphic on $\Omega \subseteq \mathbb{C}^n, n > 1$. If $\gamma \in \mathbb{C}$ is a constant and $S \equiv f^{-1}(\gamma) \neq \emptyset$, then prove that the level set S is not contained in any compact subset of Ω.

4. Use Exercise 3 to formulate and prove versions of the maximum and minimum principles for holomorphic functions of several variables.

5. Suppose that f is holomorphic on a bounded domain Ω and continuous on $\bar{\Omega}$. Use Exercise 3 or 4 to see that if γ is a constant and $f(z) = \gamma$ for all $z \in \partial\Omega$, then $f \equiv \gamma$. (This is really a fact about harmonic functions—see Section 1.3. We have simply stumbled upon a convenient proof for holomorphic functions.)

6. Notice that the exercise at the end of Section 1.1 proves that the normal (that is, uniform on compact sets) limit of holomorphic functions is holomorphic. Give another proof using Theorem 1.2.2, Corollary 1.2.4, or Theorem 1.2.5.

1.3 Basic Properties of Harmonic Functions

One of the themes of this book is that one should not look at complex analysis, either of one or of several complex variables, as a language unto itself. Rather, it is a meeting ground of several different disciplines. When one encounters a question or topic in this book, one should say "this works because holomorphic functions are harmonic" or "this works because holomorphic functions are real analytic," and so forth. Only in this fashion can one keep the hierarchy of ideas straight.

Many of the easy properties of holomorphic functions of several complex variables, especially the ones that they share with holomorphic functions of one complex variable, follow from the fact that they are harmonic. We explore harmonic functions in space in the present section.

Let $\Omega \subseteq \mathbb{R}^N$ be a domain and $f : \Omega \to \mathbb{C}$ a C^2 function. We say that f is *harmonic* if it satisfies the differential equation

$$\Delta f \equiv \sum_{j=1}^{N} \left(\frac{\partial^2}{\partial x_j^2} \right) f = 0 \qquad \text{on } \Omega.$$

(For the record, the Laplacian of differential geometry—that is, the Laplace-Beltrami operator for the Euclidean metric—is the negative of our Laplacian. In the study of eigenvalue asymptotics it is more convenient to work with the positive operator $-\Delta$.)

If $\Omega \subseteq \mathbb{C}$ and f is holomorphic on Ω, then f is harmonic because $0 = \partial\bar\partial f = (1/4)(\Delta f)dz \wedge d\bar z$. It follows that a holomorphic function of several variables is harmonic because it is holomorphic in each variable separately.

Familiarity with the basic properties of harmonic functions will be helpful for motivation when reading what follows; but for completeness we provide a self-contained treatment. As predicted at the beginning of this chapter, we begin by deriving Green's theorem from Stokes's theorem.

THEOREM 1.3.1 Let $\Omega \subseteq \mathbb{R}^N$ be a domain with C^2 boundary. Let $d\sigma$ denote area $((2N-1)$-dimensional Hausdorff) measure on $\partial\Omega$—see Appendix II. Let ν be the unit outward normal vector field on $\partial\Omega$. Then, for any functions

$u, v \in C^2(\bar{\Omega})$, we have

$$\int_{\partial\Omega} u(\nu v) - v(\nu u)\, d\sigma = \int_{\Omega} (u\Delta v - v\Delta u)\, dV.$$

Proof   Write the volume form dV as $dV = dx_1 \wedge \cdots \wedge dx_N$. By definition, $d\sigma = -\nu \lrcorner dV$, where $\lrcorner$ represents interior multiplication of the form dV by the vector field ν. (See H. Federer [1] for the concept of interior multiplication; this equality for $d\sigma$ is in fact the key to Green's theorem.) Notice that grad $u - (\nu u)\nu$ is orthogonal to ν so that $(\operatorname{grad} u - (\nu u)\nu)\lrcorner dV$ is zero when restricted to $\partial\Omega$. Therefore,

$$\int_{\partial\Omega} (v \operatorname{grad} u)\lrcorner dV = \int_{\partial\Omega} (v(\nu u)\nu)\lrcorner dV$$

or

$$\int_{\partial\Omega} (v \operatorname{grad} u)\lrcorner dV = -\int_{\partial\Omega} v(\nu u)\, d\sigma. \qquad (1.3.1.1)$$

Likewise,

$$\int_{\partial\Omega} (u \operatorname{grad} v)\lrcorner dV = -\int_{\partial\Omega} u(\nu v)\, d\sigma. \qquad (1.3.1.2)$$

We combine (1.3.1.1) and (1.3.1.2) and apply Stokes's theorem to obtain

$$\begin{aligned}
\int_{\partial\Omega} u(\nu v) - v(\nu u)\, d\sigma &= \int_{\partial\Omega} -(u \operatorname{grad} v - v \operatorname{grad} u)\lrcorner dV \\
&= \int_{\Omega} -d\left[(u \operatorname{grad} v - v \operatorname{grad} u)\lrcorner dV\right] \\
&= \int_{\Omega} \left[-du \wedge (\operatorname{grad} v \lrcorner dV) + u\Delta v\, dV\right] \\
&\quad - \int_{\Omega} \left[-dv \wedge (\operatorname{grad} u \lrcorner dV) + v\Delta u\, dV\right]. \quad (1.3.1.3)
\end{aligned}$$

Here we have used the facts that

$$-\operatorname{div}(\operatorname{grad} v)\, dV = \Delta v \cdot dV,$$

$$-\operatorname{div}(\operatorname{grad} u)\, dV = \Delta u \cdot dV,$$

and that, for any vector field X,

$$\operatorname{div} X\, dV \equiv \langle X, dV \rangle = d(X \lrcorner dV)$$

(this is essentially the definition of divergence and of interior multiplication—see Federer [1, Section 4.1.6]).

Finally,

$$du \wedge (\operatorname{grad} v \rfloor dV) = < \operatorname{grad} u, \operatorname{grad} v > dV = dv \wedge (\operatorname{grad} u \rfloor dV).$$

Therefore, (1.3.1.3) becomes

$$\int_\Omega (u \Delta v - v \Delta u) \, dV,$$

and Green's theorem is proved.   □

Although Stokes's theorem and Green's theorem are two different aspects of the same phenomenon—the fundamental theorem of calculus in several variables—they have a different flavor. We particularly wish to note that in Stokes's theorem the boundary is *oriented*, whereas in Green's theorem it is not.

Throughout this section the symbol Ω denotes a given C^2 domain, ν denotes the unit outward normal vector field to $\partial\Omega$, and $d\sigma$ is Euclidean area measure on $\partial\Omega$ (see Appendix II). The symbol ω_{N-1} denotes the σ-measure of the $(N-1)$-dimensional unit sphere in $\mathbb{R}^N$.

Many of the ensuing arguments come *grosso modo* from E. M. Stein and G. Weiss [1], which is a good source for further information on harmonic functions. See also M. Tsuji [1].

PROPOSITION 1.3.2   The fundamental solution for the Laplacian on $\mathbb{R}^N, N \geq 2$, is given by

$$\Gamma_N(x) = \Gamma(x) = \begin{cases} (2\pi)^{-1} \log |x| & \text{if} \quad N = 2, \\ (2-N)^{-1} \omega_{N-1}^{-1} |x|^{-N+2} & \text{if} \quad N > 2. \end{cases}$$

More precisely, for any $\phi \in C_c^\infty(\mathbb{R}^N)$, it holds that

$$\int (\Delta\phi)\Gamma \, dV = \phi(0).$$

Remark: This result is so important that it merits discussion. The conclusion of the proposition says that $\Delta\Gamma = \delta_0$, the Dirac delta mass, in the weak (or distribution) sense. In particular, $\Delta\Gamma(x) = 0$ when $x \neq 0$. This assertion may also be verified directly using calculus.

It is a straightforward exercise for the reader to derive from the proposition the statement that if $\phi \in C_c^\infty(\mathbb{R}^N)$, then $u = \Gamma * \phi$ satisfies $\Delta u = \phi$. Thus Γ is the kernel for a solution operator of the Laplace operator.   □

Proof of the Proposition: We do only the case $N > 2$. The case $N = 2$ has the same substance but a different form, so we leave it for the reader (see also O. Kellogg [1])

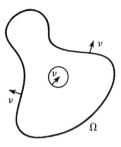

FIGURE 1.2

Fix $\phi \in C_c^\infty(\mathbb{R}^N)$ and a large ball $B = B(0, R) \subseteq \mathbb{R}^N$ such that $\operatorname{supp} \phi \subseteq B(0, R)$. Let $0 < \epsilon < R$ and apply Green's theorem on the domain $\Omega_\epsilon \equiv B(0, R)\backslash \bar{B}(0, \epsilon)$ with $u = \Gamma$ and $v = \phi$. Then, since $\Delta \Gamma(x) = 0$ for $x \neq 0$,

$$
\begin{aligned}
\int_{\Omega_\epsilon} (\Delta \phi)\Gamma \, dV &= \int_{\Omega_\epsilon} (\Delta \phi)\Gamma \, dV - \int_{\Omega_\epsilon} \phi \cdot (\Delta \Gamma) \, dV \\
&= \int_{\partial B(0,R)} \{(\nu \phi) \cdot \Gamma - \phi \cdot \nu \Gamma\} \, d\sigma \\
&\quad + \int_{\partial B(0,\epsilon)} \{(\nu \phi) \cdot \Gamma - \phi \cdot \nu \Gamma\} \, d\sigma \\
&= 0 - \phi(0) \int_{\partial B(0,\epsilon)} \nu \Gamma \, d\sigma + \left[\int_{\partial B(0,\epsilon)} (\nu \phi) \cdot \Gamma \right. \\
&\quad \left. + \int_{\partial B(0,\epsilon)} \{\phi(0) - \phi(x)\} \cdot \nu \Gamma \, d\sigma \right] \\
&= 0 - \phi(0) \int_{\partial B(0,\epsilon)} \nu \Gamma \, d\sigma + o(1) \quad \text{as } \epsilon \to 0^+.
\end{aligned}
$$

We have used here the facts that $\Gamma(x) = c \cdot |x|^{-N+2}$ and $\sigma(\partial B(0, \epsilon)) = c' \cdot \epsilon^{N-1}$. Letting $\epsilon \to 0^+$ and noting that the outward normal vector field ν to $\partial \Omega_\epsilon$ at points of $\partial B(0, \epsilon)$ is oriented *oppositely* to ν when thought of as the outward normal vector field to $\partial B(0, \epsilon)$ itself (see Figure 1.2) gives

$$
\begin{aligned}
\int (\Delta \phi)\Gamma \, dx &= \lim_{\epsilon \to 0} \phi(0)(\omega_{N-1}^{-1}) \int_{\partial B(0,\epsilon)} |x|^{-N+1} \, d\sigma(x) \\
&= \phi(0)
\end{aligned}
$$

as desired. □

LEMMA 1.3.3   If u is harmonic on $\Omega \equiv B(x_0, r) \subseteq \mathbb{R}^N$ and is C^2 on $\bar{\Omega}$, then

$$\int_{\partial B(x_0, r)} \nu u \, d\sigma = 0.$$

Proof   Apply Green's theorem with the functions u and $v \equiv 1$ on the domain Ω.   □

THEOREM 1.3.4   Let u be harmonic on a domain $\Omega \subseteq \mathbb{R}^N$. Suppose that $\bar{B}(P, r) \subseteq \Omega$. Then

$$\frac{1}{\sigma(\partial B(P, r))} \int_{\partial B(P, r)} u(x) \, d\sigma(x) = u(P).$$

Proof   We give the details for $N > 2$ only. Let $0 < \epsilon < r$ and apply Green's theorem on the region $\Omega_\epsilon \equiv B(P, r) \setminus \bar{B}(P, \epsilon)$ with u the given harmonic function and $v = \Gamma(x - P)$. The result is

$$
\begin{aligned}
0 &= \int_{\Omega_\epsilon} u \Delta v - v \Delta u \, dV \\
&= \left\{ \int_{\partial B(P, r)} u \nu \Gamma(x - P) \, d\sigma + \int_{\partial B(P, \epsilon)} u \nu \Gamma(x - P) \, d\sigma \right\} \\
&\quad - \left\{ \int_{\partial B(P, r)} (\nu u) \Gamma(x - P) \, d\sigma + \int_{\partial B(P, \epsilon)} (\nu u) \Gamma(x - P) \, d\sigma \right\}.
\end{aligned}
$$

Now the second expression in braces is equal to zero by Lemma 1.3.3, since $\Gamma(x - P)$ is constant on $\partial B(P, r)$ and on $\partial B(P, \epsilon)$. Thus we have

$$
\begin{aligned}
0 &= (-N + 2)^{-1} (\omega_{N-1})^{-1} \left[(-N + 2) r^{-N+1} \int_{\partial B(P, r)} u \, d\sigma \right. \\
&\quad \left. - (-N + 2) \epsilon^{-N+1} \int_{\partial B(P, \epsilon)} u \, d\sigma \right].
\end{aligned}
$$

The second expression inside the brackets tends, as $\epsilon \to 0^+$, to

$$\omega_{N-1}(N - 2) u(P).$$

We conclude that

$$(\omega_{N-1})^{-1} r^{-N+1} \int_{\partial B(P, r)} u(x) \, d\sigma(x) = u(P). \qquad □$$

Exercise for the Reader
Supply the details for the proofs of Proposition 1.3.2 and Theorem 1.3.4 for the case $N = 2$.

COROLLARY 1.3.5 (Maximum Principle)   Let u be a nonconstant, real-valued harmonic function on a domain $\Omega \subseteq \mathbb{R}^N$. Then for every $x \in \Omega$, it holds that

$$u(x) < \sup_{y \in \Omega} u(y).$$

Remark: There are several useful ways to restate the maximum principle. One of these is the contrapositive statement: If u is real-valued harmonic on Ω and there is an $x \in \Omega$ such that $u(x) = \sup_{y \in \Omega} u(y)$, then u is constant.
   A slightly weaker statement is this: Assume that Ω is bounded. If u is continuous on $\bar{\Omega}$ and harmonic on Ω, then $\max_{\bar{\Omega}} u = \max_{\partial \Omega} u$. This statement has a strong intuitive appeal.   □

Proof of the Corollary: Let $A = \sup_{y \in \Omega} u(y)$. We may as well suppose that A is finite or else there is nothing to prove. Let $S = \{w \in \Omega : u(w) = A\}$. We will show that S is either empty or all of Ω.
   If S is not empty, let $x \in S$. There exists a number $r_0 > 0$ such that $\bar{B}(x, r_0) \subseteq \Omega$. Then for every $0 < r < r_0$, we have from the mean value property that

$$\frac{1}{\sigma(\partial B(x,r))} \int_{\partial B(x,r)} u(y)\, d\sigma(y) = u(x) = A.$$

Since $u \leq A$, it follows that u is identically A on $B(x, r_0)$. Thus S is open. But S is clearly closed since u is continuous. Since S is not empty and (by assumption) Ω is connected, we conclude that $S = \Omega$; hence $u \equiv A$. This completes the proof.   □

COROLLARY 1.3.6   If u_1, u_2 are harmonic on $\Omega \subset\subset \mathbb{R}^N$ and continuous on $\bar{\Omega}$, and if $u_1|_{\partial \Omega} = u_2|_{\partial \Omega}$, then $u_1 \equiv u_2$ on $\bar{\Omega}$.

Remark: The corollary holds in particular for a holomorphic function on a domain Ω in $\mathbb{C}^n$ that is continuous on the closure. Compare with the exercises following Theorem 1.2.6.   □

Exercises for the Reader
1. Prove that if u is harmonic on $\Omega \subseteq \mathbb{R}^N$, $P \in \Omega$, and $\bar{B}(P,r) \subseteq \Omega$, then

$$\frac{1}{\mathrm{vol}(B(P,r))} \int_{B(P,r)} u(x)\, dV(x) = u(P).$$

2. Prove that following trivial variant of Theorem 1.3.4. Under the same hypotheses,

$$\frac{1}{\omega_{N-1}} \int_{\partial B(0,1)} u(P + ry)\, d\sigma(y) = u(P).$$

LEMMA 1.3.7   Let $\Omega \subseteq \mathbb{R}^N$ be a domain and let $u \in C^2(\Omega)$. For $x \in \Omega$, define the expression

$$\mathcal{U}(x) = \lim_{r \to 0^+} \frac{d^2}{dr^2} \frac{1}{\omega_{N-1}} \int_{\partial B(0,1)} u(x + ry)\, d\sigma(y)$$

(the limit exists because $u \in C^2$). Then

$$\mathcal{U}(x) = \frac{1}{N}(\Delta u)(x).$$

Proof   By the chain rule we have

$$\mathcal{U}(x) = \frac{1}{\omega_{N-1}} \sum_{j,k=1}^{N} \int_{\partial B(0,1)} \frac{\partial^2 u}{\partial x_j \partial x_k}(x) y_j y_k\, d\sigma(y).$$

When $j \neq k$, then oddness implies that $\int_{\partial B(0,1)} y_j y_k\, d\sigma(y) = 0$. Therefore,

$$\mathcal{U}(x) = \frac{1}{\omega_{N-1}} \sum_{j=1}^{N} \int_{\partial B(0,1)} y_j^2\, d\sigma(y) \cdot \frac{\partial^2 u(x)}{\partial x_j^2}. \tag{1.3.7.1}$$

But observe that

$$\sum_{j=1}^{N} \int_{\partial B(0,1)} y_j^2\, d\sigma(y) = \int_{\partial B(0,1)} \sum_{j=1}^{N} y_j^2\, d\sigma(y) = \omega_{N-1}$$

so that

$$\int_{\partial B(0,1)} y_j^2\, d\sigma(y) = \frac{\omega_{N-1}}{N}.$$

Hence line (1.3.7.1) becomes $\Delta u(x)/N$, as desired.   □

THEOREM 1.3.8   If u is C^2 on a domain $\Omega \subseteq \mathbb{R}^N$ and if for every $x \in \Omega$ there is a sequence $r_j \to 0$ (possibly depending on x) such that

$$\frac{1}{\omega_{N-1}} \int_{\partial B(0,1)} u(x + r_j y)\, d\sigma(y) = u(x),$$

then u is harmonic on Ω.

Proof   If we use the notation from Lemma 1.3.7, then the hypothesis of the theorem implies that $\mathcal{U}(x) \equiv 0$ on Ω or $\Delta u \equiv 0$. Hence u is harmonic. $\square$

Exercises for the Reader

1. If $\phi \in C_c^\infty(\mathbb{R}^N)$ is radial (that is, $\phi(x) = \phi(x')$ whenever $|x| = |x'|$) and $\int \phi \, dx = 1$, let $\phi_\epsilon(x) \equiv \epsilon^{-N} \phi(x/\epsilon)$. Prove that $\int \phi_\epsilon \, dx = 1$.

2. Use Exercise 1 to show that if $\Omega \subseteq \mathbb{R}^N$ is a domain and $f \in C(\Omega)$ then, defining $f_\epsilon(x) = f * \phi_\epsilon(x) = \int f(x - t)\phi_\epsilon(t) \, dt$ for ϵ sufficiently small, it holds that $f_\epsilon \to f$ uniformly on compact subsets of Ω (see Y. Katznelson [1] or E. M. Stein and G. Weiss [1] for this type of argument).

3. Let $f : \Omega \to \mathbb{C}$ be continuous on Ω and satisfy the mean value property on every sphere contained in Ω. Conclude that $f_\epsilon(x) = f(x)$ for x in a compact subset of Ω and ϵ small. Prove that f_ϵ, hence f, is harmonic.

4. If $\{u_n\}$ is a sequence of harmonic functions on $\Omega \subseteq \mathbb{R}^N$ that converges *normally* on Ω (i.e., uniformly on compacta), then the limit function u is harmonic on Ω.

5. Prove *Weyl's lemma:* If u is continuous on $\Omega \subseteq \mathbb{R}^N$ and if u is a *weak solution* of Laplace's equation in the sense that

$$\int u \Delta \phi \, dV = 0$$

for all $\phi \in C_c^\infty(\Omega)$, then u is harmonic on Ω.

6. Conclude from Exercise 5 that a continuous function $u : \Omega \to \mathbb{C}, \Omega \subseteq \mathbb{C}^N$, which is a weak solution of the Cauchy-Riemann equations, is in fact a bona fide holomorphic function.

7. The reader who knows some distribution theory can generalize Exercises 5 and 6 to the case where u is a *distribution solution* of either Δ or $\bar{\partial}$.

DEFINITION 1.3.9 (The Green's Function)   Let $\Omega \subseteq \mathbb{R}^N$ be a bounded domain with C^2 boundary. A function $G : (\Omega \times \bar{\Omega}) \setminus \{\text{diagonal}\} \to \mathbb{R}$ is the *Green's function* (for the Laplacian) on Ω if:

1. G is C^2 on $\Omega \times \Omega \setminus \{\text{diagonal}\}$ and, for any small $\epsilon > 0$, is $C^{2-\epsilon}$ up to $(\Omega \times \bar{\Omega}) \setminus \{\text{diagonal}\}$;

2. $\Delta_y G(x, y) = 0$ for $x \neq y, y \in \Omega$;

3. for each fixed $x \in \Omega$ the function $G(x, y) + \Gamma_N(y - x)$ is harmonic as a function of $y \in \Omega$ (even at the point x);

4. $G(x, y)|_{y \in \partial\Omega} = 0$ for each fixed $x \in \Omega$.

Remark: The reader should check that properties 1–4 uniquely determine the function G. Notice that we have not yet asserted that Green's functions exist; this assertion is the content of Proposition 1.3.11.

We also have not said much about the dependence of G on the x variable. We shall learn in the exercises at the end of the chapter that $G(x, y)$ is symmetric in the variables x and y.   □

THEOREM 1.3.10 (Solution of the Dirichlet Problem)   Let $\Omega \subseteq \mathbb{R}^N$ be a bounded domain with C^1 boundary. Let f be a continuous function on $\partial\Omega$. Then there exists a (unique) $F \in C(\bar{\Omega})$ such that $F|_{\partial\Omega} = f$ and F is harmonic on Ω.

Remark: In case f is real, the function F is constructed as a supremum of subharmonic functions (see Section 2.1) whose boundary values do not exceed f (this construction is called the *Perron method*). In particular, F is constructed in the abstract, not given by an explicit formula.

We shall not give a proof of Theorem 1.3.10, because the ideas involved would lead us far afield. However these ideas are important and have played a critical role in the development of analysis. We refer the reader to L. Ahlfors [1] for details of the two-dimensional case and to P. Garabedian [1] and L. Bers, F. John, and M. Schechter [1] for the higher-dimensional case.   □

Given Theorem 1.3.10, we can now see that Green's functions always exist:

PROPOSITION 1.3.11   Let $\Omega \subseteq \mathbb{R}^N$ be a bounded domain with C^2 boundary. Then Ω has a Green's function.

Proof   Fix $x \in \Omega$. Let $f_x(y) = -\Gamma(y - x)|_{y \in \partial\Omega}$. Let $F_x(y)$ be the unique solution to the Dirichlet problem with boundary data $f_x(y)$. Set $G(x, y) = -\Gamma(y - x) - F_x(y)$. Then Properties (2)–(4) of the Green's function follow immediately. The smoothness assertion (Property 1) follows from regularity properties of elliptic boundary value problems, and we shall not provide the details (see C. B. Morrey [1] or L. Bers, F. John, and M. Schechter [1] or S. Krantz [19]).   □

The reader should not be surprised to now learn that the Green's function, together with Green's formula, gives us a new integral formula.

THEOREM 1.3.12 (Poisson Integral Formula)   Let $\Omega \subseteq \mathbb{R}^N$ be a bounded domain with C^2 boundary. Let ν represent the unit outward normal vector field on $\partial\Omega$. Let the Poisson kernel on Ω be the function

$$P(x, y) = -\nu_y G(x, y).$$

If $u \in C(\bar{\Omega})$ is harmonic on Ω, then

$$u(x) = \int_{\partial\Omega} P(x, y) u(y) \, d\sigma(y) \quad \text{for all } x \in \Omega.$$

Proof   First assume that $u \in C^2(\bar{\Omega})$. Fix $x \in \Omega$ and a positive number $\epsilon <$ dist$(x, \partial\Omega)$. We apply Green's formula on $\Omega_\epsilon \equiv \Omega \setminus \bar{B}(x, \epsilon)$ with $u(y)$ the given harmonic function and $v(y) = G(x, y)$. Then

$$\int_{\partial\Omega_\epsilon} (\nu_y G(x, y)) u(y) \, d\sigma(y) - \int_{\partial\Omega_\epsilon} G(x, y)(\nu_y u(y)) \, d\sigma(y)$$

$$= \int_{\Omega_\epsilon} (\Delta_y G(x, y)) u(y) \, dV(y) - \int_{\Omega_\epsilon} G(x, y)(\Delta_y u(y)) \, dV(y).$$

Since $G(x, \cdot)$ and $u(\cdot)$ are harmonic, the right side vanishes. Using the definition of $G(x, y)$ from the proof of Proposition 1.3.11, we may rewrite the last equation as

$$\int_{\partial\Omega_\epsilon} \nu_y G(x, y) u(y) \, d\sigma(y)$$

$$= \int_{\partial\Omega} G(x, y)\nu_y u(y) \, d\sigma(y) - \int_{\partial B(x,\epsilon)} \Gamma(x - y)\nu_y u(y) \, d\sigma(y)$$

$$- \int_{\partial B(x,\epsilon)} F_x(y)\nu_y u(y) \, d\sigma(y).$$

The first term on the right vanishes, since $G(x, \cdot)|_{\partial\Omega} = 0$. The second vanishes, because $\Gamma(x - \cdot)|_{\partial B(x,\epsilon)}$ is constant and because of Lemma 1.3.3. The last term vanishes as $\epsilon \to 0^+$, since the integrand is bounded.

Thus we have

$$\int_{\partial\Omega_\epsilon} (\nu_y G(x, y)) u(y) \, d\sigma(y) = o(1)$$

or

$$\int_{\partial\Omega} (\nu_y G(x, y)) u(y) \, d\sigma(y) = - \int_{\partial B(x,\epsilon)} (\nu_y G(x, y)) u(y) \, d\sigma(y) + o(1).$$

Recall that the unit outward normal to $\partial\Omega_\epsilon$ at a point $y \in \partial B(x, \epsilon)$ is the negative of the unit outward normal to $\partial B(x, \epsilon)$ at y. With $P(x, y) = -\nu_y G(x, y)$, we may now write

$$\int_{\partial\Omega} P(x, y) u(y) \, d\sigma(y) = \int_{\partial B(x,\epsilon)} \nu_y G(x, y) u(y) \, d\sigma(y) + o(1)$$

$$= \int_{\partial B(x,\epsilon)} \nu_y \Gamma(x - y) u(y) \, d\sigma(y)$$

$$+ \int_{\partial B(x,\epsilon)} \nu_y F_x(y) u(y) \, d\sigma(y) + o(1).$$

The second term on the right side of the last equation is $o(1)$ as $\epsilon \to 0^+$ because the integrand is bounded. A now-familiar computation (see the proof of Theorem 1.1.4) then shows that the first term on the right tends to $u(x)$. Letting $\epsilon \to 0^+$, we conclude that

$$u(x) = \int_{\partial\Omega} P(x,y)u(y)\,d\sigma(y).$$

This is the Poisson integral formula when $u \in C^2(\bar{\Omega})$.

To eliminate the hypothesis that $u \in C^2(\bar{\Omega})$, we exhaust Ω by relatively compact, smoothly bounded subdomains Ω_j and apply the preceding result to u on $\bar{\Omega}_j$. A limiting argument then completes the proof. Details are left to the interested reader. $\qquad\square$

COROLLARY 1.3.13   For each fixed $y \in \partial\Omega, P(x,y)$ is harmonic in x.

Proof   Of course, this follows from the previously noted fact that $G(x,y) = G(y,x)$.

An alternate proof may be obtained by noting that $\int_{\partial\Omega} P(x,y)\phi(y)\,d\sigma(y)$ is a harmonic function of x for every $\phi \in C(\partial\Omega)$. Now choose a sequence ϕ_j such that $\phi_j\,d\sigma$ converges to δ_y in the weak$-*$ topology. [Use 1.3.10, 1.3.12, and the Ascoli-Arzela theorem.] $\qquad\square$

Exercise for the Reader
Why does a proof similar to the second proof of the corollary not suffice to imply that the Bochner-Martinelli kernel is holomorphic?

The Poisson kernel can almost never be computed explicitly. However we can compute it for the ball; later on we can exploit the explicit formula on the ball to obtain sharp estimates for the kernel on a general domain. These estimates suffice for most applications of the kernel. We begin with the following.

DEFINITION 1.3.14 (Kelvin Inversion)   Let the dimension N be at least two. The *Kelvin inversion* on the domain $\mathbb{R}^N \setminus \{0\}$ is the map

$$K : \mathbb{R}^N \setminus \{0\} \to \mathbb{R}^N \setminus \{0\}$$

given by $K(x) = x/|x|^2$. Notice that $K = K^{-1}$.

PROPOSITION 1.3.15   Let the dimension N be at least two. If Ω is an open subdomain of $\mathbb{R}^N, 0 \notin \Omega$, and if u is harmonic on Ω, then

$$x \mapsto |x|^{-N+2} \cdot (u \circ K(x))$$

is harmonic on $K(\Omega)$. Notice that in dimension two the prefactor $|x|^{-N+2}$ becomes 1.

Proof   Exercise for the reader (just compute). $\qquad\square$

THEOREM 1.3.16   The Poisson kernel for the ball $\Omega = B$ is given by

$$P(x,y) = \frac{1}{\omega_{N-1}} \frac{1 - |x|^2}{|x - y|^N}.$$

Proof   We do the details only for $N > 2$. If we can construct the Green's function for the ball, then the rest is straightforward. A glance at the proof of the existence of the Green's function shows that the main obstacle is to construct F_x for $x \in B$. Fix $x \in B$. Set

$$F_x(y) = -|y|^{-N+2} \Gamma(x - K(y)) = -\omega_{N-1}^{-1} (2 - N)^{-1} |y|^{-N+2} \left| x - \frac{y}{|y|^2} \right|^{-N+2}.$$

Now $\Gamma(x - y)$ is plainly harmonic on $\mathbb{R}^N \setminus \{x\}$; hence, $F_x(y)$ is harmonic on a neighborhood of $\bar{B} \setminus \{0\}$. It may be checked directly that, in fact, F_x is smooth at 0, so it follows from the continuity of the derivative that F_x is harmonic on a neighborhood of $\bar{B}$ (alternatively, use the identity $|y|^{-N+2} \Gamma(x - K(y)) = |x|^{-N+2} \Gamma(y - K(x))$). Since the Kelvin transform K fixes ∂B, it follows that

$$F_x(\cdot)|_{\partial B} = -\Gamma(x - \cdot)|_{\partial B}.$$

Thus F_x is the function we seek, and the Green's function for B must be

$$-\Gamma(x-y) - F_x(y) = -\omega_{N-1}^{-1}(2-N)^{-1} \left\{ |x - y|^{-N+2} - |y|^{-N+2} \left| x - \frac{y}{|y|^2} \right|^{-N+2} \right\}.$$

Finally, by Theorem 1.3.12, the Poisson kernel for the ball is

$$
\begin{aligned}
P(x,y) &= (2-N)^{-1} \omega_{N-1}^{-1} \\
&\quad \times \sum_{j=1}^{N} y_j \frac{\partial}{\partial y_j} \left\{ |x - y|^{-N+2} - |y|^{-N+2} \left| x - \frac{y}{|y|^2} \right|^{-N+2} \right\} \\
&= (2-N)^{-1} \omega_{N-1}^{-1} \sum_{j=1}^{N} y_j \left\{ \left(\frac{-N+2}{2} \right) |x - y|^{-N} \cdot 2(y_j - x_j) \right. \\
&\quad - \left(\frac{-N+2}{2} \right) |y|^{-N} 2 y_j \left| x - \frac{y}{|y|^2} \right|^{-N+2} \\
&\quad - \left(\frac{-N+2}{2} \right) |y|^{-N+2} \left| x - \frac{y}{|y|^2} \right|^{-N} \\
&\quad \left. \times \sum_{\ell=1}^{N} 2 \left(\frac{y_\ell}{|y|^2} - x_\ell \right) \left(\frac{-2 y_j y_\ell}{|y|^4} + \frac{\delta_{j\ell}}{|y|^2} \right) \right\}.
\end{aligned}
$$

When $y \in \partial B$, this yields

$$
\begin{aligned}
P(x,y) &= \omega_{N-1}^{-1} \cdot |x-y|^{-N} \Big\{ |y|^2 - x \cdot y - |y|^2 |x-y|^2 \\
&\quad - \sum_{\ell=1}^{N} \sum_{j=1}^{N} y_j (y_\ell - x_\ell)(-2 y_j y_\ell + \delta_{j\ell}) \Big\} \\
&= \omega_{N-1}^{-1} \cdot |x-y|^{-N} \{ 1 - x \cdot y - |x-y|^2 + 1 - x \cdot y \} \\
&= \omega_{N-1}^{-1} \cdot \frac{1 - |x|^2}{|x-y|^N}.
\end{aligned}
$$

This is the desired result. $\qquad\square$

Exercise for the Reader
Prove Theorem 1.3.16 for the case $N = 2$.

PROPOSITION 1.3.17 The Poisson kernel for $B \subseteq \mathbb{R}^N$ has the following properties:

1. $P(x,y) \geq 0$.

2. $\int_{\partial B} P(x,y) \, d\sigma(y) = 1$, all $x \in B$.

3. For any $\delta > 0$, any fixed $\zeta_0 \in \partial B$,

$$
\lim_{B \ni x \to \zeta_0} \int_{|\zeta_0 - y| > \delta} P(x,y) \, d\sigma(y) = 0.
$$

Proof Inequality 1 is obvious. Also, by the reproducing property of P and the fact that $P \geq 0$, we have

$$
1 = \int_{\partial B} 1 \cdot P(x,y) \, d\sigma(y) = \| P(x, \cdot) \|_{L^1(\partial B, \, d\sigma)}.
$$

As for Property 3, choose $0 < \epsilon < \delta/4$. Then $|\zeta_0 - y| > \delta$ and $|x - \zeta_0| < \min(\omega_{N-1} \epsilon^{N+1}, \delta/2)$ imply that

$$
P(x,y) = \frac{1}{\omega_{N-1}} \frac{1 - |x|^2}{|x-y|^N} \leq \frac{1}{\omega_{N-1}} \cdot \frac{2 \omega_{N-1} \epsilon^{N+1}}{(\delta/2)^N} \leq \epsilon
$$

uniformly in $|\zeta_0 - y| > \delta$. The result follows. $\qquad\square$

THEOREM 1.3.18 (Poisson Integral for the Ball)   Let $B \subseteq \mathbb{R}^N$ be the unit ball and $f \in C(\partial B)$. Then the function

$$F(x) = \begin{cases} \int_{\partial B} P(x, y) f(y) \, d\sigma(y) & \text{if } \quad x \in B \\ f(x) & \text{if } = \quad x \in \partial B \end{cases}$$

is harmonic on B and continuous on $\bar{B}$.

Proof   We already know that the Poisson kernel is harmonic in the x variable by Corollary 1.3.13 (or, with the explicit formula on the ball, one can check this directly). Therefore, by differentiating under the integral sign, F is harmonic on B. Since F is obviously continuous (indeed smooth) inside B and also is continuous on ∂B, it remains to check that if $\zeta \in \partial B$, then

$$\lim_{B \ni x \to \zeta} Pf(x) = f(\zeta).$$

Let $\epsilon > 0$. Since f is uniformly continuous on ∂B, we may find a $\delta > 0$ such that if $\xi, \zeta \in \partial B$ and $|\xi - \zeta| \leq \delta$, then $|f(\zeta) - f(\xi)| < \epsilon$. With this δ fixed, we write

$$\begin{aligned} Pf(x) - f(\zeta) &= \int_{\partial B} P(x, \xi) f(\xi) \, d\sigma(\xi) - f(\zeta) \\ &= \int_{\partial B} P(x, \xi)[f(\xi) - f(\zeta)] \, d\sigma(\xi). \end{aligned}$$

Here we have used part 2 of Proposition 1.3.17.
    But this last equals

$$\int_{\partial B \cap \{|\zeta - \xi| \leq \delta\}} P(x, \xi)[f(\xi) - f(\zeta)] \, d\sigma(\xi)$$
$$+ \int_{\partial B \cap \{|\zeta - \xi| > \delta\}} P(x, \xi)[f(\xi) - f(\zeta)] \, d\sigma(\xi) \equiv I + II.$$

Now $|I| \leq \epsilon \int_{\partial B} P(x, \xi) \, d\sigma(\xi) = \epsilon$ by the choice of δ. On the other hand,

$$|II| \leq 2 \sup_{\partial B} |f| \int_{\partial B \cap \{|\zeta - \xi| > \delta\}} P(x, \xi) \, d\sigma(\xi) \to 0$$

as $x \to \zeta$ by part 3 of Proposition 1.3.17. This is what we wished to prove. $\square$

Exercises for the Reader

1. Give another proof of Theorem 1.3.18, using Theorems 1.3.10 and 1.3.12, that involves no computation. This proof will be valid for any domain Ω with C^2 boundary.

2. Examine the proof of Theorem 1.3.18 to see that $\lim_{r \to 1^-} Pf(r\zeta)$ tends to $f(\zeta)$ *uniformly* in $\zeta \in \partial B$ when $f \in C(\partial B)$.

Remark: Let $\Omega \subseteq \mathbb{R}^N$ be any domain with C^2 boundary. It follows from the maximum principle that $G(x, y) \geq 0$. Hence, by the Hopf lemma (Exercise 22 at the end of the chapter), we conclude that $P(x, y) > 0$. Therefore, for each $x \in \Omega$, the argument in Proposition 1.3.17 shows that $\|P(x, \cdot)\|_{L^1(\partial\Omega, d\sigma)} = 1$. Thus, for $\phi \in C(\partial\Omega)$, the functional

$$\phi \mapsto \int_{\partial\Omega} P(x, y)\phi(y)\, d\sigma(y)$$

is bounded. From this, Theorem 1.3.12, and the maximum principle, we have the next result.   □

PROPOSITION 1.3.19   The Poisson kernel for a C^2 domain Ω is uniquely determined by the property that it is positive and solves the Dirichlet problem for Δ.

1.4   The Bergman Kernel

We have already noted that it is difficult to create an explicit integral formula, with holomorphic reproducing kernel, for holomorphic functions on a domain in $\mathbb{C}^n$. Although we shall carry out such a construction on an important class of domains in Chapter 5, we now examine one of several nonconstructive approaches to this problem. This circle of ideas, due to S. Bergman [1] and to G. Szegö [1] (some of the ideas presented here were anticipated by the thesis of S. Bochner [1]), will later be seen to have profound applications to the boundary regularity of holomorphic mappings.

In this section we will see some of the invariance properties of the Bergman kernel. This will lead to the definition of the Bergman metric (in which all biholomorphic mappings become isometries) and to such other canonical constructions as representative domains. The Bergman kernel has certain extremal properties that make it a powerful tool in the theory of partial differential equations (see S. Bergman and M. Schiffer [1]). Also, the form of the singularity of the Bergman kernel (calculable for some interesting classes of domains) explains many phenomena of the theory of several complex variables (see Chapter 8 for more on this matter).

1.4.1   General Properties of the Bergman Kernel

Let $\Omega \subseteq \mathbb{C}^n$ be a domain. Define the *Bergman space*

$$A^2(\Omega) = \left\{ f \text{ holomorphic on } \Omega : \int_\Omega |f(z)|^2 \, dV(z)^{1/2} \equiv \|f\|_{A^2(\Omega)} < \infty \right\}.$$

LEMMA 1.4.1   Let $K \subseteq \Omega$ be compact. There is a constant $C_K > 0$, depending on K and on n, such that

$$\sup_{z \in K} |f(z)| \leq C_K \|f\|_{A^2(\Omega)} \qquad \text{all } f \in A^2(\Omega).$$

Proof   Since K is compact, there is an $r(K) = r > 0$ so that for any $z \in K, B(z, r) \subseteq \Omega$. Therefore, for each $z \in K$ and $f \in A^2(\Omega)$, the first exercise following Corollary 1.3.6 implies that

$$\begin{aligned}
|f(z)| &= \frac{1}{V(B(z,r))} \left| \int_{B(z,r)} f(t) \, dV(t) \right| \\
&\leq (V(B(z,r)))^{-1/2} \|f\|_{L^2(B(z,r))} \\
&\leq C(n) r^{-n} \|f\|_{A^2(\Omega)} \\
&\equiv C_K \|f\|_{A^2(\Omega)}. \qquad \qquad \square
\end{aligned}$$

LEMMA 1.4.2   The space $A^2(\Omega)$ is a Hilbert space with the inner product $\langle f, g \rangle \equiv \int_\Omega f(z)\overline{g(z)} \, dV(z)$.

Proof   Everything is clear except for completeness. Let $\{f_j\} \subseteq A^2$ be a sequence that is Cauchy in norm. Since L^2 is complete, there is an L^2 limit function f. We need to see that f is holomorphic. But Lemma 1.4.1 yields that norm convergence implies normal convergence. And the last exercise in Section 1.1 or Exercise 6 at the end of Section 1.2 yields that holomorphic functions are closed under normal limits. Therefore, f is holomorphic and $A^2(\Omega)$ is complete. $\square$

LEMMA 1.4.3   For each fixed $z \in \Omega$, the functional

$$\Phi_z : f \mapsto f(z), \quad f \in A^2(\Omega)$$

is a continuous linear functional on $A^2(\Omega)$.

Proof   This is immediate from Lemma 1.4.1 if we take K to be the singleton $\{z\}$. $\square$

We may now apply the Riesz representation theorem to see that there is an element $k_z \in A^2(\Omega)$ such that the linear functional Φ_z is represented by inner product with k_z: If $f \in A^2(\Omega)$, then for all $z \in \Omega$ we have

$$f(z) = \langle f, k_z \rangle.$$

DEFINITION 1.4.4   The Bergman kernel is the function $K(z, \zeta) = \overline{k_z(\zeta)}$, $z, \zeta \in \Omega$. It has the reproducing property

$$f(z) = \int K(z, \zeta) f(\zeta)\, dV(\zeta), \quad \forall f \in A^2(\Omega).$$

PROPOSITION 1.4.5   The Bergman kernel $K(z, \zeta)$ is conjugate symmetric: $K(z, \zeta) = \overline{K(\zeta, z)}$.

Proof   By its very definition, $\overline{K(\zeta, \cdot)} \in A^2(\Omega)$ for each fixed ζ. Therefore, the reproducing property of the Bergman kernel gives

$$\int_\Omega K(z, t) \overline{K(\zeta, t)}\, dV(t) = \overline{K(\zeta, z)}.$$

On the other hand,

$$
\begin{aligned}
\int_\Omega K(z, t) \overline{K(\zeta, t)}\, dV(t) &= \overline{\int K(\zeta, t) \overline{K(z, t)}\, dV(t)}\\
&= \overline{\overline{K(z, \zeta)}} = K(z, \zeta). \qquad \square
\end{aligned}
$$

PROPOSITION 1.4.6 The Bergman kernel is uniquely determined by the properties that it is an element of $A^2(\Omega)$ in z, is conjugate symmetric, and reproduces $A^2(\Omega)$.

Proof   Let $K'(z, \zeta)$ be another such kernel. Then

$$
\begin{aligned}
K(z, \zeta) &= \overline{K(\zeta, z)} = \overline{\int K'(z, t) \overline{K(\zeta, t)}\, dV(t)}\\
&= \int K(\zeta, t) \overline{K'(z, t)}\, dV(t)\\
&= \overline{\overline{K'(z, \zeta)}} = K'(z, \zeta). \qquad \square
\end{aligned}
$$

Since $L^2(\Omega)$ is a separable Hilbert space, then so is its subspace $A^2(\Omega)$. Thus there is a complete orthonormal basis $\{\phi_j\}_{j=1}^\infty$ for $A^2(\Omega)$.

PROPOSITION 1.4.7 Let K be a compact subset of Ω. Then the series

$$\sum_{j=1}^\infty \phi_j(z) \overline{\phi_j(\zeta)}$$

sums uniformly on $K \times K$ to the Bergman kernel $K(z, \zeta)$.

Proof   By the Riesz-Fischer and Riesz representation theorems, we obtain

$$
\sup_{z \in K} \left(\sum_{j=1}^{\infty} |\phi_j(z)|^2 \right)^{1/2} = \sup_{z \in K} \left\| \{\phi_j(z)\}_{j=1}^{\infty} \right\|_{\ell^2}
$$

$$
= \sup_{\substack{\|\{a_j\}\|_{\ell^2}=1 \\ z \in K}} \left| \sum_{j=1}^{\infty} a_j \phi_j(z) \right|
$$

$$
= \sup_{\substack{\|f\|_{A^2}=1 \\ z \in K}} |f(z)| \leq C_K. \tag{1.4.7.1}
$$

In the last inequality we have used Lemma 1.4.1. Therefore,

$$
\sum_{j=1}^{\infty} \left| \phi_j(z)\overline{\phi_j(\zeta)} \right| \leq \left(\sum_{j=1}^{\infty} |\phi_j(z)|^2 \right)^{1/2} \left(\sum_{j=1}^{\infty} |\phi_j(\zeta)|^2 \right)^{1/2}
$$

and the convergence is uniform over $z, \zeta \in K$. For fixed $z \in \Omega$, (1.4.7.1) shows that $\{\phi_j(z)\}_{j=1}^{\infty} \in \ell^2$. Hence we have that $\sum \phi_j(z)\overline{\phi_j(\zeta)} \in \overline{A^2(\Omega)}$ as a function of ζ. Let the sum of the series be denoted by $K'(z, \zeta)$. Notice that K' is conjugate symmetric by its very definition. Also, for $f \in A^2(\Omega)$, we have

$$
\int K'(\cdot, \zeta)f(\zeta)\,dV(\zeta) = \sum \hat{f}(j)\phi_j(\cdot) = f(\cdot),
$$

where convergence is in the Hilbert space topology. [Here $\hat{f}(j)$ is the jth Fourier coefficient of f with respect to the basis $\{\phi_j\}$.] But Hilbert space convergence dominates pointwise convergence (Lemma 1.4.1), so

$$
f(z) = \int K'(z, \zeta)f(\zeta)\,dV(\zeta), \quad \text{all } f \in A^2(\Omega).
$$

Therefore, K' is the Bergman kernel.   □

Remark: It is worth noting explicitly that the proof of 1.4.7 shows that

$$
\sum \phi_j(z)\overline{\phi_j(\zeta)}
$$

equals the Bergman kernel $K(z, \zeta)$ *no matter what the choice* of complete orthonormal basis $\{\phi_j\}$ for $A^2(\Omega)$.   □

PROPOSITION 1.4.8 If Ω is a bounded domain in $\mathbb{C}^n$, then the mapping

$$
P : f \mapsto \int_{\Omega} K(\cdot, \zeta)f(\zeta)\,dV(\zeta)
$$

is the Hilbert space orthogonal projection of $L^2(\Omega, dV)$ onto $A^2(\Omega)$.

Proof Notice that P is idempotent and self-adjoint and that $A^2(\Omega)$ is precisely the set of elements of L^2 that are fixed by P. $\qquad\square$

DEFINITION 1.4.9 Let $\Omega \subseteq \mathbb{C}^n$ be a domain and let $f : \Omega \to \mathbb{C}^n$ be a *holomorphic mapping*; that is, $f(z) = (f_1(z), \ldots, f_n(z))$ with $f_1, \ldots, f_n$ holomorphic on Ω. Let $w_j = f_j(z), j = 1, \ldots, n$. Then the *holomorphic Jacobian matrix* of f is the matrix

$$J_{\mathbb{C}}f = \frac{\partial(w_1, \ldots, w_n)}{\partial(z_1, \ldots, z_n)}.$$

Write $z_j = x_j + iy_j, w_k = \xi_k + i\eta_k, j, k = 1, \ldots, n$. Then the *real Jacobian matrix* of f is the matrix

$$J_{\mathbb{R}}f = \frac{\partial(\xi_1, \eta_1, \ldots, \xi_n, \eta_n)}{\partial(x_1, y_1, \ldots, x_n, y_n)}.$$

PROPOSITION 1.4.10 With notation as in the definition, we have

$$\det J_{\mathbb{R}}f = |\det J_{\mathbb{C}}f|^2$$

whenever f is a holomorphic mapping.

Proof We exploit the functoriality of the Jacobian. Let $w = (w_1, \ldots, w_n) = f(z) = (f_1(z), \ldots, f_n(z))$. Write $z_j = x_j + iy_j, w_j = \xi_j + i\eta_j, j = 1, \ldots, n$. Then, by the definition of the Jacobian,

$$d\xi_1 \wedge d\eta_1 \wedge \cdots \wedge d\xi_n \wedge d\eta_n = (\det J_{\mathbb{R}}f(x, y))dx_1 \wedge dy_1 \wedge \cdots \wedge dx_n \wedge dy_n.$$

$$(1.4.10.1)$$

On the other hand,

$$\begin{aligned}
&d\xi_1 \wedge d\eta_1 \wedge \cdots \wedge d\xi_n \wedge d\eta_n \\
&= \frac{1}{(2i)^n}d\bar{w}_1 \wedge dw_1 \wedge \cdots \wedge d\bar{w}_n \wedge dw_n \\
&= \frac{1}{(2i)^n}\overline{(\det J_{\mathbb{C}}f(z))}(\det J_{\mathbb{C}}f(z))d\bar{z}_1 \wedge dz_1 \wedge \cdots \wedge d\bar{z}_n \wedge dz_n \\
&= |\det J_{\mathbb{C}}f(z)|^2 dx_1 \wedge dy_1 \wedge \cdots \wedge dx_n \wedge dy_n. \quad (1.4.10.2)
\end{aligned}$$

Equating (1.4.10.1) and (1.4.10.2) gives the result. $\qquad\square$

Exercise for the Reader

Prove Proposition 1.4.10 using only matrix theory (no differential forms). This will give rise to a great appreciation for the theory of differential forms (see L. Bers [1] for help).

Now we can prove the holomorphic implicit function theorem.

THEOREM 1.4.11   Let $f_j(w, z), j = 1, \ldots, m$ be holomorphic functions of $(w, z) = ((w_1, \ldots, w_m), (z_1, \ldots, z_n))$ near a point $(w^0, z^0) \in \mathbb{C}^m \times \mathbb{C}^n$. Assume that

$$f_j(w^0, z^0) = 0, \quad j = 1, \ldots, m,$$

and that

$$\det \left(\frac{\partial f_j}{\partial w_k} \right)_{j,k=1}^m \neq 0 \quad \text{at } (w^0, z^0).$$

Then the system of equations

$$f_j(w, z) = 0, \qquad j = 1, \ldots, m,$$

has a unique holomorphic solution $w(z)$ in a neighborhood of z^0 that satisfies $w(z^0) = w^0$.

Proof   We rewrite the system of equations as

$$\operatorname{Re} f_j(w, z) = 0, \qquad \operatorname{Im} f_j(w, z) = 0$$

for the $2m$ real variables $\operatorname{Re} w_k, \operatorname{Im} w_k, k = 1, \ldots, m$. By Proposition 1.4.10 the determinant of the Jacobian over $\mathbb{R}$ of this new system is the modulus squared of the determinant of the Jacobian over $\mathbb{C}$ of the old system. By our hypothesis, this number is nonvanishing at the point (w^0, z^0). Therefore, the classical implicit function theorem (see W. Rudin [1]) implies that there exist C^1 functions $w_k(z), k = 1, \ldots, m$, with $w(z^0) = w^0$ that solve the system. Our job is to show that these functions are, in fact, holomorphic. When properly viewed, this is purely a problem of geometric algebra

Applying exterior differentiation to the equations

$$0 = f_j(w(z), z), \qquad j = 1, \ldots, m,$$

yields that

$$0 = df_j = \sum_{k=1}^m \frac{\partial f_j}{\partial w_k} dw_k + \sum_{k=1}^n \frac{\partial f_j}{\partial z_k} dz_k.$$

There are no $d\bar{z}_j$'s and no $d\bar{w}_k$'s because the f_j's are holomorphic.

The result now follows from linear algebra only: The hypothesis on the determinant of the matrix $(\partial f_j / \partial w_k)$ implies that we can solve for dw_k in terms of dz_j. Therefore, w is a holomorphic function on z.   □

A holomorphic mapping $f : \Omega_1 \rightarrow \Omega_2$ of domains $\Omega_1 \subseteq \mathbb{C}^n, \Omega_2 \subseteq \mathbb{C}^m$ is said to be *biholomorphic* if it is one-to-one and onto and $\det J_{\mathbb{C}} f(z) \neq 0$ for every $z \in \Omega_1$.

Exercise for the Reader

Use Theorem 1.4.11 to prove that a biholomorphic mapping has a holomorphic inverse (hence the name).

Remark: It is true, but not at all obvious, that the nonvanishing of the Jacobian determinant is a superfluous condition in the definition of "biholomorphic mapping;" that is, the non-vanishing of the Jacobian follows from the univalence of the mapping. A proof of this assertion is sketched in Exercise 37 at the end of Chapter 11.   □

In what follows we denote the Bergman kernel for a given domain Ω by K_Ω.

PROPOSITION 1.4.12   Let Ω_1, Ω_2 be domains in $\mathbb{C}^n$. Let $f : \Omega_1 \rightarrow \Omega_2$ be biholomorphic. Then

$$\det J_{\mathbb{C}} f(z) K_{\Omega_2}(f(z), f(\zeta)) \det \overline{J_{\mathbb{C}} f(\zeta)} = K_{\Omega_1}(z, \zeta).$$

Proof   Let $\phi \in A^2(\Omega_1)$. Then, by change of variable,

$$\int_{\Omega_1} \det J_{\mathbb{C}} f(z) K_{\Omega_2}(f(z), f(\zeta)) \det \overline{J_{\mathbb{C}} f(\zeta)} \phi(\zeta) \, dV(\zeta)$$

$$= \int_{\Omega_2} \det J_{\mathbb{C}} f(z) K_{\Omega_2}(f(z), \tilde{\zeta}) \det \overline{J_{\mathbb{C}} f(f^{-1}(\tilde{\zeta}))} \phi(f^{-1}(\tilde{\zeta}))$$

$$\times \det J_{\mathbb{R}} f^{-1}(\tilde{\zeta}) \, dV(\tilde{\zeta}).$$

By Proposition 1.4.10 this simplifies to

$$\det J_{\mathbb{C}} f(z) \int_{\Omega_2} K_{\Omega_2}(f(z), \tilde{\zeta}) \left\{ \left(\det J_{\mathbb{C}} f(f^{-1}(\tilde{\zeta})) \right)^{-1} \phi \left(f^{-1}(\tilde{\zeta}) \right) \right\} dV(\tilde{\zeta}).$$

By change of variables, the expression in braces is an element of $A^2(\Omega_2)$. So the reproducing property of K_{Ω_2} applies, and the last line equals

$$\det J_{\mathbb{C}} f(z) \left(\det J_{\mathbb{C}} f(z) \right)^{-1} \phi \left(f^{-1}(f(z)) \right) = \phi(z).$$

By the uniqueness of the Bergman kernel, the proposition follows.   □

PROPOSITION 1.4.13   For $z \in \Omega \subset\subset \mathbb{C}^n$, it holds that $K_\Omega(z, z) > 0$.

Proof   Now

$$K_\Omega(z, z) = \sum_{j=1}^{\infty} |\phi_j(z)|^2 \geq 0.$$

If, in fact, $K(z, z) = 0$ for some z, then $\phi_j(z) = 0$ for all j; hence $f(z) = 0$ for every $f \in A^2(\Omega)$. This is absurd. $\square$

DEFINITION 1.4.14   For any $\Omega \subseteq \mathbb{C}^n$ we define a Hermitian metric on Ω by

$$g_{ij}(z) = \frac{\partial^2}{\partial z_i \partial \bar{z}_j} \log K(z, z), \quad z \in \Omega.$$

This means that the square of the length of a tangent vector $\xi = (\xi_1, \ldots, \xi_n)$ at a point $z \in \Omega$ is given by

$$|\xi|_{B,z} = \sum_{i,j} g_{ij}(z) \xi_i \bar{\xi}_j.$$

The metric that we have defined is called the *Bergman metric*.

In a Hermitian metric $\{g_{ij}\}$, the length of a C^1 curve $\gamma : [0, 1] \to \Omega$ is given by

$$\ell(\gamma) = \int_0^1 \left(\sum_{i,j} g_{i,j}(\gamma(t)) \gamma_i'(t) \overline{\gamma_j'(t)} \right)^{1/2} dt.$$

If P, Q are points of Ω, then their distance $d_\Omega(P, Q)$ in the metric is defined to be the infimum of the lengths of all piecewise C^1 curves connecting the two points.

It is not a priori obvious that the Bergman metric for a bounded domain Ω is given by a positive definite matrix at each point. We outline a proof of this fact in Exercise 39 at the end of the chapter.

PROPOSITION 1.4.15   Let $\Omega_1, \Omega_2 \subseteq \mathbb{C}^n$ be domains and let $f : \Omega_1 \to \Omega_2$ be a biholomorphic mapping. Then f induces an isometry of Bergman metrics:

$$|\xi|_{B,z} = |(J_\mathbb{C} f)\xi|_{B,f(z)}$$

for all $z \in \Omega_1, \xi \in \mathbb{C}^n$. Equivalently, f induces an isometry of Bergman distances in the sense that

$$d_{\Omega_2}(f(P), f(Q)) = d_{\Omega_1}(P, Q).$$

Proof   This is a formal exercise, but we include it for completeness:

From the definitions, it suffices to check that

$$\sum_{i,j} g_{i,j}^{\Omega_2}(f(z))\,(J_{\mathbb{C}}f(z)w)_i\,\overline{(J_{\mathbb{C}}f(z)w)}_j = \sum_{i,j} g_{ij}^{\Omega_1}(z)w_i\bar{w}_j \qquad (1.4.15.1)$$

for all $z \in \Omega, w = (w_1, \ldots, w_n) \in \mathbb{C}^n$. But by Proposition 1.4.12,

$$
\begin{aligned}
g_{ij}^{\Omega_1}(z) &= \frac{\partial^2}{\partial z_i \bar{z}_j} \log K_{\Omega_1}(z,z) \\
&= \frac{\partial^2}{\partial z_i \bar{z}_j} \log\left\{ |\det J_{\mathbb{C}}f(z)|^2 K_{\Omega_2}(f(z), f(z)) \right\} \\
&= \frac{\partial^2}{\partial z_i \bar{z}_j} \log K_{\Omega_2}(f(z), f(z)) \qquad (1.4.15.2)
\end{aligned}
$$

since $\log |\det J_{\mathbb{C}}f(z)|^2$ is locally

$$\log\left(\det J_{\mathbb{C}}f\right) + \log\left(\overline{\det J_{\mathbb{C}}f}\right) + C$$

and hence is annihilated by the mixed second derivative. But (1.4.15.2) is nothing other than

$$\sum_{\ell,m} g_{\ell,m}^{\Omega_2}(f(z)) \frac{\partial f_\ell(z)}{\partial z_i} \frac{\partial \overline{f_m(z)}}{\partial \bar{z}_j}$$

and (1.4.15.1) follows.   □

PROPOSITION 1.4.16   Let $\Omega \subset\subset \mathbb{C}^n$ be a domain. Let $z \in \Omega$. Then

$$K(z,z) = \sup_{f \in A^2(\Omega)} \frac{|f(z)|^2}{\|f\|_{A^2}^2} = \sup_{\|f\|_{A^2(\Omega)}=1} |f(z)|^2.$$

Proof   Now

$$
\begin{aligned}
K(z,z) &= \sum |\phi_j(z)|^2 \\
&= \left(\sup_{\|\{a_j\}\|_{\ell^2}=1} \left| \sum \phi_j(z)a_j \right| \right)^2 \\
&= \sup_{\|f\|_{A^2}=1} |f(z)|^2,
\end{aligned}
$$

by the Riesz-Fischer theorem. This equals

$$\sup_{f \in A^2} \frac{|f(z)|^2}{\|f\|_{A^2}^2}. \qquad \Box$$

We shall use this proposition in a moment. Meanwhile, we should like to briefly mention some open problems connected with the Bergman kernel.

The Lu Qi-Keng Conjecture

We have already noticed that $K_\Omega(z, z) > 0$, all $z \in \Omega$, any bounded Ω. It is reasonable to ask whether $K_\Omega(z, \zeta)$ is ever equal to zero. In fact, various geometric constructions connected with the Bergman metric and associated biholomorphic invariants (which involve division by K) make it particularly desirable that K be nonvanishing.

If $\Omega = D$, the unit disc, then explicit calculation (which we perform below) shows that

$$K(z, \zeta) = \frac{1}{\pi} \frac{1}{(1 - z\bar{\zeta})^2};$$

hence, $K(z, \zeta)$ is nonvanishing on $D \times D$. Proposition 1.4.12 and the Riemann mapping theorem then show that the Bergman kernel for any proper, simply connected subdomain of $\mathbb{C}$ is nonvanishing.

The Bergman kernel for the annulus was studied in M. Skwarczynski [1] and was seen to vanish at some points. It is shown in N. Suita and A. Yamada [1] that that if $\Omega \subseteq \mathbb{C}$ is a multiply connected domain with smooth boundary, then K_Ω must vanish—this is proved by an analysis of differentials on the Riemann surface consisting of the double of Ω. By using the easy fact that the Bergman kernel for a product domain is the product of the Bergman kernels (exercise), we may conclude that any domain in $\mathbb{C}^2$ of the form $A \times \Omega$, where A is multiply connected, has a Bergman kernel with zeros. The Lu Qi-Keng conjecture can be formulated as follows:

Conjecture: A topologically trivial domain in $\mathbb{C}^n$ has nonvanishing Bergman kernel.

It is known (R. E. Greene and S. G. Krantz [1, 2]) that a domain that is C^∞ sufficiently close to the ball in $\mathbb{C}^n$ has nonvanishing Bergman kernel. Also, if a domain Ω has Bergman kernel that is bounded from zero (and satisfies a modest geometric condition), then all nearby domains have Bergman kernel that is bounded from zero. Thus it came as a bit of a surprise when in H. Boas [2] it was shown that there exist topologically trivial domains—even ones with real analytic boundary and satisfying all reasonable additional geometric conditions—for which the Bergman kernel has zeros. See also J. Wiegerinck [1], where interesting ideas contributing to the solution of this problem first arose.

Exercise for the Reader

The set of smoothly bounded domains for which the Lu Qi-Keng conjecture is true is closed in the Hausdorff topology on domains.

1.4.2 Smoothness to the Boundary of K_Ω

It is of interest to know whether K_Ω is smooth on $\bar{\Omega} \times \bar{\Omega}$. We can see from the above formula for the Bergman kernel of the disc that $K_D(z, z)$ blows up as $z \to 1^-$. In fact this property of blowing up prevails at any boundary point of a domain at which there is a peaking function (apply Proposition 1.4.16 to a high power of the peaking function). The reference T. Gamelin [1] contains background information on peaking functions.

However, there is strong evidence that—as long as Ω is smoothly bounded—on compact subsets of

$$\bar{\Omega} \times \bar{\Omega} \setminus ((\partial\Omega \times \partial\Omega) \cap \{z = \zeta\}),$$

the Bergman kernel will be smooth. For strongly convex domains (all boundary curvatures are positive), this statement is true; its proof (see N. Kerzman [3]) uses deep and powerful methods of partial differential equations.

Perhaps the most central open problem in the function theory of several complex variables is to prove that a biholomorphic mapping of two smoothly bounded domains extends to a diffeomorphism of the closures (this topic is treated in detail in Chapter 11). It is known (see S. Bell and H. Boas [1]) that a sufficient condition for this problem to have an affirmative answer on a smoothly bounded domain $\Omega \subseteq \mathbb{C}^n$ is that for any multi-index α there are constants $C = C_\alpha$ and $m = m_\alpha$ such that the Bergman kernel K_Ω satisfies

$$\sup_{z \in \Omega} \left| \frac{\partial^\alpha}{\partial z^\alpha} K_\Omega(z, \zeta) \right| \leq C \cdot \delta_\Omega(\zeta)^{-m}$$

for all $\zeta \in \Omega$. Here $\delta_\Omega(w)$ denotes the distance of the point $w \in \Omega$ to the boundary of the domain.

1.4.3 Calculating the Bergman Kernel

The Bergman kernel can almost never be calculated explicitly; unless the domain Ω has a great deal of symmetry—so that a useful orthonormal basis for $A^2(\Omega)$ can be determined—there are few techniques for determining K_Ω.

In 1974 C. Fefferman [1] introduced a new technique for obtaining an asymptotic expansion for the Bergman kernel on a large class of domains. (For an alternate approach, see L. Boutet de Monvel and J. Sjöstrand [1].) This work enabled rather explicit estimations of the Bergman metric and opened up an entire branch of analysis on domains in $\mathbb{C}^n$ (see C. Fefferman [2], S. S. Chern and J. Moser [1], P. Klembeck [1], and R. E. Greene and S. G. Krantz [1–11], for example).

The Bergman theory that we have presented here would be a bit hollow if we did not at least calculate the kernel in a few instances. We complete the section by addressing that task.

Restrict attention to the ball $B \subseteq \mathbb{C}^n$. The functions z^α, α a multi-index, are each in $A^2(B)$ and are pairwise orthogonal by the symmetry of the ball. By the uniqueness of the power series expansion for an element of $A^2(B)$, the elements z^α form a complete orthonormal system on B (their closed linear span is $A^2(B)$). Setting

$$\gamma_\alpha = \int_B |z^\alpha|^2 \, dV(z),$$

we see that $\{z^\alpha / \sqrt{\gamma_\alpha}\}$ is a complete orthonornal system in $A^2(B)$. Thus, by Proposition 1.4.7,

$$K(z,\zeta) = \sum_\alpha \frac{z^\alpha \bar\zeta^\alpha}{\gamma_\alpha}.$$

If we want to calculate the Bergman kernel for the ball in closed form, we need to calculate the γ_α's. This requires some lemmas from real analysis. These lemmas will be formulated and proved on $\mathbb{R}^N$ and $B_N = \{x \in \mathbb{R}^N : |x| < 1\}$.

LEMMA 1.4.17   We have that

$$\int_{\mathbb{R}^N} e^{-\pi|x|^2} \, dx = 1.$$

Proof   The case $N = 1$ is familiar from calculus (or see E. M. Stein and G. Weiss [1]). For the N-dimensional case, write

$$\int_{\mathbb{R}^N} e^{-\pi|x|^2} \, dx = \int_{\mathbb{R}} e^{-\pi x_1^2} \, dx_1 \cdots \int_{\mathbb{R}} e^{-\pi x_N^2} \, dx_N$$

and apply the one-dimensional result.   □

Let σ be the unique rotationally invariant area measure on $S_{N-1} = \partial B_N$ (see Appendix II) and let $\omega_{N-1} = \sigma(\partial B)$.

LEMMA 1.4.18   We have

$$\omega_{N-1} = \frac{2\pi^{N/2}}{\Gamma(N/2)},$$

where

$$\Gamma(x) = \int_0^\infty t^{x-1} e^{-t} \, dt$$

is Euler's gamma function.

Proof   Introducing polar coordinates we have

$$1 = \int_{\mathbb{R}^N} e^{-\pi|x|^2}\, dx = \int_{S^{N-1}} d\sigma \int_0^\infty e^{-\pi r^2} r^{N-1}\, dr$$

or

$$\frac{1}{\omega_{N-1}} = \int_0^\infty e^{-\pi r^2} r^N \frac{dr}{r}.$$

Letting $s = r^2$ in this last integral and doing some obvious manipulations yields the result.   □

Now we return to $B \subseteq \mathbb{C}^n$. We set

$$\eta(k) = \int_{\partial B} |z_1|^{2k}\, d\sigma, \qquad N(k) = \int_B |z_1|^{2k}\, dV(z), \qquad k = 0, 1, \ldots.$$

LEMMA 1.4.19   We have

$$\eta(k) = \pi^n \frac{2(k!)}{(k+n-1)!}, \qquad N(k) = \pi^n \frac{k!}{(k+n)!}.$$

Proof   Polar coordinates show easily that $\eta(k) = 2(k+n)N(k)$. So it is enough to calculate $N(k)$. Let $z = (z_1, z_2, \ldots, z_n) = (z_1, z')$. We write

$$
\begin{aligned}
N(k) &= \int_{|z|<1} |z_1|^{2k}\, dV(z) \\
&= \int_{|z'|<1} \left(\int_{|z_1|\leq\sqrt{1-|z'|^2}} |z_1|^{2k}\, dV(z_1) \right) dV(z') \\
&= 2\pi \int_{|z'|<1} \int_0^{\sqrt{1-|z'|^2}} r^{2k} r\, dr\, dV(z') \\
&= 2\pi \int_{|z'|<1} \frac{(1-|z'|^2)^{k+1}}{2k+2}\, dV(z') \\
&= \frac{\pi}{k+1} \omega_{2n-3} \int_0^1 (1-r^2)^{k+1} r^{2n-3}\, dr \\
&= \frac{\pi}{k+1} \omega_{2n-3} \int_0^1 (1-s)^{k+1} s^{n-1} \frac{ds}{2s} \\
&= \frac{\pi}{2(k+1)} \omega_{2n-3} \beta(n-1, k+2),
\end{aligned}
$$

where β is the classical beta function of special function theory (see G. Carrier, M. Crook, and C. Pearson [1] or E. Whittaker and G. Watson [1]). By a standard identity for the beta function, we then have

$$
\begin{aligned}
N(k) &= \frac{\pi}{2(k+1)} \omega_{2n-3} \frac{\Gamma(n-1)\Gamma(k+2)}{\Gamma(n+k+1)} \\
&= \frac{\pi}{2(k+1)} \frac{2\pi^{n-1}}{\Gamma(n-1)} \frac{\Gamma(n-1)\Gamma(k+2)}{\Gamma(n+k+1)} \\
&= \frac{\pi^n k!}{(k+n)!}.
\end{aligned}
$$

This is the desired result.   □

LEMMA 1.4.20   Let $z \in B \subseteq \mathbb{C}^n$ and $0 < r < 1$. The symbol $\mathbb{1}$ denotes the point $(1, 0, \ldots, 0)$. Then

$$
K_B(z, r\mathbb{1}) = \frac{n!}{\pi^n} \frac{1}{(1 - rz_1)^{n+1}}.
$$

Proof   Refer to the formula preceding Lemma 1.4.17. Then

$$
\begin{aligned}
K_B(z, r\mathbb{1}) &= \sum_{\alpha} \frac{z^\alpha (r\mathbb{1})^\alpha}{\gamma_\alpha} = \sum_{k=0}^{\infty} \frac{z_1^k r^k}{N(k)} \\
&= \frac{1}{\pi^n} \sum_{k=0}^{\infty} (rz_1)^k \cdot \frac{(k+n)!}{k!} \\
&= \frac{n!}{\pi^n} \sum_{k=0}^{\infty} (rz_1)^k \binom{k+n}{n} \\
&= \frac{n!}{\pi^n} \cdot \frac{1}{(1 - rz_1)^{n+1}}.
\end{aligned}
$$

This is the desired result.   □

THEOREM 1.4.21   If $z, \zeta \in B$, then

$$
K_B(z, \zeta) = \frac{n!}{\pi^n} \frac{1}{(1 - z \cdot \overline{\zeta})^{n+1}},
$$

where $z \cdot \overline{\zeta} = z_1 \overline{\zeta}_1 + z_2 \overline{\zeta}_2 + \cdots + z_n \overline{\zeta}_n$.

Proof   Let $z = r\tilde{z} \in B$, where $r = |z|$ and $|\tilde{z}| = 1$. Also, fix $\zeta \in B$. Choose a unitary rotation ρ such that $\rho\tilde{z} = \mathbb{1}$. Then, by Proposition 1.4.12 and

Lemma 1.4.20 we have

$$
\begin{aligned}
K_B(z,\zeta) &= K_B(r\tilde{z},\zeta) = K(r\rho^{-1}\mathbb{1},\zeta) \\
&= K(r\,\mathbb{1},\rho\zeta) = \overline{K(\rho\zeta, r\,\mathbb{1})} \\
&= \frac{n!}{\pi^n} \cdot \frac{1}{\left(1 - r\overline{(\rho\zeta)_1}\right)^{n+1}} \\
&= \frac{n!}{\pi^n} \cdot \frac{1}{\left(1 - (r\,\mathbb{1})\cdot\overline{(\rho\zeta)}\right)^{n+1}} \\
&= \frac{n!}{\pi^n} \cdot \frac{1}{\left(1 - (r\rho^{-1}\mathbb{1})\cdot\bar{\zeta}\right)^{n+1}} \\
&= \frac{n!}{\pi^n} \cdot \frac{1}{(1 - z\cdot\bar{\zeta})^{n+1}}. \qquad\qquad \square
\end{aligned}
$$

PROPOSITION 1.4.22 The Bergman metric for the ball $B = B(0,1) \subseteq \mathbb{C}^n$ is given by

$$
g_{ij}(z) = \frac{n+1}{(1 - |z|^2)^2} \left[(1 - |z|^2)\delta_{ij} + \bar{z}_i z_j \right].
$$

Proof Since $K(z,z) = n!/(\pi^n(1 - |z|^2)^{n+1})$, this is a routine computation that we leave to the reader. $\square$

COROLLARY 1.4.23 The Bergman metric for the disc (i.e., the ball in dimension one) is

$$
g_{ij}(z) = \frac{2}{(1 - |z|^2)^2}, \qquad i = j = 1,
$$

This is the well-known Poincaré, or Poincaré-Bergman, metric.

PROPOSITION 1.4.24 The Bergman kernel for the polydisc $D^n(0,1) \subseteq \mathbb{C}^n$ is the product

$$
K(z,\zeta) = \frac{1}{\pi^n} \prod_{j=1}^{n} \frac{1}{(1 - z_j\bar{\zeta}_j)^2}.
$$

Proof This is left as an exercise for the reader. Use the uniqueness property of the Bergman kernel. $\square$

Exercise for the Reader

Calculate the Bergman metric for the polydisc.

1.4.4   The Poincaré-Bergman Metric on the Disc

If $D \subseteq \mathbb{C}$ is the unit disc, $z \in D$, then Corollary 1.4.23 shows that

$$|w|_{B,z} = \left\{ \frac{2|w|^2}{(1 - |z|^2)^2} \right\}^{1/2} = \frac{\sqrt{2}|w|}{1 - |z|^2},$$

where the subscript B indicates that we are working in the Bergman metric. We now use this formula to derive an explicit expression for the Poincaré distance from $0 \in D$ to $r + i0 \in D, 0 < r < 1$. Call this distance $d(0, r)$. Then

$$d(0, r) = \inf \left\{ \int_0^1 |\gamma'(t)|_{B, \gamma(t)} \, dt : \right.$$
$$\left. \gamma \text{ is a curve in } D, \gamma(0) = 0, \gamma(1) = r + i0 \right\}.$$

Elementary comparisons show that, among curves of the form $\psi(t) = rt + iw(t), 0 \leq t \leq 1$, the curve $\gamma(t) = tr + i0$ is the shortest in the Poincaré metric. Further elementary arguments show that a general curve of the form $\psi(t) = v(t) + iw(t)$ is always longer than some corresponding curve of the form $rt + i\tilde{w}(t)$. We leave the details of these assertions to the reader. Thus

$$d(0, r) = \int_0^1 \frac{\sqrt{2}r}{(1 - (rt)^2)} \, dt$$
$$= \sqrt{2} \int_0^r \frac{1}{1 - t^2} \, dt$$
$$= \frac{1}{\sqrt{2}} \log \left(\frac{1 + r}{1 - r} \right).$$

Since rotations are conformal maps of the disc, we may next conclude that

$$d(0, re^{i\theta}) = \frac{1}{\sqrt{2}} \log \left(\frac{1 + r}{1 - r} \right).$$

Finally, if w_1, w_2 are arbitrary, then the Möbius transformation

$$\phi : z \mapsto \frac{z - w_1}{1 - \bar{w}_1 z}$$

satisfies $\phi(w_1) = 0, \phi(w_2) = (w_2 - w_1)/(1 - \bar{w}_1 w_2)$. Then Proposition 1.4.15 yields that

$$d(w_1, w_2) = d \left(0, \frac{w_2 - w_1}{1 - \bar{w}_1 w_2} \right) = \frac{1}{\sqrt{2}} \log \left(\frac{1 + \left| \frac{w_2 - w_1}{1 - \bar{w}_1 w_2} \right|}{1 - \left| \frac{w_2 - w_1}{1 - \bar{w}_1 w_2} \right|} \right).$$

We note in passing that the expression $\rho(w_1, w_2) \equiv |(w_1 - w_2)/(1 - \bar{w}_1 w_2)|$ is called the *pseudohyperbolic distance*. It is also conformally invariant, but it does *not* arise from integrating an infinitesimal metric (i.e., lengths of tangent vectors at a point). A fuller discussion of both the Poincaré metric and the pseudohyperbolic metric on the disc may be found in J. Garnett [1] and in S. Krantz [20].

1.4.5 Appendix to Section 1.4: The Biholomorphic Inequivalence of the Ball and the Polydisc

THEOREM 1.4.25   There is no biholomorphic map of the bidisc $D^2(0,1)$ to the ball $B(0,1) \subseteq \mathbb{C}^2$.

Proof   Suppose, seeking a contradiction, that there is such a map. Since Möbius transformations act transitively on the disc, pairs of them act transitively on the bidisc. Therefore, we may compose ϕ with a self-map of the bidisc and assume that ϕ maps 0 to 0.

If $Y \in \partial B$, then the disc $d_Y = \{z \in B : z = \zeta Y, \zeta \in \mathbb{C}, |\zeta| < 1\}$ is a totally geodesic submanifold of B (informally, this means that if P, Q are points of d_Y, then the geodesic connecting them in the Riemannian manifold d_Y is the same as the geodesic connecting them in the Riemannian manifold B—see S. Kobayashi and K. Nomizu [1]). Here we work in the Bergman metric.

By our discussion in the calculation of the Poincaré metric, we may conclude that the geodesics, or paths of least length, emanating from the origin in the ball are the rays $\tau_Y : t \mapsto tY$. (This assertion may also be derived from symmetry considerations.)

Likewise, if $\alpha, \beta \in \mathbb{C}, |\alpha| = 1, |\beta| = 1$, then the disc $e_0 = \{(\zeta\alpha, \zeta\beta) : \zeta \in D\} \subseteq D^2(0,1)$ is a totally geodesic submanifold of $D^2(0,1)$. Again we may apply our discussion of the Poincaré metric on the disc to conclude that the geodesic curve emanating from the origin in the bidisc in the direction $X = (\alpha, \beta)$ is $\psi_{\alpha\beta} : t \mapsto tX$. A similar argument shows that the curve $t \mapsto (t, 0)$ is a geodesic in the bidisc.

Now if $t \mapsto tX$ is one of the above-mentioned geodesics on the bidisc, then it will be mapped under ϕ to a geodesic $t \mapsto tY$ in the ball. If $0 < t_1 < t_2 < 1$, then the points $t_1 X, t_2 X \in D^2$ will be mapped to points $t_1' Y, t_2' Y \in B$ and it must be that $0 < t_1' < t_2' < 1$, since ϕ is an isometry and hence must map the point $t_2 X$ to a point further from the origin than it maps $t_1 X$ (because $t_2 X$ is further from the origin than $t_1 X$). It follows that the limit

$$\lim_{t \to 1^-} \phi(tX)$$

exists for every choice of X and the limit lies in ∂B. After composing ϕ with a rotation we may suppose that $\{\phi(t(1,0))\}$ terminates at $(1,0)$.

Now consider the function $f(z_1, z_2) = (z_1 + 1)/2$ on B. This function has the property that $f(1,0) = 1, f$ is holomorphic on a neighborhood of $\bar{B}$, and $|f(z)| < 1$ for $z \in \bar{B} \setminus \{(1,0)\}$. For $0 < r < 1$ we invoke the mean value property for a harmonic function to write

$$\frac{1}{2\pi} \int_0^{2\pi} f \circ \phi(r, re^{i\theta})\, d\theta = f \circ \phi(r, 0). \qquad (1.4.25.1)$$

As $r \to 1^-$ the right-hand side tends to $\lim_{t \to 1^-} f(t, 0) = 1$. However, each of the paths $r \to (r, re^{i\theta})$ is a geodesic in the bidisc, as discussed above, and for different $\theta \in [0, 2\pi)$ they are distinct. Thus the curves $r \to \phi(r, re^{i\theta})$ have distinct limits in ∂B, and these limits will be different from the point $(1, 0) \in \partial B$. In particular, $\lim_{r \to 1^-} f \circ \phi(r, re^{i\theta})$ exists for each $\theta \in [0, 2\pi)$ and assumes a value of modulus strictly less than 1.

By the Lebesgue-dominated convergence theorem, we may pass to the limit as $r \to 1^-$ in the left side of (1.4.25.1) to obtain a limit that must be strictly less than one in absolute value. That is the required contradiction. □

1.5   The Szegö and Poisson-Szegö Kernels

The basic theory of the Szegö kernel is similar to that for the Bergman kernel—they are both special cases of a general theory of "Hilbert spaces with reproducing kernel" (see N. Aronszajn [1], Chang and Krantz [2]). Thus we only outline the basic steps here, leaving details to the reader.

Let $\Omega \subseteq \mathbb{C}^n$ be a bounded domain with C^2 boundary. Let $A(\Omega)$ be those functions continuous on $\bar{\Omega}$ that are holomorphic on Ω. Let $H^2(\partial\Omega)$ be the space consisting of the closure in the $L^2(\partial\Omega, d\sigma)$ topology of the restrictions to $\partial\Omega$ of elements of $A(\Omega)$. Then $H^2(\partial\Omega)$ is a proper Hilbert subspace of $L^2(\partial\Omega)$. Each element $f \in H^2(\partial\Omega)$ has a natural holomorphic extension to Ω given by its Poisson integral Pf. We shall prove in Chapter 8 that for σ-almost every $\zeta \in \partial\Omega$, it holds that

$$\lim_{\epsilon \to 0^+} f(\zeta - \epsilon\nu_\zeta) = f(\zeta).$$

Here, as usual, ν_ζ is the unit outward normal to $\partial\Omega$ at the point ζ.

For each fixed $z \in \Omega$, the functional

$$\psi_z : H^2(\partial\Omega) \ni f \mapsto Pf(z)$$

is continuous. (Why?) Let $k_z(\zeta)$ be the Hilbert space representative for the functional ψ_z. Define the Szegö kernel $S(z, \zeta)$ by the formula

$$S(z, \zeta) = \overline{k_z(\zeta)} \qquad z \in \Omega, \zeta \in \partial\Omega.$$

If $f \in H^2(\partial\Omega)$, then

$$Pf(z) = \int_{\partial\Omega} S(z,\zeta)f(\zeta)\,d\sigma(\zeta)$$

for all $z \in \Omega$. We shall not explicitly formulate and verify the various uniqueness and extremal properties for the Szegö kernel. The reader is invited to consider these topics.

Let $\{\phi_j\}_{j=1}^{\infty}$ be an orthonormal basis for $H^2(\partial\Omega)$. Define

$$S'(z,\zeta) = \sum_{j=1}^{\infty} \phi_j(z)\overline{\phi_j(\zeta)}, \qquad z,\zeta \in \Omega.$$

For convenience we tacitly identify each function with its Poisson extension to the interior of the domain. Then, for $K \subseteq \Omega$ compact, the series defining S' converges uniformly on $K \times K$. By a Riesz-Fischer argument, $S'(\cdot,\zeta)$ is the Poisson integral of an element of $H^2(\partial\Omega)$ and $S'(z,\cdot)$ is the conjugate of the Poisson integral of an element of $H^2(\partial\Omega)$. So S' extends to $(\bar{\Omega} \times \Omega) \cup (\Omega \times \bar{\Omega})$, where it is understood that all functions on the boundary are defined only almost everywhere. The kernel S' is conjugate symmetric. Also, by Riesz-Fischer theory, S' reproduces $H^2(\partial\Omega)$. Since the Szegö kernel is unique, it follows that $S = S'$.

The Szegö kernel may be thought of as representing a map

$$S : f \mapsto \int_{\partial\Omega} f(\zeta)S(\cdot,\zeta)\,d\sigma(\zeta)$$

from $L^2(\partial\Omega)$ to $H^2(\partial\Omega)$. Since $S = S'$ is self-adjoint and idempotent, it is the Hilbert space projection of $L^2(\partial\Omega)$ to $H^2(\partial\Omega)$.

The Poisson-Szegö kernel is obtained by a formal procedure from the Szegö kernel: This procedure manufactures a *positive* reproducing kernel from one that is not necessarily positive. Note in passing that, just as we argued for the Bergman kernel in the last section, $S(z,z)$ is never 0 when $z \in \Omega$.

PROPOSITION 1.5.1   Define

$$\mathcal{P}(z,\zeta) = \frac{|S(z,\zeta)|^2}{S(z,z)}, \qquad z \in \Omega,\ \zeta \in \partial\Omega.$$

Then for any $f \in A(\Omega)$ and $z \in \Omega$, it holds that

$$f(z) = \int_{\partial\Omega} f(\zeta)\mathcal{P}(z,\zeta)\,d\sigma(\zeta).$$

Proof   Fix $z \in \Omega$ and $f \in A(\Omega)$ and define

$$u(\zeta) = f(\zeta)\frac{\overline{S(z,\zeta)}}{S(z,z)}, \qquad \zeta \in \partial\Omega.$$

Then $u \in H^2(\partial\Omega)$; hence

$$\begin{aligned} f(z) &= u(z) = \int_{\partial\Omega} S(z,\zeta)u(\zeta)\,d\sigma(\zeta) \\ &= \int_{\partial\Omega} \mathcal{P}(z,\zeta)f(\zeta)\,d\sigma(\zeta). \end{aligned}$$

This is the desired formula.   □

Remark: In passing to the Poisson-Szegö kernel we gain the advantage of positivity of the kernel (for more on this circle of ideas, see Chapter 8 and also Chapter 1 of Y. Katznelson [1]). However, we lose something in that $\mathcal{P}(z,\zeta)$ is no longer holomorphic in the z variable nor conjugate holomorphic in the ζ variable. The literature on this kernel is rather sparse and there are many unresolved questions.   □

As an exercise, use the paradigm of Proposition 1.5.1 to construct a positive kernel from the Cauchy kernel on the disc (be sure to first change notation in the usual Cauchy formula so that it is written in terms of arc length measure on the boundary). What familiar kernel results?

Like the Bergman kernel, the Szegö and Poisson-Szegö kernels can almost never be explicitly computed. They can be calculated asymptotically in a number of important instances, however (see C. Fefferman [1] and L. Boutet de Monvel and J. Sjöstrand [1]). We will give explicit formulas for these kernels on the ball. The computations are similar in spirit to those in Section 1.4; fortunately, we may capitalize on much of the work done there.

LEMMA 1.5.2   The functions $\{z^\alpha\}$, where α ranges over multi-indices, are pairwise orthogonal and span $H^2(\partial B)$.

Proof   The orthogonality follows from symmetry considerations. For the completeness, notice that it suffices to see that the span of $\{z^\alpha\}$ is dense in $A(B)$ in the uniform topology on the boundary. By the Stone-Weierstrass theorem, the closed algebra generated by $\{z^\alpha\}$ and $\{\bar{z}^\alpha\}$ is all of $C(\partial B)$. But the monomials $\bar{z}^\alpha, \alpha \neq 0$, are orthogonal to $A(B)$ (use the power series expansion about the origin to see this). The claimed density follows.   □

LEMMA 1.5.3   Let $\mathbb{1} = (1,0,\ldots,0)$. Then

$$S(z,\,\mathbb{1}) = \frac{(n-1)!}{2\pi^n}\frac{1}{(1-z_1)^n}.$$

Proof   We have that

$$
\begin{aligned}
S(z, \mathbb{1}) &= \sum_{\alpha} \frac{z^{\alpha} \cdot \mathbb{1}^{\alpha}}{\|z_1^{\alpha}\|_{L^2(\partial B)}^2} = \sum_{k=0}^{\infty} \frac{z_1^k}{\eta(k)} = \frac{1}{2\pi^n} \sum_{k=0}^{\infty} \frac{z_1^k(k+n-1)!}{k!} \\
&= \frac{(n-1)!}{2\pi^n} \sum_{k=0}^{\infty} \binom{k+n-1}{n-1} z_1^k = \frac{(n-1)!}{2\pi^n} \frac{1}{(1-z_1)^n}. \qquad \square
\end{aligned}
$$

LEMMA 1.5.4   Let ρ be a unitary rotation on $\mathbb{C}^n$. For any $z \in \bar{B}, \zeta \in \partial B$, we have that $S(z, \zeta) = S(\rho z, \rho \zeta)$.

Proof   This is a standard change of variables argument and we omit it. $\square$

THEOREM 1.5.5   The Szegö kernel for the ball is

$$
S(z, \zeta) = \frac{(n-1)!}{2\pi^n} \frac{1}{(1 - z \cdot \bar{\zeta})^n}.
$$

Proof   Let $z \in B$ be arbitrary. Let ρ be the unique unitary rotation such that ρz is a multiple of $\mathbb{1}$. Then, by 1.5.4,

$$
\begin{aligned}
S(z, \zeta) &= S(\rho^{-1}\mathbb{1}, \zeta) = S(\mathbb{1}, \rho\zeta) = \overline{S(\rho\zeta, \mathbb{1})} = \frac{(n-1)!}{2\pi^n} \overline{\left(\frac{1}{1 - (\overline{\rho\zeta}) \cdot \mathbb{1}}\right)^n} \\
&= \frac{(n-1)!}{2\pi^n} \frac{1}{\left(1 - \bar{\zeta} \cdot (\rho^{-1}\mathbb{1})\right)^n} = \frac{(n-1)!}{2\pi^n} \frac{1}{(1 - z \cdot \bar{\zeta})^n}.
\end{aligned}
$$

COROLLARY 1.5.6   The Poisson-Szegö kernel for the ball is

$$
\mathcal{P}(z, \zeta) = \frac{(n-1)!}{2\pi^n} \frac{(1 - |z|^2)^n}{|1 - z \cdot \bar{\zeta}|^{2n}}.
$$

Exercise for the Reader
Calculate the Szegö and Poisson-Szegö kernel for the polydisc.

1.6 Afterword to Chapter 1

It follows from work of A. Gleason [2] and L. Bungart [1] that Cauchy-type integral formulas *that are holomorphic in the z variable* (this is crucial for constructing holomorphic functions) exist on virtually any domain. However these references tell us almost nothing about the form of the integral kernel—whether its singularity is only on the diagonal or at what rate it blows up. The Bergman and Szegö theories give other methods for producing integral formulas on a wide class of domains.

Thus we have several "canonical kernels" on domains in $\mathbb{C}^n$: the Poisson kernel, the Bergman kernel, and the Szegö kernel. Unfortunately, these kernels are virtually never computable (however see C. Fefferman [1], L. Boutet de Monvel and J. Sjöstrand [1], and L. Hua [1]). Later on we shall learn of a construction by G. M. Henkin (variants were also constructed independently by E. Ramirez [1] and H. Grauert and I. Lieb [1]) that yields rather explicitly computable kernels on a large class of domains in $\mathbb{C}^n$. For many practical applications, the Henkin kernel is just as useful as the Szegö kernel. And it turns out that the Szegö kernel may be expressed as an asymptotic expansion in terms of powers of the Henkin kernel (see N. Kerzman and E. M. Stein [2]).

The connection between canonical kernels and computable kernels has opened up new vistas and has made accessible many deep results in function theory and geometry (see N. Kerzman and E. M. Stein [1, 2], R. E. Greene and S. G. Krantz [1–11], D. H. Phong and E. M. Stein [1], C. Fefferman [1, 2], L. Boutet de Monvel and J. Sjöstrand [1], and S. Bell [3]). The subject of integral representations will be a fertile area of research for some time to come.

EXERCISES

1. Prove that if f and f^2 are real-valued harmonic on a domain $\Omega \subseteq \mathbb{R}^N$, then f is constant. What can you say for complex-valued f?

2. Let f be harmonic on $\Omega \subseteq \mathbb{R}^N$, real-valued, and nonvanishing. Prove that $\Delta|f|^p = p(p-1)|f|^{p-2}|\nabla f|^2 \geq 0$ provided that $p \geq 1$.

3. Prove that the phenomenon cited in Exercise 1 fails for pluriharmonic functions (recent private communication of P. Ahern). See Section 2.2 for the concept of pluriharmonicity.

4. Let f be holomorphic on $\Omega \subseteq \mathbb{C}^n$ and nonvanishing. Prove that $\Delta|f|^p = (p^2) \cdot |f|^{p-2}|\partial f|^2 \geq 0$ if $p > 0$. Prove that $\log|f|$ is harmonic by computing $\Delta \log|f|$.

5. *The automorphism group of the ball.* Let $B \subseteq \mathbb{C}^2$ be the unit ball. Complete the following outline to calculate the set of biholomorphic self-maps of B.
   a. Let $a \in \mathbb{C}, |a| < 1$. Then

$$\phi_a(z_1, z_2) = \left(\frac{z_1 - a}{1 - \bar{a}z_1}, \frac{\sqrt{1 - |a|^2}z_2}{1 - \bar{a}z_1} \right)$$

is a biholomorphic mapping of B to itself.
   b. If ρ is a unitary mapping of $\mathbb{C}^2$ (that is, the inverse of ρ is its conjugate transpose; equivalently, ρ preserves the Hermitian inner product on $\mathbb{C}^2$), then ρ is a biholomorphic mapping of the ball.

   c. Prove that if $\psi : B \to B$ is holomorphic and $\psi(0) = 0$, then all eigenvalues of Jac $\psi(0)$ have modulus not exceeding 1 (*Hint:* Apply the one-variable Schwarz lemma in a clever way).

   d. Apply part c to any biholomorphic mapping of B that preserves the origin, and to its inverse, to see that such a mapping has Jacobian matrix at the origin with all eigenvalues of modulus 1.

   e. Use the result of part d to see that any biholomorphic mapping of the ball that preserves the origin must be linear, indeed unitary.

   f. If now α is any biholomorphic mapping of the ball, choose a complex constant a such that $\phi_a \circ \alpha$ is a biholomorphic mapping that preserves the origin. Thus, by part e, $\phi_a \circ \alpha$ is unitary.

   g. Conclude that the mappings ϕ_a and the unitary mappings generate the automorphism group of the ball in $\mathbb{C}^2$.

6. Generalize the result of Exercise 5 to n complex dimensions.

7. Let $\Omega \subseteq \mathbb{R}^N$ be a domain. Suppose that $\partial\Omega$ is a regularly imbedded C^j manifold, $j = 1, 2, \ldots$. This means that for each $P \in \partial\Omega$ there is a neighborhood $U_P \subseteq \mathbb{R}^N$ and a C^j function $f_P : U_P \to \mathbb{R}$ with $\nabla f_P \neq 0$ and $\{x \in U_P : f_P(x) = 0\} = U_P \cap \partial\Omega$. [For a discussion of these matters see Appendix I.] Prove that there is a function $\rho : \mathbb{R}^N \to \mathbb{R}$ satisfying

   a. $\nabla\rho \neq 0$ on $\partial\Omega$;

   b. $\{x \in \mathbb{R}^N : \rho(x) < 0\} = \Omega$;

   c. ρ is C^j.

   We call ρ a *defining function* for Ω.

   Prove that if Ω has a C^j defining function, then $\partial\Omega$ is a regularly imbedded C^j submanifold of $\mathbb{R}^N$.

   Prove that both of the preceding concepts are equivalent to the following: For each $P \in \partial\Omega$ there is a neighborhood U_P, a coordinate system $t_1, \ldots, t_N$ on U_P, and a C^j function $\phi(t_1, \ldots, t_N)$ such that $\{(t_1, \ldots, t_N) \in U_P : t_N = \phi(t_1, \ldots, t_{N-1})\} = \partial\Omega \cap U_P$. This means that $\partial\Omega$ is locally the graph of a C^j function.

8. Prove that if f is harmonic on a domain $\Omega \subseteq \mathbb{R}^N$, then f is real analytic.

9. Let $\mathbf{h}^2(\partial D)$ be the space of those continuous functions that are the boundary functions of harmonic functions on the disc. Mimic the construction of the Szegö kernel to obtain a reproducing kernel. What reproducing kernel do you obtain? Why is the space $\mathbf{h}^2(\partial D)$ defined incorrectly (see B. Epstein [1, p. 63])?

10. Is there a Bergman kernel for the square integrable harmonic functions on the disc? Can you write it down explicitly? (See Ligocka [1, 2, 3] for more on these matters.)

11. Let $(X, \mu), (Y, \nu)$ be measure spaces. Suppose that $K : X \times Y \to \mathbb{C}$ satisfies

$$\int_X |K(x, y)| d\mu(x) \leq C_0, \quad \text{all } y \in Y,$$

and

$$\int_Y |K(x,y)|d\nu(y) \le C_0, \quad \text{all } x \in X,$$

for some constant C_0 that does not depend on $y \in Y, x \in X$. Show that if $f \in L^p(Y,\nu)$, then the operator

$$Tf(x) = \int K(x,y)f(y)d\nu(y)$$

satisfies

$$\|Tf\|_{L^p(X,\mu)} \le C_0\|f\|_{L^p(Y,\nu)}, \quad 1 \le p \le \infty.$$

This is called Schur's lemma. (*Hint:* Use Fubini's theorem.)

12. Use Exercise 11 and the integral formula for solutions of $\bar\partial u = f$ on $\mathbb{C}$ given in Theorem 1.1.9 to prove the following assertion. If f is C^∞ with compact support in $K \subset\subset \mathbb{C}$ and u is the unique solution to $\bar\partial u = f$ that vanishes at ∞, then

$$\|u\|_{L^p(L)} \le C(K,L)\|f\|_{L^p}$$

for any compact set $L \subseteq \mathbb{C}$.

13. Derive a formula for the Green's function and the Poisson kernel for the ball $B(x_0, r) \subseteq \mathbb{R}^N$ by using invariance properties of the Laplacian (do not imitate the proof in the text for the unit ball).

14. Use the result of Exercise 13 to prove the Harnack inequalities for a harmonic function: If u is a positive harmonic function on the ball $B(x_0, r)$, then for any $x \in B(x_0, r) \subseteq \mathbb{R}^2$, we have

$$\frac{r - |x|}{r + |x|} u(x_0) \le u(x) \le \frac{r + |x|}{r - |x|} u(x_0).$$

How do these inequalities change in dimension $N > 2$?

15. Use the result of Exercise 14 to prove Harnack's principle: If $u_1 \le u_2 \le \cdots$ are harmonic functions on a domain $\Omega \subseteq \mathbb{R}^N$, then either $u_j \nearrow +\infty$ uniformly on compact sets or the u_j converge uniformly on compacta to a harmonic function u_0.

16. Explain why, in one complex dimension, the Green's function is related to the Bergman kernel by the formula

$$K(z,w) = -\frac{2}{\pi} G_{z\bar w}(z,w).$$

Here subscripts denote derivatives. For help in this matter see B. Epstein [1]. There is an analogous formula in higher dimensions in which the Green's function is replaced by the kernel of a certain fundamental solution operator. See Harvey and Polking [1] for more on this matter.

17. Prove that if $\{f_j\}$ are harmonic and uniformly bounded on a domain Ω, then there is a normally convergent subsequence. (*Hint:* Apply Ascoli-Arzela.)

18. Use the mean value characterization of harmonic functions to prove that the Poisson kernel $P_\Omega(x,t)$ for a domain Ω must be harmonic in x.

19. Here is an alternate proof that the Poisson kernel for the ball is harmonic in the x-variable (for which I am indebted to C. Berg): Assume without loss of generality that $y = (1,0,\ldots,0)$. For $N > 2$ we write

$$\omega_{N-1} \cdot P(x,y) = -\frac{1}{|x-y|^{N-2}} - \frac{2(x-y)\cdot y}{|x-y|^N}$$
$$= \frac{-1}{|x-y|^{N-2}} + \frac{1}{N-2}\frac{\partial}{\partial x_1}\left(\frac{2}{|x-y|^{N-2}}\right).$$

20. Suppose that $\Omega \subseteq \mathbb{C}$ is a domain and $f : \Omega \to \mathbb{C}$ is conformal. Show that

$$\int_\Omega |f'(z)|^2 dV(z)$$

is precisely the area of the image of f.

21. Let $\Omega \subseteq \mathbb{R}^N$ be a bounded domain with C^2 boundary. Let $G(x,y)$ be the Green's function for the Laplacian on Ω. Prove that $G(P,Q) = G(Q,P)$, all $P \neq Q$. In particular, assuming part 1 of Definition 1.3.9, G extends to a smooth $(C^{2-\epsilon})$ function on $\bar\Omega \times \bar\Omega \setminus \{(x,x) : x \in \bar\Omega\}$ and is harmonic in each variable. (*Hint:* Fix $P,Q \in \Omega, P \neq Q$. Apply Green's theorem on

$$\Omega_\epsilon \equiv \Omega \setminus \left(\bar{B}(P,\epsilon) \cup \bar{B}(Q,\epsilon)\right),$$

ϵ small, to the functions $G(P,\cdot)$ and $G(Q,\cdot)$.)

22. Hopf's lemma, originally proved (see R. Courant and D. Hilbert [1]) for the sake of establishing the maximum principle for solutions of second-order elliptic equations, has proved to be a powerful tool in the study of functions of several complex variables. Here we state and outline a proof of this result. Although simpler proofs are available (see S. Krantz [19]), this one has the advantage of applying in rather general circumstances.

   Theorem: Let $\Omega \subseteq \mathbb{R}^N$ be a bounded domain with C^2 boundary. Let $f : \bar\Omega \to \mathbb{R}$ be harmonic and nonconstant on Ω, C^1 on $\bar\Omega$. Suppose that f assumes a (not necessarily strict) maximum at $P \in \partial\Omega$. If $\nu = \nu_P$ is the unit outward normal to $\partial\Omega$ at P, then $(\partial f/\partial\nu)(P) > 0$.

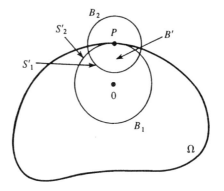

FIGURE 1.3

Outline of Proof

a. Let B_1 be a ball internally tangent to $\partial\Omega$ at P with $\partial B_1 \cap \partial\Omega = \{P\}$. Let $r > 0$ be the radius of B_1 (see Figure 1.3). Assume without loss of generality that the center of B_1 is at the origin. Let B_2 be a ball centered at P of radius $r_1 < r$. Let $B' = B_1 \cap B_2$. Notice that $\partial B' = S_1' \cup S_2'$.

b. Let $\alpha > 0$ and set $h(x) = e^{-\alpha|x|^2} - e^{-\alpha r^2}$. Then

$$\Delta h = e^{-\alpha|x|^2}\{4\alpha^2|x|^2 - 2\alpha N\}.$$

c. If $\alpha > 0$ is sufficiently large, then $\Delta h > 0$ on B'.

d. Set $v(x) \equiv f(x) + \epsilon h(x)$. If $\epsilon > 0$ is sufficiently small, then $v(x) < f(P)$ for $x \in S_1'$. Also $v(x) = f(x) < f(P)$ for $x \in S_2' \setminus \{P\}$. Use the maximum principle.

e. $\max_{x \in \bar{B}'} v(x) = f(P)$.

f. $\frac{\partial v}{\partial \nu}(P) = \frac{\partial f}{\partial \nu}(P) + \epsilon \frac{\partial h}{\partial \nu}(P) \geq 0$.

g. $\frac{\partial f}{\partial \nu}(P) > 0$.

Now suppose only that $f \in C(\bar{\Omega})$ and harmonic on Ω and that the point $P \in \partial\Omega$ is a local (not necessarily strict) maximum of f. Modify the preceding argument to prove that

$$\liminf_{\epsilon \to 0^+} \left(\frac{f(P) - f(P - \epsilon\nu)}{\epsilon} \right) > 0.$$

23. Let $f \in C_c^k(\mathbb{C}), k \geq 1$. Let u be the solution to $\bar{\partial}u = f d\bar{z}$ given by the integral formula in the text. Prove that $u \in C^k$.

24. Let $\Omega \subseteq \mathbb{C}$ be any open set. Let f be a C^k function on Ω (not necessarily compactly supported). Complete the following outline to show that there is a C^k function u on Ω with $\bar{\partial}u = f d\bar{z}$.

a. Write $\Omega = \cup_{j=1}^{\infty} K_j$, where $K_j \subset\subset K_{j+1}$ and each K_j is compact and is the closure of its interior;

b. Construct $\eta_j \in C_c^{\infty}(K_{j+1})$ such that $\eta_j \equiv 1$ on K_j.

c. Write $f = \sum f_j \equiv \eta_1 f + \sum_{j=2}^{\infty} (\eta_j - \eta_{j-1}) f$.

d. Let $u_j \in C^k(\Omega)$ with $\bar{\partial} u_j = f_j$ (see Exercise 23).

e. Notice that u_j is holomorphic on $\overset{\circ}{K}_{j-1}$ for $j = 2, 3, \dots$.

f. Apply the Runge theorem to u_j with respect to $\bar{K}_{j-1} \subset \Omega$. Find a holomorphic v_j on Ω with $\sup_{K_{j-1}} |u_j - v_j| < 2^{-j}$.

g. Let $u = \sum (u_j - v_j)$. Prove that the series converges uniformly on compacta and that $\bar{\partial} u = f d\bar{z}$.

h. Note that if $f \in C^k$, then $u \in C^k$. If $f \in C^{\infty}$, then $u \in C^{\infty}$. Where does this proof break down in $\mathbb{C}^2$?

25. Prove the classical Mittag-Leffler theorem of one complex variable, using the $\bar{\partial}$-equation, by completing the following outline (see subsection 0.3.4 for the statement of the theorem):

a. For each j, let $\Omega_j \subseteq \Omega$ be a neighborhood of z_j that does not contain any $z_k, k \neq j$. Let $\Omega_0 = \Omega \setminus \{z_j\}_{j=1}^{\infty}$.

b. There is a C^{∞} partition of unity $\{\phi_{\ell}\}_{\ell=0}^{\infty}$ subordinate to the covering $\{\Omega_{\ell}\}_{\ell=0}^{\infty}$. Say that ϕ_{ℓ} is supported on $\Omega_{j(\ell)}$.

c. Let $h_k = \sum_{\ell=0}^{\infty} \phi_{\ell}(f_k - f_{j(\ell)})$.

d. Check that

$$\psi(z) \equiv \bar{\partial} h_k(z), \quad z \in \Omega_k,$$

is a well-defined C^{∞} form on Ω.

e. Let $u \in C^{\infty}(\Omega)$ satisfy $\bar{\partial} u = \psi$ on Ω (use Exercise 24).

f. Check that

$$f(z) = f_k(z) - h_k(z) + u(z), \quad z \in \Omega_k,$$

is a well-defined meromorphic function on Ω.

26. Construct another proof of Theorem 1.2.6 (the Hartogs extension phenomenon) using the idea of the proof in the special case of the polydisc given in Section 0.3. It will only be possible to slide the contours around locally, but this and a connectedness argument (using the fact that $\Omega \setminus K$ is connected) suffices to give a proof. Note that there are delicate analytic continuation problems involved with this approach and you may have to make extra hypotheses on the boundary of the domain to make this proof work. (This attack on the problem was originated by W. F. Osgood [1].)

27. Holomorphic mappings in $\mathbb{C}^2$ are not necessarily conformal (i.e., infinitesimally angle preserving). Show that the example $F(z_1, z_2) = (z_1^2, z_1^2 + z_2)$ confirms this statement.

28. Use the classical Hurwitz theorem of one complex variable to give another proof that a holomorphic function of several variables cannot have an isolated zero.

29. Are holomorphic functions open? Are holomorphic mappings open?

30. Let $z \in \Omega_1 \subseteq \Omega_2 \subseteq \mathbb{C}^n$. Let K_j be the Bergman kernel for Ω_j. Then show that $K_2(z, z) \leq K_1(z, z)$ for all $z \in \Omega_1$. Further show that $\|K_2(z, \cdot)\|_{L^2(\Omega_2)} \leq \|K_1(z, \cdot)\|_{L^2(\Omega_1)}$.

31. Let $\Omega \subseteq \mathbb{R}^N$ be a domain and let f be harmonic on Ω. Let $P \in \Omega$ and assume that the function ϕ is compactly supported in Ω and is radially symmetric about P (that is, $\phi(x) = \phi(x')$ if $|x - P| = |x' - P|$). Further assume that $\int \phi(x) \, dx = 1$. Then prove that $f(P) = \int f(x)\phi(x) \, dx$.

32. Let $\{a_{ij}\}_{i,j=1}^n$ be complex constants such that $\sum a_{ij} w_i \bar{w}_j \geq C|w|^2$ for some $C > 0$ and all $w \in \mathbb{C}^n$. Let

$$\Omega = \left\{ z \in \mathbb{C}^n : -2 \operatorname{Re} z_1 + \sum_{i,j=1}^n a_{ij} z_i \bar{z}_j < 0 \right\}.$$

Prove that the Bergman kernel for Ω is given on the diagonal by

$$K(z, z) = \frac{n! \det (a_{ij})_{i,j=1}^n}{\pi^n (2 \operatorname{Re} z_1 - \sum_{i,j=1}^n a_{ij} z_i \bar{z}_j)^{n+1}}.$$

(*Hint:* The domain Ω is biholomorphic to the ball. See I. Graham [1] for details.)

33. Let $\Omega \subseteq \mathbb{C}^n$ be a domain. Let $\{f_j\}$ be a sequence of holomorphic functions on Ω that converges pointwise to a function f at every point of Ω. Prove that f is holomorphic on a dense open subset of Ω. (*Hint:* Let $\bar{U} \subseteq \Omega$ be the closure of any open subset U of Ω. Apply the Baire category theorem to the sets $S_M = \{z \in \bar{U} : |f_j(z)| \leq M, \text{all } j\}$.)

34. Check that $(\partial/\partial z_j)z_k \equiv \delta_{jk}$, $(\partial/\partial \bar{z}_j)\bar{z}_k \equiv \delta_{jk}$. Also $(\partial/\partial z_j)\bar{z}_k \equiv 0$ and $(\partial/\partial \bar{z}_j)z_k \equiv 0$ for every j and k. Finally, $\langle dz_j, \partial/\partial z_k \rangle \equiv \delta_{jk}$, $\langle d\bar{z}_j, \partial/\partial \bar{z}_k \rangle \equiv \delta_{jk}$, $\langle dz_j, \partial/\partial \bar{z}_k \rangle \equiv 0$, $\langle d\bar{z}_j, \partial/\partial z_k \rangle \equiv 0$ for all j, k. These equalities make the definitions of $\partial/\partial z_j, \partial/\partial \bar{z}_j, dz_j$, and $d\bar{z}_j$ seem natural.

Prove the following complex versions of the chain rule (supply the right hypotheses as well):

$$\frac{\partial}{\partial z_j}(f \circ g) = \sum_{k=1}^n \frac{\partial f}{\partial z_k} \cdot \frac{\partial g_k}{\partial z_j} + \sum_{\ell=1}^n \frac{\partial f}{\partial \bar{z}_\ell} \cdot \frac{\partial \bar{g}_\ell}{\partial z_j},$$

$$\frac{\partial}{\partial \bar{z}_j}(f \circ g) = \sum_{k=1}^n \frac{\partial f}{\partial z_k} \cdot \frac{\partial g_k}{\partial \bar{z}_j} + \sum_{\ell=1}^n \frac{\partial f}{\partial \bar{z}_\ell} \cdot \frac{\partial \bar{g}_\ell}{\partial \bar{z}_j}.$$

How do these equations simplify in the case that either f or g is holomorphic?

35. Derive the complex form of Taylor's theorem for a function f (not necessarily holomorphic) on a ball $B(z^0, r) \subseteq \mathbb{C}^n$: If $f \in C^k(B(z^0, r))$, $w \in B(z^0, r)$, then

$$f(w) \;=\; \sum_{|\alpha|+|\beta| \le k} \left(\frac{\partial}{\partial z}\right)^\alpha \left(\frac{\partial}{\partial \bar{z}}\right)^\beta f(z^0) \cdot \frac{(w - z^0)^\alpha (\bar{w} - \overline{z^0})^\beta}{\alpha!\,\beta!}$$
$$+ o(|z^0 - w|^k).$$

36. Use the notation of the Bochner-Martinelli formula. Let $\Omega \subset\subset \mathbb{C}^n$ be a domain with C^1 boundary. Let $a(\zeta) = (a_1(\zeta), \ldots, a_n(\zeta))$ with $a_j(\zeta) = p_j(\zeta)/q(\zeta), j = 1, \ldots, n$. Suppose that the a_j, p_j, q are C^1 functions on $\partial\Omega$. Let $d\sigma$ be area measure on $\partial\Omega$. Then

$$\eta(a) \wedge \omega(\zeta)\big|_{\partial\Omega} = \frac{h(\zeta)}{q(\zeta)^n} d\sigma(\zeta)$$

for some continuous h on $\partial\Omega$. (*Hint:* This is a purely algebraic fact. Write out the case $n = 2$ explicitly to see it.)

37. Complete the following outline to prove the Cauchy-Fantappiè formula:

Theorem: Let $\Omega \subset\subset \mathbb{C}^n$ be a domain with C^1 boundary. Let $w(z, \zeta) = (w_1(z, \zeta), \ldots, w_n(z, \zeta))$ be a C^1, vector-valued function on $\bar{\Omega} \times \bar{\Omega} \setminus \{\text{diagonal}\}$ that satisfies

$$\sum_{j=1}^n w_j(z, \zeta)(\zeta_j - z_j) \equiv 1.$$

Then, using the notation from Section 1.1, we have for any $f \in C^1(\bar{\Omega}) \cap \{\text{holomorphic functions on } \Omega\}$ and any $z \in \Omega$ the formula

$$f(z) = \frac{1}{nW(n)} \int_{\partial\Omega} f(\zeta)\eta(w) \wedge \omega(\zeta).$$

Proof: We may assume that $z = 0 \in \Omega$.

a. If $\alpha^1 = (a_1^1, \ldots, a_n^1), \ldots, \alpha^n = (a_1^n, \ldots, a_n^n)$ are n-tuples of C^1 functions on $\bar{\Omega}$ that satisfy $\sum_{i=1}^n a_i^j(\zeta) \cdot (\zeta_i - z_i) = 1$, let

$$B(\alpha^1, \ldots, \alpha^n) = \sum_{\sigma \in S_n} \epsilon(\sigma) a_{\sigma(1)}^1 \wedge \bar{\partial}(a_{\sigma(2)}^2) \wedge \cdots \wedge \bar{\partial}(a_{\sigma(n)}^n),$$

where S_n is the symmetric group on n letters and $\epsilon(\sigma)$ is the signature of the permutation σ. Prove that B is independent of α^1.

b. It follows that $\bar{\partial}B = 0$ on $\bar{\Omega} \setminus \{0\}$ (indeed $\bar{\partial}B$ is an expression like B with the expression $a^1_{\sigma(1)}$ replaced by $\bar{\partial}a^1_{\sigma(1)}$).

c. Use (b), especially the parenthetical remark, to prove inductively that if $\beta^1 = (b^1_1, \ldots, b^1_n), \ldots, \beta^n = (b^n_1, \ldots, b^n_n)$, then there is a form γ on $\Omega \setminus \{0\}$ such that

$$\left[B(\alpha^1, \ldots, \alpha^n) - B(\beta^1, \ldots, \beta^n) \right] \wedge \omega(\zeta) = \bar{\partial}\gamma = d\gamma.$$

d. Prove that if $\alpha^1 = \cdots = \alpha^n = (w_1, \ldots, w_n)$, then $B(\alpha^1, \ldots, \alpha^n)$ simplifies

$$B(\alpha^1, \ldots, \alpha^n) \wedge \omega(\zeta) = (n-1)! \eta(w) \wedge \omega(\zeta).$$

e. Let $\mathcal{S}$ be a small sphere of radius $\epsilon > 0$ centered at 0 such that $\mathcal{S} \subseteq \Omega$. Use part c to see that

$$\int_{\partial\Omega} f(\zeta)\eta(w) \wedge \omega(\zeta) = \int_{\mathcal{S}} f(\zeta)\eta(w) \wedge \omega(\zeta).$$

f. Now use (c) and (d) to see that

$$\int_{\mathcal{S}} f(\zeta)\eta(w) \wedge \omega(\zeta) = \int_{\mathcal{S}} f(\zeta)\eta(v) \wedge \omega(\zeta),$$

where

$$v(z, \zeta) = \frac{\bar{\zeta}_j - \bar{z}_j}{|\zeta - z|^2}.$$

(*Warning*: Be careful if you decide to apply Stokes's theorem.) From the theory of the Bochner-Martinelli kernel, we know that the last line is $n \cdot W(n) \cdot f(0)$.

In Section 9.1 we will learn a much more elegant proof of the Cauchy-Fantappiè formula, which is due to G. M. Henkin. The proof just presented is in W. Koppelman [1]. See also R. M. Range [3] for a detailed consideration of integral formulas.

38. Use a limiting argument to show that the hypotheses of the Cauchy-Fantappiè formula (the preceding exercise) may be weakened to $f \in C(\bar{\Omega})$, $w \in C(\bar{\Omega} \times \partial\Omega)$.

Prove, using only linear algebra, that if w is as in the statement of the Cauchy-Fantappiè formula, then there are functions $\psi_1, \ldots, \psi_n, \Psi$ such that $w_j = \psi_j/\Psi, j = 1, \ldots, n$ (see Exercise 36).

39. Let $\Omega_1 \subset \Omega_2 \subset \cdots \subset \mathbb{C}^n$ be bounded domains such that $\cup\Omega_j = \Omega$, also a bounded domain. Let K denote the Bergman kernel. Prove Ramadanov's theorem (I. Ramadanov [1]): $K_{\Omega_j}(\cdot, w) \to K_\Omega(\cdot, w)$ normally on Ω. (*Hint:* Use the fact that the $K_{\Omega_j}(\cdot, w)$ all represent the same linear functional.)

Show that the hypothesis of increasing union can be weakened.

40. Here is a useful way to generate an orthonormal basis for the Bergman space. Fix $z_0 \in \Omega$. Let ϕ_0 be the (unique!) element of A^2 with $\phi_0(z_0)$ real, $\|\phi_0\| = 1$, and $\phi_0(z_0)$ maximal. (Why does such a ϕ_0 exist?) Let ϕ_1 be the (unique) element of A^2 with $\phi_1(z_0) = 0, (\partial\phi_1/\partial z_1)(z_0)$ real, $\|\phi_1\| = 1$, and $(\partial\phi_1/\partial z_1)(z_0)$ maximal. (Why does such a ϕ_1 exist?) Now ϕ_1 is orthogonal to ϕ_0, or else ϕ_1 has nonzero projection on ϕ_0, leading to a contradiction. Continue this process to create an orthogonal system on Ω. Use Taylor series to see that it is complete. This circle of ideas comes from the elegant paper S. Kobayashi [1].

41. Use Exercise 40 to do the following. Let $\Omega \subseteq \mathbb{C}^n$ be a bounded domain and let (g_{ij}) be its Bergman metric. Prove that the matrix $(g_{ij}(z))$ is positive definite, each $z \in \Omega$. (*Hint:* The crucial fact is that for each $z \in \Omega$ and each j there is an element $f \in A^2(\Omega)$ such that $\partial f/\partial z_j(z) \neq 0$.)

42. Without using the calculations of Section 1.4, show that $K(0,0) = 1/V(B)$. Here K is the Bergman kernel for the ball B. Use the automorphism group of B, together with the invariance of the kernel, to calculate $K(z,z)$ for every $z \in B$. The values of K on the diagonal then completely determine $K(z,\zeta)$.

43. Let f be holomorphic on the unit bidisc. Suppose that $f(z_1, z_2) \neq 0$ if either $|z_1| < 1/2$ or $|z_2| < 1/2$. Prove that then f does not vanish on the unit ball.

44. A pair of domains (U, W) is said to exhibit *Hartogs's phenomenon* if $U \subseteq W$, $U \neq W$, and every holomorphic funtion on U analytically continues to W. Find a domain $\Omega \subseteq \mathbb{C}^2$ with the following property: There exist domains Ω_1, Ω_2 containing Ω, and distinct from Ω, such that (Ω, Ω_1) exhibits Hartogs's phenomenon, (Ω, Ω_2) exhibits Hartogs's phenomenon, but $(\Omega, \Omega_1 \cup \Omega_2)$ does not.

45. Refer to Exercise 44 for terminology. Prove that there is a domain $\Omega \subseteq \mathbb{C}^2$ that is not a domain of holomorphy, but such that there is no $\tilde{\Omega}$ properly containing Ω so that $(\Omega, \tilde{\Omega})$ exhibits Hartogs's phenomenon. (*Hint:* Think carefully about the definition of "domain of holomorphy".)

46. Let $0 < a < \infty$. Define $\Omega_a = \{(z_1, z_2) : |z_1||z_2|^a < 1\}$. Prove that Ω_a is a domain of holomorphy. Is every bounded holomorphic function on Ω_a constant? (*Hint:* Consider the case of a rational and the case of a irrational separately.)

47. Let Ω be the Hartogs triangle $\{(z_1, z_2) : |z_1| < |z_2| < 1\}$. Every analytic function on a neighborhood of $\bar{\Omega}$ continues to the bidisc. Find an orthonormal basis for $A^2(\Omega)$.

2 Subharmonicity and Its Applications

2.1 Subharmonic Functions

The analogue of the Laplacian for functions of one real variable is the differential operator d^2/dx^2. Thus the analogue of harmonic functions on $\mathbb{R}^1$ is the null space of this operator, or the linear functions.

A convex function f of a real variable is one whose graph satisfies the following property: If $(a, f(a)), (b, f(b))$ are points on the graph of f and τ is the chord connecting them, then the graph of f on the interval $[a, b]$ lies below τ. Subharmonic functions are the complex function-theoretic analogues of convex functions (with graphs of harmonic functions replacing line segments and discs replacing intervals). Although subharmonic functions are in fact more subtle than convex functions, the analogy that we have just drawn will prove useful both in formulating and in understanding the ideas connected with subharmonic function theory. We shall have a great deal more to say about convexity in Chapter 3.

Subharmonicity is a real-variable notion, so in this section all of our domains lie in $\mathbb{R}^N$. Good references for the classical theory of subharmonic functions in the complex plane are W. Hayman and P. B. Kennedy [1], W. Hayman [1], and M. Tsuji [1].

DEFINITION 2.1.1 A function $f : \Omega \to \mathbb{R} \cup \{-\infty\}$ is *upper semicontinuous* (u.s.c.) at a point $P \in \Omega$ if

$$\limsup_{\Omega \ni x \to P} f(x) \leq f(P).$$

It is upper semicontinuous on Ω if it is such at each point of Ω.

A function g on Ω is *lower semicontinuous* (l.s.c.) at $P \in \Omega$ (resp. on Ω) if $-g$ is u.s.c. at P (resp. on Ω).

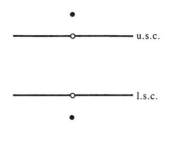

FIGURE 2.1

Exercises for the Reader

1. The real-valued function f is upper semicontinuous on Ω if and only if $\{x \in \Omega : f(x) < a\}$ is open for each $a \in \mathbb{R}$. The function g is lower semicontinuous on Ω if and only if $\{x \in \Omega : g(x) > a\}$ is open for each $a \in \mathbb{R}$. A real-valued function is continuous at a point if and only if it is both u.s.c. and l.s.c. at that point. Figure 2.1 is a useful mnemonic for distinguishing u.s.c. from l.s.c.

2. A function that is u.s.c. is bounded above on compacta. If $\{f_\alpha\}_{\alpha \in A}$ are u.s.c., then $f(x) \equiv \sup_\alpha f_\alpha(x)$ is not necessarily u.s.c., even if $f(x) < \infty$ at all points. But $g(x) = \inf_\alpha f_\alpha(x)$ *is* u.s.c. . If A is finite then f is u.s.c.

We leave the corresponding statements for l.s.c. functions for you to formulate.

PROPOSITION 2.1.2   If f is a u.s.c. function on Ω and bounded above, then there is a sequence $f_1 \geq f_2 \geq \cdots$ of continuous functions on Ω that are bounded above and that converge to f. In particular, if μ is a nonnegative finite Borel measure supported in Ω, then $\int_\Omega f d\mu$ is well defined and equals $\lim_j \int_\Omega f_j d\mu$.

Proof   Define

$$f_j(x) = \sup_{y \in \Omega}\{f(y) - j|x - y|\}, \quad x \in \Omega.$$

Clearly

$$f_1(x) \geq f_2(x) \geq \cdots \geq f(x) - j|x - x| = f(x)$$

and

$$f_j(x) \leq \sup f < \infty.$$

If we fix $\epsilon, j > 0$ and select $x_1, x_2 \in \Omega$ with $|x_1 - x_2| < \epsilon/j$, then for all $y \in \Omega$ it holds that

$$f(y) - j|x_1 - y| < f(y) - j|x_2 - y| + \epsilon.$$

Hence
$$f_j(x_1) - f_j(x_2) \leq \epsilon.$$

By symmetry,
$$f_j(x_2) - f_j(x_1) \leq \epsilon.$$

Thus each f_j is continuous.

It remains to verify that $f_j(x) \searrow f(x)$ for each x. Suppose without loss of generality that $\sup f = 1$. Fix $x \in \Omega$ and assume for simplicity that $f(x) = 0$ (the case $f(x) = -\infty$ is left for the reader). Fix $0 < \epsilon < 1$. Since f is u.s.c. there is a $\delta > 0$ such that if $|y - x| < \delta$, then $f(y) < \epsilon$. If $|y - x| \geq \delta$ and $j > 1/\delta$, then $f(y) - j|x - y| \leq 1 - 1 = 0 = f(x)$. It follows that

$$0 = f(x) \leq f_j(x) = \sup_{|y-x|<\delta} \{f(y) - j|x - y|\} \leq \epsilon.$$

Therefore, $f_j(x) \searrow f(x)$. □

DEFINITION 2.1.3   An upper semicontinuous function $f : \Omega \to \mathbb{R} \cup \{-\infty\}$ is said to be *subharmonic* if it satisfies the following condition: For every $x \in \Omega$ and $r > 0$ satisfying $\bar{B}(x, r) \subseteq \Omega$ and for every real-valued continuous function h on $\bar{B}(x, r)$ that is harmonic on $B(x, r)$ and satisfies $h \geq f$ on $\partial B(x, r)$, it holds that $h \geq f$ on $B(x, r)$.

The reason for the terminology subharmonic is manifest; refer again to the analogy with convex functions at the beginning of this section for motivation. By the maximum principle, harmonic functions are subharmonic. If both u and $-u$ are subharmonic, then u is harmonic. Our definition of subharmonic function allows the identically $-\infty$ function as an example. Some treatments rule out this function by fiat.

The following theorem will assist us in identifying and creating subharmonic functions.

THEOREM 2.1.4   Let $\Omega \subseteq \mathbb{R}^N$ and let $f : \Omega \to \mathbb{R} \cup \{-\infty\}$ be u.s.c. The following are equivalent:

1. f is subharmonic on Ω.

2. For all $\bar{B}(x, r) \subseteq \Omega$ and $P_{x,r}(\cdot, \cdot)$ the Poisson kernel for $B(x, r)$ (see Exercise 13 at the end of Chapter 1), we have

$$f(y) \leq \int_{\partial B(x,r)} P_{x,r}(y, t) f(t) \, d\sigma(t)$$

for all $y \in B(x, r)$.

3. If $x \in \Omega$ and $r > 0$ satisfy $\bar{B}(x,r) \subseteq \Omega$, then

$$f(x) \leq \frac{1}{r^{N-1}\omega_{N-1}} \int_{\partial B(x,r)} f(t) \, d\sigma(t).$$

4. If dist $(x, {}^c\Omega) > \delta > 0$ and if μ is any positive Borel measure on $[0, \delta]$, then

$$f(x) \cdot \int_0^\delta d\mu(s) \leq \frac{1}{\omega_{N-1}} \int_0^\delta \int_{\partial B} f(x + s\xi) \, d\sigma(\xi) \, d\mu(s).$$

5. For each $\delta > 0$ and x with dist $(x, {}^c\Omega) > \delta$, there exists one positive Borel measure μ on $[0, \delta]$ with $(\operatorname{supp} \mu) \cap (0, \delta] \neq \emptyset$ so that the inequality in (4) holds.

6. If $K \subset\subset \Omega$ is compact and $h \in C(K)$ is harmonic on $\overset{\circ}{K}$ and majorizes f on ∂K, then $h \geq f$ on K.

7. The function f is the limit of a decreasing sequence of subharmonic functions.

8. For all $\bar{B}(x,r) \subseteq \Omega$, we have

$$f(x) \leq \frac{1}{V(B(x,r))} \int_{B(x,r)} f(t) \, dV(t).$$

Remark: Some of the parts of the theorem are redundant, but we state each of them explicitly for convenient reference later. □

Proof of the Theorem The scheme of the proof is shown in Figure 2.2.
(4) ⟹ (5) Trivial.

(5) ⟹ (6) Let h be continuous on a compact subset K of Ω and harmonic on the interior of K, with $h \geq f$ on ∂K. If there is a $y \in \overset{\circ}{K}$ such that $w(y) \equiv f(y) - h(y) > 0$, then semicontinuity implies that w attains its maximum M over K on a relatively compact subset $L \subset\subset K$. Let $P \in L$ be a point that is nearest to ∂K. Choose $\delta > 0$ such that $\bar{B}(P, \delta) \subseteq \overset{\circ}{K}$. The extremal property of P guarantees that for $0 < s < \delta$, the sphere $\partial B(P, s)$ contains a relatively open set on which $w < M$. Thus, using the hypothesis about f and the mean value property for h, we see that

$$\int_0^\delta \int_{\partial B_{N-1}} w(P + s\xi) \, d\sigma(\xi) \, d\mu(s) < M\omega_{N-1} \int_0^\delta d\mu(s) = w(P)\omega_{N-1} \int_0^\delta d\mu(s).$$

This inequality contradicts (5).
(6) ⟹ (1) Trivial.

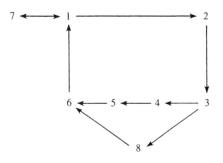

FIGURE 2.2

(1) ⇒ (2) Let f_j on $\bar{B}(x,r)$ be continuous with $f_j \searrow f$ (Proposition 2.1.2). Let h_j be the solution to the Dirichlet problem on $\bar{B}(x,r)$ with boundary data $f_j|_{\partial B(x,r)}$. Then, for $y \in B(x,r)$, it holds that

$$f(y) \le h_j(y) = \int_{\partial B(x,r)} P_{x,r}(y,t) f_j(t) \, d\sigma(t).$$

As $j \to +\infty$, the right side tends to

$$\int_{\partial B(x,r)} P_{x,r}(y,t) f(t) \, d\sigma(t).$$

(2) ⇒ (3) Set $y = x$ in (2).

(3) ⇒ (4) Integrate (3) against μ.

(1) ⇒ (7) Let $f_j = f + (1/j)$.

(7) ⇒ (1) Let $h \ge f$ on $\partial B(x,r), \bar{B}(x,r) \subseteq \Omega, h$ harmonic. Let $\epsilon > 0$. Let f_j be subharmonic, $f_j \searrow f$. Define $S_j = \{x \in \partial B(x,r) : f_j(x) \ge h(x) + \epsilon\}$. Then each S_j is compact, $S_1 \supseteq S_2 \supseteq \cdots$, and $\cap S_j = \emptyset$. Thus $S_j = \emptyset$ for j large. We conclude that $f \le f_j \le h + \epsilon$ on ∂B for j large. Since f_j is subharmonic, $f \le h + \epsilon$ on $\bar{B}(x,r)$. Letting $\epsilon \to 0^+$ now yields the result.

(3) ⇒ (8) Integrate out in the radial direction.

(8) ⇒ (6) Same as (5) ⇒ (6).   □

COROLLARY 2.1.5 Let u be subharmonic on the domain Ω and $M = \sup_\Omega u$. If $u(x) = M$ for some $x \in \Omega$ then $u \equiv M$ on Ω.

Proof   The set $S = \{x \in \Omega : u(x) = M\}$ is closed by upper semicontinuity. By part (8) of 2.1.4 and again by upper semicontinuity, S is also open. Since S is nonempty by hypothesis, it must be all of Ω.   □

COROLLARY 2.1.6 Finite sums of subharmonic functions are subharmonic.

Proof   Apply part (3) of 2.1.4.   □

COROLLARY 2.1.7 If f is subharmonic on Ω, and $\phi : \mathbb{R} \to \mathbb{R}$ is both convex and monotonically increasing, then $\phi \circ f$ is subharmonic (here $\phi(-\infty)$ is understood to equal $\lim_{t \to -\infty} \phi(t)$).

Proof   For any $x \in \Omega, r > 0$ small, we have

$$\phi \circ f(x) \;\leq\; \phi\left(\frac{1}{r^{N-1}\omega_{N-1}} \int_{\partial B(x,r)} f(t)\,d\sigma(t) \right)$$

$$\leq\; \frac{1}{r^{N-1}\omega_{N-1}} \int_{\partial B(x,r)} \phi \circ f(t)\,d\sigma(t)$$

by Jensen's inequality.   □

Exercise for the Reader
In the case that f is *harmonic*, it is necessary in Corollary 2.1.7 to assume only that ϕ is convex (not necessarily monotone increasing) to conclude that $\phi \circ f$ is subharmonic. The necessity of monotonicity for postcomposition with a convex function to preserve subsolutions is a general feature of elliptic equations.

COROLLARY 2.1.8 If each $x \in \Omega$ has a neighborhood $U_x \subseteq \Omega$ such that $f|_{U_x}$ is subharmonic on U_x, then f is subharmonic on Ω.

COROLLARY 2.1.9 If f_α are subharmonic on Ω and $f(x) \equiv \sup_\alpha f_\alpha(x)$ is u.s.c., then f is subharmonic on Ω.

Exercises for the Reader
1. In $\mathbb{R}$, harmonic functions are linear and subharmonic functions are convex. In particular, finite-valued subharmonic functions are continuous (which is not the case in higher dimensions). In dimensions at least 2, the fundamental solution Γ for the Laplacian is subharmonic (where it is understood that its value at $x = 0$ is $-\infty$). What (if anything) is the fundamental solution to the Laplacian in $\mathbb{R}^1$? Is it subharmonic? (*Hint:* You may have to use some distribution theory.)

2. If $\Omega \subseteq \mathbb{C}$ and f is holomorphic on Ω, then prove that $\log|f|$ is subharmonic. (*Hint:* Use the fact that harmonic functions are locally real parts of holomorphic functions; *do not* assume that f is nonvanishing.) Also, $|f|^p$ is subharmonic for $p > 0$.

3. Prove that if f is subharmonic on Ω, then for each $P \in \Omega$ we have

$$f(P) = \limsup_{\Omega \ni x \to P} f(x).$$

PROPOSITION 2.1.10 Assume that $\Omega \subseteq \mathbb{R}^N$ is a domain. Let $f : \Omega \to \mathbb{R} \cup \{\infty\}$ be subharmonic and *not* identically $-\infty$. Then f is locally integrable on Ω. In particular, $\mathcal{P} = \{x \in \Omega : f(x) = -\infty\}$ has zero Lebesgue measure.

Proof   Let $U = \{x \in \Omega : f$ is integrable on a neighborhood of $x\}$. If $x \in \Omega \backslash \mathcal{P}$ then, for r sufficiently small,

$$f(x) \leq \frac{1}{V(B(x,r))} \int_{B(x,r)} f(t)\, dV(t) < \infty,$$

where the right inequality follows from the fact that u.s.c. functions are bounded above on compacta. So we know that U is open and nonempty and that $^{c}U \subseteq \mathcal{P}$. But $f \equiv -\infty$ in a neighborhood of each element of ^{c}U. Hence ^{c}U is open. Since Ω is connected, and U is nonempty, we conclude that $U = \Omega$. Hence f is locally integrable on Ω. □

Exercise for the Reader
Let $\Omega \subseteq \mathbb{R}^N$ and $f : \Omega \to \mathbb{R} \cup \{\infty\}$ be subharmonic and not identically $-\infty$. If $w \in \Omega$ and $\bar{B}(w,r) \subseteq \Omega, r > 0$, then

$$\int_{\partial B(w,r)} |f(w)|\, d\sigma(w)$$

exists and is *finite*. (*Hint:* The positive part of f is easy to control. The negative part is controlled by the sub-mean value property.)

PROPOSITION 2.1.11   If f is subharmonic on Ω and not identically $-\infty$ and if $\phi \in C_c^\infty(\Omega), \phi \geq 0$, then

$$\int_\Omega f \Delta \phi \, dV(x) \geq 0.$$

Proof   The existence of the integral follows from Proposition 2.1.10. On the other hand, let $x \in \Omega, r < \text{dist}\,(x, \partial\Omega)$, and write

$$f(x) \leq \frac{1}{\omega_{N-1}} \int_{\partial B(0,1)} f(x + r\zeta)\, d\sigma(\zeta).$$

Then

$$\omega_{N-1} \int f(x)\phi(x)\, dV(x) \leq \int \phi(x) \int_{\partial B(0,1)} f(x + r\zeta)\, d\sigma(\zeta)\, dV(x)$$

$$= \int f(x) \int_{\partial B(0,1)} \phi(x - r\zeta)\, d\sigma(\zeta)\, dV(x)$$

$$= \int f(x) \int_{\partial B(0,1)} \sum_{|\alpha| \leq 2} \left(\frac{\partial}{\partial x}\right)^\alpha \phi(x) \cdot \frac{(-r\zeta)^\alpha}{\alpha!} + \mathcal{O}(r^3)\, d\sigma(\zeta)\, dV(x).$$

Now the zero-order term in the Taylor expansion on the right cancels with the far left side; also the first-order terms and the mixed second-order terms vanish by parity. If we divide through by r^2 and let $r \to 0$, then the $\mathcal{O}(r^3)$ term vanishes, and we are finally left with

$$0 \leq \int_\Omega f(x) \sum_{j=1}^{N} \frac{\partial^2 \phi}{\partial x_j^2}(x) \frac{\omega_{N-1}}{N} \, dV(x). \qquad \Box$$

COROLLARY 2.1.12   If $f \in C^2(\Omega)$ and f is subharmonic on Ω, then $\Delta f \geq 0$.

Proof   Integrate Proposition 2.1.11 by parts. $\qquad \Box$

PROPOSITION 2.1.13   If $f \in C^2(\Omega)$ and $\Delta f \geq 0$ on Ω, then f is subharmonic on Ω.

Proof   First, suppose that $\Delta f > 0$ on Ω. If f is not subharmonic, then there is a $K \subset\subset \Omega$ and a function $h \in C(K)$ that is harmonic on the interior of K such that $h \geq f$ on ∂K but $w = f - h$ is positive at some interior point of K. Let $x_0 \in \overset{\circ}{K}$ be a point at which the maximum of w on $\overset{\circ}{K}$ is attained. Since $\Delta f(x_0) > 0$, there is a contradiction (use the second-derivative test from calculus).

In the general case, apply the preceding argument to $f + \epsilon |x|^2$; then let $\epsilon \to 0^+$ and use part (7) of Theorem 2.1.4. $\qquad \Box$

COROLLARY 2.1.14   If f is harmonic on Ω, then $|f|^p$ is subharmonic for all $p \geq 1$.

Proof   According to Exercise 2 of Chapter 1, $\Delta |f|^p \geq 0$ in a neighborhood of any point where where $f \neq 0$. At points where $f(x) = 0$, part (8) of 2.1.4 is trivially satisfied.

An alternate proof may be obtained by applying Jensen's inequality to the mean value property for harmonic functions. $\qquad \Box$

COROLLARY 2.1.15   If f is holomorphic on $\Omega \subseteq \mathbb{C}^n$, then $|f|^p$ is subharmonic for all $p > 0$.

Proof   The proof is similar to the proof of the last corollary. $\qquad \Box$

Exercise for the Reader
The hypotheses of the last corollary also imply that $\log |f|$ is subharmonic.

Remark: As we shall see in Chapter 8, Corollaries 2.1.14 and 2.1.15 go a long way toward explaining the different boundary behavior of holomorphic functions and harmonic functions. $\qquad \Box$

Our next goal is to prove a converse to 2.1.11. This requires some preliminary development.

LEMMA 2.1.16 If u is locally p^{th} power integrable on Ω, $1 \leq p < \infty$, and $\phi \in C_c^\infty$, $\int \phi \, dx = 1$, then $\lim_{r \to 0^+} \int u(x - rt)\phi(t) \, dV(t) = u(x)$ in the $L^p_{\text{loc}}(\Omega)$ topology (here u is understood to be equal to 0 off Ω).

Proof   Fix a compact set $K \subset\subset \Omega$. Choose $r > 0$ so small that dist $(x, \partial\Omega) > r$ for all $x \in K$. Then

$$\int_K \left| \int u(x - rt)\phi(t) \, dV(t) - u(x) \right|^p dV(x)$$

$$= \int_K \left| \int [u(x - rt) - u(x)] \, \phi(t) \, dV(t) \right|^p dV(x)$$

$$\leq \int_{\mathbb{R}^N} \int_K |u(x - rt) - u(x)|^p \, dV(x) |\phi(t)| \, dV(t)$$

by, for instance, Jensen's inequality. The continuity of the integral implies that, for r sufficiently small, $\int_K |u(x-rt)-u(x)|^p \, dV(x) < \epsilon$, uniformly over $t \in \text{supp }\phi$ (exercise) whence the last line does not exceed $\epsilon \|\phi\|_{L^1}$.   □

Remark: See also Exercise 2 immediately following Theorem 1.3.8.   □

Let f be subharmonic on Ω and let $\bar{B}(x, r) \subseteq \Omega$. Then, on $\partial B(x, r)$, f is the limit of a decreasing sequence of continuous functions f_j (see 2.1.2). Of course we may solve the Dirichlet problem on $B(x, r)$ with boundary data f_j. Call the solution u_j. Then it is easy to see, for instance from Harnack's principle, that the u_j converge to a harmonic function u. We will refer to u as the solution of the Dirichlet problem with boundary data $f|_{\partial B(x,r)}$.

LEMMA 2.1.17 If f is subharmonic on Ω, $\bar{B}(x, r_0) \subseteq \Omega$, and $0 < r < r' \leq r_0$, then (with $B = B(0, 1)$),

$$\int_{\partial B} f(x + r\xi) \, d\sigma(\xi) \leq \int_{\partial B} f(x + r'\xi) \, d\sigma(\xi). \tag{2.1.17.1}$$

Proof   Let F be the solution to the Dirichlet problem on $B(x, r)$ with data $f|_{\partial B(x,r)}$ and define

$$\tilde{f}(t) = \begin{cases} F(t) & \text{if } |t - x| \leq r \\ f(t) & \text{if } |t - x| > r. \end{cases}$$

Then $\tilde{f}$ is subharmonic on Ω (why?), and its value at x is ω_{N-1}^{-1} times the left side of (2.1.17.1). Let G be the solution of the Dirichlet problem on $\bar{B}(x, r')$ with data $f|_{\partial B(x,r')} = \tilde{f}|_{\partial B(x,r')}$. Then G is a harmonic majorant for $\tilde{f}$ on $\bar{B}(x, r')$, and its value at x is ω_{N-1}^{-1} times the right side of (2.1.17.1).   □

THEOREM 2.1.18   If $f \in L^1_{\text{loc}}(\Omega)$ and $\int f(x)\Delta\phi(x)\,dV(x) \geq 0$ for all non-negative $\phi \in C^\infty_c(\Omega)$, then f can be corrected on a set of measure zero to be subharmonic. Indeed, if ψ is a fixed nonnegative radial function in $C^\infty_c(B(0,1))$ with $\int \psi(x)\,dV(x) = 1$, then

$$F(x) \equiv \lim_{r \to 0^+} \int f(x - rt)\psi(t)\,dV(t) \qquad \text{(in the L^1_{loc} topology)}$$

agrees with f a.e. and is subharmonic.

Proof   First suppose that $f \in C^2(\Omega)$. Then integration by parts and Proposition 2.1.13 give the result.

For the general case, replace f on

$$\Omega_\delta \equiv \{x \in \Omega : \text{dist}\,(x, \partial\Omega) > \delta\}$$

by

$$f_\delta(x) = \int f(x - \delta t)\psi(t)\,dV(t),$$

where ψ is chosen as in the statement of the theorem. Then $f_\delta \in C^\infty(\Omega_\delta)$ and f_δ satisfies the hypotheses of the theorem on Ω_δ. By the first part of the proof, f_δ is subharmonic on Ω_δ. If we can show that $\{f_\delta\}$ is a decreasing sequence as $\delta \to 0^+$, then $\lim_{\delta \to 0^+} f_\delta(x)$ exists for every x and is subharmonic by part (7) of 2.1.4. But Lemma 2.1.17 implies that

$$\int f_\delta(x - \epsilon t)\psi(t)\,dV(t)$$

decreases, for fixed δ, as $\epsilon \to 0^+$. We let $\delta \to 0^+$ in this line to obtain that f_ϵ decreases as $\epsilon \to 0^+$. Finally, since $f_\epsilon \to f$ in L^1_{loc}, it follows that $f_\epsilon \to f$ a.e. and the result is proved.   □

Remark: If G is another subharmonic function that equals f a.e., then

$$G(x) \leq \int G(x - \delta t)\psi(t)\,dV(t)$$

for each x by subharmonicity; also the lim sup of the right-hand side cannot exceed $G(x)$ by upper semicontinuity. Since this lim sup also equals $F(x)$, we conclude that there is precisely one way to correct f on a set of measure zero to make it subharmonic.   □

Exercises for the Reader

1. Check that the proof of Theorem 2.1.18 actually contains the following useful fact: Let $\Omega \subseteq \mathbb{R}^N$ and let $f : \Omega \to \mathbb{R}$ be subharmonic. For $\epsilon > 0$, let $\Omega_\epsilon \equiv \{x \in \Omega : \text{dist}(x, \partial\Omega) > \epsilon\}$. Then there is a family of functions $f_\epsilon \in C^\infty(\Omega_\epsilon)$ such that each f_ϵ is subharmonic on Ω_ϵ and $f_\epsilon \searrow f$.

2. Compare the theorem with Weyl's lemma from Chapter 1.

The following result, sometimes called Hartogs's lemma, is a key step in the proof of the separate analyticity theorem that we shall present in Section 4.

PROPOSITION 2.1.19   Let f_j be a sequence of subharmonic functions on Ω. Suppose that for each $K \subset\subset \Omega$ there is a finite $M_K > 0$ such that $f_j \leq M_K$ on K for every j. Suppose further that $\limsup_{j \to \infty} f_j(x) \leq C < \infty$ for all $x \in \Omega$. Then for each $\epsilon > 0$ and each $K \subset\subset \Omega$, there is a $J = J(\epsilon, K)$ such that

$$f_j(x) \leq C + \epsilon \qquad \text{for all } j \geq J, \text{ all } x \in K.$$

Proof   Fix $K \subset\subset \Omega$. Choose K_0 such that $K \subset\subset K_0 \subset\subset \Omega$ and K_0 is the closure of an open set. We may suppose that $f_j \leq -1$ on K_0, all j. Let $0 < r < \text{dist}(K, {}^cK_0)/3$. Choose, for $x \in K, \epsilon > 0$, a natural number j_0 so large that (by Fatou's lemma)

$$\int_{B(x,r)} f_j(t)\, dV(t) \leq V(B(0,1)) \cdot r^N \cdot (C + 2\epsilon), \qquad j \geq j_0.$$

(Note that the negativity of the functions involved reverses the usual inequality of Fatou.) Let

$$\delta = r \cdot \left[\left(\frac{C + 2\epsilon}{C + \epsilon} \right)^{1/N} - 1 \right].$$

Then $|x - x'| < \delta$ implies that

$$
\begin{aligned}
V(B)(r + \delta)^N f_j(x') &\leq \int_{B(x',r+\delta)} f_j(t)\, dV(t) \\
&\leq \int_{B(x,r)} f_j(t)\, dV(t) \\
&\leq V(B) r^N (C + 2\epsilon)
\end{aligned}
$$

or $f_j(x') \leq C + \epsilon$, all $x' \in B(x, \delta)$. Now we invoke the compactness of K to complete the proof.   □

We conclude with some results on removable sets for subharmonic functions (see also Section 2.2).

PROPOSITION 2.1.20   Let $\Omega \subseteq \mathbb{R}^N$ and $s : \Omega \to \mathbb{R} \cup \{-\infty\}$ be a subharmonic function not identically $-\infty$. Let $\mathcal{P} = \{x \in \Omega : s(x) = -\infty\}$ (we sometimes call $\mathcal{P}$ a *polar set*). Let $f \in C(\Omega)$ be subharmonic on $\Omega \setminus \mathcal{P}$. Then f is subharmonic on Ω.

Proof   Let $\Omega_0 \subset\subset \Omega$. Then s is bounded above on Ω_0, and we may assume that $s|_{\Omega_0} \leq 0$. For $\epsilon > 0$ the functions $f_\epsilon \equiv f + \epsilon s$ are subharmonic on Ω. (Exercise: This is an important trick in potential theory!) Thus if $K \subset\subset \Omega_0$ and $h \in C(K)$ is harmonic on $\overset{\circ}{K}$ and majorizes f on ∂K, it follows that

$$f_\epsilon(x) \leq f(x) \leq h(x), \quad x \in \partial K$$

whence

$$f_\epsilon(x) \leq h(x), \quad x \in K.$$

Letting $\epsilon \to 0^+$ gives

$$f(x) \leq h(x), \quad x \in K \setminus \mathcal{P}.$$

But $\mathcal{P}$ has measure zero (by 2.1.10); hence $^c\mathcal{P}$ is dense. Since f is continuous, it follows that

$$f(x) \leq h(x), \quad x \in K.$$

Hence f is subharmonic on Ω_0. But Ω_0 was an arbitrary relatively compact subset of Ω. Hence f is subharmonic on all of Ω. $\square$

Exercise for the Reader
Show that the hypothesis that f is continuous in Proposition 2.1.20 is actually necessary. In particular, show that it cannot be replaced by u.s.c. However, Proposition 2.1.21 gives a substitute result.

PROPOSITION 2.1.21   Let $\Omega \subseteq \mathbb{R}^N$ and $\mathcal{P} \subseteq \Omega$ be the polar set of a subharmonic function s on Ω. Suppose that $f : \Omega \setminus \mathcal{P} \to \mathbb{R}$ is bounded above on a neighborhood of $\mathcal{P}$ and is subharmonic. Then there exists a subharmonic function $\tilde{f}$ on all of Ω such that $\tilde{f}\big|_{\Omega \setminus \mathcal{P}} = f$.

Proof   Once again restrict attention to $\Omega_0 \subset\subset \Omega$. By the definition of u.s.c we are virtually forced to define

$$\tilde{f}(p) = \begin{cases} \limsup_{\Omega \setminus \mathcal{P} \ni x \to p} f(x) & \text{if } p \in \mathcal{P}, \\ f(p) & \text{if } p \notin \mathcal{P}. \end{cases} \tag{2.1.21.1}$$

Then $\tilde{f}$ will be u.s.c.
  Let $\epsilon > 0$. Assuming as we may that $s \leq 0$ on a neighborhood U of $\Omega_0 \cap \mathcal{P}$, we define $\tilde{f}_\epsilon = \tilde{f} + \epsilon s$. Then $\tilde{f}_\epsilon$ is subharmonic on U. Let $\bar{B}(x_0, r) \subseteq U$ and let h

be a continuous function on $\bar{B}(x_0, r)$, harmonic on the interior, that majorizes $\tilde{f}$ on $\partial B(x_0, r)$. Then

$$\tilde{f}_\epsilon(x) \leq \tilde{f}(x) \leq h(x) , \quad x \in \partial B(x_0, r) \setminus \mathcal{P}. \tag{2.1.21.2}$$

But $\tilde{f}_\epsilon = -\infty$ on $\partial B(x_0, r) \cap \mathcal{P}$ so (2.1.21.2) holds on all of $\partial B(x_0, r)$. Thus, it also holds on $B(x_0, r)$. Letting $\epsilon \to 0^+$ now yields that

$$\tilde{f}(x) \leq h(x) , \quad x \in B(x_0, r) \setminus \mathcal{P}.$$

This inequality extends to all of $B(x_0, r)$ by the definition of $\tilde{f}$. Since $B(x_0, r)$ was arbitrary in U, we conclude that $\tilde{f}$ is subharmonic U.   □

Remark (Hartogs Functions):   If $\Omega \subseteq \mathbb{C}$ and $f : \Omega \to \mathbb{C}$ is holomorphic, then $\log |f|$ is subharmonic, so that the zero set of f is polar. In one dimension this set is, of course, discrete, and we know from the Riemann removable singularities theorem that such a set is removable for locally bounded holomorphic functions. There is not such an elegant connection between pluripolar sets and removable singularities for holomorphic functions of several complex variables. That a pluripolar set is removable is true, but this is a weak implication (Cegrell [1]).

Even more is known. For, in a certain sense, the zero sets of holomorphic functions "generate" all polar sets in $\mathbb{C}$. More precisely, let $\Omega \subseteq \mathbb{C}$ and let $\mathcal{F}_\Omega$ be the smallest "cone" of functions that contains all $\log |f|$, f holomorphic, and that is closed under the following operations:

1. If $\phi_1, \phi_2 \in \mathcal{F}_\Omega$, then $\phi_1 + \phi_2 \in \mathcal{F}_\Omega$.

2. If $\phi \in \mathcal{F}_\Omega$ and $a \geq 0$, then $a\phi \in \mathcal{F}_\Omega$.

3. If $\{\phi_j\} \subseteq \mathcal{F}_\Omega, \phi_1 \geq \phi_2 \geq \cdots$, then $\lim_{j \to \infty} \phi_j \in \mathcal{F}_\Omega$.

4. If $\{\phi_j\} \subseteq \mathcal{F}_\Omega, \phi_j$ uniformly bounded above on compacta, then $\sup_j \phi_j \in \mathcal{F}_\Omega$.

5. If $\phi \in \mathcal{F}_\Omega$, then $\limsup_{z' \to z} \phi(z') \equiv \tilde{\phi}(z) \in \mathcal{F}_\Omega$.

6. If $\phi|_{\Omega'} \in \mathcal{F}_{\Omega'}$ for all $\Omega' \subset\subset \Omega$, then $\phi \in \mathcal{F}_\Omega$.

The elements of the family $\mathcal{F}_\Omega$ are called *Hartogs functions*.

It is a classical result that for $\Omega \subseteq \mathbb{C}$, the u.s.c. Hartogs functions are precisely the subharmonic functions.   □

Call a set $\mathcal{P} \subseteq \mathbb{C}$ *locally polar* if each $z \in \mathcal{P}$ has a neighborhood U_z and a subharmonic function $f_z : U_z \to \mathbb{R} \cup \{-\infty\}$ with $\mathcal{P} \cap U_z$ the polar set of f_z. It is a classical result that the locally polar sets are polar. This is proved by showing that polar sets are sets of zero logarithmic capacity (see M. Tsuji [1]). We do not investigate this matter here.

In the next section we shall comment on complex n-dimensional analogues of these ideas.

2.2   Pluriharmonic and Plurisubharmonic Functions

The function theory of several complex variables is remarkable in the range of techniques that may be profitably used to explore it: algebraic geometry, one complex variable, differential geometry, partial differential equations, harmonic analysis, and function algebras are only some of the areas that interact with the subject. Most of the important results in several complex variables bear clearly the imprint of one or more of these disciplines. But if there is one set of ideas that belongs exclusively to several complex variables, it is those centering around pluriharmonic and plurisubharmonic functions. These play a recurring role in any treatment of the subject; we record here a number of their basic properties.

Whereas the setting for the material in the last section was $\mathbb{R}^N$, the setting for the present section is $\mathbb{C}^n$. Let $a, b \in \mathbb{C}^n$. The set

$$\{a + b\zeta : \zeta \in \mathbb{C}\}$$

is called a *complex line* in $\mathbb{C}^n$. More generally, if $b^1, \ldots, b^k$ are linearly independent over $\mathbb{C}$, then the set

$$\left\{ a + \sum_{j=1}^{k} b^j \zeta_j : \zeta_1, \ldots, \zeta_k \in \mathbb{C} \right\}$$

is a k-dimensional *complex affine space* in $\mathbb{C}^n$ (it is not a complex *subspace* unless $a = 0$).

Remark: Note that *not every real two-dimensional affine space in $\mathbb{C}^n$ is a complex line.* For instance, the set $\ell = \{(x + i0, 0 + iy) : x, y \in \mathbb{R}\}$ is not a complex line in $\mathbb{C}^2$ according to our definition. This is rightly so, for the complex structures on ℓ and $\mathbb{C}^2$ are incompatible. This means the following: If $f : \mathbb{C}^2 \to \mathbb{C}$ is holomorphic, then it *does not* follow that $z = x + iy \mapsto f(x + i0, 0 + iy)$ is holomorphic. The point is that a complex line is supposed to be a (holomorphic) *complex affine imbedding* of $\mathbb{C}$ into $\mathbb{C}^n$. □

DEFINITION 2.2.1   A C^2 function $f : \Omega \to \mathbb{C}$ is said to be *pluriharmonic* if for every complex line $\ell = \{a + b\zeta\}$ the function $\zeta \mapsto f(a + b\zeta)$ is harmonic on the set $\Omega_\ell \equiv \{\zeta \in \mathbb{C} : a + b\zeta \in \Omega\}$.

Exercise for the Reader
A C^2 function f on Ω is pluriharmonic iff $(\partial^2/\partial z_j \partial \bar{z}_k) f \equiv 0$ for all $j, k = 1, \ldots, n$. This, in turn, is true iff $\partial \bar{\partial} f \equiv 0$ on Ω.

In the theory of one complex variable, harmonic functions play an important role because they are (locally) the real parts of holomorphic functions. The analogous role in several complex variables is played by pluriharmonic functions. To make this statement more precise, we first need a "Poincaré lemma."

LEMMA 2.2.2 Let $\alpha = \sum_j \alpha_j dx_j$ be a differential form with C^1 coefficients and satisfying $d\alpha = 0$ on a neighborhood of a closed box $S \subseteq \mathbb{R}^N$ with sides parallel to the axes. Then there is a function a on S satisfying $da = \alpha$.

Proof   Let $P = (p_1, \ldots, p_N) \in S$ and define, for $x \in S$,

$$a(x) = \sum_{j=1}^{N} \int_{p_j}^{x_j} \alpha_j(x_1, \ldots, x_{j-1}, t, p_{j+1}, \ldots, p_N) \, dt.$$

Then

$$
\begin{aligned}
\frac{\partial}{\partial x_k} a(x) \;=\;& \alpha_k(x_1, \ldots, x_k, p_{k+1}, \ldots, p_N) \\
& + \sum_{j=k+1}^{N} \int_{p_j}^{x_j} \frac{\partial \alpha_j}{\partial x_k}(x_1, \ldots, x_{j-1}, t, p_{j+1}, \ldots, p_N) \, dt \\
\;=\;& \alpha_k(x_1, \ldots, x_k, p_{k+1}, \ldots, p_N) \\
& + \sum_{j=k+1}^{N} \int_{p_j}^{x_j} \frac{\partial \alpha_k}{\partial x_j}(x_1, \ldots, x_{j-1}, t, p_{j+1}, \ldots, p_N) \, dt,
\end{aligned}
$$

where we have used the fact that α is d-closed. By the fundamental theorem of calculus, this last equals

$$
\begin{aligned}
& \alpha_k(x_1, \ldots, x_k, p_{k+1}, \ldots, p_N) \\
& + \sum_{j=k+1}^{N} [\alpha_k(x_1, \ldots, x_j, p_{j+1}, \ldots, p_N) - \alpha_k(x_1, \ldots, x_{j-1}, p_j, \ldots, p_N)] \\
\;=\;& \alpha_k(x_1, \ldots, x_N).
\end{aligned}
$$

That completes the proof.   □

Notice that the *proof* of the Poincaré lemma shows that if $d\alpha = 0$ and α has real coefficients, then a can be taken to be real. Now we have the following.

PROPOSITION 2.2.3 Let $D^n(P, r) \subseteq \mathbb{C}^n$ be a polydisc and assume that $f : D^n(P, r) \to \mathbb{R}$ is C^2. Then f is pluriharmonic on $D^n(P, r)$ if and only if f is the real part of a holomorphic function on $D^n(P, r)$.

Proof   The "if" part is trivial.
For "only if," notice that $\alpha \equiv i(\bar{\partial} f - \partial f)$ is real and satisfies $d\alpha = 0$. But then there exists a real function g such that $dg = \alpha$. In other words,

$d(ig) = (\partial f - \bar{\partial} f)$. Counting degrees, we see that $\bar{\partial}(ig) = -\bar{\partial} f$. It follows that $\bar{\partial}(f + ig) = \bar{\partial} f - \bar{\partial} f = 0$; hence g is the real function we seek. $\square$

Remark: If f is a function defined on a polydisc and harmonic in each variable separately, it is natural to wonder whether f is pluriharmonic. In fact the answer is "no," as the function $f(z_1, z_2) = z_1 \bar{z}_2$ demonstrates. This is a bit surprising, since the answer is affirmative when "harmonic" and "pluriharmonic" are replaced by "holomorphic" and "holomorphic."

Questions of "separate (P)," where (P) is some property, implying "joint (P)" are considered in detail in Hervé [1]. See also the recent work of Wiegerinck [2]. $\square$

A substitute result, which we present without proof, is the following (see F. Forelli [1] for details).

THEOREM 2.2.4 (Forelli)   Let $B \subseteq \mathbb{C}^n$ be the unit ball and $f \in C(\bar{B})$. Suppose that for each $b \in \partial B$ the function $z \mapsto f(z \cdot b)$ is harmonic on D, the unit disc. Suppose also that f is C^∞ in a neighborhood of the origin. Then f is pluriharmonic on B.

It is known that the hypothesis in Forelli's theorem that f be C^∞ near 0 cannot be appreciably weakened.

The proof of this theorem is clever but elementary; it uses only basic Fourier expansions of one variable. Such techniques are available only on domains—such as the ball and polydisc and, to a more limited extent, the bounded symmetric domains (see S. Helgason [1])—with a great deal of symmetry. One of the themes of this book is the following: Whereas the principal objects of study in complex analysis of one variable are functions, instead the principal objects of study in the complex analysis of several variables are domains. Every domain in complex n-dimensional space has its own geometric character that determines the function theory on that domain, and classical techniques that exploit groups of symmetries are usually neither available nor relevant.

Exercises for the Reader

1. Pluriharmonic functions are harmonic, but the converse is false.

2. If $\Omega \subseteq \mathbb{C}^n$ is a domain, $\mathcal{P}$ is its Poisson-Szegö kernel, and $f \in C(\bar{\Omega})$ is pluriharmonic on Ω, then $\mathcal{P}(f|_{\partial \Omega}) = f$. (What happens if we use the Szegö kernel instead of the Poisson-Szegö kernel?)

3. If f and f^2 are pluriharmonic, then f is either holomorphic or conjugate holomorphic (see Exercise 1 at the end of Chapter 1).

Remark: In Chapter 1 we solved the Dirichlet problem for harmonic functions on smoothly bounded domains in $\mathbb{R}^N$, in particular on the ball. Pluriharmonic functions are much more rigid objects; in fact the Dirichlet problem for these functions cannot always be solved. Let ϕ be a smooth function on the boundary of the ball B in $\mathbb{C}^2$ with the property that $\phi \equiv 1$ in a relative neighborhood of

$(1,0) \in \partial B$ and $\phi \equiv -1$ in a relative neighborhood of $(-1,0) \in \partial B$. Then any pluriharmonic function assuming ϕ as its boundary function would have to be identically equal to 1 in a neighborhood of $(1,0)$ and would have to be identically equal to -1 in a neighborhood of $(-1,0)$. Since a pluriharmonic function is real analytic, these conditions are incompatible.

In fact, there is a partial differential operator $\mathcal{L}$ on ∂B so that a smooth f on ∂B is the boundary function of a pluriharmonic function if and only if $\mathcal{L}f = 0$ (see E. Bedford [1], and E. Bedford and P. Federbush [1]). The operator $\mathcal{L}$ may be computed using just the theory of differential forms. It is remarkable that $\mathcal{L}$ is of third order. $\square$

DEFINITION 2.2.5   Let $\Omega \subseteq \mathbb{C}^n$ and let $f : \Omega \to \mathbb{R} \cup \{-\infty\}$ be u.s.c. We say that f is *plurisubharmonic* if, for each complex line $\ell = \{a + b\zeta\} \subseteq \mathbb{C}^n$, the function

$$\zeta \mapsto f(a + b\zeta)$$

is subharmonic on $\Omega_\ell \equiv \{\zeta \in \mathbb{C} : a + b\zeta \in \Omega\}$.

Remark: Because it is cumbersome to write out the word *plurisubharmonic*, a number of abbreviations for the word have come into use. Among the most common are *psh*, *plsh*, and *plush*. We shall sometimes use the first of these. $\square$

Exercise for the Reader
If $\Omega \subseteq \mathbb{C}^n$ and $f : \Omega \to \mathbb{C}$ is holomorphic, then $\log |f|$ is psh; so is $|f|^p, p > 0$. The property of plurisubharmonicity is local (see Corollary 2.1.8). A real-valued function $f \in C^2(\Omega)$ is psh iff

$$\sum_{j,k=1}^{n} \frac{\partial^2 f}{\partial z_j \partial \bar{z}_k}(z) w_j \bar{w}_k \geq 0$$

for every $z \in \Omega$ and every $w \in \mathbb{C}^n$. In other words, f is psh on Ω iff the complex Hessian of f is positive semidefinite at each point of Ω.

PROPOSITION 2.2.6   If $f : \Omega \to \mathbb{R} \cup \{-\infty\}$ is psh and $\phi : \mathbb{R} \cup \{-\infty\} \to \mathbb{R} \cup \{-\infty\}$ is convex and monotonically increasing, then $\phi \circ f$ is psh.

Proof   The proof is left as an exercise. $\square$

DEFINITION 2.2.7   A real-valued function $f \in C^2(\Omega), \Omega \subseteq \mathbb{C}^n$, is *strictly plurisubharmonic* if

$$\sum_{j,k=1}^{n} \frac{\partial^2 f}{\partial z_j \partial \bar{z}_k}(z) w_j \bar{w}_k > 0$$

for every $z \in \Omega$ and every $0 \neq w \in \mathbb{C}^n$ (see the preceding exercise for the reader for motivation).

Exercise for the Reader

With notation as in Definition 2.2.7, use the idea of homogeneity to see that if $K \subset\subset \Omega$, then there is a $C = C(K) > 0$ such that

$$\sum_{j,k=1}^{n} \frac{\partial^2 f}{\partial z_j \partial \bar{z}_k}(z) w_j \bar{w}_k \geq C|w|^2$$

for all $z \in K, w \in \mathbb{C}^n$.

PROPOSITION 2.2.8   Let $\Omega \subseteq \mathbb{C}^n$ and $f : \Omega \to \mathbb{R} \cup \{-\infty\}$ be a psh function. For $\epsilon > 0$ we set $\Omega_\epsilon = \{z \in \Omega : \operatorname{dist}(z, \partial\Omega) > \epsilon\}$. Then there is a family $f_\epsilon : \Omega \to \mathbb{R}$ such that $f_\epsilon \in C^\infty(\Omega_\epsilon)$, $f_\epsilon \searrow f$, and each f_ϵ is psh on Ω_ϵ.

Proof   Let $\phi \in C_c^\infty(\mathbb{C}^n)$ satisfy $\int \phi = 1$ and $\phi(z_1, \ldots, z_n) = \phi(|z_1|, \ldots, |z_n|)$ for all z. Assume that ϕ is supported in $B(0,1)$.
   Define

$$f_\epsilon(z) = \int f(z - \epsilon\zeta)\phi(\zeta)dV(\zeta) , \quad z \in \Omega_\epsilon.$$

Now adapt the arguments of 2.1.18.   □

Continuous plurisubharmonic functions are often called *pseudoconvex functions*.

Exercise for the Reader

If $\Omega_1, \Omega_2 \subseteq \mathbb{C}^n$, Ω_1 is bounded and $f : \Omega_2 \to \mathbb{R} \cup \{-\infty\}$ is C^2, then f is psh if and only if $f \circ \phi$ is psh for every holomorphic map $\phi : \Omega_1 \to \Omega_2$. Now prove the result assuming only that f is u.s.c. Why must Ω_1 be bounded?

The deeper properties of psh functions are beyond the scope of this book. The potential theory of psh functions is a rather well-developed subject and is intimately connected with the theory of the complex Monge-Ampère equation. Good reference for these matters are U. Cegrell [1], and M. Klimek [1]. See also the papers of E. Bedford and B. A. Taylor [2, 3, 4] and references therein. In earlier work (E. Bedford and B. A. Taylor [1]), the theory of the Dirichlet problem for psh functions is developed.

We have laid the groundwork earlier for one aspect of the potential theory of psh functions, and we should like to say a bit about it now. Call a subset $\mathcal{P} \subseteq \mathbb{C}^n$ *pluripolar* if it is the $-\infty$ set of a plurisubharmonic function. Then zero sets of holomorphic functions are obviously pluripolar. It is a result of B. Josefson [1] that locally pluripolar sets are pluripolar. In E. Bedford and B. A. Taylor [4], a capacity theory for pluripolar sets is developed that is a powerful tool for answering many natural questions about pluripolar sets. In particular, it gives another method for proving Josefson's theorem. Plurisubharmonic functions, which were first defined by Lelong, are treated in depth in the treatise L. Gruman and P. Lelong [1].

Define the Hartogs functions on a domain in $\mathbb{C}^n$ to be the smallest class of functions on Ω that contains all $\log|f|$, f holomorphic on Ω and that is closed under the operations (1)–(6) listed at the end of Section 2.1. H. Bremerman [2] showed that all psh functions are Hartogs (note that the converse is trivial) provided that Ω is a domain of holomorphy. He also showed that it is necessary for Ω to be a domain of holomorphy in order for this assertion to hold. This answered an old question of S. Bochner and W. Martin [1].

2.3   Power Series

Power series are essentially a real-variable device. The elementary results about power series representations of holomorphic functions of one or several variables are merely results about real analytic functions dressed up in complex notation. We refer the reader to the monograph S. G. Krantz and H. R. Parks [2] for a detailed treatment of the lore of real analytic functions.

In fact, there is an important point here that is valid even in the theory of one complex variable but is vital in keeping one's perspective in the function theory of several complex variables: Some facts about holomorphic functions of several variables are true merely because they are real analytic; some are true because they are harmonic; some are true because they are inherited from the function theory of one variable. Others are truly indigenous to the function theory of several complex variables. So far in this book we have encountered none of the latter type of result. These will begin to appear in Chapter 3.

The most important result of the present section is a characterization of domains of convergence of complex power series of several variables. The result contrasts with the simple result in one complex variable (all domains of convergence are discs) and gives us our first glimpse of what turns out to be the single unifying feature of the function theory of several complex variables: convexity.

DEFINITION 2.3.1   The power series $\sum_\alpha a_\alpha (x - P)^\alpha$ is said to *converge* at $x \in \mathbb{R}^N$ if some rearrangement of it converges. That is, if

$$\sum_{j=1}^{\infty} a_{\alpha(j)} (x - P)^{\alpha(j)}$$

is the rearrangement, we require that the partial sums

$$S_K \equiv \sum_{j=1}^{K} a_{\alpha(j)} (x - P)^{\alpha(j)}$$

converge numerically as $K \to \infty$.

One could worry here about different ways to define the convergence of a power series that is summed over the lattice of multi-indices. For conditional convergence, the definition that one takes for "partial sum" strongly affects the entire theory (see J. M. Ash [1] for a nice discussion of this topic). Since we will be concerned primarily with absolute convergence—which is, of course, invariant under rearrangement—these considerations are less crucial. For the record we mention that two standard methods of summing a series indexed over a lattice are to exhaust the lattice by spheres $(S_N = \sum_{|\alpha| \leq N})$ and by cubes $(T_N = \sum_{\{\alpha: |\alpha_j| \leq N, j=1,...,N\}})$. We shall not need to concern ourselves with the subtleties associated with the different methods of convergence. (However, in the theory of multiple Fourier series, there is a world of difference between the two methods— see K. M. Davis and Yang-Chun Chang [1].)

DEFINITION 2.3.2   Let $\Omega \subseteq \mathbb{R}^N$ be a domain. A function $f : \Omega \to \mathbb{C}$ is called *real analytic* if for each $P \in \Omega$, there is a neighborhood U_P of P such that for all $x \in U_P$ there is an absolutely and uniformly convergent series representation of the form

$$f(x) = \sum_\alpha a_\alpha (x - P)^\alpha.$$

Here the summation is over multi-indices α. The a_α will depend on P but, of course, not on $x \in U_P$.

Observe that polynomials are real analytic, as are the familiar transcendental functions of advanced calculus—$\sin x, \cos x, \tan x, \log |x|, \Gamma(x), \ \beta(x)$—on their respective domains. Compositions and inverses of real analytic functions are real analytic, although this is rather difficult to verify directly (see S. G. Krantz and H. R. Parks [2] for details). Simple examples of analytic functions of several variables are easily manufactured by superposition of these one-variable examples.
    The function

$$f(x) = \begin{cases} 0 & \text{if } x \leq 0, \\ e^{-1/|x|^2} & \text{if } x > 0 \end{cases}$$

is a C^∞ function that is not real analytic (why?).

Remark: If $\sum_\alpha a_\alpha (y - P)^\alpha$ converges at $y \in \mathbb{R}^N$, then there is a $C > 0$ such that $\{|a_\alpha||y - P|^\alpha\}$ is bounded by C.   □

Without loss of generality, we shall for simplicity of notation consider almost exclusively power series expanded about the origin; with this in mind we give the following definition. The result that follows it is the justification for the new terminology.

DEFINITION 2.3.3 If $x = (x_1, \ldots, x_N) \in \mathbb{R}^N$ (resp. $z = (z_1, \ldots, z_N) \in \mathbb{C}^n$), then define the *silhouette* of x (resp. the silhouette of z), denoted $s(x)$ (resp. $s(z)$) to be the set $\{(r_1 x_1, \ldots, r_N x_N) : -1 < r_j < 1, j = 1, \ldots, N\}$ (resp. $\{(\zeta_1 z_1, \ldots, \zeta_n z_n) : \zeta_j \in \mathbb{C}, |\zeta_j| < 1, j = 1, \ldots, n\}$).

Likewise, the silhouette of a set E is the union of the silhouettes of its elements.

PROPOSITION 2.3.4 (Abel's Lemma)   If $\sum_\alpha a_\alpha x^\alpha$ converges at a point $y \neq 0$, then it converges absolutely and uniformly on compact subsets of $s(y)$. Moreover, if p is a fixed polynomial of N arguments, then $\sum p(\alpha) a_\alpha x^\alpha$ converges absolutely and uniformly on compact subsets of $s(y)$.

Proof   Let $K \subset\subset s(y)$. Choose $\lambda = (\lambda_1, \ldots, \lambda_N), 0 < \lambda_j < 1$, with $|x_j| \leq \lambda_j |y_j|$, all $x \in K, j = 1, \ldots, N$. By the remark preceding the definition, there is a $C > 0$ such that $|a_\alpha y^\alpha| \leq C$, all multi-indices α. Assume that p is a monomial of degree d. Then we have, for $x \in K$,

$$\sum_\alpha |p(\alpha)||a_\alpha||x^\alpha| \;\leq\; \sum_\alpha C|p(\alpha)|\lambda^\alpha$$

$$\leq \; C \cdot C' \cdot \left[\sum_{\alpha_1} |\alpha_1 + 1|^d \lambda_1^{\alpha_1}\right] \cdots \left[\sum_{\alpha_N} |\alpha_N + 1|^d \lambda_N^{\alpha_N}\right]$$

$$< \; \infty.$$

The Weierstrass M-test completes the proof.   □

Remark: It is worth noting that the hypothesis $\sup_\alpha |a_\alpha y^\alpha| \leq C < \infty$ also implies the conclusion of the proposition.   □

PROPOSITION 2.3.5 If $f : \Omega \to \mathbb{C}$ is real analytic, then f is C^∞ and the derivatives of f on U_P (see the definition) may be obtained by termwise differentiation of the power series expansion of f about P. The derived functions are also real analytic on Ω.

Proof   Let $P \in \Omega$ and $r > 0$ be so small that the closure of $\mathcal{D}(P, r) \equiv \{x : |x_j - P_j| \leq r_j, j = 1, \ldots, N\}$ lies in U_P, where we are using the notation from the definition of real analytic. Let $Q \in \mathcal{D}(P, r)$ and select real numbers h_j such that $0 < |h_j| < r - |P_j - Q_j|, j = 1, \ldots, N$. Let $e_j = (0, \ldots, 0, 1, 0, \ldots, 0)$, with the 1 in the jth position. Then

$$\frac{f(Q + h_j e_j) - f(Q)}{h_j} = \sum_\alpha a_\alpha \cdot \frac{(Q + h_j e_j - P)^\alpha - (Q - P)^\alpha}{h_j}. \qquad (2.3.5.1)$$

The αth term in the sum does not exceed $C(|a_\alpha||\alpha||Q - P|^{|\alpha|-1})$ in absolute value. But

$$\sum_\alpha |a_\alpha||\alpha||Q - P|^{|\alpha|-1}$$

converges by the preceding proposition. By the (discrete version) of the Lebesgue dominated convergence theorem we may now conclude that the limit of (2.3.5.1) exists as $h_j \to 0$ and has the asserted limit. (All we are really doing here is passing a limit under an integral sign—a process that we have performed before—only now our measure space is discrete.)

The result for higher derivatives now follows by induction.   □

COROLLARY 2.3.6 Let $\sum_\alpha a_\alpha(x - P)^\alpha$ be a power series representation for f on $B(P,r) \subseteq \mathbb{R}^N$. Then

$$a_\alpha = \frac{1}{\alpha!} \cdot \left(\frac{\partial}{\partial x} \right)^\alpha f(P).$$

In particular, the power series expansion for a function f about a point P is unique if it exists.

Proof   The assertion for $\alpha = 0$ is obtained by setting $x = P$ in the power series expansion. The full result follows by differentiating the series (use the proposition) and setting $x = P$.   □

COROLLARY 2.3.7 If $\Omega \subseteq \mathbb{C}^n$ is a domain and $f : \Omega \to \mathbb{C}$ is holomorphic, then f is real analytic. If $P \in \Omega$, then the power series for f, expanded about P, is given by $\sum_\alpha a_\alpha(z - P)^\alpha$, where the α are multi-indices and $a_\alpha = (\partial/\partial z)^\alpha f(P)/\alpha!$.

Proof   The fact that f is real analytic follows, for instance, from the Bochner-Martinelli representation for f on any $\Omega' \subset\subset \Omega$ (or from the fact that f is harmonic)—just expand the real analytic kernel of the integral formula into a power series. The formula for the a_α follows either by writing the formula in Corollary 2.3.6 in complex notation (this is messy) or by mimicking the proofs of Proposition 2.3.5 and Corollary 2.3.6.

An alternate and somewhat more elementary proof results from using the Cauchy integral formula for a polydisc (see 1.2.2).   □

It is trivial—but should be noted—that a function defined by a uniformly convergent complex power series is of course holomorphic; this is so because it is the limit, uniformly on compact sets, of the partial sums (which are plainly holomorphic).

COROLLARY 2.3.8 Let $\Omega \subseteq \mathbb{R}^N$ be a domain and $f : \Omega \to \mathbb{C}$ be real analytic. Set $\mathcal{N} = \{x \in \Omega : f(x) = 0\}$. If $\mathcal{N}$ has nonempty interior, then $\mathcal{N} = \Omega$; i.e., $f \equiv 0$.

Proof   Let P be an interior point of $\mathcal{N}$. Then the power series expansion of f about P is identically zero. The set of all points with identically zero power series expansions is therefore open. It is also closed since f is C^∞. Since it is nonempty by hypothesis, it is all of Ω.   □

Remark: It is natural to wonder how thin the zero set of a real analytic function can be. In one real variable, if the zero set of a real analytic function has an interior limit point, then the function must be identically zero. There is an elaborate and difficult structure theorem for zero sets of real analytic functions in N dimensions (known as *real analytic varieties*) that generalizes this sharp one-dimensional result (see S. Lojaciewicz [1], and S. G. Krantz and H. R. Parks [2]).

What one can prove without much difficulty is that a real analytic variety must have N-dimensional Lebesgue measure zero. Prove this as an exercise, in contrapositive form, by induction on dimension.   □

2.3.1   Complexification

If $\sum_\alpha a_\alpha (x - P)^\alpha$ converges for $|x - P| < r$, and hence defines a real analytic function f for $|x - P| < r$, then we may define an associated *complexified function* $F(z) = \sum_\alpha a_\alpha (z - P)^\alpha$. Here we have identified the original point $P = (p_1, \ldots, p_N) \in \mathbb{R}^N$ with $P = (p_1 + i0, \ldots, p_n + i0) \in \mathbb{C}^N$. Notice that Abel's lemma implies that the new complexified series converges on the whole complex ball $B(x + i0, r)$ and defines a holomorphic function.

The notion of complexification is not simply an affectation, for complex analytic (holomorphic) functions f are easily identified by the differential condition $\bar\partial f = 0$. There is no such simple device for recognizing real analytic functions. As a simple application, to verify that the composition $f \circ g$ of two real analytic functions is real analytic it is best to complexify and then use the chain rule when applying $\partial / \partial \bar z_j$ to $F \circ G$. A more direct proof using just the definition of real analyticity is quite difficult (see S. G. Krantz and H. R. Parks [2]). One can also check that the collection of real analytic functions is closed under the arithmetic operations and under inversion. We leave details to the reader.

Exercises for the Reader

1. Let $a_0, a_1, \ldots$ be an arbitrary sequence of real numbers. Prove E. Borel's theorem: There is a C^∞ function f on $\mathbb{R}$ such that the Taylor series of f at 0 is $\sum_{j=0}^\infty a_j x^j$. More difficult is S. Besicovitch's theorem: If $\{a_j\}, \{b_j\}$ are real numbers, then then there exists an $f \in C^\infty[-1, 1]$ such that f is real analytic on $(-1, 1)$ and $f^{(j)}(-1) = a_j$, $f^{(j)}(1) = b_j$ for all j. See S. G. Krantz and H. R. Parks [2] for details.

2. Can a real analytic function be compactly supported?

3. Can a real analytic function vanish at ∞?

4. Can a real analytic function vanish exponentially at ∞?

Preliminary to our consideration of domains of convergence for power series of several variables is the next result.

LEMMA 2.3.9 (The Cauchy Estimates)   Let f be holomorphic on a neighborhood of $\bar{\mathcal{D}} \equiv \bar{D}(P_1, r_1) \times \cdots \times \bar{D}(P_n, r_n)$ and let $M = \max_{\mathcal{D}} |f|$. Then

$$\left| \left(\frac{\partial}{\partial z} \right)^\alpha f(P) \right| \leq \frac{M \cdot \alpha!}{r_1^{\alpha_1} \cdots r_n^{\alpha_n}}.$$

Proof   Write the Cauchy representation for f on the polydisc $\bar{\mathcal{D}}$, differentiate under the integral sign, and perform the most obvious estimations (as in the one-variable case).   □

LEMMA 2.3.10   Let $\sum_\alpha a_\alpha z^\alpha$ converge absolutely in a neighborhood of $w \in \mathbb{C}^n$. Then, for $\epsilon > 0$ sufficiently small,

$$|a_\alpha| \leq \frac{C_\epsilon}{\prod_{j=1}^n (|w_j| + \epsilon)^{\alpha_j}}.$$

Proof   By Abel's test, the series converges on a neighborhood of a polydisc of the form $\bar{D}(0, |w_1| + \epsilon) \times \cdots \times \bar{D}(0, |w_n| + \epsilon)$. Now apply the Cauchy estimates on this polydisc.   □

DEFINITION 2.3.11   Let $\sum_\alpha a_\alpha z^\alpha$ be a power series (as usual, for simplicity, we take $P = 0$). Let

$$\mathcal{C} = \bigcup_{r > 0} \left\{ z \in \mathbb{C}^n : \sum |a_\alpha w^\alpha| < \infty , \ \text{all } |z - w| < r \right\}.$$

We call $\mathcal{C}$ the *domain of convergence* for the power series.
    We also define

$$\mathcal{B} = \bigcup_{k=1}^\infty \left\{ z \in \mathbb{C}^n : \sup_\alpha |a_\alpha z^\alpha| \leq k \right\}.$$

Then $\mathcal{B}$ is called the *domain of boundedness* of the power series.

Easy considerations (such as the root test) show that $\mathcal{C} \subseteq \overset{\circ}{\mathcal{B}}$. However, Abel's test shows that $\overset{\circ}{\mathcal{B}} \subseteq \mathcal{C}$. Hence $\mathcal{C} = \overset{\circ}{\mathcal{B}}$.
    Not every open set can be the domain of convergence for a power series, for if $z \in \mathcal{C}$, then $(\mu_1 z_1, \ldots, \mu_n z_n) \in \mathcal{C}$ for every choice of $|\mu_j| \leq 1, j = 1, \ldots, n$. Thus we are interested in certain special sets.

DEFINITION 2.3.12   A set $S \subseteq \mathbb{C}^n$ is called *circular* (or a *circled set*) if $z \in S$ implies that $(e^{i\theta} z_1, \ldots, e^{i\theta} z_n) \in S$ for all $0 \leq \theta < 2\pi$. The set is called a *Reinhardt* *domain* if $(e^{i\theta_1}z_1,\ldots,e^{i\theta_n}z_n) \in S$ for all $0 \le \theta_j < 2\pi, j = 1,\ldots,n$. The set is called a *complete circular domain* or *complete Reinhardt* if $z \in S$ implies that $(\mu_1 z_1,\ldots,\mu_n z_n) \in S$ for all $\mu_j \in \mathbb{C}$ with $|\mu_j| \le 1, j = 1,\ldots,n$.

Warning: There is some confusion in the literature between circular and Reinhardt. Some writers do not distinguish between the two.

The silhouette of any given set is a complete circular domain. For us this will prove to be the most important example of such a domain.

DEFINITION 2.3.13   If $S \subseteq \mathbb{C}^n$, then we define

$$\log \|S\| = \{(\log|s_1|,\ldots,\log|s_n|) : s = (s_1,\ldots,s_n) \in S\}.$$

DEFINITION 2.3.14   A set $S \subseteq \mathbb{C}^n$ is said to be *logarithmically convex* if $\log \|S\| \subseteq \mathbb{R}^n$ is convex in the classical geometric sense.

Notice that it really makes sense to discuss logarithmic convexity only for a set that is Reinhardt (and, after some thought, it is only natural for a set that is complete circular—the results below will make this assertion clear). Now we have come to the main point of our work.

PROPOSITION 2.3.15   Let $\sum_\alpha a_\alpha z^\alpha$ be a power series. Then $\mathcal{C}$ is complete circular and logarithmically convex.

Proof   The complete circularity is obvious.

Let $w, w' \in \mathcal{C}, 0 \le \lambda \le 1$. Then, for $\epsilon > 0$ sufficiently small, Lemma 2.3.10 implies that

$$|a_\alpha| \le \frac{C_\epsilon}{\prod_{j=1}^n (|w_j| + \epsilon)^{\alpha_j}}$$

and

$$|a_\alpha| \le \frac{C_\epsilon}{\prod_{j=1}^n (|w_j'| + \epsilon)^{\alpha_j}}.$$

Hence, for slightly smaller ϵ,

$$|a_\alpha| \le \frac{C_\epsilon}{\prod_{j=1}^n (|w_j|^\lambda |w_j'|^{1-\lambda} + \epsilon)^{\alpha_j}}.$$

But this means that $(|w_1|^\lambda |w_1'|^{1-\lambda},\ldots,|w_n|^\lambda |w_n'|^{1-\lambda}) \in \mathcal{C}$, or

$$\lambda(\log|w_1|,\ldots,\log|w_n|) + (1-\lambda)(\log|w_1'|,\ldots,\log|w_n'|) \in \log \|\mathcal{C}\|.$$

This is the desired result.   □

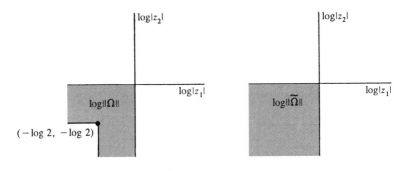

FIGURE 2.3

Remark: Abel's lemma implies that if $t \in \log \|\mathcal{C}\|$, then $t' \in \log \|\mathcal{C}\|$ for all t' satisfying $t'_j \leq t_j, j = 1, \ldots, n$. Moreover, $z \in \mathcal{C}$ if and only if $|z| \leq e^t$ componentwise for some $t \in \log \|\mathcal{C}\|$. Now the intersection of logarithmically convex sets is logarithmically convex. So if $S \subseteq \mathbb{C}^n$ is Reinhardt, then there is a smallest logarithmically convex Reinhardt domain $\tilde{S}$ containing S. Note that $\tilde{S}$ is in fact completely circular. By the proposition, any power series converging on S will actually converge on $\tilde{S}$. □

EXAMPLE   Let

$$\Omega = D(0,1) \setminus \bar{D}^2(0,1/2)$$
$$= \{z \in \mathbb{C}^2 : |z_1| < 1, 1/2 < |z_2| < 1\}$$
$$\cup \{z \in \mathbb{C}^2 : |z_2| < 1, 1/2 < |z_1| < 1\}.$$

Then Ω is Reinhardt and $\tilde{\Omega} = D^2(0,1)$. See Figure 2.3. This result is no surprise in view of Hartogs's extension phenomenon. □

PROPOSITION 2.3.16   If $\Omega \subseteq \mathbb{C}^n$ is a connected Reinhardt domain that contains 0 and $f : \Omega \to \mathbb{C}$ is holomorphic, then the power series expansion of f about 0 converges normally on Ω and hence normally on $\tilde{\Omega}$.

Proof   Write $\Omega = \cup_{j=1}^{\infty} \Omega_j$, where

$$\Omega_j \equiv \{z \in \Omega : \operatorname{dist}(z, {}^c\Omega) > |z|/j\}.$$

Notice that each Ω_j is Reinhardt, open, and contains 0. Moreover $\Omega = \cup_j \Omega_j$ and $\Omega_j \subseteq \Omega_{j+1}$ for each j. Since Ω is connected, we may replace each Ω_j by its connected component containing the origin and all the properties just listed will still hold.

Now the idea of the proof is to represent the (unique) power series expansion of f about 0 in two different ways. Fix j and $z \in \Omega_j$. Notice that the mapping

$$(\zeta_1, \ldots, \zeta_n) \mapsto f(\zeta_1 z_1, \ldots, \zeta_n z_n)$$

is well defined for $|\zeta_1| = |\zeta_2| = \cdots = |\zeta_n| = 1 + 1/j$ *by the definition of the set* Ω_j. Thus the Cauchy integral

$$f_z(w) \equiv \frac{1}{(2\pi i)^n} \int_{|\zeta_1| = 1+1/j} \cdots \int_{|\zeta_n| = 1+1/j} \frac{f(\zeta_1 z_1, \ldots, \zeta_n z_n)}{(\zeta_1 - w_1) \cdots (\zeta_n - w_n)} d\zeta_n \cdots d\zeta_1$$

defines a holomorphic function of w on $D^n(0, 1+1/j)$. If $|z|$ is small enough, then $(\zeta_1 z_1, \ldots, \zeta_n z_n) \in \Omega$ for every $\zeta \in \bar{D}^n(0, 1+1/j)$. Hence Theorem 1.2.2 yields that $f_z(1, 1, \ldots, 1) = f(z)$. By analytic continuation, $f_z(1, 1, \ldots, 1) = f(z)$ on all of Ω_j.

On the other hand, for w lying in a compact subset of $D^n(0, 1+1/j)$, we can expand the integrand of the Cauchy integral to obtain a normally convergent power series expansion for f_z about 0. The coefficient of w^α in this expansion is

$$(2\pi i)^{-n} \int_{|\zeta_1| = 1+1/j} \cdots \int_{|\zeta_n| = 1+1/j} \frac{f(\zeta_1 z_1, \ldots, \zeta_n z_n)}{\zeta_1^{\alpha_1 + 1} \cdots \zeta_n^{\alpha_n + 1}} d\zeta_n \cdots d\zeta_1$$

Now set $w = 1$, choose z small, and notice that the coefficient of w^α in fact must be $z^\alpha \partial^\alpha f(0)/\alpha!$. Thus, saying that the Taylor series for f_z converges normally on $D^n(0, 1 + 1/j)$ is just the same (by change of notation) as saying that the Taylor series expansion of f itself converges normally on Ω_j. This is what we wished to prove. $\square$

Proposition 3.4.10 will prove that every logarithmically convex complete circular domain Ω has defined on it a holomorphic function that cannot be analytically continued to any larger open set. By the preceding proposition, the domain of convergence for the power series of that function will be precisely Ω. Thus we shall learn that the Reinhardt domains that are domains of convergence are precisely the Reinhardt domains that are domains of holomorphy; these, in turn, are just the logarithmically convex Reinhardt domains.

2.4 Hartogs's Theorem on Separate Analyticity

Here we present a proof of Theorem 1.2.5—that a separately analytic function is holomorphic according to Definition 1.2.1. We again refer the reader to Hervé [1] for a detailed study of separate analyticity and related phenomena.

We begin by recalling the Baire category theorem.

THEOREM 2.4.1 Let X be a complete metric space. Then X cannot be written as a countable union of nowhere dense sets.

Proof   See S. G. Krantz [14].   □

Recall that Theorem 1.2.2 and subsequent remarks show that a function that is separately analytic and locally bounded on an open domain Ω is in fact C^∞, and hence holomorphic. So it is enough to check local boundedness.

Now the proof of Hartogs's theorem is by induction on dimension. When $n = 1$ there is, of course, nothing to prove. Assume inductively (*and for the rest of the section*) that the result has been proved in dimensions $j = 1, \ldots, n - 1$. (However, it should be mentioned that Lemma 2.4.3 stands alone and has been formulated accordingly.)

LEMMA 2.4.2   Let $\Omega \subseteq \mathbb{C}^n$ be a domain and let $f : \Omega \to \mathbb{C}$ be holomorphic in each variable separately. Let

$$\bar{\mathcal{D}} \equiv \bar{D}(z_1, r_1) \times \cdots \times \bar{D}(z_n, r_n) \subseteq \Omega$$

with $r_1, \ldots, r_n > 0$. Then there exist $\bar{D}(z_j', r_j') \subseteq \bar{D}(z_j, r_j), j = 1, \ldots, n$, with each $r_j' > 0$ and with $\bar{D}(z_n', r_n') = \bar{D}(z_n, r_n)$, such that f is bounded on

$$\bar{\mathcal{D}}' \equiv \bar{D}(z_1', r_1') \times \cdots \times \bar{D}(z_n', r_n').$$

In particular, f is C^∞ and holomorphic on $\mathcal{D}' = D(z_1', r_1') \times \cdots \times D(z_n', r_n')$.

Proof   We write $\zeta = (\zeta_1, \ldots, \zeta_n) = (\zeta', \zeta_n)$. For $M = 1, 2, \ldots$, define

$$S_M \equiv \left\{ \zeta' \in \prod_{j=1}^{n-1} \bar{D}(z_j, r_j) : |f(\zeta', \zeta_n)| \leq M, \text{ all } \zeta_n \in D(\zeta_n, r_n) \right\}.$$

Then

$$\prod_{j=1}^{n-1} \bar{D}(z_j, r_j) = \bigcup_{M=1}^{\infty} S_M.$$

By the inductive hypothesis, each $f(\cdot, \zeta_n)$ is continuous as a function of ζ', so each S_M is closed. By Baire's theorem, one of the S_M's, say S_{M_0}, has interior. The proof is completed by selecting $\bar{D}(z_1', r_1') \times \cdots \times \bar{D}(z_{n-1}', r_{n-1}') \subseteq S_{M_0}$.   □

Remark: Notice that Lemma 2.4.2 does not yet yield Hartogs's theorem. All we know so far is that every polydisc in Ω contains a polydisc (of full size in the last variable) on which f is holomorphic. Thus there may still be a closed set without interior on which f is ill behaved. This possibility is obviated by the following calculation, which is the heart of the proof.   □

LEMMA 2.4.3   Let $0 < r < R$, and suppose that $g : D^n(P, R) \to \mathbb{C}$ satisfies

1. For each $\zeta_n \in D(P_n, R)$, the function $\zeta' \mapsto g(\zeta', \zeta_n)$ is holomorphic on $D^{n-1}(P', R)$,

2. The function g is holomorphic and bounded on $\hat{D} \equiv D^{n-1}(P', r) \times D(P_n, R)$.

Then g is holomorphic on $D^n(P, R)$.

Proof   Assume for simplicity that $P = 0$. For fixed ζ_n, we expand $g(\cdot, \zeta_n)$ in a power series expansion on $D^{n-1}(0, R)$ (we use here hypothesis 1):

$$g(\zeta', \zeta_n) = \sum_{\alpha \in (\mathbb{Z}^+)^{n-1}} a_\alpha(\zeta_n)(\zeta')^\alpha. \qquad (2.4.3.1)$$

Here

$$a_\alpha(\zeta_n) = \left(\frac{\partial}{\partial \zeta'}\right)^\alpha g(0, \zeta_n)/\alpha!.$$

In particular, a_α is a holomorphic function of ζ_n by hypothesis 2.

It is our goal to see that the series (2.4.3.1) converges uniformly on compact subsets of $D^n(0, R)$. Let M be a bound for g on $\hat{D}$. Let $0 < R_1 < R_2 < R$. Now Lemma 2.3.10 yields that

$$|a_\alpha(\zeta_n)| R_2^{|\alpha|} \to 0 \qquad \text{when } |\alpha| \to \infty$$

and hypothesis (2) with the Cauchy estimates yields that

$$|a_\alpha(\zeta_n)| r^{|\alpha|} \le M \qquad \text{when } |\zeta_n| < R.$$

This last line tells us that the function

$$\zeta_n \mapsto \frac{1}{|\alpha|} \log |a_\alpha(\zeta_n)| \equiv h_\alpha(\zeta_n)$$

is subharmonic and bounded above (by $\log(M + 1)$) on the disc $\{\zeta_n : |\zeta_n| < R\}$. And the line that precedes it tells us that

$$\limsup_{|\alpha| \to \infty} h_\alpha(\zeta_n) \le \log(1/R_2).$$

Thus we may apply Hartogs's Proposition 2.1.19 to see that, for $|\alpha|$ sufficiently large,

$$\frac{1}{|\alpha|} \log |a_\alpha(\zeta_n)| \leq \log \frac{1}{R_1}, \quad |\zeta_n| < R_1.$$

In conclusion,

$$|a_\alpha(\zeta_n)| \cdot (R_1)^{|\alpha|} \leq 1, \quad \text{all } |\zeta_n| < R_1.$$

This trivially yields that the series (2.4.3.1) converges normally on $\{|\zeta_n| < R_1\}$. Since R_1 was an arbitrary positive number less than R, we have that the series (2.4.3.1) converges normally on $\{|\zeta_n| < R\}$.

We conclude that the series (2.4.3.1) converges to a harmonic, hence locally bounded, hence (by Cauchy theory) holomorpic function on $D^n(P, R)$.   □

Final Argument in the Proof of Theorem 1.2.5 Let $w \in \Omega, R > 0$ satisfy $\bar{D}^n(w, 2R) \subseteq \Omega$. Apply Lemma 2.4.2 to find a $P \in D^n(w, R)$ and $r = \min_{j=1,\ldots,n}\{r'_j\}$ so that the hypotheses of Lemma 2.4.3 are satisfied with $g = f$. Then, by 2.4.3, f is holomorphic in a neighborhood of w. Since $w \in \Omega$ was arbitrary, we are finished.   □

The proof of Hartogs's theorem that we have presented is not essentially different from the original proof of F. Hartogs [1]. In spite of the intervening years, no essentially simpler argument has been found. An old conjecture of Hervé [1] was that a separately subharmonic function is subharmonic. This implication would give an easy and natural way to see that a separately analytic function is locally bounded above, hence is holomorphic. Unfortunately, the Hervé conjecture is false, as was recently discovered by Wiegerinck [2].

It would be quite interesting to have a truly new proof of Hartogs's theorem.

EXERCISES————————————————————————

1. Prove that if $u_1, u_2 \geq 0$ on $\Omega \subseteq \mathbb{C}$ and $\log u_j$ is subharmonic, $j = 1, 2$ (where $\log 0$ is understood to be $-\infty$), then $\log(u_1 + u_2)$ is subharmonic.

2. Let $D \subseteq \mathbb{C}$ be the unit disc and, for $0 < p < \infty$, define

$$H^p(D) = \left\{ f \text{ holomorphic on } D : \sup_{0<r<1} \int_0^{2\pi} |f(re^{i\theta})|^p d\theta^{1/p} \equiv \|f\|_{H^p(D)} < \infty \right\}.$$

Prove that if $f \in H^p(D)$, then $|f|^p$ has a harmonic majorant on D in the sense that there is a harmonic function g on D with $g \geq |f|^p$. (*Hint:* Use Corollary 2.1.15 and Lemma 2.1.17).

3. Let

$$\Omega_1 = \left\{ (z_1, z_2) \in \mathbb{C}^2 : \frac{1}{4} < |z_2| < 1, |z_1| < 1 \right\}$$
$$\cup \left\{ (z_1, z_2) \in \mathbb{C}^2 : |z_2| < \frac{1}{2}, |z_1| < \frac{1}{2} \right\}.$$

Let

$$\Omega_2 = \left\{ (z_1, z_2) \in \mathbb{C}^2 : \frac{1}{4} < \sqrt{x_2^2 + y_1^2} < 1, \sqrt{x_1^2 + y_2^2} < 1 \right\}$$
$$\cup \left\{ (z_1, z_2) \in \mathbb{C}^2 : \sqrt{x_2^2 + y_1^2} < \frac{1}{2}, \sqrt{x_1^2 + y_2^2} < \frac{1}{2} \right\}.$$

Now Ω_1 is a circular domain. Compute $\log \|\Omega_1\|$ and $\log \|\tilde{\Omega}_1\|$. Although Ω_2 is not circular, can you compute the smallest Reinhardt domain that contains it? Notice the complex structure playing a role; that is, your answer should be affected by the way that Ω_2 sits in space.

4. Examine the domains of convergence for the sample series discussed in subsection 0.3.6. Verify that they are logarithmically convex and Reinhardt.

5. Suppose that $f : \mathbb{C}^n \to \mathbb{C}$ has the property that its restriction to *every* two-dimensional real affine subspace of $\mathbb{C}^n$ (*not* just the complex lines) is harmonic. Prove that f is linear. Suppose that the word harmonic is replaced by subharmonic. Then what can you say?

6. Let $E \subseteq \mathbb{R}$ be *any* closed set. Show that there is a C^∞ function $F : \mathbb{R} \to \mathbb{R}$ such that $\{x \in \mathbb{R} : f(x) = 0\} = E$. This fact generalizes to $\mathbb{R}^N, N > 1$, by way of the *Whitney decomposition* of an open set (see E. M. Stein [1]). Contrast these results with those in the text for the zero set of a real analytic function.

7. True or false? Let $\Omega \subseteq \mathbb{C}^n$ be a domain, $f : \Omega \to \mathbb{R}$ continuous. Suppose that for each compact $K \subset\subset \Omega$ and for each $h \in C(K)$ that is pluriharmonic on $\overset{\circ}{K}$, it holds that $h \geq f$ on ∂K implies $h \geq f$ on K. Then f is psh.

8. *The Poisson-Szegö kernel and the Poisson kernel.* Let $B \subseteq \mathbb{R}^N$ be the unit ball. If $f \in C^\infty(\partial B)$, then Pf, its *Poisson integral*, is in $C^\infty(\bar{B})$. This can be proved by integration by parts. The analogous result for the Poisson-Szegö kernel on $B \subseteq \mathbb{C}^n$ is false.

   a. Let $f : \partial B \to \mathbb{C}$ be given by $f(\zeta) = |\zeta_1|^2$. Calculate $\mathcal{P}f(r\mathbf{1}, 0)$ explicitly to see that, as a function of r, this function is C^1 and its first derivative is Lipschitz $1 - \epsilon$ for every $\epsilon > 0$. However, the function is not C^2.

   b. Use the fact that $\mathcal{P}$ commutes with rotations to see that if $f \in C^\infty(\partial B)$, then $\partial B \ni \zeta \mapsto \mathcal{P}f(r\zeta)$ is C^∞ for any fixed $0 < r < 1$.

c. On the ball, the Poisson-Szegö kernel solves the Dirichlet problem for the Laplace-Beltrami operator coming from the Bergman metric (see E. M. Stein [2]). This operator is given by

$$\Delta^B = \frac{4}{n+1} \cdot (1 - |z|^2) \sum_{j,k} (\delta_{j,k} - z_j \bar{z}_k) \frac{\partial^2}{\partial z_j \partial \bar{z}_k}.$$

Verify that Δ^B is elliptic on B, but not uniformly so. Compute its symbol to see this. As $|z| \to 1$, the operator degenerates.

d. It is a result of G. B. Folland [1] that a function $f \in C^\infty(\partial B)$ has Poisson-Szegö integral $\mathcal{P}f \in C^\infty(\bar{B})$ if and only if $\mathcal{P}f$ is pluriharmonic on B. Give an example of a function $f \in C^\infty(\partial B)$ for which you can verify directly that $\mathcal{P}f$ is not pluriharmonic. Verify that $\mathcal{P}(\cdot, \zeta)$ is not pluriharmonic in the first variable.

The work of C. R. Graham [1] explores in detail the nonregularity of the Poisson-Szegö operator on the ball. See also Krantz [19, Ch. 6].

9. *The generalized Cayley transform and the Siegel upper half-space.* Let $n \geq 2$ and let $B \subseteq \mathbb{C}^n$ be the unit ball. Let $\mathcal{U} = \{w \in \mathbb{C}^n : \operatorname{Im} w_1 > \sum_{j=2}^n |w_j|^2\}$. Define $\Phi(z) = (w_1, \ldots, w_n)$, where

$$w_1 = i \cdot \frac{1 - z_1}{1 + z_1}$$
$$w_j = \frac{z_j}{1 + z_1}, \quad j = 2, \ldots, n.$$

a. Verify that Φ maps B biholomorphically onto $\mathcal{U}$. The domain $\mathcal{U}$ is the standard unbounded realization of the bounded symmetric domain B. It is called a Siegel upper half-space of type II (see S. Kaneyuki [1]).

b. The set $\mathbb{C}^{n-1} \times \mathbb{R}$ can be equipped with the multiplicative structure

$$(\zeta, t) \cdot (\xi, s) = (\zeta + \xi, t + s + 2 \operatorname{Im} \zeta \cdot \bar{\xi}),$$

where $\zeta \cdot \bar{\xi} \equiv \sum_{j=1}^n \zeta_j \bar{\xi}_j$. Verify that this binary operation makes $\mathbb{C}^{n-1} \times \mathbb{R}$ into a nonabelian group (called the *Heisenberg group* and denoted by $\mathbb{H}_{n-1}$).

c. If $(t + i|z'|^2, z_2, \ldots, z_n) \in \partial\mathcal{U}$, where $z' = (z_2, \ldots, z_n)$, then identify this point with $(z', t) \in \mathbb{H}_{n-1}$. Thus $\partial\mathcal{U}$ is a group in a natural way.

d. If $w \in \mathcal{U}$, write $\rho(w) = \operatorname{Im} w_1 - |w'|^2 > 0$ and

$$w = ((\operatorname{Re} w_1 + i|w'|^2) + i(\operatorname{Im} w_1 - |w'|^2), w_2, \ldots, w_n)$$
$$\equiv (u_1, \ldots, u_n) + (i\rho(w), 0, \ldots, 0).$$

If $(z', t) = g \in \mathbb{H}_{n-1}$, then let

$$gw = g(u_1, \ldots, u_n) + (i\rho(w), 0, \ldots, 0),$$

where $g(u_1, \ldots, u_n)$ denotes the natural action of $\mathbb{H}_{n-1}$ on $\partial \mathcal{U}$. Indeed, each $g \in \mathbb{H}_{n-1}$ induces an element of $\mathrm{Aut}\, \mathcal{U}$ (the biholomorphic self-maps of $\mathcal{U}$). Denote this subgroup of $\mathrm{Aut}\, \mathcal{U}$ by N.

e. To what subgroup of $\mathrm{Aut}\, B$ does $N \subseteq \mathrm{Aut}\, \mathcal{U}$ correspond by way of the map Φ?

f. If $w = (w_1, \ldots, w_n) \in \mathcal{U}, \epsilon > 0$, define $\alpha_\epsilon(w) = (\epsilon^2 w_1, \epsilon w_2, \ldots, \epsilon w_n)$. Then $\{\alpha_\epsilon\} \subseteq \mathrm{Aut}\, \mathcal{U}$. Indeed $\{\alpha_\epsilon\}$ is a group. Call it A. To what subgroup of $\mathrm{Aut}\, B$ does it correspond?

g. Compute the subgroup K of $\mathrm{Aut}\, \mathcal{U}$ consisting of elements that fix $(i, 0, \ldots, 0)$. To what subgroup of $\mathrm{Aut}\, B$ does K correspond?

h. Show that $\mathrm{Aut}\, \mathcal{U} = K \cdot A \cdot N$. This is the *Iwasawa decomposition* of the Lie group $\mathrm{Aut}\, \mathcal{U}$ (see S. Helgason [1]).

10. How much of the construction in Exercise 9 can you carry out for the polydisc?

11. Refer to Exercise 9. Use Φ to calculate the Poisson-Szegö kernel for the Siegel upper half-space $\mathcal{U}$. Verify that the Poisson-Szegö operator on $\mathcal{U}$ can be realized as a convolution operator on the Heisenberg group (i.e., on $\partial \mathcal{U}$).

12. Let $\Omega \subset\subset \mathbb{C}^n$ be a domain with C^k boundary. Let $F : \bar{\Omega} \to \mathbb{C}^n$ be injective, holomorphic, and satisfy $|\nabla F| \geq C > 0$ on $\bar{\Omega}$. If k is large enough, $\epsilon > 0$ is small enough, and $\|F - G\|_{C^k(\bar{\Omega})} < \epsilon$ for some holomorphic G on Ω, then show that G is injective. What is the least k that suffices for this result? Is it necessary that F be holomorphic?

13. Prove that a C^∞ function f on an open interval $I \subseteq \mathbb{R}$ is real analytic if for any compact $K \subseteq I$ there exist constants C, M such that for any $k \in K$ and any positive integer j it holds that $|f^{(j)}(k)| \leq C \cdot M^k \cdot k!$. Prove the converse as well.

14. The following formula of Faà de Bruno is useful for differentiating compositions: If f, g, h are scalar-valued functions of a real variable and if $h = g \circ f$, then

$$h^{(n)}(t)$$
$$= \sum \frac{n!}{k_1! k_2! \ldots k_n!} g^{(k)}(f(t)) \left(\frac{f^{(1)}(t)}{1!} \right)^{k_1} \left(\frac{f^{(2)}(t)}{2!} \right)^{k_2} \cdots \left(\frac{f^{(n)}(t)}{n!} \right)^{k_n},$$

where $k = k_1 + k_2 + \ldots + k_n$ and the sum is taken over all $k_1, k_2, \ldots, k_n$ for which $k_1 + 2k_2 + \ldots + nk_n = n$.

Use this formula to verify directly that the composition of two real analytic functions is real analytic.

15. Use the result of Exercise 14 to prove that if $f : I \to \mathbb{R}$ is a real analytic function on an interval $I \subseteq \mathbb{R}$, if f' is nonvanishing, and if f has an inverse, then f^{-1} is real-analytic.

16. Prove that if μ is a positive, compactly supported measure of finite mass in $\mathbb{R}^N$ and Γ is the fundamental solution of the Laplacian, then

$$u(x) = \int \Gamma(x - t) \, d\mu(t)$$

defines a subharmonic function. What is the domain of u?

17. If u is a subharmonic function, then show that u is a distribution in a natural sense. Verify that Δu, interpreted in the sense of distributions, is positive and hence is a measure.

18. Let S be a countable subset of the unit disc in $\mathbb{C}$ with no interior accumulation point. Give a procedure for constructing a subharmonic function on D whose pole set is precisely S.

19. Let

$$P(z, \bar{z}) = \sum_{|\alpha| + |\beta| \leq N} a_{\alpha\beta} z^\alpha \bar{z}^\beta$$

be a polynomial in both z and $\bar{z}$. Give necessary and sufficient conditions for P to be pluriharmonic.

20. If $\mathcal{M}$ is a positive semidefinite $m \times m$ matrix and if $F = (f_1, \ldots f_m)$ is an n-tuple of holomorphic functions, then prove that

$$F \mathcal{M}^t \bar{F}$$

is plurisubharmonic.

21. What can you say about the size or geometry of the zero set of a nontrivial pluriharmonic function? (*Hint:* First consider harmonic functions in $\mathbb{R}^2$.)

22. Let f and g be pluriharmonic on a neighborhood of the closed unit ball in $\mathbb{C}^n, n > 1$. Let U be a neighborhood of the point $(1, 0, \ldots, 0) \in \partial B$. If f and g agree on $\partial B \cap U$, then prove that $f \equiv g$. Does this result still hold when $n = 1$?

23. Is there a Liouville theorem for pluriharmonic functions? For subharmonic functions? For plurisubharmonic functions?

24. A C^2, real-valued function is called convex if its real Hessian matrix is positive semidefinite at each point of its domain. Prove by an example that a subharmonic function need not be convex. Is it true that convex functions are subharmonic?

25. Suppose that $\phi : \mathbb{R} \cup \{-\infty\} \to \mathbb{R} \cup \{-\infty\}$ is a function with the property that for every subharmonic function f it holds that $\phi \circ f$ is subharmonic. What can you say about ϕ?

26. Suppose that it were the case that a function $f(z_1, \ldots, z_n)$ that is subharmonic in each complex variable separately is then subharmonic as a function on $\mathbb{C}^n \cong \mathbb{R}^{2n}$. This would provide an elegant and simple proof of the Hartogs's separate analyticity theorem. Unfortunately, the assertion about separate subharmonicity is false (see J. Wiegerinck [2]). Hervé [1] is the original source for the problem and gives background and related results.

27. Let Ω be a connected Reinhardt domain. Prove that if f is holomorphic on Ω, then f has a Laurent expansion that converges uniformly on compact subsets of Ω to f.

28. Let f be real analytic in the N-cube $(-1, 1) \times \cdots \times (-1, 1) \subseteq \mathbb{R}^N$. What sort of Cauchy estimates hold for the derivatives of f at the origin?

29. Let u be a harmonic function in $\mathbb{C}^n$ such that $z_j \cdot u$ is harmonic for $j = 1, \ldots, n$. Show that u must then be holomorphic. What other n-tuples of functions may be used instead of $(z_1, \ldots, z_n)$?

30. Let $\Omega = \{(z_1, z_2) \in \mathbb{C}^2 : |z_1| < |z_2| < 1\}$, the Hartogs triangle. Prove that there is no bounded psh function u on Ω such that $u(z)$ tends to 0 as z tends to $\partial\Omega$.

3 Convexity

3.1 Many Notions of Convexity

The concept of convexity goes back to the work of Archimedes, who used the idea in his axiomatic treatment of arc length. The notion was treated sporadically, and in an ancillary fashion, by Fermat, Cauchy, Minkowski, and others. It was not until the 1930s, however, that the first treatise on convexity (by Bonneson and Fenchel [1]) appeared. An authoritative discussion of the history of convexity can be found in Fenchel [1].

One of the most prevalent and classical definitions of convexity is as follows: A subset $S \subseteq \mathbb{R}^N$ is said to be convex if whenever $P, Q \in S$ and $0 \leq \lambda \leq 1$, then $(1 - \lambda)P + \lambda Q \in S$. In the remainder of this book we shall refer to a set or domain satisfying this condition as *geometrically convex*. From the point of view of analysis, this definition is of little use. We say this because the definition is *nonquantitative, nonlocal*, and *not formulated in the language of functions*. Put slightly differently, we have known since the time of Riemann that the most useful conditions in geometry are differential conditions. Thus we wish to find a differential characterization of convexity. We begin this chapter by relating classical notions of convexity to more analytic notions. All these ideas are properly a part of *real analysis*, so we restrict attention to $\mathbb{R}^N$.

Let $\Omega \subseteq \mathbb{R}^N$ be a domain with C^1 boundary. Let $\rho : \mathbb{R}^N \to \mathbb{R}$ be a C^1 *defining function* for Ω. Recall (Section 0.1 and Exercise 7 at the end of Chapter 1) that such a function has these properties:

1. $\Omega = \{x \in \mathbb{R}^N : \rho(x) < 0\}$

2. $^c\bar{\Omega} = \{x \in \mathbb{R}^N : \rho(x) > 0\}$

3. $\operatorname{grad} \rho(x) \neq 0 \ \ \forall x \in \partial\Omega$

If $k \geq 2$ and the boundary of Ω is a C^k manifold in the usual sense (see Appendix I), then it is straightforward to manufacture a C^1 (indeed a C^k) defining function for Ω by using the signed distance-to-the-boundary function.

DEFINITION 3.1.1   Let $\Omega \subseteq \mathbb{R}^N$ have C^1 boundary and let ρ be a C^1 defining function. Let $P \in \partial\Omega$. An N-tuple $w = (w_1, \ldots, w_N)$ of real numbers is called a *tangent vector* to $\partial\Omega$ at P if

$$\sum_{j=1}^{N}(\partial\rho/\partial x_j)(P) \cdot w_j = 0.$$

We write $w \in T_P(\partial\Omega)$.

Of course this definition makes sense only if it is independent of the choice of ρ. We shall address that issue in a moment. First, if P is a boundary point of a domain Ω with C^1 boundary, then we let ν_P denote the unit outward normal to $\partial\Omega$ at P. It should be observed that the condition defining tangent vectors simply mandates that $w \perp \nu_P$ at P. And, after all, we know from calculus that $\nabla\rho$ is the normal ν_P and that the normal is uniquely determined and independent of the choice of ρ. In principle, this settles the well-definedness issue.

However this point is so important and the point of view that we are considering is so pervasive that further discussion is warranted. The issue is this: if $\hat\rho$ is another defining function for Ω, then it should give the same tangent vectors as ρ at any point $P \in \partial\Omega$. The key to seeing that this is so is to write $\hat\rho(x) = h(x) \cdot \rho(x)$, for h a function that is nonvanishing near $\partial\Omega$. Then, for $P \in \partial\Omega$,

$$
\begin{aligned}
\sum_{j=1}^{N}(\partial\hat\rho/\partial x_j)(P) \cdot w_j &= h(P) \cdot \left(\sum_{j=1}^{N}(\partial\rho/\partial x_j)(P) \cdot w_j \right) \\
&\quad + \rho(P) \cdot \left(\sum_{j=1}^{N}(\partial h/\partial x_j)(P) \cdot w_j \right) \\
&= h(P) \cdot \left(\sum_{j=1}^{N}(\partial\rho/\partial x_j)(P) \cdot w_j \right) \\
&\quad + 0, \hspace{3cm} (3.1.2)
\end{aligned}
$$

because $\rho(P) = 0$. Thus w is a tangent vector at P vis-à-vis ρ if and only if w is a tangent vector vis-à-vis $\hat\rho$. But why does h exist?

After a change of coordinates, it is enough to assume that we are dealing with a piece of $\partial\Omega$ that is a piece of flat, $(N-1)$-dimensional real hypersurface (just use the implicit function theorem). Thus we may take $\rho(x) = x_N$ and

$P = 0$. Then any other defining function $\hat{\rho}$ for $\partial\Omega$ near P must have the Taylor expansion

$$\hat{\rho}(x) = c \cdot x_N + \mathcal{R}(x)$$

about 0. Here $\mathcal{R}$ is a remainder term satisfying $\mathcal{R}(x) = o(|x|)$. (There is no loss of generality to take $c = 1$, and we do so in what follows.) Thus we wish to define

$$h(x) = \frac{\hat{\rho}(x)}{\rho(x)} = 1 + \mathcal{S}(x).$$

Here $\mathcal{S}(x) \equiv \mathcal{R}(x)/x_N$ and $\mathcal{S}(x) = o(1)$ as $x_N \to 0$. Since this remainder term involves a derivative of $\hat{\rho}$, it is plain that h is not even differentiable. (An explicit counterexample is given by $\hat{\rho}(x) = x_N \cdot (1 + |x_N|)$.) Thus the program that we attempted in equation (3.1.2) is apparently flawed.

However an inspection of the explicit form of the remainder term $\mathcal{R}$ reveals that because $\hat{\rho}$ is constant on $\partial\Omega$, h as defined above *is* continuously differentiable *in tangential directions*. That is, for tangent vectors w (vectors that are orthogonal to ν_P), the derivative

$$\sum_j \frac{\partial h}{\partial x_j}(P) w_j$$

is defined. Thus it does indeed turn out that our definition of tangent vector is well posed when it is applied to vectors *that are already known to be tangent vectors* by the geometric definition $w \cdot \nu_P = 0$. For vectors that are *not* geometric tangent vectors, an even simpler argument shows that

$$\sum_j \frac{\partial \hat{\rho}}{\partial x_j}(P) w_j \neq 0$$

if and only if

$$\sum_j \frac{\partial \rho}{\partial x_j}(P) w_j \neq 0.$$

Thus Definition 1.1 is well posed. Questions similar to the one just discussed will come up later when we define convexity using C^2 defining functions (and also when we define the concept of pseudoconvexity). They are resolved in just the same way and we shall leave details to the reader.

The reader should check that the discussion above proves the following: If $\rho, \tilde{\rho}$ are C^k defining functions for a domain Ω, then there is a C^{k-1} function h defined near $\partial\Omega$ such that $\rho = h \cdot \tilde{\rho}$.

This somewhat protracted discussion of a small technical point seems necessary because it is recorded incorrectly in most places in the literature (including the first edition of this book).

3.1.1   The Analytic Definition of Convexity

For convenience, we restrict attention for this subsection to *bounded* domains. Many of our definitions would need to be modified and extra arguments would need to be given in proofs were we to consider unbounded domains as well.

DEFINITION 3.1.3   Let $\Omega \subset\subset \mathbb{R}^N$ be a domain with C^2 boundary and ρ a defining function for Ω. Fix a point $P \in \partial\Omega$. We say that $\partial\Omega$ is (weakly) *convex* at P if

$$\sum_{j,k=1}^{N} \frac{\partial^2 \rho}{\partial x_j \partial x_k}(P)w_j w_k \geq 0, \qquad \forall w \in T_P(\partial\Omega).$$

We say that $\partial\Omega$ is *strongly convex* at P if the inequality is strict whenever $w \neq 0$.

If $\partial\Omega$ is convex (resp. strongly convex) at each boundary point, then we say that Ω is convex (resp. strongly convex).

The quadratic form

$$\left(\frac{\partial^2 \rho}{\partial x_j \partial x_k}(P) \right)_{j,k=1}^{N}$$

is frequently called the "real Hessian" of the function ρ. This form carries considerable geometric information about the boundary of Ω. It is, of course, closely related to the second fundamental form of Riemannian geometry (see B. O'Neill [1]).

There is a technical difference between strong and strict convexity that we shall not discuss here (see L. Lempert [2] for details). It is common to use either of the words strong or strict to mean that the inequality in the last definition is strict when $w \neq 0$. The reader may wish to verify that at a strongly convex boundary point, all curvatures are positive (in fact, one may, by the positive definiteness of the matrix $(\partial^2 \rho / \partial x_j \partial x_k)$, impose a change of coordinates at P so that the boundary of Ω agrees with a ball up to and including second order at P). It is also the case that any strongly convex boundary point P is *extreme:* If $x, y \in \bar{\Omega}$ and if $P = (1 - \lambda)x + \lambda y$, some $0 < \lambda < 1$, then $x = y = P$. Although it is necessary and sufficient for strong convexity of a point P that all boundary curvatures be positive, it is only necessary that the boundary point be extreme. The point $P = (1, 0)$ in the boundary of the convex domain $\{(x_1, x_2) \in \mathbb{R}^2 : |x_1|^2 + |x_2|^4 < 1\}$ is extreme, but is not a point of strong convexity.

Now we explore our analytic notions of convexity. The first lemma is a technical one.

LEMMA 3.1.4   Let $\Omega \subseteq \mathbb{R}^N$ be strongly convex. Then there is a constant $C > 0$ and a defining function $\tilde{\rho}$ for Ω such that

$$\sum_{j,k=1}^{N} \frac{\partial^2 \tilde{\rho}}{\partial x_j \partial x_k}(P) w_j w_k \geq C|w|^2, \qquad \forall P \in \partial\Omega, w \in \mathbb{R}^N. \tag{3.1.4.1}$$

Proof   Let ρ be some fixed C^2 defining function for Ω. For $\lambda > 0$, define

$$\rho_\lambda(x) = \frac{\exp(\lambda\rho(x)) - 1}{\lambda}.$$

We shall select λ large in a moment. Let $P \in \partial\Omega$ and set

$$X = X_P = \left\{ w \in \mathbb{R}^N : |w| = 1 \text{ and } \sum_{j,k} \frac{\partial^2 \rho}{\partial x_j \partial x_k}(P) w_j w_k \leq 0 \right\}.$$

Then no element of X could be a tangent vector at P; hence $X \subseteq \{w : |w| = 1 \text{ and } \sum_j \partial\rho/\partial x_j(P)w_j \neq 0\}$. Since X is defined by a nonstrict inequality, it is closed; it is, of course, also bounded. Hence X is compact and

$$\mu \equiv \min \left\{ \left| \sum_j \partial\rho/\partial x_j(P)w_j \right| : w \in X \right\}$$

is attained and is nonzero. Define

$$\lambda = \frac{-\min_{w \in X} \sum_{j,k} \frac{\partial^2 \rho}{\partial x_j \partial x_k}(P)w_j w_k}{\mu^2} + 1.$$

Set $\tilde{\rho} = \rho_\lambda$. Then for any $w \in \mathbb{R}^N$ with $|w| = 1$, we have (since $\exp(\rho(P)) = 1$) that

$$\sum_{j,k} \frac{\partial^2 \tilde{\rho}}{\partial x_j \partial x_k}(P)w_j w_k = \sum_{j,k} \left\{ \frac{\partial^2 \rho}{\partial x_j \partial x_k}(P) + \lambda \frac{\partial\rho}{\partial x_j}(P)\frac{\partial\rho}{\partial x_k}(P) \right\} w_j w_k$$

$$= \sum_{j,k} \left\{ \frac{\partial^2 \rho}{\partial x_j \partial x_k} \right\}(P)w_j w_k + \lambda \left| \sum_j \frac{\partial\rho}{\partial x_j}(P)w_j \right|^2.$$

If $w \notin X$ then this expression is positive by definition. If $w \in X$ then the expression is positive by the choice of λ. Since $\{w \in \mathbb{R}^N : |w| = 1\}$ is compact, there is thus a $C > 0$ such that

$$\sum_{j,k} \left\{ \frac{\partial^2 \tilde{\rho}}{\partial x_j \partial x_k} \right\} (P) w_j w_k \geq C, \qquad \forall w \in \mathbb{R}^N \text{ such that } |w| = 1.$$

This establishes our inequality (3.1.4.1) for $P \in \partial\Omega$ fixed and w in the unit sphere of $\mathbb{R}^N$. For arbitrary w, we set $w = |w|\hat{w}$, with $\hat{w}$ in the unit sphere. Then (3.1.4.1) holds for $\hat{w}$. Multiplying both sides of the inequality for $\hat{w}$ by $|w|^2$ and performing some algebraic manipulations gives the result for fixed P and all $w \in \mathbb{R}^N$. (In the future we shall refer to this type of argument as a "homogeneity argument.")

Finally, notice that our estimates—in particular, the existence of C—hold uniformly over points in $\partial\Omega$ near P. Since $\partial\Omega$ is compact, we see that the constant C may be chosen uniformly over all boundary points of Ω.   □

Notice that the statement of the lemma has two important features: (1) that the constant C may be selected uniformly over the boundary and (2) that the inequality (3.1.4.1) holds for all $w \in \mathbb{R}^N$ (not just tangent vectors). In fact, it is impossible to arrange for anything like (3.1.4.1) to be true at a weakly convex point.

Our proof shows in fact that (3.1.4.1) is true not just for $P \in \partial\Omega$ but for P in a neighborhood of $\partial\Omega$. It is this sort of stability of the notion of strong convexity that makes it a more useful device than ordinary (weak) convexity.

PROPOSITION 3.1.5   If Ω is strongly convex, then Ω is geometrically convex.

Proof   We use a connectedness argument.

Clearly $\Omega \times \Omega$ is connected. Set $S = \{(P_1, P_2) \in \Omega \times \Omega : (1 - \lambda)P_1 + \lambda P_2 \in \Omega, \text{ all } 0 < \lambda < 1\}$. Then S is plainly open and nonempty.

To see that S is closed, fix a defining function $\tilde{\rho}$ for Ω as in the lemma. If S is not closed in $\Omega \times \Omega$, then there exist $P_1, P_2 \in \Omega$ such that the function

$$t \mapsto \tilde{\rho}((1 - t)P_1 + tP_2)$$

assumes an interior maximum value of 0 on $[0, 1]$. But the positive definiteness of the real Hessian of $\tilde{\rho}$ contradicts that assertion. The proof is complete.   □

We gave a special proof that strong convexity implies geometric convexity simply to illustrate the utility of the strong convexity concept. It is possible to prove that an arbitrary (weakly) convex domain is geometrically convex by showing that such a domain can be written as the increasing union of strongly convex domains. However the proof is difficult and technical (the reader interested in these matters may wish to consider them after learning the techniques in the proof of Theorem 3.3.5). We thus give another proof of this fact.

PROPOSITION 3.1.6   If Ω is (weakly) convex, then Ω is geometrically convex.

Proof   To simplify the proof we shall assume that Ω has at least C^3 boundary.
   Assume without loss of generality that $N \geq 2$ and $0 \in \Omega$. For $\epsilon > 0$, let
$\rho_\epsilon(x) = \rho(x) + \epsilon |x|^{2M}/M$ and $\Omega_\epsilon = \{x : \rho_\epsilon(x) < 0\}$. Then $\Omega_\epsilon \subseteq \Omega_{\epsilon'}$ if $\epsilon' < \epsilon$ and
$\cup_{\epsilon>0}\Omega_\epsilon = \Omega$. If $M \in \mathbb{N}$ is large and ϵ is small, then Ω_ϵ is strongly convex. By
Proposition 3.1.5, each Ω_ϵ is geometrically convex, so Ω is convex.   □

   We mention in passing that a nice treatment of convexity, from roughly
the point of view presented here, appears in V. Vladimirov [1].

PROPOSITION 3.1.7   Let $\Omega \subset\subset \mathbb{R}^N$ have C^2 boundary and be geometrically
convex. Then Ω is (weakly) convex.

Proof   Seeking a contradiction, we suppose that for some $P \in \partial\Omega$ and some
$w \in T_P(\partial\Omega)$, we have

$$\sum_{j,k} \frac{\partial^2 \rho}{\partial x_j \partial x_k}(P)w_j w_k = -2K < 0. \tag{3.1.7.1}$$

Suppose without loss of generality that coordinates have been selected in $\mathbb{R}^N$ so
that $P = 0$ and $(0, 0, \ldots, 0, 1)$ is the unit outward normal vector to $\partial\Omega$ at P.
We may further normalize the defining function ρ so that $\partial\rho/\partial x_N(0) = 1$. Let
$Q = Q^t = tw + \epsilon \cdot (0, 0, \ldots, 0, 1)$, where $\epsilon > 0$ and $t \in \mathbb{R}$. Then, by Taylor's
expansion,

$$
\begin{aligned}
\rho(Q) &= \rho(0) + \sum_{j=1}^N \frac{\partial\rho}{\partial x_j}(0)Q_j + \frac{1}{2}\sum_{j,k=1}^N \frac{\partial^2\rho}{\partial x_j \partial x_k}(0)Q_j Q_k + o(|Q|^2) \\
&= \epsilon\frac{\partial\rho}{\partial x_N}(0) + \frac{t^2}{2}\sum_{j,k=1}^N \frac{\partial^2\rho}{\partial x_j \partial x_k}(0)w_j w_k + \mathcal{O}(\epsilon^2) + o(t^2) \\
&= \epsilon - Kt^2 + \mathcal{O}(\epsilon^2) + o(t^2).
\end{aligned}
$$

Thus if $t = 0$ and $\epsilon > 0$ is small enough, then $\rho(Q) > 0$. However, for that same
value of ϵ, if $|t| > \sqrt{2\epsilon/K}$, then $\rho(Q) < 0$. This contradicts the definition of
geometric convexity.   □

Remark: The reader can already see in the proof of the proposition how useful
the quantitative version of convexity can be.
   The assumption that $\partial\Omega$ be C^2 is not very restrictive, for convex functions
of one variable are twice differentiable almost everywhere (see A. Zygmund [1]).
On the other hand, C^2 smoothness of the boundary is essential for our approach
to the subject.   □

Exercise for the Reader

If $\Omega \subseteq \mathbb{R}^N$ is a domain, then the *closed convex hull* of Ω is defined to be the closure of the set $\{\sum_{j=1}^{m} \lambda_j s_j : s_j \in \Omega, m \in \mathbb{N}, \lambda_j \geq 0, \sum \lambda_j = 1\}$.

Assume in the following problems that $\bar{\Omega} \subseteq \mathbb{R}^N$ is closed, bounded, and convex. Assume that Ω has C^2 boundary.

1. Prove that $\bar{\Omega}$ is the closed convex hull of its extreme points (this result is usually referred to as the *Krein-Milman theorem* and is true in much greater generality).

2. Let $P \in \partial\Omega$ be extreme. Let $\mathbf{p} = P + T_P(\partial\Omega)$ be the geometric tangent affine hyperplane to the boundary of Ω that passes through P. Show by an example that it is not necessarily the case that $\mathbf{p} \cap \bar{\Omega} = \{P\}$.

3. Prove that if Ω_0 is *any* bounded domain with C^2 boundary, then there is a relatively open subset U of $\partial\Omega_0$ such that U is strongly convex. (*Hint:* Fix $x_0 \in \Omega_0$ and choose $P \in \partial\Omega_0$ that is as far as possible from x_0).

4. If Ω is a convex domain, then the Minkowski functional (see S. R. Lay [1]) gives a convex defining function for Ω.

Our goal now is to pass from convexity to a complex-analytic analogue of convexity. We shall first express the differential condition for convexity in complex notation. Then we shall isolate that portion of the complexified expression that is invariant under biholomorphic mappings. This invariant version of convexity will be the focus of much of our study for the remainder of the book. Because of its centrality we have gone to extra trouble to put all these ideas into context.

Now fix $\Omega \subset\subset \mathbb{C}^n$ with C^2 boundary and assume that $\partial\Omega$ is convex at $P \in \partial\Omega$. If $w \in \mathbb{C}^n$, then the complex coordinates for w are, of course,

$$w = (w_1, \ldots, w_n) = (\xi_1 + i\eta_1, \ldots, \xi_n + i\eta_n).$$

Then it is natural to (geometrically) identify $\mathbb{C}^n$ with $\mathbb{R}^{2n}$ via the map

$$(\xi_1 + i\eta_1, \ldots, \xi_n + i\eta_n) \longleftrightarrow (\xi_1, \eta_1, \ldots, \xi_n, \eta_n).$$

Similarly, we identify $z = (z_1, \ldots, z_n) = (x_1 + iy_1, \ldots, x_n + iy_n) \in \mathbb{C}^n$ with $(x_1, y_1, \ldots, x_n, y_n) \in \mathbb{R}^{2n}$. (Strictly speaking, $\mathbb{C}^n$ is $\mathbb{R}^n \otimes_{\mathbb{R}} \mathbb{C}$. Then one equips $\mathbb{C}^n$ with a linear map J, called the *complex structure tensor*, which mediates between the algebraic operation of multiplying by i and the geometric mapping $(\xi_1, \eta_1, \ldots, \xi_n, \eta_n) \mapsto (-\eta_1, \xi_1, \ldots, -\eta_n, \xi_n)$. In this book it would be both tedious and unnatural to indulge in these niceties. In other contexts they are essential. See R. O. Wells [2] for a thorough treatment of this matter.) If ρ is a defining function for Ω that is C^2 near P, then the condition that $w \in T_P(\partial\Omega)$ is

$$\sum_j \frac{\partial\rho}{\partial x_j}\xi_j + \sum_j \frac{\partial\rho}{\partial y_j}\eta_j = 0.$$

In complex notation we may write this equation as

$$\frac{1}{2} \sum_j \left[\left(\frac{\partial}{\partial z_j} + \frac{\partial}{\partial \bar{z}_j} \right) \rho(P) \right] (w_j + \bar{w}_j)$$

$$+ \frac{1}{2} \sum_j \left[\left(\frac{1}{i} \right) \left(\frac{\partial}{\partial \bar{z}_j} - \frac{\partial}{\partial z_j} \right) \rho(P) \right] \left(\frac{1}{i} \right) (w_j - \bar{w}_j) = 0.$$

But this is the same as

$$2\mathrm{Re} \left(\sum_j \frac{\partial \rho}{\partial z_j}(P) w_j \right) = 0.$$

The space of vectors w that satisfy this last equation is not closed under multiplication by i, and hence is not a natural object of study for our purposes. Instead, we restrict attention in the following discussion to vectors $w \in \mathbb{C}^n$ that satisfy

$$\sum_j \frac{\partial \rho}{\partial z_j}(P) w_j = 0.$$

The collection of all such vectors is termed the *complex tangent space* to $\partial \Omega$ at P and is denoted by $\mathcal{T}_P(\partial \Omega)$. Clearly $\mathcal{T}_P(\partial \Omega) \subseteq T_P(\partial \Omega)$; indeed, the complex tangent space is a proper real subspace of the ordinary (real) tangent space. The reader should check that $\mathcal{T}_P(\partial \Omega)$ is the largest complex subspace of $T_P(\partial \Omega)$ in the following sense: If S is a real linear subspace of $T_P(\partial \Omega)$ that is closed under multiplication by i, then $S \subseteq \mathcal{T}_P(\partial \Omega)$. In particular, when $n = 1, \Omega \subseteq \mathbb{C}^n$, and $P \in \partial \Omega$ then $\mathcal{T}_P(\partial \Omega) = \{0\}$. At some level, this last fact explains many of the differences between the functions theories of one and several complex variables. Now we turn attention to the convexity condition.

The convexity condition on tangent vectors is

$$0 \leq \sum_{j,k=1}^n \frac{\partial^2 \rho}{\partial x_j \partial x_k}(P) \xi_j \xi_k$$

$$+ 2 \sum_{j,k=1}^n \frac{\partial^2 \rho}{\partial x_j \partial y_k}(P) \xi_j \eta_k + \sum_{j,k=1}^n \frac{\partial^2 \rho}{\partial y_j \partial y_k}(P) \eta_j \eta_k$$

$$= \frac{1}{4} \sum_{j,k=1}^n \left(\frac{\partial}{\partial z_j} + \frac{\partial}{\partial \bar{z}_j} \right) \left(\frac{\partial}{\partial z_k} + \frac{\partial}{\partial \bar{z}_k} \right) \rho(P)(w_j + \bar{w}_j)(w_k + \bar{w}_k)$$

$$+ 2 \cdot \frac{1}{4} \sum_{j,k=1}^n \left(\frac{\partial}{\partial z_j} + \frac{\partial}{\partial \bar{z}_j} \right) \left(\frac{1}{i} \right) \left(\frac{\partial}{\partial \bar{z}_k} - \frac{\partial}{\partial z_k} \right) \rho(P)$$

$$\times (w_j + \bar{w}_j) \left(\frac{1}{i}\right)(w_k - \bar{w}_k)$$

$$+ \frac{1}{4} \sum_{j,k=1}^{n} \left(\frac{1}{i}\right)\left(\frac{\partial}{\partial \bar{z}_j} - \frac{\partial}{\partial z_j}\right)\left(\frac{1}{i}\right)\left(\frac{\partial}{\partial \bar{z}_k} - \frac{\partial}{\partial z_k}\right)\rho(P)$$

$$\times \left(\frac{1}{i}\right)(w_j - \bar{w}_j)\left(\frac{1}{i}\right)(w_k - \bar{w}_k)$$

$$= \sum_{j,k=1}^{n} \frac{\partial^2 \rho}{\partial z_j \partial z_k}(P)w_j w_k + \sum_{j,k=1}^{n} \frac{\partial^2 \rho}{\partial \bar{z}_j \partial \bar{z}_k}(P)\bar{w}_j \bar{w}_k$$

$$+ 2 \sum_{j,k=1}^{n} \frac{\partial^2 \rho}{\partial z_j \partial \bar{z}_k}(P)w_j \bar{w}_k$$

$$= 2\,\mathrm{Re}\left(\sum_{j,k=1}^{n} \frac{\partial^2 \rho}{\partial z_j \partial z_k}(P)w_j w_k\right) + 2 \sum_{j,k=1}^{n} \frac{\partial^2 \rho}{\partial z_j \partial \bar{z}_k}(P)w_j \bar{w}_k.$$

(This formula could also have been derived by examining the complex form of Taylor's formula—see Exercise 35 at the end of Chapter 1.) We see that the real Hessian, when written in complex coordinates, decomposes rather naturally into two Hessian-like expressions. Our next task is to see that the first of these does not transform canonically under biholomorphic mappings but the second one does. We shall thus dub the second quadratic expression "the complex Hessian" of ρ. It will also be called the "Levi form" of the domain Ω. This form will be the object of our considerable attention for the remainder of this book.

The Riemann mapping theorem tells us, in part, that the unit disc is biholomorphic to any simply connected smoothly bounded planar domain. Since many of these domains are not convex, we see easily that biholomorphic mappings do not preserve convexity (an explicit example of this phenomenon is the mapping $\phi : D \to \phi(D), \phi(\zeta) = (\zeta + 2)^4$). We wish now to understand analytically where the failure lies. So let $\Omega \subset\subset \mathbb{C}^n$ be a convex domain with C^2 boundary. Let U be a neighborhood of $\bar{\Omega}$ and $\rho : U \to \mathbb{R}$ a defining function for Ω. Assume that $\Phi : U \to \mathbb{C}^n$ is biholomorphic onto its image and define $\Omega' = \Phi(\Omega)$. Hopf's lemma (Exercise 22 at the end of Chapter 1; the proof shows that Hopf's lemma is valid for subharmonic and hence for plurisubharmonic functions) guarantees that $\rho' \equiv \rho \circ \Phi^{-1}$ is a defining function for Ω'. Finally fix a point $P \in \partial\Omega$ and corresponding point $P' \equiv \Phi(P) \in \partial\Omega'$. If $w \in T_P(\partial\Omega)$, then

$$w' = \left(\sum_{j=1}^{n} \frac{\partial \Phi_1(P)}{\partial z_j}w_j, \ldots, \sum_{j=1}^{n} \frac{\partial \Phi_n(P)}{\partial z_j}w_j\right) \in T_{P'}(\partial\Omega').$$

Let the complex coordinates on $\Phi(U)$ be $z_1{}',\dots,z_n'$. Our task is to write the expression determining convexity,

$$2\,\mathrm{Re}\left(\sum_{j,k=1}^n \frac{\partial^2\rho}{\partial z_j\partial z_k}(P)w_jw_k\right) + 2\sum_{j,k=1}^n \frac{\partial^2\rho}{\partial z_j\partial\bar z_k}(P)w_j\bar w_k, \qquad (3.1.8)$$

in terms of the $z_j{}'$ and the $w_j{}'$. But

$$\begin{aligned}
\frac{\partial^2\rho}{\partial z_j\partial z_k}(P) &= \frac{\partial}{\partial z_j}\sum_{\ell=1}^n \frac{\partial\rho'}{\partial z_\ell'}\frac{\partial\Phi_\ell}{\partial z_k}\\
&= \sum_{\ell,m=1}^n\left\{\frac{\partial^2\rho'}{\partial z_\ell'\partial z_m'}\frac{\partial\Phi_\ell}{\partial z_k}\frac{\partial\Phi_m}{\partial z_j}\right\} + \sum_{\ell=1}^n\left\{\frac{\partial\rho'}{\partial z_\ell'}\frac{\partial^2\Phi_\ell}{\partial z_j\partial z_k}\right\},\\
\frac{\partial^2\rho}{\partial z_j\partial\bar z_k}(P) &= \frac{\partial}{\partial z_j}\sum_{\ell=1}^n \frac{\partial\rho'}{\partial\bar z_\ell'}\frac{\partial\bar\Phi_\ell}{\partial\bar z_k} = \sum_{\ell,m=1}^n \frac{\partial^2\rho'}{\partial z_m'\partial\bar z_\ell'}\frac{\partial\Phi_m}{\partial z_j}\frac{\partial\bar\Phi_\ell}{\partial\bar z_k}.
\end{aligned}$$

Therefore,

$$(3.1.8) = 2\,\mathrm{Re}\underbrace{\left\{\sum_{\ell,m=1}^n \frac{\partial^2\rho'}{\partial z_\ell'\partial z_m'}w_\ell'w_m' + \sum_{j,k=1}^n\sum_{\ell=1}^n \frac{\partial\rho'}{\partial z_\ell'}\frac{\partial^2\Phi_\ell}{\partial z_j\partial z_k}w_jw_k\right\}}_{\text{nonfunctorial}}$$

$$\underbrace{+2\sum_{\ell,m=1}^n \frac{\partial^2\rho'}{\partial z_m'\partial\bar z_\ell'}w_m'\bar w_\ell'}_{\text{functorial}}.$$

So we see that the part of the quadratic form characterizing convexity that is preserved under biholomorphic mappings is

$$\sum_{j,k=1}^n \frac{\partial^2\rho}{\partial z_j\partial\bar z_k}(P)w_j\bar w_k.$$

DEFINITION 3.1.8   Let $\Omega \subseteq \mathbb{C}^n$ be a domain with C^2 boundary and let $P \in \partial\Omega$. Let ρ be a C^2 defining function for Ω. We say that $\partial\Omega$ is *Levi pseudoconvex* at P if

$$\sum_{j,k=1}^n \frac{\partial^2\rho}{\partial z_j\partial\bar z_k}(P)w_j\bar w_k \geq 0, \qquad \forall w \in \mathcal{T}_P(\partial\Omega). \qquad (3.1.8.1)$$

The expression on the left side of (3.1.8.1) is called the *Levi form*. The point P is said to be *strongly (or strictly) Levi pseudoconvex* if the expression on the left side of (3.1.8.1) is positive whenever $w \neq 0, w \in T_P(\partial\Omega)$. A domain is called *Levi pseudoconvex* (resp. *strongly Levi pseudoconvex*) if all its boundary points are pseudoconvex (resp. strongly Levi pseudoconvex).

The reader may check that the definition of pseudoconvexity is independent of the choice of defining function for the domain in question.

The collection of Levi pseudoconvex domains is, in a local sense to be made precise later, the smallest class of domains that contains the convex domains and is closed under increasing union and biholomorphic mappings.

PROPOSITION 3.1.9   If $\Omega \subseteq \mathbb{C}^n$ is a domain with C^2 boundary and if $P \in \partial\Omega$ is a point of convexity, then P is also a point of pseudoconvexity.

Proof   Let ρ be a defining function for Ω. Let $w \in T_P(\partial\Omega)$. Then iw is also in $T_P(\partial\Omega)$. If we apply the convexity hypothesis to w at P, we obtain

$$2\operatorname{Re}\left(\sum_{j,k=1}^n \frac{\partial^2 \rho}{\partial z_j \partial z_k}(P)w_j w_k\right) + 2\sum_{j,k=1}^n \frac{\partial^2 \rho}{\partial z_j \partial \bar{z}_k}(P)w_j \bar{w}_k \geq 0.$$

However, if we apply the convexity condition to iw at P, we obtain

$$-2\operatorname{Re}\left(\sum_{j,k=1}^n \frac{\partial^2 \rho}{\partial z_j \partial z_k}(P)w_j w_k\right) + 2\sum_{j,k=1}^n \frac{\partial^2 \rho}{\partial z_j \partial \bar{z}_k}(P)w_j \bar{w}_k \geq 0.$$

Adding these two inequalities we find that

$$4\sum_{j,k=1}^n \frac{\partial^2 \rho}{\partial z_j \partial \bar{z}_k}(P)w_j \bar{w}_k \geq 0;$$

hence $\partial\Omega$ is Levi pseudoconvex at P.   □

The converse of this lemma is false. For instance, any product of smooth domains $\Omega_1 \times \Omega_2 \subseteq \mathbb{C}^2$ is Levi pseudoconvex at boundary points that are smooth (for instance, off the distinguished boundary $\partial\Omega_1 \times \partial\Omega_2$). From this observation a smooth example may be obtained simply by rounding off the product domain near its distinguished boundary. The reader should carry out the details of these remarks as an exercise.

There is no elementary geometric way to think about pseudoconvex domains. The collection of convex domains forms an important subclass, but by no means a representative subclass. As recently as 1972 it was conjectured that a pseudoconvex point $P \in \partial\Omega$ has the property that there is a holomorphic change of coordinates Φ on a neighborhood U of P such that $\Phi(U \cap \partial\Omega)$ is convex. This conjecture is false (see J. J. Kohn and L. Nirenberg [1]). In fact it is not known which pseudoconvex boundary points are "convexifiable."

The definition of Levi pseudoconvexity can be motivated by analogy with the real variable definition of convexity. However, we feel that the calculations above, which we learned from J. J. Kohn, provide the most palpable means of establishing the importance of the Levi form.

We conclude this discussion by noting that pseudoconvexity is not an interesting condition in one complex dimension because the complex tangent space to the boundary of a domain is always empty. Any domain in the complex plane is vacuously pseudoconvex.

3.1.2   Convexity with Respect to a Family of Functions

Let $\Omega \subseteq \mathbb{R}^N$ be a domain and let $\mathcal{F}$ be a family of real-valued functions on Ω (we do not assume in advance that $\mathcal{F}$ is closed under any algebraic operations, although often in practice it will be). Let K be a compact subset of Ω. Then the *convex hull of K in Ω with respect to $\mathcal{F}$* is defined to be

$$\hat{K}_{\mathcal{F}} \equiv \left\{ x \in \Omega : f(x) \leq \sup_{t \in K} f(t) \text{ for all } f \in \mathcal{F} \right\}.$$

We sometimes denote this hull by $\hat{K}$ when the family $\mathcal{F}$ is understood or when no confusion is possible. We say that Ω is *convex* with respect to $\mathcal{F}$ provided $\hat{K}_{\mathcal{F}}$ is compact in Ω whenever K is. When the functions in $\mathcal{F}$ are complex-valued, then $|f|$ replaces f in the definition of $\hat{K}_{\mathcal{F}}$.

PROPOSITION 3.1.10   Let $\Omega \subset\subset \mathbb{R}^N$ and let $\mathcal{F}$ be the family of real linear functions. Then Ω is convex with respect to $\mathcal{F}$ if and only if Ω is geometrically convex.

Proof   The proof is left as an exercise. Use the classical definition of convexity at the beginning of the section.   □

PROPOSITION 3.1.11   Let $\Omega \subset\subset \mathbb{R}^N$ be any domain. Let $\mathcal{F}$ be the family of continuous functions on Ω. Then Ω is convex with respect to $\mathcal{F}$.

Proof   If $K \subset\subset \Omega$ and $x \notin K$, then the function $F(t) = 1/(1 + |x - t|)$ is continuous on Ω. Notice that $f(x) = 1$ and $|f(k)| < 1$ for all $k \in K$. Thus $x \notin \hat{K}_{\mathcal{F}}$. Therefore, $\hat{K}_{\mathcal{F}} = K$ and Ω is convex with respect to $\mathcal{F}$.   □

PROPOSITION 3.1.12   Let $\Omega \subseteq \mathbb{C}$ be an open set and let $\mathcal{F}$ be the family of all functions holomorphic on Ω. Then Ω is convex with respect to $\mathcal{F}$.

Proof   First suppose that Ω is bounded. Let $K \subset\subset \Omega$. Let r be the Euclidean distance of K to the complement of Ω. Then $r > 0$. Suppose that $w \in \Omega$ is of distance less than r from $\partial\Omega$. Choose $w' \in \partial\Omega$ such that $|w - w'| = \text{dist}(w, {}^c\Omega)$. Then the function $f(\zeta) = 1/(\zeta - w')$ is holomorphic on Ω and $|f(w)| > \sup_{\zeta \in K} |f(\zeta)|$. Hence $w \notin \hat{K}_{\mathcal{F}}$, so $\hat{K}_{\mathcal{F}} \subset\subset \Omega$. Therefore, Ω is convex with respect to $\mathcal{F}$.

In case Ω is unbounded, we take a large disc $D(0, R)$ containing K and notice that $\hat{K}_{\mathcal{F}}$ with respect to Ω is equal to $\hat{K}_{\mathcal{F}}$ with respect to $\Omega \cap D(0, R)$, which by the first part of the proof is relatively compact. $\quad\square$

Exercise for the Reader
Prove that, in the last proposition, if the family $\mathcal{F}$ is replaced by the family $\mathcal{G}$ of bounded holomorphic functions, then not every Ω will be convex with respect to $\mathcal{G}$ (however your example will not have smooth boundary). What about the family $\mathcal{H}$ of square integrable holomorphic functions?

The reader should consider carefully why the argument in the last proposition breaks down in $\mathbb{C}^n, n \geq 2$. The reason is that, in higher dimensions, the zeros of a nonconstant holomorphic function are never isolated. The fact that holomorphic functions of one variable do have isolated zeros made it very easy to construct the required function f. But in $\mathbb{C}^2$, a holomorphic function that vanishes at $w' \in \partial\Omega$ might also perforce vanish at points inside the domain as well.

In fact, it is the notion of pseudoconvexity that helps to detect when the zeros of a holomorphic function that vanishes at a boundary point of a domain can be kept outside the domain. This is a subject that we explore in detail as the book develops.

With respect to the exercise, we note that in $\mathbb{C}^2$, there is a bounded, topologically trivial domain Ω with smooth boundary such that Ω is convex with respect to the family of all holomorphic functions but not with respect to all bounded holomorphic functions (see N. Sibony [1, 3]). A variant of this example will be considered in Exercise 7 at the end of Chapter 8.

3.1.3   Concluding Remarks

The discussion thus far in this chapter has shown that convexity for domains and convexity for functions are closely related concepts. We now develop the latter notion a bit further.

Classically, a real-valued function f on a convex domain Ω is called *convex* if, whenever $P, Q \in \Omega$ and $0 \leq \lambda \leq 1$, we have $f((1 - \lambda)P + \lambda Q) \leq (1 - \lambda)f(P) + \lambda f(Q)$. A C^2 function f is convex according to this definition if and only if the matrix $(\partial^2 f/\partial x_j \partial x_k)_{j,k=1}^N$ is positive semidefinite at each point of the domain of f. The function is convex *at a point* if this Hessian matrix is positive semidefinite at that point. It is *strongly* (or strictly) *convex* at a point of its domain if the matrix is strictly positive definite at that point. Of course, the function is called strictly convex if it is such at each point of its domain.

Now let $\Omega \subseteq \mathbb{R}^N$ be any domain. A function $\phi : \Omega \to \mathbb{R}$ is called an *exhaustion function* for Ω if, for any $c \in \mathbb{R}$, the set $\Omega_c \equiv \{x \in \Omega : \phi(x) \leq c\}$ is relatively compact in Ω. It is a fact (not easy to prove) that Ω is convex if and only if it possesses a convex exhaustion function, and that is true if and only if it possesses a strictly convex exhaustion function. In Section 3.3 we shall present the necessary tools for proving a result such as this.

We close this discussion of convexity with a geometric characterization of the property. We shall, later in the book, refer to this as the "affine sphere" characterization. First, if $\Omega \subseteq \mathbb{R}^N$ is a domain and I is a closed one-dimensional segment lying in Ω, then the boundary ∂I is the set consisting of the two end-points of I. Now the domain Ω is convex if and only if whenever $\{I_j\}_{j=1}^{\infty}$ is a collection of segments in Ω and $\{\partial I_j\}$ is relatively compact in Ω, then so is $\{I_j\}$. This is little more than a restatement of the classical definition of geometric convexity. We invite the reader to supply the details.

All our approaches to convexity will prove useful in later parts of the book. Each of them will have an analogue in the complex analytic setting, and each will be part of the arsenal of tools that we use to characterize domains of holomorphy. In fact, one of the main goals of the first five chapters of this book is to prove the following equivalence (here $\Omega \subseteq \mathbb{C}^n$ is a domain with C^2 boundary and $\mathcal{F}$ is the family of holomorphic functions on Ω):

Ω is a domain of holomorphy. $\Longleftrightarrow$

Ω is convex with respect to $\mathcal{F}$. $\Longleftrightarrow$

Ω is Levi pseudoconvex. $\Longleftrightarrow$

Ω has a C^{∞} strictly psh exhaustion function. $\Longleftrightarrow$

The equation $\bar{\partial}u = f$ can be solved on Ω for

every $\bar{\partial}$-closed (p,q) form f on Ω. $\Longleftrightarrow$

Whenever $\{\delta_j\}_{j=1}^{\infty} \subseteq \Omega$ is a family of closed analytic

discs (see Section 3.2) such that $\{\partial \delta_j\}$ is relatively

compact in Ω, then $\{\delta_j\}$ is also relatively compact

in Ω.

The hardest part of these equivalences is that a Levi pseudoconvex domain is a domain of holomorphy. This implication is known as the *Levi problem* and was solved completely for domains in $\mathbb{C}^n$, all n, only in the mid-1950s. Some generalizations of the problem to complex manifolds and analytic spaces remain open. An informative survey is Y. T. Siu [1]. We shall use the technique of the $\bar{\partial}$ equation to attack the Levi problem; that is, we first prove that the $\bar{\partial}$ equation can always be solved on a Levi pseudoconvex domain. Then we use this tool to show that a Levi pseudoconvex domain is a domain of holomorphy.

The next section collects a number of geometric properties of pseudoconvex domains. Although some of these properties are not needed for a while, it is appropriate to treat these properties all in one place.

3.2 Convexity and Pseudoconvexity

Let $\Omega \subset\subset \mathbb{C}^n$ be a domain with C^2 boundary. If $P \in \partial\Omega$, then P is a point of (weak) *Levi pseudoconvexity* if the Levi form is positive semi-definite on the space of $w \in \mathcal{T}_P(\Omega)$. Explicitly, $P \in \partial\Omega$ is a point of Levi pseudoconvexity for $\Omega = \{z \in \mathbb{C}^N : \rho(z) < 0\}$ if

$$\sum_{j,k=1}^{n} \frac{\partial^2 \rho}{\partial z_j \partial \bar{z}_k}(P) w_j \bar{w}_k \geq 0$$

for all $w \in \mathbb{C}^N$ that satisfy

$$\sum_{j=1}^{n} \frac{\partial \rho}{\partial z_j}(P) w_j = 0.$$

The point P is a point of *strong* (or *strict*) pseudoconvexity if the Levi form at P is positive definite for some choice of defining function and $w \in \mathcal{T}_P(\partial\Omega)$. One checks that these definitions are in fact independent of the choice of defining function. The domain Ω is said to be *Levi pseudoconvex* (resp. *strictly or strongly Levi pseudoconvex*) if every $P \in \partial\Omega$ is a point of Levi pseudoconvexity (resp. strict or strong Levi pseudoconvexity).

A line-by-line imitation of the proof of Lemma 3.1.4 yields the next result.

PROPOSITION 3.2.1   If Ω is strongly pseudoconvex, then Ω has a defining function $\tilde{\rho}$ such that

$$\sum_{j,k=1}^{n} \frac{\partial^2 \tilde{\rho}}{\partial z_j \partial \bar{z}_k}(P) w_j \bar{w}_k \geq C|w|^2$$

for all $P \in \partial\Omega$, all $w \in \mathbb{C}^n$.

By continuity of the second derivatives of $\tilde{\rho}$, the inequality in the proposition must in fact persist for all z in a neighborhood of $\partial\Omega$. In particular, if $P \in \partial\Omega$ is a point of strong pseudoconvexity, then so are all nearby boundary points. The analogous assertion for weakly pseudoconvex points is false.

EXAMPLE   Let $\Omega = \{(z_1, z_2) \in \mathbb{C}^2 : |z_1|^2 + |z_2|^4 < 1\}$. Then $\rho(z_1, z_2) = -1 + |z_1|^2 + |z_2|^4$ is a defining function for Ω and the Levi form applied to (w_1, w_2) is $|w_1|^2 + 4|z_2|^2 |w_2|^2$. Thus $\partial\Omega$ is strongly pseudoconvex except at boundary points where $|z_2|^2 = 0$, and the tangent vectors w satisfy $w_1 = 0$. Of course, these are just the boundary points of the form $(e^{i\theta}, 0)$. The domain is (weakly) Levi pseudoconvex at these exceptional points.   □

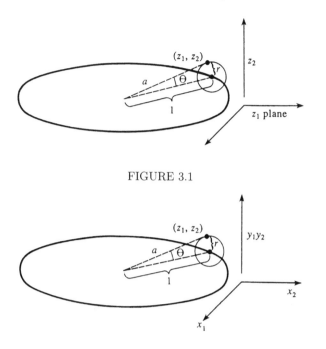

FIGURE 3.1

FIGURE 3.2

EXAMPLE   Let Ω_1 be the "solid torus" in $\mathbb{C}^2$ whose major circle of rotation (of radius 1) is independent of z_2 and whose minor circle of rotation (of radius r) is independent of z_1 (see Figure 3.1). This region is described by the equation

$$a^2 + 1^2 - 2 \cdot a \cdot 1 \cdot \cos\theta < r^2;$$

that is, the domain has defining function

$$\rho_1(z_1, z_2) = |z_1|^2 + |z_2|^2 + 1 - 2|z_1| - r^2.$$

The reader may calculate that the domain is strongly pseudoconvex for $0 < r < 1$.

On the other hand, let Ω_2 be the "solid torus" in $\mathbb{C}^2$ whose major circle of rotation (of radius 1) lies in the $x_1 - x_2$ plane and whose minor circle of rotation (of radius r) is in the $y_1 - y_2$ plane (see Figure 3.2). This region is described by the equation

$$a^2 + 1^2 - 2 \cdot a \cdot 1 \cdot \cos\theta < r^2;$$

that is, the domain has defining function

$$\rho_1(z_1, z_2) = |z_1|^2 + |z_2|^2 + 1 - 2\sqrt{x_1^2 + x_2^2} - r^2.$$

The reader may calculate the Levi form of Ω_2 and determine that Ω_2 is pseudo-convex *only if* $r \leq 1/2$. It is strongly pseudoconvex if $0 < r < 1/2$.

The two tori described here are isometric in Euclidean geometry. But they are *not* biholomorphic. They are imbedded into complex Euclidean space in such a fashion that they are different from the point of view of complex analysis. □

We see from the last example that pseudoconvexity describes something more (and less) than classical geometric properties of a domain. We will learn later that certain cohomology groups of Ω with coefficients in $\mathbb{C}$ must vanish when Ω is pseudoconvex. Also, pseudoconvexity can be characterized by a condition similar to the classical characterization of convexity (presented in Section 3.1) by families of imbedded segments; for pseudoconvexity we replace segments by holomorphically imbedded discs. However, it is important to realize that there is no simple geometric description of pseudoconvex points. Weakly pseudoconvex points are far from being well understood at this time. Matters are much clearer for strongly pseudoconvex points.

LEMMA 3.2.2 (Narasimhan)   Let $\Omega \subset\subset \mathbb{C}^n$ be a domain with C^2 boundary. Let $P \in \partial\Omega$ be a point of strong pseudoconvexity. Then there is a neighborhood $U \subseteq \mathbb{C}^n$ of P and a biholomorphic mapping Φ on U such that $\Phi(U \cap \partial\Omega)$ is strongly convex.

Proof   By Proposition 3.2.1 there is a defining function $\tilde{\rho}$ for Ω such that

$$\sum_{j,k} \frac{\partial^2 \tilde{\rho}}{\partial z_j \partial \bar{z}_k}(P) w_j \bar{w}_k \geq C|w|^2$$

for all $w \in \mathbb{C}^n$. By a rotation and translation of coordinates, we may assume that $P = 0$ and that $\nu = (1, 0, \ldots, 0)$ is the unit outward normal to $\partial\Omega$ at P. The second-order Taylor expansion of $\tilde{\rho}$ about $P = 0$ is given by

$$
\begin{aligned}
\tilde{\rho}(w) &= \tilde{\rho}(0) + \sum_{j=1}^{n} \frac{\partial \tilde{\rho}}{\partial z_j}(P) w_j + \frac{1}{2} \sum_{j,k=1}^{n} \frac{\partial^2 \tilde{\rho}}{\partial z_j \partial z_k}(P) w_j w_k \\
&\quad + \sum_{j=1}^{n} \frac{\partial \tilde{\rho}}{\partial \bar{z}_j}(P) \bar{w}_j + \frac{1}{2} \sum_{j,k=1}^{n} \frac{\partial^2 \tilde{\rho}}{\partial \bar{z}_j \partial \bar{z}_k}(P) \bar{w}_j \bar{w}_k \\
&\quad + \sum_{j,k=1}^{n} \frac{\partial^2 \tilde{\rho}}{\partial z_j \partial \bar{z}_k}(P) w_j \bar{w}_k + o(|w|^2) \\
&= 2\operatorname{Re}\left\{ \sum_{j=1}^{n} \frac{\partial \tilde{\rho}}{\partial z_j}(P) w_j + \frac{1}{2} \sum_{j,k=1}^{n} \frac{\partial^2 \tilde{\rho}}{\partial z_j \partial z_k}(P) w_j w_k \right\}
\end{aligned}
$$

$$+ \sum_{j,k=1}^{n} \frac{\partial^2 \tilde{\rho}}{\partial z_j \partial \bar{z}_k}(P) w_j \bar{w}_k + o(|w|^2)$$

$$= 2 \operatorname{Re} \left\{ w_1 + \frac{1}{2} \sum_{j,k=1}^{n} \frac{\partial^2 \tilde{\rho}}{\partial z_j \partial z_k}(P) w_j w_k \right\}$$

$$+ \sum_{j,k=1}^{n} \frac{\partial^2 \tilde{\rho}}{\partial z_j \partial \bar{z}_k}(P) w_j \bar{w}_k + o(|w|^2) \qquad (3.2.2.1)$$

by our normalization $\nu = (1, 0, \ldots, 0)$.

Define the mapping $w = (w_1, \ldots, w_n) \mapsto w' = (w'_1, \ldots, w'_n)$ by

$$w'_1 = \Phi_1(w) = w_1 + \frac{1}{2} \sum_{j,k=1}^{n} \frac{\partial^2 \tilde{\rho}}{\partial z_j \partial z_k}(P) w_j w_k$$

$$w'_2 = \Phi_2(w) = w_2$$

$$\cdot \qquad \cdot$$

$$\cdot \qquad \cdot$$

$$\cdot \qquad \cdot$$

$$w'_n = \Phi_n(w) = w_n.$$

By the implicit function theorem, we see that for w sufficiently small this is a well-defined invertible holomorphic mapping on a small neighborhood W of $P = 0$. Then equation (3.2.2.1) tells us that, in the coordinate w', the defining function becomes

$$\hat{\rho}(w') = 2 \operatorname{Re} w'_1 + \sum_{j,k=1}^{n} \frac{\partial^2 \tilde{\rho}}{\partial z'_j \partial \bar{z}'_k}(P) w'_j \bar{w}'_k + o(|w'|^2).$$

Thus the real Hessian at P of the defining function $\hat{\rho}$ is precisely the Levi form; and the latter is positive definite by our hypothesis. Hence the boundary of $\Phi(W \cap \Omega)$ is strictly convex at $\Phi(P)$. By the continuity of the second derivatives of $\hat{\rho}$, we may conclude that the boundary of $\Phi(W \cap \Omega)$ is strictly convex in a neighborhood V of $\Phi(P)$. We now select $U \subseteq W$ a neighborhood of P such that $\Phi(U) \subseteq V$ to complete the proof. □

By a very ingenious (and complicated) argument, J. E. Fornæss [1] has refined Narasimhan's lemma in the following manner.

THEOREM 3.2.3 (Fornæss)   Let $\Omega \subseteq \mathbb{C}^n$ be a strongly pseudoconvex domain with C^2 boundary. Then there is an integer $n' > n$, a strongly convex domain $\Omega' \subseteq \mathbb{C}^{n'}$, a neighborhood $\hat{\Omega}$ of $\bar{\Omega}$, and a one-to-one imbedding $\Phi : \hat{\Omega} \to \mathbb{C}^{n'}$

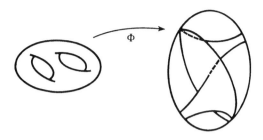

FIGURE 3.3

such that

1. $\Phi(\Omega) \subseteq \Omega'$;

2. $\Phi(\partial\Omega) \subseteq \partial\Omega'$;

3. $\Phi(\hat{\Omega} \setminus \bar{\Omega}) \subseteq \mathbb{C}^{n'} \setminus \bar{\Omega}'$;

4. $\Phi(\hat{\Omega})$ is transversal to $\partial\Omega'$.

Remark: In general, $n' >> n$ in the theorem. Sharp estimates on the size of n', in terms of the Betti numbers of Ω and other analytic data, are not known. The example that we gave above of the strongly pseudoconvex torus shows that strongly pseudoconvex domains are, in general, not strongly convex. However, Fornæss's theorem is the next best thing: a strongly pseudoconvex domain can be mapped properly into a strongly convex domain of higher dimension. Figure 3.3 suggests, roughly, what Fornæss's theorem says.

It is known (see J. Yu [1]) that if Ω has real analytic boundary, then the domain Ω' in the theorem can be taken to have real analytic boundary and the mapping Φ will extend real analytically across the boundary (see also Forstneric [1]). It is not known whether, if Ω is described by a polynomial defining function, the mapping Φ can be taken to be a polynomial. Sibony has produced an example of a smooth weakly pseudoconvex domain that cannot be mapped properly into any weakly convex domain of any dimension (even if we discard any smoothness and transversality conclusions). See N. Sibony [4] for details. It is not known which weakly pseudoconvex domains can be properly imbedded in a convex domain of some dimension.   □

DEFINITION 3.2.4   An *analytic disc* in $\mathbb{C}^n$ is a nonconstant holomorphic mapping $\phi : D \to \mathbb{C}^n$. We shall sometimes intentionally confuse the imbedding with its image (the latter is denoted by **d**). If ϕ extends continuously to $\bar{D}$ then we call $\phi(\bar{D})$ a *closed analytic disc* and $\phi(\partial D)$ the *boundary* of the analytic disc.

EXAMPLE   The analytic disc $\phi(\zeta) = (1, \zeta)$ lies entirely in the boundary of the bidisc $D \times D$. By contrast, the boundary of the ball contains no nontrivial (i.e., nonconstant) analytic discs. To see this last assertion, assume that

$\phi = (\phi_1, \phi_2) : D \to \partial B$ is an analytic disc. For simplicity, take the dimension to be two. Let $\rho(z) = -1 + |z_1|^2 + |z_2|^2$ be a defining function for B. Then $\rho \circ \phi = -1 + |\phi_1|^2 + |\phi_2|^2$ is constantly equal to 0, or $|\phi_1(\zeta)|^2 + |\phi_2(\zeta)|^2 \equiv 1$. Each function on the left side of this identity is subharmonic. By the sub-mean value property, if d is a small disc centered at ζ with radius r, then

$$1 = |\phi_1(\zeta)|^2 + |\phi_2(\zeta)|^2 \leq \frac{1}{\pi r^2} \int_d |\phi_1(\xi)|^2 + |\phi_2(\xi)|^2 \, dA(\xi) = 1.$$

Thus, in fact, we have the equality

$$|\phi_1(\zeta)|^2 + |\phi_2(\zeta)|^2 = \frac{1}{\pi r^2} \int_d |\phi_1(\xi)|^2 + |\phi_2(\xi)|^2 \, dA(\xi).$$

But also

$$|\phi_1(\zeta)|^2 \leq \frac{1}{\pi r^2} \int_d |\phi_1(\xi)|^2 \, dA(\xi)$$

and

$$|\phi_2(\zeta)|^2 \leq \frac{1}{\pi r^2} \int_d |\phi_2(\xi)|^2 \, dA(\xi).$$

It therefore must be that equality holds in these last two inequalities. But then $|\phi_1|^2$ and $|\phi_2|^2$ are harmonic. That can be true only if ϕ_1, ϕ_2 are constant.   □

Exercises for the Reader

1. Prove that the boundary of a strongly pseudoconvex domain cannot contain a nonconstant analytic disc.
   In fact, more is true: If Ω is strongly pseudoconvex, $P \in \partial\Omega$, and $\phi : D \to \bar{\Omega}$ satisfies $\phi(0) = P$, then ϕ is identically equal to P.

2. There is a precise, quantitative version of the behavior of an analytic disc at a strongly pseudoconvex point. If $P \in \partial\Omega$ is a strongly pseudoconvex point, then there is no analytic disc $\mathbf{d}$ with the property that

$$\lim_{\mathbf{d} \ni z \to P} \frac{\text{dist}(z, \partial\Omega)}{|z - P|^2} = 0.$$

This property distinguishes a weakly pseudoconvex boundary point from a strongly pseudoconvex boundary point. For example, the boundary point $(1, 0, 0)$ in the domain $\{z \in \mathbb{C}^3 : |z_1|^2 + |z_2|^2 + |z_3|^4 < 1\}$ has a zero eigenvalue of its Levi form in the direction $(0, 0, 1)$. Correspondingly, the analytic disc $\phi(\zeta) = (0, 0, \zeta)$ has order of contact with the boundary greater than 2 at the point $(0, 0, 1)$. We shall provide a detailed discussion of these ideas in Section 11.5.

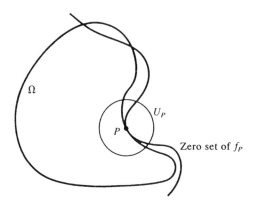

FIGURE 3.4

3.2.1   Holomorphic Support Functions

Let $\Omega \subseteq \mathbb{C}^n$ be a domain and $P \in \partial\Omega$. We say that P possesses a *holomorphic support function* for the domain Ω provided that there is a neighborhood U_P of P and a holomorphic function $f_P : U_P \to \mathbb{C}$ such that $\{z \in U_P : f_P(z) = 0\} \cap \bar{\Omega} = \{P\}$ (see Figure 3.4). Compare the notion of holomorphic support function with the classical notion of support line or support hypersurface for a convex body (see F. A. Valentine [1] or S. R. Lay [1]).

Suppose now that $P \in \partial\Omega$ is a point of strong convexity and that $T_P(\partial\Omega)$ denotes the real tangent hyperplane to $\partial\Omega$ at P. Then there is a neighborhood U_P of P such that $T_P(\partial\Omega) \cap \bar{\Omega} \cap U_P = \{P\}$ (exercise). Identify $\mathbb{C}^n$ with $\mathbb{R}^{2n}$ in the usual way. Assume, for notational simplicity, that $P = 0$. Let $(a_1, b_1, \ldots, a_n, b_n) \simeq (a_1 + ib_1, \ldots, a_n + ib_n) \equiv (\alpha_1, \ldots, \alpha_n) = \alpha$ be the unit outward normal to $\partial\Omega$ at P. Then we may think of $T_P(\partial\Omega)$ as $\{(x_1, y_1, \ldots, x_n, y_n) : \sum_{j=1}^{n} a_j x_j + b_j y_j = 0\}$. Equivalently, identifying $(x_1, y_1, \ldots, x_n, y_n)$ with $(z_1, \ldots, z_n)$, we may identify $T_P(\partial\Omega)$ with

$$\left\{ (z_1, \ldots, z_n) : \mathrm{Re} \sum_{j=1}^{n} z_j \bar{\alpha}_j = 0 \right\}.$$

Let $f(z) = \sum \bar{\alpha}_j z_j = z \cdot \bar{\alpha}$. (The notation $\langle z, \alpha \rangle$ is used in some contexts in place of $z \cdot \bar{\alpha}$.) Then f is plainly holomorphic on $\mathbb{C}^n$ and f is a support function for Ω at P, since the zero set of f lies in $T_P(\partial\Omega)$. The next proposition now follows from Narasimhan's lemma.

PROPOSITION 3.2.5   If $\Omega \subseteq \mathbb{C}^n$ is a domain and $P \in \partial\Omega$ is a point of strong pseudoconvexity, then there exists a holomorphic support function for Ω at P.

As already noted, the proposition follows immediately from Narasimhan's lemma. But the phenomenon of support functions turns out to be so important (see Chapter 5, for instance) that we now provide a separate, self-contained proof.

Proof of the Proposition Let ρ be a defining function for Ω with the property that

$$\sum_{j,k=1}^{n} \frac{\partial^2 \rho}{\partial z_j \partial \bar{z}_k}(P) w_j \bar{w}_k \geq C|w|^2$$

for all $w \in \mathbb{C}^n$. Define

$$f(z) = \sum_{j=1}^{n} \frac{\partial \rho}{\partial z_j}(P)(z_j - P_j) + \frac{1}{2} \sum_{j,k=1}^{n} \frac{\partial^2 \rho}{\partial z_j \partial z_k}(P)(z_j - P_j)(z_k - P_k).$$

The function f is obviously holomorphic. It is called the *Levi polynomial* at P. We claim that f is a support function for Ω at P. To see this, we expand ρ in a Taylor expansion about P :

$$\begin{aligned}
\rho(z) &= 2 \operatorname{Re}\left\{ \sum_{j=1}^{n} \frac{\partial \rho}{\partial z_j}(P)(z_j - P_j) \right. \\
&\quad \left. + \frac{1}{2} \sum_{j,k=1}^{n} \frac{\partial^2 \rho}{\partial z_j \partial z_k}(P)(z_j - P_j)(z_k - P_k) \right\} \\
&\quad + \sum_{j,k=1}^{n} \frac{\partial^2 \rho}{\partial z_j \partial \bar{z}_k}(P)(z_j - P_j)(\overline{z_k - P_k}) + o(|z - P|^2) \\
&= 2 \operatorname{Re} f(z) + \sum_{j,k=1}^{n} \frac{\partial^2 \rho}{\partial z_j \partial \bar{z}_k}(P)(z_j - P_j)(\overline{z_k - P_k}) + o(|z - P|^2).
\end{aligned}$$

Now let z be a point at which $f(z) = 0$. Then we find that

$$\begin{aligned}
\rho(z) &= \sum_{j,k=1}^{n} \frac{\partial^2 \rho}{\partial z_j \partial \bar{z}_k}(P)(z_j - P_j)(\overline{z_k - P_k}) + o(|z - P|^2) \\
&\geq C|z - P|^2 + o(|z - P|^2).
\end{aligned}$$

Obviously, if z is sufficiently closed to P, then we find that

$$\rho(z) \geq \frac{C}{2}|z - P|^2.$$

Thus if z is near P and $f(z) = 0$, then either $\rho(z) > 0$, which means that z lies outside $\bar{\Omega}$, or $z = P$. But this means precisely that f is a holomorphic support function for Ω at P.   □

EXAMPLE   Let $\Omega = D^3(0,1) \subseteq \mathbb{C}^3$. Then

$$\partial\Omega = (\partial D \times D \times D) \cup (D \times \partial D \times D) \cup (D \times D \times \partial D).$$

In particular, $\partial\Omega$ contains the entire bidisc $\mathbf{d} = \{(\zeta_1, \zeta_2, 1) : |\zeta_1| < 1, |\zeta_2| < 1\}$. The point $P = (0,0,1)$ is the center of $\mathbf{d}$. If there were a support function f for Ω at P, then $f|_{\mathbf{d}}$ would have an isolated zero at P. That is impossible for a function of two complex variables.

On the other hand, the function $g(z_1, z_2, z_3) = z_3 - 1$ is a *weak support function* for Ω at P in the sense that $g(P) = 0$ and $\{z : g(z) = 0\} \cap \bar{\Omega} \subseteq \partial\Omega$.

If $\Omega \subseteq \mathbb{C}^n$ is any (weakly) convex domain and $P \in \partial\Omega$, then $T_P(\partial\Omega) \cap \bar{\Omega} \subseteq \partial\Omega$. As above, a weak support function for Ω at P can therefore be constructed.   □

As recently as 1972 it was hoped that a weakly pseudoconvex domain would have at least a weak support function at each point of its boundary. These hopes were dashed by the following theorem.

THEOREM 3.2.6   Let

$$\Omega = \{(z_1, z_2) \in \mathbb{C}^2 : \operatorname{Re} z_2 + |z_1 z_2|^2 + |z_1|^8 + \frac{15}{7}|z_1|^2 \operatorname{Re} z_1^6 < 0\}.$$

Then Ω is strongly pseudoconvex at every point of $\partial\Omega$ except 0 (where it is weakly pseudoconvex). However there is no weak holomorphic support function (and hence no support function) for Ω at 0. More precisely, if f is a function holomorphic in a neighborhood U of 0 such that $f(0) = 0$, then for every neighborhood V of zero f vanishes both in $V \cap \Omega$ and in $V \cap {}^c\bar{\Omega}$.

Proof   The reader should verify the first assertion. For the second, consult J. J. Kohn and L. Nirenberg [1]. In the exercises at the end of the chapter, an argument of Hakim and Sibony is used to prove a stronger result.   □

Necessary and sufficient conditions for the existence of holomorphic support functions or of weak holomorphic support functions are not known.

3.2.2   Peaking Functions

The ideas that we have been presenting are closely related to questions about the existence of peaking functions in various function algebras on pseudoconvex domains. We briefly summarize some of these. Let $\Omega \subseteq \mathbb{C}^n$ be a pseudoconvex domain.

If $\mathcal{A}$ is any algebra of functions on $\bar{\Omega}$, we say that $P \in \bar{\Omega}$ is a *peak point* for $\mathcal{A}$ if there is a function $f \in \mathcal{A}$ satisfying $f(P) = 1$ and $|f(z)| < 1$ for all $z \in \bar{\Omega} \setminus \{P\}$. Let $\mathcal{P}(\mathcal{A})$ denote the set of all peak points for the algebra $\mathcal{A}$.

Recall that $A(\Omega)$ is the subspace of $C(\bar{\Omega})$ consisting of those functions that are holomorphic on Ω. Let $A^j(\Omega) = A(\Omega) \cap C^j(\bar{\Omega})$. The maximum principle for holomorphic functions implies that any peak point for any subalgebra of $A(\Omega)$ must lie in $\partial\Omega$. If Ω is Levi pseudoconvex with C^∞ boundary, then $\mathcal{P}(A(\Omega))$ is contained in the closure of the strongly pseudoconvex points (see Basener [1]). Also the Šilov boundary for the algebra $A(\Omega)$ is equal to the closure of the set of peak points (this follows from classical function algebra theory—see T. W. Gamelin [1]), which in turn equals the closure of the set of strongly pseudoconvex points (see Basener [1]).

The Kohn-Nirenberg domain has no peak function at 0 that extends to be holomorphic in a neighborhood of 0 (for if f were a peak function, then $f(z) - 1$ would be a holomorphic support function at 0.) Hakim and Sibony [1] showed that the same domain has no peak function at 0 for the algebra $A^8(\Omega)$. J. E. Fornæss [2] has refined the example to exhibit a domain Ω that is strongly pseudoconvex except at one boundary point P but has no peak function at P for the algebra $A^1(\Omega)$. There is no known example of a smooth, pseudoconvex boundary point that is not a peak point for the algebra $A(\Omega)$. The deepest work to date on peaking functions is E. Bedford and J. E. Fornæss [1]. See also the more recent approaches in Fornæss and Sibony [2] and in Fornæss and McNeal [1]. It is desirable to extend that work to a larger class of domains and in higher dimensions, but that program seems to be intractable.

It is reasonable to hypothesize (see the remarks in the last paragraph about the Kohn-Nirenberg domain) that if $\Omega \subseteq \mathbb{C}^n$ has C^∞ boundary and a holomorphic support function at $P \in \partial\Omega$, then there is a neighborhood U_P of P and a holomorphic function $f : U_P \cap \Omega \to \mathbb{C}$ with a smooth extension to $\overline{U_P \cap \Omega}$ so that f peaks at P. This was proved false by T. Bloom [1]. However this conjecture *is* true at a strongly pseudoconvex point.

The problem of the existence of peaking functions is still a matter of great interest, for peak functions can be used to study the boundary behavior of holomorphic mappings (see Bedford and Fornæss [2]). Recently Fornæss [4] modified the peaking function construction of Bedford and Fornæss [1] to construct reproducing formulas for holomorphic functions.

Exercise for the Reader
Show that if $\Omega \subseteq \mathbb{C}$ is a domain with boundary consisting of finitely many C^j Jordan curves, then every $P \in \partial\Omega$ is a peak point for $A^{j-1}(\Omega)$. (*Hint:* The problem is local. The Riemann mapping of a simply connected Ω to D extends $C^{j-\epsilon}$ to the boundary. See O. Kellogg [2] and, Tsuji [1].)

3.2.3   Pseudoconvex Neighborhood Systems

Strong pseudoconvexity is a more useful condition than weak pseudoconvexity for a number of reasons. Two of these are the support function and peak point phenomena already discussed. It is also important that strong pseudoconvexity is stable under C^2 perturbations of $\partial\Omega$, whereas weak pseudoconvexity is not (exercise). Indeed, if Ω is strongly pseudoconvex with defining function ρ, then, for $\epsilon_0 > 0$ small, the functions $\rho_\epsilon(z) = \rho(z) + \epsilon, 0 < \epsilon < \epsilon_0$, define an increasing family of strongly pseudoconvex domains that exhaust Ω. Likewise, the functions $\rho_\epsilon(z) = \rho(z) - \epsilon, \epsilon_0 > \epsilon > 0$, define a decreasing family of strongly pseudoconvex domains whose intersection is Ω.

We prove in the next section that every weakly pseudoconvex domain can be exhausted by an increasing union of strongly pseudoconvex domains. It was hoped that every smoothly bounded weakly pseudoconvex domain could also be written as the decreasing intersection of strongly pseudoconvex domains. This was shown not to be the case by K. Diederich and J. E. Fornæss [2] in 1976. An easy counterexample with nonsmooth boundary is given by the so-called Hartogs triangle; see Exercise 17 at the end of the chapter.

3.3   Pseudoconvexity, Plurisubharmonicity, and Analytic Discs

At this point the various components of our investigations begin to converge to a single theme. In particular, we shall directly relate the notions of pseudoconvexity, plurisubharmonicity, and domains of holomorphy. In order to effect this unity, we need a second notion of pseudoconvexity. This, in turn, requires some preliminary terminology:

DEFINITION 3.3.1   A continuous function $\mu : \mathbb{C}^n \to \mathbb{R}$ is called a *distance function* if

1. $\mu \geq 0$;
2. $\mu(z) = 0$ if and only if $z = 0$;
3. $\mu(tz) = |t|\mu(z), \forall t \in \mathbb{C}, z \in \mathbb{C}^n$.

DEFINITION 3.3.2   Let $\Omega \subseteq \mathbb{C}^n$ be a domain and μ a distance function. Define

$$\mu_\Omega(z) = \mu(z, {}^c\Omega) = \inf_{w \in {}^c\Omega} \mu(z - w).$$

If $X \subseteq \Omega$ is a set, we write

$$\mu_\Omega(X) = \inf_{x \in X} \mu_\Omega(x).$$

It is elementary to verify that the function μ_Ω is continuous. In the special case that $\mu(z) = |z|$, one checks that in fact μ_Ω must satisfy a classical Lipschitz condition with norm 1. Moreover, for this special μ, it turns out that when Ω has C^j boundary, $j \geq 2$, then μ_Ω is a C^j function near $\partial\Omega$ (see Krantz and Parks [1] or S. Gilbarg and N. Trudinger [1]). The assertion is false for $j = 1$. The matter is addressed in detail in Exercise 4 at the end of the chapter.

DEFINITION 3.3.3   Let $\Omega \subseteq \mathbb{C}^n$ be a (possibly unbounded) domain (with, possibly, unsmooth boundary). We say that Ω is *Hartogs pseudoconvex* if there is a distance function μ such that $-\log\mu_\Omega$ is plurisubharmonic on Ω.

This new definition is at first obscure. What do these different distance functions have to do with complex analysis? It turns out that they give us a flexibility that we shall need in characterizing domains of holomorphy. In practice we think of a Hartogs pseudoconvex domain as a domain that has a (strictly) plurisubharmonic exhaustion function. Theorem 3.3.5 will clarify matters. In particular, we shall prove that a given domain Ω satisfies the definition of Hartogs pseudoconvex for one distance function if and only if it does so for all distance functions.

PROPOSITION 3.3.4   Let $\Omega \subseteq \mathbb{C}$ be any planar domain. Then Ω is Hartogs pseudoconvex.

Proof   We use the Euclidean distance $\delta(z) = |z|$. Let $\bar{D}(z_0, r) \subseteq \Omega$ and let h be harmonic on a neighborhood of this closed disc. Assume that $h \geq -\log d_\Omega$ on $\partial D(z_0, r)$. Let $\tilde{h}$ be a real harmonic conjugate for h on a neighborhood of $\bar{D}(z_0, r)$. So $h + i\tilde{h}$ is holomorphic on $D(z_0, r)$ and continuous on $\bar{D}(z_0, r)$. Fix, for the moment, a point $P \in \partial\Omega$. Then

$$-\log d_\Omega(z) \;\leq\; h(z), \quad z \in \partial D(z_0, r),$$

$$\Rightarrow \left| \exp\left(-h(z) - i\tilde{h}(z) \right) \right| \;\leq\; d_\Omega(z), \quad z \in \partial D(z_0, r),$$

$$\Rightarrow \left| \frac{\exp\left(-h(z) - i\tilde{h}(z) \right)}{z - P} \right| \;\leq\; 1, \quad z \in \partial D(z_0, r).$$

But the expression in absolute value signs is holomorphic on $D(z_0, r)$ and continuous on $\bar{D}(z_0, r)$. Hence

$$\left| \frac{\exp\left(-h(z) - i\tilde{h}(z) \right)}{z - P} \right| \leq 1, \qquad \forall z \in \bar{D}(z_0, r).$$

Unwinding this inequality yields that

$$-\log|z - P| \leq h(z), \qquad \forall z \in \bar{D}(z_0, r).$$

Choosing for each $z \in D(z_0, r)$ a point $P = P_z \in \partial\Omega$ with $|z - P| = d_\Omega(z)$ now yields that

$$-\log d_\Omega(z) \le h(z) \quad \forall z \in D(z_0, r).$$

It follows that $-\log d_\Omega$ is subharmonic, hence the domain Ω is Hartogs pseudoconvex. $\square$

Exercise for the Reader
Why does the proof of Proposition 3.3.4 break down when the dimension is two or greater?

THEOREM 3.3.5   Let $\Omega \subseteq \mathbb{C}^n$ be any connected open set. The following eleven properties are then equivalent. (Note, however, that property 9 makes sense only when the boundary is C^2.)

1. $-\log \mu_\Omega$ is plurisubharmonic on Ω for *any* distance function μ.

2. Ω is Hartogs pseudoconvex.

3. There exists a continuous plurisubharmonic (i.e., pseudoconvex) function Φ on Ω such that for every $c \in \mathbb{R}$ we have $\{z \in \Omega : \Phi(z) < c\} \subset\subset \Omega$.

4. Property 3 is true for a C^∞ strictly plurisubharmonic exhaustion function Φ.

5. Ω is convex with respect to the family $P(\Omega)$ of plurisubharmonic functions on Ω.

6. Let $\{\mathbf{d}_\alpha\}_{\alpha \in A}$ be a family of closed analytic discs in Ω. If $\cup_{\alpha \in A} \partial\mathbf{d}_\alpha \subset\subset \Omega$, then $\cup_{\alpha \in A} \mathbf{d}_\alpha \subset\subset \Omega$. (This assertion is called the Kontinuitätssatz.)

7. If μ is any distance function and if $\mathbf{d} \subseteq \Omega$ is any closed analytic disc, then $\mu_\Omega(\partial\mathbf{d}) = \mu_\Omega(\mathbf{d})$.

8. Property 7 is true for just one particular distance function.

9. Ω is Levi pseudoconvex.

10. $\Omega = \cup\Omega_j$, where each Ω_j is Hartogs pseudoconvex and $\Omega_j \subset\subset \Omega_{j+1}$.

11. Property 10 is true, except that each Ω_j is a bounded, strongly Levi pseudoconvex domain with C^∞ boundary.

Remark: The scheme of the proof is shown in Figure 3.5. $\square$

Some parts of the proof are rather long. However this proof contains many of the basic techniques of the theory of several complex variables. This is material that is worth mastering. Notice that the hypothesis of C^2 boundary is used only in the proof that $(1) \Rightarrow (9) \Rightarrow (3)$. The implication $(1) \Rightarrow (3)$ is immediate for any domain Ω.

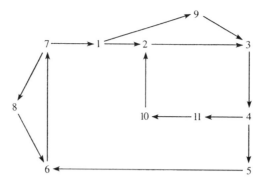

FIGURE 3.5

Proof of Theorem 3.3.5

(2) $\Rightarrow$ **(3)** If Ω is unbounded, then it is possible that $-\log \mu_\Omega(z)$ is not an exhaustion function (although, by hypothesis, it *is* psh). Thus we set $\Phi(z) = -\log \mu_\Omega(z) + |z|^2$, where μ is given by (2). Then Φ will be a psh *exhaustion*.

(9) $\Rightarrow$ **(3)** We are assuming that Ω has C^2 boundary. If (3) is false, then the Euclidean distance function $d_\Omega(z) \equiv \text{dist}(z, {}^c\Omega)$ (which is C^2 on $U \cap \bar{\Omega}$, U a tubular neighborhood of $\partial\Omega$; see Exercise 4 at the end of the chapter) has the property that $-\log \delta_\Omega$ is *not* psh.

So plurisubharmonicity of $-\log d_\Omega$ fails at some $z \in \Omega$. Since Ω has a psh exhaustion function if and only if it has one defined near $\partial\Omega$ (exercise), we may as well suppose that $z \in U \cap \Omega$. So the complex Hessian of $-\log d_\Omega$ has a negative eigenvalue at z. Quantitatively, there is a direction w so that

$$\frac{\partial^2}{\partial\zeta\partial\bar\zeta} \log d_\Omega(z + \zeta w)\bigg|_{\zeta=0} \equiv \lambda > 0.$$

To exploit this, we let $\phi(\zeta) = \log d_\Omega(z + \zeta w)$ and examine the Taylor expansion about $\zeta = 0$:

$$
\begin{aligned}
\log d_\Omega(z + \zeta w) &= \phi(\zeta) = \phi(0) + 2\,\text{Re}\left\{\frac{\partial\phi}{\partial\zeta}(0)\cdot\zeta + \frac{\partial^2\phi}{\partial\zeta^2}(0)\cdot\frac{\zeta^2}{2}\right\} \\
&\quad + \frac{\partial^2\phi}{\partial\zeta\partial\bar\zeta}(0)\cdot\zeta\cdot\bar\zeta + o(|\zeta|^2) \\
&\equiv \log d_\Omega(z) + \text{Re}\{A\zeta + B\zeta^2\} \\
&\quad + \lambda|\zeta|^2 + o(|\zeta|^2),
\end{aligned}
\tag{3.3.5.1}
$$

where A, B are defined by the last equality.

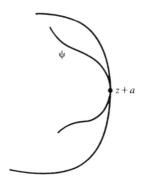

FIGURE 3.6

Now choose $a \in \mathbb{C}^n$ such that $z + a \in \partial\Omega$ and $|a| = d_\Omega(z)$. Define the function

$$\psi(\zeta) = z + \zeta w + a \exp(A\zeta + B\zeta^2), \qquad \zeta \in \mathbb{C} \text{ small},$$

and notice that $\psi(0) = z + a \in \partial\Omega$. Also, by (3.3.5.1),

$$
\begin{aligned}
d_\Omega(\psi(\zeta)) &\geq d_\Omega(z + \zeta w) - |a| \cdot |\exp(A\zeta + B\zeta^2)| \\
&\geq d_\Omega(z) \cdot |\exp(A\zeta + B\zeta^2)| \exp\left(\lambda|\zeta|^2 + o(|\zeta|^2)\right) \\
&\quad - |a| \cdot |\exp(A\zeta + B\zeta^2)| \\
&\geq |a| \cdot |\exp(A\zeta + B\zeta^2)| \left\{\exp(\lambda|\zeta|^2/2) - 1\right\} \\
&\geq 0 \qquad\qquad\qquad\qquad\qquad\qquad\qquad\qquad (3.3.5.2)
\end{aligned}
$$

if ζ is small. These estimates also show that, up to reparametrization, ψ describes an analytic disc *which is contained in* $\bar{\Omega}$ and which is internally tangent to $\partial\Omega$. The disc intersects $\partial\Omega$ at the single point $z + a$ (see Figure 3.6).

On geometrical grounds, then, $(\partial/\partial\zeta)(d_\Omega \circ \psi)(0) = 0$. For (3.3.5.2) to hold, it must then be that $(\partial^2/\partial\zeta\partial\bar\zeta)(d_\Omega \circ \psi)(0) > 0$, since the term $2\,\mathrm{Re}\,[(\partial^2/\partial\zeta^2)(d_\Omega \circ \psi)(0)\zeta^2]$ in the Taylor expansion of $d_\Omega \circ \psi$ is not of constant sign.

We have proved that the defining function

$$
\rho(z) = \begin{cases} -d_\Omega(z) & \text{if } z \in \bar\Omega \cap U, \\ d_{c\Omega}(z) & \text{if } z \in {}^c\bar\Omega \cap U \end{cases}
$$

does not satisfy the Levi pseudoconvexity condition at $z + a = \psi(0) \in \partial\Omega$ in the tangent direction $\psi'(0)$. That is a contradiction. (See the important remark at the end of the proof of the theorem.)

(3) $\Rightarrow$ (4) Let Φ be the psh function whose existence is hypothesized in (3). Let $\Omega_c \equiv \{z \in \Omega : \Phi(z) + |z|^2 < c\}, c \in \mathbb{R}$. Then each $\Omega_c \subset\subset \Omega$, by the

definition of exhaustion function, and $c' > c$ implies that $\Omega_c \subset\subset \Omega_{c'}$. Let $0 \leq \phi \in C_c^\infty(\mathbb{C}^n)$, $\int \phi = 1, \phi$ polyradial (i.e., $\phi(z_1, \dots, z_n) = \phi(|z_1|, \dots, |z_n|)$)—see Proposition 2.2.8). We may assume that ϕ is supported in $B(0,1)$. Pick $\epsilon_j > 0$ such that $\epsilon_j < \mathrm{dist}(\Omega_{j+1}, \partial\Omega)$. For $z \in \Omega_{j+1}$, set

$$\Phi_j(z) = \int_\Omega [\Phi(\zeta) + |\zeta|^2] \epsilon_j^{-2n} \phi\left((z-\zeta)/\epsilon_j\right) dV(\zeta) + |z|^2 + 1.$$

Then Φ_j is C^∞ and strictly psh on Ω_{j+1} (use Corollary 2.1.12). By the proof of Proposition 2.2.8, we know that $\Phi_j(\zeta) > \Phi(\zeta) + |\zeta|^2$ on $\bar\Omega_j$. Let $\chi \in C^\infty(\mathbb{R})$ be a convex function with $\chi(t) = 0$ when $t \leq 0$ and $\chi'(t), \chi''(t) > 0$ when $t > 0$. Observe that $\Psi_j(z) \equiv \chi(\Phi_j(z) - (j-1))$ is positive and psh on $\Omega_j \setminus \bar\Omega_{j-1}$ and is, of course, C^∞. Now we inductively construct the desired function Φ'. First, $\Phi_0 > \Phi$ on Ω_0. If a_1 is large and positive, then $\Phi_1' = \Phi_0 + a_1\Psi_1 > \Phi$ on Ω_1. Inductively, if $a_1, \dots, a_{\ell-1}$ have been chosen, select $a_\ell > 0$ such that $\Phi_\ell' = \Phi_0 + \sum_{j=1}^\ell a_j\Psi_j > \Phi$ on Ω_ℓ. Since $\Psi_{\ell+k} = 0$ on $\Omega_\ell, k > 0$, we see that $\Phi_{\ell+k}' = \Phi_{\ell+k'}'$ on Ω_ℓ for any $k, k' > 0$. So the sequence Φ_ℓ' stabilizes on compacta and $\Phi' \equiv \lim_{\ell\to\infty} \Phi_\ell'$ is a C^∞ strictly psh function that majorizes Φ. Hence Φ' is the smooth, strictly psh exhaustion function that we seek.

$(4) \Rightarrow (5)$ This is immediate from the definition of convexity with respect to a family of functions.

$(5) \Rightarrow (6)$ Let $\mathbf{d} \subseteq \Omega$ be a closed analytic disc and let $u \in P(\Omega)$. Let $\phi : \bar{D} \to \mathbf{d}$ be a parametrization of $\mathbf{d}$. Then $u \circ \phi$ is subharmonic, so for any $z \in \bar{D}$,

$$u \circ \phi(z) \leq \sup_{\zeta \in \partial D} u(\zeta).$$

It follows that, for any $p \in \mathbf{d}$,

$$u(p) \leq \sup_{\xi \in \partial\mathbf{d}} u(\xi).$$

Therefore, $\mathbf{d} \subseteq \widehat{\partial\mathbf{d}}_{P(\Omega)}$. Thus if $\{\mathbf{d}_\alpha\}_{\alpha \in A}$ is a family of closed analytic discs in Ω, then $\cup\mathbf{d}_\alpha \subseteq (\cup_\alpha\widehat{\partial\mathbf{d}_\alpha})_{P(\Omega)}$. Hence (6) holds.

$(6) \Rightarrow (7)$ If not, there is a closed analytic disc $\phi : \bar{D} \to \mathbf{d} \subseteq \Omega$ and a distance function μ such that $\mu_\Omega(\overset{\circ}{\mathbf{d}}) < \mu_\Omega(\partial\mathbf{d})$. Note that because the continuous image of a compact set is compact, $\mathbf{d}$ is both closed and bounded.

   Let $p_0 \in \overset{\circ}{\mathbf{d}}$ be the μ-nearest point to $\partial\Omega$. We may assume that $\phi(0) = p_0$. Choose $z_0 \in \partial\Omega$ so that $\mu(p_0 - z_0) = \mu_\Omega(p_0)$. It follows that the discs $\mathbf{d}_j \equiv \mathbf{d} + (1 - (1/j))(z_0 - p_0)$ satisfy $\cup\partial\mathbf{d}_j \subset\subset \Omega$ whereas $\cup\mathbf{d}_j \supseteq \{(1 - (1/j))z_0 + (1/j)p_0\} \to z_0 \in \partial\Omega$. This contradicts (6).

$(7) \Rightarrow (1)$ (This ingenious proof is due to Hartogs.) It is enough to check plurisubharmonicity at a fixed point $z_0 \in \Omega$ and for any distance function μ.

Fix a vector $a \in \mathbb{C}^n$: We must check the subharmonicity of $\psi : \zeta \mapsto -\log \mu_\Omega(z_0 + a\zeta), \zeta \in \mathbb{C}$ small. If $|a|$ is small enough, we may take $|\zeta| \leq 1$. We then show that

$$\psi(0) \leq \frac{1}{2\pi} \int_0^{2\pi} \psi(e^{i\theta}) \, d\theta.$$

Now $\psi|_{\partial D}$ is continuous. Let $\epsilon > 0$. By the Stone-Weierstrass theorem, there is a holomorphic polynomial p on $\mathbb{C}$ such that if $h = \operatorname{Re} p$, then

$$\sup_{\zeta \in \partial D} |\psi(\zeta) - h(\zeta)| < \epsilon.$$

We may assume that $h > \psi$ on ∂D. Let $b \in \mathbb{C}^n$ satisfy $\mu(b) \leq 1$. Define a closed analytic disc

$$\phi : \zeta \mapsto z_0 + \zeta a + be^{-p(\zeta)}, \qquad |\zeta| \leq 1.$$

Identifying ϕ with its image $\mathbf{d}$ as usual, our aim is to see that $\mathbf{d} \subseteq \Omega$. If we can prove this claim, then the proof of (1) is completed as follows: Since b was arbitrary, we conclude by setting $\zeta = 0$ that $z_0 + be^{-p(0)} \in \Omega$ for every choice of b a vector of μ length not exceeding 1. It follows that the ball with μ-radius $|e^{-p(0)}|$ and center z_0 is contained in Ω. Thus

$$\mu_\Omega(z_0) \geq |e^{-p(0)}| = e^{-h(0)}.$$

Equivalently,

$$\psi(0) = -\log \mu_\Omega(z_0) \leq h(0) = \frac{1}{2\pi} \int_0^{2\pi} h(e^{i\theta}) \, d\theta < \frac{1}{2\pi} \int_0^{2\pi} \psi(e^{i\theta}) \, d\theta + \epsilon.$$

Letting $\epsilon \to 0^+$ then yields the result.

It remains to check that $\mathbf{d} \subseteq \Omega$. The proof we present will seem unnecessarily messy. For it is fairly easy to see that $\partial \mathbf{d}$ lies in Ω. The trouble is that, while the spirit of (7) suggests that we may then conclude that $\mathbf{d}$ itself lies in Ω, this is not so. There are no complex analytic obstructions to this conclusion; but there *can* be topological obstructions. To show that $\mathbf{d}$ in its entirety lies in Ω, we must demonstrate that it is a continuous deformation of another disc that we know a priori lies in Ω. Thus there are some unpleasant details.

We define the family of discs

$$\mathbf{d}_\lambda : \zeta \mapsto z_0 + \zeta a + \lambda be^{-p(\zeta)}, \qquad 0 \leq \lambda \leq 1.$$

Let $S = \{\lambda : 0 \leq \lambda \leq 1 \text{ and } \mathbf{d}_\lambda \subseteq \Omega\}$. We claim that $S = [0, 1]$. Of course $\mathbf{d}_1 = \mathbf{d}$, so that will complete the proof. We use a continuity method.

First notice that, by the choice of a, $0 \in S$. Hence S is not empty.

Next, if $P \in \mathbf{d}_\lambda$, choose $\zeta_0 \in D$ so that $\mathbf{d}_\lambda(\zeta_0) = P$. If $\lambda_j \to \lambda$, then $\mathbf{d}_{\lambda_j}(\zeta_0) \equiv P_j \to P$. So the disc $\mathbf{d}_\lambda$ is the limit of the discs $\mathbf{d}_{\lambda_j}$ in a natural way. Moreover $\cup_{0 \leq \lambda \leq 1} \partial \mathbf{d}_\lambda \subset\subset \Omega$ because

$$
\begin{aligned}
\mu\left((z_0 + \zeta a) - (z_0 + \zeta a + \lambda b e^{-p(\zeta)}) \right) &= \mu(\lambda b e^{-p(\zeta)}) \\
&\leq e^{-h(\zeta)} \\
&< e^{-\psi(\zeta)} \\
&= \mu_\Omega(z_0 + \zeta a).
\end{aligned}
$$

We may not now conclude from (7) that $\cup_{0 \leq \lambda \leq 1} \partial \mathbf{d}_\lambda \subset\subset \Omega$, because the Kontinuitätssatz applies only to discs that are known a priori to lie in Ω. But we may conclude from the Kontinuitätssatz and the remarks in the present paragraph that S is closed.

Since Ω is an open domain, it is also clear that S is open.

We conclude that $S = [0, 1]$ and our proof is complete.

(1) $\Rightarrow$ (2) This is trivial.

(4) $\Rightarrow$ (11) Let Φ be as in (4). The level sets $\{z \in \Omega : \Phi(z) < c\}$ may not all have smooth boundary; at a boundary point where $\nabla \Phi$ vanishes, there could be a singularity. However Sard's theorem (J. Munkres [1]) guarantees that the set of c's for which this problem arises has measure zero in $\mathbb{R}$. Thus we let $\Omega_j = \{z \in \Omega : \Phi(z) < \lambda_j\}$, where $\lambda_j \to +\infty$ are such that each $\partial \Omega_j$ is smooth. Since Φ is strictly psh, each Ω_j is strongly pseudoconvex.

(11) $\Rightarrow$ (10) It is enough to prove that a strongly pseudoconvex domain $\mathcal{D}$ with smooth boundary is Hartogs pseudoconvex. But this follows from (9) $\Rightarrow$ (3) $\Rightarrow$ (4) $\Rightarrow$ (5) $\Rightarrow$ (6) $\Rightarrow$ (7) $\Rightarrow$ (1) $\Rightarrow$ (2) above (see Figure 3.5 to verify that we have avoided circular reasoning).

(10) $\Rightarrow$ (2) Let δ be the Euclidean distance. By (2) $\Rightarrow$ (3) $\Rightarrow$ (4) $\Rightarrow$ (5) $\Rightarrow$ (6) $\Rightarrow$ (7) $\Rightarrow$ (1) above, $-\log d_{\Omega_j}$ is psh for each j. Hence $-\log d_\Omega$ is psh and Ω is Hartogs pseudoconvex.

(7) $\Rightarrow$ (8) This is trivial.

(8) $\Rightarrow$ (6) Let μ be the distance function provided by (8). If (6) fails, then there is a sequence $\{\mathbf{d}_j\}$ of closed analytic discs lying in Ω with $\mu_\Omega(\partial \mathbf{d}_j) \geq \delta_0 > 0$, whereas $\mu_\Omega(\overset{\circ}{\mathbf{d}}_j) \to 0$. That is a contradiction.

(1) $\Rightarrow$ (9) Let δ be Euclidean distance. If $\partial \Omega$ is C^2, then $d_\Omega(\cdot)$ is C^2 at points P that are sufficiently near to $\partial \Omega$ (see Exercise 4 at the end of the chapter and Krantz and Parks [1]). Consider such a $P \in \Omega$ and $w \in \mathbb{C}^n$. By (1), $-\log d_\Omega$ is psh on Ω. Hence

$$
\sum_{j,k=1}^n \left(-d_\Omega^{-1}(P) \frac{\partial^2 d_\Omega}{\partial z_j \partial \bar{z}_k}(P) + d_\Omega^{-2}(P) \left(\frac{\partial d_\Omega}{\partial z_j}(P) \right) \left(\frac{\partial d_\Omega}{\partial \bar{z}_k}(P) \right) \right) w_j \bar{w}_k \geq 0.
$$

Multiply through by $d_\Omega(P)$, and restrict attention to w that satisfy

$$\sum (\partial d_\Omega/\partial z_j)(P)w_j = 0.$$

Letting $P \to \partial\Omega$, the inequality becomes

$$-\sum_{j,k=1}^{n} \frac{\partial^2 d_\Omega}{\partial z_j \partial \bar{z}_k}(P)w_j \bar{w}_k \geq 0, \qquad (3.3.5.3)$$

for all $P \in \partial\Omega$, all $w \in \mathbb{C}^n$ satisfying

$$\sum_j \frac{\partial d_\Omega}{\partial z_j}(P)w_j = 0.$$

But the function

$$\rho(z) = \begin{cases} -d_\Omega(z) & \text{if } z \in \bar{\Omega}, \\ d_{c_\Omega}(z) & \text{if } z \notin \bar{\Omega} \end{cases}$$

is a C^2 defining function for z near $\partial\Omega$ (which may easily be extended to all of $\mathbb{C}^n$). Thus (3.3.5.3) simply says that Ω is Levi pseudoconvex. □

Remark: The last half of the proof of (9) $\Rightarrow$ (3) explains what is geometrically significant about Levi pseudoconvexity. For it is nothing other than a classical convexity condition along complex tangential directions. This convexity condition may be computed along analytic discs that are tangent to the boundary. With this in mind, we see that Figure 3.6 already suggests that Levi pseudoconvexity does not hold—think of an appropriate submanifold of the boundary as locally the graph of a function over the analytic disc. □

Exercise for the Reader
Modify (3) $\Rightarrow$ (4) in the proof of the theorem to see that if $K \subset\subset \Omega, W$ is a relatively compact neighborhood of $\hat{K}_{P(\Omega)}$ in Ω, then the function Φ in part (4) may be constructed to have the additional property that $\Phi < 0$ on K and $\Phi > 0$ on $\Omega \setminus W$.

We now formulate some useful consequences of the theorem, some of which are restatements of the theorem and some of which go beyond the theorem. Note that we may now use the word pseudoconvex to mean *either* Hartogs's notion or Levi's notion (at least for domains with C^2 boundary).

PROPOSITION 3.3.6 Let $\Omega_j \subseteq \mathbb{C}^n$ be pseudoconvex domains. If $\Omega \equiv \cap_{j=1}^{\infty}\Omega_j$ is open and connected, then Ω is pseudoconvex.

Proof Use part 1 of the theorem together with the Euclidean distance function. □

PROPOSITION 3.3.7 If $\Omega_1 \subseteq \Omega_2 \subseteq \cdots$ are pseudoconvex domains, then $\cup_j \Omega_j$ is pseudoconvex.

Proof   The proof is left as an exercise.   □

PROPOSITION 3.3.8 Let $\Omega \subseteq \mathbb{C}^n$. Let $\Omega' \subseteq \mathbb{C}^{n'}$ be pseudoconvex, and assume that $\phi : \Omega \to \Omega'$ is a surjective (but not necessarily injective) proper holomorphic mapping. Then Ω is pseudoconvex.

Proof   Let $\Phi' : \Omega' \to \mathbb{R}$ be a pseudoconvex (continuous, plurisubharmonic) exhaustion function for Ω'. Let $\Phi \equiv \Phi' \circ \phi$. Then Φ is pseudoconvex. Let $\Omega'_c \equiv (\Phi')^{-1}\big((-\infty, c)\big)$. Then Ω'_c is relatively compact in Ω', since Φ' is an exhaustion function. But $\Omega_c \equiv \Phi^{-1}\big((-\infty, c)\big) = \phi^{-1}(\Omega'_c)$. Thus Ω_c is relatively compact in Ω by the properness of ϕ. We conclude that Ω is pseudoconvex.   □

PROPOSITION 3.3.9   Hartogs pseudoconvexity is a local property. More precisely, if $\Omega \subseteq \mathbb{C}^n$ is a domain and each $P \in \partial\Omega$ has a neighborhood U_P such that $U_P \cap \Omega$ is Hartogs pseudoconvex, then Ω is Hartogs pseudoconvex.

Proof   Since $U_P \cap \Omega$ is pseudoconvex, $-\log d_{U_P \cap \Omega}$ is psh on $U_P \cap \Omega$ (here δ is Euclidean distance). But, for z sufficiently near P, $-\log d_{U_P \cap \Omega} = -\log d_\Omega$. It follows that $-\log d_\Omega$ is psh near $\partial\Omega$, say on $\Omega \setminus F$, where F is a closed subset of Ω. Let $\phi : \mathbb{C}^n \to \mathbb{R}$ be a convex increasing function of $|z|^2$ that satisfies $\phi(z) \to \infty$ when $|z| \to \infty$ and $\phi(z) > -\log d_\Omega(z)$ for all $z \in F$. Then the function

$$\Phi(z) = \max\{\phi(z), -\log d_\Omega(z)\}$$

is continuous and plurisubharmonic on Ω and is also an exhaustion function for Ω. Thus Ω is a pseudoconvex domain.   □

Remark: If $\partial\Omega$ is C^2, then there is an alternative (but essentially equivalent) proof of the last proposition as follows: Fix $P \in \partial\Omega$. Since $U_P \cap \Omega$ is pseudoconvex, the part of its boundary that it shares with $\partial\Omega$ is Levi pseudoconvex. Hence the Levi form at P is positive semidefinite. But P was arbitrary; hence each boundary point of Ω is Levi pseudoconvex. The result follows.

It is essentially tautologous that Levi pseudoconvexity is a local property.   □

Exercise for the Reader
Check that Proposition 3.3.8 fails if the hypothesis of properness of ϕ is omitted.

3.4 Domains of Holomorphy

We now direct the machinery that has been developed to derive several characterizations of domains of holomorphy. These are presented in Theorem 3.4.5. As an immediate consequence, we shall see that every domain of holomorphy is pseudoconvex. Chapter 4 is devoted to a presentation of Hörmander's construction of solutions to the $\bar{\partial}$ equation on pseudoconvex domains. This leads in Chapter 5 to a solution of the Levi problem—that is, to see that every pseudoconvex domain is a domain of holomorphy. Thus Theorems 3.3.5 and 3.4.5, together with the solution of the Levi problem, give a total of 23 equivalent characterizations of the principal domains that are studied in the theory of several complex variables. Chapters 6 and 7 develop further cohomological and function-theoretic properties of these domains.

The reader is encouraged at this point to reread the definition of domain of holomorphy in Chapter 0. It raises many questions, because other reasonable definitions of "domain of holomorphy" are not manifestly equivalent to it. For instance, suppose that $\Omega \subseteq \mathbb{C}^n$ has the property that $\partial\Omega$ may be covered by finitely many open sets $\{U_j\}_{j=1}^M$ so that $\Omega \cap U_j$ is a domain of holomorphy, $j = 1, \ldots, M$. Is Ω then a domain of holomorphy? Suppose instead that to each $P \in \partial\Omega$ we may associate a neighborhood U_P and a holomorphic function $f_P : U_P \cap \Omega \to \mathbb{C}$ so that f_P cannot be continued analytically past P. Is Ω then a domain of holomorphy?

Fortunately, all these definitions are equivalent to the original definition of domain of holomorphy, as we shall soon see. We will ultimately learn that the property of being a domain of holomorphy is purely a local one.

In what follows, it will occasionally prove useful to allow our domains of holomorphy to be disconnected (contrary to our customary use of the word domain). We leave it to the reader to sort out this detail when appropriate. Throughout this section, the family of functions $\mathcal{F} = \mathcal{F}(\Omega)$ will denote the holomorphic functions on Ω unless explicitly mentioned otherwise.

By way of warming up to our task, let us prove a few simple assertions about $\hat{K}_{\mathcal{F}}$ when $K \subset\subset \Omega$ (note that we are assuming, in particular, that K is bounded).

LEMMA 3.4.1   The set $\hat{K}_{\mathcal{F}}$ is bounded (even if Ω is not).

Proof   The set K is bounded if and only if the holomorphic functions $f_1(z) \equiv z_1, \ldots, f_n(z) \equiv z_n$ are bounded on K. This, in turn, is true if and only if $f_1, \ldots, f_n$ are bounded on $\hat{K}_{\mathcal{F}}$; and this last is true if and only if $\hat{K}_{\mathcal{F}}$ is bounded. $\square$

LEMMA 3.4.2   The set $\hat{K}_{\mathcal{F}}$ is contained in the closed convex hull of K.

Proof   Apply the definition of $\hat{K}_{\mathcal{F}}$ to the real parts of all complex linear functionals on $\mathbb{C}^n$ (which, of course, are elements of $\mathcal{F}$). Then use the fact that $|\exp(f)| = \exp(\mathrm{Re} f)$. $\square$

LEMMA 3.4.3   Let $\mathbf{d} \subseteq \Omega$ be a closed analytic disc. Then $\mathbf{d} \subseteq \widehat{\partial \mathbf{d}}_{\mathcal{F}}$.

Proof   Let $f \in \mathcal{F}$. Let $\phi : \bar{D} \to \mathbf{d}$ be a parametrization of $\mathbf{d}$. Then $f \circ \phi$ is holomorphic on $\mathbf{d}$ and continuous on $\bar{D}$. Therefore, it assumes its maximum modulus on ∂D. It follows that

$$\sup_{z \in \mathbf{d}} |f(z)| = \sup_{z \in \partial \mathbf{d}} |f(z)|.$$ $\square$

Exercise for the Reader
Consider the Hartogs domain $\Omega = D^2(0,1) \setminus \bar{D}^2(0,1/2)$. Let $K = \{(0, 3e^{i\theta}/4) : 0 \le \theta < 2\pi\} \subset\subset \Omega$. Verify that $\hat{K}_{\mathcal{F}} = \{(0, re^{i\theta}) : 0 \le \theta < 2\pi : \frac{1}{2} < r \le \frac{3}{4}\}$, which is *not* relatively compact in Ω.

Now we turn attention to the main results of this section. First, a definition is needed. This is a localized version of the definition of domain of holomorphy.

DEFINITION 3.4.4 Let $U \subseteq \mathbb{C}^n$ be an open set. We say that $P \in \partial U$ is *essential* if there is a holomorphic function h on U such that for no connected neighborhood U_2 of P and nonempty $U_1 \subseteq U_2 \cap U$ is there an h_2 holomorphic on U_2 with $h = h_2$ on U_1.

THEOREM 3.4.5   Let $\Omega \subseteq \mathbb{C}^n$ be an open set (no smoothness of $\partial\Omega$ nor boundedness of Ω need be assumed). Let $\mathcal{F} = \mathcal{F}(\Omega)$ be the family of holomorphic functions on Ω. Then the following are equivalent. (Items appearing with a $(*)$ are cohomological assertions that will be proved in Chapters 5 and 6. We record them here for convenience.)

1. Ω is convex with respect to $\mathcal{F}$.

2. There is an $h \in \mathcal{F}$ that cannot be holomorphically continued past any $P \in \partial\Omega$ (i.e., in the definition of essential point the same function h can be used for *every* boundary point P).

3. Each $P \in \partial\Omega$ is essential (Ω is a domain of holomorphy).

4. Each $P \in \partial\Omega$ has a neighborhood U_P such that $U_P \cap \Omega$ is a domain of holomorphy.$(*)$

5. Each $P \in \partial\Omega$ has a neighborhood U_P so that $U_P \cap \Omega$ is convex with respect to $\mathcal{F}_P \equiv \{\text{holomorphic functions on } U_P \cap \Omega\}.(*)$

6. For any $f \in \mathcal{F}$, any $K \subset\subset \Omega$, and any distance function μ, the inequality

$$|f(z)| \le \mu_\Omega(z), \quad \forall z \in K$$

implies that

$$|f(z)| \le \mu_\Omega(z), \quad \forall z \in \hat{K}_{\mathcal{F}}.$$

7. For any $f \in \mathcal{F}$, any $K \subset\subset \Omega$, and any distance function μ, we have

$$\sup_{z \in K} \left\{ \frac{|f(z)|}{\mu_\Omega(z)} \right\} = \sup_{z \in \hat{K}_{\mathcal{F}}} \left\{ \frac{|f(z)|}{\mu_\Omega(z)} \right\}.$$

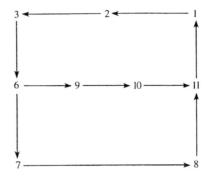

FIGURE 3.7

8. If $K \subset\subset \Omega$ then, for any distance function μ,

$$\mu_\Omega(K) = \mu_\Omega\left(\hat{K}_{\mathcal{F}}\right).$$

9. (6) is true for just one distance function μ.
10. (7) is true for just one distance function μ.
11. (8) is true for just one distance function μ.

Remark: The scheme of the proof is shown in Figure 3.7. Observe that (4) and (5) are omitted. They follow easily once the Levi problem has been solved. □

Proof of Theorem 3.4.5

(**2**) $\Rightarrow$ (**3**) This is trivial.

(**11**) $\Rightarrow$ (**1**) This is trivial. Use Lemma 3.4.1.

(**1**) $\Rightarrow$ (**2**) Choose a dense sequence $\{w_j\}_{j=1}^\infty \subseteq \Omega$ that repeats every point infinitely often. For each j, let D_j be the largest polydisc $D^n(w_j, r)$ contained in Ω. Choose a sequence $K_1 \subset\subset K_2 \subset\subset K_3 \subset\subset \cdots$ with $\cup_{j=1}^\infty K_j = \Omega$. For each j, $(\widehat{K_j})_{\mathcal{F}} \subset\subset \Omega$ by hypothesis. Thus there is a point $z_j \in D_j \setminus (\widehat{K_j})_{\mathcal{F}}$. This means that we may choose an $h_j \in \mathcal{F}$ such that $h_j(z_j) = 1, \left|h_j|_{K_j}\right| < 1$. By replacing h_j by $h_j^{M_j}, M_j$ large, we may assume that $\left|h_j|_{K_j}\right| < 2^{-j}$. Write

$$h(z) = \prod_{j=1}^\infty (1 - h_j)^j.$$

Then the product converges uniformly on each K_j, and hence normally on Ω, and the limit function h is not identically zero (this is just standard one variable theory—see, for instance, L. Ahlfors [1]). By the choice of the points w_j, every D_j contains infinitely many of the z_ℓ and hence contains points at which h vanishes to arbitrarily high order. Any analytic continuation of h to a neighborhood of a point $P \in \partial\Omega$ is a continuation to a neighborhood of some $\bar{D}_j$ and hence would necessitate that h vanish to infinite order at some point. This would imply that $h \equiv 0$, which is a clear contradiction.

(3) $\Rightarrow$ **(6)** Fix $r = (r_1, \ldots, r_n) > 0$. Define a distance function $\mu^r(z) = \max_{1 \le j \le n}\{|z_j|/r_j\}$. We first prove (6) for this distance function. Let $f \in \mathcal{F}, K \subset\subset \Omega$ satisfy $|f(z)| \le \mu_\Omega^r(z), z \in K$. We claim that for all $g \in \mathcal{F}$, all $p \in \hat{K}_\mathcal{F}$, it holds that g has a normally convergent power series expansion on

$$\{z \in \mathbb{C}^n : \mu^r(z - p) < |f(p)|\} = D^1(p_1, |f(p)| \cdot r_1) \times \cdots \times D^1(p_n, |f(p)| \cdot r_n).$$

This implies (6) for this particular distance function. For if $|f(p)| > \mu_\Omega^r(p)$ for some $p \in \hat{K}_\mathcal{F}$, then $D^1(p_1, |f(p)| \cdot r_1) \times \cdots \times D^1(p_n, |f(p)| \cdot r_n)$ has points in it that lie outside $\bar{\Omega}$ (to which every $g \in \mathcal{F}$ extends analytically!). That would contradict (3). To prove the claim, let $0 < t < 1$, let $g \in \mathcal{F}$, and let

$$S_t = \bigcup_{k \in K} \{z \in \mathbb{C}^n : \mu^r(z - k) \le t|f(k)|\}.$$

Since $S_t \subset\subset \Omega$ by the hypothesis on f, there is an $M > 0$ such that $|g| \le M$ on S_t. By Cauchy's inequalities,

$$\left|\left(\frac{\partial}{\partial z}\right)^\alpha g(k)\right| \le \frac{\alpha! M}{t^{|\alpha|} r^\alpha |f(k)|^{|\alpha|}}, \qquad \forall k \in K, \text{ all multi-indices } \alpha.$$

But then the same estimate holds on $\hat{K}_\mathcal{F}$. So the power series of g about $p \in \hat{K}_\mathcal{F}$ converges on $D^1(p_1, t|f(p)| \cdot r_1) \times \cdots \times D^1(p_n, t|f(p)| \cdot r_n)$. Since $0 < t < 1$ was arbitrary, the claim is proved. So, in the special case of distance function μ^r, the implication (3) $\Rightarrow$ (6) is proved.

Now fix any distance function μ. Define, for any $w \in \mathbb{C}^n$,

$$S_\Omega^w(z) = \sup\{r \in \mathbb{R} : z + \tau w \in \Omega, \ \forall|\tau| < r, \tau \in \mathbb{C}\}.$$

Then, trivially,

$$\mu_\Omega(z) = \inf_{\mu(w)=1} S_\Omega^w(z). \qquad (3.4.5.1)$$

If we prove (6) for S_Ω^w instead of μ_Ω, w fixed, then the full result follows from (3.4.5.1).

After a rotation and dilation, we may suppose that $w = (1, 0, \ldots, 0)$. If $k \in \mathbb{N}$, we apply the special case of (6) to the n-tuple $r^k = (1, 1/k, \ldots, 1/k)$. Notice that $\mu_\Omega^{r^k} \nearrow S_\Omega^w$ as $k \to +\infty$. Let $K \subset\subset \Omega$. Assume that $|f(z)| \le S_\Omega^w(z)$

for $z \in K$. Let $\epsilon > 0$. Define

$$A_k = \{z : |f(z)| < (1 + \epsilon)\mu_\Omega^{r^k}(z)\}.$$

Then $\{A_k\}$ is an increasing sequence of open sets whose union contains K. Thus one of the sets A_{k_0} covers K. In other words,

$$|f(z)| \leq (1 + \epsilon)\mu_\Omega^{r^{k_0}}(z), \ z \in K.$$

By what we have already proved for the μ^r,

$$|f(z)| \leq (1 + \epsilon)\mu_\Omega^{r^{k_0}}(z) \leq (1 + \epsilon)S_\Omega^w(z), \ \forall z \in \hat{K}_\mathcal{F}.$$

Letting $\epsilon \to 0^+$ yields $|f(z)| \leq S_\Omega^w(z), z \in \hat{K}_\mathcal{F}$, as desired.

$(\mathbf{6}) \Rightarrow (\mathbf{9})$ This is trivial.

$(\mathbf{9}) \Rightarrow (\mathbf{10})$ This is trivial.

$(\mathbf{10}) \Rightarrow (\mathbf{11})$ Apply (10) with $f \equiv 1$.

$(\mathbf{7}) \Rightarrow (\mathbf{8})$ Apply (7) with $f \equiv 1$.

$(\mathbf{8}) \Rightarrow (\mathbf{11})$ This is trivial.

$(\mathbf{6}) \Rightarrow (\mathbf{7})$ This is trivial.   □

3.4.1   Consequences of Theorems 3.3.5 and 3.4.5

COROLLARY 3.4.6   If $\Omega \subseteq \mathbb{C}^n$ is a domain of holomorphy, then Ω is pseudoconvex.

Proof   By part 1 of Theorem 3.4.5, Ω is holomorphically convex. It follows a fortiori that Ω is convex with respect to the family $P(\Omega)$ of all psh functions on Ω, since $|f|$ is psh whenever f is holomorphic on Ω. Thus, by part 5 of Theorem 3.3.5, Ω is pseudoconvex.   □

COROLLARY 3.4.7   Let $\{\Omega_\alpha\}_{\alpha \in A}$ be domains of holomorphy in $\mathbb{C}^n$. If $\Omega \equiv \cap_{\alpha \in A}\Omega_\alpha$ is open, then Ω is a domain of holomorphy.

Proof   Use part 8 of the theorem.   □

COROLLARY 3.4.8   If Ω is geometrically convex, then Ω is a domain of holomorphy.

Proof   Let $P \in \partial\Omega$. Let $(a_1, \ldots, a_n) \in \mathbb{C}^n$ be any unit outward normal to $\partial\Omega$ at P. Then the real tangent hyperplane to $\partial\Omega$ at P is $\left\{ z : \mathrm{Re}\left[\sum_{j=1}^{n}(z_j - P_j)\bar{a}_j \right] = 0 \right\}$. But then the function

$$f_P(z) = \frac{1}{\left(\sum_{j=1}^{n}(z_j - P_j)\bar{a}_j \right)}$$

is holomorphic on Ω and shows that P is essential. By part 3 of the theorem, that is *by definition*, Ω is a domain of holomorphy.   □

Exercise for the Reader
Construct another proof of Corollary 3.4.8 using part 1 of the Theorem.

Remark: Corollary 3.4.8 is a special case of the Levi problem. The reader should consider why this proof fails on, say, strongly pseudoconvex domains: Let $\Omega \subseteq \mathbb{C}^n$ be strongly pseudoconvex, ρ the defining function for Ω given by Proposition 3.2.1. For $P \in \partial\Omega$,

$$\begin{aligned} \rho(z) \quad = \quad & \rho(P) + 2\,\mathrm{Re}\left\{ \sum_{j=1}^{n} \frac{\partial\rho}{\partial z_j}(P)(z_j - P_j) \right. \\ & \left. + \frac{1}{2}\sum_{j,k=1}^{n} \frac{\partial^2\rho}{\partial z_j \partial z_k}(P)(z_j - P_j)(z_k - P_k) \right\} \\ & + \sum_{j,k=1}^{n} \frac{\partial^2\rho}{\partial z_j \partial\bar{z}_k}(P)(z_j - P_j)(\bar{z}_k - \bar{P}_k) + o(|z - P|^2). \end{aligned}$$

Define

$$L_P(t) = \sum_{j=1}^{n} \frac{\partial\rho}{\partial z_j}(P)(z_j - P_j) + \frac{1}{2}\sum_{j,k=1}^{n} \frac{\partial^2\rho}{\partial z_j \partial z_k}(P)(z_j - P_j)(z_k - P_k).$$

The function $L_P(z)$ is called the *Levi polynomial* at P. For $|z - P|$ sufficiently small, $z \in \bar{\Omega}$, we have that $L_P(z) = 0$ if and only if $z = P$. For $L_P(z) = 0$ means that $\rho(z) \geq C|z - P|^2 + o(|z - P|^2)$. Hence L_P is an *ersatz* for f_P in the proof of Corollary 3.4.8 *near* P. In short, P is "locally essential." It requires powerful additional machinery to conclude from this that P is (globally) essential.   □

3.4.2   Consequences of the Levi Problem

Assume for the moment that we have proved that pseudoconvex domains are domains of holomorphy (we *did* prove the converse of this statement in Corollary 3.4.6). Then we may quickly dispatch several interesting questions:

$\big((3)$ of $(3.4.5)\big) \Leftrightarrow \big((4)$ of $(3.4.5)\big)$ It is enough to show that $(4) \Rightarrow (3)$. By Corollary 3.4.6, each $U_P \cap \Omega$ is pseudoconvex. By Proposition 3.3.9, it follows that Ω is pseudoconvex. By the Levi problem, Ω is a domain of holomorphy.   □

$\big((4)$ of $(3.4.5)\big) \Leftrightarrow \big((5)$ of $(3.4.5)\big)$ This is obvious.   □

THEOREM 3.4.9 (Behnke-Stein)   Let $\Omega_1 \subseteq \Omega_2 \subseteq \cdots$ be domains of holomorphy. Then $\Omega \equiv \cup_j \Omega_j$ is a domain of holomorphy.

Proof   Each Ω_j is pseudoconvex (by 3.4.6); hence, by Proposition 3.3.7, Ω is pseudoconvex. By the Levi problem, Ω is a domain of holomorphy.   □

Remark: It is possible, but rather difficult, to prove the Behnke-Stein theorem directly, without any reference to the Levi problem. Classically, the Levi problem was solved for strongly pseudoconvex domains and then the fact that any weakly pseudoconvex domain is the increasing union of strongly pseudoconvex domains, together with Behnke-Stein, was used to complete the argument. See L. Bers [1] for a treatment of this approach.   □

3.4.3   The Levi Problem for Complete Circular Domains

We have just seen what a useful tool the solution of the Levi problem can be. Our full solution of the Levi problem comes in Chapter 5. We conclude the present section by solving the Levi problem for complete circular domains.

PROPOSITION 3.4.10   If $\Omega \subseteq \mathbb{C}^n$ is a logarithmically convex complete circular domain, then Ω is a domain of holomorphy.

Proof   Let $K \subset\subset \Omega$. For each $P \in K$, there is a neighborhood U_P of P and a $\zeta^P \in \Omega$ such that $|z_j| \leq |\zeta_j^P|$ for all $z \in U_P$, all $j = 1, \ldots, n$. By the compactness of K, we may choose $\zeta^1, \ldots, \zeta^k \in \Omega$ such that

$$K \subseteq \bigcup_{\ell=1}^{k} \{z : |z_j| \leq |\zeta_j^\ell|, j = 1, \ldots, n\}.$$

We may assume that $\zeta_j^\ell \neq 0$ for all j, ℓ. Let $\mathcal{F}$ be the family of holomorphic functions on Ω. Now let $z \in \hat{K}_{\mathcal{F}}$ be arbitrary. Assume without loss of generality (since the domain is circular) that there is an integer $m, 1 \leq m \leq n$, such that

$z_1, \ldots, z_m \neq 0$ while $z_{m+1}, \ldots, z_n = 0$. Then, by definition of $\hat{K}_{\mathcal{F}}$,

$$|z_1^{\alpha_1} \cdots z_m^{\alpha_m}| \leq \sup_{\xi \in K} |\xi_1^{\alpha_1} \cdots \xi_m^{\alpha_m}| \leq \max_{1 \leq \ell \leq k} |(\zeta_1^{\ell})^{\alpha_1} \cdots (\zeta_m^{\ell})^{\alpha_m}| \qquad (3.4.10.1)$$

for any multi-index α. Define $\mu_i = \alpha_i / |\alpha|$. Then (3.4.10.1) says that

$$\sum_{i=1}^{m} \mu_i \log |z_i| \leq \max_{1 \leq \ell \leq k} \sum_{i=1}^{m} \mu_i \log |\zeta_i^{\ell}| \qquad (3.4.10.2)$$

for any rationals $0 \leq \mu_i \leq 1$ with $\sum_{i=1}^{m} \mu_i = 1$. It follows that (3.4.10.2) holds for any real numbers $0 \leq \mu_i \leq 1$ with $\sum_i \mu_i = 1$. Therefore, $(\log |z_1|, \ldots, \log |z_m|)$ is in the convex hull of

$$\{(t_1, \ldots, t_m) \in \mathbb{R}^m : \text{for some } \ell, 0 \leq t_i \leq \log |\zeta_i^{\ell}|, i = 1, \ldots, m\}.$$

Therefore,

$$
\begin{aligned}
(\log |z_1|, \ldots, \log |z_m|) \quad &\in \quad \text{convex hull } \{(t_1, \ldots, t_m) : t_i \leq \log |\zeta_i^{\ell}|, \\
&\qquad\qquad i = 1, \ldots, m, \text{some } \ell\} \\
&\equiv \quad L \\
&\subset\subset \quad \log \|\Omega\| \cap (\text{the } t_1, \ldots, t_m \text{ plane}),
\end{aligned}
$$

because $\log \|\Omega\|$ is convex. Hence z lies in a complete circular, relatively compact subset $\tilde{L}$ of Ω (the "circularization" of L). But z was an arbitrary element of $\hat{K}_{\mathcal{F}}$, so we conclude that $\hat{K}_{\mathcal{F}} \subset\subset \Omega$. $\qquad \square$

Remark: Proposition 3.4.10 completes the discussion, begun in Section 2.3, of complete circular domains. The logarithmically convex ones are, on the one hand, the smallest domains on which power series converge. On the other hand, by Proposition 3.4.10, they are also the largest. $\qquad \square$

3.5 Examples of Domains of Holomorphy and the Edge-of-the-Wedge Theorem

3.5.1 Analytic Polyhedra

Variants of this notion are also called *Weil domains*. Let $W \subseteq \mathbb{C}^n$ be an open set that is homeomorphic to the ball. Let $f_1, \ldots, f_k$ be holomorphic functions on W. We define

$$\Omega = \Omega(f_1, \ldots, f_k) \equiv \{z \in W : |f_j(z)| < 1, j = 1, \ldots, k\}.$$

If $\Omega \subset\subset W$, then we call Ω an analytic polyhedron. Of course, we require that $\Omega \subset\subset W$ so that the boundary of Ω is given just by the analytic functions $\{f_j\}$ and not by the boundary of W; the requirement that W be homeomorphic to a ball is imposed for similar reasons. An analytic polyhedron is a domain of holomorphy because if $P \in \partial\Omega$, then, for some $1 \leq j \leq k$, the function $(f_j(P) - f_j(z))^{-1}$ is holomorphic on Ω and singular at P. In the early days of the function theory of several complex variables, analytic polyhedra received considerable attention. The classical proof of the Behnke-Stein theorem provides a method for exhausting any domain of holomorphy by analytic polyhedra. See L. Bers [1].

3.5.2 Tube Domains

Tube domains are the simplest cases of Siegel domains, which we shall encounter later in this book. Let $\omega \subseteq \mathbb{R}^n$ be an open set. The tube domain associated to ω is the set

$$T_\omega \equiv \{z \in \mathbb{C}^n : \operatorname{Re} z \in \omega\}.$$

THEOREM 3.5.1   Let $T_\omega \subseteq \mathbb{C}^n$ be a tube domain. The following are equivalent:

1. T_ω is pseudoconvex;
2. T_ω is geometrically convex;
3. T_ω is a domain of holomorphy.

Proof   First notice that T_ω is convex if and only if ω is convex.

Observe that (2) $\Rightarrow$ (3) is Corollary 3.4.8 and that (3) $\Rightarrow$ (1) is Corollary 3.4.6. To prove (1) $\Rightarrow$ (2) it is enough, by the "affine sphere" characterization convexity in Section 3.1, to prove that if I is a closed segment in ω and δ the Euclidean distance function, then $d_\omega(I) = d_\omega(\partial I)$. Let x^1, x^2 be the endpoints of I, and consider the analytic disc $\mathbf{d}$ given by

$$
\begin{aligned}
\phi : D &\rightarrow T_\omega \\
\zeta &\mapsto ((1-\zeta)/2)x^1 + ((1+\zeta)/2)x^2.
\end{aligned}
$$

Since T_ω is a tube domain, this disc lies in T_ω. Let $\rho = -\log d_{T_\omega}$; since T_ω is pseudoconvex, then ρ is a psh exhaustion function for T_ω. If for some $t \in (-1,1)$ it held that $[(1-t)/2]x^1 + [(1+t)/2]x^2$ were nearer to $\partial\omega$ than are the points x^1, x^2, it would follow that the subharmonic function $u(\zeta) = \rho \circ \phi(\zeta)$ took an interior maximum. Thus u would be constant and the elements of I would all be at the same distance from ∂T_ω. This contradiction establishes the result.   □

Remark: Theorem 3.5.1 solves the Levi problem for tube domains and gives a simplified version of the sort of characterization that we seek for arbitrary domains of holomorphy.   □

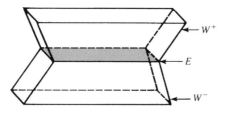

FIGURE 3.8

The simple proof of the characterization of tube domains that are domains of holomorphy is facilitated by the free mobility in Ω in the imaginary directions. Philosophically related to this result is the edge-of-the-wedge theorem, which also deals with domains that are less restrictive in the imaginary variables. The result was originally discovered in the context of verification of the dispersion relations in quantum field theory (N. N. Bogoliubov and D. V. Shirkov [1], and N. N. Bogoliubov and V. Vladimirov [1]). In more recent times it has been studied for its intrinsic merit and for applications to partial differential equations (E. Bedford [3], W. Rudin [2], and V. Guilleman, M. Kashiwara, and T. Kawai [1]) and to other areas of mathematical physics (see V. Vladimirov [1]). For the introductory purposes here, we present only a very simple version of the theorem.

DEFINITION 3.5.2   An *open cone* in $\mathbb{R}^N$ is an open set $V \subseteq \mathbb{R}^N$ such that $y \in V$ implies $ty \in V$ for all $t > 0$.

EXAMPLE   Let $\mathcal{O} \subseteq \mathbb{R}^N$ be any open set. Define $V_\mathcal{O} \equiv \{tx : x \in \mathcal{O}, t > 0\}$. Then $V_\mathcal{O} \subseteq \mathbb{R}^N$ is an open cone.   □

In harmonic analysis, convex open cones that are closed under addition play an important role because they are, for many purposes, the right generalization of the upper half-space. For instance, harmonic analysis on the forward light cone (from relativity theory) is a matter of considerable interest (see E. M. Stein and G. Weiss [1] and references therein).

If $V \subseteq \mathbb{R}^N$ is an open cone, $R > 0$, let $U = V \cap B(0, R)$. Let $E \subseteq \mathbb{R}^N$ be *any* nonempty open set. Define $W^+ \subseteq \mathbb{C}^n, W^- \subseteq \mathbb{C}^n$ by

$$W^+ = E + iU, \qquad W^- = E - iU$$

(see Figure 3.8). We shall call $W^+ \cup W^- \cup E$ a *truncated wedge domain*. In a certain sense, which can only be approximately indicated in the figure, E is the "edge" of the wedges W^+, W^-. It is convenient in what follows to identify $\{x + i0 : x \in E\}$ with E, and we do so without further comment.

EXAMPLE   Let $E \subseteq \mathbb{R}^2$ be open and bounded. Define $V = \{(y_1, y_2) \in \mathbb{R}^2 : y_1 > 0, y_2 > 0\}$. Let $U = V \cap B(0, 1)$. Then $W^+ \cup W^- \cup E$ is a truncated wedge domain. Further note that $W^+ \cap W^- = \emptyset, \overline{W^+} \cap \overline{W^-} = \bar{E}$, and $W^+ \cup W^- \cup E$ is *not* open.   □

THEOREM 3.5.3   Let $n \geq 2$. With the notation just given, there exists an open set $\Omega \subseteq \mathbb{C}^n$ such that $\Omega \supseteq (W^+ \cup W^- \cup E)$ and such that the following property holds: If $f : W^+ \cup W^- \cup E \to \mathbb{C}$ is continuous and f is holomorphic on $W^+ \cup W^-$, then there is a holomorphic F on Ω such that $F|_{W^+ \cup W^- \cup E} = f$.

The theorem is a remarkable analytic continuation statement. It says, in effect, that there is an open "halo" about the set E to which all analytic functions on $W^+ \cup W^- \cup E$ extend.

Proof of the Theorem (W. Rudin [2]) Let $x \in E$. We may assume that $x = 0$. After composition with a real linear transformation with real coefficients, we may assume that $V \supseteq \{(y_1, \ldots, y_n) \in \mathbb{R}^n : y_j > 0, j = 1, \ldots, n\}$. After a translation and change of scale, we may also suppose that $R > 6\sqrt{n}$ and that $E \subseteq \{x \in \mathbb{R}^n : |x_j| < 6, j = 1, \ldots, n\}$. So it is enough to prove the following.

CLAIM: Let

$$E = \{x \in \mathbb{R}^n : |x_j| < 6, \ j = 1, \ldots, n\},$$

$$U = \{y \in \mathbb{R}^n : 0 < y_j < 6, \ j = 1, \ldots, n\},$$

$$W^+ = E + iU \ , W^- = E - iU.$$

Let $\Omega = D^n(0, 1)$. If $f : W^+ \cup W^- \cup E \to \mathbb{C}$ is continuous and if f is holomorphic on $W^+ \cup W^-$, then there is a holomorphic F on Ω with $F|_{W^+ \cup W^- \cup E} = f$.

The idea in the proof of the claim is to map $D^n(0, 1)$ into $W^+ \cup W^-$ and then to compose f with this map. Since there are obvious topological obstructions to carrying out this program, we instead construct approximate maps; then the average of these approximate mappings gives the result that we desire.
Let $c = \sqrt{2} - 1$ and define

$$\phi : \bar{D}^2(0, 1) \ \to \ \mathbb{C}$$
$$(w, \lambda) \ \mapsto \ \frac{w + \lambda/c}{1 + c\lambda w}.$$

Then

$$\phi(w, \lambda) = \frac{w + \lambda/c + |\lambda|^2 \bar{w} + c\bar{\lambda}|w|^2}{|1 + c\lambda w|^2};$$

hence

$$\mathrm{Im}\phi(w, \lambda) = \frac{(1 - |\lambda|^2)\mathrm{Im}(cw) + (1 - |cw|^2)\mathrm{Im}\lambda}{c|1 + c\lambda w|^2}.$$

Notice that

1. $\operatorname{sgn}(\operatorname{Im}\phi) = \operatorname{sgn}(\operatorname{Im}\lambda)$ if $|\lambda| = 1$;

2. $\operatorname{sgn}(\operatorname{Im}\phi) = \operatorname{sgn}(\operatorname{Im}\lambda)$ if $w \in \mathbb{R}$;

3. $\phi(w, 0) = w$;

4. $|\phi(w, \lambda)| \le (1 + 1/c)(1 - c) < 6$.
   (Here the *sign* of a nonzero complex number z is given by $\operatorname{sgn} z = z/|z|$.) It
   follows that the function

$$\begin{aligned} \Phi : D^n(0,1) \times \bar{D} &\rightarrow \mathbb{C}^n \\ (z, \lambda) &\mapsto (\phi(z_1, \lambda), \dots, \phi(z_n, \lambda)) \end{aligned}$$

satisfies

5. $\Phi(z, e^{i\theta}) \in W^+$ if $0 < \theta < \pi$.

6. $\Phi(z, e^{i\theta}) \in W^-$ if $\pi < \theta < 2\pi$.

7. $\Phi(z, e^{i\theta}) \in E$ if $\theta = 0$ or $\theta = \pi$.

In short, $\Phi(z, e^{i\theta}) \in \operatorname{Dom} f$ for all $0 \le \theta < 2\pi$, all $z \in D^n(0,1)$. So we may define

$$F(z) = \frac{1}{2\pi} \int_0^{2\pi} f(\Phi(z, e^{i\theta})) \, d\theta, \qquad z \in D^n(0,1).$$

We claim that this is the F we seek (clearly F is unique if it exists).

First, F is holomorphic by an application of Morera's theorem. Next,
we note that for $(x_1, \dots, x_n) \equiv x \in E \cap D^n(0,1)$, the function $f(\Phi(x, \cdot))$ is
continuous on $\bar{D}$ and holomorphic on $\{\lambda \in D : \operatorname{Im}\lambda > 0\}$ and on $\{\lambda \in D : \operatorname{Im}\lambda < 0\}$. Again, by Morera, $f(\Phi(x, \cdot))$ is holomorphic on all of D. It follows that

$$F(x) = \frac{1}{2\pi} \int_0^{2\pi} f(\Phi(x, e^{i\theta})) \, d\theta = f(\Phi(x, 0)) = f(x);$$

hence $F = f$ on $E \cap D^n(0,1)$.

Now we are done because $E \cap D^n(0,1)$ is a totally real submanifold of
$D^n(0,1)$ (see Exercise 1 at the end of Chapter 8). In the present context we can
be more explicit: If $x + iy \in \mathbb{C}^n$ is fixed, $|x + iy| < 1/2, y > 0$, then the functions

$$\xi \mapsto f(x + \xi y)$$

$$\xi \mapsto F(x + \xi y)$$

are holomorphic for $\xi \in \mathbb{C}$ small and agree when ξ is real. It follows that these two functions are identical. Since this argument applies to all $x + iy \in B(0, 1/2), y > 0$, it follows that $F \equiv f$ on $W^+ \cup W^- \cup E$. $\qquad\square$

Remark: Given U and E as in the edge-of-the-wedge theorem, the preceding proof gives little information about the optimal size of Ω. Indeed, little is know about this matter. $\qquad\square$

There are many different proofs of the edge-of-the-wedge theorem. The most classical ones use Fourier analysis and are closely related to the Paley-Wiener theory (see E. M. Stein and G. Weiss [1]). A new proof due to J. Korevaar and J. Wiegerinck [1] provides potential techniques for estimating the size of Ω. The continuity of the function f at E is much stronger than is necessary. Only a mild growth condition is required. The theory of hyperfunctions was inspired in part by the edge-of-the-wedge theorem.

Exercise for the Reader
In the proof of Theorem 3.5.3, give an alternate argument that F is holomorphic by passing the $\bar{\partial}$ operator under the integral sign. You will need a subtle limiting argument.

EXERCISES——————————————————————————————————

1. Let $\Omega \subseteq \mathbb{C}^n$ be a domain of holomorphy and let $\Omega_0 \subseteq \mathbb{C}^k$ be another domain of holomorphy. Assume that $f : \Omega \to \mathbb{C}^k$ is a holomorphic mapping. Prove that $f^{-1}(\Omega_0) \cap \Omega$ is a domain of holomorphy.

2. Let $\Omega \subseteq \mathbb{R}^3$ be a domain. Prove that Ω is convex if and only if for every hyperplane $\mathbf{h} \subseteq \mathbb{R}^3$ it holds that either $\partial\Omega \cap \mathbf{h} = \emptyset$ or $\partial\Omega \cap \mathbf{h}$ is a Jordan curve.

3. Let $\epsilon > 0$ and consider the curve $\gamma = \gamma_\epsilon : t \mapsto (t, |t|^{2-\epsilon}) \subseteq \mathbb{R}^2$. Then γ is not C^2, but it is C^1 and γ' satisfies a classical Lipschitz condition of order $1 - \epsilon$. If $A = (0, a), a > 0$, is a point on the y-axis, then prove that A has *two* nearest points on γ. Conclude that the function $\delta_\gamma(x) \equiv \text{dist}(x, \gamma)$ cannot be differentiable in any open neighborhood of the image of γ.

4. Let $\Omega \subset\subset \mathbb{R}^{N+1}$ have C^k boundary $M, k \geq 2$. Let

$$\delta_M^*(x) = \begin{cases} -\delta_\Omega(x) & \text{if } x \in \bar{\Omega} \\ \delta_\Omega(x) & \text{if } x \notin \bar{\Omega} \end{cases}$$

Complete the following outline to prove that there is a neighborhood U of M on which δ_M^* is C^k.

a. It suffices to prove the result in a neighborhood of a fixed point of M and we may take that point to be the origin. Near 0, we take M to have the form

$$M = \{(t_1,\ldots,t_N,f(t_1,\ldots,t_N)) : (t_1,\ldots,t_N) \in W\},$$

some $W \subseteq \mathbb{R}^N$ open. We assume that f is C^k and that $\partial f/\partial t_j(0) = 0$, $j = 1,\ldots,N$. Assume that $t_{N+1} > f(t_1,\ldots,t_N)$ corresponds to $t \in \Omega$.

b. If $(x,y) = ((x_1,\ldots,x_N),y) \in \Omega$ is near 0, we know by the implicit function theorem that (x,y) has a unique nearest point $(t_1,\ldots,t_N,f(t_1,\ldots,t_N))$ in M, $t_j = t_j(x,y), j = 1,\ldots,N$. By calculus, it holds for $j = 1,\ldots,N$ that

$$0 = 2(t_j - x_j) + 2(f(t) - y)\frac{\partial f}{\partial t_j}. \tag{1}$$

Therefore, we may write

$$\delta_M^*(x,y) = (f(t) - y)\left(1 + \sum_{j=1}^{N}\left(\frac{\partial f}{\partial t_j}\right)^2\right)^{1/2}. \tag{2}$$

c. Differentiate (1) to obtain

$$0 = \frac{\partial t_j}{\partial x_i} - \delta_{ij} + \sum_{q=1}^{N}\frac{\partial f}{\partial t_j}\frac{\partial f}{\partial t_q}\frac{\partial t_q}{\partial x_i} + \sum_{q=1}^{N}(f(t) - y)\frac{\partial^2 f}{\partial t_q \partial t_j}\frac{\partial t_q}{\partial x_i} \tag{3}$$

$$0 = \frac{\partial t_j}{\partial y} + \sum_{q=1}^{N}\frac{\partial f}{\partial t_j}\frac{\partial f}{\partial t_q}\frac{\partial t_q}{\partial y} + \sum_{q=1}^{N}(f(t) - y)\frac{\partial^2 f}{\partial t_q \partial t_j}\frac{\partial t_q}{\partial y} - \frac{\partial f}{\partial t_j} \tag{4}$$

d. Compute from (3) and (2) that

$$\frac{\partial}{\partial x_i}\delta_M^* = \left(1 + \sum_{j=1}^{N}\left(\frac{\partial f}{\partial t_j}\right)^2\right)^{-1/2} \cdot \frac{\partial f}{\partial t_i}$$

so that $\frac{\partial}{\partial x_i}\delta_M^*$ is C^{k-1}.

e. Compute from (4) and (2) that

$$\frac{\partial}{\partial y}\delta_M^* = -\left(1 + \sum_{j=1}^{N}\left(\frac{\partial f}{\partial t_j}\right)^2\right)^{-1/2},$$

so that $\frac{\partial}{\partial y}\delta_M^*$ is C^{k-1}.

Where was the hypothesis that $k \geq 2$ used? Where was the restriction to a small neighborhood of M used? See Krantz and Parks [1] for more on this matter.

5. Let $\Omega \subseteq \mathbb{C}^N$ be a pseudoconvex domain. Let $u : \Omega \to \mathbb{R}$ be a pseudoconvex (i.e., continuous psh) function on Ω. Let $\Omega' \subset\subset \Omega$. Prove that there is a sequence u_j of C^∞ psh functions on Ω such that $u_j \searrow u$ uniformly on Ω'.

6. Strengthen the conclusion of Exercise 5 by showing that the u_j may be taken to be *strictly* psh.

7. Strengthen the conclusion of Exercise 6 by showing that if Ω is pseudoconvex and if $\Omega' \subset\subset \Omega_1 \subset\subset \Omega_2 \subset\subset \cdots \Omega$ and if M_j are positive numbers, then the u_j may be taken to satisfy $u_j \geq M_j$ on $\Omega \setminus \Omega_j$.

8. Strengthen the conclusion of Exercise 7 by showing that the if M_j are positive numbers and $K \in \mathbb{N}$ is fixed, then the u_j may be taken to satisfy

   a. $\left| \left(\frac{\partial}{\partial z} \right)^\alpha \left(\frac{\partial}{\partial \bar{z}} \right)^\beta u_j(z) \right| \geq M_j$ for all $z \in \Omega \setminus \Omega_j$ and all multi-indices α, β with $|\alpha| + |\beta| \leq K$;

   b. $\sum_{\ell,m} \frac{\partial}{\partial z_\ell} \frac{\partial}{\partial \bar{z}_m} u_j(z) w_\ell \bar{w}_m \geq M_j |w|^2$, all $z \in \Omega \setminus \Omega_j$, all $w \in \mathbb{C}^n$.

9. Let $\Omega \subseteq \mathbb{C}^n$ be a domain of holomorphy. Let $K \subset\subset \Omega$. Show that there is an analytic polyhedron A such that $K \subset\subset A \subset\subset \Omega$. Conclude that any domain of holomorphy can be written as the increasing union of analytic polyhedra. Show that the hypothesis that Ω be a domain of holomorphy is necessary.

10. Let $\phi : \mathbb{R}^+ \to \mathbb{R}$ be any continuous function. Prove that there is a C^∞ convex function f that satisfies $f \geq \phi$. If ϕ is increasing, show that f may be taken to be increasing. Indeed, if $\psi_1, \psi_2 : \mathbb{R}^+ \to \mathbb{R}$ are continuous, it may be arranged that $f'' \geq \psi_1, f' \geq \psi_2$ and $f \geq \phi$.

11. Let $\phi : \mathbb{R}^+ \to \mathbb{R}$ be a positive continuous function. Prove that there is a positive convex function f on $\mathbb{R}$ that satisfies $0 < f < \phi$.

12. Complete the following outline to obtain an elementary proof of a special case of the edge-of-the-wedge theorem (see E. Bedford [2]). Let

$$U^+ = \{(y_1, y_2) \in \mathbb{R}^2 : y_1 > 0, y_2 > 0\}$$

$$U^- = \{(y_1, y_2) \in \mathbb{R}^2 : y_1 < 0, y_2 < 0\}$$

and

$$W^+ = \{z \in \mathbb{C}^2 : \mathrm{Im}\, z \in U^+, |z| < 3\}$$

$$W^- = \{z \in \mathbb{C}^2 : \mathrm{Im}\, z \in U^-, |z| < 3\}.$$

Let $f : \mathbb{C}^2 \to \mathbb{C}$ be C^2 with support in $B(0,3)$, and suppose that $f\big|_{(W^+ \cup W^-) \cap B(0,2)}$ is holomorphic. Then $f\big|_{W^+ \cup W^-}$ has a holomorphic extension to all of $B(0,1)$.

a. Solve the $\bar{\partial}$ problems

$$\bar{\partial}h_1 = \begin{cases} \bar{\partial}f & \text{if } \mathrm{Im}z_1 > 0, \\ 0 & \text{otherwise,} \end{cases}$$

$$\bar{\partial}h_2 = \begin{cases} \bar{\partial}f & \text{if } \mathrm{Im}z_1 < 0, \\ 0 & \text{otherwise.} \end{cases}$$

b. Let $X^+ = \{(z_1, z_2) : \mathrm{Im}z_1 > 0, \mathrm{Im}z_2 < 0\}, X^- = \{(z_1, z_2) : \mathrm{Im}z_1 < 0, \mathrm{Im}z_2 > 0\}$. By Theorem 3.5.1, $h_1\big|_{X^+}$ and $h_2\big|_{X^-}$ extend holomorphically to all of $B(0,1)$. Call the extensions H_1 and H_2, respectively.

c. The desired extension of f is $f + (H_1 - h_1) + (H_2 - h_2)$.

d. Extend the theorem to $\mathbb{C}^n, n > 2$, by considering two-dimensional slices.

13. Let

$$\Omega = \{(z, w) \in \mathbb{C}^2 : \rho(z, w) \equiv \mathrm{Re}w + |zw|^2 + |z|^8 + \frac{15}{7}|z|^2\mathrm{Re}z^6 < 0\}.$$

Then the domain Ω is pseudoconvex. Indeed, it is strongly pseudoconvex except at $(0,0)$. Suppose that $f \in C^8(\bar{\Omega})$ is holomorphic on Ω and satisfies $f(0) = 1$ and $|f(z, w)| < 1$ when $(z, w) \neq (0, 0)$. Derive a contradiction, à la Hakim and Sibony [1], by completing the following outline.

a. Let $\tilde{f}$ denote a C^8 extension of f to a neighborhood of $\bar{\Omega}$. Write $\tilde{f} = f_1 + if_2, w = u + iv$. By Hopf's lemma, $(\partial f/\partial u)(0) > 0$. So $(\partial f/\partial w)(0) \neq 0$.

b.

$$0 \neq \left|\frac{\partial\tilde{f}}{\partial w}(0)\right|^2 = \det\begin{vmatrix} \partial f_1/\partial u & \partial f_2/\partial u \\ \partial f_1/\partial v & \partial f_2/\partial v \end{vmatrix}$$

c. There is a small neighborhood U of 0 such that $\mathcal{Z} = \{(z, w) \in U : \tilde{f}(z, w) = 1\}$ is a two-dimensional real manifold and there is a C^8 function $h(z)$ such that $\tilde{f}(z, h(z)) = 1$.

d. The equation

$$\left(\frac{\partial}{\partial z}\right)^\alpha \left(\frac{\partial}{\partial \bar{z}}\right)^\beta h(0) = 0$$

holds for all multi-indices α, β with $\beta \neq 0, |\alpha| + |\beta| \leq 8$. (*Hint:* Begin by computing $(\partial/\partial\bar{z})\tilde{f}(z, h(z))(0)$. Now use induction.)

e. Conclude that the kth order Taylor expansion of h for all $k \leq 8$ is

$$h(z) = p(z) + o(|z|^k),$$

where p is a pure holomorphic polynomial of degree not exceeding k.

f. It follows that

$$\rho(z, p(z)) = q_s(z) + o(|z|^s),$$

where q_s is the homogeneous polynomial of least degree s in the Taylor expansion of $\rho(z, p(z))$ about 0 and $s \leq k$.

g. It holds that

$$\rho(z, p(z)) = \rho(z, h(z)) + o(|z|^k).$$

h. Finally,

$$0 \leq \lim_{\substack{\lambda > 0 \\ \lambda \to 0}} \frac{\rho(\lambda z, h(\lambda z))}{\lambda^s} = \frac{q_s(\lambda z)}{\lambda^s} = q_s(z).$$

i. Show that (h) cannot hold by writing $p(z) = cz^r + o(|z|^r)$ and considering separately the cases (i) $r < 8$, (ii) $r = 8$, and (iii) $r > 8$. (*Hint:* In case i, $q_s(z) = \mathrm{Re}cz^r$; in case ii, $q_s(z) = \mathrm{Re}cz^8 + |z|^8 + (15/7)|z|^2\mathrm{Re}z^6$; in case iii, $q_s(z) = |z|^8 + (15/7)|z|^2\mathrm{Re}z^6$).

Remark: The domain Ω is the one constructed by Kohn and Nirenberg [1]. They used quite a different argument to show the weaker result that there is no peak function at 0 that is *holomorphic* in a neighborhood of 0.

14. *The tomato can principle.* Let $\Omega \subseteq \mathbb{C}^n$ have C^2 boundary, $P \in \partial\Omega$, and suppose that the Levi form at P has a negative eigenvalue. Prove that every holomorphic f on Ω extends analytically to a neighborhood of P. (*Hint:* Use analytic discs.)

15. Complete the following argument to produce the construction (R. M. Range [4]) of a bounded psh exhaustion function on any weakly pseudoconvex domain with C^3 boundary. (This was originally done by Diederich and Fornæss [1] for domains with C^2 boundary. With the extra smoothness hypothesis, Range's argument is simpler. By an extremely ingenious argument, N. Kerzman and J. P. Rosay [1] have produced bounded psh exhaustion functions on domains with C^1 boundary.)

a. Let W be a tubular neighborhood of $\partial\Omega$ and let $\pi : W \to \partial\Omega$ be Euclidean orthogonal projection. Let $\mathcal{L}_\rho(p; \xi)$ denote the Levi form defined with respect to a defining function ρ at a point p in the direction ξ. Now let $p \in W \cap \Omega$. If $a \in \mathbb{C}^n$, then we may decompose a into complex tangential and complex normal components, $a = a|_p = a_p^t + a_p^n$, with regard to the associated boundary point $\pi(p)$. Fix a C^3 defining function ρ for Ω.

b. The function $\mathcal{L}_\rho(p; a_p^t)$ is of class C^1 in p. Thus

$$\mathcal{L}_\rho(p; a_p^t) - \mathcal{L}_\rho(\pi(p); a_{\pi(p)}^t) = \mathcal{O}(|\rho(p)|)|a|^2.$$

c. We conclude that

$$\mathcal{L}_\rho(p, a^t) \geq c|\rho(p)||a|^2$$

for $p \in \Omega \cap W$ and $a \in \mathbb{C}^n$.

d. We may calculate that

$$\mathcal{L}_\rho(p; a) \geq -A|\rho(p)||a|^2 - A|a|_p|\langle \partial \rho_p, a\rangle| \tag{1}$$

for $p \in W \cap \Omega, a \in \mathbb{C}^n$, and some $A > 0$.

e. Set $r(z) = -(-\rho e^{-K|z|^2})^\eta$, where K and η will be selected momentarily. For convenience, set $\psi(z) = |z|^2$.

f. One may calculate that

$$
\begin{aligned}
\mathcal{L}_r(p; a) =\ & \eta(-\rho)^{\eta-2}e^{-\eta K\psi}\Bigg\{ K\rho^2\big[\mathcal{L}_\psi(p; a) - \eta K|\langle \partial\psi, a\rangle|^2\big] \\
& +(-\rho)\big[\mathcal{L}_\rho(p; a) - 2\eta K\operatorname{Re}\langle \partial\rho, a\rangle\overline{\langle \partial\psi, a\rangle}\big] \\
& +(1-\eta)|\langle \partial\rho, a\rangle|^2\Bigg\}.
\end{aligned}
\tag{2}
$$

g. Let the expression in braces in part f be denoted by $D(a)$. We want to select η, K so that $D(a) > 0$ for $a \neq 0$. To this end, note that there is an $A_1 > 0$ such that $\mathcal{L}_\psi(p; a) \geq A_1|a|^2$. By (1) and (2), we may choose $0 < \eta \leq \eta_0(K)$ such that

$$D(a) \geq K\rho^2\big[A_1 - A_1/2\big]|a|^2 + (-\rho)\big[\rho A|a|^2 - A_2|\langle \partial\rho, a\rangle||t|\big] + \frac{1}{2}|\langle \partial\rho, a\rangle|^2.$$

h. Since

$$|\rho|A_2|\langle \partial\rho, a\rangle||a| \leq \frac{1}{4}|\langle \partial\rho, a\rangle|^2 + A_3\rho^2|a|^2,$$

we may conclude that

$$D(a) \geq \rho^2\big[KA_1/2 - A_4\big]|a|^2.$$

In this estimate, all the A's are positive and independent of K, η, a. Clearly if K is chosen large enough then we are done.

16. Define the Hartogs triangle

$$\Omega = \{(z, w) \in \mathbb{C}^2 : |z| < |w| < 1\}.$$

Suppose that ϕ is a bounded psh exhaustion function on Ω (refer to the preceding exercise for terminology). Show that $\phi(0, w)$ must be constant, thus obtaining a contradiction.

17. Let $\Omega = \{(z, w) \in \mathbb{C}^2 : |z| < |w| < 1\}$. Then Ω is a domain of holomorphy. But if U is a sufficiently small neighborhood of $\bar\Omega$, then U is not a domain of holomorphy (for any function holomorphic on U extends analytically to $D^2(0, 1)$). Thus Ω does not have a neighborhood basis consisting of domains of holomorphy. Prove these statements.

There also exists a smoothly bounded domain of holomorphy without a pseudoconvex neighborhood basis. The construction of this "worm domain" is due to K. Diederich and J. E. Fornæss [2]. Although the idea behind their construction is similar in spirit to the Hartogs triangle above, it is much more complicated.

18. Fatou and Bieberbach, in the early part of this century, constructed examples of entire mappings $f : \mathbb{C}^2 \to \mathbb{C}^2$ that are one-to-one, have constant nonzero Jacobian determinant, and yet are not onto (see S. Bochner and W. Martin [1] for details). Indeed, the range of the Fatou-Bieberbach mapping omits a nontrivial open set. Notice that, by the Riemann mapping theorem and Picard's theorem, such a result is impossible in one complex dimension.

More recently J. P. Rosay and W. Rudin [1] and also Dixon and Esterle [1] have constructed remarkably pathological holomorphic mappings of $\mathbb{C}^2$ into $\mathbb{C}^2$ whose images omit large sets. The method in these papers is to iterate a given holomorphic mapping whose Jacobian at the origin has been normalized so that the iterates converge. See Section 11.6 for more on this approach.

A Fatou-Bieberbach mapping $f : \mathbb{C}^2 \to \mathbb{C}^2$ exhibits $\mathbb{C}^2$ as a proper submanifold of itself. Put another way, it provides an extension of the complex manifold $\mathbb{C}^2$. By iterating the map, we may obtain countably many extensions of the complex manifolds $\mathbb{C}^2$. The extended manifold is called "long $\mathbb{C}^2$." It is not known whether long $\mathbb{C}^2$ is biholomorphic to $\mathbb{C}^2$ itself. Is there a "long $\mathbb{C}^1$"? Why or why not?

19. Let $\Omega \subset\subset \mathbb{C}^n$ be strongly pseudoconvex with C^k boundary, $k \geq 2$. Prove that there is a C^k defining function $\rho : \mathbb{C}^n \to \mathbb{R}$ for Ω so that ρ is strictly psh on a neighborhood of $\bar\Omega$. (*Hint:* First examine the analogue of this question for convex domains in $\mathbb{R}^1$ and $\mathbb{R}^2$.)

20. This problem is based on a construction in Basener [1].
    a. Let $\Omega \subset\subset \mathbb{C}^n$ be a domain with C^2 boundary. Suppose that $\Omega \subseteq B(0, r)$, some $r > 0$, and that $P \in \partial\Omega \cap \partial B(0, r)$. Then P is a point of strong convexity for Ω.
    b. The strongly pseudoconvex analogue for (a) is as follows. Let $\Omega \subset\subset \mathbb{C}^n$ have C^2 boundary. Let $P \in \partial\Omega$ and $U \subseteq \mathbb{C}^n$ be a neighborhood of P. Let $v \in C^2(U)$ be a real-valued strictly psh function. Suppose that $v(P) = 0$ and that $v\big|_{\Omega \cap U} < 0$. Then P is a point of strict pseudoconvexity for Ω.
    c. Let $\Omega \subset\subset \mathbb{C}^n$ be a C^2 domain. Let $P \in \partial\Omega$. Let $U \subset\subset U_1$ be open neighborhoods of P in $\mathbb{C}^n$. Suppose that u is a psh function on U_1 such

that

$$\sup_{U_1 \setminus U} u < \sup_U u.$$

Then $U \cap \partial\Omega$ contains a strictly pseudoconvex point of $\partial\Omega$. To see this, we may assume that u is strictly psh. Choose a constant c with

$$\sup_{U_1 \setminus U} u < c < \sup_U u.$$

Shrinking U, U_1 if necessary, we may suppose that

$$\rho(z) = \begin{cases} -\delta_\Omega(z) & \text{if } z \in U_1 \cap \Omega, \\ \delta_\Omega(z) & \text{if } z \in U_1 \cap {}^c\Omega \end{cases}$$

is a C^2 defining function for Ω on U_1. Choose $\epsilon_0 > 0$ as large as possible so that there is a $z_0 \in \{\rho(z) = -\epsilon_0\}$ with $u(z_0) = c$. Let $w_0 \in \partial\Omega \cap U$ be the unique nearest point to z_0 in $\partial\Omega$. (Why may we assume that $w_0 \in U$?) Let $\nu = \nu_{w_0}$ be the unit outward normal to $\partial\Omega$ at w_0. Define $\tilde{u}(z) = u(z - \epsilon_0 \nu) - c$. Then $\tilde{u}(w_0) = 0$ and $\tilde{u} < 0$ on Ω near w_0. Now apply (b).

   d. Use (c) to show that the set of peak points for $A(\Omega)$, $\Omega \subset\subset \mathbb{C}^n$ with C^2 boundary, is contained in the closure of the strongly pseudoconvex points.

   e. The Šilov boundary for $A(\Omega)$ is the closure of the peak points (see T. W. Gamelin [1]). Therefore, you can now explicitly identify the Šilov boundary of any $\Omega \subset\subset \mathbb{C}^n$ with C^2 boundary as the closure of the strongly pseudoconvex boundary points.

21. Let $f : \mathbb{R}^N \to \mathbb{R}$ be convex. Assume that f has a minimum at 0. Let $x \in \mathbb{R}^N, x \neq 0$. Prove that the normal to the level curve of f at x cannot have negative inner product with $\overrightarrow{0x}$.

22. Refine Narashimhan's lemma in the following manner. Let $\Omega \subset\subset \mathbb{C}^n$ be a domain with C^2 boundary. Let $P \in \partial\Omega$ be a point of strict pseudoconvexity. There is a neighborhood U of P and a biholomorphic change of coordinates $z \leftrightarrow w$ so that, in the new coordinates,

$$\Omega \cap U = \{w \in U : 2\operatorname{Re} w_n < -|w|^2 + o(|w|^2)\}.$$

This says, in effect, that a strongly pseudoconvex point agrees with the ball (in suitable local coordinates) to second order. Fefferman's lemma (C. Fefferman [1]) shows, in fact, that it can be arranged for the agreement to be to fourth order. Powerful use of this fact can be made in geometric/analytic arguments.

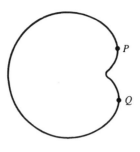

FIGURE 3.9

23. Let $\Omega \subset\subset \mathbb{R}^N$ be a domain with C^2 boundary. If Ω is strongly convex, prove that whenever I is a closed segment in $\bar{\Omega}$, then $\overset{\circ}{I} \subseteq \Omega$. Prove that the converse is false. What is the complex-analytic analogue of this assertion?

24. Let $\Omega_1, \Omega_2 \subset\subset \mathbb{C}^n$ be domains with C^2 boundary. Let $f : \Omega_1 \to \Omega_2$ be biholomorphic. Assume that f, f^{-1} extend C^1 to the boundary. Prove that $|\det \mathrm{Jac}_{\mathbb{C}} f(z)| \geq c > 0$, all $z \in \bar{\Omega}$. Now let $f : \Omega_1 \to \Omega_2$ be proper holomorphic (not biholomorphic) and assume that Ω_1, Ω_2 are strongly pseudoconvex and that f extends C^2 to $\bar{\Omega}_1$. Let $\tilde{\rho}$ be a C^2 strictly plurisubharmonic defining function for Ω_2 (we take $\tilde{\rho}$ to be defined and psh on a neighborhood of $\bar{\Omega}_2$). Prove that, for $A > 0$ sufficiently large, $\rho = \exp(A \cdot \tilde{\rho} \circ f) - 1$ is a similar such function for Ω_1 (you will need to use Hopf's lemma to see that $\nabla \rho \neq 0$ near Ω_1). However, the complex Hessian of ρ is divisible by $|\nabla f|^2$. Thus $|\nabla f| \geq c > 0$ on $\bar{\Omega}_1$. J. E. Fornæss [3] has proved a similar result under the hypothesis that Ω_1, Ω_2 are weakly pseudoconvex.

25. Let $\Omega \subset\subset \mathbb{R}^2$ have C^∞ boundary. Then the convex hull of Ω need not have C^∞ boundary. (*Hint:* Look at a disc in $\mathbb{R}^2$ with a smooth dent as shown in Figure 3.9.)

26. Let $\Omega \subset\subset \mathbb{C}^n$ be a domain with C^2 boundary, $n \geq 3$. Show that Ω is pseudoconvex if and only if its intersection with every complex hyperplane is either empty or pseudoconvex. Why is the case $n = 2$ different?

27. Let $\Omega \subseteq \mathbb{C}^n$ be a domain. Let $P \in \partial\Omega$. We say that P is *locally essential* if there is a neighborhood Ω_P of P and a holomorphic function u on $\Omega_P \cap \Omega$ such that for *no* connected neighborhood Ω_2 of P and nonempty $\Omega_1 \subseteq \Omega_P \cap \Omega_2$ is there a u_2 holomorphic on Ω_2 with $u = u_2$ on Ω_1. If Ω in addition has C^2 boundary and every $P \in \partial\Omega$ is locally essential, prove that Ω is Levi pseudoconvex. What if Ω does not have C^2 boundary?

28. Give a proof that the ball $B = \{z \in \mathbb{C}^2 : |z| < 1\}$ is not biholomorphic to the half-space $H = \{z \in \mathbb{C}^2 : \mathrm{Re} z_1 > 0\}$ via a biholomorphism that is C^2 to the boundary. Can you prove the result without assuming that the biholomorphism is C^2?

29. Which tube domains are strongly pseudoconvex?

30. Prove that the unit ball in $\mathbb{C}^2$ is biholomorphic to a tube domain over the paraboloid

$$\{x = (x_1, x_2) : x_1 - x_2^2 > 0\}.$$

See P. Yang [1] for more on this matter. However the ball is *not* biholomorphic to a tube domain over a cone. (This last assertion follows from the classification theory of Siegel domains—see Kaneyuki [1]—but can also be seen by more elementary means.)

31. True or false: The complement of a pseudoconvex domain is pseudoconvex. Is this *ever* true? Under what conditions on the boundary?

32. Consider the domain

$$\Omega_\lambda \equiv \{z \in \mathbb{C}^2 : \operatorname{Re} z_1 + |z_2|^4 + 2\lambda \operatorname{Re} z_2 \bar{z}_2^3 < 0\}.$$

Here λ is a real parameter. Prove that if λ is small, then Ω_λ is strongly pseudoconvex. For larger λ the domain is not pseudoconvex; in particular, the hull of holomorphy of the domain is strictly larger than Ω. If λ is large enough, then the hull of holomorphy is all of space.

33. Prove that a domain Ω is convex if and only if the function $-\log \delta_\Omega$ is a convex function on Ω (here δ_Ω is Euclidean distance to the boundary).

34. Compute the Levi form for the surfaces defined by
    a. $\rho(z) = 2 \operatorname{Re} z_n + \sum_{j=1}^{n-1} |z_j|^{2p_j}, \quad p_j \in \mathbb{N}$;
    b. $\rho(z) = 2 \operatorname{Re} z_n + \sum_{j=1}^{N} |f_j(z)|^2, \quad f_j$ holomorphic;
    c. $\rho(z) = 2 \operatorname{Re} z_2 + |z_1|^8 + k|z_1|^6 \operatorname{Re}(z_1{}^2)$.
    In part c, what range of k will make the surface pseudoconvex near the origin? In part b, if $N = n - 1$ and the f_j are independent of z_n, then find a simple formula for the determinant of the Levi form.

35. Let M be a real hypersurface in $\mathbb{C}^n$ and assume that the Levi form has rank *at least* q at each point of M. Prove that M contains no $(n - q)$-dimensional complex analytic manifold. Prove by example that if the rank of the Levi form is $q - 1$, then the conclusion is false.

36. Give a suitable definition for the "polynomial hull" of a set. If K is a compact subset of the unit disc in $\mathbb{C}$, then what is its polynomial hull?

4 Hörmander's Solution of the $\bar{\partial}$ Equation

4.1 Generalities about the $\bar{\partial}$ Problem

Let $\Omega \subseteq \mathbb{C}^n$ be a domain, $f = \sum_{j=1}^{n} f_j d\bar{z}_j$ a $(0,1)$ form on Ω. The equation $\bar{\partial}u = f$, that is, $\partial u/\partial \bar{z}_j = f_j, j = 1,\ldots,n$, is *over determined* (i.e., there are more equations than unknowns) as soon as $n > 1$. This motivates, at least intuitively, the compatibility condition $\bar{\partial}f = 0$. It also suggests why it may be difficult to prove existence and regularity theorems (the rigorous treatment of what constitutes an over-determined system is rather elaborate and is best formulated in terms of sheaf cohomology—see D. C. Spencer [1]).

D. C. Spencer introduced the formalism of the $\bar{\partial}$-Neumann problem that demonstrates why the geometry of $\partial\Omega$ plays an essential role in solving the $\bar{\partial}$ problem. C. B. Morrey [1] further developed this theory but did not find the right condition on the boundary to solve the problem. J. J. Kohn discovered how to exploit strong pseudoconvexity to solve the equation and prove regularity (J. J. Kohn [1], G. B. Folland and J. J. Kohn [1], and S. G. Krantz [19]). Lars Hörmander [1] (and Andreotti and Vesentini [1]) discovered how to suppress the boundary phenomena by working with weighted Sobolev spaces. The initial choice of weights takes into account the shape of $\partial\Omega$—but once they are introduced, the result is that one is doing analysis on a Riemannian manifold in which the boundary is at infinite distance from any interior point (i.e., the manifold is *complete*). As a result, the boundary no longer plays a role. Therefore, the method of weights applies directly to all weakly pseudoconvex domains.

The techniques due to Kohn are more natural from the point of view of partial differential equations—especially in view of the well-known elliptic theory of the Laplacian. Also Kohn's methods are more difficult and technical than are those of Hörmander. But they yield much more information.

A leisurely treatment of Kohn's approach is given in S. G. Krantz [19], together with detailed background material. Although Hörmander's techniques appear at first to be rather *ad hoc*, they yield the information we desire rather quickly. Also they are rather flexible: In the hands of Skoda (H. Skoda [1, 2]) and others they have led to a body of powerful tools. In order to get to the solution of the Levi problem most directly, we shall use Hörmander's methods. (It should be mentioned that, for sheer brevity, the method of R. M. Range [3] for solving the Levi problem is the best. But it is also an unusual combination of classical and modern methods that, for our purposes, is less instructive.)

In Section 4.2 we review some basic notions connected with the theory of unbounded operators on a Hilbert space. We will then be able to fit the generalities of this section into a theoretical framework and to reduce solving the $\bar{\partial}$ problem to the problem of proving an estimate.

4.2   Unbounded Operators on a Hilbert Space

Let $\mathcal{H}^1, \mathcal{H}^2$ be complex Hilbert spaces, $\mathcal{E} \subseteq \mathcal{H}^1$ a subspace, and $T : \mathcal{E} \to \mathcal{H}^2$ a *bounded linear operator*. Then T extends naturally to $\bar{\mathcal{E}}$, and $\bar{\mathcal{E}}$ is a Hilbert space. In practice, then, it is most convenient to study T as an operator on $\bar{\mathcal{E}}$. This simple idea makes the basic theory of bounded Hilbert space operators rather easy. The theory of bounded operators is suitable for the study of integral operators such as those given by convolution with an L^1 function.

Most integral operators are distributions and so are automatically continuous in a suitable sense on the topological vector space C_c^∞. Many are bounded on some Hilbert space (usually a Sobolev space), although this always requires some work to see.

The typical unbounded Hilbert space operator is a differential operator. Take $\mathcal{H}^1 = \mathcal{H}^2 = L^2(\mathbb{R})$. Consider $T = d/dx$ on $\mathcal{H}^1$. Unfortunately, T is not defined on all of $\mathcal{H}^1$. Also, there exist $f \in L^2$ such that Tf is defined (in a suitable sense) but $Tf \notin L^2$. However, on the dense subset $C_c^\infty \subseteq L^2$, T is well defined and if $f \in C_c^\infty$, then $Tf \in C_c^\infty \subseteq L^2$. Everywhere-defined, unbounded Hilbert space operators do not occur in nature. More precisely, a theorem due to J. D. M. Wright [1] says in effect that an everywhere-defined unbounded Hilbert space operator cannot be constructed without making use of the axiom of choice. On the other hand, we may usually take our T to be at least densely defined by merely shrinking $\mathcal{H}^1$. Since all of our operators T in this chapter will be differential operators and all our Hilbert spaces Sobolev spaces (to be defined later), then T will be defined on the dense subspace C_c^∞.

Analysis of unbounded, densely defined operators is subtle, although not always hard. Assertions that would be obvious for bounded operators sometimes require tedious checking in the unbounded operator setting. An operator that is unbounded may still be closed or may still have a bounded inverse. The adjoint of an unbounded operator is defined only on a subspace of the full Hilbert space $\mathcal{H}^2$. All these phenomena will be encountered below in a rather explicit setting.

In what follows, let $\mathcal{H}^j$ be complex Hilbert spaces with inner products $\langle\,,\,\rangle_j$ and norms $\|\quad\|_j, j = 1, 2$. Let $\mathcal{D} \subseteq \mathcal{H}^1$ be a dense subspace and $T : \mathcal{D} \to \mathcal{H}^2$ a linear operator that we do *not* assume to be bounded. It is sometimes convenient, for emphasis, to write $\mathcal{D}_T$ for the domain of T instead of $\mathcal{D}$.

DEFINITION 4.2.1   We say that T is *closed* if its graph

$$\mathcal{G}_T = \{(x, Tx) : x \in \mathcal{D}\} \subseteq \mathcal{H}^1 \times \mathcal{H}^2$$

is a closed set.

DEFINITION 4.2.2   Let $\psi \in \mathcal{H}^2$. We say that $\psi \in \mathcal{D}_{T^*}$ (ψ is in the domain of the *adjoint* T^* of T) if there is a constant $C = C(\psi) > 0$ such that

$$|\langle T\phi, \psi\rangle_2| \le C\|\phi\|_1, \quad \text{all } \phi \in \mathcal{D}_T. \qquad (4.2.2.1)$$

Remark: Notice that if T is a bounded operator, then the adjoint of T always exists and is defined on all of $\mathcal{H}^2$. Thus an inequality of the form (4.2.2.1) holds trivially for all elements of $\mathcal{H}^2$.   □

The last definition makes sense because of the next proposition.

PROPOSITION 4.2.3   If $\psi \in \mathcal{D}_{T^*}$, then there exists a unique element $T^*\psi \in \mathcal{H}^1$ such that

$$\langle\phi, T^*\psi\rangle_1 = \langle T\phi, \psi\rangle_2, \text{ all } \phi \in \mathcal{D}_T.$$

Proof   The linear functional $\phi \mapsto \langle T\phi, \psi\rangle_2$ is densely defined and bounded (by (4.2.2.1)), so it extends uniquely to a bounded linear functional on all of $\mathcal{H}^1$. By the Riesz representation theorem, there is a unique element $\sigma \in \mathcal{H}^1$ such that the functional is given by pairing with σ. Set $T^*\psi = \sigma$.   □

Thus we have a new linear operator $T^* : \mathcal{D}_{T^*} \to \mathcal{H}^1$.

LEMMA 4.2.4   The space $\mathcal{H}^1 \times \mathcal{H}^2$ is a Hilbert space when equipped with the inner product

$$\langle\langle(h_1, h_2), (h_1', h_2')\rangle\rangle = \langle h_1, h_1'\rangle_1 + \langle h_2, h_2'\rangle_2.$$

Proof   The proof is left as an exercise.   □

PROPOSITION 4.2.5   If $\mathcal{H}$ is any Hilbert space and $\mathcal{E} \subseteq \mathcal{H}$ any subset, then $\mathcal{E}^\perp \equiv \{h \in \mathcal{H} : \langle h, e\rangle = 0 \ \forall \ e \in \mathcal{E}\}$ is closed.

Proof   The proof is left as an exercise.   □

DEFINITION 4.2.6   Let $J : \mathcal{H}^2 \times \mathcal{H}^1 \to \mathcal{H}^1 \times \mathcal{H}^2$ be given by $J(h_2, h_1) = (-h_1, h_2)$.

Remark: J and J^{-1} take closed spaces to closed spaces.   □

PROPOSITION 4.2.7   If $T : \mathcal{H}^1 \to \mathcal{H}^2$ is a linear operator, then

$$(\mathcal{G}_T)^\perp = J(\mathcal{G}_{T^*}).$$

Proof   Let $(-T^*y, y) \in J(\mathcal{G}_{T^*}), (x, Tx) \in \mathcal{G}_T$. Then

$$\langle\langle(-T^*y, y), (x, Tx)\rangle\rangle \equiv \langle -T^*y, x\rangle_1 + \langle y, Tx\rangle_2.$$

But $y \in \mathcal{D}_{T^*}$, of course, so this last line is

$$-\langle y, Tx\rangle_2 + \langle y, Tx\rangle_2 = 0,$$

whence $J(\mathcal{G}_{T^*}) \subseteq (\mathcal{G}_T)^\perp$.

For the converse inclusion, let $(a, b) \in (\mathcal{G}_T)^\perp$. So for any $x \in \mathcal{D}_T$, we have

$$0 = \langle\langle(a, b), (x, Tx)\rangle\rangle \equiv \langle a, x\rangle_1 + \langle b, Tx\rangle_2. \qquad (4.2.7.1)$$

Therefore,

$$|\langle b, Tx\rangle_2| = |\langle a, x\rangle_1| \leq \|a\|_1 \|x\|_1$$

so that $b \in \mathcal{D}_{T^*}$. Thus (4.2.7.1) may be rewritten as

$$0 = \langle a + T^*b, x\rangle_1, \qquad \text{all } x \in \mathcal{D}_T.$$

Since $\mathcal{D}_T$ is dense in $\mathcal{H}^1$, we have that $a + T^*b = 0$ or $T^*b = -a$. Hence $(a, b) \in J(\mathcal{G}_{T^*})$.   □

COROLLARY 4.2.8   The operator T^* is closed.

Proof   Apply 4.2.5 and 4.2.7.   □

Exercises for the Reader

1. Let $T : \mathcal{H}^1 \to \mathcal{H}^2$ be any linear operator. Then T^* exists and is well defined if $\mathcal{D}_T$ is dense in $\mathcal{H}^1$. Moreover, T admits a closed linear extension if and only if $T^{**} = (T^*)^*$ exists. This, in turn, is true if and only if $\mathcal{D}_{T^*}$ is dense in $\mathcal{H}^2$. Under these circumstances, T^{**} is a closed extension of T. If T is closed, then $T^{**} = T$.

2. If $y \perp (\text{Range } T)$, then $y \in \mathcal{D}_{T^*}$ and $\overline{T^*y = 0}$. If T is closed, if $\mathcal{D}_{T^*}$ is dense, and if $x \perp (\text{Kernel } T)$ then $x \in \overline{(\text{Range } T^*)}$. If $y \perp (\text{Kernel } T^*)$ then $y \in \overline{\text{Range } T}$.

Let $\bar{\partial}_{p,q}$ denote the $\bar{\partial}$ operator on (p,q) forms. For smooth forms, we know that $\text{Range}\,(\bar{\partial}_{p,q}) \subseteq \text{Kernel}(\bar{\partial}_{p,q+1})$. Our principal goal is to see that, when Ω is pseudoconvex, this inclusion is really an equality—that is, we want to see that every $\bar{\partial}$-closed form is $\bar{\partial}$-exact. We shall formulate our problem in the context of suitable Hilbert spaces so that we may utilize the following lemma.

LEMMA 4.2.9 Assume that $T : \mathcal{H}^1 \to \mathcal{H}^2$ is closed and densely defined. Let $F \subseteq \mathcal{H}^2$ be a closed subspace, and suppose that $F \supseteq \text{Range}\,T$. Suppose also that $\mathcal{D}_{T^*}$ is dense. Then $F = \text{Range}\,T$ if and only if there is a constant $C > 0$ such that

$$\|y\|_{\mathcal{H}^2} \leq C\|T^*y\|_{\mathcal{H}^1}, \qquad \text{all } y \in F \cap \mathcal{D}_{T^*}. \tag{4.2.9.1}$$

Proof   Assume that (4.2.9.1) holds. Let $z \in F$. We construct a bounded linear functional φ on $\mathcal{H}^1$ whose representative x satisfies $Tx = z$. If $w \in T^*(F \cap \mathcal{D}_{T^*})$, $w = T^*y$, define $\varphi(w) = \langle y, z \rangle_2$. Note that the choice of y is unique by (4.2.9.1). Thus φ is bounded in norm, on the domain $T^*(F \cap \mathcal{D}_{T^*})$, by $C\|z\|_2$ by (4.2.9.1). Apply the Hahn-Banach theorem to extend φ to all of $\mathcal{H}^2$. Then φ has a unique representative x. So

$$\langle y, z \rangle_2 = \varphi(w) = \langle w, x \rangle_1 = \langle T^*y, x \rangle_1$$

for all $y \in \mathcal{D}_{T^*} \cap F$ and hence for all $y \in \mathcal{D}_{T^{**}}$. It follows that $x \in \mathcal{D}_{T^*} = \mathcal{D}_T$ whence $\langle y, z \rangle_2 = \langle y, Tx \rangle_2$ for all $y \in \mathcal{D}_{T^*} \cap F$, hence $(z - Tx) \perp (\mathcal{D}_{T^*} \cap f)$, so $(z - Tx) \perp F$ (remember that $\mathcal{D}_{T^*}$ is dense), so $z = Tx$.

Conversely, if $F = \text{Range}\,T$, then we assign to each $z \in F$ an $x_z \in \mathcal{H}^1$ with $Tx_z = z$. Then, for any $f \in F \cap \mathcal{D}_{T^*}$, we have

$$|\langle f, z \rangle_2| = |\langle f, Tx_z \rangle_2| = |\langle T^*f, x_z \rangle_1| \leq \|T^*f\|_1\|x_z\|_1.$$

The inequality holds trivially for $z \in F^\perp$ since the left side is then 0. Thinking of the set $F \cap \mathcal{D}_{T^*} \cap \{f \in \mathcal{H}^2 : \|T^*f\|_1 \leq 1\}$ as linear functionals on $\mathcal{H}^2$, we conclude by the uniform boundedness principle that (4.2.9.1) holds. $\square$

4.3   The Formalism of the $\bar{\partial}$ Problem

Fix $\Omega \subseteq \mathbb{C}^n$. Let $0 \leq p, q \leq n$ and let $L^2_{(p,q)}(\Omega)$ denote the (p,q) forms on Ω with $L^2(\Omega)$ coefficients. If f is such a form, we write

$$f = \sum_{|\alpha|=p, |\beta|=q}' f_{\alpha\beta} \, dz^\alpha \wedge d\bar{z}^\beta;$$

here for specificity the sum $\sum'$ is taken *only* over increasing multi-indices α, β (here a multi-index α is said to be increasing if $\alpha_1 \leq \alpha_2 \leq \cdots \leq \alpha_n$). Let $|f|^2 = \sum'_{|\alpha|=p,|\beta|=q} |f_{\alpha\beta}|^2$. If $g = \sum'_{|\alpha|=p,|\beta|=q} g_{\alpha\beta} dz^\alpha \wedge d\bar{z}^\beta$ is another such form, we let

$$\langle f, g \rangle = \sum'_{\alpha,\beta} \int f_{\alpha\beta} \overline{g_{\alpha\beta}} dV,$$

$$\|f\|^2 = \langle f, f \rangle.$$

The inner product turns $L^2_{(p,q)}$ into a Hilbert space.

Although we use $\sum'_{\alpha,\beta}$ as much as possible, we also have occasion to use the full sum $1/p!1/q! \sum_{\alpha,\beta} f_{\alpha\beta} dz^\alpha d\bar{z}^\beta$. In this case the consistent policy is to require that if $\tilde{\alpha}, \tilde{\beta}$ are permutations of α, β, then $f_{\alpha\beta} = \epsilon^\alpha_{\tilde{\alpha}} \epsilon^\beta_{\tilde{\beta}} f_{\tilde{\alpha}\tilde{\beta}}$, where $\epsilon^\alpha_{\tilde{\alpha}}, \epsilon^\beta_{\tilde{\beta}}$ are, respectively, the *signs* of the permutations in S_n that take α to $\tilde{\alpha}$ and β to $\tilde{\beta}$. Since we must do some rather precise calculating and counting, these notations will prove essential.

We use the notation D and C^∞_c interchangeably. The former is most amenable to additional subscripts and superscripts. Since forms with C^∞_c coefficients are dense in $L^2_{(p,q)}$ (equivalently, $D_{(p,q)}$ is dense in $L^2_{(p,q)}$), we may think of $\bar{\partial}$ as a densely defined operator on $L^2_{(p,q)}$. However, we also want it to be *closed*, so we extend the definition of the operator $\bar{\partial}$ by means of the concept of *weak derivative*.

DEFINITION 4.3.1   Fix a domain $\Omega \subseteq \mathbb{C}^n$. Let $D = (\partial/\partial z)^\alpha (\partial/\partial \bar{z})^\beta$ be a differential monomial. Let $f, g \in L^1(\Omega)$. We say that $Df = g$ in the *weak sense* (or *weakly* or *in the sense of distributions*) if for all $\phi \in C^\infty_c$ we have $\int f D\phi \, dV = (-1)^{|\alpha|+|\beta|} \int g\phi \, dV$.

Remark: The definition is motivated by integration by parts (which is valid in the special case that f is smooth). This explains the signature.

Note that there is an $f \in L^\infty(\Omega)$ such that for no $g \in L^1(\Omega)$ does it hold that $\partial f/\partial z_1 = g$ in the weak sense.   □

Now the weak form of exterior differentiation when applied to *functions* is easy: Let f be an L^1 function on Ω and let $g = \sum_j g_j d\bar{z}_j$ be a $(0,1)$ form on Ω with L^1 coefficients. We say that $\bar{\partial} f = g$ *in the weak sense* if $\partial f/\partial \bar{z}_j = g_j$ in the weak sense for each j. Of course, ∂f in the weak sense is defined similarly.

For exterior differentiation of forms of degree at least one, matters are a bit more tedious. If $f = \sum'_{\alpha,\beta} f_{\alpha\beta} dz^\alpha \wedge d\bar{z}^\beta$ is a form with *smooth* coefficients, then recall that

$$
\begin{aligned}
\bar{\partial} f &= {\sum_{j,\alpha,\beta}}' \frac{\partial f_{\alpha\beta}}{\partial \bar{z}_j} d\bar{z}_j \wedge dz^\alpha \wedge d\bar{z}^\beta \\
&= {\sum_{j,\alpha,\beta}}' \frac{\partial f_{\alpha\beta}}{\partial \bar{z}_j} (-1)^{|\alpha|} dz^\alpha \wedge d\bar{z}_j \wedge d\bar{z}^\beta \\
&= {\sum_\alpha}' {\sum_\gamma}' {\sum_{j,\beta}}' \frac{\partial f_{\alpha\beta}}{\partial \bar{z}_j} \epsilon_{j\beta}^\gamma (-1)^{|\alpha|} dz^\alpha \wedge d\bar{z}^\gamma .
\end{aligned}
$$

Here $\epsilon_{j\beta}^\gamma$ is the sign of the permutation taking $j\beta$ to γ—in other words, we have grouped together all terms that have the same set of differentials. The point of this calculation is that when we are defining the concept of the weak exterior ($\bar{\partial}$) derivative of a form, we must treat (for γ fixed)

$$
{\sum_{j,\beta}}' \frac{\partial f_{\alpha\beta}}{\partial \bar{z}_j} \epsilon_{j\beta}^\gamma (-1)^{|\alpha|} \, dz^\alpha \wedge d\bar{z}^\gamma
$$

as a unit. Our definition of weak $\bar{\partial}$ derivative is then as follows.

Let $f = {\sum}'_{\alpha,\beta} f_{\alpha\beta} dz^\alpha \wedge d\bar{z}^\beta$ be a form with L^1 coefficients on Ω. Let $g = {\sum}'_{\mu,\eta} g_{\mu\eta} dz^\mu \wedge d\bar{z}^\eta$ be another such form. We say that $\bar{\partial} f = g$ *in the weak sense* on Ω if for each fixed α, γ and $\phi \in C_c^\infty(\Omega)$ we have

$$
-\int_\Omega (-1)^{|\alpha|} {\sum_{j,\beta}}' f_{\alpha\beta} \frac{\partial \phi}{\partial \bar{z}_j} \epsilon_{j\beta}^\gamma = \int_\Omega g_{\alpha\gamma} \phi .
$$

Of course the weak operator ∂ on forms is defined similarly. It is a formal exercise to check that the weakly defined operator $\bar{\partial}$ is a closed operator from forms with L^p coefficients to forms with L^p coefficients, $1 \leq p \leq \infty$.

We may now think of the operator

$$
\bar{\partial}_{p,q} : L^2_{(p,q)} \to L^2_{(p,q+1)}
$$

as a closed, densely defined, unbounded linear operator. Notice that

$$
\bar{\partial}^*_{p,q} : L^2_{(p,q+1)} \to L^2_{(p,q)} .
$$

It is natural (see J. J. Kohn [1], G. B. Folland and J. J. Kohn [1], and S. G. Krantz [19]) to consider the associated operator

$$
\Box = \bar{\partial}_{p,q} \bar{\partial}^*_{p,q} + \bar{\partial}^*_{p,q+1} \bar{\partial}_{p,q+1} : L^2_{(p,q+1)} \to L^2_{(p,q+1)},
$$

which is defined on a subspace of $L^2_{(p,q+1)}$ that contains the space of forms with C^∞_c coefficients.

The operator $\square$ has the virtue of being second order, self-adjoint, and elliptic. The disadvantage is that the operators $\bar{\partial}^*_{p,q}$ and $\bar{\partial}^*_{p,q+1}$ are not known explicitly. If one restricts attention to forms with C^∞_c coefficients, then it is a simple matter to compute the *formal adjoints* of $\bar{\partial}_{p,q}$ and $\bar{\partial}_{p,q+1}$ by integration by parts (we compute some formal adjoints later). What is subtle is to see how these formal adjoints are related to the Hilbert space adjoints $\bar{\partial}^*_{p,q}$ and $\bar{\partial}^*_{p,q+1}$. This is again reckoned by integration by parts on *noncompactly supported forms*, and the process gives rise to the $\bar{\partial}$-Neumann boundary conditions. It is these nonelliptic boundary conditions that account for all the subtlety of the $\bar{\partial}$ problem. In particular, one needs to determine explicitly *when* a form is in the domain of the adjoint operator. When it *is* in the domain, then the adjoint is computed according to the formula for the formal adjoint. All of these matters are considered in detail in the books by G. B. Folland and J. J. Kohn [1] and S. G. Krantz [19].

However we wish to avoid the subtleties associated to adjoints that were discussed in the last paragraph. To do so, we work in the $L^2(\Omega, \phi)$ spaces of Hörmander, where $\phi : \Omega \to \mathbb{R}$ is an appropriate smooth function. Here $f \in L^2(\Omega, \phi)$ if and only if

$$\|f\|^2_\phi \equiv \int |f|^2 e^{-\phi}\, dV < \infty.$$

The inner product on $L^2(\Omega, \phi)$ is of course

$$\langle f, g \rangle_\phi = \int f\bar{g}e^{-\phi}dV.$$

As usual, the inner product and norm are extended by linearity to forms. In practice, ϕ will be positive and blow up near $\partial\Omega$—indeed, ϕ will be manufactured from the psh exhaustion functions that we studied in Chapter 3.

4.4 Some Computations

To enable us to focus in the next section on the simplicity and elegance of Hörmander's ideas, we isolate in the present section the routine computations. Here is the setup: $\Omega \subseteq \mathbb{C}^n$ is a domain. Consider the complex

$$L^2_{(p,q)}(\Omega, \phi_1) \xrightarrow{\bar{\partial}_{p,q}} L^2_{(p,q+1)}(\Omega, \phi_2) \xrightarrow{\bar{\partial}_{p,q+1}} L^2_{(p,q+2)}(\Omega, \phi_3).$$

The weights ϕ_1, ϕ_2, ϕ_3 are to be chosen at the propitious moment. To simplify notation, we set $T = \bar{\partial}_{p,q}, S = \bar{\partial}_{p,q+1}, \mathcal{H}^1 = L^2_{(p,q)}(\Omega, \phi_1), \mathcal{H}^2 = L^2_{(p,q+1)}(\Omega, \phi_2),$ $\mathcal{H}^3 = L^2_{(p,q+2)}(\Omega, \phi_3), F = \operatorname{Ker} S \subseteq \mathcal{H}^2$. We let

$$\langle \ , \ \rangle_j = \langle \ , \ \rangle_{\phi_j} = \langle \ , \ \rangle_{\mathcal{H}^j}$$

be the inner product on $\mathcal{H}^j$ and

$$\| \ \|_j = \| \ \|_{\phi_j} = \| \ \|_{\mathcal{H}^j}$$

the corresponding norm. Notice that the statement $f \in \mathcal{D}_T$ means, of course, that $f \in L^2_{(p,q)}(\Omega, \phi_1)$ but also that $\bar{\partial} f$ exists (in the weak sense) *and* that $\bar{\partial} f \in L^2_{(p,q+1)}(\Omega, \phi_2)$. A similar statement holds for the meaning of $g \in \mathcal{D}_S$.

The reader should verify the following simple facts:

1. If $\phi \in C(\Omega)$, then $L^2(\Omega, \phi) \subseteq L^2_{\text{loc}}(\Omega)$. (Here, for $1 \leq p < \infty, L^p_{\text{loc}}(\Omega)$ consists of those measurable functions that are p^{th} power integrable on compact sets.)

2. If $f \in L^2_{\text{loc}}(\Omega)$, then there exists a continuous $\phi : \Omega \to \mathbb{R}$ such that $f \in L^2(\Omega, \phi)$.

3. The analogue of (2), with $\phi \in C^\infty(\Omega)$, holds.

4. If $\phi \in C(\Omega)$, then $C^\infty_c(\Omega)$ is dense in $L^2(\Omega, \phi)$.

5. If $\phi \in C_c(\Omega)$, then $L^2(\Omega, \phi) = L^2(\Omega)$ with comparable norms.

Now, for Ω pseudoconvex and an appropriate choice of $\phi_j, j = 1, 2, 3$, we will prove the following:

MAIN GOAL

$$\|f\|^2_2 \leq C^2 \left\{ \|T^* f\|^2_1 + \|S f\|^2_3 \right\}, \quad \text{all } f \in \mathcal{D}_{T^*} \cap \mathcal{D}_S. \tag{4.4.1}$$

Notice that when $f \in \operatorname{Ker} S$, then equation (4.4.1) reduces to (4.2.9.1). Thus we can obtain the desired result that Range $T = F$; that is, all $\bar{\partial}$-closed forms are $\bar{\partial}$-exact. The inequality displayed in (4.4.1) is a chain homotopy type of condition—it is frequently advantageous to prove a symmetric statement like this one rather than an asymmetric one. The main purpose of this section is to reduce the **MAIN GOAL** to a simpler assertion involving only forms with C^∞_c coefficients. That will be the **MODIFIED MAIN GOAL**.

LEMMA 4.4.2 Let $\eta \in C^\infty_c(\Omega), f \in \mathcal{D}_S$. Then $\eta f \in \mathcal{D}_S$.

Proof   That $\eta f \in \mathcal{H}^2$ if $f \in \mathcal{H}^2$ is clear. Assume first that $\partial f_{\alpha\beta}/\partial \bar{z}_j$ exists in the weak sense. If $\phi \in C_c^\infty(\Omega)$ is arbitrary, then

$$
\begin{aligned}
\int (\eta f_{\alpha\beta}) \frac{\partial \phi}{\partial \bar{z}_j} dV &= \int f_{\alpha\beta} \left(\eta \frac{\partial \phi}{\partial \bar{z}_j} \right) dV \\
&= \int f_{\alpha\beta} \frac{\partial}{\partial \bar{z}_j}(\eta\phi) dV - \int f_{\alpha\beta} \left(\phi \frac{\partial \eta}{\partial \bar{z}_j} \right) dV \\
&= -\int \frac{\partial f_{\alpha\beta}}{\partial \bar{z}_j} \eta\phi \, dV - \int f_{\alpha\beta} \frac{\partial \eta}{\partial \bar{z}_j} \phi \, dV \\
&= -\int \left\{ \frac{\partial f_{\alpha\beta}}{\partial \bar{z}_j} \eta + f_{\alpha\beta} \frac{\partial \eta}{\partial \bar{z}_j} \right\} \phi \, dV
\end{aligned}
$$

or $(\partial/\partial \bar{z}_j)(\eta f_{\alpha\beta}) = (\partial f_{\alpha\beta}/\partial \bar{z}_j)\eta + f_{\alpha\beta}(\partial \eta/\partial \bar{z}_j)$ in the weak sense. For the general case, group terms as in Section 4.3. □

LEMMA 4.4.3   Let $\eta \in C_c^\infty(\Omega), f \in \mathcal{D}_S$. Then $S(\eta f) - \eta \cdot Sf = \bar{\partial}\eta \wedge f$.

Proof   By the proof of the last lemma, $S(\eta f) = \bar{\partial}\eta \wedge f + \eta \wedge \bar{\partial}f$. Also $\eta Sf = \eta\bar{\partial}f$. So the result follows. □

LEMMA 4.4.4   If $f \in \mathcal{D}_{T^*}$ and $\eta \in C_c^\infty(\Omega)$, then $\eta f \in \mathcal{D}_{T^*}$.

Proof   For simplicity assume that $p = q = 0$. If $u \in \mathcal{D}_T$, then

$$
\begin{aligned}
\langle \eta f, Tu \rangle_2 &= \langle f, \bar{\eta}Tu \rangle_2 = \langle f, T(\bar{\eta}u) \rangle_2 + \langle f, \bar{\eta}Tu - T(\bar{\eta}u) \rangle_2 \\
&= \langle T^*f, \bar{\eta}u \rangle_1 + \langle f, \bar{\eta}Tu - T(\bar{\eta}u) \rangle_2 \\
&= \langle \eta T^*f, u \rangle_1 + \langle f, -\bar{\partial}\bar{\eta} \wedge u \rangle_2.
\end{aligned}
$$

Thus

$$
|\langle \eta f, Tu \rangle_2| \leq C \left\{ \|\eta T^*f\|_1 + \|f\|_2 \|\bar{\partial}\bar{\eta}\|_{L^\infty} \right\} \|u\|_1,
$$

so $\eta f \in \mathcal{D}_{T^*}$. (Here we have used property 5 from the beginning of the section to estimate the second term—notice that η has compact support.) □

The only way that we can achieve our goals is to pass from abstractions about unbounded operators to specific facts about $\bar{\partial}$ acting on the spaces $\mathcal{H}^j$.

LEMMA 4.4.5   Let $K_0 \subset\subset K_1 \subset\subset \cdots \subset\subset \Omega$ satisfy $\cup_j K_j = \Omega$. Let $\eta_j \in C_c^\infty(\Omega)$ satisfy $\eta_j = 1$ on K_{j-1}, $\text{supp}\,\eta_j \subseteq K_j, 0 \leq \eta_j \leq 1, j = 1, 2, \ldots$. Then there exists a function $\psi \in C^\infty(\Omega)$ such that

$$
\sum_{k=1}^n \left| \frac{\partial \eta_\ell}{\partial \bar{z}_k} \right|^2 \leq e^\psi, \qquad \text{all } \ell = 1, 2, \ldots.
$$

Proof   The assertion is equivalent with

$$\left|\bar{\partial}\eta_\ell\right|^2 \le e^\psi.$$

For each $j \in \mathbb{N}$, the left-hand side is nonzero on K_j only for $\ell = 1, 2, \ldots, j$. Hence one can build ψ inductively on j to satisfy the conditions

$$e^\psi \ge \|\eta_1\|_{C^1} \qquad \text{on} \quad K_1,$$

$$e^\psi \ge \sum_{\ell=1}^{j} \|\eta_\ell\|_{C^1} \qquad \text{on } K_j \setminus K_{j-1}.$$

In this fashion we clearly obtain a function $\psi \in C(\Omega)$ with the desired properties. By the techniques of Exercises 5 through 8 at the end of Chapter 3, we obtain a function $\psi \in C^\infty(\Omega)$ that satisfies the desired conclusions.   □

Now we specialize further and declare the form of the weight functions ϕ_1, ϕ_2, ϕ_3.

DEFINITION 4.4.6   With η_j fixed as in Lemma 4.4.5, let ψ satisfy the conclusions of that lemma. Let $\phi \in C^\infty(\Omega)$ (We will choose a particular ϕ later in our applications.) Define

$$\phi_1 = \phi - 2\psi, \qquad \phi_2 = \phi - \psi, \qquad \phi_3 = \phi.$$

It is immediate that

$$e^{-\phi_3}\left|\bar{\partial}\eta_\ell\right|^2 \le e^{-\phi_2} \tag{4.4.6.1}$$

and

$$e^{-\phi_2}\left|\bar{\partial}\eta_\ell\right|^2 \le e^{-\phi_1} \tag{4.4.6.2}$$

for all $\ell = 1, 2, \ldots$.

In what follows, properties (4.4.6.1) and (4.4.6.2) are crucial. They will be used frequently without comment.

LEMMA 4.4.7   If $f \in \mathcal{D}_S$, then as $\ell \to \infty$,

$$S(\eta_\ell f) - \eta_\ell S f \to 0 \qquad \text{in } \mathcal{H}^3.$$

Proof   Now

$$\|S(\eta_\ell f) - \eta_\ell(Sf)\|_3^2 = \int_\Omega |\bar{\partial}\eta_\ell \wedge f|^2 e^{-\phi_3}\, dV$$

$$\leq C \cdot \int_{K_\ell \setminus K_{\ell-1}} |f|^2 |\bar{\partial}\eta_\ell|^2 e^{-\phi_3}\, dV$$

$$\leq C \cdot \int_{K_\ell \setminus K_{\ell-1}} |f|^2 e^{-\phi_2}\, dV \to 0. \qquad \square$$

LEMMA 4.4.8   If $f \in \mathcal{D}_{T^*}$, then

$$T^* \eta_\ell f - \eta_\ell T^* f \to 0 \quad \text{in} \quad \mathcal{H}^1.$$

Proof   Use the definition of T^* to write out and simplify the left-hand side. Then imitate the proof of the preceding lemma. $\qquad \square$

COROLLARY 4.4.9   If $f \in \mathcal{D}_{T^*} \cap \mathcal{D}_S$, then $\eta_\ell f \to f$ in the graph norm

$$\|f\|_{\mathcal{G}}^2 \equiv \|f\|_2^2 + \|T^* f\|_1^2 + \|Sf\|_3^2.$$

Proof   Use Lemmas 4.4.7 and 4.4.8 and the Lebesgue dominated convergence theorem (this last for the $\|\ \|_2$ term). $\qquad \square$

LEMMA 4.4.10   If $f \in \mathcal{D}_S$ satisfies supp $f \subset\subset \Omega$, then there exist forms $f_\delta \in D_{(p,q+1)}, 0 < \delta < 1$, such that $f_\delta \to f$ in $\mathcal{H}^2$ and $Sf_\delta \to Sf$ in $\mathcal{H}^3$ as $\delta \to 0^+$.

Proof   Let $\Phi \in C_c^\infty(\mathbb{C}^n), \text{supp}\,\Phi \subseteq B(0,1)$, and $\int \Phi dV = 1$. Let $\Phi_\delta(z) = \delta^{-2n}\Phi(z/\delta)$. If $f = \sum' f_{\alpha\beta}dz^\alpha \wedge d\bar{z}^\beta$, let $f_\delta = \sum'(f_{\alpha\beta} * \Phi_\delta)dz^\alpha \wedge d\bar{z}^\beta$ for $\delta << \text{dist}\,(\text{supp}\,f, {}^c\Omega)$. Then supp $f_\delta \subset\subset \Omega$ and $f_\delta \to f$ in L^2 by Lemma 2.1.16. Hence $f_\delta \to f$ in $\mathcal{H}^2$. Notice also that

$$Sf_\delta = \sum_j \left(\frac{\partial}{\partial\bar{z}_j} \sum_{\alpha,\beta}' f_{\alpha\beta} \right) * \Phi_\delta\, d\bar{z}_j \wedge dz^\alpha \wedge d\bar{z}^\beta \to Sf$$

in $\mathcal{H}^3$ as desired. $\qquad \square$

We would like to prove a result analogous to Lemma 4.4.10 for T^*. If $f \in \mathcal{D}_{T^*}$, it will still be the case that $f_\delta \to f$ in $\mathcal{H}^2$. What needs some work is seeing that $T^* f_\delta \to T^* f$ since T^* is not a constant-coefficient operator.

LEMMA 4.4.11   Let

$$f = \sum_{\substack{|\alpha|=p \\ |\beta|=q+1}}' f_{\alpha\beta}\, dz^\alpha \wedge d\bar{z}^\beta \in \mathcal{D}_{T^*}.$$

Then

$$T^* f = (-1)^{p-1} \sideset{}{'}\sum_{\substack{|\alpha|=p \\ |\gamma|=q}} \sum_{j=1}^{n} e^{\phi_1} \left\{ \frac{\partial(e^{-\phi_2} f_{\alpha,j\gamma})}{\partial z_j} \right\} dz^\alpha \wedge d\bar{z}^\gamma.$$

Here

$$f_{\alpha,j\gamma} = \sideset{}{'}\sum_{|\beta|=q+1} f_{\alpha\beta} \epsilon^\beta_{j\gamma},$$

where $\epsilon^\beta_{j\gamma}$ is the sign of the permutation taking β to $j\gamma$ (if there is one) and is 0 otherwise.

Proof We must compute. Let

$$u = \sideset{}{'}\sum_{\substack{|\alpha|=p \\ |\gamma|=q}} u_{\alpha\gamma} dz^\alpha \wedge d\bar{z}^\gamma \in D_{(p,q)}(\Omega).$$

Then

$$\int_\Omega \sideset{}{'}\sum_{\alpha,\gamma} (T^* f)_{\alpha\gamma} \bar{u}_{\alpha\gamma} e^{-\phi_1} \, dV$$

$$\equiv \langle T^* f, u \rangle_1$$
$$= \langle f, Tu \rangle_2 \quad (\text{since } u \in D_{(p,q)} \subseteq \mathcal{D}_T)$$
$$= (-1)^p \int_\Omega \sideset{}{'}\sum_{\alpha,\beta,\gamma} \sum_{j=1}^n f_{\alpha\beta} \epsilon^\beta_{j\gamma} \overline{\left(\frac{\partial u_{\alpha\gamma}}{\partial \bar{z}_j} \right)} e^{-\phi_2} \, dV$$
$$= (-1)^p \sideset{}{'}\sum_{\alpha,\gamma} \sum_{j=1}^n \int_\Omega f_{\alpha,j\gamma} \overline{\left(\frac{\partial u_{\alpha\gamma}}{\partial \bar{z}_j} \right)} e^{-\phi_2} \, dV$$
$$= \int_\Omega \sideset{}{'}\sum_{\alpha,\gamma} \left\{ (-1)^{p-1} \sum_{j=1}^n e^{\phi_1} \frac{\partial}{\partial z_j} \left(e^{-\phi_2} f_{\alpha,j\gamma} \right) \right\} \bar{u}_{\alpha\gamma} e^{-\phi_1} \, dV.$$

Since $D_{(p,q)}(\Omega)$ is dense in $L^2_{(p,q)}(\Omega)$, the result follows. □

COROLLARY 4.4.12 We may write $e^{\phi_2-\phi_1} T^*$ as a constant-coefficient first-order partial differential operator $\mathcal{T}^*$ plus a zero-order operator $\mathcal{A}$ (where by a "zero-order" operator we mean one that is given by multiplication by a C^∞ function).

Proof   We write

$$
\begin{aligned}
e^{\phi_2-\phi_1} T^* f &= (-1)^{p-1} \sideset{}{'}\sum_{\substack{|\alpha|=p \\ |\gamma|=q}} \sum_{j=1}^n \frac{\partial f_{\alpha,j\gamma}}{\partial z_j} \, dz^\alpha \wedge d\bar{z}^\gamma \\
&\quad - (-1)^{p-1} \sideset{}{'}\sum_{\substack{|\alpha|=p \\ |\gamma|=q}} \sum_{j=1}^n \left(\frac{\partial \phi_2}{\partial z_j} \right) f_{\alpha,j\gamma} \, dz^\alpha \wedge d\bar{z}^\gamma \\
&\equiv \ \mathcal{T}^* f + \mathcal{A} f. \qquad\qquad\qquad\qquad\qquad\qquad \square
\end{aligned}
$$

COROLLARY 4.4.13   With notation as in Lemma 4.4.10, if $f \in \mathcal{D}_{T^*}$ and supp $f \subset\subset \Omega$, then $T^* f_\delta \to T^* f$ in $\mathcal{H}^1$.

Proof   Using some algebra, we see that

$$
(\mathcal{T}^* + \mathcal{A}) f_\delta = ((\mathcal{T}^* + \mathcal{A}) f) * \Phi_\delta + \mathcal{A}(f * \Phi_\delta) - (\mathcal{A} f) * \Phi_\delta,
$$

which tends to $(\mathcal{T}^* + \mathcal{A}) f$ in the $L^2_{(p,q)}(\Omega)$ topology. Thus $e^{\phi_1-\phi_2}(\mathcal{T}^* + \mathcal{A}) f_\delta \to e^{\phi_1-\phi_2}(\mathcal{T}^* + \mathcal{A}) f$ in $\mathcal{H}^1$ or $T^* f_\delta \to T^* f$ in $\mathcal{H}^1$, as desired. $\qquad \square$

The next corollary summarizes the point of our calculations.

COROLLARY 4.4.14   The space $D_{(p,q+1)}$ is dense in $\mathcal{D}_{T^*} \cap \mathcal{D}_S$ in the graph norm $\|f\|_{\mathcal{G}} = \|T^* f\|_1 + \|S f\|_3 + \|f\|_2$.

Proof   Let $f \in \mathcal{D}_{T^*} \cap \mathcal{D}_S$. Let $\epsilon > 0$. By Corollary 4.4.9, there is an $\ell > 0$ such that $\|\eta_\ell f - f\|_{\mathcal{G}} < \epsilon/2$. Now by Lemma 4.4.10, Corollary 4.4.13, and Lemma 2.1.16, there is a $\delta > 0$ such that $\|(\eta_\ell f)_\delta - \eta_\ell f\|_{\mathcal{G}} < \epsilon/2$. Then $\|(\eta_\ell f)_\delta - f\|_{\mathcal{G}} < \epsilon$, as desired. $\qquad \square$

Now the reward for our work is that in order for us to prove that Range $T = F$, we no longer have to prove the **MAIN GOAL**. Indeed, the following (simpler) modified main goal implies the **MAIN GOAL**.

MODIFIED MAIN GOAL

$$
\|f\|_2^2 \le C^2 \{ \|T^* f\|_1^2 + \|S f\|_3^2 \}, \qquad \text{all } f \in D_{(p,q+1)}(\Omega).
$$

This is the same estimate as in the **MAIN GOAL**, but because of our approximation lemmas we now need only check it for forms with coefficients in C_c^∞. This reduction will yield tremendous simplifications for us.

Let us conclude this section by isolating a few more easy, but tedious, calculations.

LEMMA 4.4.15   Let $f \in D_{(p,q+1)}$,

$$f = \sum_{\substack{|\alpha|=p \\ |\beta|=q+1}}{}' f_{\alpha\beta} dz^\alpha \wedge d\bar{z}^\beta.$$

Then

$$|\bar{\partial}f|^2 = \sum_{\alpha,\beta}{}' \sum_j \left|\frac{\partial f_{\alpha\beta}}{\partial \bar{z}_j}\right|^2 - \sum_{\alpha,\gamma}{}' \sum_{j,k=1}^n \left(\frac{\partial f_{\alpha,j\gamma}}{\partial \bar{z}_k}\right) \overline{\left(\frac{\partial f_{\alpha,k\gamma}}{\partial \bar{z}_j}\right)}.$$

Proof   This is nearly obvious, so the following collection of calculations should be considered *pro forma*.
First,

$$\bar{\partial}f = \sum_{|\alpha|=p}{}' \sum_{|\beta|=q+1}{}' \sum_{j=1}^n \frac{\partial f_{\alpha\beta}}{\partial \bar{z}_j} d\bar{z}_j \wedge dz^\alpha \wedge d\bar{z}^\beta$$

so that

$$|\bar{\partial}f|^2 = \sum_{\alpha,\beta,\mu}{}' \sum_{b,m=1}^n \frac{\overline{\partial f_{\alpha\beta}}}{\partial \bar{z}_b} \frac{\partial f_{\alpha\mu}}{\partial \bar{z}_m} \epsilon_{m\mu}^{b\beta}.$$

Here we recall that $\epsilon_{m\mu}^{b\beta}$ is the sign of the permutation taking $b\beta$ to $m\mu$, if one exists, and is zero otherwise. The last line is

$$\sum_{\alpha,\beta,\mu}{}' \sum_{b=m} + \sum_{\alpha,\beta,\mu}{}' \sum_{b \neq m} \equiv A + B.$$

Now $b = m$ implies that $\beta = \mu$ (otherwise the term is zero), so

$$A = \sum_{\alpha,\beta}{}' \sum_{b \notin \beta} \left|\frac{\partial f_{\alpha\beta}}{\partial \bar{z}_b}\right|^2. \tag{4.4.15.1}$$

For the case $b \neq m$, the nonzero terms must have $m \in \beta$ and $b \in \mu$. Moreover, deletion of m from β or of b from μ must leave the same q^{th} order multi-index ξ. If we remember that $f_{\alpha,j\gamma}$ is defined to be $\epsilon_{j\gamma}^\beta \cdot f_{\alpha,\beta}$, where β is the index equivalent to $j\gamma$ but with increasingly ordered entries, then

$$\begin{aligned} B &= \sum_{\alpha,\beta,\mu}{}' \sum_{b \neq m} \frac{\overline{\partial f_{\alpha\beta}}}{\partial \bar{z}_b} \frac{\partial f_{\alpha\mu}}{\partial \bar{z}_m} \epsilon_{m\xi}^\beta \epsilon_\mu^{b\xi} \\ &= -\sum_{\alpha,\xi}{}' \sum_{b \neq m} \frac{\partial f_{\alpha,b\xi}}{\partial \bar{z}_m} \frac{\overline{\partial f_{\alpha,m\xi}}}{\partial \bar{z}_b}. \tag{4.4.15.2} \end{aligned}$$

Note that we sum over ξ because it depends on β and μ and these no longer appear in the sum. Combining (4.4.15.1) and (4.4.15.2) yields the desired result.   □

DEFINITION 4.4.16   For any $g \in C^1(\Omega)$, let

$$\delta_j g = e^\phi \frac{\partial}{\partial z_j}(g e^{-\phi}) = \frac{\partial g}{\partial z_j} - g \frac{\partial \phi}{\partial z_j}.$$

Exercises for the Reader

1. The operator δ_j is the formal adjoint of the operator $-\partial/\partial \bar{z}_j$ on the space $L^2(\Omega, \phi)$.

2. If $f \in D_{(p,q+1)}$, then

$$
\begin{aligned}
e^\psi T^* f &= (-1)^{p-1} \sideset{}{'}\sum_{\alpha,\gamma} \sum_{j=1}^{n} \delta_j f_{\alpha,j\gamma}\, dz^\alpha \wedge d\bar{z}^\gamma \\
&\quad + (-1)^{p-1} \sideset{}{'}\sum_{\alpha,\gamma} \sum_{j=1}^{n} f_{\alpha,j\gamma} \frac{\partial \psi}{\partial z_j}\, dz^\alpha \wedge d\bar{z}^\gamma.
\end{aligned}
$$

3.

$$\delta_j \frac{\partial}{\partial \bar{z}_k} - \frac{\partial}{\partial \bar{z}_k} \delta_j = \frac{\partial^2 \phi}{\partial z_j \partial \bar{z}_k}.$$

It is Exercise 3 that brings the Levi form into play.

4.5   An Existence Theorem for the $\bar{\partial}$ Operator

It is important for us not to get lost in the details. So far we have done the following:

1. We have reformulated the problem of solving $\bar{\partial} u = f$, f a $\bar{\partial}$-closed $(p, q+1)$ form, as a statement about unbounded Hilbert space operators; specifically, we wish to show (in our notation) that Range $T = F$.

2. We have seen that showing Range $T = F$ is equivalent to proving the estimate

$$\|f\|_2^2 \le C^2(\|T^* f\|_1^2 + \|Sf\|_3^2), \qquad \text{all } f \in \mathcal{D}_S \cap \mathcal{D}_{T^*}.$$

3. We have seen that the inequality in (2) is equivalent to

$$\|f\|_2^2 \le C^2(\|T^*f\|_1^2 + \|Sf\|_3^2), \qquad \text{all } f \in D_{(p,q+1)}$$

provided that the weights that define our Hilbert spaces satisfy the following: $\phi_1 = \phi - 2\psi, \phi_2 = \phi - \psi, \phi_3 = \phi, \phi$ is any C^∞ function, and ψ is chosen according to Lemma 4.4.5.

There is a subtle point here that is not generally noted explicitly in the literature—namely, given a $(p, q+1)$ form f that has L^2_{loc} coefficients, we need to see that ϕ, ψ can be chosen so that both $f \in L^2_{(p,q+1)}(\Omega, \phi_2)$ and property 3 holds. This is realized, finally, in Corollary 4.5.3. Meanwhile, what remains to be seen is that on any pseudoconvex domain we can choose ϕ such that estimate 3 holds (we worry about the integrability of f later). Notice that, up until now, we have not used the pseudoconvexity of the domain. The required property of ϕ is contained in the next lemma.

LEMMA 4.5.1   Let $\Omega \subseteq \mathbb{C}^n$ be pseudoconvex (Ω need not have smooth boundary nor be bounded). Then there exists a C^∞ exhaustion function $\phi : \Omega \to \mathbb{R}$ such that

$$\sum_{j,k=1}^n \frac{\partial^2 \phi}{\partial z_j \partial \bar{z}_k} w_j \bar{w}_k \ge 2\left(|\bar\partial\psi|^2 + e^\psi\right)\sum_{j=1}^n |w_j|^2, \qquad \text{all } w \in \mathbb{C}^n.$$

Remark: Lemma 4.5.1 says that ϕ is a strictly psh exhaustion function on Ω such that the eigenvalues of its complex Hessian are bounded below in a specified manner (note that the properties of ψ specified in Lemma 4.4.5 imply that these eigenvalues will in fact blow up at the boundary of the domain). The reader who has done the exercises in Chapter 3 will not be surprised that such a ϕ may be constructed. $\square$

We leave it as an easy exercise, after you have read the proof of the lemma, to verify the following slight strengthening of the lemma. Let κ be a given continuous function on Ω. There is a function ϕ satisfying the conclusions of the lemma and the additional conclusion that $\phi \ge \kappa$ on Ω.

Proof of the Lemma: Use part 4 of Theorem 3.3.5 to select a positive, C^∞ exhaustion function $p : \Omega \to \mathbb{R}$ for Ω that satisfies (for a positive continuous function $m(z)$)

$$\sum_{j,k=1}^n \frac{\partial^2 p}{\partial z_j \partial \bar{z}_k}(z) w_j w_k \ge m(z)|w|^2, \qquad \text{all } z \in \Omega, \text{ all } w \in \mathbb{C}^n.$$

Let

$$\Omega_t = \{z \in \Omega : p(z) \le t\}, \quad t \in \mathbb{R}.$$

Then $\Omega_t \subset\subset \Omega$, all t. Choose $\chi : \mathbb{R}^+ \to \mathbb{R}$ a C^∞ increasing convex function satisfying

$$\chi'(t) \geq \sup_{z \in \Omega_t} \{2\left(|\bar{\partial}\psi(z)|^2 + e^\psi\right)/m(z)\}, \qquad \text{all } t \in \mathbb{R}$$

(see Exercise 11 at the end of Chapter 3). Then $\phi = \chi \circ p$ does the job.   □

THEOREM 4.5.2   Let $\Omega \subseteq \mathbb{C}^n$ be pseudoconvex (neither boundedness nor boundary smoothness is assumed). If $\phi \in C^\infty(\Omega)$ is chosen as in Lemma 4.5.1 and ψ as in Lemma 4.4.5 and if we let $\phi_1 = \phi - 2\psi, \phi_2 = \phi - \psi, \phi_3 = \phi$, then

$$\|f\|_2^2 \leq C^2 \left(\|T^*f\|_1^2 + \|Sf\|_3^2\right), \qquad \text{all } f \in D_{(p,q+1)}.$$

The constant C may be taken to be 1. As a consequence, for every $f \in L^2_{(p,q+1)}(\Omega, \phi_2)$ with $\bar{\partial}f = 0$ (in the weak sense), there is a $u \in L^2_{(p,q)}(\Omega, \phi_1)$ with $\bar{\partial}u = f$.

Proof   Let $f \in D_{(p,q+1)}$. Then, by Exercise 2 at the end of Section 4.4, we obtain

$$T^*f = (-1)^{p-1} e^{-\psi} \sum_{\alpha,\gamma}{}' \sum_{j=1}^n \delta_j f_{\alpha,j\gamma} \, dz^\alpha \wedge d\bar{z}^\gamma$$

$$+ (-1)^{p-1} e^{-\psi} \sum_{\alpha,\gamma}{}' \sum_{j=1}^n f_{\alpha,j\gamma} \frac{\partial\psi}{\partial z_j} \, dz^\alpha \wedge d\bar{z}^\gamma \equiv A + B.$$

Using the trivial algebraic fact that

$$z = x + y \quad \text{implies} \quad 2|z|^2 \geq |x|^2 - 2|y|^2,$$

we have

$$2\|T^*f\|_1^2 \geq \|A\|_1^2 - 2\|B\|_1^2$$

or

$$\begin{aligned}
2\|T^*f\|_1^2 &\geq \int_\Omega \sum_{\alpha,\gamma}{}' \sum_{j,k=1}^n \delta_j f_{\alpha,j\gamma} \overline{\delta_k f_{\alpha,k\gamma}} e^{-\phi} \, dV \\
&\quad -2 \int_\Omega |f|^2 |\partial\psi|^2 e^{-\phi} \, dV \\
&= -\int_\Omega \sum_{\alpha,\gamma}{}' \sum_{j,k=1}^n \left(\frac{\partial}{\partial\bar{z}_k}\delta_j f_{\alpha,j\gamma}\right) \overline{f_{\alpha,k\gamma}} e^{-\phi} \, dV \\
&\quad -2 \int_\Omega |f|^2 |\partial\psi|^2 e^{-\phi} \, dV
\end{aligned} \tag{4.5.2.1}$$

by Exercise 1 at the end of Section 4.4. Lemma 4.4.15 now gives

$$
\begin{aligned}
\|Sf\|_3^2 &= \int_\Omega \sideset{}{'}\sum_{\alpha,\beta} \sum_{j=1}^n \left|\frac{\partial f_{\alpha,\beta}}{\partial \bar{z}_j}\right|^2 e^{-\phi}\,dV \\
&\quad - \int_\Omega \sideset{}{'}\sum_{\alpha,\gamma} \sum_{j,k=1}^n \left(\frac{\partial f_{\alpha,j\gamma}}{\partial \bar{z}_k}\right)\left(\overline{\frac{\partial f_{\alpha,k\gamma}}{\partial \bar{z}_j}}\right) e^{-\phi}\,dV \\
&= \int_\Omega \sideset{}{'}\sum_{\alpha,\beta} \sum_{j=1}^n \left|\frac{\partial f_{\alpha,\beta}}{\partial \bar{z}_j}\right|^2 e^{-\phi}\,dV \\
&\quad + \int_\Omega \sideset{}{'}\sum_{\alpha,\gamma} \sum_{j,k=1}^n \left(\delta_j \frac{\partial}{\partial \bar{z}_k} f_{\alpha,j\gamma}\right)\overline{f_{\alpha,k\gamma}} e^{-\phi}\,dV
\end{aligned}
\tag{4.5.2.2}
$$

by Exercise 1 at the end of Section 4.4.

Now (4.5.2.1) and (4.5.2.2) together give

$$
\begin{aligned}
&2\|T^*f\|_1^2 + \|Sf\|_3^2 \\
&\geq \int_\Omega \sideset{}{'}\sum_{\alpha,\gamma} \sum_{j,k=1}^n \left\{\left(\delta_j \frac{\partial}{\partial \bar{z}_k} - \frac{\partial}{\partial \bar{z}_k}\delta_j\right) f_{\alpha,j\gamma}\right\}\overline{f_{\alpha,k\gamma}} e^{-\phi}\,dV \\
&\quad + \int_\Omega \sideset{}{'}\sum_{\alpha,\beta} \left|\frac{\partial f_{\alpha,\beta}}{\partial \bar{z}_j}\right|^2 e^{-\phi}\,dV \\
&\quad - 2\int_\Omega |f|^2 |\partial\psi|^2 e^{-\phi}\,dV.
\end{aligned}
\tag{4.5.2.3}
$$

By Exercise 3 at the end of Section 4.4, this is equal

$$
\begin{aligned}
&\int_\Omega \sideset{}{'}\sum_{\alpha,\gamma} \sum_{j,k=1}^n \left\{\frac{\partial^2\phi}{\partial z_j \partial \bar{z}_k}\right\} f_{\alpha,j\gamma}\overline{f_{\alpha,k\gamma}} e^{-\phi}\,dV \\
&\quad + \int_\Omega \sideset{}{'}\sum_{\alpha,\beta} \sum_{j=1}^n \left|\frac{\partial f_{\alpha,\beta}}{\partial \bar{z}_j}\right|^2 e^{-\phi}\,dV \\
&\quad - 2\int_\Omega |f|^2 |\partial\psi|^2 e^{-\phi}\,dV \\
&\geq \int_\Omega \sideset{}{'}\sum_{\alpha,\gamma} \sum_{j=1}^n 2(|\partial\psi|^2 + e^\psi)|f_{\alpha,j\gamma}|^2 e^{-\phi}\,dV \\
&\quad - 2\int_\Omega |f|^2 |\partial\psi|^2 e^{-\phi}\,dV.
\end{aligned}
\tag{4.5.2.4}
$$

Here we have used the estimate from below on the eigenvalues of the complex Hessian of ϕ given by Lemma 4.5.1 and we have simply discarded the second term on the right. Now we combine the first and last terms to obtain

$$2\|T^*f\|_1^2 + \|Sf\|_3^2 \;\geq\; 2\int \sum_{\alpha,\gamma}{}' \sum_{j=1}^{n} |f_{\alpha,j\gamma}|^2 e^{-\phi_2}\, dV$$
$$= \; 2\|f\|_2^2. \qquad \Box$$

COROLLARY 4.5.3   Let f be a $(p, q+1)$ form on Ω with L^2_{loc} coefficients satisfying $\bar{\partial}f = 0$ in the weak sense. Then there is a (p,q) form u on Ω with L^2_{loc} coefficients satisfying $\bar{\partial}u = f$ in the weak sense.

Proof   There is a $\tilde{\phi} : \Omega \to \mathbb{R}$ such that the coefficients of f are in $L^2(\Omega, \tilde{\phi})$. Choose $\phi : \Omega \to \mathbb{R}$ such that ϕ satisfies the conclusion of Lemma 4.5.1 and $\phi - \psi \geq \tilde{\phi}$ (note that this is just a matter of making ϕ large enough). Then the coefficients of f are also in $L^2(\Omega, \phi_2)$, where $\phi_2 = \phi - \psi$. Therefore Theorem 4.5.2 applies and there is a (p,q) form u with coefficients in $L^2(\Omega, \phi_1), \phi_1 = \phi - 2\psi$, satisfying $\bar{\partial}u = f$. But then the coefficients of u are also in L^2_{loc}. That is the desired conclusion. $\qquad \Box$

Remark: Hörmander [3] has refined the corollary to show that if Ω is bounded and pseudoconvex and f has L^2 coefficients, then there is a solution u with L^2 coefficients. Moreover, he has shown that the solution u may be chosen to satisfy an estimate of the form

$$\|u\|_{L^2} \leq C\|f\|_{L^2},$$

where the constant C depends *only* on the diameter of Ω and the dimension n. The analogue of this statement with L^2 replaced by L^∞ is known to be false (see N. Sibony [2]). In fact Sibony produces domains in which there are no L^∞ estimates whatever. In N. Sibony [6], he produces counterexamples for every C^k norm. The domains in these examples are smoothly bounded and are strongly pseudoconvex *except at one point*. $\qquad \Box$

4.6   A Regularity Theorem for the $\bar{\partial}$ Operator

The first point to understand about regularity for the $\bar{\partial}$ operator is that $\bar{\partial}$ has a large kernel. Therefore, for a given f, the equation $\bar{\partial}u = f$ has an entire coset of this kernel as its solution space. Not all these solutions will be smooth when f is.

EXAMPLE   Let $\Omega = \{z \in \mathbb{C}^2 : |z| < 1\}$. Define $f = (2\bar{z}_1 + 2\bar{z}_2)\, d\bar{z}_1 \wedge d\bar{z}_2$. Then f is a $\bar{\partial}$-closed $(0,2)$ form on Ω with C^∞ coefficients. We now construct a nonsmooth solution to $\bar{\partial}u = f$:

Let

$$\tilde{u}_1(z) = \begin{cases} 0 & \text{if } \operatorname{Re} z_2 \leq 0, \\ 1 & \text{if } \operatorname{Re} z_2 > 0, \end{cases}$$

$$\tilde{u}_2(z) = \begin{cases} 0 & \text{if } \operatorname{Re} z_1 \leq 0, \\ 1 & \text{if } \operatorname{Re} z_1 > 0, \end{cases}$$

Define

$$u(z) = \left(-\bar{z}_2^2 + \tilde{u}_1 \cdot \frac{\partial \tilde{u}_2}{\partial \bar{z}_1}\right) d\bar{z}_1 + \left(\bar{z}_1^2 + \tilde{u}_2 \cdot \frac{\partial \tilde{u}_1}{\partial \bar{z}_2}\right) d\bar{z}_2.$$

We calculate that $\bar{\partial}u = f$, but u does not even have continuous coefficients. On the other hand,

$$v(z) = -\bar{z}_2^2 \, d\bar{z}_1 + \bar{z}_1^2 \, d\bar{z}_2$$

and

$$w(z) = -2\bar{z}_1\bar{z}_2 \, d\bar{z}_1 + 2\bar{z}_1\bar{z}_2 \, d\bar{z}_2$$

satisfy $\bar{\partial}v = \bar{\partial}w = f$ and both v and w have smooth coefficients.

Thus both v and w lie in the same coset of Ker $\bar{\partial}$ as does u. In some sense, v and w are "good" solutions of the $\bar{\partial}$ equation, whereas u is a "bad" solution. We must develop a method for choosing a good solution. □

EXAMPLE   Let f be a $\bar{\partial}$-closed $(0,1)$ form on a domain $\Omega \subseteq \mathbb{C}^n$. Let u, v be functions that satisfy $\bar{\partial}u = f, \bar{\partial}v = f$. Then $u - v$ is holomorphic, hence smooth. This shows that if one solution of a $\bar{\partial}$ problem, with data a $(0,1)$ form, is smooth, then all solutions are. *This assertion holds even when the derivatives are interpreted in the weak sense,* as we shall see below.

This is a special case of standard interior regularity theory for elliptic partial differential operators. The $\bar{\partial}$ operator is elliptic on functions but not on forms. □

The idea for choosing the right solution to the $\bar{\partial}$ equation is due to J. J. Kohn, although it has old roots in Hodge theory. Let $\mathcal{K} \subseteq \mathcal{H}^1$ be the kernel of T. Then $\mathcal{K}$ is closed in $\mathcal{H}^1$ (exercise—remember to use weak derivatives!). Let $\mathcal{P} : \mathcal{H}^1 \to \mathcal{K}$ be the Hilbert space projection. If $f \in \mathcal{H}^2$ satisfies $Sf = 0$, then let u be some solution to $Tu = f$. Then $\tilde{u} = u - \mathcal{P}u$ is also a solution, and $\tilde{u} \perp \mathcal{K}$. If $\tilde{u}^*$ is another solution that is orthogonal to $\mathcal{K}$, then $\tilde{u} - \tilde{u}^* \in \mathcal{K}$ and $\tilde{u} - \tilde{u}^* \perp \mathcal{K}$. It follows that $\tilde{u} = \tilde{u}^*$. The unique solution in $\mathcal{K}^\perp$ to the equation $\bar{\partial}u = f$ is often called the *canonical solution* or the *Kohn solution* (especially when the Hilbert space in question is classical $L^2(\Omega)$). The condition $\tilde{u} \perp \mathcal{K}$ provides an additional differential equation that will enable us to prove regularity theorems.

We will need a device for mediating between functions with L^2 derivatives (which, as we have seen, are convenient) and classical C^k functions (which are more useful in applications because they are more meaningful). The standard tool for this purpose is the Sobolev theory.

LEMMA 4.6.1   Let $g \in C_c^\infty(\mathbb{R}^N)$. Then, for any $x \in \mathbb{R}^N$,

$$|g(x)| \leq \int_{\mathbb{R}^N} \left| \frac{\partial^N}{\partial t_1 \cdots \partial t_N} g(t) \right| dV(t).$$

Proof   We write

$$
\begin{aligned}
|g(x)| &= \left| \int_{-\infty}^{x_1} \int_{-\infty}^{x_2} \cdots \int_{-\infty}^{x_N} \frac{\partial^N}{\partial t_1 \partial t_2 \cdots \partial t_N} g(t_1, \ldots, t_N) \, dt_N \cdots dt_2 \, dt_1 \right| \\
&\leq \int_{\mathbb{R}^N} \left| \frac{\partial^N}{\partial t_1 \partial t_2 \cdots \partial t_N} g(t) \right| dV(t). \qquad \Box
\end{aligned}
$$

LEMMA 4.6.2   Let $g : \mathbb{R}^N \to \mathbb{C}$ have compact support and satisfy $(\partial/\partial x)^\alpha g \in L^2$, all $|\alpha| \leq N + 1$. Here derivatives are interpreted in the weak sense. Then, after correction on a set of measure zero, g is continuous.

Proof   Let $\Phi \in C_c^\infty(\mathbb{R}^N), \Phi \geq 0, \int \Phi = 1$. Let $\Phi_\epsilon(x) = \epsilon^{-N} \Phi(x/\epsilon)$ and $g_\epsilon \equiv g * \Phi_\epsilon$. For each α with $|\alpha| \leq N + 1$, it follows that

$$\left\| \left(\frac{\partial}{\partial x} \right)^\alpha g_\epsilon \right\|_{L^1} \leq C \cdot \left\| \left(\frac{\partial}{\partial x} \right)^\alpha g_\epsilon \right\|_{L^2} \leq C \cdot \left\| \left(\frac{\partial}{\partial x} \right)^\alpha g \right\|_{L^2}, \qquad \text{all } \epsilon > 0.$$

But then, by the previous lemma,

$$\left\| \frac{\partial}{\partial x_j} g_\epsilon \right\|_{L^\infty} \leq C, \qquad \text{all } j = 1, \ldots, N, \quad \text{all } \epsilon > 0.$$

So the functions $\{g_\epsilon\}_{0 < \epsilon < 1}$ form an equicontinuous family on some large compact set that contains the support of all the functions $g_\epsilon, 0 < \epsilon < 1$. Therefore, there is a subsequence g_{ϵ_j} that converges uniformly to a continuous function $\tilde{g}$. On the other hand, $g_\epsilon \to g$ in L^2. It follows that $\tilde{g} = g$ almost everywhere. $\qquad \Box$

THEOREM 4.6.3 (Sobolev)   Fix a nonnegative integer k. Let $g : \mathbb{R}^N \to \mathbb{C}$ have compact support and satisfy $(\partial/\partial x)^\alpha g \in L^2$, all $|\alpha| \leq N + k + 1$. Then, after correction on a set of measure zero, $g \in C^k(\mathbb{R}^N)$.

Proof   Apply induction to the preceding lemma. $\qquad \Box$

Remark: The spaces of functions used in Theorem 4.6.3, called *Sobolev spaces*, have been the subject of intense study (see R. Adams [1] and, E. M. Stein [1]). There is a theory not only for L^2 derivatives but for L^p derivatives, $1 \le p \le \infty$. There are variants of Theorem 4.6.3 for fractional derivatives (which yield sharper results) and results *up to* the boundary of a domain. When the boundary of the domain is rough, then the geometry of the boundary plays a crucial role in the conclusions of the Sobolev theorem; see the optimal results of P. W. Jones [1] and Exercises 7 and 8 at the end of the chapter. □

COROLLARY 4.6.4 Let $\Omega \subseteq \mathbb{R}^N$ be any open set. Let $g : \Omega \subseteq \mathbb{C}$ satisfy $(\partial/\partial x)^\alpha g \in L^2_{\text{loc}}$ for *all* multi-indices α. Then $g \in C^\infty(\Omega)$ after correction on a set of measure zero.

Proof Let $\eta \in C^\infty_c(\Omega)$ be fixed. Apply Theorem 4.6.3 to ηg with $k = 0$. Then, after correction on a set of measure zero, ηg is continuous. Inductively apply the theorem to ηg for each k. (No corrections on sets of measure zero are necessary after the first one; why?) Thus $\eta g \in C^k$ for all k, so $\eta g \in C^\infty(\Omega)$. Since η was arbitrary, $g \in C^\infty(\Omega)$ after correction on a set of measure zero. □

What we do now is a prototype for a great deal of the regularity theory of partial differential equations; namely, we find a device for passing from information about the L^2 norms of $T^* f$ and Sf to information about the L^2 norm of any term $(\partial/\partial z_j)f_{\alpha\beta}$ or $(\partial/\partial \bar{z}_j)f_{\alpha\beta}$ (in effect, we have to untangle the differential forms). Then the Sobolev theorem will give us the desired regularity results.

DEFINITION 4.6.5 Let $\Omega \subseteq \mathbb{R}^N$ be any open set. For $0 \le s \in \mathbb{Z}$, let $W^s(\Omega)$ (the Sobolev space of order s) be those functions f with $(\partial/\partial x)^\alpha f \in L^2(\Omega)$ for all multi-indices α with $|\alpha| \le s$. As usual, all derivatives are in the weak sense. Define W^s_{loc} likewise.

Let $W^s_{(p,q)}(\Omega), W^s_{(p,q)}(\Omega, \text{loc})$ be the corresponding spaces of forms. If $f = \sum_{\alpha,\beta} f_{\alpha\beta}\, dz^\alpha \wedge d\bar{z}^\beta \in W^s_{(p,q)}(\Omega)$, let

$$\|f\|^2_{W^s(\Omega)} \equiv \sum_{\alpha,\beta} \sum_{|\gamma|\le s} \left\| \left(\frac{\partial}{\partial x}\right)^\gamma f_{\alpha\beta} \right\|^2_{L^2(\Omega)}.$$

We leave it to the reader to formulate the analogue of the definition of Sobolev space in complex notation. One simply requires that *all* derivatives, in both $\partial/\partial z_j$ and $\partial/\partial \bar{z}_k$, of order not exceeding s lie in L^2.

LEMMA 4.6.6 For $f \in C^\infty_c(\mathbb{C}^n), j = 1, \dots, n$, we have

$$\left\| \frac{\partial f}{\partial z_j} \right\|_{L^2} = \left\| \frac{\partial f}{\partial \bar{z}_j} \right\|_{L^2}.$$

Proof Integrate by parts. □

COROLLARY 4.6.7   If $f \in L^2$ has compact support and if $\partial f / \partial \bar{z}_j \in L^2$, $j = 1, \ldots, n$, then $f \in W^1$ and

$$\left\| \frac{\partial f}{\partial z_j} \right\|_{L^2} = \left\| \frac{\partial f}{\partial \bar{z}_j} \right\|_{L^2}, \quad \text{all } j.$$

Proof   Apply the preceding lemma and the usual approximation arguments.   □

LEMMA 4.6.8   Let $f \in L^2_{(p,q+1)}(\Omega)$. Define

$$T^* f = (-1)^{p-1} \sum_{\alpha, \gamma} {}' \sum_{j=1}^{n} \frac{\partial f_{\alpha,j\gamma}}{\partial z_j} \, dz^\alpha \wedge d\bar{z}^\gamma$$

(as in Corollary 4.4.12). Suppose that f has compact support in Ω. If $\bar{\partial} f \in L^2_{(p,q+2)}$ and $T^* f \in L^2_{(p,q)}$ (derivatives are computed in the weak sense), then $f \in W^1_{(p,q+1)}(\mathbb{C}^n)$.

Proof   By the proof of Lemma 4.6.2, it suffices to prove an a priori estimate for elements of $D_{(p,q+1)}$. But lines (4.5.2.3) and (4.5.2.4) are formal calculations that apply for any choice of ϕ and ψ. They say for the case $\phi = \psi = 0$ that

$$\|f\|_{L^2}^2 + 2\|T^* f\|_{L^2}^2 + \|\bar{\partial} f\|_{L^2}^2 \geq \|f\|_{L^2}^2 + \sum_{\alpha, \beta} \sum_{j=1}^{n} \left\| \frac{\partial f_{\alpha\beta}}{\partial \bar{z}_j} \right\|_{L^2}^2.$$

Now Corollary 4.6.7 says that the right-hand side is equal

$$\|f\|_{L^2}^2 + \frac{1}{2} \sum_{\alpha, \beta} \sum_{j=1}^{n} \left\{ \left\| \frac{\partial f_{\alpha\beta}}{\partial \bar{z}_j} \right\|_{L^2}^2 + \left\| \frac{\partial f_{\alpha\beta}}{\partial z_j} \right\|_{L^2}^2 \right\} \geq \frac{1}{2} \|f\|_{W^1}^2.   \qquad □$$

THEOREM 4.6.9   Let $\Omega \subseteq \mathbb{C}^n$ be a pseudoconvex open set (no boundedness or smoothness of $\partial \Omega$ is assumed). Let $0 \leq s \in \mathbb{Z}$, and let $f \in W^s_{(p,q+1)}(\Omega, \mathrm{loc})$ satisfy $\bar{\partial} f = 0$ weakly. Then the canonical solution u to $\bar{\partial} u = f$ satisfies $u \in W^{s+1}_{(p,q)}(\Omega, \mathrm{loc})$.

Proof   Choose ϕ, ψ according to the proof of 4.5.3. Define ϕ_1, ϕ_2, ϕ_3—and the corresponding Hilbert spaces—as usual. Let u be the canonical solution (in the Hilbert space $\mathcal{H}^1$) to $\bar{\partial} u = f$; this solution is guaranteed to exist by Theorem 4.5.2. Assume for the moment that $q \geq 1$. Then $u \perp \mathrm{Ker}\, T$, so that u lies in the closure of $\mathrm{Range}\, T^*$. But $\mathrm{Range}\, T^*$ is closed by equation (4.2.9.1) and by Theorem 4.5.2. Thus there is a $v \in L^2_{(p,q+1)}(\Omega, \mathrm{loc})$ with $T^* v = u$. Now if we let

$$U = \bar{\partial}_{p,q-1} : L^2_{(p,q-1)}(\Omega, \phi_0) \xrightarrow{\bar{\partial}} L^2_{(p,q)}(\Omega, \phi_1)$$

for some weight function ϕ_0, then $U^*u = U^*T^*v = 0$. Writing $e^{\phi_1 - \phi_0}U^* = \mathcal{U}^* + \mathcal{B}^*$ just as in Corollary 4.4.12, where $\mathcal{U}^*$ has constant coefficients and $\mathcal{B}^*$ is given by multiplication by smooth functions, we find that

$$\mathcal{U}^*u = -\mathcal{B}^*u. \tag{4.6.9.1}$$

As in Corollary 4.4.12, we obtain

$$\mathcal{U}^*u = (-1)^{p-1} \sideset{}{'}\sum_{\substack{|\alpha|=p \\ |\gamma|=q-1}} \sum_{j=1}^n \frac{\partial u_{\alpha,j\gamma}}{\partial z_j}\, dz^\alpha \wedge d\bar{z}^\gamma.$$

Let $0 \le \theta \le s$ and suppose that we have proved that $u \in W^\theta_{(p,q)}(\Omega, \mathrm{loc})$ (we have already done the step $\theta = 0$ in 4.5.2, 4.2.9). Let $\eta \in C^\infty_c(\Omega)$ be fixed. Then $\bar{\partial}(\eta u) = \bar{\partial}\eta \wedge u + \eta \wedge \bar{\partial}u \in W^\theta_{(p,q+1)}(\Omega)$ and

$$
\begin{aligned}
\mathcal{U}^*(\eta u) &= \eta\mathcal{U}^*u + \text{(terms involving differentiation of } \eta \text{ only)} \\
&= -\eta\mathcal{B}^*u + \text{(terms involving differentiation of } \eta \text{ only)} \\
&\in W^\theta_{(p,q-1)}(\Omega).
\end{aligned}
$$

If μ, ν are multi-indices with $|\mu| + |\nu| \le \theta$, then it follows inductively that

$$\bar{\partial}\left\{ \left(\frac{\partial}{\partial z}\right)^\mu \left(\frac{\partial}{\partial\bar{z}}\right)^\nu (\eta u) \right\} \in L^2_{(p,q+1)}(\Omega)$$

and

$$\mathcal{U}^*\left\{ \left(\frac{\partial}{\partial z}\right)^\mu \left(\frac{\partial}{\partial\bar{z}}\right)^\nu (\eta u) \right\} \in L^2_{(p,q-1)}(\Omega).$$

(Here a *differential operator* applied to a form is just applied to the coefficients.) By Lemma 4.6.8 with $\mathcal{T}^*$ replaced by $\mathcal{U}^*$, we conclude that

$$\left(\frac{\partial}{\partial z}\right)^\mu \left(\frac{\partial}{\partial\bar{z}}\right)^\nu (\eta u) \in W^1_{(p,q)}(\Omega).$$

Hence $u \in W^{\theta+1}_{(p,q)}(\Omega, \mathrm{loc})$. By induction on θ, we conclude that

$$u \in W^{s+1}_{(p,q)}(\Omega, \mathrm{loc}).$$

For the case $q = 0$, the operator U makes no sense. But it is not needed. We still have $\bar{\partial}(\eta u) \in W^\theta_{(p,q+1)}$, whence $(\partial/\partial\bar{z}_j)(\eta u) \in W^\theta, j = 1,\ldots,n$. By Lemma 4.6.6, $(\partial/\partial z_j)(\eta u) \in W^\theta, j = 1,\ldots,n$, so $\eta u \in W^{\theta+1}$. The argument is then completed as before. $\square$

COROLLARY 4.6.10   If $\Omega \subseteq \mathbb{C}^n$ is pseudoconvex and f is a $(p, q+1)$ form on Ω with C^∞ coefficients and satisfying $\bar{\partial} f = 0$, then there is a (p, q) form u on Ω with C^∞ coefficients satisfying $\bar{\partial} u = f$.

Proof   Apply the Sobolev theorem.   □

Remark: Notice that there would be a problem with attempting to derive Corollary 4.6.10 from Theorem 4.6.9 if we had to choose a different solution u for every s. But the canonical solution works for every s. (*Note:* Even if there were a different solution for every s, Hörmander has developed a technique for manufacturing a C^∞ solution. We do not discuss this method now, but refer the reader to Exercise 13 at the end of Chapter 10; see also J. J. Kohn [4].)   □

An immediate consequence of the proof of Corollary 4.6.10 is the following.

THEOREM 4.6.11   If $\Omega \subseteq \mathbb{C}^n$ is any pseudoconvex domain (no boundary smoothness or boundedness is required), if $0 \leq s \in \mathbb{Z}$, and if $f \in W^s_{(p,1)}(\Omega, \mathrm{loc})$ satisfies $\bar{\partial} f = 0$ weakly, then every solution u of $\bar{\partial} u = f$ satisfies $u \in W^{s+1}_{(p,0)}(\Omega, \mathrm{loc})$.

Remark: Notice that the theorem asserts that for f a $(0,1)$ form there is no need to choose a "good" solution to obtain a smooth one. Let us understand why this is so. Suppose that $u \in L^2_{\mathrm{loc}}(\Omega)$ and that $\bar{\partial} u = 0$ in the weak sense. Then $\partial u / \partial \bar{z}_j = 0, j = 1, \ldots, n$, in the weak sense. Let $\eta \in C^\infty_c(\Omega)$ be fixed. Then

$$\frac{\partial}{\partial \bar{z}_j}(\eta u) = \eta \frac{\partial u}{\partial \bar{z}_j} + u \frac{\partial \eta}{\partial \bar{z}_j} = u \frac{\partial \eta}{\partial \bar{z}_j}$$

in the weak sense. Moreover, the function on the right-hand side of the equation is in $L^2(\mathbb{C}^n)$. By Corollary 4.6.7, it follows that $(\partial / \partial z_j)(\eta u) \in L^2(\mathbb{C}^n)$ as well. Inductively, we may prove that

$$\left(\frac{\partial}{\partial z} \right)^\alpha \left(\frac{\partial}{\partial \bar{z}} \right)^\beta (\eta u) \in L^2(\mathbb{C}^n)$$

for all multi-indices α, β. So $\eta u \in C^\infty$ (after correction on a set of measure zero) by Sobolev's theorem. Since η was arbitrary, we conclude that $u \in C^\infty$. So u is a holomorphic function in the sense of Chapter 1—that is, it satisfies $\bar{\partial} u = 0$ in the strong sense. In short, there are no "fake" holomorphic functions. (An alternate way to see this is to note that a weakly holomorphic function is also weakly harmonic and then to invoke Weyl's lemma; see M. Tsuji [1]).

The preceding paragraph fleshes out our remarks in the second example of this section: Two (weak) solutions of the same equation $\bar{\partial} u = f$, for f a $\bar{\partial}$-closed $(0,1)$ form, must differ by a weakly holomorphic function. But we now know that a weakly holomorphic function is perforce holomorphic in the classical sense, hence smooth. Thus all solutions to the $\bar{\partial}$ equation are smooth if one is.   □

Let

$$M = M_{(p,q)} : L^2_{(p,q+1)}(\Omega, \phi_2) \cap F \to \mathcal{K}^\perp \cap L^2_{(p,q)}(\Omega, \phi_1)$$

be the operator assigning to each $f \in F = \operatorname{Ker} \bar{\partial}_{(p,q+1)}$ the canonical solution to $\bar{\partial}u = f$ that we have just constructed.

COROLLARY 4.6.12 The operator M is continuous.

Proof   Since the domain $\mathcal{D}$ and range $\mathcal{R}$ of M are Banach spaces, we may apply the closed graph theorem. For simplicity, let us prove the result when $p = q = 0$. Let $f^k \in \mathcal{D}, f^k = \sum_{j=1}^n f^k_j d\bar{z}_j$. Suppose that $f^k \to f = \sum_{j=1}^n f_j d\bar{z}_j$ and $Mf^k \to g$. Let $\phi \in C^\infty_c(\Omega)$. Then

$$\int_\Omega g \frac{\partial}{\partial \bar{z}_j} \phi \, dV = \lim_{k \to \infty} \int_\Omega Mf^k \frac{\partial}{\partial \bar{z}_j} \phi \, dV = -\lim_{k \to \infty} \int_\Omega f^k_j \phi \, dV = -\int_\Omega f_j \phi \, dV.$$

Hence $\bar{\partial}g = f$ in the weak sense. Also, the fact that $Mf^k \in \mathcal{K}^\perp$ for all k implies that $g \in \mathcal{K}^\perp$. So g is the canonical solution to $\bar{\partial}g = f$. By uniqueness, $g = Mf$. We conclude that M is continuous.   □

Exercises for the Reader

1. Fix $\eta \in C^\infty_c(\Omega)$ and $s \in \{0, 1, \dots\}$. Let $W^s(\Omega, \phi_2)$ be the weighted Sobolev space of functions with weak derivatives up to order s in $L^2(\Omega, \phi_2)$. With the natural inner product, $W^s(\Omega, \phi_2)$ is a Hilbert space. Define $W^s_{(p,q)}(\Omega, \phi_2)$ analogously. Let

$$M^\eta_s : W^s_{(p,q+1)}(\Omega, \phi_2) \cap F \to \mathcal{K}^\perp \cap W^{s+1}_{(p,q)}(\Omega, \phi_1) \qquad (4.6.13)$$

be the operator assigning to each f in the domain the form ηu, where u is the unique canonical solution to $\bar{\partial}u = f$. Prove that M^η_s is continuous.

2. Examine the proofs in this section to see that Corollary 4.6.12 and Exercise 1 can be obtained directly from the estimates, without recourse to the closed graph theorem.

EXERCISES

1. Prove that the operators M^η_s in line (4.6.13) are compact in the W^s topology.

2. If $f \in L^1_{\text{loc}}(\mathbb{R}^N)$ and the weak derivatives $\partial f / \partial x_j = 0$ almost everywhere, $j = 1, \dots, N$, then prove that f is constant after correction on a set of measure zero.

3. *Distributions.* Let $D = C_c^\infty(\mathbb{R}^N)$. If $K \subset\subset \mathbb{R}^N$ and α is a multi-index, then we let $\rho_{K,\alpha}$ on D be the seminorm

$$\rho_{K,\alpha}(f) = \sup_K \left| \left(\frac{\partial}{\partial x} \right)^\alpha f \right|.$$

A sequence $\{\phi_j\}_{j=1}^\infty \subseteq D$ is said to converge to a limit function $\phi \in D$ if all ϕ_j are supported in a common compact set and $\lim_{j\to\infty} \rho_{K,\alpha}(\phi_j - \phi) = 0$ for every K, α.

Prove each of the following statements:

a. With the topology described above, the space D becomes a topological vector space.

b. The dual D' is called the space of distributions. Write down the definition of continuity of a distribution.

c. If β is a multi-index, then the differential operator $f \mapsto (\partial/\partial x)^\beta f(0)$ is a distribution.

d. If $g \in L^1_{\text{loc}}(\mathbb{R}^N)$, then the operator $f \mapsto \int f g dV$ is a distribution.

e. If μ is a finite Borel measure on $\mathbb{R}^N$, then $f \mapsto \int f d\mu$ is a distribution.

f. If $T \in D'$ and K is a fixed compact set, then the restriction of T to those f that are supported in K depends on only finitely many seminorms; that is, there is a number $M > 0$ such that

$$|Tf| \le C \cdot \sum_{|\alpha| \le M} \|f\|_{\rho_{K,\alpha}}, \quad \text{all } f \in C_c^\infty(K).$$

g. In general, $T \in D'$ will not depend only on seminorms on a fixed compact set.

4. *The Sobolev spaces by way of the Fourier transform.* For $f \in L^1(\mathbb{R}^N)$, let $\hat{f}(\xi) = \int f(t) e^{i\xi t} dt$. Verify each of the following assertions.

a. If $f, g \in L^1(\mathbb{R}^N)$, then $\widehat{f * g}(\xi) = \hat{f} \cdot \hat{g}(\xi)$.

b. If $f \in L^1(\mathbb{R}^N)$, let $\tilde{f}(x) = f(-x)$. Then $\hat{\tilde{f}}(\xi) = \overline{\hat{f}(\xi)}$ and $\hat{\bar{f}} = \overline{\hat{f}}(\xi)$.

c. If $f \in D(\mathbb{R}^N)$, then $|\hat{f}(\xi)| \le C \cdot (1 + |\xi|)^{-N-1}$. (*Hint:* Integrate by parts.) Therefore, $\hat{f} \in L^1(\mathbb{R}^N)$.

d. If $f \in D(\mathbb{R}^N)$, then $f(0) = c_N \int \hat{f}(\xi) d\xi$.

e. Apply (d) to $f(x) = g * \tilde{\bar{g}}$, any $g \in D$. Then

$$\int |g(t)|^2 dt = c_N \int |\hat{g}(\xi)|^2 d\xi.$$

f. Extend (e) to all of $g \in L^2(\mathbb{R}^N)$.

g. If $f \in D, \alpha$ is a multi-index, then

$$\left(\left(\frac{\partial}{\partial x}\right)^\alpha f\right)\hat{}(\xi) = (-i\xi)^\alpha \hat{f}(\xi).$$

h. If $r > 0$, then we define the Sobolev space

$$W^r(\mathbb{R}^N) = \left\{ f \in L^2(\mathbb{R}^N) : \int |\hat{f}(\xi)|^2 (1 + |\xi|^2)^r dV(\xi) \equiv \|f\|_{W^r}^2 < \infty \right\}.$$

Let $r \in \{0, 1, \ldots, \}$. Then $f \in W^r$ if and only if for all multi-indices α with $|\alpha| \le r$, we have that $(\partial/\partial x)^\alpha f$ exists weakly and is in L^2.

5. This problem is needed in order to make the next one rigorous. Let $\mathcal{S}(\mathbb{R}^N)$, the *Schwartz space*, consist of those $f \in C^\infty(\mathbb{R}^N)$ satisfying

$$\rho_{\alpha,\beta}(f) = \sup_x \left| x^\alpha \left(\frac{\partial}{\partial x}\right)^\beta f \right| < \infty$$

for all multi-indices α, β. Prove each of the following statements.

a. The seminorms $\rho_{\alpha,\beta}$ make $\mathcal{S}$ into a Frechet space. Its dual $\mathcal{S}'$ is called the *space of Schwartz distributions*. Write the condition describing the continuity of a Schwartz distribution.

b. If $f \in \mathcal{S}$, then $((ix)^\alpha f)\hat{} = (\partial/\partial\xi)^\alpha \hat{f}(\xi)$ and $((\partial/\partial x)^\alpha f)\hat{} = (-i\xi)^\alpha \hat{f}$ for any multi-index α. Use this fact, together with Exercise 4(d), to conclude that $\hat{}$ maps $\mathcal{S}$ to $\mathcal{S}$ both injectively and surjectively.

c. If $\mu \in \mathcal{S}'$ then define $\hat{\mu}$ by $\hat{\mu}(f) = \mu(\hat{f})$ for all $f \in \mathcal{S}$. What property of the Fourier transform can you derive from Exercise 4 that justifies this definition?

d. If $\mu \in \mathcal{S}'$ and $f \in \mathcal{S}$, then define $\mu * f$ to be the function given by $(\mu * f)(x) = \mu(\tau_x \tilde{f})$, where $(\tau_x(g))(t) \equiv g(t-x)$ and $\tilde{}$ is as in Exercise 4. What property of convolution of functions motivates this definition?

e. If $\mu \in \mathcal{S}'$ and $f \in \mathcal{S}$, then what can you say about $\widehat{\mu * f}$?

6. Discover the fundamental solution for the Laplacian using the Fourier transform.

a. If $\delta > 0, f \in \mathcal{S}(\mathbb{R}^N)$, then let $\alpha_\delta f(x) = f(\delta x)$.

b. $\widehat{\alpha_\delta f}(x) = \delta^{-N} \hat{f}(x/\delta) \equiv \alpha^\delta \hat{f}(x)$.

c. If ρ is a rotation of $\mathbb{R}^N$, then $\widehat{f \circ \rho}(\xi) = \hat{f} \circ \rho(\xi)$, all $f \in \mathcal{S}$.

d. If $f \in D$, then $\widehat{\Delta f} = -|\xi|^2 \hat{f}$.

e. Assuming that $f = \Delta f * \Gamma$ for some function Γ to be determined, then $\hat{\Gamma}(\xi) = -1/|\xi|^2$.

f. The function Γ is rotationally invariant and $\alpha_\delta \Gamma(x) = \delta^{-N+2}\Gamma(x)$.

g. $\Gamma(x) = c_N |x|^{-N+2}$.

7. *The Sobolev imbedding theorem on domains.* Let $\Omega \subseteq \mathbb{R}^N$. Let

$$W^s(\Omega) = \left\{ f \in L^2(\Omega) : \left(\frac{\partial}{\partial x} \right)^\alpha f \in L^2(\Omega), \text{ all } |\alpha| \leq s \right\}.$$

Here, as usual, the derivatives are interpreted in the weak sense. We wish to consider a Sobolev imbedding theorem for these Sobolev spaces. However, geometric conditions must be imposed on $\partial\Omega$ (see Exercise 8).

Optimal conditions on $\partial\Omega$ are rather complicated. But suppose that Ω is bounded and that $\partial\Omega$ is $C^k, k > s$. Prove that the imbedding theorem as stated in Section 4.6 then prevails for our new Sobolev spaces. (*Hint:* After a partition of unity, the problem is a local one. But then a coordinate change maps $\partial\Omega$ locally to $\{(x_1, \ldots, x_N) : x_N = 0\}$. So it is enough to do the problem on the Euclidean upper half-space in a neighborhood of 0, say on $\tilde{\Omega} = B \cap \{x \in \mathbb{R}^N : x_N > 0\}$. Show that the functions that extend to be C^k on B are dense in $W^s(\tilde{\Omega})$—use convolution and dilation. Then compute an a priori estimate for C^∞ functions.) See R. Adams [1] for more on this matter.

8. Here is a version of the classical Sobolev imbedding theorem on domains. Let $\Omega \subset\subset \mathbb{R}^N$ have C^1 boundary. Let $L_k^p(\Omega) = \{f \in L^p(\Omega) : (\partial/\partial x)^\alpha f \in L^p, \text{all } |\alpha| \leq k\}$. Here, as usual, derivatives are all in the weak sense. If $1/q = 1/p - k/N$ and $1 \leq p < N/k$, then $L_k^p(\Omega) \subseteq L^q(\Omega)$ with continuous inclusion. If $p > N/k$, then $L_k^p(\Omega) \subseteq \Lambda_\alpha^{loc}(\Omega)$, where $\alpha = k - N/p$. See R. Adams [1] for a detailed treatment of these assertions. Parts a–d show that these hypotheses are necessary.

a. Let $\Omega = \{|x| < 1\} \subseteq \mathbb{R}^N$. Let $f(x) = |x|^{(-N/p)+1} / \left(\log(1/|x|) \right)^{N+1}$. Assume that $N > p$. Then $f \in L_1^p(\Omega)$, but $f \notin L^q(\Omega)$ if $q > Np/(N-p), 1 \leq p < \infty$.

b. Generalize (a) to the case $k > 1$. What about $p = \infty$?

c. Some form of boundary regularity is necessary. Let $\Omega = \{(x_1, x_2) \in \mathbb{R}^2 : |x_1| < |x_2|^{1+\epsilon} < 1, x_2 > 0\}$, and let $f(x) = |x|^{-\epsilon/4}$. Then $f \in L_1^{2+\epsilon/4}(\Omega)$, but $f \notin L^\infty(\Omega)$.

d. Generalize (c) to $\mathbb{R}^N$.

9. Verify that the Laplacian, acting on $C_c^\infty(\mathbb{R}^N)$, is self-adjoint. Verify that $-\Delta$ is a positive operator in the sense that $\langle -\Delta\phi, \phi \rangle \geq 0$ for all $\phi \in C_c^\infty$. Use integration by parts on this inner product to prove the a priori estimate

$$\|\nabla\phi\|_{L^2}^2 \leq C \cdot \|\phi\|_{L^2}^2 + C \cdot \|\Delta\phi\|_{L^2}^2.$$

Conclude that

$$\|f\|_{W^1}^2 \leq C\|f\|_{L^2}^2 + C\|\Delta f\|_{L^2}^2, \qquad \text{all } f \in W^2.$$

10. Use the Fourier transform definition of the Sobolev spaces $W^s(\mathbb{R}^N)$ (see Exercise 4) to prove that if $s_1 > s_2$, then W^{s_1} is compact in W^{s_2} in the sense that the natural inclusion operator $i : W^{s_1} \to W^{s_2}$ is a compact operator. Equivalently, if $\{f_j\} \subseteq W^{s_2}$ is bounded in the W^{s_1} topology, then it has a convergent subsequence in W^{s_2}. The analogous result for Lipschitz spaces is that if $\alpha > \beta$, then $\mathrm{Lip}_\alpha(\mathrm{T})$ is compact in $\mathrm{Lip}_\beta(\mathrm{T})$. Here $\mathrm{T} = \mathbb{R}/\{2\pi\mathbb{Z}\}$. For more on this, see Chapter 8.

11. On elements of the Schwartz space, the Laplacian Δ is equivalent to the Fourier multiplier $-|\xi|^2$. More generally, if $L(D) = \sum_\alpha a_\alpha \cdot (\partial/\partial x)^\alpha$ is a *linear, constant coefficient*, partial differential operator, then $L(D)$ corresponds to the Fourier multiplier $\sum_\alpha a_\alpha(-i\xi)^\alpha$. It is a powerful idea of Mikhlin, Calderón, Zygmund, Kohn, Nirenberg, Hörmander, and others to reverse this process and concentrate on the Fourier multiplier as the principal tool. We allow the multiplier to depend both on x (the space variable) and on ξ (the Fourier transform or frequency variable).

Define a smooth function $p(x, \xi)$ on $\mathbb{R}^N \times (\mathbb{R}^N \setminus \{0\})$ to be a *symbol* of class $m \in \mathbb{Z}$ (denoted $p \in S^m$) if p is compactly supported in x and

$$\left| (\partial/\partial x)^\alpha \, (\partial/\partial \xi)^\beta \, p(x, \xi) \right| \leq C_{\alpha,\beta} \cdot (1 + |\xi|)^{m-|\beta|}$$

for any multi-indices α, β. To such a symbol p is associated a *pseudodifferential operator*

$$T_p \phi(x) = \int_{\mathbb{R}^N} p(x, \xi) e^{-ix \cdot \xi} \hat{\phi}(\xi) \, d\xi.$$

The principal results of an elementary calculus of pseudodifferential operators amount to proving that calculus on the operator level (with the operators T_p) is equivalent (up to acceptable error terms) with calculus on the symbol level (with the symbols p). More precisely, we have

(i) $(T_p)^* = T_{\bar{p}} +$ (negligible error),
(ii) $T_p \circ T_q = T_{pq} +$ (negligible error).

By the phrase *negligible error* we mean here an operator with symbol that lies in S^r with $r < m$ (case i) or $r < m + n$ (case ii). This calculus enables one to construct, in a natural fashion, parametrices (approximate inverses) for a large collection of partial differential operators (including elliptic ones). It is hard work to make all this precise, and we do not attempt to do so (good references are M. E. Taylor [1], F. Treves [1], and S. Krantz [19]). Instead, we include a few simple properties of the symbolic calculus, which the reader may verify, and give an application to the regularity theory of the Laplacian. It is of historical interest to note that Kohn and Nirenberg [2] introduced the first modern calculus of pseudodifferential operators; they were motivated in their work by considerations connected with the $\bar{\partial}$ problem. See G. B. Folland and J. J. Kohn [1] and S. G. Krantz [19] for more on these matters.

a. Let $p \in S^m$, some $m \in \mathbb{Z}$. Then, for x fixed, the operator $\phi \mapsto T_p \phi(x)$ is a distribution.

b. If $L = \sum_\alpha a_\alpha(x) \cdot (\partial/\partial x)^\alpha$ is a linear partial differential operator of degree m with coefficients $a_\alpha \in C_c^\infty(\mathbb{R}^N)$, then L is a pseudodifferential operator. What is its symbol? In what symbol class S^m does it lie?

c. Let $p(x,\xi) \in C_c^\infty(\mathbb{R}^N \times \mathbb{R}^N)$. Then $p \in \cap_{m=-\infty}^\infty S^m$.

d. If L_1 and L_2 are partial differential operators of degrees m_1 and m_2 respectively (as in (b)), then $[L_1, L_2]$ is of degree not exceeding $m_1 + m_2 - 1$. A similar result holds for T_{p_1}, T_{p_2} in S^{m_1}, S^{m_2}. but this is more difficult to prove. It also holds that $T_{p_1} \circ T_{p_2} \in S^{m_1+m_2}$.

e. Prove that if $p \in S^m$, then $T_p : W^s \to W^{s-m}$, every $s \in \mathbb{Z}$.

f. Fix a $\phi \in C_c^\infty$. Define $L = \phi(x) \cdot \Delta$. The symbol of L is $p(x,\xi) = -\phi(x)|\xi|^2$. Assume that $\phi(x) = 1$ when $|x| \le 1, \phi(x) = 0$ when $|x| \ge 2$. Now write

$$
\begin{aligned}
p(x,\xi) &= \phi(\xi) \cdot p(x,\xi) + (1 - \phi(\xi)) \cdot p(x,\xi) \\
&= p_1(x,\xi) + p_2(x,\xi).
\end{aligned}
$$

Then $p_1 \in S^m$, every m, while $p_2 \in S^2$. Define

$$
q(x,\xi) = \phi(2x)(1 - \phi(\xi/2))/p(x,\xi).
$$

Then the operator T_q is a parametrix for L in the following sense. First write

$$
T_p T_q = T_{(p_1+p_2)} T_q = T_{p_1} T_q + T_{p_2} T_q \equiv E + F.
$$

Now $E \in S^m$ for every m. Also, by (ii),

$$
\begin{aligned}
T_{p_2} T_q &= T_{p_2 q} + \mathcal{E} = T_{\phi(2x)(1-\phi(\xi/2))} + \mathcal{E} \\
&= T_{\phi(2x)} + \left(-T_{\phi(2x)\phi(\xi/2)}\right) + \mathcal{E} \\
&= \phi(2x) \cdot I + \mathcal{E}
\end{aligned}
$$

Here $\mathcal{E}$ is a "negligible error" whose meaning changes from line to line and is, we shall assume without proof, in S^{-1}. Therefore, we have determined that

$$
LT_q = T_p T_q = \phi(2x) \cdot I + \mathcal{E},
$$

where $\mathcal{E} \in S^{-1}$. Use a similar argument to show that $T_q L = \phi(2x) \cdot I + \mathcal{E}'$, some $\mathcal{E}' \in S^{-1}$. Note that $q \in S^{-2}$.

g. We apply the results of (f) as follows.   Let $\eta \in C_c^\infty$ with supp $\eta \subseteq \{x : \phi(2x) = 1\}$. Assume that $L\eta = \psi$. We wish to estimate the Sobolev norms of η in terms of the Sobolev norms of ψ. So we write

$$
\begin{aligned}
\|\eta\|_{W^2} &= \|\phi(2x)\eta\|_{W^2} \\
&= \|\phi(2x)I\eta\|_{W^2} \\
&= \|T_q L\eta - \mathcal{E}'\eta\|_{W^2} \\
&\leq \|T_q\psi\|_{W^2} + \|\mathcal{E}'\eta\|_{W^2} \\
&\leq C\|\psi\|_{W^0} + C\|\eta\|_{W^1} \\
&\leq C\|\psi\|_{W^0} + C\|\eta\|_{W^0} + (1/2)\|\eta\|_{W^2}.
\end{aligned}
$$

Thus

$$\|\eta\|_{W^2} \leq C\|\psi\|_{W^0} + C\|\eta\|_{W^0}.$$

Iterate this device to prove

$$\|\eta\|_{W^{s+2}} \leq C\left\{\|\psi\|_{W^s} + \|\eta\|_{W^0}\right\}, \qquad \text{any } s \in \mathbb{Z}.$$

Use this result to say something about regularity for Δ.
h. Give two reasons why the naive theory of pseudodifferential operators presented above is useless for solving the $\bar\partial$ problem on a pseudoconvex domain.

12. J. J. Kohn [3] used the method of weight functions to prove the following result: Let $\Omega \subset\subset \mathbb{C}^n$ be a pseudoconvex domain with C^∞ boundary. Let f be a $\bar\partial$-closed $(0,1)$ form on Ω with coefficients that extend to be C^∞ on $\bar\Omega$. Then there is a $u \in C^\infty(\bar\Omega)$ such that $\bar\partial u = f$. Kohn's theorem is difficult and we shall not prove it. Let us take Kohn's result for granted.

Suppose that $P \in \partial\Omega$ and U is a neighborhood of P. Let $g : U \cap \Omega \to \mathbb{C}$ be holomorphic and satisfy (a) g extends to be C^∞ on $\overline{U \cap \Omega}$, (b) $g(P) = 0$, (c) (Image g) omits a sector in $\mathbb{C}$. Use Kohn's result to construct a (global) peaking function for $A(\Omega)$ at P. Will your peaking function be C^∞ on $\bar\Omega$? Will it satisfy a Lipschitz condition?

13. Let f be a nonnegative singular measure supported on the unit circle in $\mathbb{C}$. Suppose that f has total mass 1 and that f is rotationally invariant. Compute explicitly a solution to $\bar\partial u = f\,d\bar z$. Is u in any Sobolev class near the unit circle? How does u behave away from the circle?

14. *The Lewy-Pincuk reflection principle* (H. Lewy [1] and S. Pincuk [1]). The proof of Schwarz's reflection principle is so simple that it obscures the essential geometry of the reflection. The following generalization is harder to prove but is, in the end, more enlightening.

Theorem: Let $\Omega_1, \Omega_2 \subset\subset \mathbb{C}^n$ be strongly pseudoconvex domains with real analytic boundaries. Let $P \in \partial\Omega_1$. Let $U \subseteq \mathbb{C}^n$ be a neighborhood of P. Suppose that $f : \overline{U \cap \Omega_1} \to \mathbb{C}^n$ is C^1 and univalent and that $f|_{U\cap\Omega_1}$ is holomorphic. Finally, suppose that $f(U\cap\partial\Omega_1) \subseteq \partial\Omega_2$. Then f extends holomorphically to a neighborhood of $U \cap \partial\Omega_1$.

Complete the following outline to obtain a proof of this theorem.

a. We may as well suppose from the outset that $\partial\Omega_1 \cap U$ is strongly convex.

b. If $w, \zeta \in \mathbb{C}^n$, then define the complex line

$$\ell^w(\zeta) = \{\zeta + w\lambda : \lambda \in \mathbb{C}\}.$$

Let

$$\gamma^w(\zeta) = \ell^w(\zeta) \cap \partial\Omega_1.$$

For $\epsilon > 0, \rho_j$ a defining function for Ω_j, define

$$
\begin{aligned}
\Omega^-(\epsilon, w, \zeta) &= \{z \in \ell^w(\zeta) : -\epsilon < \rho_1(z) < 0\}, \\
\Omega^+(\epsilon, w, \zeta) &= \{z \in \ell^w(\zeta) : 0 < \rho_1(z) < \epsilon\}, \\
\Omega(\epsilon, w, \zeta) &= \{z \in \ell^w(\zeta) : -\epsilon < \rho_1(z) < \epsilon\}.
\end{aligned}
$$

The idea is to choose w, ζ^0 so that $\gamma^w(\zeta)$ is a closed real analytic curve in $\partial\Omega_1$ for all ζ sufficiently close to ζ^0. *Suppose hereafter that $\zeta_0 = P$ and w is chosen in this way.* Then we extend f in the classical fashion, across each $\gamma^w(\zeta)$, from $\Omega^-(\epsilon, w, \zeta)$ to $\Omega^+(\epsilon, w, \zeta)$.

c. There is a neighborhood $V \subseteq \mathbb{C}^n$ of ζ^0 such that the following holds: If $\epsilon > 0$, then there is a $\delta > 0$ such that for all $\zeta \in V$ and g holomorphic on $\Omega^-(\epsilon, w, \zeta)$ and continuous up to $\gamma^w(\zeta)$, there is a function G conjugate holomorphic on $\Omega^+(\epsilon, w, \zeta)$ and continuous up to $\gamma^w(\zeta)$ such that $G|_{\gamma^w(\zeta)} = g|_{\gamma^w(\zeta)}$.

d. Let ϕ be real analytic on $\partial\Omega_1$. There is an $\epsilon > 0$ and a neighborhood $W \subseteq \mathbb{C}^n$ of ζ^0 such that, for any $\zeta \in W$, it holds that $\phi|_{\gamma^w(\zeta)}$ extends holomorphically to $\Omega(\epsilon, w, \zeta)$.

e. If $z \in U\cap\partial\Omega_1$, then $\rho_2 \circ f(z) = 0$. Assume that $\partial\rho_1/\partial z_n \neq 0$ on $U\cap\partial\Omega_1$. Define $T_j = (\partial\rho_1/\partial z_n)\partial/\partial z_j - (\partial\rho_1/\partial z_j)\partial/\partial z_n$ on $\partial\Omega_1, j = 1, \ldots, n-1$. Then $T_j(\rho_2 \circ f) = 0$ on $U\cap\partial\Omega_1$.

f. The result of (e) may be rewritten as

$$\sum_{k=1}^{n} \frac{\partial\rho_2}{\partial w_k}(f(z)) \, T_j f_k(z) = 0, \quad j = 1, \ldots, n-1, \; z \in \partial\Omega_1.$$

g. Let w be the complex coordinate on Ω_2. Writing $w_k = u_k + iv_k$, $k = 1, \ldots, n$, and assuming as we may that $P = f(P) = 0$, we may expand

$$\rho_2(u + iv) = \rho_2(w) = \sum_{\alpha,\beta} a_{\alpha\beta} u^\alpha v^\beta$$

in a neighborhood of $f(P)$.

h. Write $u_k = (w_k + \bar{w}_k)/2$, $v_k = (w_k - \bar{w}_k)/(2i)$ and note that this transforms the expansion in (g) to one of the form

$$\mathcal{P}(w, \bar{w}) = \sum_{\alpha,\beta} a'_{\alpha\beta} w^\alpha \bar{w}^\beta.$$

Now formally define

$$\mathcal{P}(w, \mu) = \sum a'_{\alpha\beta} w^\alpha \mu^\beta, \quad w \in \mathbb{C}^n, \quad \mu \in \mathbb{C}^n.$$

i. Rewrite the n equations in (e) as

$$\mathcal{P}\left(f(z), \overline{f(z)}\right) = 0$$

$$\sum_{k=1}^{n} \frac{\partial \mathcal{P}}{\partial w_k}\left(f(z), \overline{f(z)}\right) T_j f_k(z), \quad j = 1, \ldots, n-1.$$

j. Consider the matrix $\mathcal{C} = (c_{ij})_{i,j=1}^n$ given by the condition

$$c_{ij} = \sum_{k=1}^{n} \frac{\partial^2 \mathcal{P}}{\partial w_k \partial \mu_j}(0,0) T_i f_k(0), \quad i = 1, \ldots, n-1,$$

$$c_{nj} = \frac{\partial \mathcal{P}}{\partial \mu_j}(0,0).$$

Refer to Exercise 37 at the end of Chapter 11 to notice that if f is nonconstant (which we may as well assume), then $\det J_\mathbb{C} f(0) \neq 0$. As a result, the vectors $T_i f(0), i = 1, \ldots, n-1$ span the complex tangent space to $\partial\Omega_1$ at 0. Since the Levi form is positive definite, the first $(n-1)$ rows of $\mathcal{C}$ are linearly independent at 0. Since the last row is a normal vector, $\mathcal{C}$ has rank n.

k. Parts i and j imply that we may use the implicit function theorem to solve for $\overline{f_1(z)}, \ldots, \overline{f_n(z)}$ as holomorphic functions of the n^2 variables $f_1(z), \ldots, f_n(z)$ and $T_j f_k, j = 1, \ldots, n-1, k = 1, \ldots n$. More precisely, write

$$Tf(z) = (T_1 f_1(z), \ldots, T_1 f_n(z), \ldots, T_{n-1} f_1(z), \ldots, T_{n-1} f_n(z)).$$

Then we can find holomorphic function h_i of n^2 complex variables such that

$$\tilde{f}_1(z) = h_1 \left(f(z), Tf(z) \right)$$

$$\vdots$$

$$\tilde{f}_n(z) = h_n \left(f(z), Tf(z) \right)$$

whenever $(f(z), Tf(z))$ lies in a neighborhood N of $(0, Tf(0))$.

l. If w is fixed appropriately and ζ is sufficiently near P, then we may apply (d) to find $\epsilon > 0$ and a neighborhood W of 0 in $\mathbb{C}^n$ such that for $\zeta \in W$, the $T_i f_k|_{\partial\Omega_1}$ extend holomorphically to $\Omega^-(\epsilon, w, \zeta)$. We may also assume that $h_k(f, Tf)$ extends holomorphically to $\Omega^-(\epsilon, w, \zeta)$.

m. By (c), the $h_k(f, Tf)$ extend conjugate holomorphically to $\Omega^+(\epsilon, w, \zeta)$, some $\epsilon > 0$.

n. Shrinking W if necessary, relation (k) is satisfied on $\partial\Omega_1 \cap U$ so that $f_k|_{\gamma^w(\zeta)}$ can be holomorphically continued to $\Omega(\delta, w, \zeta), k = 1, \ldots, n$, some $\delta > 0$.

o. Now conclude that f holomorphically continues to some small neighborhood of P. Complete the proof in an obvious way. (See also Exercise 20 at the end of Chapter 5.)

p. Notice that the injectivity of f was needed only to see that $J_{\mathbb{C}} f(0)$ has rank n. This turns out to be true even if f is not univalent. Try to prove this. Use the technique of Exercise 37 at the end of Chapter 11.

15. Apply the result of the preceding exercise to prove the following.

Theorem: If Ω_1, Ω_2 are strongly pseudoconvex domains with real analytic boundaries and $f : \Omega_1 \to \Omega_2$ is biholomorphic, then f extends holomorphically to a neighborhood of $\bar{\Omega}_1$.

(*Hint:* Assume Fefferman's theorem (11.4.3), which states that f and f^{-1} extend smoothly to $\bar{\Omega}_1, \bar{\Omega}_2$ respectively.)

16. Let $\Omega \subset\subset \mathbb{C}^n$ have C^k boundary. Let $f \in C^{k+1}(\Omega)$. If $(\partial/\partial x)^\alpha f$ is bounded for every $|\alpha| \leq k+1$, then f has a C^k extension to $\bar{\Omega}$. Can you obtain the same result with a weaker hypothesis than the existence and boundedness of the $(k+1)^{\text{st}}$ derivatives?

17. *The tangential Cauchy-Riemann equations and the Bochner extension phenomenon.* Let $\Omega \subset\subset \mathbb{C}^n$ be a domain, $n > 1$, with C^1 boundary and defining function ρ. Let f be a C^1 function on $\partial\Omega$. We say that f satisfies the *tangential Cauchy-Riemann equations* if there is a C^1 extension F of f to a neighborhood of $\partial\Omega$ such that $\bar{\partial}F \wedge \bar{\partial}\rho = 0$ on $\partial\Omega$. (Note: This *ad hoc* definition is equivalent to several other more intrinsic ones—see G. B. Folland and J. J. Kohn [1] and A. Boggess [1].)

  a. Show that the preceding definition is unambiguous (i.e., is independent of the choice of F).

  b. Show that f satisfies the tangential Cauchy-Riemann equations if and only if, on $\partial\Omega$,

$$\sum_j \alpha_j \frac{\partial F}{\partial \bar{z}_j} = 0 \quad \text{whenever} \quad \sum_j \alpha_j \frac{\partial \rho}{\partial \bar{z}_j} = 0.$$

  Here F is *any* C^1 extension of f.

  c. Let Ω now have C^4 boundary, and suppose that $f \in C^4(\partial\Omega)$ satisfies the tangential Cauchy-Riemann equations. Let F be a C^4 extension of f to $\bar{\Omega}$. Then there exist $\alpha \in C^3(\bar{\Omega}), \beta \in C^2_{(0,1)}(\bar{\Omega})$ with $\bar{\partial}F = \alpha\bar{\partial}\rho + \beta \cdot \rho$ on $\bar{\Omega}$. (Why is β only C^2?)

  d. Notice that $\bar{\partial}(F - \alpha\rho) = \gamma \cdot \rho$, some $\gamma \in C^2_{(0,1)}(\bar{\Omega})$.

  e. Compute that $0 = \bar{\partial}(\rho\gamma) = \bar{\partial}\rho \wedge \gamma + \rho\bar{\partial}\gamma$. Therefore, $\bar{\partial}\rho \wedge \gamma = 0$ on $\partial\Omega$. So $\gamma = \mu\bar{\partial}\rho + \rho\eta$, some $\mu \in C^2(\bar{\Omega}), \eta \in C^1_{(0,1)}(\bar{\Omega})$.

  f. Define $\Phi = f - \alpha\rho - \mu\rho^2/2$. Then Φ agrees with f on $\partial\Omega$, $\Phi \in C^2(\bar{\Omega})$ and $\bar{\partial}\Phi = O(\rho^2)$.

   Theorem: Let $\Omega \subset\subset \mathbb{C}^n$ have C^4 boundary. Assume that ${}^c\bar{\Omega}$ is connected. Let $f \in C^4(\partial\Omega)$ satisfy the tangential Cauchy-Riemann equations. Then there is a function $\hat{f} \in C^1(\hat{\Omega})$ that is holomorphic on Ω and such that $\hat{f}\big|_{\partial\Omega} = f$.

  g. To prove the theorem, choose Φ as in (f). Let

$$\omega(z) = \begin{cases} \bar{\partial}\Phi(z) & \text{if } z \in \bar{\Omega}, \\ 0 & \text{if } z \notin \bar{\Omega}. \end{cases}$$

  Then $\omega \in C^1_c(\mathbb{C}^n)$ and $\bar{\partial}\omega = 0$. So there is a function $u \in C^1_c(\mathbb{C}^n)$ with $\bar{\partial}u = \omega$, $u \equiv 0$ off Ω. (Careful—what hypothesis do we need to use here?) Now set $\hat{f} = \Phi - u$.

  h. The hypotheses of this theorem may be considerably weakened (see A. Boggess [1]). Even in the proof just outlined (due to Hörmander [3]), the hypothesis may be weakened from C^4 to C^2. There are local versions of the extension theorem, but they require extra hypotheses and are harder to prove. Many problems remain open.

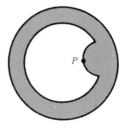

FIGURE 4.1

18. Let $\Omega \subseteq \mathbb{C}^2$ be given by $\Omega = B((0,0),6) \setminus \{\bar{B}((0,0),5) \setminus B((5,0),1)$, as indicated by Figure 4.1. Show that, on $\Omega, \bar{\partial}$ on functions does not have closed range. (*Hint:* Construct a peaking function at $P = (3,0)$.) Details of this construction may be found in S. G. Krantz [15].

19. Let $\Omega \subseteq \mathbb{C}^n$ be any domain. Let f be a $\bar{\partial}$-closed $(0,1)$ form on Ω with C^k coefficients. Assume that $u \in L^2_{\text{loc}}(\Omega)$ is a function on Ω that satisfies $\bar{\partial}u = f$ in the weak sense. Prove that u is C^k on Ω. (*Hint:* Exploit the trivial solution to the $\bar{\partial}$ equation for compactly supported forms that was developed in Chapter 1 together with the fact proved in Section 4.6 that there are no "fake" holomorphic functions.)

5 Solution of the Levi Problem and Other Applications of $\bar{\partial}$ Techniques

5.1 An Extension Problem

Let $\Omega \subseteq \mathbb{C}^n$ be a domain. Let $\omega = \{z \in \Omega : z_n = 0\}$. Then ω can be identified in a natural way with a subset of $\mathbb{C}^{n-1}$. With this in mind, let $f : \omega \to \mathbb{C}$ be holomorphic (notice that ω may be disconnected). Does there exist an $F : \Omega \to \mathbb{C}$ such that F is holomorphic and $F|_{\omega} = f$? In case the setup is that shown in Figure 5.1, then the solution is trivially $F(z_1, \ldots, z_n) = f(z_1, \ldots, z_{n-1}, 0)$. However, if the setup is that shown in Figure 5.2, then the problem is nontrivial. It turns out that if Ω is pseudoconvex, then we can always solve the problem. First, let us look at a critical example to see that for general domains we may not be able to solve the problem.

EXAMPLE   Let $\Omega = B(0,1) \setminus \bar{B}(0, 1/2) \subseteq \mathbb{C}^2$. Then $\omega = \Omega \cap \{(z_1, z_2) : z_2 = 0\} = \{(z_1, 0) : 1/2 < |z_1| < 1\}$. Let $f : \omega \to \mathbb{C}$ be given by $f(z_1, 0) = 1/(z_1 - 1/2)$. Then f is holomorphic on ω. Suppose that F is a holomorphic extension of f to all of Ω. By Hartogs's phenomenon, F has a holomorphic continuation to $B(0,1)$. In particular, this F will be bounded in every neighborhood of the point $(1/2, 0)$. Therefore F cannot agree with f on ω.   □

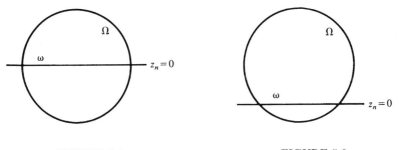

FIGURE 5.1   FIGURE 5.2

Now we can appreciate the next theorem.

THEOREM 5.1.1   Let $\Omega \subseteq \mathbb{C}^n$ be pseudoconvex (no assumptions about boundary smoothness, or even boundedness, need be made). Let $\omega = \Omega \cap \{(z_1, \ldots, z_n) : z_n = 0\}$. Let $f : \omega \to \mathbb{C}$ satisfy the property that the map

$$(z_1, \ldots, z_{n-1}) \mapsto f(z_1, \ldots, z_{n-1}, 0)$$

is holomorphic on $\tilde{\omega} = \{(z_1, \ldots, z_{n-1}) \in \mathbb{C}^{n-1} : (z_1, \ldots, z_{n-1}, 0) \in \omega\}$. Then there is a holomorphic $F : \Omega \to \mathbb{C}$ such that $F|_\omega = f$. Indeed there is a linear operator

$$\mathcal{E}_{\omega,\Omega} : \{\text{holomorphic functions on } \omega\} \to \{\text{holomorphic functions on } \Omega\}$$

such that $(\mathcal{E}_{\omega,\Omega}f)|_\omega = f$. The operator is continuous in the topology of normal convergence.

Proof   Let $\pi : \mathbb{C}^n \to \mathbb{C}^n$ be the Euclidean projection $(z_1, \ldots, z_n) \mapsto (z_1, \ldots, z_{n-1}, 0)$. Let $\mathcal{B} = \{z \in \Omega : \pi z \notin \omega\}$. Then $\mathcal{B}$ and ω are *relatively closed* disjoint subsets of Ω. Hence there is a function $\Psi : \Omega \to [0,1], \Psi \in C^\infty(\Omega)$, such that $\Psi \equiv 1$ on a relative neighborhood of ω and $\Psi \equiv 0$ on $\mathcal{B}$. (This last assertion is intuitively nonobvious. It is a version of the C^∞ Urysohn lemma, for which see M. Hirsch [1]. It is also a good exercise for the reader to construct Ψ by hand.) Set

$$F(z) = \Psi(z) \cdot f(\pi(z)) + z_n \cdot v(z),$$

where v is an unknown function to be determined.

Notice that $f(\pi(z))$ is well defined on supp Ψ. We wish to select $v \in C^\infty(\Omega)$ so that $\bar{\partial}F = 0$. Then the function F defined by the displayed equation will be the function that we seek.

Thus we require that

$$\bar{\partial}v(z) = \frac{\left(-\bar{\partial}\Psi(z)\right) \cdot f\left(\pi(z)\right)}{z_n}. \tag{5.1.1.1}$$

Now the right side of this equation is C^∞, since $\bar{\partial}\Psi \equiv 0$ on a neighborhood of ω. Also, by inspection, the right side is annihilated by the $\bar{\partial}$ operator (remember that $\bar{\partial}^2 = 0$). By Corollary 4.6.12, there exists a $v \in C^\infty(\Omega)$ that satisfies (5.1.1.1). Therefore, the extension F exists and is holomorphic.

Note, finally, that since the $\bar{\partial}$ solution operator that we constructed in Chapter 4 is linear, it follows that F depends linearly on f.

Because the solution operator for the $\bar{\partial}$ equation that we constructed in Chapter 4 is bounded in a weighted L^2 space, it is straightforward to check that the operator $\mathcal{E}_{\omega,\Omega}$ is bounded in L^2 of any compact set. But then the Cauchy estimates show that the operator is bounded in the topology of normal convergence.   □

THEOREM 5.1.2 (Solution of the Levi Problem)   Let $\Omega \subseteq \mathbb{C}^n$ be pseudoconvex (no assumptions about boundary smoothness or boundedness are necessary). Then Ω is a domain of holomorphy.

Proof   The proof is by induction on dimension. For the case $n = 1$, the result was proved in Chapter 0. Assume the result to be proved in dimension $n-1$, and let $\Omega \subseteq \mathbb{C}^n$. Then for a dense open set of $P \in \partial\Omega$, the following construction is valid: There is a vector $w \in \mathbb{C}^n$ such that $\mathbf{h} \equiv \{z \in \mathbb{C}^n : \sum_{j=1}^n (z_j - P_j)\bar{w}_j = 0\}$ satisfies $\mathbf{h} \cap \Omega \neq \emptyset$ and $P \in \partial(\mathbf{h}\cap\Omega)$ (here we are thinking of $\mathbf{h}\cap\Omega$ as an $(n-1)$-dimensional complex manifold. In fact, if $q \in \Omega$ is near the boundary, then let B_q be the largest open ball centered at q and lying in Ω. Choose $P \in \bar{B}_q \cap \partial\Omega$. Then taking w to be any vector that is Hermitian orthogonal to the vector determined by the points q and P will do. If $x \in \partial\Omega$ and $B(x,r)$ is a small ball, then choosing $q \in B(x,r)\cap\Omega$ results in $P \in B(x,r)\cap\partial\Omega$. So the set of P's for which this geometric construction works is dense in $\partial\Omega$.

After a change of coordinates, we may assume that $P = 0$ and $\mathbf{h} = \{z \in \mathbb{C}^n : z_n = 0\}$. Then let $\omega = \Omega \cap \mathbf{h}$. Clearly ω is pseudoconvex (exercise). Therefore, by the inductive hypothesis, there is a holomorphic function f on ω that is singular at P. Apply Theorem 5.1.1 to obtain a holomorphic function F on Ω such that $F|_\omega = f$. A fortiori, F is singular at P. Thus P is an essential point of $\partial\Omega$. Since these P are dense in $\partial\Omega$, it follows by definition that every point of $\partial\Omega$ is essential.

We conclude from part 3 of Theorem 3.4.5 that Ω is a domain of holomorphy.   □

We conclude the section with a theorem that summarizes many of the key ideas that have been considered so far in this text. It is an important exercise for the reader to verify that all these statements have been proved.

THEOREM 5.1.3   Let $\Omega \subseteq \mathbb{C}^n$ be a domain. The following are equivalent:

(1) Ω is pseudoconvex.

(2) The equation $\bar{\partial}u = f$ always has a solution $u \in C_{(p,q)}^\infty(\Omega)$ for any form $f \in C_{(p,q+1)}^\infty(\Omega)$ with $\bar{\partial}f = 0, q = 0, 1, \ldots, n - 1$.

(3) Ω is a domain of holomorphy.

All the parts of the theorem have been stated explicitly in the text except for (2) $\Rightarrow$ (3). The proof of that assertion is a variant of the proof of 5.1.2. Details may also be found in Hörmander [3, p. 88].

5.2 Singular Functions on Strongly Pseudoconvex Domains

In 1970, independently and nearly simultaneously, G. M. Henkin and E. Ramirez discovered how to construct a smoothly varying holomorphic separating function on a strongly pseudoconvex domain. They were able to exploit this construction along with the Cauchy-Fantappiè formalism to obtain integral representation formulas *with holomorphic kernels* on such domains. These formulas have a wealth of function-theoretic consequences including (nearly explicit) integral solution formulas for the $\bar{\partial}$ equation. (It should be mentioned that Kerzman, Øvrelid, and Grauert-Lieb also obtained such formulas.) In a long series of papers by many authors, the latter formulas have been studied and estimates in many classical function spaces—L^p, Lipschitz, and C^k, for instance—have been obtained for solutions to the $\bar{\partial}$ problem. These estimates, in turn, can be applied to the study of problems such as the extension problem of Section 5.1 and to problems of function theory.

In the present section we shall present the construction, essentially in the style of Henkin, of smoothly varying holomorphic separating functions on a strongly pseudoconvex domain. We apply them first to construct smoothly varying *peaking* functions for the algebra $A(\Omega)$. In the next section we derive integral reproducing formulas for holomorphic functions.

Fix a nonnegative integer k and a strongly pseudoconvex domain $\Omega \subset\subset \mathbb{C}^n$ with C^{k+3} boundary. Let $\rho : \mathbb{C}^n \to \mathbb{R}$ be a C^{k+3} defining function for Ω with the property that (by Proposition 3.2.1) it is *strictly psh* in a neighborhood of $\partial\Omega$. According to the remark following Corollary 3.4.8, the function

$$L : \mathbb{C}^n \times \mathbb{C}^n \to \mathbb{C}$$

given by

$$
\begin{aligned}
L_P(z) \;=\; & L(z, P) \equiv \rho(P) + \sum_{j=1}^{n} \frac{\partial \rho}{\partial z_j}(P)(z_j - P_j) \\
& + \frac{1}{2} \sum_{j,k=1}^{n} \frac{\partial^2 \rho(P)}{\partial z_j \partial z_k}(z_j - P_j)(z_k - P_k)
\end{aligned}
$$

satisfies the following properties:

(5.2.1) For each $P \in \mathbb{C}^n$, the function $z \mapsto L(z, P)$ is holomorphic (indeed, it is a polynomial).

(5.2.2) For each $z \in \mathbb{C}^n$, the function $P \mapsto L(z, P)$ is C^{k+1}.

(5.2.3) For each $P \in \partial\Omega$, there is a neighborhood U_P such that if $z \in \bar{\Omega} \cap \{w \in U_P : L_P(w) = 0\}$, then $z = P$.

Our goal is to remove the need to restrict in (5.2.3) to a small neighborhood of $P \in \partial\Omega$ while preserving properties (5.2.1) and (5.2.2). We proceed through a sequence of lemmas. Following Henkin, we use the notation

$$\begin{aligned}\Omega_\delta &= \{z \in \mathbb{C}^n : \rho(z) < \delta\};\\ U_\delta &= \{z \in \mathbb{C}^n : |\rho(z)| < \delta\}, \quad \delta > 0.\end{aligned}$$

Further, let us fix the following constants:

(5.2.4) Choose $\delta > 0$ and $\gamma > 0$ such that

$$\sum_{j,k=1}^{n} \frac{\partial^2 \rho}{\partial z_j \partial \bar{z}_k}(P) w_j \bar{w}_k \geq \gamma |w|^2, \qquad \text{all } P \in U_\delta, \text{ all } w \in \mathbb{C}^n.$$

(5.2.5) Shrinking δ if necessary, we may select $\kappa > 0$ so that

$$|\operatorname{grad} \rho(z)| \geq \kappa \quad \text{for all } z \in U_\delta.$$

(5.2.6) With δ as above, let

$$K = \sum_{|\alpha|+|\beta|\leq 3} \left\| \left(\frac{\partial}{\partial z}\right)^\alpha \left(\frac{\partial}{\partial \bar{z}}\right)^\beta \rho(z) \right\|_{L^\infty(U_\delta)}.$$

LEMMA 5.2.7 There is a $\lambda > 0$ such that if $P \in \partial\Omega$ and $|z - P| < \lambda$, then

$$2 \operatorname{Re} L_P(z) \leq \rho(z) - \gamma |z - P|^2/2.$$

Proof Let $\lambda = \frac{1}{2}\gamma/(K+1)$. If $|z - P| < \lambda$, then

$$\rho(z) = \rho(P) + 2 \operatorname{Re} L_P(z) + \sum_{j,k=1}^{n} \frac{\partial^2 \rho}{\partial z_j \partial \bar{z}_k}(P)(z_j - P_j)(\bar{z}_j - \bar{P}_k) + R_P(z),$$

where R_P is the remainder term for Taylor's formula. Therefore,

$$\begin{aligned}2 \operatorname{Re} L_P(z) &\leq \rho(z) - \sum_{j,k=1}^{n} \frac{\partial^2 \rho}{\partial z_j \partial \bar{z}_k}(P)(z_j - P_j)(\bar{z}_j - \bar{P}_k) + |R_P(z)|\\ &\leq \rho(z) - \gamma|z - P|^2 + K|z - P|^3\\ &\leq \rho(z) - \gamma|z - P|^2/2. \qquad \square\end{aligned}$$

LEMMA 5.2.8 Let $\epsilon = \gamma\lambda^2/20$. If $P \in \partial\Omega, z \in \Omega_\epsilon, \lambda/3 \leq |z - P| \leq 2\lambda/3$, then

$$\operatorname{Re} L_P(z) < 0.$$

Proof   With z as in the hypotheses, we have by Lemma 5.2.7 that

$$2 \operatorname{Re} L_P(z) \leq \epsilon - \gamma \frac{(\lambda/3)^2}{2} = \gamma \frac{\lambda^2}{20} - \gamma \frac{\lambda^2}{18} < 0. \qquad \square$$

We may assume that $\epsilon < \lambda < \delta < 1$ (where δ is as in (5.2.4) and (5.2.5)). Let $\eta : \mathbb{R} \to [0, 1]$ be a C^∞ function that satisfies $\eta(x) = 0$ for $x \geq 2\lambda/3$ and $\eta(x) = 1$ for $x \leq \lambda/3$.

LEMMA 5.2.9   Fix $P \in \partial\Omega$. The $(0, 1)$ form

$$f_P(z) = \begin{cases} -\bar{\partial}_z \{\eta(|z - P|)\} \cdot \log L_P(z) & \text{if } |z - P| < \lambda, z \in \Omega_\epsilon \\ 0 & \text{if } |z - P| \geq \lambda, z \in \Omega_\epsilon \end{cases}$$

is well defined (if we take the principal branch for logarithm) and has C^∞ coefficients for $z \in \Omega_\epsilon$. If z is fixed, then $f_P(z)$ depends C^k on P. Finally, $\bar{\partial}_z f_P(z) = 0$ on Ω_ϵ. (One may note that this construction is valid even for P sufficiently near $\partial\Omega$.)

Proof   On $\operatorname{supp}\{\bar{\partial}\eta(|z - P|)\}$ we have $\lambda/3 \leq |z - P| \leq 2\lambda/3$, so if z is also in Ω_ϵ, then Lemma 5.2.8 applies and $\operatorname{Re} L_P(z) < 0$. Therefore, $\log L_P(z)$ makes sense. It follows that the form has C^∞ coefficients for $|z - P| < \lambda$. When $|z - P| > 2\lambda/3$, we have $\bar{\partial}_z\{\eta(|z - P|)\} \equiv 0$, so that $f_P(z)$ is smooth. Since $\log L_P(z)$ is holomorphic on $\operatorname{supp} \bar{\partial}_z\eta(|z - P|)$, it follows that $\bar{\partial} f_P(z) = 0$ on all of Ω_ϵ. The fact that $f_P(z)$ depends C^k on P is clear, since $L_P(z)$ does. $\square$

LEMMA 5.2.10   There is a C^∞ function u_P on Ω_ϵ such that $\bar{\partial} u_P = f_P$.

Proof   Since $\epsilon < \delta$, we know that $\rho_\epsilon(z) \equiv \rho(z) - \epsilon$ is a defining function for Ω_ϵ; hence Ω_ϵ is strongly pseudoconvex. By Corollary 4.6.12 and Lemma 5.2.9, such a function u_P must exist. $\square$

We now define

$$\Phi(z, P) = \begin{cases} [\exp u_P(z)] \cdot L_P(z) & \text{if } |z - P| < \lambda/3, \\ \exp[u_P(z) + \eta(|z - P|) \log L_P(z)] & \text{if } \lambda/3 \leq |z - P| < \lambda, \\ \exp(u_P(z)) & \text{if } \lambda \leq |z - P|. \end{cases}$$

Notice that Φ is unambiguously defined. To study the properties of Φ, we require two technical lemmas.

LEMMA 5.2.11   If $U \subseteq \mathbb{C}^n$ is any open set and $K \subset\subset U$, then any $u \in C^1(U)$ satisfies

$$\sup_K |u| \leq C \left(\|u\|_{L^2(U)} + \|\bar{\partial} u\|_{L^\infty(U)} \right).$$

Here the constant C depends on U and K but not on u.

Proof   Let $V \subset\subset U$ be a C^1 domain such that $K \subset\subset V$. Choose $\eta \in C_c^\infty(V)$ such that $\eta \equiv 1$ on K. Apply the full Bochner-Martinelli formula to the function ηu on V. Then the boundary term (the first term of the Bochner-Martinelli formula) vanishes, and the desired estimate follows directly from the integrability of the kernel in the remaining term.

The reader who knows something about partial differential equations will note that this result also follows from the uniform ellipticity of the $\bar\partial$ operator on compact subsets of U. □

Exercise for the Reader
Construct an elementary proof, from first principles, of Lemma 5.2.11.

COROLLARY 5.2.12   Let $\Omega \subset\subset \mathbb{C}^n$ be pseudoconvex and $K \subset\subset \Omega$. Let f be a $\bar\partial$-closed $(0,1)$ form on Ω with C^1 coefficients. If $u = Mf$ is the Hörmander solution to $\bar\partial u = f$, then we have

$$\|u\|_{L^\infty(K)} \le C\left(\|f\|_{L^\infty(\Omega)}\right),$$

where C depends only on K and Ω (and not on f or u).

Proof   Let $K \subset\subset \Omega' \subset\subset \Omega$. Recall Hörmander's formalism:

$$T = \bar\partial_{(0,0)} : L^2_{(0,0)}(\Omega, \phi_1) \to L^2_{(0,1)}(\Omega, \phi_2).$$

Note that we had some flexibility in choosing the weights. In particular, we may assume that ϕ_2 blows up at every boundary point. Then

$$
\begin{aligned}
\sup_K |u| &\le C\left(\|u\|_{L^2(\Omega')} + \|\bar\partial u\|_{L^\infty_{(0,1)}(\Omega')}\right) \quad \text{(by 5.2.11)}\\
&\le C'\left(\|u\|_{L^2(\Omega,\phi_1)} + \|f\|_{L^\infty_{(0,1)}(\Omega')}\right)\\
&\le C''\left(\|f\|_{L^2_{(0,1)}(\Omega,\phi_2)} + \|f\|_{L^\infty_{(0,1)}(\Omega')}\right)\\
&\le C'''\left(\|f\|_{L^\infty_{(0,1)}(\Omega)}\right)
\end{aligned}
$$

since $e^{-\phi_2}$ is bounded. □

PROPOSITION 5.2.13   Assume once more that $\Omega \subset\subset \mathbb{C}^n$ has C^{k+3} boundary. Then $\Phi(\cdot, P)$ is holomorphic on Ω_ϵ. Also there is a $C > 0$, independent of P, such that for all $z \in \Omega_{\epsilon/2}$ we have

$$\text{if} \quad |z - P| < \lambda/3, \quad \text{then } |\Phi(z,P)| \ge C|L_P(z)|; \quad (5.2.13.1)$$
$$\text{if} \quad |z - P| \ge \lambda/3, \quad \text{then } |\Phi(z,P)| \ge C. \quad (5.2.13.2)$$

Proof   If $|z - P| \geq 2\lambda/3$, then $\Phi(z, P) = \exp u_P(z)$ and $\bar{\partial}_z \Phi(z, P) = (\exp u_P(z)) \cdot \bar{\partial} u_P(z) = (\exp u_P(z)) \cdot f_P(z) = 0$ by construction. If $\lambda/3 \leq |z - P| < 2\lambda/3$, then

$$
\begin{aligned}
\bar{\partial}_z \Phi(z, P) &= \exp\left[u_P(z) + \eta(|z - P|)\log L_P(z)\right] \\
&\quad \cdot \left\{\bar{\partial}\left[u_P(z) + \eta(|z - P|)\log L_P(z)\right]\right\} \\
&= \exp\left[u_P(z) + \eta(|z - P|)\log L_P(z)\right] \\
&\quad \cdot \left\{f_P(z) + \bar{\partial}\eta(|z - P|) \cdot \log L_P(z)\right\}
\end{aligned}
$$

since $\log L_P(z)$ is holomorphic when $\lambda/3 \leq |z - P| < 2\lambda/3$. The last line is 0 by definition of f_P. The calculation for $|z - P| < \lambda/3$ is trivial. Hence we find that Φ_P is holomorphic in the z variable, $z \in \Omega_\epsilon$.

For the estimate, notice that f_P is bounded on Ω_ϵ, uniformly in P, so that u_P is bounded on $\bar{\Omega}_{\epsilon/2}$ uniformly in P (by Corollary 5.2.12). So there is a $C' > 0$ such that $|\exp u_P(z)| \geq C'$. Thus

$$
|\Phi(z, P)| = |\exp u_P(z)| \geq C' \quad \text{if } |z - P| \geq 2\lambda/3
$$

and

$$
|\Phi(z, P)| = |\exp u_P(z)| \cdot |L_P(z)| \geq C'|L_P(z)| \quad \text{if } |z - P| \leq \lambda/3.
$$

For $\lambda/3 \leq |z - P| \leq 2\lambda/3$, we have by Lemmas 5.2.7 and 5.2.8 that

$$
\begin{aligned}
\mathrm{Re} L_P(z) &\leq \frac{\epsilon}{2} - \gamma\frac{(\lambda/3)^2}{4} \\
&= \frac{\gamma\lambda^2}{40} - \gamma\frac{(\lambda/3)^2}{4} \\
&= -\frac{\gamma\lambda^2}{360}.
\end{aligned}
$$

Thus

$$
|L_P(z)| \geq \frac{11\gamma\lambda^2}{360}.
$$

We conclude that, for $\lambda/3 \leq |z - P| \leq 2\lambda/3$,

$$
|\Phi(z, P)| \geq |\exp u_P(z)| \cdot |L_P(z)| \geq C''\frac{\gamma\lambda^2}{360}. \qquad \square
$$

Now we would like to consider the smooth dependence of Φ on P. The subtlety is that our construction of Φ_P involved solving $\bar{\partial} u_P = f_P$, so, in principle, it appears that we must check the smooth dependence of Hörmander's solution operator on parameters. In fact this type of smooth dependence has been checked for various solutions of the $\bar{\partial}$ problem (see R. E. Greene and S. G. Krantz [2]). But, by using a little functional analysis, we may avoid such difficult calculations.

Now fix $z \in \Omega$. Let $\theta \in C_c^\infty(\Omega_\epsilon)$ satisfy $\theta(z) = 1$. Let $s > 2n$. Let M_s^θ be the right inverse to $\bar{\partial}_{0,0}$ (the Hörmander solution operator) for the pseudoconvex domain Ω_ϵ followed by θ. Let notation be as in 5.2.7 through 5.2.13. Let μ be in the dual space of $W^s(\Omega, \phi_1)$ (naturally, this dual space is just $W^s(\Omega, \phi_1)$ itself). Then

$$\langle \mu, M_s^\theta f_P \rangle = \langle (M_s^\theta)^* \mu, f_P \rangle,$$

which depends C^k on P because f_P does.

PROPOSITION 5.2.14   The function $\Phi(z, P)$ depends in a C^k fashion on P for fixed $z \in \Omega_{\epsilon/2}$.

Proof   Fix $s > 2n$ and let $\mu = e_z$ be the point evaluation functional on $C(\Omega) \supseteq W^s(\Omega, \phi_1)$. Then, by the preceding discussion,

$$\langle e_z, M_s^\theta f_P \rangle = \left(M_s^\theta f_P \right)(z) = u_P(z)$$

depends C^k on P. Therefore, $\Phi(z, P)$ itself depends C^k on P. $\square$

Now we may use Henkin's *construction* of the function Φ to obtain the main result of the present section.

THEOREM 5.2.15   Let $\Omega \subseteq \mathbb{C}^n$ be strongly pseudoconvex with C^{k+3} boundary. There is a function

$$\mathcal{P} : \bar{\Omega} \times \partial\Omega \to \mathbb{C}$$

such that

1. $\mathcal{P}(\cdot, P)$ is holomorphic on $\Omega_{\epsilon/2}$ for each fixed $P \in \partial\Omega$,

2. $\mathcal{P}(z, \cdot)$ is C^k for each $z \in \Omega_{\epsilon/2}$,

3. For each $P \in \partial\Omega$, the function $z \mapsto \mathcal{P}(z, P)$ *peaks* at P in the sense that $\mathcal{P}(P, P) = 1$ and $|\mathcal{P}(z, P)| < 1$ for all $z \in \bar{\Omega} \setminus \{P\}$.
   One may note that, in fact, $\mathcal{P}$ is defined on $\Omega_{\epsilon/2} \times U_\delta$.

Proof   Examine Lemmas 5.2.9 and 5.2.10. Now u_P and $\eta(|z - P|) \log L_P(z)$ both have bounded imaginary parts on $\Omega_{\epsilon/2}$. Therefore, there is a small constant m such that $m\{\operatorname{Im} u_P + \operatorname{Im} [\eta(|z - P|) \log L_P(z)]\} > 0$ and

$$m \left\{ |\operatorname{Im} u_P| + |\operatorname{Im} [\eta(|z - P|) \log L_P(z)]| \right\} < \frac{\pi}{2}.$$

Let

$$\Psi(z, P) = \exp\left\{mu_P(z) + m\eta(|z - P|)\log L_P(z)\right\}, \quad z \in \Omega_{\epsilon/2}.$$

Then $\Psi(P, P) = 0$. Moreover $\Psi(\cdot, P)$ is holomorphic on $\Omega_{\epsilon/2}$. Finally, $\Psi(z, \cdot)$ is C^k just as in Proposition 5.2.14. Therefore

$$\mathcal{P}(z, P) \equiv \exp\left\{-\Psi(z, P)\right\}$$

has all the desired properties.   □

Exercises for the Reader

1. Here is an important alternate approach to the Henkin function $\Phi(z, P)$ that was developed by John Fornæss (see J. E. Fornæss [1]). Let $\Omega \subset\subset \mathbb{C}^n$ be a strongly *convex* domain with C^2 boundary and $\rho : \mathbb{C}^n \to \mathbb{R}$ a C^2 defining function. Then

$$S(z, P) = \sum_{j=1}^{n} \frac{\partial \rho}{\partial z_j}(P)(z_j - P_j)$$

has all the properties of the Levi polynomial *and* of the Henkin function Φ. Moreover, the function $\mathcal{P}(z, P) = \exp\{-S(z, P)\}$ gives the continuously varying peaking functions of Theorem 5.2.15.

2. Now apply the Fornæss imbedding theorem (Theorem 3.2.3) to see that Φ may be obtained for any strongly pseudoconvex domain from the S in the preceding paragraph.

We went to considerable extra trouble in the preceding paragraphs to ensure that peaking functions not only exist on strongly pseudoconvex domains, but they vary with the boundary point in a regular fashion. In light of this, the following result is of some interest. Let X be a compact metric space, and let $\mathcal{A}$ be a closed subalgebra of $C(X)$. Call $P \in X$ a *peak point* for $\mathcal{A}$ if there is an $f \in \mathcal{A}$ such that $f(P) = 1$ while $|f(x)| < 1$ for all $x \in X \setminus \{P\}$. Let $\mathcal{P}$ be the set of all peak points for $\mathcal{A}$. Then we have the following.

THEOREM 5.2.16 (Fornæss/Krantz [1])   With notation as above, there is a continuous function $\Phi : \mathcal{P} \to \mathcal{A}$ such that $\Phi(P)$ peaks at P for every $P \in \mathcal{P}$.

So the fact that peak functions vary continuously comes "for free." Under what circumstances the *smooth* varying of the peak functions comes for free is not well understood at this writing.

5.3 Hefer's Lemma and Henkin's Integral Representation

Let $\Omega \subseteq \mathbb{C}$ and $z^0 \in \Omega$. We know that when $f : \Omega \to \mathbb{C}$ is holomorphic, $f(z^0) = 0$, and f is not identically zero, then the zero of f can be factored out:

$$f(z) = (z - z^0)^k \cdot \tilde{f}(z)$$

for some $k \in \{0, 1, \ldots\}$. It is clear that no such simple factorization can be performed in $\mathbb{C}^n$; the simple example consisting of the family of functions $F^w(z) = \sum_{j=1}^n z_j w_j, w \in \mathbb{C}^n$, shows that there is no *single* polynomial factor that will contain the zeros of all holomorphic functions that vanish at $z^0 = 0$. (After all, the zero set of a holomorphic polynomial is, generically, an $(n-1)$-dimensional complex variety; we shall prove this assertion in detail in Chapter 7.) The correct alternative has been suggested in the material on factoring holomorphic functions in Section 0.3. We now prove the result stated there without the sup norm estimates. The sup norm estimates are derived in Chapter 10 when we have some additional machinery available.

PROPOSITION 5.3.1 Let $\Omega \subseteq \mathbb{C}^n$ be pseudoconvex. Let $\Omega_k = \Omega \cap \{z \in \mathbb{C}^n : z_1, \ldots, z_k = 0\}, k = 1, \ldots, n$. Let $A_k(\Omega) = \{f \text{ holomorphic on } \Omega : f|_{\Omega_k} = 0\}$. Then there are linear operators

$$Q_i^k : A_k(\Omega) \to \{f \text{ holomorphic on } \Omega\}, \quad i = 1, \ldots, k,$$

such that

$$f(z) = \sum_{i=1}^k z_i \cdot (Q_i^k f)(z)$$

for all $f \in A_k(\Omega)$.

Remark: We are primarily interested in the proposition when $k = n$. However, the proof is by induction on k. We shall give a classical proof using Theorem 5.1.1. In Chapter 6 we shall give another proof using sheaf cohomology.   □

Proof of the Proposition If $k = 1$ and n is arbitrary, then the result follows by setting $Q_1 f(z) = f(z)/z_1$. Now suppose that the result has been proved for $k = K - 1$ and for any n.

Let $\tilde{\Omega} = \{z \in \Omega : z_K = 0\}$. Let $f \in A_K(\Omega)$. Then $\tilde{f} \equiv f|_{\tilde{\Omega}} \in A_{K-1}(\tilde{\Omega})$. Therefore, by the inductive hypothesis,

$$\tilde{f}(z) = \sum_{i=1}^{K-1} z_i \cdot \left(\tilde{Q}_i^{K-1}\tilde{f}\right)(z), \quad z \in \tilde{\Omega},$$

where $\tilde{Q}_i^{K-1}$ are the operators assumed to exist on $\tilde{\Omega}$ for $K - 1, n - 1$.

Now we apply Theorem 5.1.1. Indeed, we let

$$Q_K^K(f)(z) = \frac{f(z) - \sum_{i=1}^{K-1} z_i \left(\mathcal{E}_{\tilde{\Omega},\Omega} \tilde{Q}_i^{K-1} \tilde{f}(z) \right)}{z_K}.$$

This is well defined and holomorphic on Ω, since the expression in the numerator vanishes on $\tilde{\Omega}$. Also, let

$$Q_i^K f(z) = \mathcal{E}_{\tilde{\Omega},\Omega} \tilde{Q}_i^{K-1} \tilde{f}(z), \quad i = 1, \ldots, K - 1.$$

By algebra, $f(z) = \sum_{i=1}^{K} z_i \cdot Q_i^K f(z)$, all $z \in \Omega$. The induction is now complete. □

COROLLARY 5.3.2   Let $\Omega \subseteq \mathbb{C}^n$ be pseudoconvex. Then there are continuous linear operators

$$T_i : \{\text{holomorphic functions on } \Omega\} \to \{\text{holomorphic functions on } \Omega \times \Omega\}$$

such that for any holomorphic $f : \Omega \to \mathbb{C}$ we have

$$f(z) - f(w) = \sum_{i=1}^{n} (z_i - w_i) T_i f(z, w), \quad \text{all } z, w \in \Omega.$$

Proof   Apply Proposition 5.3.1 to the function $F(z, w) = f(z) - f(w)$ on the domain $\Omega \times \Omega$ with coordinates

$$\begin{aligned}
z_1' &= z_1 - w_1 \\
&\vdots \\
z_n' &= z_n - w_n \\
z_{n+1}' &= z_1 \\
&\vdots \\
z_{2n}' &= z_n.
\end{aligned}$$

The continuity will follow from the closed graph theorem. □

PROPOSITION 5.3.3 (Hefer's Lemma)   Let $\Omega \subseteq \mathbb{C}^n$ be strongly pseudoconvex with C^4 boundary. Let $\Phi : \Omega_{\epsilon/2} \times \partial\Omega \to \mathbb{C}$ be the C^1 singular function constructed in Section 5.2. Then we may write

$$\Phi(z,\zeta) = \sum_{i=1}^{n} (\zeta_i - z_i) \cdot P_i(z,\zeta), \quad z \in \Omega_{\epsilon/2}, \zeta \in \partial\Omega,$$

where each P_i is holomorphic in $z \in \Omega_{\epsilon/2}$ and C^1 in $\zeta \in \partial\Omega$.

Proof   Fix $\zeta \in \partial\Omega$. Apply Corollary 5.3.2 to the function $\Phi_\zeta(\cdot) = \Phi(\cdot,\zeta)$ on $\Omega_{\epsilon/2}$. So

$$\Phi(z,\zeta) - \Phi(w,\zeta) = \sum_{i=1}^{n} (z_i - w_i)[(T_i\Phi_\zeta)(z,w)].$$

Since this is true for all $w \in \Omega_{\epsilon/2}$, we may set $w = \zeta \in \partial\Omega$ to obtain

$$\Phi(z,\zeta) = \sum_{i=1}^{n} (z_i - \zeta_i)\left[(T_i\Phi_\zeta)(z,\zeta)\right] \equiv \sum_{i=1}^{n} (\zeta_i - z_i)P_i(z,\zeta).$$

It remains to check that P_i is C^1 in ζ. For this, it is enough to verify that $(T_i\Phi_\zeta)(z,w)$ is C^1 in ζ. But, just as in the proof of Proposition 5.2.14, we let $e_{(z,w)}$ be the point evaluation functional on $\Omega \times \Omega$ and observe that

$$(T_i\Phi_\zeta)(z,w) = \langle e_{(z,w)}, T_i\Phi_\zeta \rangle = \langle T_i^* e_{(z,w)}, \Phi_\zeta \rangle.$$

The last expression is C^1 in ζ by Proposition 5.2.14. $\square$

THEOREM 5.3.4 (Henkin [2])   Let $\Omega \subseteq \mathbb{C}^n$ be a strongly pseudoconvex domain with C^4 boundary. Let $\Phi : \Omega_{\epsilon/2} \times \partial\Omega \to \mathbb{C}$ be the Henkin singular function. Define

$$w_i(z,\zeta) = \frac{P_i(z,\zeta)}{\Phi(z,\zeta)}, \qquad i = 1,\dots,n.$$

Here $P_i(z,\zeta)$ are as in Proposition 5.3.3. Just as in Section 1.1, let

$$\eta(w) = \sum_{i=1}^{n} (-1)^{i+1} w_i dw_1 \wedge \cdots \wedge dw_{i-1} \wedge dw_{i+1} \wedge \cdots \wedge dw_n$$

and

$$\omega(\zeta) = d\zeta_1 \wedge \cdots \wedge d\zeta_n.$$

Then for any $f \in C^1(\bar{\Omega}) \cap \{\text{holomorphic functions on } \Omega\}$, we have the integral representation

$$f(z) = \int_{\partial\Omega} f(\zeta)\eta(w) \wedge \omega(\zeta).$$

Proof   The functions w_i satisfy

$$\sum_{i=1}^{n} w_i(z,\zeta)(\zeta_i - z_i) \equiv 1, \quad z \in \Omega, \quad \zeta \in \partial\Omega.$$

Now apply the Cauchy-Fantappiè formula (Exercise 37 at the end of Chapter 1).   □

COROLLARY 5.3.5   With notation as in the Theorem, we have

$$f(z) = \int_{\partial\Omega} f(\zeta) \frac{K(z,\zeta)}{\Phi^n(z,\zeta)} d\sigma(\zeta), \qquad (5.3.5.1)$$

where $K : \Omega_{\epsilon/2} \times \partial\Omega$ is holomorphic in z and continuous in ζ. In fact, $K(z,\zeta)d\sigma(\zeta) = \eta(z) \wedge \omega(\zeta)$.

Proof   Apply Exercise 36 at the end of Chapter 1.   □

Remark: A thorough treatment of the Cauchy-Fantappiè formalism and of integral formulas appears in R. M. Range [3].   □

It is straightforward but tedious to see, by a limiting argument, that both the theorem and the corollary may be extended to functions f that are only continuous on $\bar{\Omega}$ and holomorphic on Ω.

What is crucial about formula (5.3.5.1) in the corollary is that the right-hand side is holomorphic no matter what f is put into it. Indeed,

$$\int_{\partial\Omega} \frac{K(z,\zeta)}{\Phi^n(z,\zeta)} d\mu(\zeta)$$

is holomorphic in z for any finite Borel measure μ. In this respect the Henkin kernel is very much like the classical Cauchy kernel of one variable (except that the kernel *depends on the domain*). The property of creating, as well as reproducing, holomorphic functions distinguishes Henkin's kernel from that of Bochner-Martinelli. We exploit this distinction decisively in Chapter 10 when we construct Henkin's solution to the $\bar{\partial}$-problem.

The problem of constructing integral formulas for domains of various kinds has become a small industry. For domains of general type, techniques are as yet unavailable for producing integral formulas of the type described above (however, work of Gleason [2] and Bungart [1] guarantees abstractly that such integral formulas exist). Since the existence of separating functions is closely related to the existence of the Henkin Φ function, we realize that a function Φ as constructed above *cannot exist* on the Kohn-Nirenberg domain. In spite of this, J. E. Fornæss [4] has produced (with an ingenious argument) an alternate type of integral formula on a collection of domains that includes the Kohn-Nirenberg domain.

In an arbitrary function algebra, any multiplicative homomorphism (such as point evaluation at an interior point) can be represented by integration agains a complex measure supported on the Šilov boundary (see T. W. Gamelin [1, 2]). In the case of a smoothly bounded domain Ω, the Šilov boundary for $A(\Omega)$ is contained in the closure of the strongly pseudoconvex points in $\partial\Omega$ (see H. Rossi [1]). In general, this could be a proper subset of the boundary. (However for a bounded pseudoconvex domain with real analytic boundary or a domain of finite type—see Section 11.5—the closure of the strongly pseudoconvex points is indeed the entire boundary.) It would require new ideas to generate a useful integral integral kernel supported on only a part of the boundary (although for heartening progress see Hatziafratis [1] and Bonneau [1]).

On the other hand, for domains with sufficient symmetry, there is a rich theory. For a domain with a transitive group of biholomorphic mappings, one has an integral formula (at least in principle) as soon as one has written down a single representing measure for a single point (why?). However (see Section 11.3), the only strongly pseudoconvex domain with transitive automorphism group is the ball. So there is no simple way to generate integral formulas on all strongly pseudoconvex domains.

Some complete circular domains have nice integral formulas (B. A. Fuks [1]). Furthermore, a Weil domain (a sort of analytic polyhedron) will have integral formulas (see G. M. Henkin [1]). Integral formulas for certain domains with only piecewise smooth boundaries have been obtained (see, for example, R. M. Range and Y. T. Siu [1]).

We conclude this section with a simple, though technical, application of Corollary 5.3.5.

THEOREM 5.3.6 (Localization of Singularities)   Let $\Omega \subset\subset \mathbb{C}^n$ be a strongly pseudoconvex domain with C^4 boundary. Let $\{U_j\}_{j=1}^{\infty}$ be open sets in $\mathbb{C}^n$ that cover $\partial\Omega$. Let f be holomorphic on Ω and and continuous on $\bar{\Omega}$. Then the function f may be written $f = f_1 + \cdots + f_k$, where each f_j is holomorphic on a neighborhood of $\overline{\Omega \setminus U_j}$.

Proof   Fix $\{\phi_j\}_{j=1}^{\infty}$, a partition of unity on $\partial\Omega$ subordinate to $\{U_j\}$. By the corollary and the remark following it, we may write

$$
\begin{aligned}
f(z) &= \int_{\partial\Omega} f(\zeta) \frac{K(z,\zeta)}{\Phi^n(z,\zeta)} \, d\sigma(\zeta) \\
&= \sum_{j=1}^{k} \int_{\partial\Omega} [\phi_j(\zeta)f(\zeta)] \frac{K(z,\zeta)}{\Phi^n(z,\zeta)} \, d\sigma(\zeta) \\
&\equiv \sum_{j=1}^{k} f_j(z).
\end{aligned}
$$

Since $K(\cdot,\zeta), \Phi(\cdot,\zeta)$ are holomorphic on a neighborhood of $\bar{\Omega}$ and $\Phi(\cdot,\zeta)$ vanishes only at ζ, the result follows. $\qquad\qquad\square$

5.4   Approximation Problems

The basic approximation theorems of one complex variable are striking in their elegance and universality. Runge's theorem allows one to approximate a function holomorphic in a neighborhood of a compact set by a rational function with poles in the complement. Mergelyan's theorem allows approximation of a function continuous on a compact set and holomorphic on its interior by *polynomials* provided only that the complement of the compact set has just one connected component. In both cases the hypotheses are purely topological. Such simplicity does not obtain in several complex variables, as the following example shows.

EXAMPLE   Let $\Omega = B \subseteq \mathbb{C}^n$. Let $U = \{(z_1, z_2) : \frac{1}{4} < |z_1| < \frac{1}{2}, |z_2| < \frac{1}{2}\}$. Notice that Ω is a domain of holomorphy and that U is also a domain of holomorphy, since it is a product domain. Let $f(z_1, z_2) = 1/z_1$. Then f is holomorphic on a neighborhood of U. However, it is not possible to approximate f uniformly on U by F holomorphic on Ω.

For suppose that $F : \Omega \to \mathbb{C}$ with $\sup_U |F - f| < \frac{1}{2}$. Then

$$
1 = \left| \frac{1}{2\pi i} \int_{|z_1|=3/8} f(z_1, 0) - F(z_1, 0) dz_1 \right| \leq \frac{1}{2}
$$

which is a contradiction. It is worth noting that if we adjoin to U a single disc—say we replace U by $U' = U \cup \{(z_1, z_2) : |z_1| \leq \frac{1}{2}, |z_2| = 0\}$—then the example fails by Hartogs's phenomenon. In fact, it *will* be the case that functions holomorphic on a neighborhood of U' *can* be approximated on U' by functions holomorphic on Ω. This will also be the case for the set $U'' = \bar{B}(0, \frac{1}{2}) \setminus B(0, \frac{1}{4})$.

Finally, notice that, since U is pseudoconvex, we can (by part 11 of 3.3.5) write U as an increasing union of strongly pseudoconvex domains U_j. For j sufficiently large, U_j will be homeomorphic to U, and the approximation property will not hold for U_j. Thus we have obtained a smooth, strongly pseudoconvex counterexample.

So something other than topology alone or pseudoconvexity alone is controlling this problem. $\square$

In fact the "Runge approximation property" in several complex variables is not completely understood. However, the machinery of Hörmander, which we used to solve the $\bar{\partial}$ problem, gives a nice method for deriving an approximation theorem. To begin, we need an easy Hilbert space lemma.

LEMMA 5.4.1 Let $\mathcal{H}^1, \mathcal{H}^2$ be Hilbert spaces and $T : \mathcal{H}^1 \to \mathcal{H}^2$ a linear operator that is closed and densely defined. Suppose that $F \subseteq \mathcal{H}^2$ is a closed subspace such that $F \supseteq \operatorname{Range} T$. If

$$\|f\|_{\mathcal{H}^2} \le C_0 \|T^* f\|_{\mathcal{H}^1}, \quad \text{all} \ f \in \mathcal{D}_{T^*} \cap F, \qquad (5.4.1.1)$$

then for every $v \in (\operatorname{Ker} T)^{\perp} \subseteq \mathcal{H}^1$ there is an $f \in \mathcal{D}_{T^*}$ that satisfies $T^* f = v$ and

$$\|f\|_{\mathcal{H}^2} \le C_0 \|v\|_{\mathcal{H}^1}.$$

It should be noted that the constant C_0 in the conclusion is the same as the constant C_0 in the hypothesis.

Proof We know that v is in the closure of $\operatorname{Range} T^*$. Likewise, $F^{\perp} = \operatorname{Ker} T^*$. Hence

$$T^* : F \cap \mathcal{D}_{T^*} \to \operatorname{Range} T^*$$

has the same range as $T^* : \mathcal{D}_{T^*} \to \operatorname{Range} T^*$. Since equation (5.4.1.1) says that this operator has closed range, we find that there must be an element $f \in F \cap \mathcal{D}_{T^*}$ such that $T^* f = v$. The desired estimate follows now from (5.4.1.1). $\square$

THEOREM 5.4.2 Let $\Omega \subseteq \mathbb{C}^n$ be pseudoconvex. Let p be a strictly psh exhaustion function for Ω. For $c \in \mathbb{R}$ we set $K_c = \{z \in \Omega : p(z) \le c\} \subset\subset \Omega$. If f is holomorphic on a neighborhood of K_c and if $\epsilon > 0$, then there is an F holomorphic on Ω such that $\|f - F\|_{L^2(K_c)} < \epsilon$.

Proof By the Hahn-Banach theorem, it is necessary and sufficient to show that if μ is a linear functional on $L^2(K_c)$ that annihilates the restrictions to K_c of all holomorphic functions on Ω, then it annihilates all functions that are holomorphic only in a neighborhood of K_c.

All such functionals are given by integration against some $\bar{v} \, dV, v \in L^2(K_c)$. Fix one such. We may assume that $c = 0$ and we drop the subscript. Define

v to be equal to 0 on $\Omega \setminus K$. Using the language of Chapter 4, we let $F = \operatorname{Ker} S = \operatorname{Range} T$, so that inequality (5.4.1.1) is satisfied. Assume without loss of generality that $\|v\| = 1$.

Our hypothesis says that $e^{\phi_1} v \in (\operatorname{Ker} T)^{\perp} \subseteq L^2(\Omega, \phi_1)$. By Lemma 5.4.1, there is an $f \in \mathcal{D}_{T^*}$ with $T^* f = e^{\phi_1} v$ and $\|f\|_{\phi_2} \leq C_0 \|e^{\phi_1} v\|_{\phi_1}$. Write $f = \sum_j f_j d\bar{z}_j$. Then Lemma 4.4.11 yields that

$$e^{\phi_1} v = -e^{\phi_1} \sum_j \frac{\partial(e^{-\phi_2} f_j)}{\partial z_j}$$

or

$$v = \delta(e^{-\phi_2} f)$$

where δ is the formal adjoint of the operator $\bar{\partial} : L^2(\Omega) \to L^2_{(0,1)}(\Omega)$ (exercise). Denote $e^{-\phi_2} f$ by g.

The construction we have just performed is valid for any $\phi_1 = 2\psi - \phi, \phi_2 = \psi - \phi$. Here ψ is as in Lemma 4.4.5 and ϕ is a strictly psh exhaustion function, $\phi = \chi \circ p$ as in Lemma 4.5.1. Examine the proof of Lemma 4.5.1. We may suppose that χ strictly increases to ∞ and that $\chi(0) = 0$. Let $\eta : \mathbb{R} \to \mathbb{R}$ be C^{∞}, identically 1 for $x \leq 0$ and strictly convex and strictly increasing to ∞ for $x > 0$. Let $\chi_n(x) = \eta^n(x) \cdot \chi(x)$. Then $\chi_n(x) = \chi_m(x)$, all $x \leq 0$, all $n, m \geq 1$. Also $\chi_{n+1}(x) \geq \chi_n(x)$, all $x \geq 0$, all $n \geq 1$. Finally, for any $x > 0$, we have $\chi_n(x) \nearrow \infty$, as $n \to \infty$. We may repeat the construction of the preceding paragraph for ϕ_1^n, ϕ_2^n manufactured from each of these χ_n. Thus we obtain for each n a $(0,1)$ form g_n with $v = \delta(g_n)$ and

$$\left\| g_n e^{\phi_2^n} \right\|_{\phi_2^n} \leq C_0 \left\| e^{\phi_1^n} v \right\|_{\phi_1^n}.$$

It is important to note that the right-hand side of this inequality does not depend on n. For v is supported on the set where all the ϕ_1^n agree (remember that $K = \{z : p(z) \leq 0\}$). Also, the constant C_0 comes from Lemma 5.4.1, which is the same constant as the one in Lemma 4.2.9. In our application of Lemma 4.2.9 (namely, Theorem 4.5.2) we showed that this constant can be taken to be 1 provided only that χ is large enough. So we have

$$\delta g_n = v, \quad \|g_n e^{\phi_2^n}\|_{\phi_2^n} \leq 1, \quad n = 1, 2, \ldots .$$

Since $L^2(\Omega, \phi_2^n) \subseteq L^2(\Omega, \phi_2^1)$, all $n \geq 1$, we have that $\{g_n\} \subseteq L^2_{(0,1)}(\Omega, \phi_2^1)$ is a bounded subset. Let g_0 be a weak accumulation point. So some subsequence $g_{n_j} \to g_0$ weakly. It follows that

$$\left\| g_0 e^{\phi_2^n} \right\|_{\phi_2^n} \leq 1, \quad n = 1, 2, \ldots . \tag{5.4.2.1}$$

But $\phi_2^n \nearrow \infty$ off K; hence $g_0 = 0$ a.e. off K (otherwise (5.4.2.1) could not hold). Let $\mu \in C_c^\infty(\Omega)$. The assertion that $g_{n_j} \to g_0$ weakly implies that $\delta g_{n_j} \to \delta g_0$ in the weak sense; that is,

$$\int \mu \bar{v} \, dV = \int \mu \overline{\delta g_{n_j}} \, dV \to \int \mu \overline{\delta g_0} \, dV = \int \left(\sum_1^n \frac{\partial \mu}{\partial \bar{z}_k} \cdot (\overline{g_0})_k \right) dV. \quad (5.4.2.2)$$

Since g_0, v are supported on K, this identity persists for any μ that is C^∞ on a neighborhood of K. In particular, it holds for any μ holomorphic on a neighborhood of K. For such a μ, the right-hand side of (5.4.2.2) is 0, as we set out to prove.   □

COROLLARY 5.4.3   Let $\Omega \subseteq \mathbb{C}^n$ be pseudoconvex, and suppose that K is a compact subset of Ω. Assume that $\hat{K}_{P(\Omega)} = K$ (here $P(\Omega)$ is the space of psh functions on Ω). Then any function holomorphic on a neighborhood of K can be approximated uniformly on K by functions holomorphic on Ω.

Proof   Let $K \subset\subset \omega \subset\subset \Omega$ with ω open. By the exercises at the end of Chapter 3, there is a strictly psh exhaustion function p on Ω such that $p < 0$ on K and $p > 0$ off ω. If f is holomorphic on ω, then f can be approximated in L^2 norm on $K_0 \equiv \{z : p(z) < 0\}$ and $K \subset\subset K_0$. By Lemma 1.4.1, f can therefore be approximated uniformly on K.   □

Remark: Notice that the last result does not contradict the example at the beginning of the section. For the set K in the example does not equal its hull with respect to $P(\Omega)$.

When $K \subset\subset \Omega$ satisfies the conclusions of Corollary 5.4.3, then we say that K satisfies the Runge property with respect to Ω (more concisely, K is *Runge* in Ω).   □

EXERCISES

1. Prove that the following four assertions are equivalent for pseudoconvex domains $\Omega_1 \subset\subset \Omega_2 \subseteq \mathbb{C}^n$ (here $\mathcal{F}(\Omega)$ denotes the space of holomorphic functions on Ω).

    a. If $K \subset\subset \Omega_1$ and $f \in \mathcal{F}(\Omega_1)$, then f can be approximated, uniformly on K, by elements of $\mathcal{F}(\Omega_2)$.

    b. If $K \subset\subset \Omega_1$, we have $\hat{K}_{\mathcal{F}(\Omega_1)} = \hat{K}_{\mathcal{F}(\Omega_2)}$.

    c. If $K \subset\subset \Omega_1$, we have $\hat{K}_{\mathcal{F}(\Omega_2)} \cap \Omega_1 = \hat{K}_{\mathcal{F}(\Omega_1)}$.

    d. If $K \subset\subset \Omega_1$, we have $\hat{K}_{\mathcal{F}(\Omega_2)} \cap \Omega_1 \subset\subset \Omega_1$.

(*Hint:* The difficult step is (d)$\Rightarrow$(a). Use the Runge approximation property on the function

$$\tilde{f}(z) = \begin{cases} f & \text{if} \quad z \in \hat{K}_{\mathcal{F}(\Omega_2)} \cap \Omega_1, \\ 1 & \text{if} \quad z \in \hat{K}_{\mathcal{F}(\Omega_2)} \setminus \Omega_1.) \end{cases}$$

2. Let $P(\Omega)$ be the psh functions on Ω and $\mathcal{F}(\Omega)$ be the holomorphic functions on Ω. Assume that $\Omega \subseteq \mathbb{C}^n$ is pseudoconvex. If $K \subset\subset \Omega$, then prove that

$$\hat{K}_{P(\Omega)} = \hat{K}_{\mathcal{F}(\Omega)}.$$

3. A *Hartogs domain* in $\mathbb{C}^n$ is a domain of the form

$$\{(z_1, \ldots, z_n) : |z_n| < |f(z_1, \ldots, z_{n-1})|\},$$

where f is holomorphic. Prove that a Hartogs domain is a domain of holomorphy.

4. When doing function theory on manifolds, it is useful to have a coordinate-free definition of the Levi form. Let $\Omega \subseteq \mathbb{C}^n$ be a domain with C^2 boundary and ρ a C^2 defining function. Let $P \in \partial\Omega$. We say that $L = \sum_j a_j(\partial/\partial z_j)$ is in $T_{1,0}^P(\partial\Omega)$ if $(L\rho)(P) = 0$. If $L, M \in T_{1,0}^P(\partial\Omega)$, define the Levi form $\mathcal{L}$ by

$$\mathcal{L}(L, \bar{M})\big|_P = \langle \partial\bar{\partial}\rho, L \wedge \bar{M}\rangle\big|_P,$$

where $\langle\ ,\ \rangle$ represents the pairing between vectors and covectors (see Appendix IV). Show that this definition is consistent with the definition in the text. Show that the rank of $\mathcal{L}$ at a point and the dimensions of the subspaces of $T_{1,0}^P(\partial\Omega)$ on which $\mathcal{L}$ is positive definite or negative definite are independent of the choice of ρ.

5. Let notation be as in Exercise 4. Let $L, M \in T_{1,0}^P(\partial\Omega)$. Extend L, M to smooth vector fields in a neigborhood of P; denote these vector fields by $\tilde{L}, \tilde{M}$, respectively. Then verify that the formula

$$\mathcal{L}(L, \bar{M}) = \langle [\tilde{L}, \overline{\tilde{M}}], \partial\rho\rangle \tag{$*$}$$

where $[\tilde{L}, \overline{\tilde{M}}] = \tilde{L}\overline{\tilde{M}} - \overline{\tilde{M}}\tilde{L}$ and $\langle\ ,\ \rangle$ represents the pairing between vectors and covectors. (*Hint:* First verify Cartan's formula

$$\langle A \wedge B, d\theta\rangle = A\langle B, \theta\rangle - B\langle A, \theta\rangle - \langle [A, B], \theta\rangle, \tag{$**$}$$

where A, B are 1-vectors and θ is a 1-covector: do so by introducing local coordinates. Formula $(**)$ is often used to give a coordinate-free definition of the exterior differential d.) Verify that this formula is independent of the choice of $\tilde{L}, \tilde{M}$.

Suppose that $\partial\Omega$ is strongly pseudoconvex at $P \in \partial\Omega$. What does $(*)$ say about the normal component of $[\tilde{L}, \overline{\tilde{M}}]$?

6. Let $\Omega \subset\subset \mathbb{C}^n$ be strongly pseudoconvex with C^4 boundary. For the sake of this problem, it is useful to define (as we did in Section 1.5) $H^2(\partial\Omega)$ to be the L^2 closure of the restriction to $\partial\Omega$ of all elements of $A(\Omega)$. (For smooth pseudoconvex domains this is equivalent to the definition in Chapter 8—see F. Beatrous [1].) Let $\phi \in L^\infty(\partial\Omega)$. Let S be the Szegö projection for Ω. The *Toeplitz operator* with symbol ϕ is the operator

$$T_\phi : H^2(\partial\Omega) \ni f \mapsto S(\phi f) \in H^2(\partial\Omega).$$

Since S is a projection on L^2, T_ϕ is well defined. Many function-theoretic problems may be studied by way of Toeplitz operators (see, e.g., A. M. Davie and N. Jewell [1], N. Jewell and S. G. Krantz [1], and R. Douglas [1]). Fill in the following outline to prove that $\|T_\phi\|_{(L^2,L^2)} = \|\phi\|_\infty$.

a. Trivially, $\|T_\phi\|_{(L^2,L^2)} \le \|\phi\|_\infty$.
   (Steps b through e prove that T_ϕ invertible implies ϕ is invertible in L^∞.)

b. If T_ϕ is invertible, then for some $\epsilon > 0$,

$$\|\phi f\|_2 \ge \epsilon\|f\|_2, \quad \text{all } f \in H^2(\partial\Omega).$$

c. Let $\mathcal{P}(z, P)$ be the continuously varying peaking function of Theorem 5.2.15. Apply (b) to the function

$$f(z) = [\mathcal{P}(z, P)]^k, \quad k \in \mathbb{N}, P \in \partial\Omega \text{ fixed.}$$

Obtain

$$\int_{\partial\Omega} |\mathcal{P}(z,P)|^{2k}|\phi(z)|^2\,d\sigma(z) \ge \epsilon^2 \int_{\partial\Omega} |\mathcal{P}(z,P)|^{2k}\,d\sigma(z).$$

d. If $g \in C(\partial\Omega), g \ge 0$, is arbitrary, then integrate both sides of (c) against g and apply Fubini to obtain

$$\frac{\int_{\partial\Omega} |\mathcal{P}(z,P)|^{2k}g(P)\,d\sigma(P)|\phi(z)|^2\,d\sigma(z)}{\int_{\partial\Omega} |\mathcal{P}(z,P)|^{2k}\,d\sigma(P)}$$

$$\ge \epsilon^2 \frac{\int_{\partial\Omega} |\mathcal{P}(z,P)|^{2k}g(P)\,d\sigma(P)\,d\sigma(z)}{\int_{\partial\Omega} |\mathcal{P}(z,P)|^{2k}\,d\sigma(P)}.$$

e. Let $k \to \infty$ and obtain

$$\int_{\partial\Omega} |\phi(z)|^2 g(z) \, d\sigma(z) \geq \epsilon^2 \int_{\partial\Omega} g(z) \, d\sigma(z),$$

whence $|\phi| \geq \epsilon$ a.e.; therefore, ϕ is invertible in L^∞.

f. Conclude that $\|T_\phi\|_{(L^2, L^2)} = \|\phi\|_\infty$. (*Hint:* Look at $\lambda I - T_\phi$ and $\lambda - \phi$, $\lambda \in \mathbb{C}$.)

g. Conclude that the spectral radius of T_ϕ is $\|\phi\|_\infty$ (use the spectral radius theorem—W. Rudin [9]). More precisely, the spectrum of T_ϕ equals the essential range of ϕ.

h. Where was the continuity of $\mathcal{P}(z, P)$ used?

Not a great deal is known about Toeplitz operators of several complex variables. Perhaps two of the deepest pieces of work are U. Venugopalkrishna [1] and L. Boutet de Monvel [1]. Many basic questions remain open.

7. *Peak sets and peak interpolation sets.* The results of this problem are true on strictly pseudoconvex domains (some of them are true in still greater generality). We give some hints for the proof on the ball in $\mathbb{C}^2$, some further hints for the proof on a strongly *convex* domain, and invite the reader to fill in as many details as desired and to apply the Fornæss imbedding theorem to extend the result to the strictly pseudoconvex case (see W. Rudin [5] for further details).

If $P \in \partial B \subseteq \mathbb{C}^2, P = (z_1, z_2) \approx (x_1, y_1, x_2, y_2)$, let ν_P be the vector $(x_1, y_1, x_2,$
$y_2) \approx (x_1 + iy_1, x_2 + iy_2)$ and η_P the vector $\langle -y_1, x_1, -y_2, x_2 \rangle \approx (-y_1 + ix_1, -y_2 + ix_2)$. Throughout this problem, $\langle z, w \rangle$ denotes the inner product of complex vectors that is given by $\sum z_j \bar{w}_j$.

Let $\gamma : [0, 1] \to \partial B$ be a C^1 curve. We assume that γ is complex tangential, that is,

$$\langle \dot{\gamma}(t), \eta_{\gamma(t)} \rangle \equiv \sum_{j=1}^{n} \dot{\gamma}_j(t) (\overline{\eta_{\gamma(t)}}) = 0$$

for all $t \in [0, 1]$. Abusing notation, we also let the symbol γ denote $\{\gamma(t) : 0 \leq t \leq 1\}$. Let $f : \gamma \to \mathbb{C}$ be continuous. We may assume, after reparametrization, that $|\dot{\gamma}(t)| = 1$, all t. Define

$$
\begin{aligned}
g(t) &= \int_{-\infty}^{\infty} \left(1 + \frac{1}{2}|\dot{\gamma}(t)s|^2\right)^{-1} ds \\
&= \frac{\pi\sqrt{2}}{|\dot{\gamma}(t)|} \\
&= \pi\sqrt{2}, \quad 0 \leq t \leq 1.
\end{aligned}
$$

Let $\delta > 0$ and, noting that $\mathrm{Re}\langle \gamma(t) - z, \eta_{\gamma(t)}\rangle \geq 0$, let

$$h_\delta(z) = \frac{1}{\pi\sqrt{2}} \int_0^1 \frac{\delta f(t)\, dt}{\delta^2 + \langle \gamma(t) - z, \eta_{\gamma(t)}\rangle}.$$

a. Prove that $|h_\delta(z)| \leq C$, all $0 < \delta < 1$, all $z \in \bar{B}$, and $h_\delta \in A(B)$.
b. Prove that $\lim_{\delta \to 0^+} h_\delta(z) = 0, z \in \bar{D} \setminus \gamma$.
c. Prove that $\lim_{\delta \to 0^+} h_\delta(\gamma(s)) = f(\gamma(s))$, all $0 \leq s \leq 1$.
d. If μ is a Borel measure on $\partial B, \mu \perp A(B)$, conclude from (a), (b), and (c) that $\mu|_\gamma = 0$.
e. Apply E. Bishop's peak interpolation theorem (T. W. Gamelin [1]) to conclude that there is an $F \in A(B)$ such that $F|_\gamma = f$ and $|F(z)| < \sup_\gamma |f|$, all $z \in \bar{B} \setminus \gamma$.

Assume now that B is replaced by $\Omega \subset\subset \mathbb{C}^2$ strongly convex with C^2 boundary. Let ρ be a C^2 defining function for Ω. Replace ν_P by $((\partial\rho/\partial\bar{z}_1)(P), \ldots, (\partial\rho/\partial\bar{z}_n)(P)), \eta_P$ by $i\nu_P$.
f. Verify that

$$2\,\mathrm{Re}\langle w - z, \nu_w \rangle \geq C|w - z|^2,$$

all $w \in \partial\Omega$, all $z \in \bar{\Omega}$.
g. Let $\Psi(t) = \nu_{\gamma(t)}, 0 \leq t \leq 1$. Then

$$\langle \dot{\gamma}(t), \dot{\Psi}(t)\rangle = Q_{\gamma(t)}\left(\dot{\gamma}(t)\right), \quad 0 \leq t \leq 1,$$

where

$$Q_P(w) = \sum_{j,k} \frac{\partial^2\rho}{\partial z_j \partial z_k}(P)w_j w_k + \sum_{j,k} \frac{\partial^2\rho}{\partial z_j \partial \bar{z}_k}(P)w_j \bar{w}_k,$$

all $P \in \partial\Omega$, all $w \in \mathbb{C}^n$.
h. We have

$$\lim_{\delta \to 0^+} \left\langle \frac{\gamma(t + \delta a) - \gamma(t + \delta b)}{\delta^2}, \nu_{\gamma(t+\delta a)} \right\rangle = \frac{1}{2}Q_{\gamma(t)}\left((b - a)\dot{\gamma}(t)\right)$$

where $0 \leq a, b \in \mathbb{R}$.
i. Let

$$g(t) = \int_{-\infty}^\infty \left(1 + \frac{1}{2}Q_{\gamma(t)}\left(\dot{\gamma}(t)s\right)\right)^{-1} ds.$$

Then $g(t) \neq 0$, all $0 \leq t \leq 1$.

j. Define, for $f \in C(\gamma), z \in \bar{\Omega}$,

$$h_\delta(z) = \int_0^1 \frac{\delta \cdot (f(t)/g(t))\, dt}{\delta^2 + \langle \gamma(t) - z, \nu_{\gamma(t)} \rangle}.$$

Then $h_\delta \in A(\Omega)$.
  k. Apply the change of variable $t \mapsto y + \delta t$ in (j), where the integrand is defined to be 0 off $[0, 1]$. [Use (h) to verify (a), (b), and (c) in the present context.]
  l. Notice that (d) and (e) follow from (j) in the present setting.
  m. Apply the Fornæss imbedding theorem to derive the general case.
  n. For the case $\Omega \subseteq \mathbb{C}^n, n > 2$, the curve γ may be replaced by more general manifolds and the proof becomes slightly more difficult (see W. Rudin [5]).
  o. For the case $f \equiv 1$, part e [resp. (k), (l)] yields a function in $A(\Omega)$ that is $\equiv 1$ on γ and has modulus < 1 elsewhere, thus proving that γ is a peak set.

8. The purpose of this exercise is to acquaint the reader with some of the subtleties of approximation theory. For further details, see L. Hörmander and J. Wermer [1], L. Nirenberg and R. O. Wells [1], Wells [1], E. Bedford and J. E. Fornæss [3], and references cited therein.

Throughout this exercise, if M is a complex manifold and $U \subseteq M$ is relatively open, then $\mathcal{O}(U)$ is the algebra of holomorphic functions on U equipped with the compact-open topology. If $K \subseteq U$ is closed, let $\mathcal{O}(K)$ be those functions holomorphic on some neighborhood (in U) of K. Topologize $\mathcal{O}(K)$ by the compact-open topology. Let $C(K)$ be the bounded, continuous functions on K. We consider the restriction mappings

$$\mathcal{O}(M) \xrightarrow{r} \mathcal{O}(U) \xrightarrow{s} \mathcal{O}(K) \xrightarrow{t} C(K),$$

where r, s are restriction and t is injection. In this subject we determine whether r, s, t are surjective or have dense image.
  a. First consider the case $M = \mathbb{C}^1$. Prove that U is simply connected if and only if r has dense range. Let $U = \mathbb{C} \setminus \{0\}, K = \{z \in \mathbb{C} : |z| = 1\}$. Then s has dense range, and t is surjective. If $K \subseteq \mathbb{C}$ is any compact subset without interior such that $\mathbb{C} \setminus K$ has finitely many components, then $\mathcal{O}(K)$ is dense in $C(K)$. Finally, if $\mathbb{C} \supseteq M \underset{\neq}{\supseteq} U$, then r is not surjective.
  b. Let $M = \mathbb{C}^2, U = \{(z_1, z_2) : |z_1|^2 + |z_2|^2 < 1\}, K = \{(z_1, 0) \in U : |z_1| \leq 1\}$. Then K has no interior in $M, {}^c K$ is connected, but t does not have dense range. However, r and s have dense range.
  c. If $K = \partial B \subseteq \mathbb{C}^2$, then the restriction $\mathcal{O}(\bar{B}) \to \mathcal{O}(K)$ has dense range. However, $t : \mathcal{O}(K) \to C(K)$ does not.

d. If $K = \{(x_1 + i0, x_2 + i0) : |x_1| \leq 1, |x_2| \leq 1\}$, then the restriction $\mathcal{O}(\bar{D}^2(0,1)) \to \mathcal{O}(K)$ has dense range.

e. Let

$$S = \left\{(x_1 + iy_1, x_2 + iy_2, x_3 + iy_3) \in \mathbb{C}^3 : \right.$$

$$\left. x_1^2 + y_1^2 + x_2^2 = 1, y_2 = x_3 = y_3 = 0 \right\},$$

$$S' = \left\{(x_1 + iy_1, x_2 + iy_2, x_3 + iy_3) \in \mathbb{C}^3 : \right.$$

$$\left. x_1^2 + x_2^2 + x_3^2 = 1, y_1 = y_2 = y_3 = 0 \right\}.$$

Then S, S' are both 2-spheres, but $\mathcal{O}(S') \overset{t}{\to} C(S')$ has dense range, whereas $\mathcal{O}(S) \overset{t}{\to} C(S)$ does not.

f. Let

$$\begin{aligned}
T &= \{z \in \mathbb{C}^2 : |z_1| = |z_2| = 1\}; \\
T' &= \left\{(x_1 + iy_1, x_2 + iy_2) \in \mathbb{C}^2 : \right. \\
& \qquad \left. \frac{3}{8} + x_1^2 + y_1^2 + x_2^2 = \sqrt{y_1^2 + x_2^2}, y_2 = 0 \right\}; \\
T'' &= \left\{(x_1 + iy_1, x_2 + iy_2) \in \mathbb{C}^2 : \right. \\
& \qquad \left. \frac{3}{8} + x_1^2 + y_1^2 + x_2^2 = \sqrt{x_1^2 + y_1^2}, y_2 = 0 \right\}.
\end{aligned}$$

Then T, T', T'' are toruses (T' and T'' are isometric). The torus T' lies in $\mathbb{C} \times \mathbb{R}$, has 0 as center of symmetry, the x_1-axis as axis of symmetry, major radius 1, and minor radius $\frac{1}{2}$. The torus T'' is the same, except that the x_2-axis is the axis of symmetry. Prove that the restriction map

$$\mathcal{O}\left((\mathbb{C} \setminus \{0\}) \times (\mathbb{C} \setminus \{0\})\right) \to C(T)$$

has dense range. This is false if T is replaced by T' or T''.

g. Let $\Omega = \{(z_1, z_2) \in \mathbb{C}^2 : |z_2| < |z_1| < 1\}$ be the Hartogs triangle. Let $f(z_1, z_2) = z_2^2/z_1$. Then $f \in A(\Omega) \equiv C(\bar{\Omega}) \cap \{$holomorphic functions on $\Omega\}$. However, the restriction map $\mathcal{O}(\bar{\Omega}) \to A(\Omega)$ does not have f in the closure of its image.

Most of the preceding examples come from Wells [1].

9. *The Oka map.* Let $K \subseteq \mathbb{C}^n$ be compact. Suppose that $K \times \bar{D}$ has a neighborhood basis of domains of holomorphy. Let ϕ be holomorphic on a neighborhood ω of K. Define

$$K_\phi = \{z \in K : |\phi(z)| \leq 1\}.$$

Let f be a $\bar{\partial}$-closed (p,q) form with C^∞ coefficients on a neighborhood U of K_ϕ. Define the Oka map

$$\Theta : \omega \;\to\; \mathbb{C}^{n+1}$$
$$z \;\mapsto\; (z, \phi(z)).$$

Our goal is to write $f = \Theta^* F$ for some smooth $\bar{\partial}$-closed (p,q) form F on a neighborhood of $K \times \bar{D}$.

a. Let $\pi : \mathbb{C}^{n+1} \to \mathbb{C}^n$ be projection on the first n coordinates. Then $\pi^* f \in C^\infty_{(p,q)}(\pi^{-1}U)$ and $\bar{\partial}\pi^* f = 0$ on $\pi^{-1}(U) \supseteq \Theta(K_\phi)$.

b. Let $\psi \in C^\infty_c(\pi^{-1}U)$, $\psi \equiv 1$ on a neighborhood of $K_\phi \times \bar{D}(0,1)$, and let

$$Q(z_1, \ldots, z_{n+1}) = z_{n+1} - \phi(z_1, \ldots, z_n).$$

Then there is a smooth (p,q) form G on a neighborhood of $K \times \bar{D}$ such that

$$F = \psi\pi^* f - QG$$

is $\bar{\partial}$-closed. This F does the job.

c. Let $\phi_1, \ldots, \phi_m$ be holomorphic on a neighborhood ω of $\bar{D}^n(0,r)$. Let

$$\Theta : \omega \;\to\; \mathbb{C}^{n+m}$$
$$z \;\mapsto\; (z, \phi_1(z), \ldots, \phi_m(z))$$

be the Oka map. Define

$$K = \left\{ z \in \bar{D}^n(0,r) : |\phi_j(z)| \le 1, \;\; j = 1, \ldots, m \right\}.$$

If f is a $\bar{\partial}$-closed (p,q) form with C^∞ coefficients on a neighborhood U of K, then there is a smooth $\bar{\partial}$-closed (p,q) form F on a neighborhood of $\bar{D}^n(0,r) \times \bar{D}^m(0,1)$ such that $f = \Phi^* F$ on a neighborhood of K. (*Hint:* Use induction on the first part of the problem.)

10. *Polynomial convexity.* Let $\mathcal{P}$ denote the space of holomorphic polynomials on $\mathbb{C}^n$. If $S \subseteq \mathbb{C}^n$, then $\hat{S}_{\mathcal{P}} = \hat{S}$ is defined in the usual way (see Section 3.1). The set S is said to be *polynomially convex* if $\hat{S} = S$. Let $P_1, \ldots, P_m \in \mathcal{P}$, and define

$$\Gamma = \{z \in \mathbb{C}^n : |P_j(z)| \le 1, j = 1, \ldots, m\}.$$

Following the language of Section 3.5, we call Γ a *polynomial polyhedron*.

a. The set Γ is polynomially convex and holomorphically convex.

b. If $K \subseteq \mathbb{C}^n$ is compact and polynomially convex and if ω is a neighborhood of K, then there is a polynomial polyhedron Γ with $K \subset\subset \Gamma \subset\subset \omega$.

c. Let K, ω, Γ be as in (b). Let f be holomorphic on ω. Apply the theory of the Oka map developed in Exercise 9 to find an F with $f = F(z, P_1(z), \ldots, P_m(z))$ on a neighborhood of Γ. Show that F may be taken to be defined on a polydisc containing Γ.

d. Expand F in a power series, and conclude that f may be uniformly approximated on K by polynomials.

11. Let $\Omega \subseteq \mathbb{C}^n$ be a domain of holomorphy. We say that Ω is *Runge* if each holomorphic function on Ω can be approximated by polynomials, uniformly on compact subsets of Ω. Let $\mathcal{F} = \mathcal{F}(\Omega)$ denote the holomorphic functions on Ω, p the polynomials. Prove that the following are equivalent:

a. Ω is Runge;

b. $\forall K \subset\subset \Omega, \hat{K}_{\mathcal{P}} = \hat{K}_{\mathcal{F}}$;

c. $\forall K \subset\subset \Omega, \hat{K}_{\mathcal{P}} \cap \Omega = \hat{K}_{\mathcal{F}}$;

d. $\forall K \subset\subset \Omega, \hat{K}_{\mathcal{P}} \cap \Omega \subset\subset \Omega$.

(*Hint:* Refer to Exercise 1. Use the results of Exercise 10 as well.)

12. Which open sets in $\mathbb{C}^1$ are Runge? (See Exercise 11.)

13. Let $\Omega \subset\subset \mathbb{C}^n$ be a complete circular domain. Prove that Ω is Runge.

14. *An application of the Koszul complex.* Let $\Omega \subset\subset \mathbb{C}^n$ be a domain of holomorphy. Let $\delta > 0$, and let $f_1, \ldots, f_k$ be holomorphic functions on Ω such that $\sum_{j=1}^{k} |f_j(z)| \geq \delta$ for all $z \in \Omega$. Then there are holomorphic functions g_j on Ω such that $\sum f_j g_j \equiv 1$ on Ω.

To prove this, let $U_j = \{z \in \Omega : |f_j(z)| > \delta/(2k)\}$. Then the sets U_j cover Ω. Let $\{\phi_j\}$ be a partition of unity subordinate to the cover $\{U_j\}$. We assume that the problem has a solution of the form

$$g_j = \frac{\phi_j}{f_j} + \sum_i v_{ji} f_i$$

where the v_{ji} are unknown functions that satisfy $v_{ij} = -v_{ji}$ and $v_{jj} = 0$ for all $i, j = 1, \ldots, k$. (This form of the g_j is motivated by a device of homological algebra known as the Koszul complex.) Functions g_j of this form automatically satisfy $\sum_j f_j g_j \equiv 1$. Solve a $\bar{\partial}$ problem to produce the v_{ji}.

15. The following is essentially the simplest known proof that B_n is not biholomorphic to the polydisc D^n when $n \geq 2$.

a. If there is a biholomorphic $F : B_n \to D^n$, then we may assume that $F(0) = 0$.

b. If $\Omega \subset\subset \mathbb{C}^n$ is a domain and $0 \in \Omega$, then we define

$$U(\Omega) = \{\xi \in \mathbb{C}^n : \text{there is a holomorphic } \phi : D \to \Omega$$

$$\text{with } \phi(0) = 0 \text{ and } \phi'(0) = \xi\}.$$

  c. The sets $U(B_n)$ and $U(D^n)$ can be determined explicitly using the Schwarz lemma.
  d. It must be that $J_{\mathbb{C}}F : U(B_n) \to U(D^n)$ is injective, surjective, and linear. This is impossible. See Krantz [20] for details of this proof.

16. Let $\Omega \subset\subset \mathbb{C}^n$ be a domain. Let $P \in \partial\Omega$ and $U \subseteq \mathbb{C}^n$ a neighborhood of P such that $U \cap \partial\Omega$ is strongly pseudoconvex. Let $f : U \to \mathbb{C}$ be holomorphic, and suppose that $F(P) = 0$. Prove that it cannot be the case that $\{z \in U : f(z) = 0\} \subseteq \bar{\Omega} \cap U$. (*Hint:* Apply the tomato can principle to $1/f$ on $U \setminus \bar{\Omega}$.)

17. Let $\Omega \subset\subset \mathbb{C}^n$ be open. Let $\mathcal{O}(\Omega)$ be the space of holomorphic functions on Ω. Equip $\mathcal{O}$ with the topology of uniform convergence on compact sets. Let $\mathrm{Hom}\,(\mathcal{O}(\Omega), \mathbb{C})$ be the continuous, nonzero, multiplicative functionals on $\mathcal{O}(\Omega)$. Topologize $\mathrm{Hom}\,(\mathcal{O}(\Omega), \mathbb{C})$ with the weak-* topology—that is, a subbasis for the topology on $\mathrm{Hom}\,(\mathcal{O}(\Omega), \mathbb{C})$ is given by the sets

$$S_{\alpha, f, \epsilon} = \{\beta \in \mathrm{Hom}\,(\mathcal{O}(\Omega), \mathbb{C}) : |(\alpha - \beta)f| < \epsilon\},$$

any $\alpha \in \mathrm{Hom}\,(\mathcal{O}(\Omega), \mathbb{C}), f \in \mathcal{O}(\Omega), \epsilon > 0$. Notice that if $x \in \Omega$, then

$$\mathcal{E}_x : \mathcal{O}(\Omega) \ni f \mapsto f(x)$$

  is an element of $\mathrm{Hom}\,(\mathcal{O}(\Omega))$. Let the *envelope of holomorphy* $E(\Omega)$ of Ω be the connected component of $H(\mathcal{O}(\Omega), \mathbb{C})$ that contains $\{\mathcal{E}_x\}_{x \in \Omega}$. We say that Ω is *holomorphically convex* if $E(\Omega) = \{\mathcal{E}_x\}_{x \in \Omega}$. This exercise considers envelopes of holomorphy and holomorphic convexity.
  a. If $\alpha \in \mathrm{Hom}\,(\mathcal{O}(\Omega), \mathbb{C})$, then there is a compact $K \subset\subset \Omega$ such that $|\alpha(f)| \leq C\|f\|_K$ for all $f \in \mathcal{O}(\Omega)$. Here $\|f\|_K \equiv \sup_{z \in K} |f(z)|$.
  b. If Ω is not pseudoconvex, then Ω is not holomorphically convex. More precisely, there is an $x \notin \Omega$ such that $\mathcal{E}_x \in E(\Omega)$.
  c. If Ω is pseudoconvex, let $\alpha \in E(\Omega)$. The coordinate functions $z_j : \Omega \to \mathbb{C}$ are in $\mathcal{O}(\Omega)$. Let $a_j = \alpha(z_j)$. If the functions $z_1 - a_1, \ldots, z_n - a_n$ have no common zero on Ω, then there exist $g_1, \ldots, g_n \in \mathcal{O}(\Omega)$ with $\sum_j g_j(z) \cdot (z_j - a_j) \equiv 1$ on Ω (here we are using Exercise 14). It follows that $\{f \in \mathcal{O}(\Omega) : \alpha(f) = 0\} = \mathcal{O}(\Omega)$, which is impossible. Thus $a \equiv (a_1, \ldots, a_n) \in \Omega$. If there is an $f \in \mathcal{O}(\Omega)$ such that $\alpha(f) \neq \mathcal{E}_a(f)$, then the functions $f(z) - \alpha(f), z_1 - a_1, \ldots, z_n - a_n$ have no common zero on Ω. Again, it follows that $\alpha(1) = 0$, so $\mathrm{Ker}\,\alpha = \mathcal{O}(\Omega)$, which is impossible.
  d. Parts b and c show that Ω is holomorphically convex if and only if Ω is pseudoconvex. The novelty is that the approach is function-algebraic.
  e. Let $K \subseteq \mathbb{C}^n$ be compact. Define $\mathcal{O}(K)$ as in Exercise 8 (i.e., $\mathcal{O}(K)$ is the direct limit of $\mathcal{O}(\Omega)$ for $\Omega \supseteq K$). Show that every multiplicative homomorphism on $\mathcal{O}(K)$ is continuous. (*Hint:* If not, then there is an $f \in \mathcal{O}(K)$ with $\|f\| < 1$ but $\alpha(f) = 1$. Then $1 - f$ is invertible in $\mathcal{O}(K)$, whereas $\alpha(1 - f)$ is not invertible in $\mathbb{C}$.)

f. Does e still hold if K is replaced by Ω open and $\mathcal{O}(\Omega)$ is topologized as in the beginning of this exercise?

g. Show that if $K \subseteq \mathbb{C}^n$ is compact, $K = \cap \Omega_j$, and each Ω_j is pseudoconvex, then K is holomorphically convex.

h. Show that

$$K = \{z \in \mathbb{C}^2 : |z_1|^2 + |z_1|^{-2} + |\mathrm{Re}z_2|^2 = 4, \mathrm{Im}z_2 = 0\}$$

is not holomorphically convex. Indeed, any function holomorphic in a neighborhood of K extends to

$$\tilde{K} = \{z \in \mathbb{C}^2 : |z_1|^2 + |z_2|^{-2} + |\mathrm{Re}z_2|^2 \leq 4, \mathrm{Im}z_2 = 0\}.$$

i. It is known (M. Freeman and R. Harvey [1]) that the set K in (h) is locally holomorphically convex in the sense that for each $P \in K$ there is an $r > 0$ such that $\{z \in K : |z - P| \leq r\}$ is holomorphically convex.

j. Let $\Omega \subset\subset \mathbb{C}^n$ be open. Let $\{z_j\} \subseteq \Omega, z_j \to P \in \partial\Omega$. Let $\mathcal{J} = \{f \in \mathcal{O}(\Omega) : f$ vanishes eventually on $\{z_j\}\}$. Then $\mathcal{J}$ is an ideal in the algebra $\mathcal{O}$, but $\mathcal{J}$ is not finitely generated. (Part of this problem is to figure out what "eventually" must mean.)

18. Let $\Omega \subset\subset \mathbb{C}^n$ be smoothly bounded. Show that there exists at least one peak function in $A(\Omega)$. (*Hint:* Choose $P \in \Omega$ that is furthest from 0. This is a very special case of a theorem due to H. G. Dales [1].)

19. *Two classical definitions of analytic convexity.* Let $\Omega \subseteq \mathbb{C}^n$ be a connected open set. Let $a \in \partial\Omega$. We say that Ω is *analytically convex in the sense of Hartogs* at the point a if the following two conditions are satisfied:

a. Whenever $\epsilon > 0$ and $\{z \in \mathbb{C}^n : |z_1 - a_1| = \epsilon, z_2 = a_2, \ldots, z_n = a_n\}$ is contained in Ω, then there is a $\delta > 0$ such that whenever $|b_2 - a_2| < \delta, \ldots, |b_n - a_n| < \delta$, then the closed analytic disc $\{z \in \mathbb{C}^n : |z_1 - a_1| \leq \epsilon, z_2 = b_2, \ldots, z_n = b_n\}$ will contain points not belonging to the region Ω.

b. Property (a) is preserved under biholomorphic coordinate changes near (a).

Figure out an invariant formulation of this definition that avoids the need for (b). We call Ω analytically convex in the sense of Hartogs if each boundary point is. Show that an open set is analytically convex in the sense of Hartogs if and only if it is a domain of holomorphy.

Now let $\Omega \subset\subset \mathbb{C}^n$ be a domain with C^2 boundary. We say that Ω is *analytically convex in the sense of Levi* at a point $P \in \partial\Omega$ if whenever U is a neighborhood of P and M is an $(n-1)$-dimensional complex manifold passing through P, then $M \cap (U \setminus (\Omega \cup \{P\})) \neq \emptyset$. Prove that $P \in \partial\Omega$ is analytically convex in the sense of Levi if and only if P is a point of Levi pseudoconvexity as defined in Chapter 3. With the Levi problem solved, we now know that a C^2 bounded domain Ω is analytically convex in the sense

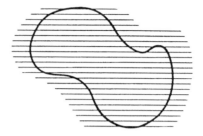

FIGURE 5.3

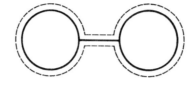

FIGURE 5.4

of Hartogs if and only if it is analytically convex in the sense of Levi. This, in turn, is true if and only if the domain is pseudoconvex in one of the two senses defined in this book.

20. Let $\Omega \subset\subset \mathbb{C}^n$ be a domain. For each $\zeta' = (\zeta_2, \ldots, \zeta_n) \in \mathbb{C}^{n-1}$, let $\ell_{\zeta'} = \{z \in \mathbb{C}^n : z_2 = \zeta_2, \ldots, z_n = \zeta_n\}$, and let $\Omega_{\zeta'} = \ell_{\zeta'} \cap \Omega$. For $r > 0$ let $\ell_{\zeta'}^r = \{z \in \ell_{\zeta'} : \operatorname{dist}(z, \bar{\Omega}) < r\}$. Let $f : \Omega \to \mathbb{C}$ be holomorphic. Suppose that for each ζ' with $\Omega_{\zeta'} \neq \emptyset$ it holds that $f|_{\Omega_{\zeta'}}$ analytically continues to $\ell_{\zeta'}^r$. Prove that f analytically continues (in the n-variable sense) to $\cup\{\ell_{\zeta'}^r : \ell_{\zeta'}^r \cap \Omega \neq \emptyset\}$. See Figure 5.3.

21. *The barbell domain.* Let $\mathcal{D} \subseteq \mathbb{C}^2$ be given by

$$\mathcal{D} = \bar{B}(2, 1) \cup \bar{B}(-2, 1) \cup \{(x_1 + i0, 0) : -2 \leq x_1 \leq 2\}.$$

Show that for every $\epsilon > 0$, there is a C^2 bounded domain $\Omega \supseteq \mathcal{D}$ such that $\operatorname{dist}(\mathcal{D}, {}^c\Omega) < \epsilon$ and Ω is strongly pseudoconvex (see Figure 5.4).

6 Cousin Problems, Cohomology, and Sheaves

6.1 Cousin Problems

We shall motivate the form of the Cousin problems by first looking at a little combinatorial topology.

EXAMPLE   Let

$$\Omega^0 = \{z \in \mathbb{C} : 1 < |z| < 2\}$$
$$\Omega^1 = \{z \in \mathbb{C} : |z| < 2\}.$$

Consider the open coverings of Ω^0 and of Ω^1 illustrated in Figure 6.1.

We shall play the following game with these coverings: To each nonempty intersection $U_i \cap U_j$ we assign an integer g_{ij}, subject to the rules

$$g_{ij} = -g_{ji} \tag{6.1.1}$$

$$g_{ij} + g_{jk} + g_{ki} = 0 \qquad \text{whenever} \qquad U_i \cap U_j \cap U_k \neq \emptyset \tag{6.1.2}$$

(We similarly assign integers h_{ij} to each nonempty intersection $V_i \cap V_j$.)

The question that we wish to consider is this: Will it always be possible to choose integers $g_i, i = 1, 2, 3, \ldots$, such that $g_{ij} = g_j - g_i$ for all i, j [resp. will it always be possible to choose integers $h_i, i = 1, 2, 3, \ldots$ such that $h_{ij} = h_j - h_i$]? As motivation for this question, the reader should consider the problem of choosing a branch for $\log z$, z in the complex plane.

On the annulus, with cover $\{U_i\}$, the answer to our question is decidedly no. For a counterexample, let $g_{12} = 1, g_{23} = 10$, and $g_{31} = 100$. Define the other g_{ij} according to (6.1.1). Observe that (6.1.2) is vacuous. If g_1, g_2, g_3 are chosen so that $g_2 - g_1 = 1$ and $g_3 - g_2 = 10$, then it must be that $g_3 - g_1 = 11$, and that is inconsistent with $g_{13} = -100$.

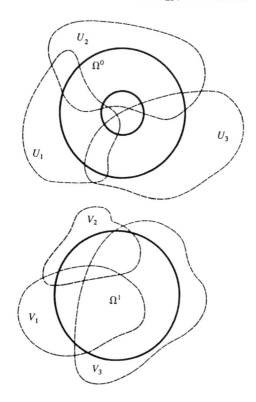

FIGURE 6.1

By contrast, our problem can always be solved on the disc Ω^1 with cover $\{V_i\}$ because the (now nonvacuous) condition (6.1.2) guarantees that the number g_{13} will be compatible with the conditions forced by $g_{12} = g_2 - g_1$ and $g_{23} = g_3 - g_2$. The contradiction that occurred on the annulus cannot arise.   □

The game we have been describing is an algebraic/combinatorial device for detecting the fact that the annulus has a hole and the disc does not. The subject of algebraic topology (especially cohomology theory) has its roots in considerations such as these (see Lefschetz [3]). The next example is, from our perspective, a more natural and interesting one containing the same thread of ideas.

EXAMPLE   (Another proof of Hefer's lemma 5.3.3 in $\mathbb{C}^2$) Let $\Omega \subseteq \mathbb{C}^2$ be pseudoconvex. Assume that $0 \in \Omega$. If $f : \Omega \to \mathbb{C}$ is a holomorphic function that satisfies $f(0) = 0$, then there are holomorphic functions f_1, f_2 on Ω such that $f(z) = z_1 f_1(z) + z_2 f_2(z)$ on Ω.

Proof   We first solve the problem in the "real variable" category using a partition of unity. Then we pass back to holomorphic function theory by correcting with the $\bar{\partial}$ operator.

By hypothesis, the local power series expansion of f about 0 has no constant term. Thus there are holomorphic functions $f_1^{(0)}$ and $f_2^{(0)}$ on a neighborhood U_0 of 0 such that $f(z) = z_1 f_1^{(0)} + z_2 f_2^{(0)}$ on U_0.

If $P \neq 0$, then either $P_1 \neq 0$ or $P_2 \neq 0$; hence either $f(z) = z_1 f_1^{(P)} + z_2 \cdot 0$ or $f(z) = z_1 \cdot 0 + z_2 f_2^{(P)}$ on a neighborhood U_P of P, with $f_j^{(P)}$ holomorphic on U_P.

We pass to a locally finite refinement $\{U_j\}$ of the covering $\{U_P\}$ of Ω with a corresponding pair $f_1^{(j)}, f_2^{(j)}$ of holomorphic functions on U_j such that $f = z_1 f_1^{(j)} + z_2 f_2^{(j)}$. For every pair j, k such that $U_j \cap U_k \neq \emptyset$, we define $g_{jk}^1 = f_1^k - f_1^j$. Observe that

$$g_{jk}^1 = -g_{kj}^1$$

and

$$g_{jk}^1 + g_{k\ell}^1 + g_{\ell j}^1 = 0 \qquad \text{on } U_j \cap U_k \cap U_\ell.$$

Now let $\{\phi_i\}$ be a C^∞ partition of unity on Ω that is subordinate to the covering $\{U_i\}$. For $z \in U_i$, define $h_1^i = \sum_k \phi_k g_{ki}^1$. Notice that, on $U_i \cap U_j$, we have

$$h_1^j - h_1^i = \sum \phi_k \left(g_{kj}^1 - g_{ki}^1 \right) = \sum_k \phi_k(g_{ij}^1) = g_{ij}^1 = f_1^j - f_1^i; \qquad (6.1.3)$$

hence

$$\bar{\partial} h_1^j - \bar{\partial} h_1^i = 0 \qquad \text{on } U_i \cap U_j. \qquad (6.1.4)$$

Thus it is consistent to define a $(0, 1)$ form

$$\alpha_1(z) \equiv \bar{\partial} h_1^i(z) \qquad \text{when } z \in U_i.$$

By inspection, α_1 is a $\bar{\partial}$-closed $(0, 1)$ on Ω with smooth coefficients. Because Ω is pseudoconvex, we may apply Corollary 4.6.12 to conclude that there is a C^∞ function h_1 on Ω satisfying $\bar{\partial} h_1 = \alpha_1$ on Ω.

We will exploit the simplicity of two-dimensional complex space in order to construct a companion function h_2. Set $h_2^i = \sum_k \phi_k g_{ki}^2$, where, of course, $g_{ki}^2 = f_2^i - f_2^k$. Then

$$0 = z_1 h_1^i + z_2 h_2^i \qquad \text{on } U_i \qquad (6.1.5)$$

by construction. Hence

$$h_2^i = -\frac{z_1 h_1^i}{z_2}$$

is a well-defined C^∞ function on U_i (the right side is C^∞ since the left is). Using (6.1.3) again we see that

$$\alpha_2(z) \equiv \frac{-z_1 \bar\partial h_1^i(z)}{z_2}, \quad z \in U_i,$$

is a well-defined, $\bar\partial$-closed (0,1) form with smooth coefficients. Moreover, the function

$$h_2(z) = \frac{-z_1 h_1(z)}{z_2} \tag{6.1.6}$$

satisfies $\bar\partial h_2 = \alpha_2$ on Ω. By Corollary 4.6.12 we know that h_2 is smooth.

We are almost ready to define f_1 and f_2. First, set $g_1^i = h_1^i - h_1$ and $g_2^i = h_2^i - h_2$ on U_i. Then each g_ℓ^i is holomorphic on U_i, $l = 1, 2$. Formula (6.1.3) tells us that, on $U_i \cap U_j$, we have

$$f_1^i - g_1^i = f_1^i - (h_1^i - h_1) = f_1^j - (h_1^j - h_1) = f_1^j - g_1^j.$$

Thus

$$f_1(z) \equiv f_1^i(z) - g_1^i(z), \quad z \in U_i,$$

gives a well-defined holomorphic function on Ω. Similarly,

$$f_2^i - g_2^i = f_2^i - (h_2^i - h_2) = f_2^j - (h_2^j - h_2) = f_2^j - g_2^j$$

on $U_i \cap U_j$; hence

$$f_2(z) \equiv f_2^i(z) - g_2^i(z), \quad z \in U_i$$

is a well-defined holomorphic function on Ω.

Finally, for $z \in U_i$ we use (6.1.5) and (6.1.6) to see that

$$\begin{aligned}
z_1 f_1(z) + z_2 f_2(z) &= z_1(f_1^i(z) - g_1^i(z)) + z_2(f_2^i(z) - g_2^i(z)) \\
&= z_1\{h_1(z) + f_1^i(z) - h_1^i(z)\} \\
&\quad + z_2\{h_2(z) + f_2^i(z) - h_2^i(z)\} \\
&= \left(-z_1 h_1^i(z) - z_2 h_2^i(z)\right) \\
&\quad + \left(z_1 f_1^i(z) + z_2 f_2^i(z)\right) \\
&\quad + \left(z_1 h_1(z) + z_2 h_2(z)\right) \\
&= 0 + f(z) + 0
\end{aligned}$$

as desired. $\qquad\square$

Inspired by this example, we now set up a conceptual framework that will make an argument like this one rather natural. Our first step in this program is to consider the Cousin problems. We enunciate them as follows:

FIRST COUSIN PROBLEM   Let $\Omega \subseteq \mathbb{C}^n$ be a domain. Let $\{U_i\}$ be an open covering of Ω. Suppose that for each U_j, U_k with nonempty intersection there is a holomorphic $g_{jk} : U_j \cap U_k \to \mathbb{C}$ satisfying

    a. $g_{jk} = -g_{kj}$;

    b. $g_{jk} + g_{k\ell} + g_{\ell j} = 0$ on $U_j \cap U_k \cap U_\ell$.

Problem: Find holomorphic functions g_j on U_j such that $g_{jk} = g_k - g_j$ on $U_j \cap U_k$ whenever this intersection is nonempty.

We also shall consider a multiplicative version of this problem.

SECOND COUSIN PROBLEM   Let $\Omega \subseteq \mathbb{C}^n$ be a domain. Let $\{U_i\}$ be an open covering of Ω. Suppose that for each U_j, U_k with nonempty intersection there is a *nonvanishing* holomorphic g_{jk} on $U_j \cap U_k$ such that

    a. $g_{jk} \cdot g_{kj} = 1$;

    b. $g_{jk} \cdot g_{k\ell} \cdot g_{\ell j} = 1$ on $U_j \cap U_k \cap U_\ell$.

Problem: Find nonvanishing holomorphic functions g_j on U_j such that $g_{jk} = g_k/g_j$ on $U_j \cap U_k$ whenever $U_j \cap U_k \neq \emptyset$.

PROPOSITION 6.1.7   The analogue of the first Cousin problem for C^∞ (rather than holomorphic) functions always has a solution.

Proof   Let $\{\phi_i\}$ be a C^∞ partition of unity on Ω subordinate to the covering $\{U_i\}$. For each i, define

$$g_i(z) = \sum_k \phi_k(z) g_{ki}(z), \qquad z \in U_i.$$

Then on $U_i \cap U_j$ we have

$$
\begin{aligned}
g_j(z) - g_i(z) &= \sum_k \phi_k(z) \{ g_{kj}(z) - g_{ki}(z) \} \\
&= \sum_k \phi_k(z) g_{ij}(z) \\
&= g_{ij}(z). \qquad \square
\end{aligned}
$$

PROPOSITION 6.1.8   If $\Omega \subseteq \mathbb{C}^n$ is a domain of holomorphy (equivalently, Ω is pseudoconvex), then the first Cousin problem can always be solved on Ω.

Proof   Let $\{\phi_i\}$ be a partition of unity subordinate to $\{U_i\}$. Define

$$h_i = \sum_k \phi_k g_{ki} \qquad \text{on } U_i.$$

Of course, h_i may not be holomorphic. However, on $U_i \cap U_j$ we have

$$
\begin{aligned}
\bar{\partial}(h_j - h_i) &= \bar{\partial} \sum_k \phi_k(g_{kj} - g_{ki}) \\
&= \bar{\partial} \sum_k \phi_k(g_{ij}) \\
&= \bar{\partial} g_{ij} \\
&= 0.
\end{aligned}
$$

Therefore, the form

$$
f \equiv \bar{\partial} h_j \qquad \text{on } U_j
$$

is well defined, $\bar{\partial}$-closed, and C^∞ on Ω. By Corollary 4.6.12, there is a $u \in C^\infty(\Omega)$ such that $\bar{\partial} u = f$. Set

$$
g_j = h_j - u \qquad \text{on } U_j.
$$

Then on $U_i \cap U_j$ we have

$$
g_j - g_i = h_j - h_i = g_{ij}.
$$

Also, for each i, $\bar{\partial} g_i = \bar{\partial} h_i - \bar{\partial} u = 0$ on U_i. Hence each g_i is holomorphic and we are done.   □

LEMMA 6.1.9   Let $\Omega \subseteq \mathbb{R}^N$ be simply connected. Let $f : \Omega \to \mathbb{C}$ be continuous and nonvanishing. Then there is a continuous g on Ω such that $\exp g = f$. If f is C^k, any $k \geq 0$, then g is C^k.

Proof   By continuity, each $P \in \Omega$ has a neighborhood U_P on which f has a continuous logarithm. Fix $P_0 \in \Omega$ and let $\gamma : [0,1] \to \Omega$ be a continuous Jordan curve with $\gamma(0) = P_0, \gamma(1) = P_0$. By a continuity argument, $f \circ \gamma(t)$ has a continuous logarithm for $0 \leq t < 1$ (that is, the set $S = \{s \in [0,1) : \log(f \circ \gamma(t))$ is well defined and continuous for $0 \leq t \leq s\}$ is open, closed, and nonempty). Seeking a contradiction, we suppose that $\lim_{t \to 1^-} \log f \circ \gamma(t) \neq \log f \circ \gamma(0)$. Let $u(s,t)$ be a fixed-point homotopy of the curve γ with the constant curve at P_0. Thus

- u is continuous on $[0,1] \times [0,1]$;
- $u(0,t) = \gamma(t)$, all $t \in [0,1]$;
- $u(s,0) = u(s,1) = P_0$, all $s \in [0,1]$;
- $u(1,t) \equiv P_0$, all $t \in [0,1]$.

The function

$$\rho(s) = \frac{1}{2\pi i} \left\{ \lim_{t \to 1^-} \log f(u(s,t)) - \log f(u(s,0)) \right\}$$

is then a continuous, integer-valued function of s that satisfies $\rho(0) = 0$ and $\rho(1) \neq 0$. This is a contradiction. Thus, in fact, $\lim_{t \to 1^-} \log f \circ \gamma(t) = \log f \circ \gamma(0)$.

Now a standard argument shows that the set of all $P \in \Omega$ to which $\log f$ can be unambiguously propagated from the initial point P_0 is both open and closed. That completes the proof in the continuous, or C^0, case.

The C^k result follows by applying implicit differentiation to the equation $\exp g = f$.   □

LEMMA 6.1.10   Let $\Omega \subseteq \mathbb{C}^n$ be simply connected and let $f : \Omega \to \mathbb{C}$ be a holomorphic and nonvanishing function. Then there is a holomorpic function g on Ω such that $\exp g = f$.

Proof   By the preceding lemma, there is a C^1 function g that satisfies $\exp g = f$. But then g satisfies the Cauchy-Riemann equations, as we see by implicit differentiation.   □

PROPOSITION 6.1.11   Let $\Omega \subseteq \mathbb{C}^n$ be a domain of holomorphy (equivalently, Ω is pseudoconvex). Let $\{g_{ij}\}$ be a set of holomorphic Cousin II data for the covering $\{U_i\}$. If there exist nonvanishing, continuous functions $g'_i : U_i \to \mathbb{C}$ such that $g_{ij} = g'_j g'^{-1}_i$ on $U_i \cap U_j$, then there exist nonvanishing, holomorphic g_i such that $g_{ij} = g_j g^{-1}_i$. That is, the second Cousin problem can be solved on Ω if and only if it can be solved just continuously.

Remark: The hypothesis that the second Cousin problem can be solved just continuously is, in fact, a specific topological assumption. We shall say more about this hypothesis later.

The "Oka Principle" is the assertion that, on a domain of holomorphy, what can be done continuously can also be done holomorphically.   □

Proof of the Proposition   We first consider a special case.
Part I: Assume that each U_i is a polydisc. Let g'_i be as in the hypotheses. By Lemma 6.1.9, we may write $g'_i = \exp h'_i$ on U_i, with h'_i continuous (this is where we use the hypothesis that each U_i is a polydisc, hence topologically trivial). Let $h_{ij} = h'_j - h'_i$. Then $g_{ij} = \exp h_{ij}$, and since h_{ij} is a continuous logarithm for g_{ij} on $U_i \cap U_j$, then it is also a holomorphic logarithm.

Then $\{h_{ij}\}$ is a set of holomorphic Cousin I data for the cover U_i. So there exist holomorphic $h_i : U_i \to \mathbb{C}$ such that $h_{ij} = h_j - h_i$ on $U_i \cap U_j$. But then the functions $g_i \equiv \exp h_i$ are obviously holomorphic and nonvanishing and solve the Cousin II problem for the data $\{g_{ij}\}$: to wit,

$$g_j g^{-1}_i = \exp(h_j - h_i) = \exp h_{ij} = \exp(h'_j - h'_i) = g'_j g'^{-1}_i = g_{ij}.$$

Part II: In case the U_i are not all polydiscs (hence not necessarily all topolog-ically trivial), we let $\{\tilde{U}_j\}$ be a refinement of the covering $\{U_i\}$ such that each $\tilde{U}_j$ is a polydisc. Let $\rho : \mathbb{N} \to \mathbb{N}$ satisfy $\tilde{U}_i \subseteq U_{\rho(i)}$ for each i (ρ is called an *affinity function*). Define

$$\tilde{g}_{ij} : \tilde{U}_i \cap \tilde{U}_j \to \mathbb{C}$$

by $\tilde{g}_{ij}(z) = g_{\rho(i)\rho(j)}(z)$ (this function is understood to be 0 if $\tilde{U}_i \cap \tilde{U}_j = \emptyset$). Then $\{\tilde{g}_{ij}\}$ is a set of holomorphic Cousin II data for the covering $\{\tilde{U}_i\}$. Let $\{\tilde{g}_i\}$ be nonvanishing holomorphic functions that solve this Cousin II problem (as provided by part I of the proof).

Now for any i, j, k and $z \in \tilde{U}_i \cap \tilde{U}_j \cap \tilde{U}_k$, we have

$$\begin{aligned}
\tilde{g}_k \tilde{g}_j^{-1} g_{\rho(k),i} g_{i,\rho(j)}(z) &= \tilde{g}_k \tilde{g}_j^{-1} g_{\rho(k)\rho(j)}(z) \\
&= \tilde{g}_k \tilde{g}_j^{-1} \tilde{g}_{kj}(z) \\
&= 1.
\end{aligned}$$

Therefore,

$$\tilde{g}_k g_{\rho(k),i} = \tilde{g}_j g_{\rho(j),i}(z) \qquad \text{on } \tilde{U}_i \cap \tilde{U}_j \cap \tilde{U}_k.$$

As a result,

$$g_i(z) \equiv \tilde{g}_k g_{\rho(k),i}(z) \qquad \text{on } \tilde{U}_i \cap \tilde{U}_k$$

gives a well-defined nonvanishing holomorphic function on $\tilde{U}_i$.

Finally, for any i, j, k and $z \in \tilde{U}_k \cap \tilde{U}_i \cap \tilde{U}_j$, we have

$$g_j g_i^{-1}(z) = \tilde{g}_k g_{\rho(k),j}(z) \tilde{g}_k^{-1} g_{\rho(k),i}^{-1}(z) = g_{ij}(z).$$

This is the desired result.   □

EXAMPLE (Oka)   There exists a domain of holomorphy on which Cousin II cannot always be solved (either continuously or holomorphically). In particular, it is not the case that Cousin I ⇒ Cousin II.

Proof   Let $\Omega = \{(z_1, z_2) \in \mathbb{C}^2 : \frac{3}{4} < |z_j| < \frac{5}{4}, j = 1, 2\}$. This is a domain of holomorphy, since it is a product domain. Let $\mathcal{P} = \{(z_1, z_2) : z_1 - z_2 - 1 = 0\}$ and $\omega = \mathcal{P} \cap \Omega$. Notice that ω is relatively closed in Ω. Writing $z_j = x_j + iy_j$, $j = 1, 2$, we have the following simple facts:

1. $\omega \cap \{(z_1, z_2) : y_1 = 0\} = \emptyset$. This is so because $y_1 = y_2$ and if both of these numbers are zero then $3/4 < |x_j| < 5/4$, $j = 1, 2$; this contradicts $x_1 - x_2 - 1 = 0$.

2. It follows from (1) that

$$\begin{aligned} \omega &= (\omega \cap \{z \in \Omega : y_1, y_2 > 0\}) \cup (\omega \cap \{z \in \Omega : y_1, y_2 < 0\}) \\ &\equiv \omega_1 \cup \omega_2. \end{aligned}$$

Let

$$U_1 = \{z \in \Omega : y_1, y_2 > 0\}$$

and

$$U_2 = \Omega \setminus \omega_1.$$

Then $\{U_j\}_{j=1,2}$ is an open covering of Ω. Notice that on $U_1 \cap U_2$ we have that $g_{21}(z) \equiv z_1 - z_2 - 1$ is never zero. All the Cousin II compatibility conditions are vacuously satisfied for $g_{12} = 1/g_{21}$.

Assume, seeking a contradiction, that Cousin II can be solved on Ω. Let $g_j : U_j \to \mathbb{C}, j = 1, 2$, be nonvanishing holomorphic functions satisfying $g_{12} = g_2/g_1$ on $U_1 \cap U_2$. Thus

$$\frac{g_2}{g_1} = \frac{1}{z_1 - z_2 - 1} \qquad \text{on } U_1 \cap U_2.$$

Hence the function

$$F(z) = \begin{cases} (z_1 - z_2 - 1)/g_1 & \text{if } z \in U_1 \\ 1/g_2 & \text{if } z \in U_2 \end{cases}$$

is a well-defined holomorphic function on Ω.

Now consider

$$g(\alpha, \beta) \equiv F(e^{i\alpha}, e^{i\beta}), \qquad 0 \le \alpha, \beta < 2\pi.$$

Then g can vanish only for $(e^{i\alpha}, e^{i\beta}) \in U_1$, and then only at the point $(e^{\pi i/3}, e^{2\pi i/3})$.

We wish to obtain now a contradiction from consideration of winding numbers. The argument is homological in nature. The idea is to show that one of the two functions, g_1 or g_2, must share the zeros of the function $z_1 - z_2 - 1$. The argument will take place on the unit torus in $\mathbb{C}^2$ and *thus is purely real variable in nature*.

In the α-β plane, consider the curve Γ_1 shown in Figure 6.2. We claim that $\arg g$ is well defined and smooth on Γ_1, since g is periodic in each variable. More precisely, any $2\pi k$ increment in the argument that is introduced in the lower part of Γ_1 (resp. the right part of Γ_1) is canceled by a corresponding increment on the upper part of Γ_1 (resp. the left part of Γ_1).

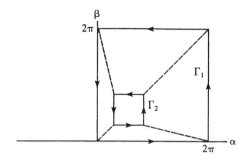

FIGURE 6.2

With Γ_2 as in Figure 6.2, no matter how small Γ_2, we thus have (with the cycles indicated by the dotted lines and the orientations as given):

$$0 = \int_{\Gamma_1} d\log g = \int_{\Gamma_2} d\log g.$$

On the other hand,

$$F(z_1, z_2) = h(z_1, z_2) \cdot (z_1 - z_2 - 1) \qquad \text{near } \Gamma_2$$

for some invertible holomorphic h, *by its very definition*. Hence

$$0 = \int_{\Gamma_2} d\log g = \int_{\Gamma_2} d\log(e^{i\alpha} - e^{i\beta} - 1). \tag{6.1.12}$$

(*Note:* It does not matter in this step that h is holomorphic—just that it is smooth and nonvanishing.)

By a trivial change of variable, this last line says that

$$0 = \int_{\Gamma_2'} d\log(e^{i\alpha' + \pi i/3} - e^{i\beta' + 2\pi i/3} - 1) \equiv \int_{\Gamma_2'} d\log(g'(\alpha', \beta')).$$

Here $\alpha' = \alpha - \pi/3$, $\beta' = \beta - 2\pi/3$, and Γ_2' is the corresponding translated cycle. Observe that the function $g'(\alpha', \beta') \equiv e^{i\alpha' + \pi i/3} - e^{i\beta' + 2\pi i/3} - 1$ has a zero at the point $(0,0)$.

Set $u + iv = g'(\alpha', \beta')$. The Jacobian determinant of this change of coordinates is

$$\det \begin{pmatrix} -\sin\left(\alpha' + \frac{\pi}{3}\right) & \sin\left(\beta' + \frac{2\pi}{3}\right) \\ \cos\left(\alpha' + \frac{\pi}{3}\right) & -\cos\left(\beta' + \frac{2\pi}{3}\right) \end{pmatrix} = \sin\left(\alpha' - \beta' - \frac{\pi}{3}\right).$$

If Γ_2 is sufficiently small, then this Jacobian determinant is as near to the constant $\sin(-\pi/3) = -\sqrt{3}/2$ as we please. Since

$$\int_{\Gamma_2'} d\log(u + iv) \neq 0,$$

then we may conclude that

$$\int_{\Gamma_2'} d\log(g'(\alpha', \beta')) \neq 0.$$

This inequality contradicts (6.1.12).   □

Corollary to the Example   The Cousin II problem in the Oka example cannot even be solved continuously (this is implicit in the proof but also follows from Proposition 6.1.11).

Cousin II can be solved continuously on any domain of holomorphy Ω that is homeomorphic to the ball. This assertion may be verified by inspection of the discussion in this section. A more precise criterion for continuous solvability will be derived in ensuing sections.

The next result is a typical application of Cousin II.

THEOREM 6.1.13   Let $\Omega \subseteq \mathbb{C}^n$ be a domain on which Cousin II is always solvable. Let $M \subseteq \Omega$ be a properly imbedded submanifold with the property that, for each $m \in M$, there is a neighborhood U_m of m in Ω and a holomorphic function $f_m : U_m \to \mathbb{C}$ such that $\{z \in U_m : f_m(z) = 0\} = M \cap U_m$ and $\partial f_m \neq 0$ on U_m. Then there is a holomorphic f on all of Ω such that $M = \{z \in \Omega : f(z) = 0\}$.

Proof   Let $\{U_i\}$ be a locally finite subcover of M subordinate to $\{U_m\}_{m \in M}$. Let $U_0 = \Omega \setminus M$. Then U_0 is open in Ω. Let f_i be the given holomorphic function on U_i and let f_0 on U_0 be identically equal to 1.

Define Cousin II data

$$g_{ij} = \frac{f_j}{f_i} \qquad \text{on } U_i \cap U_j. \tag{6.1.13.1}$$

Since Cousin II is solvable on Ω, there exist nonvanishing holomorphic g_i on U_i such that

$$g_{ij} = \frac{g_j}{g_i} \qquad \text{on } U_i \cap U_j. \tag{6.1.13.2}$$

Define

$$f(z) = \frac{f_i(z)}{g_i(z)} \qquad \text{on } U_i.$$

Then (6.1.13.1) and (6.1.13.2) imply that f is well defined and holomorphic on Ω. Also, $M = \{z \in \Omega : f(z) = 0\}$.   □

Exercises for the Reader

1. What do the first and second Cousin problems have to do with the theorems of Weierstrass and Mittag-Leffler in the function theory of one complex variable?

2. Review the second example and find the Cousin I problem lurking in the background. The method of that example used the *proof* of Cousin I rather than Cousin I itself. Indeed, it is a bit tricky to derive the result of the second example directly from Proposition 6.1.8. In the next section we will derive a more flexible tool—sheaf cohomology—that effectively subsumes the Cousin problems and makes such results easy.

6.2 Sheaves

Let $\mathcal{F}$ be a topological space, X be a Hausdorff space, and $\pi : \mathcal{F} \to X$ be a continuous mapping satisfying

(6.2.1) π is surjective;

(6.2.2) If $f \in \mathcal{F}$, then there is a neighborhood $W \subseteq \mathcal{F}$ of f such that $\pi|_W$ is a homeomorphism.

Then the triple $(\mathcal{F}, X, \pi)$ is called a *sheaf*. The letter $\mathcal{F}$ alludes to the French *fasceau* (sheaf), honoring the French mathematicians Leray and Serre, who developed the idea. K. Oka and H. Cartan first used sheaves in the theory of several complex variables. The map π is called the sheaf *projection*. If $x \in X$, then the inverse image $\mathcal{F}_x \equiv \pi^{-1}(\{x\})$ is called the *stalk* over x. If $U \subseteq X$ and $\mu : U \to \mathcal{F}$ is a continuous map that satisfies $\pi \circ \mu(x) = x$ for all $x \in U$, then μ is called a *section* of $\mathcal{F}$ over U. The set of all sections of $\mathcal{F}$ over U is denoted by $\Gamma(U, \mathcal{F})$. Frequently the stalks are each equipped with an algebraic structure such as group, ring, or module. In such a case we speak of a sheaf of groups, rings, or modules.

We shall see that sheaves are a natural vehicle for passing from local information to global information. Local statements are usually formulated in terms of elements of $\Gamma(U, \mathcal{F}), U$ a small open set in X. Global statements are generally seen to be assertions about $\Gamma(X, \mathcal{F})$. There are several important elementary examples that should be kept in mind for the rest of the chapter.

EXAMPLE   Let $X = M$ be a manifold. Define $\mathcal{F} = \mathbb{Z} \times M$ with the obvious topology. Let $\pi : \mathcal{F} \to M$ be given by $\pi(z, m) = m$. This is a *trivial*, or *product*, sheaf. If $U \subseteq M$, then $\mu : U \to \mathcal{F}$ given by $\mu(m) = (0, m)$ is a section of $\mathcal{F}$ over U. Of course, $\mathcal{F}$ is a sheaf of rings.   □

EXAMPLE   Let $X = M$ be a paracompact manifold and $m_0 \in M$. Define

$$\mathcal{F} = (\mathbb{Z} \times \{m_0\}) \cup (\{0\} \times (M \setminus \{m_0\})).$$

Topologize $\mathcal{F}$ as follows:

(i) For $(0, m) \in \mathcal{F}, m \neq m_0$, we let U be a neighborhoof of m in M that is disjoint from m_0. Then $\{0\} \times U$ is a neighborhood of $(0, m)$ in M.

(ii) The only neighborhood of $(k, m_0) \in \mathcal{F}$ is the singleton $\{(k, m_0)\}$ itself.

Using the sets described in (i) and (ii) as a subbase, we may generate a topology for $\mathcal{F}$. Of course the projection is given by $\pi(k, m) = m$. Then π is surjective and is a local homeomorphism. This sheaf is known as the *skyscraper sheaf*.   □

EXAMPLE   Let M be a paracompact C^∞ manifold. If $m \in M$ and $U \subseteq M$ is a neighborhood of m, then we define $C^\infty(m, U)$ to be the collection of all scalar-valued C^∞ functions defined on U. Let

$$C^\infty(m) = \bigcup_{U \ni m} C^\infty(m, U).$$

If $f, g \in C^\infty(m)$, then we define the relation $f \sim g$ if $f \equiv g$ on some small neighborhood of m. This is clearly an equivalence relation. The set $C_m^\infty \equiv C^\infty(m)/\sim$ is called *the ring of germs of C^∞ functions at m*.

If $f \in C^\infty(m)$ then we denote the residue class of f in C_m^∞ by either $\gamma_m f$ or $[f]_m$, depending on the context. Our sheaf $\mathcal{F}$ will be

$$\mathcal{F} = \bigcup_{m \in M} C_m^\infty.$$

This is the *sheaf of germs of C^∞ functions* on M.

Topologize $\mathcal{F}$ as follows: If $F \in \mathcal{F}$, let f be a representative of F. Thus f is defined on an open $W \subseteq M$. Let an open neighborhood of F in $\mathcal{F}$ be given by

$$\{f, W\} \equiv \{\gamma_x f : x \in W\}.$$

The set of all such $\{f, W\}$ generates the coarsest topology under which the projection $\pi : [f]_x \mapsto x$ is a local homeomorphism.

If a C^∞ function g on an open set $U \subseteq M$ is fixed, then the map

$$U \ni x \mapsto \gamma_x g \in \mathcal{F}$$

is a section of $\mathcal{F}$ over U. (Note, in particular, that a globally defined C^∞ function on M is just an element of $\Gamma(X, \mathcal{F})$.) Conversely, any section of $\mathcal{F}$ over an open set $U \subseteq M$ is induced by a C^∞ function on U.

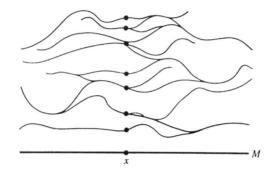

FIGURE 6.3

The given topology on $\mathcal{F}$ is not Hausdorff. To see this, take $M = \mathbb{R}$ and $m = 0$ and consider the two sheaf elements induced by the identically 0 function and by the function

$$f(x) = \begin{cases} 0 & \text{if } x \le 0, \\ e^{-1/x^2} & \text{if } x > 0. \end{cases} \qquad \square$$

Remark: Naively, a sheaf looks like Figure 6.3. In the last example, each C^∞ function on an open set $U \subseteq M$ corresponds to a "layer" in $\mathcal{F}$. For each $x \in M$, the stalk over x corresponds to the germs of all C^∞ functions at x. $\square$

Exercise for the Reader
Describe C_x^∞ in terms of formal power series. As you do this, keep in mind the theorem of E. Borel: that *any* formal power series about x is the Taylor series of some genuine C^∞ function defined in a neighborhood of x. Of course, the series will not, in general, converge to that C^∞ function.

The last example goes through in just the same way if we replace the C^∞ functions by either the C^k functions or the differential forms of degree m (or bidegree (p, q) on a complex manifold) or the real analytic functions (when M is a real analytic manifold) or the holomorphic functions (when M is a complex analytic manifold). The sheaf of germs of holomorphic functions *is* Hausdorff (exercise).

EXAMPLE   The constant sheaf $\mathbb{C} \times X$ is a sheaf over X if and only if $\mathbb{C}$ is equipped with the discrete topology (exercise). $\square$

Given a sheaf $(\mathcal{F}, X, \pi)$ of abelian groups (with group operations written *additively*) and a countable open covering $\mathcal{U} = \{U_i\}_{i=1}^\infty$ of X, we construct *cohomology groups* as follows. First, if $r \in \{0, 1, 2, \dots\}$, then an r-*cochain* with respect to $\mathcal{U}$ is a function f that assigns to each $(i_0, \dots, i_r) \in \mathbb{N}^{r+1}$ an element $f(i_0, \dots, i_r) \in \Gamma(U_{i_0} \cap \dots \cap U_{i_r}, \mathcal{F})$, subject to the condition that f is an alternating function of its indices. Let $C^r(\mathcal{U}, \mathcal{F})$ denote the set of all r-cochains of $\mathcal{F}$ with respect to the cover $\mathcal{U}$. Notice that C^r is an abelian group in a natural way. (Although we use the notation C^r in other contexts to denote spaces of smooth functions, there is little danger of confusion.)

Define the *coboundary operator*

$$\delta : C^r \to C^{r+1}$$

by

$$(\delta f)(i_0, \ldots, i_{r+1}) = \sum_{j=0}^{r+1} (-1)^j f\left(i_0, \ldots, \hat{i}_j, \ldots, i_{r+1}\right).$$

Here $\hat{\ }$ denotes omission.

LEMMA 6.2.3 We have

$$\delta^2 = 0.$$

Proof Observe that $\delta^2 f$ is a sum of terms

$$f(i_0, \ldots, \hat{i}_j, \ldots, \hat{i}_k, \ldots, i_{r+2})$$

with $j < k$. When i_j is removed first, the signature is $(-1)^{j+(k-1)}$. If i_k is removed first then the signature is $(-1)^{k+j}$. By cancellation, the sum is zero. □

Define $Z^r = Z^r(\mathcal{U}, \mathcal{F})$ to be those $f \in C^r$ such that $\delta f = 0$. Define $B^r = B^r(\mathcal{U}, \mathcal{F})$ to be those $g \in C^r$ such that $g = \delta f$ for some $f \in C^{r-1}$, $r \geq 1$. Then Z^r and B^r are subgroups of C^r. We call B^r the group of *coboundaries* and Z^r the group of *cocycles*.

Notice that B^r is a subgroup of Z^r. Thus the quotient

$$H^r(\mathcal{U}, \mathcal{F}) = H^r \equiv Z^r / B^r, \qquad r \geq 1,$$

is well defined and is called the rth *cohomology group of X* (with respect to the covering $\mathcal{U}$) *with coefficients in the sheaf $\mathcal{F}$*. By convention, since B^0 must clearly be $\emptyset$, we set $H^0 = Z^0$.

In more sophisticated language, the lemma tells us that we have the *semi-exact sequence*

$$C^0 \xrightarrow{\delta} C^1 \xrightarrow{\delta} C^2 \xrightarrow{\delta} \cdots$$

where, at each C^r, $B^r = \mathrm{Im}\,(\delta|_{C^{r-1}}) \subseteq \mathrm{Ker}\,(\delta|_{C^r}) = Z^r$. We measure the extent to which this sequence fails to be exact at C^r by way of the cohomology group H^r.

EXAMPLE Let Ω be the annulus in $\mathbb{C}$ with $\mathcal{U} = \{U_1, U_2, U_3\}$ the covering indicated in the first example of Section 6.1. Let $\mathcal{F} = \mathbb{Z} \times \Omega$. Then $H^0(\mathcal{U}, \mathcal{F}) \equiv Z^0$ consists of those f on $I = \{1, 2, 3\}$ such that $f(i_0) - f(i_1) = 0$ on $U_{i_0} \cap U_{i_1}$ for all i_0, i_1. That is, f assigns to each U_i the same integer. In brief, $H^0 = Z^0 \cong \Gamma(\Omega, \mathcal{F}) \cong \mathbb{Z}$.

Now Z^1 consists of those f on $I \times I$ such that $f(i_1, i_2) - f(i_0, i_2) + f(i_0, i_1) = 0$ on $U_{i_0} \cap U_{i_1} \cap U_{i_2}$. (*Note:* This is vacuously satisfied since $U_1 \cap U_2 \cap U_3 = \emptyset$.) Thus $Z^1 = C^1$.

On the other hand, $Z^1 \ni f = \delta g$ for some $g \in C^0$ if and only if $f(i_0, i_1) = g(i_1) - g(i_0)$. (This simply says that the analogue of Cousin I can be solved for the sheaf $\mathcal{F}$.) In summary, $H^1 \equiv Z^1/B^1 = C^1/B^1$ counts the number of unsolvable Cousin I problems with data in the group $\mathbb{Z}$.

The 1-cocycle

$$f(1,2) = 1, f(1,3) = 0, f(2,3) = 0$$

is clearly not a coboundary. Also the 1-cocycle

$$g(1,2) = 0, g(1,3) = 1, g(2,3) = 0$$

is not a coboundary. Finally, the 1-cocycle

$$h(1,2) = 0, h(1,3) = 0, h(2,3) = 1$$

is not a coboundary. However, notice that both $g + f$ and $h + f$ are coboundaries. Since f, g, h clearly generate all of Z^1 as a group, it follows that the residue of f alone in H^1 generates H^1. Thus H^1 is cyclic. Indeed, $H^1 \equiv \mathbb{Z}$.

Finally, $Z^r = B^r = C^r$ for $r \geq 2$ for vacuous reasons, hence $H^r = 0$ for $r \geq 2$.

Thus we see that the annulus Ω has a "one-dimensional" hole, and it is detected by $H^1(\mathcal{U}, \mathcal{F})$. The annulus has no holes of dimension two or higher. □

EXAMPLE   Let $\Omega, \{U_i\}$ be as in the last example. However, let $\mathcal{F}$ be the sheaf of germs of C^k functions. Then Z^0 consists of those $f(i)$ such that $f(i_0) = f(i_1)$ on $U_{i_0} \cap U_{i_1}$. We therefore see that $Z^0 = C^k(\Omega)$. Just as in Section 6.1, we may always use partitions of unity to solve Cousin I for C^k functions. Thus $H^1(\Omega, \mathcal{F}) = 0$. This is a manifestation of the fact that the sheaf of germs of C^k functions (or C^∞ functions) is *fine* (see E. Spanier [1]).= For trivial reasons, $H^r = 0$ when $r \geq 2$.   □

EXAMPLE   Let $\Omega, \mathcal{U}$ be as above. Let $\mathcal{F}$ be the sheaf of germs of holomorphic functions on Ω. As in the preceding example, $H^0(\Omega, \mathcal{F})$ is isomorphic to the group of holomorphic functions on Ω. By Proposition 6.1.8, Cousin I can always be solved on Ω, so $H^1 = 0$. Finally, $H^r = 0, r \geq 2$, for trivial reasons.   □

EXAMPLE   Let $\Omega = \{z \in \mathbb{C} : |z| < 2\}$ and let $\mathcal{V}$ be the covering (of Ω^1) as in the first example of Section 6.1. Let $\mathcal{F} = \mathbb{Z} \times \Omega$ as in the example following 6.2.3. Then $H^0 = Z^0 = \mathbb{Z}$, as in the example above for the annulus.

Now $f \in C^1$ satisfies $\delta f = 0$ precisely when $f(2,3) - f(1,3) + f(1,2) = 0$ on $V_1 \cap V_2 \cap V_3$ (this condition is no longer vacuous). Also, $f = \delta g$ precisely when $f(i_0, i_1) = g(i_1) - g(i_0)$. As discussed in Section 6.1, it follows that $\delta f = 0$ implies that $f = \delta g$ for some $g \in C^0$. Hence $H^1 = A^1/B^1 = 0$. For trivial reasons, $H^r = 0$ when $r \geq 2$. □

Exercise for the Reader
    Let $\Omega \subseteq \mathbb{C}^n$ be an open set and $\mathcal{U} = \{U_j\}$ an open covering of Ω. Let $(\mathcal{F}, \Omega, \pi)$ be the sheaf of germs of holomorphic functions. Then $H^1(\mathcal{U}, \mathcal{F}) = 0$ if and only if every Cousin I problem for the covering $\mathcal{U}$ can be solved.

Return now to the general setting of arbitrary Hausdorff X, sheaf of abelian groups $\mathcal{F}$, and covering $\mathcal{U}$. We need to see to what extent the cohomoloy groups depend on the covering $\mathcal{U}$. The hope is that, if the covering becomes sufficiently fine, then the cohomology groups are independent of the choice of covering. When formulated suitably carefully, this hope is fulfilled.
    If $\mathcal{V} = \{V_i\}$ is a refinement of the covering $\mathcal{U}$, then we need to relate the cohomology computed with respect to $\mathcal{V}$ to that computed with respect to $\mathcal{U}$. In the end we hope to see that the cohomology stabilizes under the direct limit process for coverings.
    To make the preceding paragraphs precise, we let $\mathcal{U} = \{U_j\}_{j \in J}$ be an open covering for X and $\mathcal{V} = \{V_i\}_{i \in I}$ be an open covering that refines $\mathcal{U}$. Let $\sigma : I \to J$ satisfy $V_i \subseteq U_{\sigma(i)}$ for all $i \in I$. We call σ an *affinity function*. Then σ induces a map σ^* of $C^r(\mathcal{U}, \mathcal{F})$ to $C^r(\mathcal{V}, \mathcal{F})$ as follows: Let $f \in C^r(\mathcal{U}, \mathcal{F})$. Let $(i_0, \ldots, i_r) \in \mathbb{N}^{r+1}$. Define $\sigma^* f \in C^r(\mathcal{V}, \mathcal{F})$ by

$$(\sigma^* f)(i_0, \ldots, i_r) = f(\sigma(i_0), \ldots, \sigma(i_r))|_{V_{i_0} \cap \cdots \cap V_{i_r}}.$$

Exercise for the Reader
Verify the following simple facts about σ^* :

1. σ^* is a homomorphism;

2. $\delta \sigma^* = \sigma^* \delta$;

3. $\sigma^*(Z^r) \subseteq Z^r$ and $\sigma^*(B^r) \subseteq B^r$;

4. By (b), the induced map $\sigma^* : H^r(\mathcal{U}, \mathcal{F}) \to H^r(\mathcal{V}, \mathcal{F})$ is well defined.

We call two r-cocycles with respect to a fixed covering *cohomologous* if they differ by a coboundary (i.e., if they are representatives of the same cohomology class in H^r). The next lemma is technical but essential.

LEMMA 6.2.4   Let $\mathcal{U}$ be a covering of X and $\mathcal{V}$ be a refinement of $\mathcal{U}$. Let σ and σ' be two affinity functions for $\mathcal{U}$ and $\mathcal{V}$. Then $\sigma^* = \sigma'^*$ on H^r.

Proof   We proceed by means of a standard device of homological algebra known as a *chain homotopy*. Since the assertion is obvious for $r = 0$, we may assume that $r \geq 1$. Let $\Theta : C^r(\mathcal{U}, \mathcal{F}) \to C^{r-1}(\mathcal{V}, \mathcal{F})$ be given by

$$(\Theta f)(i_0, \ldots, i_{r-1})$$
$$= \sum_{\ell=0}^{r-1} (-1)^l f(\sigma(i_0), \ldots, \sigma(i_{\ell-1}), \sigma(i_\ell), \sigma'(i_\ell), \sigma'(i_{\ell+1}), \ldots, \sigma'(i_{r-1})).$$

(Notice that Θ is a sort of adjoint for δ.) We claim that

$$(\sigma')^* f - \sigma^* f = \Theta \delta f + \delta \Theta f. \tag{6.2.4.1}$$

This is the chain homotopy condition. Assuming that we have proved (6.2.4.1), notice that if $f \in C^r(\mathcal{U}, \mathcal{F}), \delta f = 0$, then $(\sigma')^* f - \sigma^* f = \delta \Theta f$. Thus $(\sigma')^* f$ and $\sigma^* f$ are cohomologous. The result follows.

It remains to prove (6.2.4.1). We compute that

$$\delta(\Theta f)(i_0, \ldots, i_r) = \sum_{k=0}^{r} (-1)^k (\Theta f)(i_0, \ldots, \hat{i}_k, \ldots, i_r)$$

$$= \sum_{k=0}^{r} (-1)^k \left\{ \sum_{\ell=0}^{k-1} (-1)^\ell f\left(\sigma(i_0), \ldots, \sigma(i_\ell), \sigma'(i_\ell), \ldots, \widehat{\sigma'(i_k)}, \ldots, \sigma'(i_r) \right) \right.$$

$$\left. + \sum_{\ell=k}^{r-1} (-1)^\ell f\left(\sigma(i_0), \ldots, \widehat{\sigma(i_k)}, \ldots, \sigma(i_{\ell+1}), \sigma'(i_{\ell+1}), \ldots, \sigma'(i_r) \right) \right\}$$

$$= \sum_{k=0}^{r} \sum_{\ell=0}^{k-1} (-1)^{k+\ell} f\left(\sigma(i_0), \ldots, \sigma(i_\ell), \sigma'(i_\ell), \ldots, \widehat{\sigma'(i_k)}, \ldots, \sigma'(i_r) \right)$$

$$- \sum_{k=0}^{r} \sum_{p=k+1}^{r} (-1)^{p+\ell} f\left(\sigma(i_0), \ldots, \widehat{\sigma(i_k)}, \ldots, \sigma(i_p), \sigma'(i_p), \ldots, \sigma'(i_r) \right).$$

On the other hand,

$$\Theta(\delta f)(i_0, \ldots, i_r)$$
$$= \sum_{\ell=0}^{r} (-1)^\ell \delta f(\sigma(i_0), \ldots, \sigma(i_\ell), \sigma'(i_\ell), \ldots, \sigma'(i_r))$$

$$= \sum_{\ell=0}^{r}(-1)^{\ell}\left\{\sum_{k=0}^{\ell}(-1)^k f\left(\sigma(i_0),\ldots,\widehat{\sigma(i_k)},\ldots,\right.\right.$$

$$\sigma(i_\ell),\sigma'(i_\ell),\ldots,\sigma'(i_r))$$

$$+\sum_{k=\ell+1}^{r+1}(-1)^k f\left(\sigma(i_0),\ldots,\sigma(i_\ell),\sigma'(i_\ell),\ldots,\right.$$

$$\left.\left.\sigma'(\widehat{i_{k-1}}),\ldots,\sigma'(i_r))\right\}\right.$$

$$= \sum_{\ell=0}^{r}\sum_{k=0}^{\ell}(-1)^{\ell+k} f\left(\sigma(i_0),\ldots,\widehat{\sigma(i_k)},\ldots,\sigma(i_\ell),\sigma'(i_\ell),\ldots,\sigma'(i_r)\right)$$

$$-\sum_{\ell=0}^{r}\sum_{p=\ell}^{r}(-1)^{p+\ell} f\left(\sigma(i_0),\ldots,\sigma(i_\ell),\sigma'(i_\ell),\ldots,\widehat{\sigma'(i_p)},\ldots,\sigma'(i_r)\right)$$

$$= \sum_{\ell=0}^{r}\sum_{k=0}^{\ell-1}(-1)^{\ell+k} f\left(\sigma(i_0),\ldots,\widehat{\sigma(i_k)},\ldots,\sigma(i_\ell),\sigma'(i_\ell),\ldots,\sigma'(i_r)\right)$$

$$-\sum_{\ell=0}^{r}\sum_{p=\ell+1}^{r}(-1)^{p+\ell} f\left(\sigma(i_0),\ldots,\sigma(i_\ell),\sigma'(i_\ell),\ldots,\widehat{\sigma'(i_p)},\ldots,\sigma'(i_r)\right)$$

$$+\left\{\sum_{\ell=0}^{r} f\left(\sigma(i_0),\ldots,\sigma(i_{\ell-1}),\sigma'(i_\ell),\ldots,\sigma'(i_r)\right)\right.$$

$$\left.-\sum_{\ell=0}^{r} f\left(\sigma(i_0),\ldots,\sigma(i_\ell),\sigma'(i_{\ell+1}),\ldots,\sigma'(i_r)\right)\right\}$$

$$= -\delta(\Theta f)(i_0,\ldots,i_r) + \{f(\sigma'(i_0),\ldots,\sigma'(i_r)) - f(\sigma(i_0),\ldots,\sigma(i_r))\}.$$

Here the first part of the last equality comes from renaming indices. Moreover, the equality of the expressions in braces holds because the series cancel. Now the last line is

$$= -\delta(\Theta f)(i_0,\ldots,i_r) + \{(\sigma')^* f(i_0,\ldots,i_r) - \sigma^* f(i_0,\ldots,i_r)\}.$$

It follows that

$$(\Theta\delta)f(i_0,\ldots,i_r) + (\delta\Theta)f(i_0,\ldots,i_r) = (\sigma')^* f(i_0,\ldots,i_r) - \sigma^* f(i_0,\ldots,i_r)$$

as desired. $\qquad\square$

If $f \in Z^r(\mathcal{U},\mathcal{F})$, then it is convenient to adopt the standard notation $[f]$ to denote the equivalence class of f in $H^r = Z^r/B^r$. The lemma says that $\sigma^*([f]) = \sigma'^*([f])$ for all $[f] \in H^r(\mathcal{U},\mathcal{F})$ and any two affinity functions σ, σ'. If $\mathcal{V}$ is a refinement of $\mathcal{U}$, then f induces a well-defined cohomology class with respect to $\mathcal{V}$ (namely, $\sigma^* f$ for *any* affinity function σ). Let

$$\mathcal{Z}^r(X,\mathcal{F}) \equiv \bigcup_{\text{coverings } \mathcal{U} \text{ of } X} Z^r(\mathcal{U},\mathcal{F}).$$

Let $f, g \in \mathcal{Z}^r$—say that $f \in Z^r(\mathcal{U},\mathcal{F}), g \in Z^r(\mathcal{V},\mathcal{F})$. We write $f \sim g$ provided that there is a common refinement $\mathcal{W}$ such that the induced cohomology classes of f and g with respect to $\mathcal{W}$ are equal. Then $\sim$ is an equivalence relation (exercise). The collection of equivalence classes is denoted $H^r(X,\mathcal{F})$, the rth cohomology of X with coefficients in the sheaf $\mathcal{F}$. In an obvious fashion, $H^r(X,\mathcal{F})$ is an abelian group. If $[f] \in H^r(\mathcal{U},\mathcal{F})$, let $[[f]]$ denote the residue class of $[f]$ in $H^r(X,\mathcal{F})$. Thus $[f] \mapsto [[f]]$ is a homomorphism of $H^r(\mathcal{U},\mathcal{F})$ into $H^r(X,\mathcal{F})$ for each r and each covering $\mathcal{U}$. The group $H^r(X,\mathcal{F})$ is the *direct limit* of the groups $H^r(\mathcal{U},\mathcal{F})$.

Each element of $H^r(X,\mathcal{F})$ comes from some covering $\mathcal{U}$. Thus the elements of $H^0(X,\mathcal{F})$ are in one-to-one correspondence with the global sections of $\mathcal{F}$. To understand $H^1(X,\mathcal{F})$, we need the next lemma.

LEMMA 6.2.5   Let $\mathcal{U}$ be an open covering of X and $\mathcal{V}$ a covering that is a refinement of $\mathcal{U}$. Let σ be an affinity function for $\mathcal{U}$ and $\mathcal{V}$. Then $\sigma^* : H^1(\mathcal{U},\mathcal{F}) \to H^1(\mathcal{V},\mathcal{F})$ is injective.

COROLLARY 6.2.6   For any covering $\mathcal{U}$ of X it holds that the map $H^1(\mathcal{U},\mathcal{F}) \to H^1(X,\mathcal{F})$ is injective.

Proof of Lemma 6.2.5   It suffices to show that $\operatorname{Ker} \sigma^* = [0]$. For this, we must prove that if $c \in Z^1(\mathcal{U},\mathcal{F})$ satisfies $\sigma^* c \in B^1(\mathcal{V},\mathcal{F})$, then $c \in B^1(\mathcal{U},\mathcal{F})$. To see this, notice that the hypothesis $\sigma^* c \in B^1(\mathcal{V},\mathcal{F})$ implies that $\sigma^* c = \delta\gamma$ for some $\gamma \in C^0(\mathcal{V},\mathcal{F})$. With the notation $\mathcal{V} = \{V_i\}_{i \in I}, \mathcal{U} = \{U_j\}_{j \in J}$, we have for all $m, n \in I$ that

$$\gamma_n - \gamma_m = c_{\sigma(m)\sigma(n)} = c_{\sigma(m)j} - c_{\sigma(n)j} \quad \text{on} \quad U_j \cap V_m \cap V_n$$

(here we have used the condition that $\delta c = 0$). In other words,

$$\gamma_n + c_{\sigma(n)j} = \gamma_m + c_{\sigma(m)j} \quad \text{on} \quad U_j \cap V_m \cap V_n.$$

Hence the section $e_j \in \Gamma(U_j,\mathcal{F})$ given by

$$e_j = \gamma_n + c_{\sigma(n)j} \quad \text{on} \quad U_j \cap V_n$$

is well defined. We now compute, on $U_k \cap U_\ell \cap V_n$,

$$e_\ell - e_k = (\gamma_n + c_{\sigma(n)\ell}) - (\gamma_n + c_{\sigma(n)k}) = c_{k\sigma(n)} + c_{\sigma(n)\ell} = c_{k\ell}.$$

Hence $c = \delta e$ or $c \in B^1(\mathcal{U},\mathcal{F})$.   □

It follows from Corollary 6.2.6 that $H^1(X, \mathcal{F}) = 0$ if and only if $H^1(\mathcal{U}, \mathcal{F}) = 0$ for some covering $\mathcal{U}$. Therefore Cousin I for the sheaf $\mathcal{F}$ is always solvable if and only if $H^1(X, \mathcal{F}) = 0$.

Unfortunately, there are no such simple descriptions of $H^p(X, \mathcal{F})$ in terms of $H^p(\mathcal{U}, \mathcal{F})$ for $p \geq 2$. On the other hand, the following important result of J. Leray often makes it unnecessary to pass to the direct limit in order to calculate with cohomology.

THEOREM 6.2.7   Let X be a paracompact manifold, $\mathcal{F}$ a sheaf over X, and $\mathcal{U} = \{U_i\}_{i \in I}$ an open covering on X such that $H^p(U_{i_0} \cap \cdots \cap U_{i_k}, \mathcal{F}) = 0$ for every $p \geq 1$ and every choice of $i_0, \ldots, i_k$. Then $H^p(\mathcal{U}, \mathcal{F}) = H^p(X, \mathcal{F})$ for all $p \geq 0$.

We do not supply a proof of this theorem but instead refer the reader to Gunning [1, p. 44]. It should be noted that, in many applications with X an open domain in Euclidean space, a covering by balls or polydiscs will plainly satisfy the hypotheses of Leray's theorem. So the theorem is not difficult to use.

Our goal now is to develop a little homological algebraic machinery so that we can prove the Dolbeault theorem to the effect that cohomology with coefficients in the sheaf of germs of holomorphic functions is equivalent to the cohomology arising from the $\bar{\partial}$ complex. We begin with the following exercise.

Exercise for the Reader
Let X be a C^∞ paracompact manifold. Let $\mathcal{E}$ be the sheaf of germs of C^∞ functions on Ω. Let $\mathcal{F}$ be a sheaf of $\mathcal{E}$-modules (i.e., each $\mathcal{F}_x$ is an $\mathcal{E}$-module). Then $H^p(\mathcal{U}, \mathcal{F}) = 0$ for every open covering $\mathcal{U} = \{U_i\}_{i \in I}$ and every $p > 0$. (*Hint:* This will be similar to the solution of Cousin I for C^∞ functions. Let $\{\phi_i\}_{i \in I}$ be a partition of unity subordinate to $\mathcal{U}$. If $c \in Z^p(\mathcal{U}, \mathcal{F})$ and $s = (s_0, s_1, \ldots, s_{p-1}) \in I^p$, then set $e(i) = \sum_k \phi_k c_{k,i}$. Then $\delta e = c$.)

The *métier* of homological algebra is exact sequences. If A, B, C are abelian groups and

$$A \overset{\phi}{\to} B \overset{\psi}{\to} C$$

are homomorphisms, then we say that this sequence is *exact* at B if Image $\phi = $ Ker ψ. If $C = \{0\}$ then "exactness" at B means that ϕ is surjective. If $A = \{0\}$, then exactness at B means that ψ is injective.

If $\mathcal{F}, \mathcal{G}, \mathcal{H}$ are sheaves (of abelian groups) over X and

$$0 \to \mathcal{F} \overset{\phi}{\to} \mathcal{G} \overset{\psi}{\to} \mathcal{H} \to 0$$

are homomorphisms (on each stalk), then "exactness" is defined in an obvious manner. These homomorphisms trivially induce homomorphisms

$$0 \to C^p(\mathcal{U}, \mathcal{F}) \overset{\phi^*}{\to} C^p(\mathcal{U}, \mathcal{G}) \overset{\psi^*}{\to} C^p(\mathcal{U}, \mathcal{H}),$$

any $p > 0$, any open covering $\mathcal{U} = \{U_i\}_{i \in I}$. The induced sequence is also exact (exercise), except that the final map may not be surjective (which is why we have omitted the final 0).

EXAMPLE   Let $\mathcal{F}$ be the constant sheaf $2\pi i \mathbb{Z} \times X$, where $X = \{z \in \mathbb{C} : 1 < |z| < 2\}$. Let $\mathcal{G}$ be the sheaf of germs of C^∞ functions over X. Let $\mathcal{H}$ be the sheaf of germs of nonvanishing C^∞ functions (under multiplication). In the diagram

$$0 \to \mathcal{F} \xrightarrow{\phi} \mathcal{G} \xrightarrow{\psi} \mathcal{H} \to 0 \qquad (6.2.8)$$

we take ϕ to be the trivial injection and ψ the exponential map. Then the sequence is exact (because we check the assertion on the stalk level). But the induced exact sequence

$$0 \to C^0(\mathcal{U}, \mathcal{F}) \xrightarrow{\phi^*} C^0(\mathcal{U}, \mathcal{G}) \xrightarrow{\psi^*} C^0(\mathcal{U}, \mathcal{H}) \qquad (6.2.9)$$

is *not* surjective at the final stage for the covering $\mathcal{U} = \{X\}$ (or for *any* covering) because not every C^∞ function on the space X has a well-defined C^∞ logarithm. □

This failure of exactness is a source of both difficulty and rich theory. Indeed, the failure of an algebraic property of groups (in this case the property of exactness) can often be measured by a *resolution* of groups for which the property holds. In the present case, the resolution is given by an exact sequence of cohomology groups. We construct the resolution as follows.

Instead of (6.2.9) we write

$$0 \to C^p(\mathcal{U}, \mathcal{F}) \xrightarrow{\phi^*} C^p(\mathcal{U}, \mathcal{G}) \xrightarrow{\psi^*} C_0^p(\mathcal{U}, \mathcal{H}) \to 0, \qquad (6.2.10)$$

where C_0^p is precisely the subgroup of C^p that is the image of the map ψ^* induced by ψ. We have the commutative diagram given in Figure 6.4. The phrase *commutative diagram* means, for instance, that $\phi^* \circ \delta$ (where $\delta : C^{p-1}(\mathcal{U}, \mathcal{F}) \to C^p(\mathcal{U}, \mathcal{F})$ and $\phi^* : C^p(\mathcal{U}, \mathcal{F}) \to C^p(\mathcal{U}, \mathcal{G})$) equals $\delta \circ \phi^*$ (where $\delta : C^{p-1}(\mathcal{U}, \mathcal{G}) \to C^p(\mathcal{U}, \mathcal{G})$ and $\phi^* : C^{p-1}(\mathcal{U}, \mathcal{F}) \to C^{p-1}(\mathcal{U}, \mathcal{G})$) and likewise for all the other "boxes" in Figure 6.4. (The only possible way that one can write about this subject is to force the reader to verify the details.)

It follows that there are induced maps

$$0 \to Z^p(\mathcal{U}, \mathcal{F}) \to Z^p(\mathcal{U}, \mathcal{G}) \to Z_0^p(\mathcal{U}, \mathcal{H}) \to 0$$

and

$$0 \to B^p(\mathcal{U}, \mathcal{F}) \to B^p(\mathcal{U}, \mathcal{G}) \to B_0^p(\mathcal{U}, \mathcal{H}) \to 0.$$

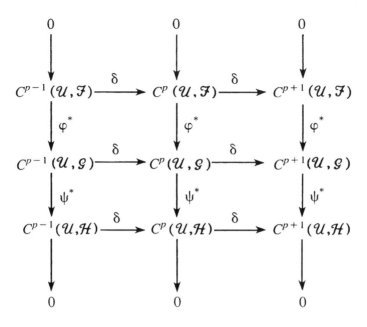

FIGURE 6.4

Hence the maps

$$0 \to \quad H^0(\mathcal{U}, \mathcal{F}) \quad \to H^0(\mathcal{U}, \mathcal{G}) \to H^0_0(\mathcal{U}, \mathcal{H})$$
$$H^1(\mathcal{U}, \mathcal{F}) \quad \to H^1(\mathcal{U}, \mathcal{G}) \to H^1_0(\mathcal{U}, \mathcal{H})$$
$$\vdots \qquad\qquad\qquad (6.2.11)$$

are well defined. We wish to construct homomorphisms that connect up the rows.

PROPOSITION 6.2.12 The commutative diagram (Figure 6.4) induces a natural homomorphism

$$\delta^* : H^p_0(\mathcal{U}, \mathcal{H}) \to H^{p+1}(\mathcal{U}, \mathcal{F})$$

for every $p \geq 0$.

Proof   Let $[h] \in H^p_0(\mathcal{U}, \mathcal{H})$, and let h be a representative. Since ψ^* is surjective, there is a $g \in Z^p(\mathcal{U}, \mathcal{G})$ such that $\psi^* g = h$. We claim that $[(\phi^*)^{-1}\delta g]$ makes sense and is uniquely determined as an element of $H^{p+1}(\mathcal{U}, \mathcal{F})$ by $[h]$.

Now $\psi^* \delta g = \delta \psi^* g = \delta h = 0$, whence, by exactness, $\delta g \in \phi^*(C^{p+1}(\mathcal{U}, \mathcal{F}))$. So our expression $[(\phi^*)^{-1}\delta g]$ makes sense but may not be uniquely determined. If $\delta g = \phi^* f$, then $\phi^* \delta f = \delta \phi^* f = \delta \delta g = 0$. Thus the injectivity of ϕ^* implies that $\delta f = 0$. So $\delta g \in \phi^*(Z^{p+1}(\mathcal{U}, \mathcal{F}))$.

It remains to check that when $[h] = 0$, that is, $h \in B_0^p(\mathcal{U}, \mathcal{H})$, then $(\phi^*)^{-1}\delta g \in B^{p+1}(\mathcal{U}, \mathcal{F})$. Now $h \in B_0^p(\mathcal{U}, \mathcal{H})$ implies that $h = \delta k$, some $k \in C_0^{p-1}(\mathcal{U}, \mathcal{H})$. The surjectivity of ψ^* yields an $m \in C_0^{p-1}(\mathcal{U}, \mathcal{G})$ with $\psi^* m = k$. But then $\delta m - g$ (with g as in the first paragraph) satisfies

$$\psi^*(\delta m - g) = \psi^* \delta m - \psi^* g = \delta \psi^* m - h = \delta k - h = 0.$$

By exactness, we have $\delta m - g \in \phi^*(C^p(\mathcal{U}, \mathcal{F}))$; say that $\delta m - g = \phi^* \mu$, some $\mu \in C^p(\mathcal{U}, \mathcal{F})$. Then

$$-\delta g = \delta(\delta m - g) = \delta \phi^* \mu = \phi^* \delta \mu \in \phi^*(B^{p+1}(\mathcal{U}, \mathcal{F})). \qquad \square$$

THEOREM 6.2.13 (The Snake Lemma)   The sequence

$$0 \;\to\; H^0(\mathcal{U}, \mathcal{F}) \xrightarrow{\phi^*} H^0(\mathcal{U}, \mathcal{G}) \xrightarrow{\psi^*} H_0^0(\mathcal{U}, \mathcal{H}) \xrightarrow{\delta^*} H^1(\mathcal{U}, \mathcal{F})$$
$$\xrightarrow{\phi^*} H^1(\mathcal{U}, \mathcal{G}) \xrightarrow{\psi^*} H_0^1(\mathcal{U}, \mathcal{H}) \xrightarrow{\delta^*} H^2(\mathcal{U}, \mathcal{F}) \to \cdots$$

is exact.

Proof
Exactness at $H^p(\mathcal{U}, \mathcal{F})$:   Reviewing the proof of Proposition 6.2.12, we note that the image of δ^* in $H^p(\mathcal{U}, \mathcal{F})$ is precisely the set of cohomology classes determined by $Z^p(\mathcal{U}, \mathcal{F}) \cap (\phi^*)^{-1}\{B^p(\mathcal{U}, \mathcal{G})\}$. On the other hand, the kernel of ϕ^* in $H^p(\mathcal{U}, \mathcal{F})$ must be the subgroup of $(\phi^*)^{-1}\{B^p(\mathcal{U}, \mathcal{G})\}$ that lies in $Z^p(\mathcal{U}, \mathcal{F})$. (Note that exactness at the stage $H^0(\mathcal{U}, \mathcal{F})$ is trivial.)

Exactness at $H^p(\mathcal{U}, \mathcal{G})$:   That Image $\phi^* \subseteq \operatorname{Ker} \psi^*$ is immediate. Conversely, if $[g] \in \operatorname{Ker} \psi^*$, then let $g \in Z^p$ be a representative for $[g]$. Then $\psi^* g \in B_0^p(\mathcal{U}, \mathcal{H})$, so $\psi^* g = \delta f$ for some $f \in C_0^{p-1}(\mathcal{U}, \mathcal{H})$. Referring again to Figure 6.4, we see that since ψ is surjective, there is an $e \in C^{p-1}(\mathcal{U}, \mathcal{G})$ with $\psi^* e = f$. But then

$$\psi^*(g - \delta e) = \psi^* g - \psi^* \delta e = \delta f - \delta \psi^* e = \delta f - \delta f = 0.$$

Therefore, $g - \delta e \in \operatorname{Ker} \psi^*$ *on the cochain level.* Thus there is an $h \in C^p(\mathcal{U}, \mathcal{F})$ with $\phi^* h = g - \delta e$. Also, $\phi^* \delta h = \delta \phi^* h = \delta(g - \delta e) = 0 - 0 = 0$. Since ϕ^* is injective, we conclude that $\delta h = 0$, hence $h \in Z^p$. Finally, $\phi^*[h] = [\phi^* h] = [g - \delta e] = [g]$, as desired.

Exactness at $H^p(\mathcal{U}, \mathcal{H})$:   If $[g] \in H^p(\mathcal{U}, \mathcal{G})$, then $\delta^* \psi^*[g] = (\phi^*)^{-1}\delta[g]$ by definition of δ^*. But $\delta[g] = 0$ and ϕ^* is univalent. Hence $\delta^* \psi^*[g] = 0$.
    Now suppose that $[h] \in H^p(\mathcal{U}, \mathcal{H})$ satisfies $\delta^*[h] = 0$. This means, according to the definition of δ^*, that there is a $g \in C^p(\mathcal{U}, \mathcal{G})$ with $\psi^* g = h$ and $\delta^*[h] = (\phi^*)^{-1}\delta[g] = 0$. Set $x = (\phi^*)^{-1}\delta g$. Then x is a coboundary, so $x = \delta x_1$ for some $x_1 \in C^p(\mathcal{U}, \mathcal{F})$. Let $f = g - \phi^* x_1$. Then $\delta f = \delta g - \phi^* \delta x_1 = \delta g - \phi^* x = 0$. Hence $f \in Z^p(\mathcal{U}, \mathcal{G})$. Finally, $\psi^*[f] = [\psi^* g - \psi^* \phi^* x_1] = [\psi^* g] = [h]$, as desired. $\qquad \square$

In the ensuing sections we apply the long exact cohomology sequence of Theorem 6.2.13. However, we first wish to eliminate the need for the special groups H_0^p. If $\mathcal{U} = \{U_j\}_{j \in J}$ is an open covering of X and $\mathcal{V} = \{V_i\}_{i \in I}$ is another open covering that refines $\mathcal{U}$, let $\sigma : I \to J$ be an affinity function for these coverings. Then, as above, σ induces a homomorphism

$$\sigma^* : H_0^p(\mathcal{U}, \mathcal{H}) \to H_0^p(\mathcal{V}, \mathcal{H})$$

(where we continue to use the notation of the Snake Lemma).

LEMMA 6.2.14 Assume that X is a paracompact manifold. Let $\mathcal{U} = \{U_j\}_{j \in J}$ be an open covering of X and $g \in C^p(\mathcal{U}, \mathcal{H})$. Then there is a refinement $\mathcal{V} = \{V_i\}_{i \in I}$ of $\mathcal{U}$ and an affinity function $\sigma : I \to J$ such that $\sigma^* g \in C_0^p(\mathcal{V}, \mathcal{H})$.

Proof   By passing to a refinement, we may suppose that $\mathcal{U}$ is locally finite. Also we may choose open sets $\{W_j\}_{j \in J}$ such that $\cup_j W_j = X$ and $\overline{W}_j \subseteq U_j$, each j. Now we claim that each $x \in X$ has an open neighborhood V_x satisfying the following conditions:

(6.2.14.1) If $s \in J^{p+1}$ and $x \in U_{s_0} \cap \cdots \cap U_{s_p} \equiv U_s$, then $V_x \subseteq U_s$ and there is a $\mu \in \Gamma(V_x, \mathcal{G})$ such that $\psi \circ \mu = g$.

(6.2.14.2) If $x \in W_j$, then $V_x \subseteq W_j$.

(6.2.14.3) If $V_x \cap W_j \neq \emptyset$, then $V_x \subseteq U_j$.

To see that such V_x exist, note that the fact that the manifold is normal and that $\mathcal{U}$ is locally finite makes all the assertions but the second part of (6.2.14.1) trivial. But the latter follows from part (6.2.2) of the definition of sheaf. For each $x \in X$, let $\rho(x) \in J$ be such that $x \in W_{\rho(x)}$. By (6.2.14.2), we know that $V_x \subseteq W_{\rho(x)}$. If $x_0, \ldots, x_p \in X$ and $V_{x_0} \cap \cdots \cap V_{x_p} \neq \emptyset$, then $V_{x_0} \cap W_{\rho(x_k)} \neq \emptyset$ for all $k = 0, \ldots, p$. Hence, by (6.2.14.3), $V_{x_0} \subseteq U_{\rho(x_0)} \cap \cdots \cap U_{\rho(x_p)}$. It follows from (6.2.14.1) that $g_{\rho(x_0) \cdots \rho(x_p)} \in \Gamma(U_{\rho(x_0)} \cap \cdots \cap U_{\rho(x_p)}, \mathcal{H}) \subseteq \Gamma(V_{x_0} \cap \cdots \cap V_{x_p}, \mathcal{H})$ satisfies $g_{\rho(x_0) \cdots \rho(x_p)} = \psi \circ \mu$ for some $\mu \in \Gamma(V_{x_0} \cap \cdots V_{x_p}, \mathcal{G})$. Now let $\mathcal{V} = \{V_i\}$ be a locally finite refinement of the covering $\{V_x\}_{x \in X}$ of X. If $\sigma : I \to J$ is an affinity function such that $x \in V_i$ implies $\rho(x) = \sigma(i)$, then, by definition, $\sigma^* g \in C_0^p(\mathcal{V}, \mathcal{H})$.   □

COROLLARY 6.2.15 If X is a paracompact manifold, then the canonical map λ of the direct limit of $H_0^p(\mathcal{U}, \mathcal{H})$ into $H^p(X, \mathcal{H})$ is a surjective isomorphism.

Proof   Denote the direct limit of $H_0^p(\mathcal{U}, \mathcal{H})$ by $H_0^p(X, \mathcal{H})$. The surjectivity is immediate from Lemma 6.2.14. For the injectivity, it suffices to observe that if $\mathcal{U}$ is an open covering of X, then

$$C_0^p(\mathcal{U}, \mathcal{H}) \cap B^p(\mathcal{U}, \mathcal{H}) = B_0^p(\mathcal{U}, \mathcal{H}).$$ □

THEOREM 6.2.16   If X is a paracompact manifold and $0 \to \mathcal{F} \to \mathcal{G} \to \mathcal{H} \to 0$ is an exact sequence of sheaves of abelian groups over X, then there is a long exact cohomology sequence:

$$
\begin{aligned}
0 \ \to \ & H^0(X,\mathcal{F}) \to H^0(X,\mathcal{G}) \to H^0(X,\mathcal{H}) \to H^1(X,\mathcal{F}) \\
\to \ & H^1(X,\mathcal{G}) \to H^1(X,\mathcal{H}) \to \cdots
\end{aligned}
$$

Proof   Combine Theorem 6.2.13 and Corollary 6.2.15.   □

Remark: Notice that in Theorem 6.2.16 we did not specify what the maps are in the exact sequence. In applications of exact sequences, it does not matter what the specific maps are. What is important is the relationship specified by the exactness of the sequence.   □

6.3   Dolbeault Isomorphism

Let $\Omega \subseteq \mathbb{C}^n$ be a domain. Let $\wedge^{p,q}(\Omega)$ denote the module of (p,q) forms on Ω with C^∞ coefficients. Consider the long semiexact sequence

$$
\wedge^{0,0} \overset{\bar\partial_{0,0}}{\to} \wedge^{0,1} \overset{\bar\partial_{0,1}}{\to} \wedge^{0,2} \overset{\bar\partial_{0,2}}{\to} \cdots.
$$

This is the *Dolbeault complex*. The associated cohomology classes are

$$
\begin{aligned}
H_D^p(\Omega) \ &= \ \operatorname{Ker}\bar\partial_{0,p}/\operatorname{Image}\bar\partial_{0,p-1}, \qquad p > 0; \\
H_D^0(\Omega) \ &= \ \operatorname{Ker}\bar\partial_{0,0}, \qquad p = 0.
\end{aligned}
$$

Now let $\mathcal{O}$ be the sheaf of germs of holomorphic functions on Ω. We wish to relate $H^q(\Omega,\mathcal{O})$ and $H_D^q(\Omega)$.

THEOREM 6.3.1 (Dolbeault)   Let $\Omega \subseteq \mathbb{C}^n$ be an open set and $\mathcal{U}$ an open covering of Ω *by domains of holomorphy.* Then $H^p(\Omega,\mathcal{O}) = H^p(\mathcal{U},\mathcal{O}) \cong H_D^p(\Omega)$, every $p \geq 0$.

Remark: This theorem says two things: (1) that two different cohomology theories are the same and (2) that a covering consisting of cohomologically trivial open sets (whose intersections are *also* cohomologically trivial—recall Corollary 3.4.7) already contains all the cohomological information about Ω. Notice that when $p = 0$ then the theorem trivially holds because, by *fiat*, all three cohomology groups equal $\Gamma(\Omega,\mathcal{O})$.   □

Proof of Theorem 6.3.1   Let $\mathcal{E}^q$ be the sheaf of germs of $(0, q)$ forms with C^∞ coefficients and $\mathcal{Z}^q$ the subsheaf consisting of germs of $\bar{\partial}$-closed forms. Now the sequence

$$0 \to \mathcal{Z}^q \xrightarrow{i} \mathcal{E}^q \xrightarrow{\bar{\partial}} \mathcal{Z}^{q+1} \to 0$$

is exact. This assertion is trivial at $\mathcal{Z}^q$. At $\mathcal{E}^q$ and $\mathcal{Z}^{q+1}$ the exactness follows because for each $x \in \Omega$ and neighborhood ω of x, there is a polydisc D^n such that $x \in D^n \subseteq \omega$. And the $\bar{\partial}$ problem is solvable on polydiscs subject to the usual compatibility conditions (this is just the Poincaré lemma). Now, as usual, we have the exact sequence

$$0 \to C^p(\mathcal{U}, \mathcal{Z}^q) \to C^p(\mathcal{U}, \mathcal{E}^q) \to C_0^p(\mathcal{U}, \mathcal{Z}^{q+1}) \to 0.$$

However, because each U_i is a domain of holomorphy, because intersections of domains of holomorphy are domains of holomorphy, and because the $\bar{\partial}$ problem can always be solved on domains of holomorphy, we may conclude that $C_0^p(\mathcal{U}, \mathcal{Z}^{q+1}) = C^p(\mathcal{U}, \mathcal{Z}^{q+1})$. Hence

$$0 \to C^p(\mathcal{U}, \mathcal{Z}^q) \to C^p(\mathcal{U}, \mathcal{E}^q) \to C^p(\mathcal{U}, \mathcal{Z}^{q+1}) \to 0.$$

is exact. There follows the long exact cohomology sequence

$$\begin{aligned}
0 \quad &\to \Gamma(\Omega, \mathcal{Z}^q) \to \Gamma(\Omega, \mathcal{E}^q) \to \Gamma(\Omega, \mathcal{Z}^{q+1}) \\
&\to H^1(\mathcal{U}, \mathcal{Z}^q) \to H^1(\mathcal{U}, \mathcal{E}^q) \to H^1(\mathcal{U}, \mathcal{Z}^{q+1}) \\
&\to H^2(\mathcal{U}, \mathcal{Z}^q) \to H^2(\mathcal{U}, \mathcal{E}^q) \to \cdots .
\end{aligned} \tag{6.3.1.1}$$

To exploit (6.3.1.1), we notice that $H^p(\mathcal{U}, \mathcal{E}^q) = 0$, every $p > 0$, by the exercise for the reader following 6.2.7. (Note here that $\mathcal{E}^q$ is a sheaf of C^∞ modules, whereas $\mathcal{Z}^q$ is definitely not; this observation arises repeatedly in the subject and causes a great deal of trouble.) We conclude that

$$0 \to H^p(\mathcal{U}, \mathcal{Z}^{q+1}) \to H^{p+1}(\mathcal{U}, \mathcal{Z}^q) \to 0$$

is exact, all $p > 0$. In other words,

$$H^p(\mathcal{U}, \mathcal{Z}^{q+1}) \cong H^{p+1}(\mathcal{U}, \mathcal{Z}^q), \qquad \text{all } p > 0. \tag{6.3.1.2}$$

Likewise,

$$\Gamma(\Omega, \mathcal{E}^q) \xrightarrow{\bar{\partial}^*} \Gamma(\Omega, \mathcal{Z}^{q+1}) \xrightarrow{\delta^*} H^1(\mathcal{U}, \mathcal{Z}^q) \to 0$$

is exact. By elementary algebra,

$$H^1(\mathcal{U}, \mathcal{Z}^q) \cong \Gamma(\Omega, \mathcal{Z}^{q+1})/\bar{\partial}^*(\Omega, \mathcal{E}^q) \equiv H_D^{q+1}(\Omega). \qquad (6.3.1.3)$$

Thus we obtain

$$\begin{aligned} H^p(\mathcal{U}, \mathcal{O}) &\equiv H^p(\mathcal{U}, \mathcal{Z}^0) \cong H^{p-1}(\mathcal{U}, \mathcal{Z}^1) \cong \cdots \\ &\cong H^1(\mathcal{U}, \mathcal{Z}^{p-1}), \end{aligned}$$

where we have used (6.3.1.2) iteratively. Now by (6.3.1.3), the last line is isomorphic to $H_D^p(\Omega)$. It remains to notice that every open covering of Ω has a refinement by domains of holomorphy (for example, polydiscs) so that $H^p(\Omega, \mathcal{O}) \cong H_D^p(\Omega)$, as desired. □

COROLLARY 6.3.2 If Ω is a domain of holomorphy, then $H^p(\Omega, \mathcal{O}) = 0$ for all $p > 0$. Indeed, $H^p(\mathcal{U}, \Omega) = 0$ for any $p > 0$ and any covering $\mathcal{U}$ consisting of domains of holomorphy.

Proof   Apply 4.6.12 to the theorem. □

Remark: Recall that $H^0(\Omega, \mathcal{O})$ is canonically isomorphic to the space of all holomorphic functions on Ω. So of course it will not be 0. □

The argument that we have used to prove Dolbeault's theorem was devised by A. Weil to give a proof of de Rham's theorem. We now formulate and prove both the real and complex versions of that theorem. Just as with the theorem of Dolbeault, any cohomology isomorphism of this sort must begin with a local triviality statement.

LEMMA 6.3.3 (Poincaré Lemma)   Let $U \subseteq \mathbb{R}^N$ be a rectangular open N-cell:

$$U = \{x \in \mathbb{R}^N : a_i < x_i < b_i : i = 1, \dots, N\}. \qquad (6.3.3.1)$$

Let γ be a d-closed $(k+1)$-form on $U, k \geq 0$. Then there exists a k-form u on U such that $du = \gamma$.

Proof   The argument is elementary and well known. But it is a classic, so we include it for completeness.

The proof is by induction on m, where the form γ is hypothesized to include only the differentials $dx_1, \dots, dx_m$. For the case $m = 1$, we have $\gamma(x) = g(x)\,dx_1$ for some smooth function g; the condition $d\gamma = 0$ means that $\partial g/\partial x_j = 0, j = 2, \dots, N$. We apply the fundamental theorem of calculus to the function g in the x_1 variable to obtain a function u on U with $\partial u/\partial x_1 = g$ and $\partial u/\partial x_j = 0, j = 2, \dots, N$. Then $du = \gamma$ as desired.

Now assume the result for $m - 1$. There is no loss of generality in assuming that we are dealing with a form of pure degree $k+1$. If γ is a $(k+1)$-form involving only the differentials $dx_1, \ldots, dx_m$, then we may write $\gamma = \alpha \wedge dx_m + \beta$, where α is a form of degree k, β is a form of degree $(k+1)$, and neither α nor β involves the differentials $dx_m, \ldots, dx_N$. By hypothesis,

$$0 = d\gamma = d\alpha \wedge dx_m + d\beta.$$

Notice that $d\beta$ is a sum of terms, each of which involves either a subset of $\{dx_1, \ldots, dx_m\}$ or a subset of $\{dx_1, \ldots, dx_{m-1}, dx_{m+1}, \ldots, dx_N\}$. In particular, no term of $d\beta$ involves both dx_m and dx_j for some $j > m$. It follows that $d\alpha \wedge dx_m$ has the same property so that $d\alpha$ has no terms involving $dx_j, j > m$. Since α itself has no such term, we conclude that the coefficients of α do not depend on $x_{m+1}, \ldots, x_N$.

Write $\alpha = \sum \alpha_\mu dx^\mu$. For each μ, let α_μ^* be an antiderivative of α_μ in the variable x_m. Set $\alpha^* = \sum_\mu \alpha_\mu^* dx^\mu$. Notice that we may assume that α_μ^* does not depend on $x_{m+1}, \ldots, x_n$ for every μ. In conclusion,

$$
\begin{aligned}
d\alpha^* \;=\; & \sum_{j=1}^{m-1} \sum_\mu \frac{\partial \alpha_\mu^*}{\partial x_j} dx_j \wedge dx^\mu + dx_m \wedge \alpha \\
& + \sum_{j=m+1}^{N} \sum_\mu \frac{\partial \alpha_\mu^*}{\partial x_j} dx_j \wedge dx^\mu \\
=\; & I + II + III.
\end{aligned}
$$

Observe that $III = 0$. Hence

$$d\alpha^* = dx_m \wedge \alpha + (\text{terms involving only } dx_1, \ldots, dx_{m-1}).$$

Thus $\gamma + (-1)^{k+1} d\alpha^*$ involves only $dx_1, \ldots, dx_{m-1}$. And $d(\gamma + (-1)^{k+1} d\alpha^*) = 0$. Finally, $\gamma + (-1)^{k+1} d\alpha^*$ is a form of pure degree $k+1$. So the inductive hypothesis applies and there is a k-form u^* such that $du^* = \gamma + (-1)^{k+1} d\alpha^*$. But then $u = u^* - (-1)^{k+1} \alpha^*$ satisfies $du = \gamma$ as desired. This completes the induction.   □

Let γ be a $(p, 0)$ form on $U \subseteq \mathbb{C}^n$. Call γ a *holomorphic p-form* if its coefficients are all holomorphic functions.

LEMMA 6.3.4   Let $U \subseteq \mathbb{C}^n$ be a polydisc. Let γ be a $\bar{\partial}$-closed holomorphic p-form on $U, 0 < p \leq n$. Then there is a holomorphic $(p-1)$-form u on U such that $\bar{\partial} u = \gamma$.

Proof   The proof is left as an exercise. Mimic the proof of Lemma 6.3.3. Replace the fundamental theorem of calculus with the standard construction of primitives using line integrals (see Lemma 2.2.2).   □

Let us call a covering $\mathcal{U} = \{U_j\}$ of $\Omega \subseteq \mathbb{R}^N$ *d-simple* if for each $U_{i_1}, \ldots, U_{i_q} \in$ $\mathcal{U}$ and each d-closed k-form γ on $U_{i_1} \cap \cdots \cap U_{i_q}$ with $k \geq 1$ there is a $(k-1)$-form u on $U_{i_1} \cap \cdots \cap U_{i_q}$ such that $du = \gamma$ on Ω. (Similarly, we call a covering $\mathcal{U} = \{U_j\}$ of $\Omega \subseteq \mathbb{C}^n$ *∂-simple* if for each $U_{i_1}, \ldots, U_{i_q} \in \mathcal{U}$ and each ∂-closed holomorphic p-form γ on $U_{i_1} \cap \cdots \cap U_{i_q}$ with $p \geq 1$ there is a holomorphic $(p-1)$-form u on $U_{i_1} \cap \cdots \cap U_{i_q}$ such that $\partial u = \gamma$ on Ω.)

Now if $\Omega \subseteq \mathbb{R}^N$ is a domain, we let $\mathcal{D}^k$ be the sheaf of germs of k-forms over Ω and $\mathcal{Y}^k$ the subsheaf of d-closed forms. (Notice that $\mathcal{Y}^0$ is the constant sheaf $\mathbb{C} \times \Omega$.) We have the short exact sequence

$$0 \to \mathcal{Y}^k \xrightarrow{i} \mathcal{D}^k \xrightarrow{d} \mathcal{Y}^{k+1} \to 0$$

where exactness at $\mathcal{D}^k, \mathcal{Y}^{k+1}$ follow by Lemma 6.3.3. We also have the de Rham complex

$$\wedge^0(\Omega) \xrightarrow{d_0} \wedge^1(\Omega) \xrightarrow{d_1} \wedge^2(\Omega) \to \cdots$$

where $\wedge^k(\Omega)$ denotes the space of k-forms on Ω with C^∞ coefficients. We may thereby define the de Rham cohomology groups

$$H_{DR}^p(\Omega) \equiv \operatorname{Ker} d_p/\operatorname{Image} d_{p-1}, \quad p > 0,$$
$$H_{DR}^0(\Omega) \equiv \operatorname{Ker} d_0 \equiv \mathbb{C}.$$

Now the de Rham theorem relates the de Rham cohomology to $H^p(\Omega, \mathcal{Y}^0) = H^p(\Omega, \mathbb{C})$.

THEOREM 6.3.5 (Real de Rham Theorem)   If $\Omega \subseteq \mathbb{R}^N$ is an open set and if $\mathcal{U} = \{U_j\}_{j \in J}$ is a d-simple covering of Ω (e.g., if the U_i are rectangular cells as in (6.3.3.1)), then for $p \geq 0$ we have

$$H_{DR}^p(\Omega) = H^p(\Omega, \mathbb{C}) = H^p(\mathcal{U}, \mathbb{C}).$$

Proof   The proof is left as an exercise. Imitate the proof of Theorem 6.3.1.   □

Notice that Theorem 6.3.5 imposes no special hypotheses on the domains being studied. This theorem is a theorem of real analysis/topology; therefore domains are (locally) indistinguishable.

Finally, we formulate the complex de Rham theorem. Denote the holomorphic p-forms on $\Omega \subseteq \mathbb{C}^n$ by $\wedge_{\mathcal{H}}^p(\Omega)$. Note that $\wedge_{\mathcal{H}}^0(\Omega)$ is simply the set of all holomorphic functions on Ω. Let $\mathcal{D}_{\mathcal{H}}^p$ be the *sheaf of germs* of holomorphic p-forms on Ω. Let $\mathcal{Y}_{\mathcal{H}}^p \subseteq \mathcal{D}_{\mathcal{H}}^p$ be the subsheaf of germs of ∂-closed holomorphic p-forms on Ω. Then the sequence

$$0 \to \mathcal{Y}_{\mathcal{H}}^p \xrightarrow{i} \mathcal{D}_{\mathcal{H}}^p \xrightarrow{\partial} \mathcal{Y}_{\mathcal{H}}^{p+1} \to 0$$

is exact by Lemma 6.3.4. We also have the holomorphic de Rham complex

$$\bigwedge_{\mathcal{H}}^0 \overset{\partial_0}{\to} \bigwedge_{\mathcal{H}}^1 \overset{\partial_1}{\to} \bigwedge_{\mathcal{H}}^2 \to \cdots$$

and may define thereby the holomorphic de Rham cohomology groups

$$
\begin{aligned}
H^p_{DR\mathcal{H}}(\Omega) &= \operatorname{Ker}\partial_p/\operatorname{Image}\partial_{p-1} \quad, \quad p > 0, \\
H^0_{DR\mathcal{H}}(\Omega) &= \operatorname{Ker}\partial_0 \cong \mathbb{C}.
\end{aligned}
$$

Now a line-by-line repetition of the proofs of the preceding two theorems yields the *holomorphic de Rham theorem*.

THEOREM 6.3.6 Let $\Omega \subseteq \mathbb{C}^n$ be an open set such that for every $p \geq 0$ and every $k > 0$ we have $H^k(\Omega, \mathcal{D}^p_{\mathcal{H}}) = 0$. Let $\mathcal{U} = \{U_j\}_{j \in J}$ be a ∂-simple open covering of Ω. Then, for $k \geq 0$, we have

$$H^k_{DR\mathcal{H}}(\Omega) \cong H^k(\Omega, \mathbb{C}) \cong H^k(\mathcal{U}, \mathbb{C}).$$

COROLLARY 6.3.7 If Ω satisfies the hypotheses of Theorem 6.3.6, then $H^k(\Omega, \mathbb{C}) = 0, k > n$.

Proof Trivially, $H^k_{DR\mathcal{H}}(\Omega) = 0$.   □

COROLLARY 6.3.8 If $\Omega \subseteq \mathbb{C}^n$ is a domain of holomorphy and $\mathcal{U}$ is a ∂-simple open covering of Ω, then

$$H^k_{DR\mathcal{H}}(\Omega) \cong H^k(\Omega, \mathbb{C}) \cong H^k(\mathcal{U}, \mathbb{C}).$$

Proof By Theorem 6.3.6, we need prove only that $H^k(\Omega, \mathcal{D}^p_{\mathcal{H}}) = 0$, for all $k > 0, p \geq 0$. But this is equivalent to the assertion that $H^k(\Omega, \mathcal{O}^r) = 0$, all $k > 0, r \geq 0$. And the latter assertion follows just as in the proof of the Dolbeault theorem.   □

Here is an alternative approach to the de Rham theory that is of independent interest.

LEMMA 6.3.9 Let $\Omega \subseteq \mathbb{C}^n$ be a domain of holomorphy. Let f be a differential form on Ω with $df \in \wedge^{(p+1,0)}$. Then f is cohomologous to a form f^* of type $(p, 0)$; that is, $f - f^* = dg$ for some differential form g of degree $p - 1$. In particular, $df = df^* \in \wedge^{(p+1,0)}$ so that $\bar{\partial}f^* = \bar{\partial}f = 0$.

Proof Write $f = \sum_{j=0}^p f_{p-j,j}$, each $f_{p-j,j}$ a C^∞ form of type $(p - j, j)$. We prove the result by induction on the complexity of f. First, if $f = f_{p,0}$, then we are done trivially. Now suppose that $k \leq p$ and that we have proved the result for f of the form $f = \sum_{j=0}^{k-1} f_{p-j,j}$ (in what follows we shall refer to such expressions as having "order of complexity" $(k-1)$). We then prove the result for forms $\sum_{j=0}^{k} f_{p-j,j}$. By hypothesis,

$$\bigwedge^{(p+1,0)} \ni df = \partial f + \bar{\partial} f = \left(\partial f + \bar{\partial} \sum_{j=0}^{k-1} f_{p-j,j} \right) + \bar{\partial} f_{p-k,k}.$$

The last term here is the only one of type $(p-k, k+1)$, so it must be 0. Since Ω is a domain of holomorphy, it follows that there is a $u \in \wedge^{p-k,k-1}$ such that $\bar{\partial} u = f_{p-k,k}$. Hence

$$f - du = \sum_{j=0}^{k} f_{p-j,j} - \partial u - \bar{\partial} u = \sum_{j=0}^{k-1} f_{p-j,j} - \partial u.$$

We may apply the inductive hypothesis to $f - du$, since $d(f - du) = df \in \wedge^{(p+1,0)}$ and $f - du$ is of order of complexity $k - 1$. Thus $f - du$ is cohomologous to some $f^* \in \wedge^{(p,0)}$. That is, $(f - du) - f^* = dv$, for some v. In conclusion, $f - f^* = d(u + v)$, as desired. $\qquad\square$

It follows from the lemma that when Ω is a domain of holomorphy, then each cohomology class in $H^p_{DR}(\Omega)$, $p \geq 0$, has a representative of the form $f \in \wedge^{(p,0)}(\Omega)$. (Results of this nature are fundamental to the cohomology induced by differential forms. The celebrated Hodge conjecture concerns harmonic representatives for cohomology in the middle dimension.) Moreover, if $p \geq 1$ and f is cohomologous to 0—that is, $f = dg$—then g may be chosen to be in $\wedge^{(p-1,0)}$. Since $d = \partial + \bar{\partial}$, we may suppose that $\bar{\partial} g = 0$, that is, that g has holomorphic coefficients.

Likewise, since $df = 0$, we have $\bar{\partial} f = 0$; hence f has holomorphic coefficients. In conclusion,

$$H^p_{DR\mathcal{H}}(\Omega) = H^p_{DR}(\Omega), \qquad \text{all } p > 0.$$

Therefore, using 6.3.6, we have

THEOREM 6.3.10 If $\Omega \subseteq \mathbb{C}^n$ is a domain of holomorphy and if $\mathcal{U} = \{U_j\}$ is a ∂-simple open covering of Ω (e.g., by polydiscs), then for $k > 0$ we have

$$H^k_{DR\mathcal{H}}(\Omega) \cong H^k_{DR}(\Omega) = H^k(\Omega, \mathbb{C}) \cong H^k(\mathcal{U}, \mathbb{C}).$$

6.4   Algebraic Properties of the Ring of Germs of Holomorphic Functions

Let $0 \in \mathbb{C}^n$ and let $\mathcal{O}_0$ be the ring of germs of holomorphic functions at 0. We wish to explore the structure of this ring. We assume familiarity with the following standard ring-theoretic terminology: *ideal, maximal ideal, integral domain, unit, prime* (irreducible) *element of a ring, reducible element of a ring, principal ideal domain* (PID), *Noetherian ring, Artinian ring, valuation,* and *unique factorization domain* (UFD).

A ring is called *local* if it has a unique maximal ideal. Two elements α, β of a ring R are called *equivalent* if there is a unit $u \in R$ such that $\alpha = u\beta$. All of this terminology is discussed in detail in Herstein [1], M. Atiyah and I. MacDonald [1], S. Lang [1], and B. L. van der Waerden [1].

The following lemmas may be considered exercises or may be found in the aforementioned references. Let R be a ring.

LEMMA 6.4.1 (Gauss's Lemma)   Let R be an integral domain and let k be its quotient field. If $p \in R[z]$ is irreducible over $R[z]$, then p is irreducible over $k[z]$.

Proof   See Lang [1, p. 127].   □

LEMMA 6.4.2   If R is a UFD, then $R[z]$ is also a UFD.

Notice that if $0 \in \mathbb{C}^n$, then $\mathcal{O}_0$ is an integral domain. A germ $\gamma \in \mathcal{O}_0$ is a unit if and only if some representative g for γ satisfies $g(0) \neq 0$; an equivalent condition is that the constant term in the power series expansion for g about 0 not be 0. The condition holds for one representative if and only if it holds for all.

In particular, any proper ideal $\mathcal{V} \subseteq \mathcal{O}_0$ consists only of elements that vanish at 0. So the unique maximal ideal in $\mathcal{O}_0$ is $\{[g] \in \mathcal{O}_0 : g(0) = 0\}$. We thus see that $\mathcal{O}_0$ is a local ring. Although $\mathcal{O}_0$ is not a PID (consider the ideal generated by $[z_1], [z_2]$), it does turn out to be Noetherian. It is not Artinian (the descending chain consisting of the principal ideals $\langle [z_1] \rangle, \langle [z_1^2] \rangle, \langle [z_1^3] \rangle, \ldots$ does not stabilize). The ring $\mathcal{O}_0$ does turn out to be a UFD, as we shall see later.

The Weierstrass preparation theorem and the Weierstrass division theorem are the starting point for any study of the algebraic properties of the ring of germs of holomorphic functions. The preparation theorem says that each germ of a holomorphic function is in fact a monic polynomial of one variable (with coefficients in the other variables) times a unit. The division theorem says that, subject to appropriate normalizations, $\mathcal{O}_0$ is a division ring.

DEFINITION 6.4.3   Let $f = \sum_\alpha a_\alpha z^\alpha$ be holomorphic in a neighborhood of 0 in $\mathbb{C}^n$. Define ord f, the *order* of f, to be the least $|\alpha|$ for which $a_\alpha \neq 0$.

Exercise for the Reader
The function ord is a valuation on the ring $\mathcal{O}_0$.

If ord $f = k$, then there is a nonsingular linear change of coordinates so that, in the new coordinates, the coefficient of z_n^k is 1 (exercise). When f is of this form, it is said to be *normalized* (with respect to the variable z_n) *of order* k.

DEFINITION 6.4.4   A function f, holomorphic in a neighborhood of $0 \in \mathbb{C}^n$, is called a *Weierstrass polynomial* of degree k if, using the notation $z' = (z_1, \dots, z_{n-1})$, we have

$$f(z_1, \dots, z_n) = f(z', z_n) = z_n^k + a_{k-1}(z')z_n^{k-1} + \cdots + a_1(z')z_n + a_0(z'),$$

where the a_j are holomorphic functions (not necessarily polynomials) in a neighborhood of $0 \in \mathbb{C}^{n-1}$ and $a_j(0) = 0$ for $j = 0, \dots, k-1$.

It will be useful to let $\mathcal{O}_0'$ denote the ring of germs of holomorphic functions in the variables $z_1, \dots, z_{n-1}$. So Weierstrass polynomial P is an element of $\mathcal{O}_0'[z_n]$.

THEOREM 6.4.5 (The Weierstrass Preparation Theorem)   Let f be holomorphic in a neighborhood of $0 \in \mathbb{C}^n$. Assume that f is normalized of order m. Then in a small neighborhood of 0, the function f may be written in a unique way as

$$f(z) = u(z) \cdot W(z),$$

where u is a unit in $\mathcal{O}_0$ and

$$W(z) = z_n^m + a_{n-1}(z')z_n^{m-1} + \cdots + a_0(z')$$

is a Weierstrass polynomial.

Proof (Siegel)   We may assume that $m \geq 1$. Notice then that $f(0,0) = 0$. Choose $r > 0$ such that $f(0, re^{i\theta}) \neq 0$ for $0 \leq \theta \leq 2\pi$. By continuity we may choose a number $h > 0$ such that $f(z', z_n) \neq 0$ for $|z'| \leq h, |z_n| = r$. The function

$$N(z') = \frac{1}{2\pi i} \oint_{|z_n|=r} \frac{\partial f(z', z_n)/\partial z_n}{f(z', z_n)} dz_n$$

counts the number of zeros (counting multiplicity) of $f(z', \cdot)$. Since N is a continuous, integer-valued function for $|z'| \leq h$, it is constant. Say that $N(z') \equiv \ell$ when $|z'| \leq h$. For each such z', let $\alpha_1(z'), \cdots, \alpha_m(z')$ denote the zeros of $f(z', \cdot)$, counting multiplicities. Shrinking r if necessary, we may assume that $\alpha_j(0) = 0$, all j, and that $\ell = m$.

If ϕ is a holomorphic function of a single complex variable in a neighborhood of $\bar{D}(0,r)$, then it holds that

$$J_\phi(z') \equiv \sum_{j=1}^{m} \phi(\alpha_j(z')) = \frac{1}{2\pi i} \oint_{|z_n|=r} \phi(z_n) \cdot \frac{\partial f(z', z_n)/\partial z_n}{f(z', z_n)} dz_n. \qquad (6.4.5.1)$$

For the functions $\phi(w) \equiv w^k, k = 1, \ldots, m$, we obtain the following symmetric functions of the roots α_j :

$$J_k(z') = J_{w^k}(z') \equiv \sum_{j=1}^{m} (\alpha_j(z'))^k.$$

Then the coefficients of the polynomial

$$W(z', z_n) = W_{z'}(z_n) \equiv \prod_{j=1}^{m} (z_n - \alpha_j(z')),$$

whose roots are $\alpha_1(z'), \cdots, \alpha_m(z')$, are polynomial combinations of the functions $J_1(z'), \ldots, J_m(z')$. Notice that, by (6.4.5.1), these are holomorphic functions of z' for $|z'| < m$. Also, when $z' = 0$, then $\alpha_1 = \cdots = \alpha_m = 0$. Hence $J_1(0) = \cdots = J_m(0) = 0$. Thus W is a Weierstrass polynomial.

Define $u(z', z_n) = f(z', z_n)/W(z', z_n)$. For each fixed z', this is a holomorphic function in z_n except possibly at $\alpha_1(z'), \ldots, \alpha_m(z')$. But these singularities are removable by construction. So $u(z', \cdot)$ is holomorphic, each z'. Thus, for any $|z_n| < r$,

$$u(z', z_n) = \frac{1}{2\pi i} \oint_{|\zeta|=r} \frac{u(z', \zeta)}{\zeta - z_n} d\zeta.$$

This formula makes it clear that u is holomorphic in $|z'| < h$ (for f and W are nonvanishing when $|z_n| = r, |z'| \leq h$.) It follows from Hartogs's theorem that u is holomorphic. It is a unit by construction. So we have the decomposition

$$f = u \cdot W.$$

For uniqueness, suppose that

$$u_1 W_1 = u_2 W_2. \qquad (6.4.5.2)$$

Then, setting $z' = 0$, we obtain

$$u_1(0, z_n) z_n^m = u_2(0, z_n) z_n^m$$

or

$$u_1(0, z_n) = u_2(0, z_n).$$

Differentiating both sides of (6.4.5.2) with respect to $z_j, j = 1, \ldots, n-1$, and setting $z' = 0$, yields

$$\frac{\partial u_1}{\partial z_j}(0, z_n) z_n^m + u_1(0, z_n)\frac{\partial W_1}{\partial z_j}(0, z_n) = \frac{\partial u_2}{\partial z_j}(0, z_n) z_n^m + u_2(0, z_n)\frac{\partial W_2}{\partial z_j}(0, z_n).$$

For z_n small, counting degrees in z_n, we see that

$$\frac{\partial u_1}{\partial z_j}(0, z_n) = \frac{\partial u_2}{\partial z_j}(0, z_n), \qquad j = 1, \ldots, n-1.$$

Continuing, we conclude that $u_1(\cdot, z_n)$ and $u_2(\cdot, z_n)$ have the same power series expansion about 0, so that $u_1 \equiv u_2$. We conclude that $W_1 \equiv W_2$. □

THEOREM 6.4.6 (The Weierstrass Division Theorem)   Let W be a Weierstrass polynomial of degree k. Let f be holomorphic in a neighborhood of $0 \in \mathbb{C}^n$. Then, in a small neighborhood of 0, the function f may be written in a unique fashion as

$$f = W \cdot q + r$$

where r is a polynomial in z_n of degree less than k with coefficients that are functions of $z' = (z_1, \ldots, z_{n-1})$ and q is holomorphic. (*Note that, in general, r will not be a Weierstrass polynomial.*)

Proof   First we consider uniqueness: If

$$q_1 \cdot W + r_1 = q_2 \cdot W + r_2 = f,$$

then

$$r_1 - r_2 = (q_2 - q_1) \cdot W.$$

It follows that $r_1 - r_2$ is a polynomial of degree not exceeding $k-1$ in z_n with k zeros (counting multiplicity) in z_n. Thus $r_1 - r_2 \equiv 0$. Therefore, $q_1 \equiv q_2$.

For existence, we note that there is a change of coordinates such that both W and f are normalized at 0 and W is monic (exercise). Say that f has degree m as a polynomial in z_n.

If $m < k$, then we may write

$$f = W \cdot 0 + f$$

and we are done. If $m \geq k$, then we may apply the Euclidean algorithm for polynomials of one variable to write $f = W \cdot q + r$, with the degree of r (as a polynomial in r_n) not exceeding $k - 1$. $\qquad \square$

Remark: It is also possible to construct q and r using Cauchy integrals, much as in the proof of the preparation theorem. We leave details to the reader, or see L. Bers [1]. A unified treatment of these two theorems is in S. Lojaciewicz [2]. $\qquad \square$

LEMMA 6.4.7 Let f, g, W be holomorphic near $0 \in \mathbb{C}^n$. Assume that f is a polynomial in z_n with coefficients that are holomorphic functions of z' and that W is a Weierstrass polynomial. If

$$f = g \cdot W,$$

then g is also a polynomial in z_n with coefficients that are polynomials in z'.

Proof This is a special case of the uniqueness portion of the division theorem. $\qquad \square$

Exercise for the Reader
The lemma is false in general if W is not a Weierstrass polynomial.

LEMMA 6.4.8 Let f, g, W be holomorphic near $0 \in \mathbb{C}^n$ and suppose that W is a Weierstrass polynomial and that f, g are polynomials in z_n with coefficients that are holomorphic functions of z'. If $W = f \cdot g$, then (up to an invertible factor that is a holomorphic function of z' only and independent of z_n) both f and g are Weierstrass polynomials.

Proof Let the degrees of W, f, g (in the z_n variable) be α, β, γ, respectively. Write $z_n^\alpha = z_n^{\beta + \gamma} = W(0, z_n) = f(0, z_n) \cdot g(0, z_n)$, whence

$$1 = \left(\frac{f(0, z_n)}{z_n^\beta} \right) \left(\frac{g(0, z_n)}{z_n^\gamma} \right).$$

It follows that f, g have leading coefficients that are units and that depend only on z'; also the lower order coefficients vanish when $z' = 0$. Therefore, f, g are Weierstrass polynomials, up to the aforementioned units. $\qquad \square$

PROPOSITION 6.4.9 The ring $\mathcal{O}_0$ is a UFD.

Proof We proceed by induction on dimension. For $n = 1$, the theorem is trivial: If $[f] \in \mathcal{O}_0$ has order k, we write $f(z) = z^k \cdot g$, where $g(0) \neq 0$, so that g is invertible.

Now assume the result to be proved in $\mathcal{O}_0^{n-1}$, the ring of germs of holomorphic functions at $0 \in \mathbb{C}^{n-1}$. Then it also follows (see Lang [1]) that $\mathcal{O}_0^{n-1}[z_n]$ is a UFD. We work, without further comment, with representatives of germs. Let $f \in \mathcal{O}_0^n$. We may assume that f is normalized of order k. Then $f = u \cdot W$ by the

Weierstrass preparation theorem. Note that $W \in \mathcal{O}_0^{n-1}[z_n]$. We claim that f is irreducible in $\mathcal{O}_0^n$ if and only if W is irreducible in $\mathcal{O}_0^{n-1}[z_n]$.

Assuming the claim, we use the inductive hypothesis and Lemma 6.4.2 to write $W = W_1 \cdots W_r$, where the W_i are irreducible elements of $\mathcal{O}_0^{n-1}[z_n]$. By Lemma 6.4.8, the W_i are Weierstrass. Therefore, $f = u \cdot W_1 \cdots W_r$. If f could also be written as $f = V_1 \cdots V_\ell$, then we apply Theorem 6.4.5 to each V_i to obtain $f = u' \cdot W_1' \cdots W_\ell'$, where u' is a unit and each W_i' is a Weierstrass polynomial. Since there is only one way to write f as a unit times a Weierstrass polynomial, we conclude that

$$W_1 \cdots W_r = W_1' \cdots W_\ell'.$$

By the inductive hypothesis,

$$\{W_1, \ldots, W_r\} = \{W_1', \ldots, W_\ell'\}.$$

Hence we are finished modulo the claim.

The claim follows, by taking contrapositives, from the Weierstrass preparation theorem (exercise). □

PROPOSITION 6.4.10   The ring $\mathcal{O}_0$ is Noetherian.

Proof   We proceed by induction on dimension. When $n = 1$, then all ideals are principal so the result is trivial.

Assume that the result has been proved for $\mathcal{O}_0^{n-1}$. If $\mathcal{I}$ is a nonzero ideal in $\mathcal{O}_0^n$, then, after a change of coordinates, we may suppose that $\mathcal{I}$ contains an element α that is normalized of order k with respect to z_n. (Now we use the letter α to also denote a representative for α and do likewise with other ring elements below without comment.) Dividing out by a unit, we may suppose that α is a Weierstrass polynomial.

If $\beta \in \mathcal{I}$ is arbitrary, then we apply the Weierstrass division theorem to obtain

$$\beta = q \cdot \alpha + r.$$

Here the degree of r as a polynomial in z_n is less than k. Let

$$M = \{h \in \mathcal{I} : \text{the degree of } h \text{ in } z_n < k\}.$$

Then M is a module over $\mathcal{O}_0^{n-1}$. Therefore, M is finitely generated (an immediate application of the Hilbert basis theorem). Let $\gamma_1, \ldots, \gamma_r$ be generators for M. Then $\gamma_1, \ldots, \gamma_r, \alpha$ generate $\mathcal{I}$. □

Let f, g be holomorphic on an open set $U \subseteq \mathbb{C}^n$. If $z \in U$, then $\gamma_z f, \gamma_z g$ are their respective germs at z. We say that $\gamma_z f, \gamma_z g$ are *relatively prime* if they are relatively prime *as elements of the ring* $\mathcal{O}_z$ (equivalently, if their power series expansions about the point z have no common factor that is given by a convergent power series about the point z).

PROPOSITION 6.4.11 Suppose that $0 \in U \subseteq \mathbb{C}^n$, that f and g are holomorphic on U, and that $\gamma_0 f, \gamma_0 g$ are relatively prime in $\mathcal{O}_0$. Then $\gamma_z f$ and $\gamma_z g$ are relatively prime in $\mathcal{O}_z$ for all z sufficiently close to 0.

Proof   We may assume that f and g are Weierstrass polynomials. Letting $\mathcal{O}'_0$ be the ring of germs in $z_1, \ldots, z_{n-1}$, we have by Lemma 6.4.8 that $\gamma_0 f$ and $\gamma_0 g$ are relatively prime in $\mathcal{O}'_0[z_n]$ (note that this observation is not trivial: The ring $\mathcal{O}'_0[z_n]$ is smaller than $\mathcal{O}_0$, but it also has fewer units). If $\mathcal{M}'_0$ is the quotient field of $\mathcal{O}'_0$, then Gauss's lemma (Lemma 6.4.1) implies that $\gamma_0 f$ and $\gamma_0 g$ are relatively prime in $\mathcal{M}'_0[z_n]$. But this means, by definition, that the ideal generated by $\gamma_0 f$ and $\gamma_0 g$ in $\mathcal{M}'_0[z_n]$ is all of $\mathcal{M}'_0[z_n]$ (note that $\mathcal{M}'_0[z_n]$ is a PID, so that if $\langle \gamma_0 f, \gamma_0 g \rangle$ were not the entire ring, then its generator would be a common factor). Thus there are $\tilde{f}_1, \tilde{g}_2 \in \mathcal{M}'_0[z_n]$ with

$$1 = \tilde{f}_1 \cdot f + \tilde{g}_1 \cdot g$$

near 0. Clearing denominators yields

$$h(z') = f_1 \cdot f + g_1 \cdot g,$$

where h is the least common denominator. But any common factor p of $\gamma_z f, \gamma_z g$ would necessarily be normalized with respect to z_n (since f and g are) and so p may be taken to be a Weierstrass polynomial. Since it follows that p divides $h(z')$ at z, it must therefore be a unit. We conclude that $\gamma_z f$ and $\gamma_z g$ are relatively prime.   □

6.5   Sheaf of Divisors, Chern Classes, and the Obstruction to Solving Cousin II

Let $\Omega \subseteq \mathbb{C}^n$ be a domain. For each $z \in \Omega$, let $\mathcal{O}_z$ be the ring of germs of holomorphic functions at z. Then $\mathcal{O}_z$ is a commutative integral domain, so we may consider the quotient field $\mathcal{M}_z$. We let $\mathcal{M} = \cup_{z \in \Omega} \mathcal{M}_z$ and define $\pi : \mathcal{M} \to \Omega$ by $\pi([\alpha/\beta]) = z$ for $[\alpha/\beta] \in \mathcal{M}_z$. To make this a sheaf under *addition and multiplication*, we need to topologize $\mathcal{M}$. If $q \in \mathcal{M}_{z_0}$, then by the fact that $\mathcal{O}$ is a UFD we may choose f and g holomorphic near z_0 that are coprime at z_0 with $q = [f]/[g]$. Then, by Proposition 6.4.11, there is a neighborhood U of z_0 on which f and g are still coprime. Define a neighborhood of q in $\mathcal{M}$ by

$$\{\gamma_y f / \gamma_y g : y \in U\}$$

where, as usual, $\gamma_y f, \gamma_y g$ represent the equivalence classes in $\mathcal{O}_y$ of f, g respectively.

We call $\mathcal{M}_z$ the *ring of germs of meromorphic functions* at z and $(\mathcal{M}, \Omega, \pi)$ the sheaf of germs of meromorphic functions on Ω. A *meromorphic function* on an open set $U \subseteq \Omega$ is an element of $\Gamma(U, \mathcal{M})$. We now list a number of properties of meromorphic functions that the reader is invited to verify (assume that U is connected):

(6.5.1) The meromorphic functions form a field.

(6.5.2) Near a point, a meromorphic function is given by the quotient of two holomorphic functions.

(6.5.3) If f is meromorphic, $f = f_1/f_2$ near z_0, and $f_2(z_0) \neq 0$, then f is holomorphic near z_0.

With notation as in (6.5.3), and f_1, f_2 coprime at z_0 (and hence *near* z_0), we call z_0 *regular* if $f_2(z_0) \neq 0$, a *pole* if $f_2(z_0) = 0$ and $f_1(z_0) \neq 0$, and *indeterminate* if $f_2(z_0) = f_1(z_0) = 0$. The following properties are easy exercises (here the dimension n is at least 2):

(6.5.4) The set $\mathcal{R}$ of regular points is open. Also, $f|_{\mathcal{R}}$ is holomorphic.

(6.5.5) Poles of f are never isolated (because the zeros of a holomorphic function are never isolated). If $\mathcal{P}$ denotes the set of poles of f, then $\mathcal{P}$ is never open, but $\mathcal{P} \cup \mathcal{R}$ is open.

(6.5.6) If $p \in \mathcal{P}$ and $M > 0$, then there is a neighborhood U of p such that $|f| > M$ at points of $U \setminus \{p\}$ on which f is defined.

(6.5.7) An indeterminate point may or may not be the limit of other indeterminate points. However, if p is indeterminate, then either there are $p_j \in \mathcal{P}$ with $p_j \to p$ or there are $r_j \in \mathcal{R}$ with $r_j \to p$. If f_1, f_2 are coprime, then both eventualities occur.

(6.5.8) If p is an indeterminate point and U is any neighborhood of p, then $f|_{U \setminus \{p\}}$ assumes all complex values (assuming that the dimension is at least two).

We let $\mathcal{O} \subseteq \mathcal{M}$ denote the subsheaf of germs of holomorphic functions.

Exercise for the Reader

Let $\Omega \subseteq \mathbb{C}^n$ be a domain. Then Cousin I is always solvable on Ω if and only if every section of $\mathcal{M}/\mathcal{O} \equiv \cup_x \mathcal{M}_x/\mathcal{O}_x$ is the image under the quotient map of a section of $\mathcal{M}$ (see Figure 6.5).

Let $\mathcal{M}^* \subseteq \mathcal{M}$ be the sheaf obtained by removing the zero section. Let $\mathcal{O}^* \subseteq \mathcal{M}^*$ be the invertible holomorphic elements. Then we may define the sheaf $\mathcal{M}^*/\mathcal{O}^* = \cup_x \mathcal{M}_x^*/\mathcal{O}_x^*$. Here the quotient is taken in the *multiplicative group* $\mathcal{M}^*$. The sheaf $\mathcal{M}^*/\mathcal{O}^*$ is called the *sheaf of germs of divisors* on Ω. A section of $\mathcal{M}^*/\mathcal{O}^*$ is called a *divisor*. A divisor is called *integral* if the germ

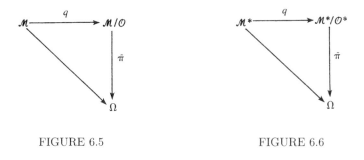

FIGURE 6.5   FIGURE 6.6

at each point is the germ of a holomorphic function. If $U \subseteq X$ is open and $\alpha \in \Gamma(U, \mathcal{M}^*/\mathcal{O}^*)$ is a divisor, then we call α *principal* if it can be pulled back to $\mathcal{M}^*$. That is, α is principal if there is a section $\alpha' \in \Gamma(U, \mathcal{M}^*)$ such that $q \circ \alpha' = \alpha$ (see Figure 6.6).

The following is a form of Cousin II that is pertinent to the study of meromorphic functions.

COUSIN II FOR MEROMORPHIC FUNCTIONS Let $\Omega \subseteq \mathbb{C}^n$ be a domain. Let $\mathcal{U} = \{U_i\}_{i \in I}$ be an open covering, and let F_i be meromorphic functions on U_i, each i, such that F_i/F_j is holomorphic on U_i/U_j. Does there exist a meromorphic F on Ω such that F/F_i is holomorphic on U_i for each i? (Denote this problem by the name "meromorphic Cousin II.")

Exercises for the Reader

1. Cousin II implies meromorphic Cousin II. The converse is also true but is rather more difficult to see (cf. Hörmander [3, p. 138]).

2. Meromorphic Cousin II is always solvable on Ω if and only if every divisor is principal. (*Hint:* Call two sets of Cousin data $\{(F_i, U_i)\}_{i \in I}$ and $\{(G_j, U_j)\}_{j \in J}$ equivalent if $\{(F_i, U_i), (G_j, V_j) : i \in I, j \in J\}$ is a set of Cousin data. The equivalence classes are sections of $\mathcal{M}^*/\mathcal{O}^*$.)

We now can prove a necessary and sufficient condition for solving meromorphic Cousin II, says that not all divisors need be checked.

PROPOSITION 6.5.9 Every integral divisor on Ω is principal if and only if meromorphic Cousin II is always solvable on Ω.

Proof   If meromorphic Cousin II is always solvable, then we have already observed that every divisor is principal.

Conversely, suppose that every integral divisor on Ω is principal. We will show that every divisor is a quotient of integral divisors, which is an even stronger assertion than what is required (and has important consequences, such as Theorem 6.5.11). Thus let α be a divisor. Let $p \in \Omega$ and let F_p be a meromorphic function on a neighborhood N_p of p that represents the germ $\gamma_p \alpha$. Shrinking N_p if necessary, we write $F_p = A_p/B_p$ on N_p, where A_p, B_p are holomorphic and coprime. Let $q \in N_p$ and, similarly, decompose a representative F_q for $\gamma_q \alpha$ as

$F_q = A_q/B_q$. We claim that $A_q/A_p \in \mathcal{O}$ on the intersection of their domains of definition and likewise for B_q/B_p (abbreviate these assertions with the notation $A_q \sim A_p$ and $B_q \sim B_p$). This will prove the result, for then $p \mapsto [A_p]$ and $p \mapsto [B_p]$ are the integral divisors that we seek.

To prove the claim, we note that

$$\frac{A_p}{B_p} \sim \frac{A_q}{B_q} \quad \text{in a neighborhood of } q;$$

hence

$$A_p B_q \sim A_q B_p \quad \text{in a neighborhood of } q.$$

So

$$A_p | A_q \,,\, A_q | A_p \,,\, B_p | B_q \,,\, B_q | B_p \quad \text{in a neighborhood of } q.$$

Hence $A_p \sim A_q, B_p \sim B_q$, as desired.   □

COROLLARY 6.5.10   On any $\Omega \subseteq \mathbb{C}^n$, every divisor is a quotient of integral divisors.

Proof   This follows from inspection of the proof of Proposition 6.5.9.   □

THEOREM 6.5.11 (The Strong Poincaré Problem)   Let $\Omega \subseteq \mathbb{C}^n$ be a domain in which meromorphic Cousin II is always solvable. Let α be a meromorphic function on Ω. Then there are holomorphic functions A and B on Ω such that A and B are coprime at every point of Ω and $\alpha = A/B$.

Proof   By Corollary 6.5.10, we may write $[\alpha] = a/b$, where a and b are integral divisors (careful—an integral divisor is *not* a holomorphic function). By Proposition 6.5.9, a and b are, in fact, principal. Hence there are elements $\mathcal{A}$ and $\mathcal{B}$ of $\Gamma(\Omega, \mathcal{M}^*)$ that induce a and b, respectively. Then $\alpha/(\mathcal{A}/\mathcal{B})$ is locally a unit. So it defines on all of Ω a nonvanishing holomorphic function β. Thus $\alpha = \mathcal{A}\beta/\mathcal{B}$ is the decomposition we seek, for $\mathcal{A}$ and $\mathcal{B}$ are holomorphic. (Exercise: $\mathcal{A}\beta$ and $\mathcal{B}$ are also coprime at each point.)   □

Given the material in Section 6.1 and in the present section, we now have more than sufficient reason to compute the obstruction to solving meromorphic Cousin II. The long exact cohomology sequence renders the computation easy. Let $\Omega \subseteq \mathbb{C}^n$ be a domain, and consider the exact sequence

$$0 \to \mathcal{O}^* \xrightarrow{i} \mathcal{M}^* \xrightarrow{q} \mathcal{M}^*/\mathcal{O}^* \to 0,$$

where i is inclusion and q is the canonical quotient map. The long exact cohomology sequence gives

$$
\begin{aligned}
0 \;\to\; & H^0(\Omega, \mathcal{O}^*) \xrightarrow{i^*} H^0(\Omega, \mathcal{M}^*) \xrightarrow{q^*} H^0(\Omega, \mathcal{M}^*/\mathcal{O}^*) \\
\xrightarrow{\delta^*}\; & H^1(\Omega, \mathcal{O}^*) \xrightarrow{i^*} H^1(\Omega, \mathcal{M}^*) \xrightarrow{q^*} H^1(\Omega, \mathcal{M}^*/\mathcal{O}^*) \\
\xrightarrow{\delta^*}\; & H^2(\Omega, \mathcal{O}^*) \to \cdots
\end{aligned}
\tag{6.5.12}
$$

Notice that, by the exercises for the reader preceding Proposition 6.5.9, the first map q^* takes each element of

$$
H^0(\Omega, \mathcal{M}^*) = \Gamma(\Omega, \mathcal{M}^*) = \{\text{meromorphic functions}\}
$$

into the equivalence class of meromorphic Cousin II problems that it solves. By exactness, if $\beta \in H^0(\Omega, \mathcal{M}^*/\mathcal{O}^*) = \Gamma(\Omega, \mathcal{M}^*/\mathcal{O}^*)$ satisfies $\delta^*\beta = 0$, then it is the image of a $u \in \Gamma(\Omega, \mathcal{M}^*)$. The converse holds as well. So the set of solvable meromorphic Cousin II problems is precisely the kernel of δ^* on $H^0(\Omega, \mathcal{M}^*/\mathcal{O}^*)$. Therefore, it behooves us to study the image group $H^1(\Omega, \mathcal{O}^*)$. We need another exact sequence.

Consider

$$
0 \to \mathbb{Z} \xrightarrow{i} \mathcal{O} \xrightarrow{\widetilde{\exp}} \mathcal{O}^* \to 0,
$$

where $\mathbb{Z}$ is thought of as the subsheaf of germs of constant integer-valued holomorphic functions on $\mathcal{O}$. Here $\widetilde{\exp}$ is the map $z \mapsto e^{2\pi i z}$. Exactness at $\mathcal{O}^*$ holds because nonvanishing holomorphic functions always have local logarithms. By the long exact cohomology sequence, we have

$$
\begin{aligned}
0 \;\to\; & H^0(\Omega, \mathbb{Z}) \xrightarrow{i^*} H^0(\Omega, \mathcal{O}) \xrightarrow{\widetilde{\exp}^*} H^0(\Omega, \mathcal{O}^*) \\
\xrightarrow{'\delta^*}\; & H^1(\Omega, \mathbb{Z}) \xrightarrow{i^*} H^1(\Omega, \mathcal{O}) \xrightarrow{\widetilde{\exp}^*} H^1(\Omega, \mathcal{O}^*) \\
\xrightarrow{'\delta^*}\; & H^2(\Omega, \mathbb{Z}) \xrightarrow{i^*} \cdots
\end{aligned}
$$

where we use the notation $'\delta^*$ to distinguish this map from δ^* of (6.5.12). Now the composite map $C = {}'\delta^* \circ \delta^*$,

$$
H^0(\Omega, \mathcal{M}^*/\mathcal{O}^*) \xrightarrow{\delta^*} H^1(\Omega, \mathcal{O}^*) \xrightarrow{'\delta^*} H^2(\Omega, \mathbb{Z})
$$

is called the *Chern map*. It assigns to each set of meromorphic Cousin II data $\beta \in H^0(\Omega, \mathcal{M}^*/\mathcal{O}^*)$ its *Chern class* $C(\beta)$ in $H^2(\Omega, \mathbb{Z})$.

THEOREM 6.5.13 (Oka/Serre)   Let $\Omega \subseteq \mathbb{C}^n$ be a domain. Let β be a divisor on Ω. Then the following properties hold:

(6.5.13.1) If β is a principal divisor on Ω, then $C(\beta) = 0$.

(6.5.13.2) If $H^1(\Omega, \mathcal{O}) = 0$ (in particular, if Ω is a domain of holomorphy) and if $C(\beta) = 0$, then β is principal.

Proof   If β is principal, then $\delta^*\beta = 0$, hence $C(\beta) = 0$.
If $H^1(\Omega, \mathcal{O}) = 0$, then

$$H^1(\Omega, \mathcal{O}^*) \overset{'\delta^*}{\to} H^2(\Omega, \mathbb{Z})$$

is injective. Therefore, $C = {}'\delta^* \circ \delta^*\beta = 0$ implies that $\delta^*\beta = 0$. In conclusion, β is principal.   □

COROLLARY 6.5.14   Let $\Omega \subseteq \mathbb{C}^n$ be a domain. If $H^1(\Omega, \mathcal{O}) = H^2(\Omega, \mathbb{Z}) = 0$, then meromorphic Cousin II is always solvable on Ω.

EXERCISES_____

1. Complete the following outline to relate the cohomology of slices of a domain to the cohomology of the domain. (Although we shall not develop these ideas here, it is worth noting that this provides yet another of the classical approaches to the Levi problem (see L. Bers [1] for details). It also explains, by way of the Dolbeault isomorphisms, why the property of being a domain of holomorphy is equivalent to solvability of the problem $\bar{\partial} u = \alpha$ for α a $\bar{\partial}$-closed (p, q) form, *any* q, not just for $q = 1$.)
   Throughout, $\Omega \subseteq \mathbb{C}^n$ is an open set and $\omega = \{z_n = 0\} \cap \Omega \neq \emptyset$. Let $\mathcal{U} = \{U_i\}$ be a covering of Ω by polydiscs, and assume that

   $$H^r(\Omega, \mathcal{O}) = H^{r+1}(\Omega, \mathcal{O}) = 0.$$

   a. Let $f \in Z^r(\omega \cap \mathcal{U}, \mathcal{O})$ on ω. Then f extends to an element $\tilde{f}$ of $C^r(\mathcal{U}, \mathcal{O})$ on Ω.
   b. It holds that $\delta \tilde{f} = z_n \tilde{g}$, some $\tilde{g} \in C^{r+1}(\mathcal{U}, \mathcal{O})$ on Ω.
   c. It follows that $\tilde{g} \in Z^{r+1}(\mathcal{U}, \mathcal{O})$ on Ω.
   d. There is an $\tilde{h} \in C^r(\mathcal{U}, \mathcal{O})$ on Ω such that $\delta \tilde{h} = \tilde{g}$.
   e. The cochain $\tilde{f} - z_n \tilde{h}$ is an r-cocycle.
   f. There is an $\tilde{F} \in C^{r-1}(\mathcal{U}, \mathcal{O})$ on Ω with $\delta \tilde{F} = \tilde{f} - z_n \tilde{h}$.
   g. $F \equiv \tilde{F}\big|_\omega$ satisfies $\delta F = f$.
   h. $H^r(\omega, \mathcal{O}) = 0$.

2. Complete the following outline to prove *Oka's extension theorem:* Let $\Omega \subseteq \mathbb{C}^n$ be an open set with $H^1(\Omega, \mathcal{O}) = 0$. Let $\omega = \Omega \cap \{z : z_n = 0\}$. If $f : \omega \to \mathbb{C}$

is holomorphic, then there is an $F : \Omega \to \mathbb{C}$ holomorphic such that $F|_\omega = f$. (This is a refinement of Theorem 5.1.1; the proof of that theorem can be related to the proof that we now present by using Dolbeault's isomorphisms.)

a. Let $\mathcal{U} = \{U_i\}_{i=1}^\infty$ be an open covering of ω by polydiscs, and let $U_0 = \Omega \backslash \omega$. On each $U_i, i \geq 1$, there is a holomorphic F_i with $F_i|_{U_i \cap \omega} = f$. Let $F_0 \equiv 1$ on U_0.

b. Define functions g_{ij} on $U_i \cap U_j$ by

$$g_{ij}(z) = \begin{cases} \frac{F_j(z) - F_i(z)}{z_n} & \text{if } i, j \neq 0 \\ F_0 - \frac{F_i(z)}{z_n} & \text{if } i \neq 0, j = 0 \end{cases}$$

Then the g_{ij} are a set of holomorphic Cousin I data for the covering $\{U_i\}_{i=0}^\infty$.

c. Let $\{g_i\}$ be a holomorphic solution to this Cousin I problem, and define

$$F(z) = \begin{cases} z_n \cdot \left(\frac{F_i}{z_n} - g_i \right) & \text{if } z \in U_i , \ i \neq 0 \\ z_n \cdot (1 - g_0) & \text{if } z \in U_0 \end{cases}$$

Then F is holomorphic on Ω and $F|_\omega = f$.

3. Use Exercises 1 and 2 to show that if $\Omega \subseteq \mathbb{C}^n$ is open and $H^1(\Omega, \mathcal{O}) = \cdots = H^n(\Omega, \mathcal{O}) = 0$, then Ω is a domain of holomorphy. (*Hint:* Imitate the proof of Theorem 5.1.2).

4. Verify that the following statements about an open set $\Omega \subseteq \mathbb{C}^n$ are equivalent:

a. Ω is a domain of holomorphy.

b. Ω is pseudoconvex.

c. $H^r(\Omega, \mathcal{O}) = 0, r = 1, \ldots, n$.

d. For all $f \in C_{(p,q)}^\infty(\Omega), 1 \leq q \leq n$, with $\bar{\partial} f = 0$, there is a $u \in C_{(p,q-1)}^\infty(\Omega)$ such that $\bar{\partial} u = f$.

e. For every complex affine subspace p in $\mathbb{C}^n$, if $f : p \cap \Omega \to \mathbb{C}$ is holomorphic, then there is a holomorphic F on Ω with $F|_{p \cap \Omega} = f$.

5. *Cohomology of complex projective space*

a. Define $\mathbb{CP}^n$, *complex projective space*, to be the set of all complex lines through 0 in $\mathbb{C}^{n+1}$. Give $\mathbb{CP}^n$ the structure of a complex manifold by introducing *homogeneous coordinates* as follows:

$$\mathbb{CP}^n = \left(\mathbb{C}^{n+1} \backslash \{0\} \right) / \sim,$$

where $z \sim w$ if $z = \lambda w$, some $\lambda \in \mathbb{C}$. Let the notation $[z_1, \ldots, z_{n+1}]$ be used to denote the equivalence classes. Let

$$U_j = \{ [z_1, \ldots, z_{n+1}] : z_j \neq 0 \}, \quad j = 1, \ldots, n+1.$$

Define $\phi_j : U_j \to \mathbb{C}^n$ by

$$\phi_j([z_1, \ldots, z_{n+1}]) = \left(z_1/z_j, \ldots, \widehat{z_j/z_j}, \ldots, z_{n+1}/z_j\right).$$

Then ϕ_j is a homeomorphism of topological spaces, and each $\phi_j \circ \phi_i^{-1}$ is a biholomorphic map on subsets of $\mathbb{C}^n$. If $U \subseteq \mathbb{CP}^n$ is open, $f : U \to \mathbb{C}$, then we say that f is holomorphic if $f \circ \phi_j^{-1}$ is holomorphic in the classical sense, $j = 1, \ldots, n+1$.

b. The manifold $\mathbb{CP}^n$ is compact. If $f : \mathbb{CP}^n \to \mathbb{C}$ is holomorphic, then f is constant (use the maximum principle). Therefore, $H^0(\mathbb{CP}^n, \mathcal{O}) \cong \mathbb{C}$.

c. Trivially, $H^p(\mathbb{CP}^n, \mathcal{O}) = 0$ if $p > n$ (use Dolbeault's theorem). Restrict attention to $\mathbb{CP}^1$. We now compute cohomology using the covering $\mathcal{U} = \{U_1, U_2\}$. Notice that $U_1 \cap U_2 = \{[z_1, z_2] : z_1 \neq 0, z_2 \neq 0\} = \mathbb{C} \setminus \{0\}$. Then $H^p(U_1 \cap U_2, \mathcal{O}) = 0$, all $p > 0$ (use the Dolbeault theorem). So Leray's theorem applies and $H^p(\mathbb{CP}^1, \mathcal{O}) \cong H^p(\mathcal{U}, \mathcal{O})$.

Compute $H^p(\mathcal{U}, \mathcal{O})$ as follows: Suppose that f is a 1-cochain for the cover $\mathcal{U}$. Then

$$f = \sum_{j=-\infty}^{\infty} a_j z_1^j = \sum_{j=-\infty}^{\infty} a_j z_2^{-j}.$$

Write f explicitly as $f_1 - f_2$, where $\{f_j\}$ is a 0-cochain on $\{U_j\}$. Therefore, $H^1(\mathcal{U}, \mathcal{O}) = 0$.

d. Now compute that $H^0(\mathbb{CP}^1, \mathcal{D}_{\mathcal{H}}^0) \cong \mathbb{C}, H^1(\mathbb{CP}^1, \mathcal{D}_{\mathcal{H}}^1) \cong \mathbb{C}$.

6. Let $\Omega = \mathbb{C}^2 \setminus \{0\}$. Prove that $\mathcal{O}(\Omega) = \mathcal{O}(\mathbb{C}^2)$. Let $U_1 = \{z_1 \neq 0\}, U_2 = \{z_2 \neq 0\}, \mathcal{U} = \{U_1, U_2\}$. Then the cover $\mathcal{U}$ is *acyclic* for $\mathcal{O}$ (that is, the cohomology, for $p > 1$, of $U_i \cap U_j$ is trivial). Calculate that $\dim H^1(\mathbb{C}^2 \setminus \{0\}, \mathcal{O}) = \infty$ by noticing that $\delta C^0(\mathcal{U}, \mathcal{O})$ contains no Laurent series with terms $z_1^m \cdot z_2^n$ and either m or n negative. Of course, $H^p(\mathbb{C}^2 \setminus \{0\}, \mathcal{O}) = 0$ if $p > 2$. What can you say about $H^2(\mathbb{C}^2 \setminus \{0\}, \mathcal{O})$?

7. Let $\mathbb{C}^*$ denote $\mathbb{C} \setminus \{0\}$. Let $k, \ell \in \mathbb{N}$. Prove that $H^p(\mathbb{C}^k \times (\mathbb{C}^*)^\ell, \mathcal{O}) = 0$ for all $p > 0$.

8. *Simplicial cohomology* We now discuss a number of assertions about the cohomology theory generated by simplices in space (this theory is closely related to, but not the same as, *singular cohomology*). The interested reader should verify all assertions.

Let

$$\Delta_k = \{x \in \mathbb{R}^{k+1} : x_i \geq 0, i = 1, \ldots, k+1, \sum x_i = 1\}.$$

See Figure 6.7 for the case $k = 2$.

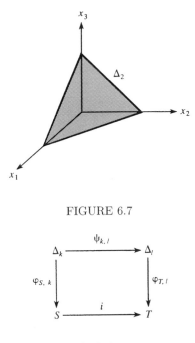

FIGURE 6.7

FIGURE 6.8

A set $S \subseteq \mathbb{R}^N$ is a k-simplex if there is a surjective *affine imbedding* $\phi_{S,k} : \Delta_k \hookrightarrow S$. If $S, T \subseteq \mathbb{R}^N$ are simplices of dimensions $k < \ell$, then S is said to be a *subsimplex* of T if $\phi_{S,k}, \phi_{T,\ell}$ can be chosen so that the diagram in Figure 6.8 commutes. Here i is injection and $\psi_{k,\ell}$ is given by

$$\psi_{k,\ell} : \mathbb{R}^{k+1} \quad \to \quad \mathbb{R}^{\ell+1}$$
$$(x_1, \ldots, x_{k+1}) \quad \mapsto \quad (x_1, \ldots, x_{k+1}, 0, \ldots, 0).$$

A *simplicial complex* $K \subseteq \mathbb{R}^N$ is a finite union $\cup_{j=1}^{k} S_j$ of simplices of various dimensions subject to the condition that if $j \neq \ell$, then either $S_j \cap S_\ell \neq \emptyset$ or $S_j \cap S_\ell$ is a subsimplex of both S_j and S_ℓ (see Figure 6.9).

If $S \subseteq \mathbb{R}^N$ is a simplex, then its 0-subsimplices are called *vertices* and its 1-subsimplices are called *edges*. A p-simplex is completely determined by its $(p+1)$ vertices. Therefore, if S is a simplex and $\nu_{\alpha_0}, \ldots, \nu_{\alpha_p}$ are its vertices, we may denote S by $\langle \nu_{\alpha_0}, \ldots, \nu_{\alpha_p} \rangle$.

A p-simplicial complex is one for which each S_j is a p-simplex. If K is a p-simplicial complex, then a *p-simplicial cochain* on K is a function that assigns to each p-subsimplex of K an integer. The function should be alternating in the vertices. Denote the collection of p-simplicial cochains by $C^p(K, \mathbb{Z})$. The coboundary operator $\delta : C^p(K, \mathbb{Z}) \to C^{p+1}(K, \mathbb{Z})$ is defined

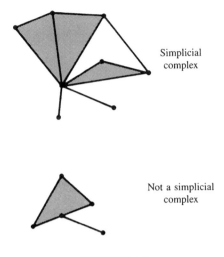

Simplicial
complex

Not a simplicial
complex

FIGURE 6.9

by

$$(\delta\eta)\left(\langle\nu_{\alpha_0},\dots,\nu_{\alpha_{p+1}}\rangle\right) = \sum_j (-1)^{j+1}\eta\left(\langle\nu_{\alpha_o},\dots,\hat{\nu}_{\alpha_j},\dots,\nu_{\alpha_{p+1}}\rangle\right)$$

for $\eta \in C^p(K,\mathbb{Z})$. Then $\delta^2 = 0$, so

$$\cdots \to C^{p-1}(K,\mathcal{S}) \xrightarrow{\delta} C^p(K,\mathbb{Z}) \xrightarrow{\delta} C^{p+1}(K,\mathbb{Z}) \to \cdots$$

is a complex, and we may form the associated cohomology groups $H_S^p(K,\mathbb{Z})$. If ν_α is a vertex of K, then we define $St(\nu_\alpha)$, the "star of ν_α," to be the union of all the simplices in K that have ν_α as a vertex. The collection $\mathcal{U} = \{St(\nu_\alpha) : \nu_\alpha \in K\}$ is a finite covering of K such that the intersection of any $(p+1)$ elements of $\mathcal{U}$ is either a p-simplex or is empty. If we ignore the fact that the elements of $\mathcal{U}$ are not open (which fact causes no difficulties), then we may consider $C^p(\mathcal{U},\mathbb{Z})$. If $\eta \in C^p(\mathcal{U},\mathbb{Z})$, and if $S = \langle\nu_{\alpha_0},\dots,\nu_{\alpha_p}\rangle$ is a p-simplex in K, then we may define $(\phi\eta)(S) = \eta(\alpha_0,\dots,\alpha_p)$. Then

$$\phi : C^p(\mathcal{U},\mathbb{Z}) \to C^p(K,\mathbb{Z}).$$

Moreover, $\phi\delta = \delta\phi$. This equality leads to an isomorphism $\phi^* : H^p(\mathcal{U},\mathbb{Z}) \to H_S^p(K,\mathbb{Z})$. Since K can be subdivided into finer and finer subsimplices without altering $H^p(K,\mathbb{Z})$, it follows that

$$H^p(K,\mathbb{Z}) \cong H_S^p(K,\mathbb{Z}).$$

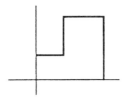

FIGURE 6.10

If K, K' are p-simplicial complexes and $\phi : K \to K'$ is a univalent, surjective simplicial mapping (how should this last concept be defined?), then ϕ^* (resp. $(\phi^{-1})^* = (\phi^*)^{-1}$) is a group isomorphism (see P. Griffiths and J. Harris [1] and Lefschetz [2] for further details).

The reader should check that there is no loss of generality to compute p-cohomology of p-simplicial complexes only.

9. Let $[f], [g]$ be relatively prime germs of holomorphic functions at $0 \in \mathbb{C}^2$. If $[h]$ is another germ, then let $m_{[f]}([h]) = [f][h] = [fh]$ and likewise for $m_{[g]}$. Then the sequence

$$0 \to \mathcal{O} \xrightarrow{m_{[f]} \times m_{[g]}} \mathcal{O} \times \mathcal{O} \xrightarrow{p} \mathcal{O} \to 0,$$

where $p([h], [k]) = [gh] - [fk]$, is exact. If f, g represent $[f], [g]$ on a small neighborhood U of 0 and if $V = \{z \in U : f(z) = g(z) = 0\}$, then we have the exact cohomology sequence

$$\Gamma(U \setminus V, \mathcal{O}) \xrightarrow{\delta^*} H^1(U \setminus V, \mathcal{O}) \xrightarrow{(m_{[f]} \times m_{[g]})^*} H^1(U \setminus V, \mathcal{O}^2).$$

Prove that the image of the cohomology class represented by 1 in $\Gamma(U \setminus V, \mathcal{O})$ is the cohomology class represented by the Bochner-Martinelli form $(\bar{f} d\bar{g} - \bar{g} d\bar{f})/(f\bar{f} + g\bar{g})^2$ in $H^1(U \setminus V, \mathcal{O})$ under the Dolbeault isomorphism. (*Hint:* Show that $\dim(\Gamma(U \setminus V, \mathcal{O})/\mathrm{Ker}\,\delta^*) = 1$. So the kernel of $(m_{[f]} \times m_{[g]})^*$ is one dimensional. The problem may also be done by direct computation.)

10. Let $a = (a_1, \ldots, a_n), 0 < a_j < 1, j = 1, \ldots, n$. Let $2 \leq k \leq n$, and define $D_k \equiv \{z \in D^n(0,1) : |z_1| \leq a_1, a_k \leq |z_k| < 1\}$. Let $\Omega = D^n(0,1) \setminus \cup_{k=2}^n D_k$. Then (D^n, Ω) is called a *simple Euclidean Hartogs figure* (see Figure 6.10). Prove that Ω is Reinhardt. Prove also that the logarithmically convex hull of Ω is D^n. If $F : D^n \to \tilde{D} \subseteq \mathbb{C}^n$ is biholomorphic, then let $\tilde{\Omega} = F(\Omega)$. Then $(\tilde{D}, \tilde{\Omega})$ is called a *simple Hartogs figure*. Give a proof of the tomato can principle (Exercise 14 at the end of Chapter 3), also called the *Kugelsatz*, using simple Hartogs figures.

Let $\Omega \subset\subset \mathbb{C}^n$ be pseudoconvex; prove that if $(\tilde{D}, \tilde{\Omega})$ is a simple Hartogs figure and $\tilde{\Omega} \subseteq \Omega$, then $\tilde{D} \subseteq \Omega$. Prove that this property characterizes pseudoconvex domains (see H. Grauert and K. Fritzsche [1] for details on these matters). A generalized Euclidean Hartogs figure is defined as follows: Fix $a = (a_1, \ldots, a_n)$ as before. Choose $1 \le k \le n - 1$. Let

$$\Omega \;=\; \{z \in D^n(0,1) : |z_j| < a_j, k < j \le n\}$$
$$\cup \left(\bigcup_{j=1}^{k} \{z \in D^n(0,1) : a_j < |z_j| < 1\} \right).$$

Then (D^n, Ω) is a *generalized Euclidean Hartogs figure* (of type k). Of course, a *generalized Hartogs figure* is the biholomorphic image of a generalized Euclidean Hartogs figure. Show that every simple Hartogs figure is a generalized Hartogs figure.

For some purposes, generalized Hartogs figures play the same role vis-à-vis the geometry of several complex variables as do the Eilenberg-Maclane spaces vis-à-vis algebraic topology. That is, for each $1 \le q \le n - 1$, the generalized Euclidean Hartogs figure (D^n, Ω) of type $k = q$ has $H^p(\Omega, \mathcal{O}) \ne 0$ only for $p = 0$ or $p = q$. It is rather tricky to prove this in general (see A. Andreotti and H. Grauert [1]). However the following special cases are accessible:

a. The cohomology of a Hartogs figure is independent of the choice of a. So we take $a_1 = a_2 = \cdots = a_n$ for the remainder of the problem.

b. Let $\mathbb{C}^n$ be fixed, $1 \le k \le n - 1$. Then the generalized Euclidean Hartogs figure Ω of type k can be written as the union of $k + 1$ product domains. This will be a Leray cover. Therefore, $H^p(\Omega, \mathcal{O}) = 0$ when $p > k$.

c. With notation as in (b), show $H^1(\Omega, \mathcal{O}) \ne 0$ when $n = 3, k = 1$.

11. Let $\Omega = \{z \in \mathbb{C}^n : 1 < |z| < 2\}$. Calculate $H^p(\Omega, \mathcal{O})$.

12. Let $\Omega \subseteq \mathbb{R}^N$ be a domain. Formulate a Cousin I problem for continuous integer-valued functions (use the first example in Section 6.1 as a guide). Prove that if Cousin I is always solvable, then every curl-free vector field is a gradient.

13. Use what you learned in the preceding problem to prove that if $\Omega \subseteq \mathbb{C}^n$ is a domain on which the first Cousin problem is always solvable (the usual one for holomorphic functions), then the equation $\bar{\partial}u = f, f$ a $\bar{\partial}$-closed $(0,1)$ form, is always solvable. Do this from first principles, using a Poincaré lemma for the local result. *Do not* quote one of our cohomology isomorphisms.

14. Let $\Omega \subseteq \mathbb{C}^n$ be a domain for which $H^p(\Omega, \mathbb{Z}) = 0$, some $p > 0$. Prove that, as a consequence, $H^p(\Omega, \mathbb{C}) = 0$. (*Hint:* The right proof is just one line.)

7 The Zero Set of a Holomorphic Function

7.1 Coherent Analytic Sheaves

Although Theorems A and B of Cartan are absolutely essential to the modern theory of several complex variables, they are not germane to the principal theme of this book. Therefore, we present these two results, together with the Oka coherence theorem, but we do not prove them. (Comprehensive treatments may be found in L. Hörmander [3] and in R. C. Gunning [3].) Section 7.2 contains a number of applications of these results. Our intent with this superficial introduction to coherent analytic sheaves is to acquaint the reader with the significance of this language and methodology without burdening him or her with the details of the (rather difficult) proofs.

Loosely speaking, the theorems of Cartan are vector-valued versions of the results of Chapter 6. So the proofs consist largely of elaborate inductive arguments. They contains no fundamentally new ideas: indeed, the concept of a sheaf, together with Cartan's theorems, constitutes an effective language. Without the language, the theorems cannot be formulated; with the language, new insights are obtained.

The central notion in Sections 7.1 and 7.2 is the concept of *coherence*. We proceed now with the necessary definitions.

DEFINITION 7.1.1   Let $\Omega \subseteq \mathbb{C}^n$ be an open set. A sheaf of $\mathcal{O}$-modules $(\mathcal{F}, \Omega, \pi)$ is called an *analytic sheaf.*

DEFINITION 7.1.2   An analytic sheaf $(\mathcal{F}, \Omega, \pi)$ is said to be *locally finitely generated* if for each $z \in \Omega$ there is a neighborhood U of z in Ω and finitely many $f_1, \ldots, f_k \in \Gamma(U, \mathcal{F})$ such that $\{[f_j]_\zeta\}_{j=1}^k$ generates $\mathcal{F}_\zeta$ as an $\mathcal{O}_\zeta$-module for each $\zeta \in U$.

EXAMPLE   Let $\Omega \subseteq \mathbb{C}^2$ be open. Let $\mathcal{F} \subseteq \mathcal{O}$ be the sheaf of germs that vanish on $\omega = \Omega \cap \{z \in \mathbb{C}^2 : z_2 = 0\}$. (*Note:* If $z \notin \omega$, then $\mathcal{F}_z \equiv \mathcal{O}$.) Then (exercise) $\mathcal{F}$ is a locally finitely generated analytic sheaf.   □

EXAMPLE   Let $U \subset\subset \Omega \subset\subset \mathbb{C}^n$. Let $\mathcal{F} \subseteq \mathcal{O}$ be the subsheaf of germs vanishing on $\Omega \setminus U$. Then $\mathcal{F}$ is not locally finitely generated at any point of $\partial(\bar{U})$.   □

EXAMPLE   The sheaf of germs of C^∞ functions over Ω, considered as an $\mathcal{O}$-module, is *not* locally finitely generated.   □

The next lemma makes locally finitely generated sheaves easier to handle in practice.

LEMMA 7.1.3   Suppose that $(\mathcal{F}, \Omega, \pi)$ is a locally finitely generated analytic sheaf, that $z \in \Omega$, that U is a neighborhood of z in Ω, and that $f_1, \ldots, f_k \in \Gamma(U, \mathcal{F})$ have the property that $[f_1]_z, \ldots, [f_k]_z$ generate $\mathcal{F}_z$. Then there is a (possibly smaller) neighborhood U' of z such that $[f_1]_\zeta, \ldots, [f_k]_\zeta$ generate $\mathcal{F}_\zeta$ for all $\zeta \in U'$.

Proof   By the definition of "locally finitely generated," there exist $g_1, \ldots, g_r \in \Gamma(\tilde{U}, \mathcal{F})$, some neighborhood $\tilde{U}$ of z, such that $[g_1]_\zeta, \ldots, [g_r]_\zeta$ generate $\mathcal{F}_\zeta$ for each $\zeta \in \tilde{U}$. By hypothesis, there exist $a_{ij}(\cdot)$ holomorphic in a neighborhood of z such that

$$[g_i]_z = \sum_{j=1}^{k} [a_{ij}]_z \, [f_j]_z, \qquad i = 1, \ldots, r.$$

By the definition of germ, this identity persists for g_i, f_i, a_{ij} in a neighborhood U' of z. The conclusion follows.   □

DEFINITION 7.1.4   If $(\mathcal{F}, \Omega, \pi)$ is an analytic sheaf, $U \subseteq \Omega$ is an open set, and $f_1, \ldots, f_k \in \Gamma(U, \mathcal{F})$, then the kernel of the homormorphism

$$\phi_{(f)} : \mathcal{O}^k \quad \to \quad \mathcal{O}$$
$$([g_1]_z, \ldots, [g_k]_z) \quad \mapsto \quad \sum_j [f_j g_j]_z$$

is called the (sub) sheaf $\mathcal{R}(f_1, \ldots, f_k) \subseteq \mathcal{O}^k$ of *relations* among $f_1, \ldots, f_k$.

DEFINITION 7.1.5   Let $(\mathcal{F}, \Omega, \pi)$ be an analytic sheaf. We say that $\mathcal{F}$ is *coherent* if

(7.1.5.1) $\mathcal{F}$ is locally finitely generated;

(7.1.5.2) For each open $U \subseteq \Omega$ and each $f_1, \ldots, f_k \in \Gamma(U, \mathcal{F})$, the subsheaf $\mathcal{R}(f_1, \ldots, f_k)$ is locally finitely generated.

THEOREM 7.1.6 (Theorem A of Cartan)   Let $\Omega \subseteq \mathbb{C}^n$ be pseudocon-vex and let $(\mathcal{F}, \Omega, \pi)$ be a coherent analytic sheaf over Ω. Then for each $z \in \Omega$, the $\mathcal{O}_z$-module $\mathcal{F}_z$ is generated by finitely many germs at z of elements of $\Gamma(\Omega, \mathcal{F})$.

THEOREM 7.1.7 (Theorem B of Cartan)   Let $\Omega \subseteq \mathbb{C}^n$ be pseudo-convex and let $(\mathcal{F}, \Omega, \pi)$ be a coherent analytic sheaf over Ω. Then, for $r > 0$, $H^r(\Omega, \mathcal{F}) = 0$.

THEOREM 7.1.8 (Oka Coherence Theorem)   Every locally finitely gen-erated subsheaf of $\mathcal{O}^k$ is coherent.

Theorem A says that a coherent analytic sheaf has plenty of global sections (so $H^0(\Omega, \mathcal{F})$ is large). Theorem B says, among other things, that first Cousin problems with values in $\mathcal{F}$ can always be solved. Oka's theorem provides the most important examples of coherent analytic sheaves. Notice that the definition of coherence is local; thus it is not surprising that Oka's theorem has nothing to do with pseudoconvexity.

7.2 Applications of the Cartan and Oka Theorems: The Structure of Ideals

Some of the applications given in this section are repetitions of results that we have already established. Others are definitely new. The point is to see that Cartan's theorems gather many of the ideas in this subject under one great umbrella.

DEFINITION 7.2.1 A paracompact Hausdorff space M is called a *complex man-ifold of dimension* k if there is a covering $\mathcal{U} = \{U_i\}$ of M by open sets (we call $\mathcal{U}$ an *atlas*) and continuous maps $\phi_i : U_i \to W_i \subseteq \mathbb{C}^k$ such that each ϕ_i is one-to-one and onto and, for all i, j,

$$\phi_j \circ \phi_i^{-1} : \phi_i(U_i \cap U_j) \to \phi_j(U_i \cap U_j)$$

is holomorphic. We say that $f : M \to \mathbb{C}$ is *holomorphic* if $f \circ \phi_i^{-1} : W_i \to \mathbb{C}$ is holomorphic for each i.

DEFINITION 7.2.2   Let $\Omega \subseteq \mathbb{C}^n$ be an open set and $M \subseteq \Omega$ a k-dimensional complex manifold. We say that M is *locally regularly presented* if each $m \in M$ has a neighborhood U_m and holomorphic functions $f_1, \ldots, f_{n-k}$ such that (i) the matrix $(\partial f_i / \partial z_j)$ is of maximal rank on U_m and (ii) $M \cap U_m = \{z \in U_m : f_1(z) = \cdots = f_{n-k}(z) = 0\}$.

Remark: The concept of "regularly presented" addresses the phenomenon that an abstractly presented smooth manifold can be badly imbedded into space. For example, the mapping $\zeta \mapsto (\zeta, |\zeta|)$ badly imbeds the manifold $\mathbb{C}$ into $\mathbb{C}^{2}$ so that the image is not regularly presented.

We will also need the concept of a "proper" mapping. A map $f : X \to Y$ of topological spaces is called *proper* if $f^{-1}(K)$ is compact in X whenever K is compact in Y. When X, Y are manifolds in space, then the concept of "proper" has the following intuitively appealing interpretation: If $x_j \in X$ and x_j tend to the boundary of X, then $f(x_j)$ tend to the boundary of Y. A manifold M is properly imbedded in $\Omega \subseteq \mathbb{C}^n$ if the inclusion mapping $i : M \to \Omega$ is a proper map.   □

DEFINITION 7.2.3   Let $\Omega \subseteq \mathbb{C}^n$ be open and $M \subseteq \Omega$ a properly imbedded, locally regularly presented complex manifold of dimension k. The *ideal sheaf* $\mathcal{J} = \mathcal{J}(M)$ of M is defined as follows: If $z \notin M$, then $\mathcal{J}_z = \mathcal{O}_z$; if $z \in M$ then $\mathcal{J}_z \subseteq \mathcal{O}_z$ is the submodule of all germs of functions vanishing on M.

Exercise for the Reader
The sheaf $\mathcal{J}(M)$ is locally finitely generated. (*Hint:* After a change of coordinates, the functions $f_1, \ldots, f_{n-k}$ may be taken to be the coordinate functions $z_{k+1}, \ldots, z_n$.)

THEOREM 7.2.4   Let $\Omega \subseteq \mathbb{C}^n$ be pseudoconvex and $M \subseteq \Omega$ a locally regularly imbedded k-dimensional complex manifold. Then there exist holomorphic functions $\{f_i\}_{i=1}^{\infty}$ on Ω such that $M = \{z \in \Omega : f_i(z) = 0, \forall i\}$.

Proof   The sheaf $\mathcal{J}(M)$ is a locally finitely generated subsheaf of $\mathcal{O}$. By Oka's theorem, it is coherent. If $p \notin M$, then $\mathcal{O}_p$ contains the germ $[1]$, and Theorem A now guarantees that there is an element $f_p \in \Gamma(\Omega, \mathcal{J})$ such that $[f_p]_p \neq 0$. This inequality persists in a neighborhood U_p of p. Now countably many of the U_p cover $\Omega \setminus M$. The corresponding f_p do the job.   □

THEOREM 7.2.5   Let $(\mathcal{F}, \Omega, \pi)$ be a coherent analytic sheaf, and let Ω be pseudoconvex. Suppose that there exist finitely many $f_1, \ldots, f_k \in \Gamma(\Omega, \mathcal{F})$ such that $[f_1]_z, \ldots, [f_k]_z$ generate $\mathcal{F}_z$, each $z \in \Omega$. Then for each $g \in \Gamma(\Omega, \mathcal{F})$, there exist $g_1, \ldots, g_k \in \Gamma(\Omega, \mathcal{O})$ with $g = f_1 g_1 + \cdots + f_k g_k$.

Proof   Consider the exact sheaf sequence

$$0 \to \mathcal{R} \xrightarrow{i} \mathcal{O}^k \xrightarrow{\psi} \mathcal{F} \to 0,$$

where $[\psi([(g_1, \ldots, g_k)])]_z = \sum_j [f_j g_j]_z$, $\mathcal{R}$ is the sheaf of relations, and i is the inclusion mapping. The sheaf $\mathcal{R}$ is coherent so, by Theorem B, $H^1(\Omega, \mathcal{R}) = 0$.

Now the long exact cohomology sequence gives

$$H^0(\Omega, \mathcal{O}^k) \xrightarrow{\psi^*} H^0(\Omega, \mathcal{F}) \xrightarrow{\delta^*} H^1(\Omega, \mathcal{R}) = 0.$$

Therefore, ψ^* is surjective as desired.   □

COROLLARY 7.2.6   If $f_1, \ldots, f_k$ are holomorphic functions on a pseudoconvex $\Omega \subseteq \mathbb{C}^n$ such that $\{f_j\}$ have no common zeros, then there exist holomorphic functions $g_1, \ldots, g_k$ on Ω satisfying $\sum_j f_j g_j \equiv 1$.

Proof   The functions $f_1, \ldots, f_k$ satisfy the hypotheses of the theorem and the function $g \equiv 1 \in \Gamma(\Omega, \mathcal{O})$. So the theorem applies.   □

Remark: The corona problem consists in proving the corollary with the additional hypothesis that the f_j each be bounded (and $\sum |f_j|$ be bounded from zero) and the additional conclusion that the g_j each be bounded. This problem is unsolved in $\mathbb{C}^n, n > 1$, both for the ball and the polydisc. There are some pseudoconvex sets (not smoothly bounded) for which it is elementary to see that the corona problem fails (see Exercise 7 at the end of Chapter 8). In fact, by more sophisticated techniques, one may construct a smooth domain that is strictly pseudoconvex except at one boundary point and for which the corona problem fails (see N. Sibony [3]). See also Exercise 14 at the end of Chapter 5.

It is curious that there are no known domains in $\mathbb{C}^1$ for which the corona problem is known to fail (for the state of the art, see J. Garnett and Peter W. Jones [1], P. Jones and T. Wolff [1], and L. Carleson [2]). There are also no known domains in $\mathbb{C}^n, n > 1$, for which the corona problem is known to have an affirmative solution. The corona problem has a positive solution for closed Riemann surfaces of finite genus (E. L. Stout [2]). However, there are Riemann surfaces of infinite genus for which the corona problem fails (see T. W. Gamelin [2]).   □

THEOREM 7.2.7   Let Ω be a domain of holomorphy. Then Cousin I is always solvable on Ω.

Proof   The sheaf $\mathcal{O}$ is coherent by Oka's theorem. By Theorem B, $H^1(\Omega, \mathcal{O}) = 0$. Therefore, Cousin I is always solvable on Ω.   □

THEOREM 7.2.8   Let Ω be a domain of holomorphy. Let $M \subseteq \Omega$ be a properly imbedded, locally regularly presented manifold. Let $f : M \to \mathbb{C}$ be holomorphic. Then there is a holomorphic F on Ω such that $F|_M = f$.

Proof   Let $\mathcal{J}$ be the ideal sheaf of M. Consider the exact sequence

$$0 \to \mathcal{J} \xrightarrow{i} \mathcal{O} \xrightarrow{q} \mathcal{O}/\mathcal{J} \to 0,$$

where q is the quotient mapping. Observe that $\mathcal{O}/\mathcal{J} \cong \mathcal{O}(M)$ (exercise). The long exact cohomology sequence yields

$$H^0(\Omega, \mathcal{O}) \xrightarrow{q^*} H^0(\Omega, \mathcal{O}(M)) \xrightarrow{\delta^*} H^1(\Omega, \mathcal{J}).$$

But $H^1(\Omega, \mathcal{J}) = 0$ by Theorem B since $\mathcal{J}$ is coherent. Therefore, q^* is surjective.
□

Remark: Of course, the last two theorems are not new to us. We present them to show how some of our old ideas fit elegantly into the present new framework.
□

THEOREM 7.2.9   Let $\Omega \subseteq \mathbb{C}^n$ be pseudoconvex, and assume that $0 \in \Omega$. Let $f : \Omega \to \mathbb{C}$ be a holomorphic function such that $f(0) = 0$. Then there exist $f_1, \ldots, f_n$ holomorphic on Ω such that $z_1 f_1 + \cdots z_n f_n = f$ on Ω.

Proof   Let $\mathcal{J}_0$ be the ideal sheaf of $\{0\}$. Then $\mathcal{J}$ is coherent. Moreover, it is straightforward to check that the functions $f_j(z) = z_j, j = 1, \ldots, n$ have the property that $[f_1]_z, \ldots, [f_n]_z$ generate $(\mathcal{J}_0)_z$ for each $z \in \Omega$ (for the origin, use the local power series expansion of f). Now Theorem 7.2.5 applies to give the desired conclusion.
□

The reader will want to compare the elegant proof of 7.2.9 with the long (and by now rather clumsy) proof of the weaker result given in the second Example of section 6.1.

7.3   Zeros of One Holomorphic Function

In Section 7.2 we considered some properties of the ideal of functions vanishing on a given set. Conversely, it is of interest to consider the zero set of a fixed holomorphic function. What topological restrictions are there on such sets? What metric restrictions?

Recall that in one complex variable the zero set of a holomorphic function must be discrete; and any discrete set is the zero set of some holomorphic function. In several complex variables, the zero set of a holomorphic function is never discrete; indeed there can be no isolated zeros. What we shall learn in this section is that these two apparently disparate facts can be reconciled and can both be seen to be special cases of a single geometric/analytic fact. We begin with a qualitative result that is a consequence of the Weierstrass preparation theorem.

Recall the concept of the *resultant* (van der Waerden [1, p. 103]): If $f = a_0 + a_1 z + \cdots + a_j z^j$ and $g = b_0 + b_1 z + \cdots + b_k z^k$ are polynomials of one complex variable with complex coefficients, then we define

$$
R(f,g) \equiv \det \left[
\begin{array}{ccccccccc}
a_0 & a_1 & \cdot & & \cdot & \cdot & a_n & & \\
 & a_0 & a_1 & \cdot & \cdot & & & a_n & \\
 & & \cdot & & & & & & \\
 & 0 & & \cdot & & & & & \\
 & & & & a_0 & a_1 & \cdot & \cdot & a_n \\
b_0 & b_1 & \cdot & & \cdot & b_n & & & \\
 & b_0 & b_1 & \cdot & \cdot & & b_n & & \\
 & & \cdot & & & & & 0 & \\
 & 0 & & \cdot & & & & & \\
 & & & & b_0 & b_1 & \cdot & \cdot & b_n
\end{array}
\right] \left.\vphantom{\begin{array}{c} \\ \\ \\ \\ \\ \\ \\ \\ \\ \end{array}}\right\} j+k
$$

$$\underbrace{\hspace{5cm}}_{j+k}$$

Then f and g have a common root if and only if $R(f,g) = 0$. In particular, f has a multiple root if and only if $R(f,f') = 0$.

Recall also the notion of *topological dimension* (W. Hurewicz and H. Wallman [1]):

1. The set $\emptyset$ has dimension -1.

2. The set $S \subseteq \mathbb{R}^N$ has dimension less than or equal to k if each $x \in S$ has a neighborhood basis of open sets U_j such that $\dim((\partial U_j) \cap S) \leq k - 1$.

3. The set S has dimension k if k is the least nonnegative integer such that S has dimension less than or equal to k but not less than or equal to $k - 1$.

Exercise for the Reader

A k cell in $\mathbb{R}^N$ has dimension k. The Cantor set in $\mathbb{R}^1$ has dimension 0. The dimension of the set of irrational numbers in $\mathbb{R}^1$ is 0. Dimension is not quite subadditive. In fact, $\dim(A \cup B) \leq \dim A + \dim B + 1$.

PROPOSITION 7.3.1 Let f be a holomorphic function that is normalized of order k on $D^n(0,r)$. Assume that f is an irreducible element in the ring of holomorphic functions on $D^n(0,r)$. Write $z = (z_1, \ldots, z_{n-1}, z_n) \equiv (z', z_n)$, and let D' denote $D^{n-1}(0,r)$ in the variable z'. Let $\mathcal{N} = f^{-1}(\{0\})$. Then, shrinking, r if necessary, there is a set $E \subseteq D'$ such that E has topological dimension less than or equal to $2n - 4$ and the map

$$
\begin{aligned}
\pi : \mathcal{N} \setminus \pi^{-1}(E) &\to D' \setminus E \\
z &\mapsto z'
\end{aligned}
$$

is a k-sheeted analytic covering. In particular, π is locally biholomorphic on $\mathcal{N} \setminus \pi^{-1}(E)$, and the topological dimension of $\mathcal{N}$ is $2n - 2$.

Remark: This theorem is in the vein of an entire class of theorems known as "resolution of singularities" theorems. Hironaka's celebrated theorem (see H. Hironaka [1]) gives a very general framework for creating theorems of this sort. A self-contained discussion of resolution of singularities theorems can be found in S. G. Krantz and H. R. Parks [2]. The discussion here is much more *ad hoc* and elementary. □

Proof of Proposition 7.3.1 The proof is by induction on n. For the case $n = 1$, the result is trivial. In this case, E is empty (since f is irreducible), $\mathcal{N}$ has topological dimension 0, and D' is a singleton.

If the result has been proved on $D^{n-1}(0, r)$, let f be normalized of order k on $D^n(0, r)$. Shrinking r if necessary, we write $f = u \cdot W$ with u a unit and W a Weierstrass polynomial. Write

$$W(z) = z_n^k + a_{k-1}(z')z_n^{k-1} + \cdots + a_0(z').$$

Let $R(z') = R(W(z', z_n), (\partial/\partial z_n)W(z', z_n))$. Then $R(z')$ is a holomorphic function on $D'(0, r)$ that is not identically zero. Let $E \subseteq D'(0, r)$ be the zero set of R. Shrinking r if necessary, we have by induction that the topological dimension of E does not exceed $2n - 4$. Finally, $E \subseteq D'(0, r)$ is closed. If $z' \in D'(0, r) \setminus E$, then $W(z', \cdot)$ has k distinct zeros. By the implicit function theorem, $\mathcal{N} \setminus \pi^{-1}(E)$ is a complex $(n-1)$-dimensional manifold, and π has the desired properties. □

Thus the zero set of a holomorphic function is locally an $(n-1)$-dimensional complex manifold, except on a branching set of complex dimension $n - 2$. It is possible to inductively apply this analysis and obtain a stratification of the zero set in terms of complex manifolds of dimensions $(n-1), (n-2), \ldots, 0$ (see R. C. Gunning [2]). We shall not pursue this matter further.

COROLLARY 7.3.2 Let $\Omega \subseteq \mathbb{C}^n$ be open. Let f be a meromorphic function on Ω that is regular except on a set $A \subseteq \Omega$ of topological dimension less than or equal to $2n - 3$. Then f is holomorphic.

Proof Let $P \in A$. Express $f = g/h$ on a neighborhood U of P, where g and h are holomorphic. Then the set $\{z \in U : h(z) = 0\}$ has topological dimension less than or equal to $2n - 3$, so it must be empty. □

We next establish a connection between the zeros of holomorphic functions and removable singularities through (a generalization of) the Riemann removable singularities theorem.

THEOREM 7.3.3 Let $\Omega \subseteq \mathbb{C}^n$ be open, with g holomorphic on Ω. Assume that g is not identically zero, and let $A = \{z \in \Omega : g(z) = 0\}$. Let f be holomorphic on $\Omega \setminus A$. Suppose that for each $a \in A$ there is a neighborhood U_a with $f|_{U_a \setminus A}$ bounded. Then there is an F holomorphic on Ω such that $F|_{\Omega \setminus A} = f$.

Proof   For the case $n = 1$, the result is well known and proved as follows: The points of A are isolated in Ω. Let a be one such point. The Laurent series for f at a is $\sum_{k=-\infty}^{\infty} f_k z^k$ with

$$f_k = \frac{1}{2\pi i} \oint_{|\zeta-a|=r} \frac{f(\zeta)}{(\zeta-a)^{k+1}} d\zeta,$$

r small. But f bounded near a implies $|f_k| \leq M \cdot r^{-k}$. When $k < 0$ and $r \to 0$, this yields $f_k = 0$. Therefore, $F(z) = \sum_{k=0}^{\infty} f_k z^k$ is the desired extension near a.

For general n, let $a \in A$. Suppose without loss of generality that $a = 0$ and that f is normalized of order k. For each fixed z' small, the case $n = 1$ now applies to the function $z_n \mapsto f(z', z_n)$, which is holomorphic and bounded for small z_n except on the finite set $\{z_n : g(z', z_n) = 0\}$.   □

In the classical theory, very precise information about zero sets is obtained when growth restrictions are imposed on the holomorphic function in question. Let $B \subseteq \mathbb{C}^n$ be the ball. For $0 < p < \infty$, let

$$H^p(B) = \left\{ f \text{ holomorphic on } B : \sup_{0<r<1} \int_{\partial B} |f(r\zeta)|^p d\sigma(\zeta)^{1/p} \equiv \|f\|_{H^p} < \infty \right\}.$$

Also set

$$H^\infty(B) = \left\{ f \text{ holomorphic on } B : \sup_{z\in B} |f(z)| \equiv \|f\|_{H^\infty(B)} < \infty \right\}$$

and define the Nevanlinna class

$$N(B) = \left\{ f \text{ holomorphic on } B : \sup_{0<r<1} \int_{\partial B} \log^+ |f(r\zeta)| d\sigma(\zeta) < \infty \right\},$$

where $\log^+ u = \max(0, \log u), u \geq 0$. Trivially, $H^p(B) \subseteq N(B)$, all $0 < p \leq \infty$.

In Section 8.1, we learn that if $\{z_j\}_{j=1}^{\infty}$ is the zero set of an $f \in N(D), 0 < p \leq \infty$, then $\sum_{j=1}^{\infty} 1 - |z_j| < \infty$. Conversely, if $\sum_j 1 - |z_j| < \infty$, then there is an $f \in H^\infty$ (which can be written explicitly) such that $\{z_j\}$ is the zero set of f. Even more can be said: $f \in N(D)$ if and only if $f = g/h$ where $g, h \in H^\infty$ and h is nonvanishing (see Garnett [1] for details).

It is a remarkable discovery of W. Rudin [3] that the zero sets of $H^p(B)$ when $B \subseteq \mathbb{C}^n, n \geq 2$, are all different.

THEOREM 7.3.4   Let $B \subseteq \mathbb{C}^n, n \geq 2$. Let $0 < p_0 < p_1 < \infty$. There is a set $Z \subseteq B$ such that Z is the zero set of an $f \in H^{p_0}(B)$ but not the zero set (with the same multiplicities) of any function $f \in H^{p_1}(B)$. More precisely, there is an $f \in H^{p_0}$ such that if $h \in H^{p_1}$ satisfies h/f is holomorphic, then $h \equiv 0$.

At the same time, Rudin [7, Section 7.3.6] has proved that the zero sets of $H^p(B)$ when $n \geq 2$ are not characterized by a simple "mass distribution" condition.

THEOREM 7.3.5   Let $0 < p < \infty$. Let $\mathbf{d}$ be the analytic disc in $\mathbb{C}^2$ given by $\{(z_1, 0) : |z_1| < 1\}$. There is a set $S_p \subseteq B \subseteq \mathbb{C}^2$ such that S_p is not the zero set of any $f \in H^p(B)$, but $S_p \cup \mathbf{d}$ is the zero set of an $f \in A(B)$.

The reader who wishes to explore this subject more thoroughly should consult H. Skoda [4], G. M. Henkin [6], and N. Th. Varopoulos [3]. In the first two of these, necessary and sufficient conditions are derived for a set S in a strongly pseudoconvex domain $\Omega \subseteq \mathbb{C}^n$ to be the zero set of an f in the Nevanlinna class. The third paper gives sufficient (but by no means necessary) conditions on a set $S \subseteq \Omega$ to be the zero set of a function f in some $H^p, p > 0$ (but p cannot be identified in any simple fashion from the proof). It is an important and deep problem to give explicit geometric characterizations of zero sets of functions in the H^p classes—even on the ball.

Even though they are formulated only on the ball, Theorems 7.3.4 and 7.3.5 are of fundamental importance in the function theory of several complex variables. Many of the beautiful insights in the books W. Rudin [7, 10], formulated and proved on the ball and/or the polydisc, have never been developed or explored on more general domains. Several important ideas lay dormant here.

We prove Theorems 7.3.4 and 7.3.5 in Sections 7.4 and 7.5 respectively.

7.4   Zero Sets for Different H^p Spaces Are Different

We prove Theorem 7.3.4 by means of a long sequence of lemmas. For simplicity, we work in $\mathbb{C}^2$ only. The proof follows rather closely the one in W. Rudin [3]. Throughout, $B \subseteq \mathbb{C}^2$ is the unit ball, $S \equiv \partial B$ is the boundary of the ball, and $D \subseteq \mathbb{C}$ is the unit disc. Also $\langle z, w \rangle \equiv z \cdot \bar{w} = z_1 \bar{w}_1 + z_2 \bar{w}_2$ whenever $z, w \in \mathbb{C}^2$.

LEMMA 7.4.1   There is a countable set $E = \{e_j\} \subseteq \partial D$ such that no point of E is a limit point of E and $E' = \{z \in \partial D : z$ is a limit point of $E\}$ has the property that if p is a holomorphic polynomial of one complex variable that vanishes on E' then $p \equiv 0$.

Proof   Let $E = \cup_{j=1}^{\infty} \cup_{k=1}^{\infty} \{\exp(i/j + 100^{-k}i/j^2)\}$, where $i = \sqrt{-1}$. Then $E' = \{1\} \cup \{\exp(i/j)\}_{j=1}^{\infty}$ and $E' \cap E = \emptyset$. Since a holomorphic polynomial has only finitely many roots, we are done.   □

LEMMA 7.4.2   There is a collection $\{X_i\}_{i=1}^{\infty}$ of relatively open circled subsets of S such that if $Q(z_1, z_2)$ is a homogeneous polynomial (not identically 0), then there exists an infinite set $J \subseteq \mathbb{N}$ and an $\eta > 0$ such that $|Q(z)| > \eta$ for all $z \in X_j$ and $j \in J$.

Proof   Let $E = \{e_j\}_{j=1}^{\infty}$ be as in Lemma 7.4.1. For each $j \in \mathbb{N}$, define

$$\Gamma_j = \{(z_1, z_2) \in S : z_1/z_2 = e_j\}.$$

Then each Γ_j is a circled set and

$$\Gamma_j \cap \bigcup_{i \neq j} \Gamma_i = \emptyset, \qquad j \in \mathbb{N}.$$

Let

$$G = \{(z_1, z_2) \in S : z_1/z_2 \in E'\}.$$

Of course G is also circled. Now there exist relatively open circled sets $X_i \subseteq S$ such that $\Gamma_i \subseteq X_i$ and every relatively open circled $U \subseteq S$ with $G \cap U \neq \emptyset$ satisfies $X_j \subseteq U$ for all j in some infinite $J \subseteq \mathbb{N}$.

Now if Q is as in the statement of the lemma, then $Q(z) \neq 0$ for some $z \in G$ (by Lemma 7.4.1), whence $|Q(w)| > \eta > 0$ for all w in some circled neighborhood U of z in S (this is where the homogeneity of Q is used). Thus $|Q| > \eta$ on X_j for all j in the infinite set J provided at the end of the last paragraph. $\square$

Fix $0 < p_0 < p_1 < \infty$, and define

$$\Phi_j(t) = \begin{cases} (1+t)^{p_j} - 1 & \text{if } t \geq 0, \\ 0 & \text{if } t < 0, \end{cases}$$

$j = 0, 1$. Notice that a holomorphic f on B is in H^{p_j} if and only if

$$\sup_{0 < r < 1} \int_{\partial B} \Phi_j\left(|f(r\zeta)|\right) d\sigma(\zeta) < \infty.$$

LEMMA 7.4.3   Let $0 < p_0 < p_1 < \infty$. There exist $\zeta_j \in X_j, k_j \in \mathbb{N}, t_j > j + 4$, and $a_j > 0$ such that the functions

$$F_j(z) \equiv a_j \langle z, \zeta_j \rangle^{k_j}$$

and the sets

$$S_j \equiv \{z \in S : |F_j(z)| > t_j\}$$

satisfy, for all $j \in \mathbb{N}$,

(7.4.3.1) $\int_S \Phi_0(|F_j(z)|) d\sigma(z) = 2/j^2$.
(7.4.3.2) $\int_{S_j} \Phi_0(|F_j(z)|) d\sigma(z) > 1/j^2$.
(7.4.3.3) $|F_j(z)| < 2^{-j}$ if $z \in (S \setminus X_j) \cup \{|z| < 1 - 1/j\}$.
(7.4.3.4) $\int_{S_j} \Phi_1(|F_j(z)|) d\sigma(z) > j$.

Proof   Choose $\zeta_j \in X_j$ at random. Choose $t_j > j + 4$ such that

$$\Phi_1(t) > j^3 \Phi_0(t), \qquad \text{all } t > t_j. \tag{7.4.3.5}$$

Now fix j. For $k = 1, 2, \dots$, choose $c_k > 0$ such that

$$\int_S \Phi_0 \left(c_k |\langle z, \zeta_j \rangle|^k \right) d\sigma(z) = 2/j^2.$$

If $0 < \beta < 1$, then

$$\Phi_0(c_k \beta^k) \cdot \sigma\{|\langle z, \zeta_j \rangle| > \beta\} \le \int_{\{|\langle z, \zeta_j \rangle| > \beta\}} \Phi_0 \left(c_k |\langle z, \zeta_j \rangle|^k \right) d\sigma \le 2/j^2.$$

Therefore, $\{c_k \beta^k\}_{k=1}^{\infty}$ is bounded. Let $\alpha_j = \max\{|\langle z, \zeta_j \rangle| : z \in S \backslash X_j\} < 1$. If j is fixed and $1 > \beta > \max(1 - 1/j, \alpha_j)$, then we have $c_k \alpha_j^k \to 0$ and $c_k(1 - 1/j)^k \to 0$. Hence (7.4.3.3) holds when $k_j = k(j)$ is large enough and $a_j = c_{k(j)}$. The Lebesgue dominated convergence theorem may then be applied to yield

$$\int_S \Phi_0 \left(c_k |\langle z, \zeta_j \rangle|^k \right) \cdot \chi_{S \backslash S_j}(z) \, d\sigma \to 0$$

as $k \to \infty$. It follows that (7.4.3.2) is true if $k = k(j)$ is large enough. Finally, (7.4.3.2) and (7.4.3.5) imply (7.4.3.4).

We have guaranteed the choice of $k_j = k(j)$ and $a_j = c_{k_j}$, thus finishing the proof. $\square$

LEMMA 7.4.4   The series $f(z) = \sum_{j=0}^{\infty} F_j(z)$ defines an element of $H^{p_0}(B)$.

Proof   The estimate (7.4.3.3) guarantees that the series converges normally. So f is a well-defined holomorphic function. If $N \in \mathbb{N}$, let $f_N = \sum_{j=0}^{N} F_j(z)$. Then, for $0 < r < 1$, we have that $\Phi_0(|f_N(\zeta)| - 1)$ is subharmonic. Hence

$$\int_S \Phi_0(|f_N(r\zeta)| - 1) \, d\sigma(\zeta) \le \int_S \Phi_0(|f_N(\zeta)| - 1) \, d\sigma(\zeta).$$

But $\Phi_0(|f_N(\zeta)| - 1) = 0$ if $\zeta \in S \setminus \cup X_j$ by (7.4.3.3) and the definition of Φ_0. Hence

$$\int_S \Phi_0(|f_N(r\zeta)| - 1) \, d\sigma(\zeta) \le \sum_j \int_{X_j} \Phi_0(|f_N(\zeta)| - 1) \, d\sigma(\zeta).$$

This last is

$$\le \sum_j \int_{X_j} \Phi_0(|F_j(\zeta)|) \, d\sigma(\zeta)$$

by (7.4.3.3) and the disjointedness of the X_j's. Now (7.4.3.1) gives

$$\int_S \Phi_0(|f_N(r\zeta)| - 1)d\sigma(\zeta) \leq 4,$$

and Fatou's lemma yields

$$\int_S \Phi_0(|f(r\zeta)| - 1)d\sigma(\zeta) \leq 4.$$

It follows that $f \in H^p(B)$. □

LEMMA 7.4.5 Let $h \in H^{p_1}(B)$ satisfy $h/f \equiv g$ is holomorphic on B. Assume that h is not identically 0. Write

$$g = \sum_{j=m}^{\infty} G_j(z),$$

where each G_j is a holomorphic homogeneous polynomial of degree j and G_m (for some m) is not the zero function. Choose, according to Lemma 7.4.2, an infinite $J \subseteq \mathbb{N}$ and a number η such that $1 > \eta > 0$ and the homogeneous polynomial G_m satisfies $|G_m| > \eta$ on $\cup_{j \in J} X_j$.
    Now fix $j \in J$. Choose $0 < r < 1$ such that $r^m > 1/2$ and $(1 - r^{k_j})\|F_j\|_\infty < 2^{-j}$. Let $\lambda \in \mathbb{C}, |\lambda| = r$, and $w \in \mathcal{S}_j$. Then the following hold:

(7.4.5.1) $|f(\lambda w)| > (1 + |F_j(w)|)/2$.

(7.4.5.2) $\frac{1}{2\pi} \int_{-\pi}^{\pi} \log |2\eta^{-1} g(re^{i\theta}w)|\, d\theta > 0$.

Proof   Note that F_i is a homogeneous polynomial of degree k_i. Now

$$|f(\lambda w)| = \left| \sum_{i=1}^{\infty} F_i(w)\lambda^{k_i} \right| \geq |F_j(w)|r^{k_j} - (1 - 2^{-j})$$

by (7.4.3.3), the disjointness of the X_i, and the choice of λ. Now this is greater than or equals

$$|F_j(w)| - 1 + [2^{-j} - (1 - r^{k_j})|F_j(w)|] \quad > \quad |F_j(w)| - 1$$
$$\geq \quad \frac{1}{2}(1 + |F_j(w)|)$$

by the choice of w and the fact that $t_j \geq 4$. This establishes (7.4.5.1).
    For (7.4.5.2), we use the plurisubharmonicity of $\log |g|$ to notice that the function $\phi : \lambda \mapsto \log |2\eta^{-1}\lambda^{-m}g(\lambda w)|$ is subharmonic. Notice that, by (7.4.3.3),

$\mathcal{S}_j \subseteq X_j$. So

$$
\begin{aligned}
\log 2 \le \log |2\eta^{-1} G_m(w)| \;=\; & \phi(0) \le \frac{1}{2\pi} \int_{-\pi}^{\pi} \phi(e^{i\theta})\, d\theta \\
=\; & \log(r^{-m}) + \frac{1}{2\pi} \int_{-\pi}^{\pi} \log |2\eta^{-1} g(re^{i\theta})|\, d\theta \\
<\; & \log 2 + \frac{1}{2\pi} \int_{-\pi}^{\pi} \log |2\eta^{-1} g(re^{i\theta})|\, d\theta
\end{aligned}
$$

as desired. $\qquad\square$

LEMMA 7.4.6 With notation as in Lemma 7.4.5, it holds that

$$
\frac{1}{2\pi} \int_{-\pi}^{\pi} d\theta \int_{\mathcal{S}_j} \Phi_1(|h(re^{i\theta}w)|)\, d\sigma(w) > \left(\frac{\eta}{4}\right)^{p_1} \cdot j. \tag{7.4.6.1}
$$

Proof By Jensen's inequality,

$$
\frac{1}{2\pi} \int_{-\pi}^{\pi} \Phi_1\left((4/\eta)|h(re^{i\theta}w)|\right) d\theta \ge \phi_1\left(\frac{1}{2\pi} \int_{-\pi}^{\pi} \log |(4/\eta)h(re^{i\theta}w)|d\theta\right)
$$

where $\phi_1(s) \equiv \exp(p_1 s) - 1$. Now by the relation $h = fg$ and (7.4.5.1), the last line exceeds

$$
\phi_1\left(\frac{1}{2\pi} \int_{-\pi}^{\pi} \log |(2/\eta)g(re^{i\theta}w)(1 + |F_j(w)|)|\, d\theta\right).
$$

But (7.4.5.2) yields that this is greater than or equal

$$
\phi_1\left(\frac{1}{2\pi} \int_{-\pi}^{\pi} \log(1 + |F_j(w)|)d\theta\right) = \phi_1(\log(1 + |F_j(w)|)).
$$

The inequality we have obtained holds for each $w \in \mathcal{S}_j$. Integrating out over this set yields

$$
\frac{1}{2\pi} \int_{-\pi}^{\pi} \int_{\mathcal{S}_j} \Phi_1\left((4/\eta)|h(re^{i\theta}w)|\right) d\sigma(w)\, d\theta \ge \int_{\mathcal{S}_j} \phi_1(\log(1 + |F_j(w)|))\, d\sigma(w),
$$

which by (7.4.3.4) exceeds j. $\qquad\square$

Conclusion of the Proof of the Theorem Since the set $\mathcal{S}_j$ is circular, the inner integral in (7.4.6.1) does not depend on θ. So we have

$$
\int_S \Phi_1(|h(rw)|)\, d\sigma(w) \ge \int_{\mathcal{S}_j} \Phi_1(|h(rw)|)\, d\sigma(w) > (\eta/4)^{p_1} \cdot j.
$$

Since, by Lemma 7.4.5, j may be made as large as we please by letting $r \to 1^-$, we obtain

$$\sup_{0<r<1} \int_S \Phi_1(|h(rw)|)\, d\sigma(w) = \infty.$$

So $\|h\|_{H^{p_1}} = \infty$. This, together with the choice of h in Lemma 7.4.5, is the contrapositive statement of our theorem. □

Exercise for the Reader
What modifications need to be made in the proof of the theorem to construct an $f \in H^{p_0}(B)$ whose zero set is not the zero set of any $h \in H^p(B)$ for any $p > p_0$? (suggested by R. Diaz)

7.5   Zero Sets of H^p Functions Do Not Have a Simple Mass Distribution Characterization

We prove Theorem 7.3.5 by way of a sequence of lemmas. We work on the unit ball $B \subseteq \mathbb{C}^2$. General notation is the same as in Section 7.4. Let $0 < p < \infty$ and let $\mathbf{d}$ be fixed as in the statement of the theorem.

LEMMA 7.5.1   Let $f \in H^p(B)$. Then

$$\int_{\mathbf{d}} |f(x_1 + iy_1, 0)|^p\, dx_1\, dy_1 < \infty.$$

Proof   Let $\frac{1}{2} < r < 1$. By hypothesis,

$$
\begin{aligned}
\infty &> C \\
&\geq \int_{|\zeta|=r} |f(\zeta)|^p\, d\sigma \\
&\geq C' \int_{|x_1|^2+|y_1|^2<r^2} \int_0^{2\pi} \left| f\left(x_1 + iy_1, \sqrt{r^2 - x_1^2 - y_1^2} \cdot e^{i\theta}\right) \right|^p d\theta\, dx_1\, dy_1 \\
&\geq C'' \int_{|x_1|^2+|y_1|^2<r^2} |f(x_1 + iy_1, 0)|^p\, dx_1\, dy_1
\end{aligned}
$$

by plurisubharmonicity. Now let $r \to 1^-$. □

Remark: This observation, that the restriction of an H^p function on B to $\mathbf{d}$ is in the corresponding "Bergman space" on $\mathbf{d}$, is due to Rudin. It has proved to be of great utility in function theory (see, e.g., J. Cima and S. G. Krantz [1]). □

LEMMA 7.5.2   Let $h : D \to \mathbb{C}$ be holomorphic. Assume that $h(0) = 1$. If $0 < t < 1$, then define

$$n(t) = \text{the number of zeros of } h \text{ in } \bar{D}(0,t).$$

Then

$$\int_0^r n(t)\, dt \leq \frac{1}{2\pi} \int_{-\pi}^{\pi} \log |h(re^{i\theta})|\, d\theta. \tag{7.5.2.1}$$

Proof   Let $z_1, \ldots, z_k$ be the zeros of h in $\bar{D}(0,r)$. Assume that there are no zeros of h on $\partial D(0,r)$. Let $\phi_k(z) = (z/r - z_k/r)/(1 - \bar{z}_k z/r^2)$. Then $|\phi_k(z)| = 1 \Leftrightarrow |z| = r$, $\phi_k : D(0,r) \to D(0,1)$ biholomorphically, and $\phi_k(z_k) = 0$. Let

$$\tilde{h}(z) = \frac{h(z)}{\phi_1(z) \cdots \phi_k(z)}.$$

Then $\tilde{h}$ is holomorphic and zero-free on $\bar{D}(0,r)$. So $\log |\tilde{h}|$ is harmonic there. It follows, for $0 < r < 1$, that

$$\frac{1}{2\pi} \int_0^{2\pi} \log |\tilde{h}(re^{i\theta})|\, d\theta = \log |\tilde{h}(0)| = -\sum_{j=1}^{k} \log |z_j|$$

$$= \int_0^r \frac{n(t)}{t}\, dt \geq \int_0^r n(t)\, dt. \qquad \square$$

DEFINITION 7.5.3   Let $4/p < m \in \mathbb{N}$. Define

$$T_k = \{(1 - 2^{-k})e^{i\theta} : \theta = 2\pi j/(m2^k)\,, 1 \leq j \leq m2^k\};$$
$$Z_k = \{(z_1, z_2) \in B : z_1 \in T_k\}.$$

Then Z_k is a union of $m2^k$ (vertical) analytic discs. Let $Z = \cup_{k=1}^{\infty} Z_k$.

LEMMA 7.5.4   Suppose that $g : B \to \mathbb{C}$ is holomorphic, g vanishes on Z, and $g(0) \neq 0$. Then $g \notin H^p(B)$.

Proof   Define $h : D \to \mathbb{C}$ by $h(z) = g(z,0)$. It is enough to show, by Lemma 7.5.1, that

$$\int_D |h(x + iy)|^p\, dx\, dy = \infty.$$

Let $n(t)$ be as in Lemma 7.5.2. We may assume that $h(0) = g(0) = 1$. By (7.5.2.1) and Jensen's inequality, we have

$$\exp \int_0^r pn(t)\, dt \leq \frac{1}{2\pi} \int_{-\pi}^{\pi} |h(re^{i\theta})|^p\, d\theta.$$

So

$$
\int_0^1 \left\{ \exp \int_0^r pn(t)\, dt \right\} r\, dr \;\le\; \frac{1}{2\pi} \int_{-\pi}^{\pi} \int_0^1 |h(re^{i\theta})|^p r\, dr\, d\theta
$$

$$
= \int_D |h(x+iy)|^p\, dx\, dy. \tag{7.5.4.1}
$$

If $1/2 < t < 1$, then there is a k such that $1 - 2^{-k} < t \le 1 - 2^{-k-1}$. But then

$$
pn(t) \ge pm2^k > \frac{4}{1 - (1 - 2^{-k})} \ge \frac{2}{1-t}.
$$

So the left side of (7.5.4.1) exceeds

$$
\int_{1/2}^1 \left\{ \exp \int_{1/2}^r \frac{2}{1-t}\, dt \right\} r\, dr = \int_{1/2}^1 \frac{r}{(2(1-r))^2}\, dr = \infty. \qquad \square
$$

LEMMA 7.5.5   Define

$$
f(z_1, z_2) = \prod_{k=1}^\infty \left\{ 1 - \left(\frac{z_1}{1 - 2^{-k}} \right)^{m2^k} \right\}.
$$

Then the infinite product converges normally on B so that f is holomorphic on B. Also,

$$
|f(z)| \le \frac{K}{(1 - |z|)^{2m+1}}.
$$

Proof   Notice that $(1 - 2^{-k})^{2^k} \ge \frac{1}{4}$. Therefore,

$$
|f(z)| < \prod_{k=1}^\infty \left\{ 1 + 4^m |z_1|^{2^k} \right\}.
$$

Gathering the right side in powers of $|z_1|$ yields

$$
|f(z)| \le 1 + \sum_{j=1}^\infty 4^{m\epsilon(j)} |z_1|^j
$$

where $\epsilon(j)$ is the number of terms in the binary expansion of j as $2^{k_1} + 2^{k_2} + \cdots + 2^{k_\ell}$. In particular, $\epsilon(j) \le \log_2 j$. Therefore, $4^{m\epsilon(j)} \le j^{2m}$. Thus

$$
|f(z)| \le 1 + \sum_{j=1}^\infty j^{2m} |z_1|^j \le \frac{K}{(1 - |z_1|)^{2m+1}},
$$

where $K = K(m)$. (*Exercise:* First do the case $m = 0$, and then take derivatives.)
□

LEMMA 7.5.6   With notation as in Lemma 7.5.5, the function

$$F(z_1, z_2) = z_2^{4m+3} f(z)$$

satisfies $F \in A(B)$ and $F^{-1}(0) = \mathbf{d} \cup Z$.

Proof   Since $|z_2| \leq 2\sqrt{1 - |z_1|}$ on B, the first assertion follows from Lemma 7.5.5. The second assertion is obvious.
□

The theorem follows from Lemmas 7.5.4 and 7.5.6.

EXERCISES

1. In Section 8.1 we shall prove the classical result that if $f \in H^p(D)$, then we may write $f = B \cdot F$, where $B \in H^\infty(D)$ and $F \in H^p(D), F$ is nonvanishing, and $\|F\|_{H^p(D)} \cong \|f\|_{H^p(D)}$. Use Theorem 7.3.4 to prove that such a decomposition is impossible for functions in $H^p(B), B$ the ball in $\mathbb{C}^2$. The impossibility of such a decomposition is also explored in L. A. Rubel and A. Shields [1], and an alternative decomposition is proposed there.

2. Let $B \subseteq \mathbb{C}^n$ be the ball. Let $f_1, f_2, \ldots$ be holomorphic functions with $f_j(0) = 0$, all j. Suppose that for no neighborhood U of 0 and for no j, k is it the case that $f_j = h \cdot f_k$ for some holomorphic h on U. Prove that there does not exist a neighborhood V of 0 and finitely many nonconstant $g_1, \ldots, g_m$ holomorphic on V such that $\cup_{j=1}^m g_j^{-1}(0) \supseteq \cup_{j=1}^\infty f_j^{-1}(0)$.

3. Prove the following generalization to domains in $\mathbb{C}^n$ of Hurwitz's classical theorem: Let $\Omega \subseteq \mathbb{C}^n$ be open. Let f_j be holomorphic functions on Ω for $j = 1, 2, \ldots$. Suppose that f is holomorphic on Ω and that $f_j \to f$ normally. If each f_j is zero-free, then prove that either f is zero-free or $f \equiv 0$ on Ω.

4. Let $U \subseteq \mathbb{R}^N$ be open. Let $v : U \to \mathbb{R}^N$ be a smooth vector field. That is, v assigns to each point x of U a vector $v(x)$. Let $P \in U$. Prove that there are $\epsilon_0, \epsilon_1 > 0$ and a unique curve $\gamma : (-\epsilon_0, \epsilon_1) \to U$ such that $\gamma(0) = P$ and $\gamma'(t) = v(\gamma(t))$, all $t \in (-\epsilon_0, \epsilon_1)$. (*Hint:* Use Picard's existence and uniqueness theorem for first-order ODE's.) We call γ an *integral curve* for the vector field v. Prove that there are well-defined maximal ϵ_0, ϵ_1 such that γ exists on $(-\epsilon_0, \epsilon_1)$ (ϵ_0, ϵ_1 could be ∞). Prove that if $\|v(x)\| \geq \lambda > 0$ and $\epsilon_1 < \infty$, then it cannot be that $\gamma(\epsilon_1) \in U$; that is, γ cannot terminate in U. Show by an example, however, that γ *can* intersect itself. Prove that the set of all maximal γ in U provides a local foliation of U in a neighborhood of any P where $v(P) \neq 0$ (the term *foliation* is defined in Section 1.1.1).

5. Let $B \subseteq \mathbb{C}^2 \cong \mathbb{R}^4$ be the ball. Define two smooth vector fields on ∂B as follows:

$$v_1(x_1, y_1, x_2, y_2) = (y_2, x_2, -y_1, -x_1)$$
$$v_2(x_1, y_1, x_2, y_2) = (-x_2, y_2, x_1, -y_1).$$

Prove that there exists no 2-manifold $M \subseteq \partial B$, even locally, such that $T_P(M)$ is spanned by $v_1(P), v_2(P)$ at each point of M. (*Hint:* If there were such a manifold, then the integral curves of v_1, v_2 would have to lie in M).

6. Prove the following version of the Campbell-Baker-Hausdorff formula for two vector fields X, Y on an open set $U \subseteq \mathbb{R}^N$. Let $P \exp sX \exp tY$ denote the point obtained by beginning at P, traveling s units along the unique integral curve of X, and then traveling t units along the unique integral curve of Y. The purpose of the theorem is to compare $P \exp sX \exp tY$ and $P \exp tY \exp sX$.

Theorem We have, for $|s|$ and $|t|$ small,

$$P \exp sX \exp tY = P \exp tY \exp sX \exp st[X,Y] \exp \left(O(s^3 + t^3) \right).$$

(*Hint:* Recall that $[X, Y] \equiv XY - YX$. Use Taylor series to prove the theorem.) Can you compute the third-order term exactly?

7. Let us recall Frobenius's integrability theorem.

Theorem Let $U \subseteq \mathbb{R}^N$ be an open set. Let $V_1, \ldots, V_k, 1 \leq k \leq N$, be smooth vector fields on U. The following two conditions are equivalent:
   a. For each $P \in U$, there is a smooth k-manifold $M(P)$ such that $P \in M(P) \subseteq U$ and $T_P(M(P))$ is spanned by $V_1(P), \ldots, V_k(P)$. The manifolds $M(P)$ foliate U (see Section 11.1 for definitions).
   b. If $\mathcal{L}$ is the module over $C^\infty(U)$ of vector fields spanned by all

$$[V_i, V_j], [V_i, [V_j, V_k]], \ldots,$$

then $\mathcal{L}$ is spanned at each point $P \in U$ by $V_1(P), \ldots, V_k(P)$.

The manifold $M(P)$ is called an *integral manifold* of $V_1, \ldots, V_k$. The proof of $(b) \Rightarrow (a)$ in Frobenius's theorem is rather involved. However, $(a) \Rightarrow (b)$ is straightforward. We refer the reader to Chevalley [1, p. 94] for details. As an exercise, the reader may do the following: Let $P \in U$. Assume (b). Define

$$M(P) = \{\text{all those points in } U \text{ that can be connected to } P$$
$$\text{by a sequence of finitely many integral curves}$$
$$\text{of } V_1, \ldots, V_k\}.$$

Assuming that $M(P)$ is indeed a smooth k-dimensional manifold, prove that $T_Q(M(P))$ is spanned by $V_1(Q), \ldots V_k(Q)$, each $Q \in M(P)$.

8. Let $\Omega \subseteq \mathbb{C}^n$ be a domain with C^2 boundary. Let $P \in \partial\Omega$. Let ρ be a defining function for Ω, and let $T_{1,0}^P(\partial\Omega) = \{w \in \mathbb{C}^n : \sum_{j=1}^n (\partial\rho/\partial z_j)(P)w_j = 0\}$. This is just the space on which the Levi form is defined. Show that $T_{1,0}^P$ is well defined independent of the choice of ρ. We say that the Levi form has *rank* $k, 0 \le k \le n-1$, at P if

$$
\begin{aligned}
&\mathcal{N}_P(L) \\
&\equiv \left\{ q \in T_{1,0}^P(\partial\Omega) : \sum_{j,k=1}^n \frac{\partial^2\rho}{\partial z_j \partial\bar{z}_k}(P)q_j\bar{w}_k = 0 \quad \text{for all} \quad w \in T_{1,0}^P(\partial\Omega) \right\}
\end{aligned}
$$

has complex dimension $n-1-k$. Prove that if the Levi form has rank k at P, then the Levi form has rank *at least* k for all $z \in \partial\Omega$ that are sufficiently near P.

9. *Foliations by complex analytic varieties.* (Refer for background to Exercises 7 and 8.)

   Theorem Let $\Omega \subseteq \mathbb{C}^n$ be a domain with C^∞ boundary. Let $U \subseteq \partial\Omega$ be a relatively open subset. Suppose that the Levi form has constant rank k on $U, 0 \le k \le n-1$. Then there is a foliation of U by $(n-k-1)$-dimensional complex analytic varieties (see Section 11.1 for the definition).

   Prove this result by completing the following outline to verify the hypotheses of the Frobenius integrability theorem (Exercise 7 above):

   a. If $S, T \in T_{1,0}(U)$ are smooth vector fields, then $[S, T] \in T_{1,0}(U)$. (*Hint:* Apply $[S, T]$ to ρ, where ρ is a smooth defining function.)

   b. If $S, T \in T_{1,0}(U)$, write $S = \sum_j s^j(\partial/\partial z_j), T = \sum_j t^j(\partial/\partial z_j)$. Suppose that $S(P) \in \mathcal{N}_P(L)$, all $P \in U$, and that $N = \sum_j n^j(\partial/\partial z_j) \in T_{1,0}(U)$ is arbitrary. Then

   $$
   \sum_{j,k} \rho_{z_j \bar{z}_k} s^j \bar{n}^k = 0 \quad \text{on } U.
   $$

   c. Apply T to the last identity to obtain

   $$
   \sum_{i,j,k} \left\{ \rho_{z_j \bar{z}_k z_i} t^i s^j \bar{n}^k + \rho_{z_j \bar{z}_k} t^i s_{z_i}^j \bar{n}^k + \rho_{z_j \bar{z}_k} t^i s^j \bar{n}_{z_i}^k = 0 \right\}
   $$

   on U.

   d. Apply T to the identity

   $$
   \sum_k \rho_{\bar{z}_k} \bar{n}^k = 0 \tag{$*$}
   $$

to obtain

$$\sum_{i,k}\left\{\rho_{\bar{z}_k z_i}t^i\bar{n}^k + \rho_{\bar{z}_k}t^i\bar{n}^k_{z_i}\right\} = 0 \qquad \text{on } U.$$

e. If $T \in \mathcal{N}_P(L)$, all $P \in U$, it follows that

$$\sum_{i,k}\rho_{\bar{z}_k}t^i\bar{n}^k_{z_i} = 0 \qquad \text{on } U.$$

Apply S to this equality to obtain

$$\sum_{j,k}\left\{\rho_{\bar{z}_k z_j}s^j t^i\bar{n}^k_{z_i} + \rho_{\bar{z}_k}s^j t^i_{z_j}\bar{n}^k_{z_i} + \rho_{\bar{z}_k}s^j t^i\bar{n}^k_{z_i z_j}\right\} = 0 \qquad \text{on } U.$$

f. From (c) and (e) we infer that

$$\sum_{i,j,k}\left\{\rho_{z_j\bar{z}_k z_i}t^i s^j\bar{n}^k + \rho_{z_j\bar{z}_k}t^i s^j_{z_i}\bar{n}^k - \rho_{\bar{z}_k}s^j t^i_{z_j}\bar{n}^k_{z_i} - \rho_{\bar{z}_k}s^j t^i\bar{n}^k_{z_i z_j}\right\} = 0$$

on U.

g. Reversing the roles of S and T gives

$$\sum_{i,j,k}\left\{\rho_{z_j\bar{z}_k z_i}s^i t^j\bar{n}^k + \rho_{z_j\bar{z}_k}s^i t^j_{z_i}\bar{n}^k - \rho_{\bar{z}_k}t^j s^i_{z_j}\bar{n}^k_{z_i} - \rho_{\bar{z}_k}s^i t^j\bar{n}^k_{z_i z_j}\right\} = 0$$

on U.

h. Subtract (f) from (g) to get

$$\sum_{j,k}\rho_{z_j\bar{z}_k}\left(s^i t^j_{z_i} - t^i s^j_{z_i}\right)\bar{n}^k + \sum_{j,k}\left(s^j t^i_{z_j} - t^j s^i_{z_j}\right)\rho_{\bar{z}_k}\bar{n}^k_{z_i} = 0 \qquad \text{on } U$$

or

$$\sum_{j,k}\rho_{z_j\bar{z}_k}\left(s^i t^j_{z_i} - t^i s^j_{z_i}\right)\bar{n}^k + \sum_{j,k}\left(s^j t^i_{z_j} - t^j s^i_{z_j}\right)\rho_{z_i\bar{z}_k}\bar{n}^k = 0 \qquad \text{on } U.$$

This last follows by applying $ST - TS$ to $(*)$.

i. Write $[S,T] = \sum_j u_j(\partial/\partial z_j)$. Conclude from (h) that

$$2\sum_k \rho_{z_j\bar{z}_k}u_j\bar{n}^k = 0;$$

hence, $[S,T] \in \mathcal{N}_L(U) \subseteq \mathcal{N}_L(U) \oplus \bar{\mathcal{N}}_L(U)$. Likewise, $[\bar{S},T] \in \mathcal{N}_L(U) \oplus \bar{\mathcal{N}}_L(U)$.

j. Perform a similar calculation to see that

$$[S, \bar{T}] \in \mathcal{N}_L(U) \oplus \bar{\mathcal{N}}_L(U).$$

k. Apply the Frobenius theorem to conclude the proof of the foliation theorem.

(I learned these calculations from J. J. Kohn.)

10. Use Cartan's formula (Exercise 5 at the end of Chapter 5) to give a second proof of the foliation theorem presented in Exercise 9. See M. Freeman [1] for details.

11. Prove that Theorem 7.2.4 is false if Ω is not pseudoconvex.

12. Prove that Theorem 7.2.8 is false if Ω is not pseudoconvex.

13. Let P be an entire function on $\mathbb{C}^n$. Then the components of the interior of $\{x : f(x + iy) \neq 0 \text{ for all } y \in \mathbb{R}^n\}$ are all convex. (*Hint:* There is a tube domain lurking around.)

14. *Stein manifolds.* For many purposes, the most natural arena for complex function theory of several variables is something more general than domains in $\mathbb{C}^n$. Let M be a complex manifold of dimension n and $\mathcal{F} = \mathcal{F}(M)$ the family of holomorphic functions on M. We say that M is a Stein manifold if:

i. $\hat{K}_{\mathcal{F}} \subset\subset M$ whenever $K \subset\subset M$.

ii. For any $z, w \in M, z \neq w$, there is an $f \in \mathcal{F}$ such that $f(z) \neq f(w)$.

iii. For each $z \in M$, there exist $f_1, \ldots, f_n \in \mathcal{F}$ that form a holomorphic coordinate system at z.

Prove the following facts about Stein manifolds:

a. If $\Omega \subset\subset \mathbb{C}^n$ is a domain of holomorphy, then Ω is a Stein manifold.

b. If M is a Stein manifold and $M' \subseteq M$ is a properly regularly imbedded complex analytic submanifold, then M' is Stein.

c. If M is Stein, then M has a strictly plurisubharmonic exhaustion function. (*Hint:* You will have to think carefully about Sections 3.2 and 3.3. You will have to define plurisubharmonic. Notice that if $f_1, \ldots, f_k \in \mathcal{F}$, then $|f_1|^2 + \cdots + |f_k|^2$ is plurisubharmonic.)

d. It is a deep theorem due to Grauert [2] that if M is a complex manifold that possesses a strictly plurisubharmonic exhaustion function, then M is Stein. Notice that only conditions (i) and (iii) are used to prove (c). Therefore it follows from Grauert's theorem that condition (ii) is superfluous in the definition of Stein manifold.

15. *Imbedding of Stein manifolds.* One of the reasons that Stein manifolds are so natural is that they support plenty of nonconstant holomorphic functions. This point is driven home by the following remarkable result due to E. Bishop [2] and R. Narasimhan [2].

Theorem A complex analytic manifold M of dimension n is Stein if and only if there is an (injective) holomorphic imbedding $f : M \to \mathbb{C}^{2n+1}$ that is both proper and regular (i.e., the rank of f is maximal at each point of M).

The "proper" part of this theorem is very difficult, and we shall not discuss it. However the reader may complete the following outline to prove the other parts. We use the notation of the preceding exercise. Notice that the "if" part is easy. Now consider the converse.

a. Let K be a compact subset of M. Choose $f_1, \ldots, f_k \in \mathcal{F}$ such that for each $z \in K$ there are $f_{i_1}, \ldots, f_{i_n}$ that form a coordinate system near z.

b. Let $\Delta = \{(z, z) \in K \times K\}$ be the diagonal. There is a neighborhood U of Δ such that if $(z, w) \in U$ and $f_j(z) = f_j(w), j = 1, \ldots, k$, then $z = w$. Now choose $f_{k+1}, \ldots, f_N \in \mathcal{F}$ such that if $(z, w) \in K \times K \setminus U$ and $f_j(z) = f_j(w)$ for $k + 1 \leq j \leq N$, then $z = w$. Then the map $f = (f_1, \ldots, f_N)$ gives a regular (injective) imbedding of M into $\mathbb{C}^N$. The remainder of the argument consists of reducing N to $2n + 1$.

c. If K is a compact subset of M and if $N > n$, then $f(K)$ has Lebesgue $\mathcal{L}^{2N}$ measure 0 in $\mathbb{C}^N$.

d. If $N \geq 2n, K$ is compact in M, and $(f_1, \ldots, f_{N+1})$ is a regular map on K, then for $\mathcal{L}^{2N}$ almost every $a \in \mathbb{C}^n$, it holds that the map

$$(f_1 - a_1 f_{N+1}, \ldots, f_N - a_N f_{N+1})$$

is regular on K. To see this, assume that K is contained in a coordinate chart with coordinates $z_1, \ldots, z_n$. The problem is equivalent to showing that for almost every a the matrix

$$\left(\frac{\partial f_j}{\partial z_k} - a_j \frac{\partial f_{N+1}}{\partial z_k} \right)_{\substack{j=1,\ldots,N \\ k=1,\ldots,n}}$$

has rank n at each point of K. By setting $a_{N+1} = 1, \mu = \sum_{k=1}^n \lambda_k \partial f_{N+1}/\partial z_k$, we may rewrite this condition as

$$\left(\frac{\partial f_j}{\partial z_k} \right) \lambda = \mu a \qquad \text{implies} \quad \lambda = 0.$$

To prove this, it suffices to choose a that is not in the range of the map

$$\mathbb{C}^n \times K \ni (\lambda, z) \to \left(\frac{\partial f_j}{\partial z_k} \right)(z)(\lambda) \in \mathbb{C}^{n+1}.$$

Use (c) to see that this obtains for $\mathcal{L}^{2N}$ almost every a.

e. If $(f_1, \ldots, f_{N+1})$ is a regular, injective map on a compact $K \subseteq M$, $N \geq 2n + 1$, then for $\mathcal{L}^{2N}$ almost every $a \in \mathbb{C}^N$, it holds that

$$(f_1 - a_1 f_{N+1}, \ldots, f_N - a_N f_{N+1})$$

is regular injective on K. To see this, note that (d) already provides the regularity for almost every a. It is sufficient now to show that we can choose a with $a_{N+1} = 1$ and so that

$$f(z) - f(w) = \alpha \cdot a$$

and

$$\alpha = f_{N+1}(z) - f_{N+1}(w)$$

implies that $\alpha = 0$.

But the range of the map

$$\mathbb{C} \times K \times K \ni (\mu, z, w) \to \mu \cdot (f(z) - f(w)) \in \mathbb{C}^{N+1}$$

is a set of zero measure since $N + 1 > 2n + 1$.

f. To prove the theorem, let $K \subset\subset M$. Let $N \geq 2n + 1$. We show that $S = \{f \in \mathcal{F}^N : f$ is not regular on $K\}$ is of the first category in $\mathcal{F}^N$. Clearly, S is closed (in the topology of normal convergence on $\mathcal{F}^N$). Now, to see that S has no interior, let $f \in S$. Let $g = (g_1, \ldots, g_Q)$ be a regular map on K. Apply (e) repeatedly to the map $(f, g) : M \to \mathbb{C}^{N+Q}$ to see that there are arbitrarily small complex numbers $\{a_{jk}\}, j = 1, \ldots, N, k = 1, \ldots, Q$, such that $f' \equiv f + (a_{ij})g$ is regular on K. Hence f is not in the interior of S.

16. If $\Omega \subset\subset \mathbb{C}^n$ is a domain, then one would like to speak of the "largest" $\Omega^* \supseteq \Omega$ to which all f holomorphic on Ω can be analytically continued. One would like Ω^* to be unique and to be a domain of holomorphy. For the case $\Omega = B(0, 2) \setminus \bar{B}(0, 1)$, then $\Omega^* = B(0, 2)$ easily. But for the theory to work for *any* Ω, we must allow Ω to be a manifold.

a. Consider the subdomain of $\mathbb{C}$ indicated in Figure 7.1.

Consider the functional element given by the principal branch of $\sqrt{z}$ on the shaded portion of the domain. It extends analytically to all of Ω. Now the resulting functional element on the dotted region continues to all of the first quadrant. Also the functional element on the shaded region continues to the first quadrant. But the two continuations are different. This ambiguity is resolved by passing to the Riemann surface for $\sqrt{z}$.

A theory of *Riemann spread domains* over $\mathbb{C}^n$, analogous to the one-variable theory, is developed in the remainder of this exercise. Throughout, Ω and Ω^* are connected complex manifolds. Also, $\mathcal{F}(\Omega), \mathcal{F}(\Omega^*)$ are the families of holomorphic functions on these manifolds, equipped with

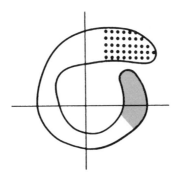

FIGURE 7.1

the topology of normal convergence. Call Ω^* a *holomorphic extension* of Ω if (i) Ω is an open subset of Ω^*, (ii) each holomorphic coordinate chart on Ω is a fortiori a coordinate chart on Ω^* (i.e., the complex structures are consistent), and (iii) for each $f \in \mathcal{F}_\Omega$, there is a (unique) $f^* \in \mathcal{F}_{\Omega^*}$ with $f^*|_\Omega = f$.

b. If Ω^* is a holomorphic extension of Ω, then for every $K^* \subset\subset \Omega^*$, there is a $K \subset\subset \Omega$ such that $K^* \subseteq \hat{K}_{\mathcal{F}(\Omega^*)}$. To see this, apply the open mapping theorem to the restriction mapping $\mathcal{F}(\Omega^*) \to \mathcal{F}(\Omega)$. Given K^*, conclude that there is a $K \subset\subset \Omega$ such that

$$\|f^*\|_{K^*} \leq C\|f^*\|_K \quad , \qquad \text{all} \quad f^* \in \mathcal{F}(\Omega^*).$$

By taking powers of f^*, conclude that C may be taken to be 1.

c. If Ω is Stein and Ω^* is a holomorphic extension of Ω, then $\Omega^* = \Omega$.

d. If Ω^* extends Ω and Ω^* is Stein, then Ω^* is maximal. More precisely, suppose that Ω^* and Ω^{**} extend Ω. If Ω^* is Stein and Ω^{**} satisfies axioms (ii) and (iii) (in Exercise 14) for a Stein manifold, then there is a holomorphic imbedding of Ω^{**} into Ω^* that restricts to the identity on Ω. To see this, let $E^* : \mathcal{F}(\Omega) \to \mathcal{F}(\Omega^*), E^{**} : \mathcal{F}(\Omega) \to \mathcal{F}(\Omega^{**})$ be given by analytic continuation. If $z^{**} \in \Omega^{**}$, $z^* \in \Omega^*$, and $(E^{**}f)(z^{**}) = (E^*f)(z^*)$ for every $f \in \mathcal{F}(\Omega)$, then define $\phi(z^{**}) = z^*$. It follows that ϕ is defined, continuous, and injective on a closed subset of Ω^{**} (you will need to use (b)). If $z^{**} \in \Omega^{**}, z^* \in \Omega^*$, choose $f^1, \ldots, f^n \in \mathcal{F}(\Omega)$ such that $E^*f^1, \ldots, E^*f^n$ give local coordinates at z^* and choose $g^1, \ldots, g^n \in \mathcal{F}(\Omega)$ such that $E^{**}g^1, \ldots, E^{**}g^n$ give local coordinates at z^{**}. There are constants a, b such that $\{h^j = af^j + bg^j\}_{j=1}^n$ give local coordinates at both z^* and z^{**}. Now let

$$S = \{z \in \text{domain } \phi : \text{for all } f \in \mathcal{F}(\Omega), E^{**}f \text{ and } E^*f \text{ have the same}$$
$$\text{formal power series in } \{h^j\}_{j=1}^n \text{ at } z \text{ and } \phi(z), \text{respectively}\}.$$

Then S is open and closed as a subst of Ω^{**}. So $S = \Omega^{**}$.

e. Let hypotheses be as in (d). If Ω^{**} is Stein, then ϕ is surjective. (*Hint:* Ω^* extends $\phi(\Omega^{**})$.)

f. We now know that Ω has at most one holomorphic extension that is Stein. It is maximal. When such an extension exists, it is called an *envelope of holomorphy*. What does this definition have to do with Exercise 11 at the end of Chapter 5?

g. If Ω is a complex manifold of dimension n, if $\mathcal{F}(\Omega)$ separates points on Ω, and if there is a holomorphic imbedding $\phi : \Omega \to \mathbb{C}^n$ that has maximal rank at every point, then Ω is called a Riemann domain spread over $\mathbb{C}^n$. For instance, the Riemann surface for $\sqrt{z}$ is a Riemann domain. It is a difficult theorem (whose proof we omit) that every Riemann domain has an envelope of holomorphy that is a Riemann spread domain. Compute the envelope of holomorphy of the domain $\mathbb{C}^2$ given by

$$B(0,1) \setminus \{(z_1, z_2) : \mathrm{Re}z_1 > 0, \mathrm{Im}z_1 = 0, z_2 = 0\}.$$

Compute the envelope of holomorphy of the domain

$$\{(z_1, z_2) : 1 < |z_2| < 2, |z_1| < 2\} \cup \{(z_1, z_2) : |z_1| < 1, |z_2| < 2\}.$$

17. Let $[f], [g]$ be nonunits in $\mathcal{O}_0$ in $\mathbb{C}^n$. Then $[f]$ and $[g]$ are relatively prime if and only if the topological dimension of the intersection of the zero sets of f and g is less than or equal to $2n - 4$.

18. Let f be a nonconstant holomorphic function on a domain Ω. Let $\mathcal{Z} = \{z \in \Omega : f(z) = 0\}$. Prove that $\Omega \setminus \mathcal{Z}$ is connected.

19. Let f be holomorphic in a neighborhood of $(\partial D \times \partial D) \cup (D \times \{1\}) \subseteq \mathbb{C}^2$. Does f continue analytically to a neighborhood of $\bar{D}$? Does your answer change if instead we hypothesize that f is holomorphic in a neighborhood of $(\partial D \times \partial D) \cup (D \times \{1\}) \cup (\{1\} \times D)$?

20. Let Ω be the Hartogs triangle $\{(z_1, z_2) : |z_1| < |z_2| < 1\}$. Prove that for each $j = 1, 2, \ldots$ there is a function $f \in C^j(\bar{\Omega})$, holomorphic on Ω, such that f cannot be analytically continued to any larger domain. Prove that if C^j is replaced by C^∞, then f continues analytically to the bidisc.

21. A set $S \subseteq \mathbb{C}^n$ is called a *set of uniqueness* if any holomorphic function on $\mathbb{C}^n$ that vanishes on S must be identically zero. Which of the following are sets of uniqueness in $\mathbb{C}^2$?

a. $S = \mathbb{Z} \times \mathbb{Z} \subseteq \mathbb{C} \times \mathbb{C}$
b. $S = \mathbb{C} \times \{0\}$
c. $S = \{(1/n, 1/m) : n \in \mathbb{N}, m \in \mathbb{N}\}$
d. $S = \{(1/n, e^{-n}) : n = 1, 2, 3, \ldots\}$
e. $S = \{(x_1 + i0, x_2 + i0)\}$

f. Call a manifold $W \subseteq \mathbb{C}^n$ *totally real* if the tangent space at each point of W contains no complex line. If S is a totally real, real analytic, n-dimensional manifold, then prove that S is a set of uniqueness.

g. Let $M_1, M_2, \ldots$ be disjoint, connected, closed complex manifolds in $\mathbb{C}^2$. Show that $\cup M_j$ is a set of uniqueness if and only if there is a bounded sequence $\{z_j\}$ with $z_j \in M_j$.

8 Some Harmonic Analysis

8.1 Review of the Classical Theory of H^p Spaces on the Disc

Throughout this section we let $D \subseteq \mathbb{C}$ denote the unit disc. Let $0 < p < \infty$. We define

$$H^p(D) = \left\{ f \text{ holomorphic on } D : \sup_{0<r<1} \left[\frac{1}{2\pi} \int_0^{2\pi} |f(re^{i\theta})|^p \, d\theta \right]^{1/p} \right.$$
$$\left. \equiv \|f\|_{H^p} < \infty \right\}.$$

Also define

$$H^\infty(D) = \left\{ f \text{ holomorphic on } D : \sup_D |f| \equiv \|f\|_{H^p} < \infty \right\}.$$

The fundamental result in the subject of H^p, or *Hardy*, spaces (and also the fundamental result of this section) is that if $f \in H^p(D)$ then the limit

$$\lim_{r \to 1^-} f(re^{i\theta}) \equiv \tilde{f}(e^{i\theta})$$

exists for almost every $\theta \in [0, 2\pi)$. For $1 \leq p \leq \infty$, the function f can be recovered from $\tilde{f}$ by way of the Cauchy or Poisson integral formulas; for $p < 1$ this "recovery" process is more subtle and must proceed by way of distributions. Once this pointwise boundary limit result is established, then an enormous and rich mathematical structure unfolds (see Y. Katznelson [1], K. Hoffman [1], and J. Garnett [1]).

Recall from Chapter 1 that the Poisson kernel for the disc is

$$P_r(e^{i\theta}) = \frac{1}{2\pi} \frac{1 - r^2}{1 - 2r\cos\theta + r^2}.$$

Let

$$\mathbf{h}^p(D) = \left\{ f \text{ harmonic on } D : \sup_{0<r<1} \left[\frac{1}{2\pi} \int_0^{2\pi} |f(re^{i\theta})|^p \, d\theta \right]^{1/p} \equiv \|f\|_{\mathbf{h}^p} < \infty \right\}$$

and

$$\mathbf{h}^\infty(D) = \left\{ f \text{ harmonic on } D : \sup_D |f| \equiv \|f\|_{H^p} < \infty \right\}.$$

Throughout this section, arithmetic and measure theory on $[0, 2\pi)$ (equivalently on ∂D) is done by identifying $[0, 2\pi)$ with $\mathbb{R}/2\pi\mathbb{Z}$. See Y. Katznelson [1] for more on this identification procedure.

PROPOSITION 8.1.1   Let $1 < p \leq \infty$ and $f \in \mathbf{h}^p(D)$. Then there is an $\tilde{f} \in L^p(\partial D)$ such that

$$f(re^{i\theta}) = \int_0^{2\pi} \tilde{f}(e^{i\psi}) P_r(e^{i(\theta - \psi)}) \, d\psi.$$

Proof   Define $f_r(e^{i\theta}) = f(re^{i\theta}), 0 < r < 1$. Then $\{f_r\}_{0<r<1}$ is a bounded subset of $(L^{p'}(\partial D))^*, p' = p/(p-1)$. By the Banach-Alaoglu theorem (see W. Rudin [11]), there is a subsequence f_{r_j} that converges weak-$*$ to some $\tilde{f}$ in $L^p(\partial D)$. For any $0 < r < 1$, let $r < r_j < 1$. Then

$$f(re^{i\theta}) = f_{r_j}\left((r/r_j)e^{i\theta} \right) = \int_0^{2\pi} f_{r_j}(e^{i\psi}) P_{r/r_j}\left(e^{i(\theta - \psi)} \right) d\psi$$

because $f_{r_j} \in C(\bar{D})$. Now $P_{r/r_j} \in C(\partial D) \subseteq L^{p'}(\partial D)$. Thus the right-hand side of the last equation is

$$\int_0^{2\pi} f_{r_j}(e^{i\psi}) P_r(e^{i(\theta - \psi)}) d\psi + \int_0^{2\pi} f_{r_j}(e^{i\psi}) \left[P_{r/r_j}(e^{i(\theta - \psi)}) - P_r(e^{i(\theta - \psi)}) \right] d\psi.$$

As $j \to \infty$, the second integral vanishes (because the expression in brackets converges uniformly to 0) and the first integral tends to

$$\int_0^{2\pi} \tilde{f}(e^{i\psi}) P_r(e^{i(\theta - \psi)}) \, d\psi.$$

This is the desired result.   □

Remark: It is easy to see that the proof breaks down for $p = 1$ since L^1 is not the dual of any Banach space. This breakdown is not merely ostensible: The harmonic function

$$f(re^{i\theta}) = P_r(e^{i\theta})$$

satisfies

$$\sup_{0<r<1} \frac{1}{2\pi} \int_0^{2\pi} |f(re^{i\theta})| \, d\theta < \infty,$$

but the Dirac δ mass is the only measure of which f is the Poisson integral.   □

Exercise for the Reader

If $f \in \mathbf{h}^1$, then there is a Borel measure μ_f on ∂D such that $f(re^{i\theta}) = P_r(\mu_f)(e^{i\theta})$.

PROPOSITION 8.1.2 Let $f \in L^p(\partial D), 1 \le p < \infty$. Then $\lim_{r \to 1^-} P_r f = f$ in the L^p norm.

Remark: The result is false for $p = \infty$ if f is discontinuous. The correct analogue in the uniform case is that if $f \in C(\partial D)$ then $P_r f \to f$ uniformly.

As an exercise, consider a Borel measure μ on ∂D. Show that its Poisson integral converges in the weak-* topology to μ.   □

Proof of Proposition 8.1.2 If $f \in C(\partial D)$, then the result is clear by the solution of the Dirichlet problem. If $f \in L^p(\partial D)$ is arbitrary, let $\epsilon > 0$ and choose $g \in C(\partial D)$ such that $\|f - g\|_{L^p} < \epsilon$. Then

$$
\begin{aligned}
\|P_r f - f\|_{L^p} &\le \|P_r(f - g)\|_{L^p} + \|P_r g - g\|_{L^p} + \|g - f\|_{L^p} \\
&\le \|P_r\|_{L^1} \|f - g\|_{L^p} + \|P_r g - g\|_{L^p} + \epsilon \\
&\le \epsilon + o(1) + \epsilon
\end{aligned}
$$

as $r \to 1^-$.   □

PROPOSITION 8.1.3 Let $K \subseteq \partial D$ be compact, and let $\{I_\alpha\}_{\alpha \in A}$ be a covering of K by open intervals. Then there is a subcovering $\{I_{\alpha_j}\}_{j=1}^M$ such that every point of K is contained in at least one but not more than two of the I_{α_j}'s. (We call such a subcover a cover of *valence two*.)

Proof Since K is compact, $\{I_\alpha\}_{\alpha \in A}$ has a finite subcollection $\{I_{\alpha_j}\}_{j=1}^J$ that still covers K. Now, in sequence, discard any I_{α_j} that is contained in the union of the other intervals.   □

DEFINITION 8.1.4   If $f \in L^1(\partial D)$, let

$$Mf(\theta) = \sup_{R>0} \frac{1}{2R} \int_{-R}^{R} |f(e^{i(\theta - \psi)})| \, d\psi.$$

The function Mf is called the *Hardy-Littlewood maximal function* of f.

DEFINITION 8.1.5   Let (X, μ) be a measure space and let $f : X \to \mathbb{C}$ be measurable. We say that f is of *weak type* $p, 0 < p < \infty$, if $\mu\{x : |f(x)| > \lambda\} \leq C/\lambda^p$, all $0 < \lambda < \infty$. The space weak type ∞ is defined to be L^∞.

LEMMA 8.1.6 (Chebycheff's Inequality)   If $f \in L^p(X, d\mu)$, then f is weak type $p, 1 \leq p < \infty$.

Proof   Let $\lambda > 0$. Then

$$\mu\{x : |f(x)| > \lambda\} \leq \int_{\{x : |f(x)| > \lambda\}} |f(x)|^p / \lambda^p \, d\mu(x) \leq \lambda^{-p} \|f\|_{L^p}^p. \qquad \Box$$

Exercise for the Reader
There exist functions that are of weak type p but not in $L^p, 1 \leq p < \infty$.

DEFINITION 8.1.7   An operator $T : L^p(X, d\mu) \to \{\text{measurable functions}\}$ is said to be of *weak type* $(p, p), 0 < p < \infty$, if

$$\mu\{x : |Tf(x)| > \lambda\} \leq C\|f\|_{L^p}^p / \lambda^p, \qquad \text{all } f \in L^p, \ \lambda > 0.$$

PROPOSITION 8.1.8   The operator M is of weak type $(1, 1)$.

Proof   Let $\lambda > 0$. Set $S_\lambda = \{\theta : |Mf(e^{i\theta})| > \lambda\}$. Let $K \subseteq S_\lambda$ be a compact subset with $2m(K) \geq m(S_\lambda)$. For each $k \in K$, there is an interval $I_k \ni k$ with $|I_k|^{-1} \int_{I_k} |f(e^{i\psi})| \, d\psi > \lambda$. Then $\{I_k\}_{k \in K}$ is an open cover of K. By Proposition 8.1.3, there is a subcover $\{I_{k_j}\}_{j=1}^M$ of K of valence not exceeding 2. Then

$$
\begin{aligned}
m(S_\lambda) \leq 2m(K) \ &\leq \ 2m\left(\bigcup_{j=1}^M I_{k_j}\right) \leq 2\sum_{j=1}^M m(I_{k_j}) \\
&\leq \ \sum_{j=1}^M \frac{2}{\lambda} \int_{I_{k_j}} |f(e^{i\psi})| \, d\psi \\
&\leq \ \frac{4}{\lambda}\|f\|_{L^1}. \qquad \Box
\end{aligned}
$$

DEFINITION 8.1.9   If $e^{i\theta} \in \partial D, 1 < \alpha < \infty$, then define the *Stolz region* (or *nontangential approach region* or *cone*) with vertex $e^{i\theta}$ and aperture α to be

$$\Gamma_\alpha(e^{i\theta}) = \{z \in D : |z - e^{i\theta}| < \alpha(1 - |z|)\}.$$

Exercise for the Reader

Draw a careful sketch of $\Gamma_\alpha(e^{i\theta})$. Save it for reference as you read the rest of the chapter.

PROPOSITION 8.1.10   If $e^{i\theta} \in \partial D$, $1 < \alpha < \infty$, then there is a constant $C_\alpha > 0$ such that if $f \in L^1(\partial D)$, then

$$\sup_{re^{i\phi}\in\Gamma_\alpha(e^{i\theta})} |P_r f(e^{i\phi})| \le C_\alpha M f(e^{i\theta}).$$

Proof   For $re^{i\phi} \in \Gamma_\alpha(e^{i\theta})$, we have

$$|\theta - \phi| \le 2\alpha(1 - r).$$

Therefore, for $1/\alpha \le r < 1$, we obtain

$$
\begin{aligned}
|P_r f(e^{i\phi})| &= \left| \frac{1}{2\pi} \int_0^{2\pi} f(e^{i(\phi-\psi)}) \frac{1-r^2}{1 - 2r\cos\psi + r^2} d\psi \right| \\
&= \left| \frac{1}{2\pi} \int_0^{2\pi} f(e^{i(\phi-\psi)}) \frac{1-r^2}{(1-r)^2 + 2r(1-\cos\psi)} d\psi \right| \\
&\le \frac{4}{2\pi} \sum_{j=0}^{\log_2(\pi/\alpha(1-r))} \\
&\quad \times \int_{S_j} |f(e^{i(\phi-\psi)})| \frac{1-r^2}{(1-r)^2 + 2r(2^{j-1}\alpha(1-r))^2} d\psi \\
&\quad + \frac{1}{2\pi} \int_{|\psi|<\alpha(1-r)} |f(e^{i(\phi-\psi)})| \frac{1-r^2}{(1-r)^2} d\psi,
\end{aligned}
$$

where $S_j = \{\psi : 2^j \alpha(1-r) \le |\psi| < 2^{j+1}\alpha(1-r)\}$. Now this is less than or equal to

$$
\begin{aligned}
&\frac{4\alpha}{4\pi\alpha^2} \sum_{j=0}^{\infty} \frac{1}{2^{2j-2}(1-r)} \int_{|\psi|<(2+2^{j+1})\alpha(1-r)} |f(e^{i(\theta-\psi)})| \, d\psi \\
&\quad + \frac{2}{2\pi} \frac{1}{1-r} \int_{|\psi|<3\alpha(1-r)} |f(e^{i(\theta-\psi)})| d\psi \\
&\le \frac{32}{\pi} \sum_{j=0}^{\infty} 2^{-j} \left[\frac{1}{2\alpha(2+2^{j+1})(1-r)} \int_{|\psi|<(2+2^{j+1})\alpha(1-r)} |f(e^{i(\theta-\psi)})| \, d\psi \right] \\
&\quad + \frac{6\alpha}{\pi} \frac{1}{2\cdot 3\alpha(1-r)} \int_{|\psi|<3\alpha(1-r)} |f(e^{i(\theta-\psi)})| \, d\psi
\end{aligned}
$$

$$\leq \frac{32}{\pi} \cdot \sum_{j=0}^{\infty} 2^{-j} M f(\theta) + \frac{6\alpha}{\pi} M f(\theta)$$

$$\leq \frac{64}{\pi} M f(\theta) + \frac{6\alpha}{\pi} M f(\theta).$$

If $0 < r \leq 1/\alpha$, then

$$|P_r f(\phi)| \leq \frac{1}{2\pi} \int_0^{2\pi} |f(e^{i(\phi-\psi)})| (2\alpha/(\alpha-1)) \, d\psi$$

$$\leq \frac{2\alpha}{\alpha-1} M f(\theta). \qquad \square$$

THEOREM 8.1.11 Let $f \in \mathbf{h}^p(D)$ and $1 < p \leq \infty$. Let $\tilde{f}$ be as in Proposition 8.1.1 and $1 < \alpha < \infty$. Then

$$\lim_{\Gamma_\alpha(e^{i\theta}) \ni z \to e^{i\theta}} f(z) = \tilde{f}(e^{i\theta}), \quad \text{a.e. } e^{i\theta} \in \partial D.$$

Proof   It suffices to handle the case $p < \infty$ and f real-valued. If $\epsilon > 0$ then choose $g \in C(\partial D)$ real-valued so that $\|\tilde{f} - g\|_{L^p(\partial D)} < \epsilon^2$. We know by the theory of the Dirichlet problem that

$$\lim_{\Gamma_\alpha(e^{i\theta}) \ni z \to e^{i\theta}} g(z) = g(e^{i\theta}), \quad \text{all } e^{i\theta} \in \partial D, \qquad (8.1.11.1)$$

where $g(re^{i\theta}) \equiv P_r g(e^{i\theta})$. Therefore,

$$m\{e^{i\theta} : \limsup_{\Gamma_\alpha(e^{i\theta}) \ni z \to e^{i\theta}} |f(z) - \tilde{f}(e^{i\theta})| > \epsilon\}$$

$$\leq m\{e^{i\theta} : \limsup_{\Gamma_\alpha(e^{i\theta}) \ni z \to e^{i\theta}} |f(z) - g(z)| > \epsilon/3\}$$

$$+ m\{e^{i\theta} : \limsup_{\Gamma_\alpha(e^{i\theta}) \ni z \to e^{i\theta}} |g(z) - g(e^{i\theta})| > \epsilon/3\}$$

$$+ m\{e^{i\theta} : |g(e^{i\theta}) - \tilde{f}(e^{i\theta})| > \epsilon/3\}$$

$$\leq \{e^{i\theta} : C_\alpha M(\tilde{f} - g) > \epsilon/3\} + 0 + (3\|g - \tilde{f}\|_{L^p}/\epsilon)^p.$$

In the last estimate we used Lemma 8.1.6 and (8.1.11.1). Now the last line is majorized by

$$C'_\alpha \|\tilde{f} - g\|_{L^1}/(\epsilon/3) + 3^p \epsilon^p \leq C''_\alpha \|\tilde{f} - g\|_{L^p}/(\epsilon/3) + 3^p \epsilon^p$$

$$\leq C'''_\alpha \epsilon.$$

It follows that

$$\lim_{\Gamma_\alpha(e^{i\theta}) \ni z \to e^{i\theta}} f(z) = \tilde{f}(e^{i\theta}), \qquad \text{a.e.} \quad e^{i\theta} \in \partial D. \qquad \square$$

The informal statement of Theorem 8.1.11 is that f has nontangential boundary limits almost everywhere.

Theorem 8.1.11 contains essentially all that can be said about the boundary behavior of harmonic functions. Why are holomorphic functions better? The classical way of answering this question (which will, in the long run, be seen to be misleading) is to use Blaschke factorization.

DEFINITION 8.1.12   If $a \in \mathbb{C}, |a| < 1$, then the *Blaschke factor* at a is

$$B_a(z) = \frac{z - a}{1 - \bar{a}z}.$$

It is elementary to verify that B_a is holomorphic on a neighborhood of $\bar{D}$ and that $|B_a(e^{i\theta})| = 1$ for all θ.

LEMMA 8.1.13 If $0 < r < 1$ and f is holomorphic on a neighborhood of $\bar{D}(0,r)$, let $p_1, \ldots, p_k$ be the zeroes of f (listed with multiplicity) in $D(0,r)$. Assume that $f(0) \neq 0$ and that $f(re^{it}) \neq 0$, all t. Then

$$\log|f(0)| + \log \prod_{j=1}^{k} r|p_j|^{-1} = \frac{1}{2\pi} \int_0^{2\pi} \log|f(re^{it})|\, dt.$$

Proof   The function

$$F(z) = \frac{f(z)}{\prod_{j=1}^{k} B_{p_j/r}(z/r)}$$

is holomorphic on a neighborhood of $\bar{D}(0,r)$; hence $\log|F|$ is harmonic on a neighborhood of $\bar{D}(0,r)$. Thus

$$\log|F(0)| = \frac{1}{2\pi} \int_0^{2\pi} \log|F(re^{it})|\, dt$$

$$= \frac{1}{2\pi} \int_0^{2\pi} \log|f(re^{it})|\, dt$$

or

$$\log|f(0)| + \log \prod_{j=1}^{k} \frac{r}{|p_j|} = \frac{1}{2\pi} \int_0^{2\pi} \log|f(re^{it})|\, dt. \qquad \square$$

Notice that, by the continuity of the integral, Lemma 8.1.13 holds even if f has zeros on $\{re^{it}\}$. The reader should also notice that the proof of the lemma is essentially the same as the proof of Lemma 7.5.2.

COROLLARY 8.1.14 If f is holomorphic in a neighborhood of $\bar{D}(0,r)$, then

$$\log|f(0)| \le \frac{1}{2\pi}\int_0^{2\pi}\log|f(re^{it})|\,dt.$$

Proof   The term $\log\prod_{j=1}^k|r/p_j|$ in Lemma 8.1.13 is positive. □

COROLLARY 8.1.15 If f is holomorphic on D, $f(0)\ne 0$ and $\{p_1,p_2,\ldots\}$ are the zeros of f counting multiplicities, then

$$\log|f(0)| + \log\prod_{j=1}^\infty\frac{1}{|p_j|} \le \sup_{0<r<1}\frac{1}{2\pi}\int_0^{2\pi}\log^+|f(re^{it})|\,dt.$$

Proof   Apply Lemma 8.1.13, letting $r\to 1^-$. □

COROLLARY 8.1.16 If $f\in H^p(D), 0<p\le\infty$, and $\{p_1,p_2,\ldots\}$ are the zeros of f counting multiplicities, then $\sum_{j=1}^\infty(1-|p_j|)<\infty$.

Proof   Since f vanishes to finite order k at 0, we may replace f by $f(z)/z^k$ and assume that $f(0)\ne 0$. It follows from Corollary 8.1.15 that

$$\log\prod_{j=1}^\infty\left\{\frac{1}{|p_j|}\right\}<\infty$$

or $\prod(1/|p_j|)$ converges; hence $\prod|p_j|$ converges. So $\sum_j(1-|p_j|)<\infty$. □

PROPOSITION 8.1.17 If $\{p_1,p_2,\ldots\}\subseteq D$ satisfy $\sum_j(1-|p_j|)<\infty, p_j\ne 0$ for all j, then

$$\prod_{j=1}^\infty\frac{-\bar{p}_j}{|p_j|}B_{p_j}(z)$$

converges normally on D.

Proof   Restrict attention to $|z|\le r<1$. Then the assertion that the infinite product converges uniformly on this disc is equivalent to the assertion that

$$\sum_j\left|1+\frac{\bar{p}_j}{|p_j|}B_{p_j}(z)\right|$$

converges uniformly. But

$$
\begin{aligned}
\left| 1 + \frac{\bar{p}_j}{|p_j|} B_{p_j}(z) \right| &= \left| \frac{|p_j| - |p_j|\bar{p}_j z + \bar{p}_j z - |p_j|^2}{|p_j|(1 - z\bar{p}_j)} \right| \\
&= \left| \frac{(|p_j| + z\bar{p}_j)(1 - |p_j|)}{|p_j|(1 - z\bar{p}_j)} \right| \\
&\leq \frac{(1 + r)(1 - |p_j|)}{1 - r},
\end{aligned}
$$

so the convergence is uniform.   □

DEFINITION 8.1.18   Let $0 < p \leq \infty$ and $f \in H^p(D)$. Let $\{p_1, p_2, \dots\}$ be the zeros of f counted according to multiplicities. Let

$$
B(z) = \prod_{j=1}^{\infty} \frac{-\bar{p}_j}{|p_j|} B_{p_j}(z)
$$

(where each $p_j = 0$ is understood to give rise to a factor of z). Then B is a well-defined holomorphic function on D by Proposition 8.1.17. Let $F(z) = f(z)/B(z)$. By the Riemann removable singularities theorem, F is a well-defined, nonvanishing holomorphic function on D. The representation $f = F \cdot B$ is called the *canonical factorization* of f.

Exercise for the Reader
All the assertions of Definition 8.1.18 hold for $f \in N(D)$, the Nevanlinna class.

PROPOSITION 8.1.19   Let $f \in H^p(D), 0 < p \leq \infty$, and let $f = F \cdot B$ be its canonical factorization. Then $F \in H^p(D)$ and $\|F\|_{H^p(D)} = \|f\|_{H^p(D)}$.

Proof   Trivially, $|F| = |f/B| \geq |f|$, so $\|F\|_{H^p} \geq \|f\|_{H^p}$. If $N = 1, 2, \dots$, let

$$
B_N(z) = \prod_{j=1}^{N} \frac{-\bar{p}_j}{|p_j|} B_{p_j}(z)
$$

(where the factors corresponding to $p_j = 0$ are just z).
Let $F_N = f/B_N$. Since $|B_N(e^{it})| = 1$, all t, it holds that $\|F_N\|_{H^p} = \|f\|_{H^p}$ (use Lemma 2.1.17 and the fact that $B_N(re^{it}) \to B_N(e^{it})$ uniformly in t as $r \to 1^-$.) If $0 < r < 1$, then

$$
\begin{aligned}
\left[\frac{1}{2\pi} \int_0^{2\pi} |F(re^{it})|^p \, dt \right]^{1/p} &= \lim_{N \to \infty} \left[\frac{1}{2\pi} \int_0^{2\pi} |F_N(re^{it})|^p \, dt \right]^{1/p} \\
&\leq \lim_{N \to \infty} \|F_N\|_{H^p} = \|f\|_{H^p}.
\end{aligned}
$$

Therefore, $\|F\|_{H^p} \leq \|f\|_{H^p}$. □

COROLLARY 8.1.20   If $\{p_1, p_2, \ldots\}$ is a sequence of points in D satisfying $\sum_j (1 - |p_j|) < \infty$ and if $B(z) = \prod_j (-\bar{p}_j/|p_j|) B_{p_j}(z)$ is the corresponding Blaschke product, then

$$\lim_{r \to 1^-} B(re^{it})$$

exists and has modulus 1 almost everywhere.

Proof   The conclusion that the limit exists follows from Theorem 8.1.11 and the fact that $B \in H^\infty$. For the other assertion, note that the canonical factorization for B is $B = 1 \cdot B$. Therefore, by Proposition 8.1.19,

$$\left[\frac{1}{2\pi} \int |\tilde{B}(e^{it})|^2 dt \right]^{1/2} = \|B\|_{H^2} = \|1\|_{H^2} = 1;$$

hence $|\tilde{B}(e^{it})| = 1$ almost everywhere. □

THEOREM 8.1.21   If $f \in H^p(D), 0 < p \leq \infty$, and $1 < \alpha < \infty$, then

$$\lim_{\Gamma_\alpha(D) \ni z \to e^{i\theta}} f(z)$$

exists for almost every $e^{i\theta} \in \partial D$ and equals $\tilde{f}(e^{i\theta})$. Also, $\tilde{f} \in L^p(\partial D)$ and

$$\|\tilde{f}\|_{L^p} = \|f\|_{H^p} \equiv \sup_{0 < r < 1} \left[\frac{1}{2\pi} \int_0^{2\pi} |f(re^{i\theta})|^p d\theta \right]^{1/p}.$$

Proof   By Definition 8.1.18, write $f = B \cdot F$, where F has no zeros and B is a Blaschke product. Then $F^{p/2}$ is a well-defined H^2 function and thus has the appropriate boundary values almost everywhere. A fortiori, F has nontangential boundary limits almost everywhere. Since $B \in H^\infty$, B has nontangential boundary limits almost everywhere. It follows that f does as well. The final assertion follows from the corresponding fact for H^2 functions (exercise). □

8.2 Three Propositions about the Poisson Kernel

The crux of our arguments in Section 8.1 was Proposition 8.1.10, in which the Poisson integral was majorized by the Hardy-Littlewood maximal function. In this estimation, the explicit form of the Poisson kernel for the disc was exploited. If we wish to use a similar program to study the boundary behavior of harmonic and holomorphic functions on general domains in $\mathbb{R}^N$ and $\mathbb{C}^n$, then we must again estimate the Poisson integral by a maximal function.

However there is no hope of obtaining an explicit formula for the Poisson kernel of an arbitrary smoothly bounded domain. In this section we shall instead obtain some rather sharp estimates that will suffice for our purposes. Although estimates of this nature have been part of the folklore of partial differential equations for some time, there is no treatment in print.

We shall present here a classical treatment that was developed by N. Kerzman [4]. Kerzman's methods have the virtue of being self-contained (in the sense that they use only facts that we developed in Chapter 1) and giving precisely the estimates that we need. We present the proofs of Propositions 8.2.2 and 8.2.6 with the generous permission of Norberto Kerzman.

Recall from Section 1.3 that the Poisson kernel for a C^2 domain $\Omega \subseteq \mathbb{R}^N$ is given by $P(x,y) = -\nu_y G(x,y), x \in \Omega, y \in \partial\Omega$. Here ν_y is the unit outward normal vector field to $\partial\Omega$ at y, and $G(x,y)$ is the Green's function for Ω. Recall that, for $N > 2$, we have $G(x,y) = c_N|x-y|^{-N+2} - F_x(y)$, where F depends in a $C^{2-\epsilon}$ fashion on x and y jointly and F is harmonic in y. By Exercise 20 at the end of Chapter 1, G is $C^{2-\epsilon}$ on $\bar{\Omega} \times \bar{\Omega} \setminus \{\text{diagonal}\}$ and $G(x,y) = G(y,x)$. It follows that $P(x,y)$ behaves qualitatively like $|x-y|^{-N+1}$. (These observations persist in $\mathbb{R}^2$ by a slightly different argument.) In this section we use the method of comparison domains to refine this qualitative estimate on the Poisson kernel.

We begin with a geometric fact.

GEOMETRIC FACT   Let $\Omega \subset\subset \mathbb{R}^N$ have C^2 boundary. There are numbers r, $\tilde{r} > 0$ such that for each $y \in \partial\Omega$ there are balls $B(c_y,r) \equiv B_y \subseteq \Omega$ and $B(\tilde{c}_y,\tilde{r}) \equiv \tilde{B}_y \subseteq {}^c\bar{\Omega}$ that satisfy

i. $\bar{B}(\tilde{c}_y,\tilde{r}) \cap \bar{\Omega} = \{y\}$;

ii. $\bar{B}(c_y,r) \cap \overline{{}^c\Omega} = \{y\}$.

Let us indicate why these balls exist. Fix $P \in \partial\Omega$. Applying the implicit function theorem to the mapping

$$\partial\Omega \times (-1,1) \quad \rightarrow \quad \mathbb{R}^N,$$
$$(\zeta,t) \quad \mapsto \quad \zeta + t\nu_\zeta,$$

at the point $(P,0)$, we find a neighborhood U_P of the point $(P,0)$ on which the mapping is one-to-one. By the compactness of the boundary, there is thus a neighborhood U of $\partial\Omega$ such that each $x \in U$ has a unique nearest point in $\partial\Omega$. It further follows that there is an $\epsilon > 0$ such that if ζ_1, ζ_2 are distinct points of $\partial\Omega$ then $I_1 = \{\zeta_1 + t\nu_{\zeta_1} : |t| < 2\epsilon\}$ and $I_2 = \{\zeta_2 + t\nu_{\zeta_2} : |t| < 2\epsilon\}$ are disjoint sets (that is, the normal bundle is locally trivial in a natural way). From this it follows that if $y \in \partial\Omega$, then we may take $c_y = y - \epsilon\nu_y, \tilde{c}_y = y + \epsilon\nu_y$, and $r = \tilde{r} = \epsilon$. Refer to Figure 8.1.

We may assume in what follows that $r = \tilde{r} < \text{diam}\,\Omega/2$.

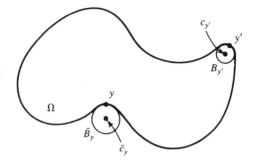

FIGURE 8.1

We now consider estimates for $P_\Omega(x,y)$. Results are stated for all dimensions $N \geq 2$ but, for convenience, are proved only for $N \geq 3$. The basic technique is to compare the Green's function and Poisson kernel for the given domain Ω with the corresponding functions for the internally and externally tangent balls $B(c_y, r)$ and $B(\tilde{c}_y, \tilde{r})$ (and also for their complements).

PROPOSITION 8.2.1   Let $\Omega \subseteq \mathbb{R}^N$ be a domain with C^2 boundary. Let $P = P_\Omega : \Omega \times \partial\Omega \to \mathbb{R}$ be its Poisson kernel. Then for each $x \in \Omega$, there is a positive constant C_x such that

$$0 < C_x \leq P(x,y) \leq \frac{C}{|x-y|^{N-1}} \leq \frac{C}{\delta(x)^{N-1}}.$$

Here $\delta(x) = \text{dist}\,(x, \partial\Omega)$.

Proof   Recall that $G(x,y) = 0$ for $x \in \Omega, y \in \partial\Omega$. By the maximum principle applied to $-G(x, \cdot)$ on $\Omega \setminus B(x, \epsilon)$, we have that $G > 0$ on Ω. By the Hopf lemma, $-\nu_y G(x,y) > 0$ when $y \in \partial\Omega$. Since $\partial\Omega$ is compact, the bound from below on $P(x,y)$ is established.

Now for fixed $x \in \Omega, y \in \partial\Omega$, we know that $G(x, \cdot)$ equals 0 on $\partial\Omega$, whereas $G_{^c\tilde{B}_y}(x, \cdot) \geq 0$ on $\partial\Omega$. Since $G_{^c\tilde{B}_y}(x, \cdot) - G_\Omega(x, \cdot)$ is harmonic on Ω, it follows that

$$\begin{aligned}
G_{^c\tilde{B}_y}(x,t) &\geq G_\Omega(x,t), \qquad t \in \bar{\Omega}, \\
G_{^c\tilde{B}_y}(x,y) &= G_\Omega(x,y) = 0.
\end{aligned}$$

Therefore,

$$P_{^c\tilde{B}_y}(x,y) = -\nu_y G_{^c\tilde{B}_y}(x,y) \geq -\nu_y G_\Omega(x,y) = P_\Omega(x,y). \tag{8.2.1.1}$$

The Poisson kernel for $\tilde{B}_y$ is

$$P_{\tilde{B}_y}(x,t) = \frac{1}{\omega_{N-1}\tilde{r}} \cdot \frac{\tilde{r}^2 - |x - \tilde{c}_y|^2}{|x - t|^N}.$$

By Kelvin reflection, the Poisson kernel for ${}^c\tilde{B}_y$ is

$$P_{{}^c\tilde{B}_y}(x,t) = \frac{1}{\omega_{N-1}\tilde{r}} \cdot \frac{|x - \tilde{c}_y|^2 - \tilde{r}^2}{|x - t|^N}.$$

Thus (8.2.1.1) gives

$$\begin{aligned}
P_\Omega(x,y) &\leq \frac{1}{\omega_{N-1}\tilde{r}} \cdot \frac{|x - \tilde{c}_y|^2 - \tilde{r}^2}{|x - y|^N} \\
&= \frac{1}{\omega_{N-1}\tilde{r}} \cdot \frac{(|x - \tilde{c}_y| - |\tilde{c}_y - y|)(|x - \tilde{c}_y| + \tilde{r})}{|x - y|^N} \\
&\leq \frac{2}{\omega_{N-1}\tilde{r}} \cdot \frac{|x - y|(\operatorname{diam}\Omega + \tilde{r})}{|x - y|^N} \\
&= C|x - y|^{-N+1} \leq C\delta(x)^{-N+1}. \qquad \square
\end{aligned}$$

Exercise for the Reader
Use the explicit formula for the Poisson kernel of $B(0,R) \subseteq \mathbb{R}^N$ to prove that if u is a nonnegative harmonic function on $B(0,R)$ and $\xi \in B(0,R)$, then:

(i) $u(\xi) \leq 2(R/d(\xi))^{N-1} \cdot u(0)$;

(ii) $u(\xi) \geq [d(\xi)/(2^N R)] \cdot u(0)$, where $d(\xi) = \operatorname{dist}(\xi, {}^cB(0,R))$

These first two inequalities are essentially the Harnack inequalities of Chapter 1. Apply inequality ii to $G_\Omega(x,\cdot)$ on a ball B_y^* internally tangent to $\partial\Omega$ at $y \in \partial\Omega$ with $x \notin B_y^*$ to obtain

(iii) $G_\Omega(x,t) \geq C_x[d(t)]G_\Omega(x,c_y)$, $t \in B_y^*$, $x \in \Omega$;

Conclude that

(iv) $-\frac{\partial}{\partial\nu_y}G_\Omega(x,y) \geq C_x G_\Omega(x,c_y) > 0$,

so we have found a second proof that

(v) $P_\Omega(x,y) \geq C_x' > 0$.

Parts i–iv also have independent interest, as we shall see.

If $x \in \Omega, \delta(x) < r$, let $\pi x \in \partial\Omega$ be the nearest point to x. Apply (ii) to $P_\Omega(\cdot,y)$ on $B_{\pi x}$ to obtain

(vi) $P_\Omega(x,y) \geq C\delta(x) \cdot P_\Omega(c_{\pi x}, y)$ provided $\delta(x) < r$;

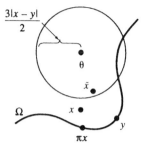

FIGURE 8.2

Conclude from (vi) and the fact that $\{c_{\pi x}\} \subset\subset \Omega$ that

(vii) $P_\Omega(x,y) \geq C'\delta(x)$ for all $x \in \Omega$.

For the case $\delta(x) < r$ and $y = \pi x$, we may apply the argument in the proof of Proposition 8.2.1 to compare $P_\Omega(x,y)$ with $P_{B_y}(x,y)$ to obtain

(viii) $P_\Omega(x,y) \geq c|x-y|^{-N+1}$ when $y = \pi x$.

With the results of the exercises in hand, it is now easy to prove the next proposition:

PROPOSITION 8.2.2 If $\Omega \subset\subset \mathbb{R}^n$ is a domain with C^2 boundary, then

$$P_\Omega(x,y) \geq C \cdot \frac{\delta(x)}{|x-y|^N}.$$

Proof　Choose $0 < \eta < r/2$ so small that the following is true: Let $x \in \Omega, y \in \partial\Omega, |x-y| < \eta$, and let $\theta = \theta(x,y) = \pi x - 2|x-y| \cdot \nu_{\pi x}$. Then there is an $\tilde{x} \in B(\theta, 3|x-y|/2)$ such that $|\tilde{x} - y| = \delta(\tilde{x})$ (see Figure 8.2).

Consider $y \in \partial\Omega$ to be fixed. To prove the result, we need only consider x such that $|x-y| < \eta$ (otherwise (vii) above applies). Notice that

$$
\begin{aligned}
|\tilde{x}-y| &\leq |\tilde{x}-\theta| + |\theta - x| + |x-y| \\
&\leq 3\frac{|x-y|}{2} + 2|x-y| + |x-y| \leq 5|x-y|.
\end{aligned}
\tag{8.2.2.1}
$$

Applying (ii) to $P_\Omega(\cdot,y)$ on $B(\theta, 2|x-y|)$ and then (i) to $P_\Omega(\cdot,y)$ on $B(\theta, 2|x-y|)$ gives

$$
\begin{aligned}
P_\Omega(x,y) &\geq C\frac{\delta(x)}{2|x-y|} \cdot P_\Omega(\theta,y) \\
&\geq C'\frac{\delta(x)}{|x-y|}\left(\frac{|x-y|/2}{2|x-y|}\right)^{N-1} P_\Omega(\tilde{x},y) = C''\delta(x)\frac{P_\Omega(\tilde{x},y)}{|x-y|}.
\end{aligned}
$$

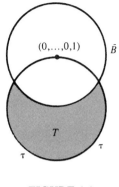

FIGURE 8.3

But (viii), together with (8.2.2.1), shows that this majorizes $C'''\delta(x)/|x - y|^N$. □

The next estimate for P_Ω (Proposition 8.2.6) is the most useful and the deepest. In order to prove it we must first consider an auxiliary function to which we will compare it.

LEMMA 8.2.3   Let $B \subseteq \mathbb{R}^N$ be the unit ball. Let $\tilde{B} = B\,((0,\ldots,0,1),1)$ and $T = B \cap {}^c\tilde{B}$. Let $\tau = \partial B \setminus \tilde{B}$. There is a nonnegative harmonic function H on a neighborhood of $\bar{T}$ and a $c > 0$ such that

(8.2.3.1) $H(0) = 0$;

(8.2.3.2) $H(x) > 0$ if $x \in \bar{T} \setminus \{0\}$;

(8.2.3.3) $H \geq 1$ on τ;

(8.2.3.4) $H(0,\ldots,0,-t) \leq ct$, all $0 \leq t \leq 1$.

Proof   Let H be a large positive multiple of the Kelvin transform in $\tilde{B}$ of the function $x \mapsto x_N$ (see Figure 8.3). A moment's reflection shows that this H works (*do not calculate*—think geometrically). □

Remark: If $0 < \alpha$, let T_α be the region obtained from T by dilation of order α (i.e., T_α is obtained from balls of radius α). Then $H_\alpha(x) = H(x/\alpha)$ has the same properties relative to T_α as H does relative to T. □

LEMMA 8.2.4   There is a constant $C = C(\Omega, N)$ such that $G_\Omega(x,y) \leq C|x - y|^{-N+2}$.

Proof   Let $\epsilon > 0$. Both $G_\Omega(x,y) = -c_N|x - y|^{-N+2} - F_x(y)$ and $-2c_N|x - y|^{-N+2}$ are harmonic on $\Omega \setminus B(x,\epsilon)$, and the latter majorizes the former on $\partial(\Omega \setminus B(x,\epsilon))$ if ϵ is small. (Remember how F_x was constructed!) By the maximum principle, the majorization persists on $\Omega \setminus B(x,\epsilon)$, hence on Ω. □

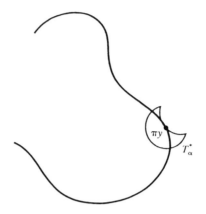

FIGURE 8.4

LEMMA 8.2.5 We have

$$G_\Omega(x, y) \leq C(\Omega, N) \cdot \frac{\delta(y)}{|x - y|^{N-1}}, \qquad (8.2.5.1)$$

$$G_\Omega(x, y) \leq C(\Omega, N) \cdot \frac{\delta(x)}{|x - y|^{N-1}}. \qquad (8.2.5.2)$$

Proof We first prove (8.2.5.1). Recall that $\tilde{r} < \operatorname{diam}\Omega/2$. If $\delta(y) \geq \tilde{r}|x - y|/(8\operatorname{diam}\Omega)$, then there is nothing to prove by Lemma 8.2.4.

If $\delta(y) < \tilde{r}|x - y|/(8\operatorname{diam}\Omega)$, let $\alpha = \tilde{r}|x - y|/(4\operatorname{diam}\Omega) > 2\delta(y)$. Rotate and translate T_α so that 0 is sent to πy (the projection of y to $\partial\Omega$) and the resulting region T_α^* is tangent to $\partial\Omega$ at πy (see Figure 8.4). Observe in the figure that Ω and T_α^* have the same inward normal at πy.

Let $t \in \partial T_\alpha^* \cap \Omega$. Then $|x - t| \geq |x - y| - |y - t| \geq |x - y| - 2\alpha \geq \frac{|x-y|}{2}$. Therefore,

$$G_\Omega(x, t) \leq C|x - t|^{-N+2} \leq C|x - y|^{-N+2}. \qquad (8.2.5.3)$$

Let H_α^* be the harmonic function on T_α^* whose existence is guaranteed by Lemma 8.2.3. By (8.2.3.3) and (8.2.5.3),

$$H_\alpha^*(t)C|x - y|^{-N+2} \geq G_\Omega(x, t), \qquad t \in \partial T_\alpha^* \cap \Omega. \qquad (8.2.5.4)$$

But, trivially,

$$H_\alpha^*(t)C|x - y|^{-N+2} \geq 0 = G_\Omega(x, t), \qquad t \in T_\alpha^* \cap \partial\Omega. \qquad (8.2.5.5)$$

By the maximum principle,

$$H_\alpha^*(t)C|x-y|^{-N+2} \geq G_\Omega(x,t), \qquad \text{all } t \in T_\alpha^* \cap \Omega. \qquad (8.2.5.6)$$

Therefore,

$$H_\alpha^*(y)C|x-y|^{-N+2} \geq G_\Omega(x,y).$$

Now (8.2.3.4) yields that

$$C''\left(\delta(y)/\alpha\right)|x-y|^{-N+2} \geq G_\Omega(x,y); \qquad (8.2.5.7)$$

hence the choice of α implies

$$C''\delta(y)|x-y|^{-N+1} \geq G_\Omega(x,y).$$

Now (8.2.5.2) follows from (8.2.5.1) by symmetry.   □

The final estimate on $P_\Omega(x,y)$ is the following.

PROPOSITION 8.2.6 If $\Omega \subset\subset \mathbb{R}^N$ is a domain with C^2 boundary, then

$$P_\Omega(x,y) \leq C \frac{\delta(x)}{|x-y|^N}.$$

Proof   We exploit Lemma 8.2.5. We prove that

$$G_\Omega(x,y) \leq C \frac{\delta(x)\delta(y)}{|x-y|^N}. \qquad (8.2.6.1)$$

For the case $\delta(x) \geq \tilde{r}|x-y|/(8\,\mathrm{diam}\,\Omega)$, there is nothing to prove. So we may assume that $\delta(x) < \tilde{r}|x-y|/(8\,\mathrm{diam}\,\Omega)$.

Use the notation of the proof of Lemma 8.2.5. We do a comparison on T_α^*. The argument used to prove (8.2.5.7) now shows that (8.2.6.1) holds because, with (8.2.5.2) replacing (8.2.5.3), we obtain

$$G_\Omega(x,t) \leq C \cdot \frac{\delta(x)H_\alpha^*(t)}{|x-t|^{N-1}}, \qquad t \in \partial(T_\alpha^* \cap \Omega)$$

instead of (8.2.5.6).

By the maximum principle, (8.2.6.1) follows. Using the fact that

$$G_\Omega(x,\cdot)|_{\partial\Omega} = 0$$

and differentiating, the proposition follows.   □

8.3 Subharmonicity, Harmonic Majorants, and Boundary Values

Let $\Omega \subseteq \mathbb{C}^N$ be a domain and $f : \Omega \to \mathbb{R}$ a function. The function f is said to have a *harmonic majorant* if there is a (necessarily) nonnegative harmonic u on Ω with $|f| \leq u$. We are interested in harmonic majorants for subharmonic functions.

PROPOSITION 8.3.1   If $f : B \to \mathbb{R}^+$ is subharmonic, then f has a harmonic majorant if and only if

$$\sup_{0 < r < 1} \int_{\partial B} f(r\zeta)\, d\sigma(\zeta) < \infty.$$

Proof   Let u be a harmonic majorant for f. Then

$$\int_{\partial B} f(r\zeta)\, d\sigma(\zeta) \leq \int_{\partial B} u(r\zeta)\, d\sigma(\zeta) = \omega_{N-1} \cdot u(0) \equiv C < \infty,$$

as claimed.

Conversely, if f satisfies

$$\sup_{r} \int_{\partial B} f(r\zeta)\, d\sigma(\zeta) < \infty,$$

then the functions $f_r : \partial B \to \mathbb{C}$ given by $f_r(\zeta) = f(r\zeta)$ form a bounded subset of $L^1(\partial B) \subseteq \mathcal{M}(\partial B)$. Let $\tilde{f} \in \mathcal{M}(\partial B)$ be a weak-$*$ accumulation point of the functions f_r. Then $F(r\zeta) \equiv P\tilde{f}(r\zeta)$ is harmonic on B, and for any $x \in B$ and $1 > r > |x|$ we have

$$
\begin{aligned}
0 \leq f(x) \;\; &\leq \;\; \int P(x/r, \zeta) f_r(\zeta) d\sigma(\zeta) \\
&\to \;\; \int P(x, \zeta) d\tilde{f}(\zeta) = F(x) \qquad \text{as } r \to 1^-. \qquad \square
\end{aligned}
$$

A consequence of Proposition 8.3.1 is that not all subharmonic functions have harmonic majorants. Harmonic majorants play a significant role in the theory of boundary behavior of harmonic and holomorphic functions. Proposition 8.3.1 suggests why growth conditions may, therefore, play a role. The fact that f harmonic implies $|f|^p$ subharmonic only for $p \geq 1$, whereas f holomorphic implies $|f|^p$ subharmonic for $p > 0$ suggests that we may expect different behavior for harmonic and for holomorphic functions.

By way of putting these remarks in perspective and generalizing Proposition 8.3.1, we consider $\mathbf{h}^p(\Omega)$ (resp. $H^p(\Omega)$), with Ω any smoothly bounded domain in $\mathbb{R}^N$ (resp. $\mathbb{C}^n$). First we require some preliminary groundwork.

Let $\Omega \subset\subset \mathbb{R}^N$ be a domain with C^2 boundary. Let $\phi : \mathbb{R} \to [0,1]$ be a C^∞ function supported in $[-2,2]$ with $\phi \equiv 1$ on $[-1,1]$. Then Exercise 4 at the end of Chapter 3 implies, with $\delta_\Omega(x) \equiv \text{dist}\,(x, \partial\Omega)$ and $\epsilon_0 > 0$ sufficiently small, that

$$\rho(x) = \begin{cases} -\phi(|x|/\epsilon_0)\delta_\Omega(|x|) - (1 - \phi(|x|/\epsilon_0)) & \text{if } x \in \bar{\Omega}, \\ \phi(|x|/\epsilon_0)\delta_\Omega(|x|) + (1 - \phi(|x|/\epsilon_0)) & \text{if } x \notin \bar{\Omega} \end{cases}$$

is a C^2 defining function for Ω. The implicit function theorem implies that if $0 < \epsilon < \epsilon_0$, then $\partial\Omega_\epsilon \equiv \{x \in \Omega : \rho(x) = -\epsilon\}$ is a C^2 manifold that bounds $\Omega_\epsilon \equiv \{x \in \Omega : \rho_\epsilon(x) \equiv \rho(x) + \epsilon < 0\}$. Now let $d\sigma_\epsilon$ denote area measure on $\partial\Omega_\epsilon$. Then it is natural to let

$$\begin{aligned} \mathbf{h}^p(\Omega) &= \{f \text{ harmonic on } \Omega : \sup_{0 < \epsilon < \epsilon_0} \int_{\partial\Omega_\epsilon} |f(\zeta)|^p d\sigma_\epsilon(\zeta)^{1/p} \\ &\equiv \|f\|_{\mathbf{h}^p(\Omega)} < \infty\}, \qquad 0 < p < \infty, \\ \mathbf{h}^\infty(\Omega) &= \{f \text{ harmonic on } \Omega : \sup_{x \in \Omega} |f(x)| \equiv \|f\|_{\mathbf{h}^\infty} < \infty\}. \end{aligned}$$

In case Ω is a subdomain of *complex space*, we define

$$H^p(\Omega) = \mathbf{h}^p(\Omega) \cap \{\text{holomorphic functions}\}, \qquad 0 < p \leq \infty.$$

The next lemma serves to free the definitions from their somewhat artificial dependence on δ_Ω and ρ.

LEMMA 8.3.2 (Stein)   Let ρ_1, ρ_2 be two C^2 defining functions for a domain $\Omega \subseteq \mathbb{R}^N$. For $\epsilon > 0$ small and $i = 1, 2$, let

$$\begin{aligned} \Omega_\epsilon^i &= \{x \in \Omega : \rho_i(x) < -\epsilon\}, \\ \partial\Omega_\epsilon^i &= \{x \in \Omega : \rho_i(x) = -\epsilon\}. \end{aligned}$$

Let σ_ϵ^i be area measure on $\partial\Omega_\epsilon^i$. Then for f harmonic on Ω we have

$$\sup_{\epsilon > 0} \int_{\partial\Omega_\epsilon^1} |f(\zeta)|^p d\sigma_\epsilon^1(\zeta) < \infty$$

if and only if

$$\sup_{\epsilon > 0} \int_{\partial\Omega_\epsilon^2} |f(\zeta)|^p d\sigma_\epsilon^2(\zeta) < \infty.$$

(*Note: Since f is bounded on compact sets—equivalently, the supremum is only of interest as $\epsilon \to 0$—there is no ambiguity in this assertion.*)

Proof   By definition of defining function, $\operatorname{grad} \rho_i \neq 0$ on $\partial\Omega$. Since $\partial\Omega$ is compact, we may choose $\epsilon_0 > 0$ so small that there is a constant $\lambda, 0 < \lambda < 1$, with $0 < \lambda \leq |\operatorname{grad} \rho_i(x)| < 1/\lambda$ whenever $x \in \Omega, \delta_\Omega(x) < \epsilon_0$. If $0 < \epsilon < \epsilon_0$, then notice that for $x \in \partial\Omega_\epsilon^2$ we have

$$B(x, \lambda\epsilon/2) \subseteq \Omega$$

and, what is stronger,

$$B(x, \lambda\epsilon/2) \subseteq \left\{ t : -3\epsilon/\lambda^2 < \rho_1(t) < -\lambda^2 \cdot \epsilon/3 \right\} \equiv S(\epsilon). \tag{8.3.2.1}$$

Therefore,

$$|f(x)|^p \leq \frac{1}{V(B(x, \lambda\epsilon/2))} \int_{B(x,\lambda\epsilon/2)} |f(t)|^p \, dV(t).$$

As a result,

$$
\begin{aligned}
\int_{\partial\Omega_\epsilon^2} |f(x)|^p \, d\sigma_\epsilon^2 &\leq C\epsilon^{-N} \int_{\partial\Omega_\epsilon^2} \int_{B(x,\lambda\epsilon/2)} |f(t)|^p \, dV(t) d\sigma_\epsilon^2(x) \\
&= C\epsilon^{-N} \int_{S(\epsilon)} \int_{\partial\Omega_\epsilon^2} \chi_{B(x,\lambda\epsilon/2)}(t) |f(t)|^p \, d\sigma_\epsilon^2(x) \, dV(t) \\
&\leq C\epsilon^{-N} \int_{S(\epsilon)} |f(t)|^p \int_{\partial\Omega_\epsilon^2 \cap B(t,\lambda\epsilon/2)} d\sigma_\epsilon^2(x) \, dV(t) \\
&\leq C'\epsilon^{-N}\epsilon^{N-1} \int_{S(\epsilon)} |f(t)|^p \, dV(t) \\
&\leq C'' \sup_\epsilon \int_{\partial\Omega_\epsilon^1} |f(t)|^p \, d\sigma_\epsilon^1(t).
\end{aligned}
$$

Of course, the reverse inequality follows by symmetry.   □

One technical difficulty that we face on an arbitrary Ω is that the device of considering the dilated functions $f_r(\zeta) = f(r\zeta)$ as harmonic functions on $\bar\Omega$ is no longer available. However this notion is an unnecessary crutch, and it is well to be rid of it. As a substitute, we cover Ω by finitely many domains $\Omega_1, \ldots, \Omega_k$ with the following properties:

(8.3.3) $\Omega = \cup_j \Omega_j$.

(8.3.4) For each j, the set $\partial\Omega \cap \partial\Omega_j$ is an $(N-1)$-dimensional manifold with boundary.

(8.3.5) There is an $\epsilon_0 > 0$ and a vector ν_j transversal to $\partial\Omega \cap \partial\Omega_j$ and pointing out of Ω such that $\Omega_j - \epsilon\nu_j \equiv \{z - \epsilon\nu_j : z \in \Omega_j\} \subset\subset \Omega$, all $0 < \epsilon < \epsilon_0$.

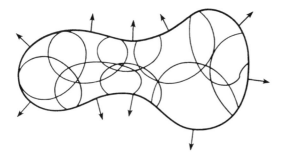

FIGURE 8.5

We leave the detailed verification of the existence of the sets Ω_j satisfying (8.3.3)–(8.3.5) as an exercise. See Figure 8.5 for an illustration of these ideas.

For a general C^2 bounded domain, the substitute for dilation will be to fix $j \in \{1, \ldots, k\}$ and consider the translated functions $f_\epsilon(x) = f(x - \epsilon\nu_j)$, $f_\epsilon : \Omega_j \to \mathbb{C}$, as $\epsilon \to 0^+$.

THEOREM 8.3.6   Let $\Omega \subseteq \mathbb{R}^N$ be a domain and f harmonic on Ω. Let $1 \leq p < \infty$. The following are equivalent:

(8.3.6.1) $f \in \mathbf{h}^p(\Omega)$.

(8.3.6.2) If $p > 1$, then there is an $\tilde{f} \in L^p(\partial\Omega)$ such that

$$f(x) = \int_{\partial\Omega} P(x, y)\tilde{f}(y)\, d\sigma(y)$$

(resp. if $p=1$, then there is a $\mu \in \mathcal{M}(\partial\Omega)$ such that $f(x)=\int_{\partial\Omega} P(x,y)d\mu(y)$). Moreover, $\|f\|_{\mathbf{h}^p} \cong \|\tilde{f}\|_{L^p}$.

(8.3.6.3) $|f|^p$ has a harmonic majorant on Ω.

Proof   **(2)** $\Rightarrow$ **(3)** If $p > 1$, let

$$h(x) = \int_{\partial\Omega} P(x, y)|\tilde{f}(y)|^p\, d\sigma(y).$$

Then, treating $P(x, \cdot)\, d\sigma$ as a positive measure of total mass 1, we have

$$\begin{aligned}
|f(x)|^p \quad &= \quad \left|\int_{\partial\Omega} \tilde{f}(y)P(x, y)\, d\sigma(y)\right|^p \\
&\overset{\text{(Jensen)}}{\leq} \quad \int_{\partial\Omega} |\tilde{f}(y)|^p P(x, y)\, d\sigma(y) \equiv h(x).
\end{aligned}$$

The proof for $p = 1$ is similar.

(3) $\Rightarrow$ (1) If $\epsilon > 0$ is small, $x_0 \in \Omega$ is fixed, and G_Ω is the Green's function for Ω, then $G_\Omega(x_0, \cdot)$ has nonvanishing gradient near $\partial\Omega$ (use Hopf's lemma). Therefore,

$$\tilde{\Omega}_\epsilon \equiv \{x \in \Omega : -G_\Omega(x, \cdot) < -\epsilon\}$$

are well-defined domains for ϵ small. Moreover (*check the proof*), the Poisson kernel for $\tilde{\Omega}_\epsilon$ is $P_\epsilon(x, y) = -\nu_y^\epsilon G_\Omega(x, y)$. Here ν_y^ϵ is the normal to $\partial\tilde{\Omega}_\epsilon$ at $y \in \partial\tilde{\Omega}_\epsilon$. Assume that $\epsilon > 0$ is so small that $x_0 \in \tilde{\Omega}_\epsilon$. So if h is the harmonic majorant for $|f|^p$, then

$$h(x_0) = \int_{\partial\tilde{\Omega}_\epsilon} -\nu_y^\epsilon G_\Omega(x_0, y) h(y) \, d\sigma(y). \qquad (8.3.6.4)$$

Let $\pi_\epsilon : \partial\tilde{\Omega}_\epsilon \to \partial\Omega$ be normal projection for ϵ small. Then

$$-\nu_y^\epsilon G_\Omega(x_0, \pi_\epsilon^{-1}(\,\cdot\,)) \to -\nu_y G_\Omega(x_0, \,\cdot\,)$$

uniformly on $\partial\Omega$ as $\epsilon \to 0^+$. By 8.2.1, $-\nu_y G_\Omega(x_0, \,\cdot\,) \geq c_{x_0} > 0$ for some constant c_{x_0}. Thus $-\nu_y^\epsilon G_\Omega(x_0, \pi_\epsilon^{-1}(\cdot))$ are all bounded below by $c_{x_0}/2$ if ϵ is small enough. As a result, (8.3.6.4) yields

$$\int_{\partial\tilde{\Omega}_\epsilon} h(y) \, d\sigma(y) \leq 2h(x_0)/c_{x_0}$$

for $\epsilon > 0$ small. In conclusion,

$$\int_{\partial\tilde{\Omega}_\epsilon} |f(y)|^p \, d\sigma(y) \leq 2h(x_0)/c_{x_0}.$$

(1) $\Rightarrow$ (2) Let Ω_j be as in (8.3.3) through (8.3.5). Fix j. Define on Ω_j the functions $f_\epsilon(x) = f(x - \epsilon\nu_j), 0 < \epsilon < \epsilon_0$. Then the hypothesis and (a small modification of) Lemma 8.3.2 show that $\{f_\epsilon\}$ forms a bounded subset of $L^p(\partial\Omega_j)$. If $p > 1$, let $\tilde{f}_j \in L^p(\partial\Omega_j)$ be a weak-$*$ accumulation point (for the case $p = 1$, replace $\tilde{f}_j$ by a Borel measure $\tilde{\mu}_j$). The crucial observation at this point is that f is the Poisson integral of $\tilde{f}_j$ on Ω_j. Therefore, f on Ω_j is completely determined by $\tilde{f}_j$ and conversely (see also the exercises at the end of the section). A moment's reflection now shows that $\tilde{f}_j = \tilde{f}_k$ almost everywhere $[d\sigma]$ in $\partial\Omega_j \cap \partial\Omega_k \cap \partial\Omega$ so that $\tilde{f} \equiv \tilde{f}_j$ on $\partial\Omega_j \cap \partial\Omega$ is well defined. By appealing to a partition of unity on $\partial\Omega$ that is subordinate to the open cover induced by the (relative) interiors of the sets $\partial\Omega_j \cap \partial\Omega$, we see that $f_\epsilon = f \circ \pi_\epsilon^{-1}$ converges weak-$*$ to $\tilde{f}$ on $\partial\Omega$ when $p > 1$ (resp. $f_\epsilon \to \tilde{\mu}$ weak-$*$ when $p = 1$).

Referring to the proof of (3) $\Rightarrow$ (1) we write, for $x_0 \in \Omega$ fixed,

$$
\begin{aligned}
f(x_0) &= \int_{\partial\Omega_\epsilon} -\nu_y^\epsilon G_\Omega(x_0, y) f(y)\, d\sigma_\epsilon(y) \\
&= \int_{\partial\Omega} -\nu_y^\epsilon G_\Omega\left(x_0, \pi_\epsilon^{-1}(y)\right) f\left(\pi_\epsilon^{-1}(y)\right) \mathcal{J}^\epsilon(y)\, d\sigma(y),
\end{aligned}
$$

where $\mathcal{J}^\epsilon$ is the Jacobian of the mapping $\pi_\epsilon^{-1} : \partial\Omega \to \partial\Omega_\epsilon$. The fact that $\partial\Omega$ is C^2 combined with previous observations implies that the last line tends to

$$
\int_{\partial\Omega} -\nu_y G_\Omega(x_0, y) \tilde{f}(y)\, d\sigma(y) = \int_{\partial\Omega} P_\Omega(x_0, y) \tilde{f}(y)\, d\sigma(y)
$$

$$
\left(\text{resp.} \int_{\partial\Omega} -\nu_y G_\Omega(x_0, y)\, d\tilde{\mu}(y) = \int_{\partial\Omega} P_\Omega(x_0, y)\, d\tilde{\mu}(y) \right)
$$

as $\epsilon \to 0^+$. $\qquad\square$

Exercises for the Reader

1. Prove the last statement in Theorem 8.3.6.

2. Imitate the proof of Theorem 8.3.6 to show that if u is continuous and *subharmonic* on Ω and if

$$
\sup_\epsilon \int_{\partial\Omega_\epsilon} |u(\zeta)|^p\, d\sigma(\zeta) < \infty, \qquad p \geq 1,
$$

then u has a harmonic majorant h. If $p > 1$, then h is the Poisson integral of an L^p function $\tilde{h}$ on $\partial\Omega$. If $p = 1$, then h is the Poisson integral of a Borel measure $\tilde{\mu}$ on $\partial\Omega$.

3. Let $\Omega \subseteq \mathbb{R}^N$ be a domain with C^2 boundary and let ρ be a C^2 defining function for Ω. Define $\Omega_\epsilon = \{x \in \Omega : \rho(x) < -\epsilon\}, 0 < \epsilon < \epsilon_0$. Let $\partial\Omega_\epsilon$ and $d\sigma_\epsilon$ be as usual. Let $\pi_\epsilon : \partial\Omega_\epsilon \to \partial\Omega$ be orthogonal projection. Let $f \in L^p(\partial\Omega), 1 \leq p < \infty$. Define

$$
F(x) = \int_{\partial\Omega} P_\Omega(x, y) f(y)\, d\sigma(y).
$$

a. Prove that $\int_{\partial\Omega} P_\Omega(x, y)\, d\sigma(y) = 1$, any $x \in \Omega$.
b. There is a $C > 0$ such that, for any $y \in \partial\Omega$,

$$
\int_{\partial\Omega_\epsilon} P_\Omega(x, y)\, d\sigma_\epsilon(x) \leq C, \qquad \text{any}\quad 0 < \epsilon < \epsilon_0.
$$

c. There is a $C' > 0$ such that

$$\int_{\partial\Omega_\epsilon} |F(x)|^p \, d\sigma_\epsilon \leq C', \qquad \text{any} \quad 0 < \epsilon < \epsilon_0.$$

d. If $\phi \in C(\Omega)$ satisfies $\|\phi - f\|_{L^p(\partial\Omega)} < \eta$ and

$$G(x) \equiv \int_{\partial\Omega} P_\Omega(x, y) \left(\phi(y) - f(y)\right) d\sigma(y),$$

then

$$\int_{\partial\Omega_\epsilon} |G(x)|^p \, d\sigma_\epsilon(x) \leq C'\eta^p, \qquad \text{any} \quad 0 < \epsilon < \epsilon_0.$$

e. With ϕ as in part **d** and $\Phi(x) = \int_{\partial\Omega} P_\Omega(x,y)\phi(y)\,d\sigma(y)$, then $\left(\Phi|_{\partial\Omega_\epsilon}\right) \circ \pi_\epsilon^{-1} \to \phi$ uniformly on $\partial\Omega$.

f. Imitate the proof of Proposition 8.1.2 to see that $F \circ \pi_\epsilon^{-1} \to f$ in the $L^p(\partial\Omega)$ norm.

8.4 Pointwise Convergence for Harmonic Functions on Domains in $\mathbb{R}^N$

Let $\Omega \subseteq \mathbb{R}^N$ be a domain. For $P \in \partial\Omega, \alpha > 1$, we define

$$\Gamma_\alpha(P) = \{x \in \Omega : |x - P| < \alpha\delta_\Omega(x)\}.$$

This is the N-dimensional analogue of the Stolz region considered in Section 8.1. Draw a sketch of $\Gamma_\alpha(P)$. Now our theorem is as follows:

THEOREM 8.4.1 Let $\Omega \subset\subset \mathbb{R}^N$ be a domain with C^2 boundary. Let $\alpha > 1$. If $1 < p \leq \infty$ and $f \in \mathbf{h}^p$, then

$$\lim_{\Gamma_\alpha(P) \ni x \to P} f(x) \equiv \tilde{f}(P) \qquad \text{exists for almost every } P \in \partial\Omega.$$

Moreover,

$$\|\tilde{f}\|_{L^p(\partial\Omega)} \cong \|f\|_{\mathbf{h}^p(\Omega)}.$$

Proof   We may as well assume that $p < \infty$. We already know from (8.3.6.2) that there exists an $\tilde{f} \in L^p(\partial\Omega), \|\tilde{f}\|_{L^p} \cong \|f\|_{\mathbf{h}^p}$, such that $f = P\tilde{f}$. It remains to show that $\tilde{f}$ satisfies the conclusions of the present theorem. This will follow just as in the proof of Theorem 8.1.11 as soon as we prove two things: First,

$$\sup_{x \in \Gamma_\alpha(P)} |P\tilde{f}(x)| \leq C_\alpha M_1 \tilde{f}(P), \tag{8.4.1.1}$$

where

$$M_1 \tilde{f}(P) \equiv \sup_{R>0} \frac{1}{\sigma(B(P,R) \cap \partial\Omega)} \int_{B(P,R) \cap \partial\Omega} |f(t)| d\sigma(t).$$

Second,

$$\sigma\{y \in \partial\Omega : M_1 \tilde{f}(y) > \lambda\} \leq C\frac{\|\tilde{f}\|_{L^1(\partial\Omega)}}{\lambda}, \qquad \text{all} \quad \lambda > 0. \tag{8.4.1.2}$$

Now (8.4.1.1) is proved just as in Proposition 8.1.10. It is necessary to use the estimate given in Proposition 8.2.6. On the other hand, (8.4.1.2) is not obvious; we supply a proof in the paragraphs that follow. $\qquad\square$

LEMMA 8.4.2 (Wiener)   Let $K \subseteq \mathbb{R}^N$ be a compact set that is covered by the open balls $\{B_\alpha\}_{\alpha \in A}, B_\alpha = B(c_\alpha, r_\alpha)$. There is a subcover $B_{\alpha_1}, B_{\alpha_2}, \ldots, B_{\alpha_m}$, consisting of pairwise disjoint balls, such that

$$\bigcup_{j=1}^{m} B(c_{\alpha_j}, 3r_{\alpha_j}) \supseteq K.$$

Proof   Since K is compact, we may immediately assume that there are only finitely many B_α. Let B_{α_1} be the ball in this collection that has the greatest radius (this ball may not be unique). Let B_{α_2} be the ball that has greatest radius and is also disjoint from B_{α_1}. At the jth step choose the (not necessarily unique) ball of greatest radius that is disjoint from $B_{\alpha_1}, \ldots, B_{\alpha_{j-1}}$. Continue. The process ends in finitely many steps. We claim that the B_{α_j} chosen in this fashion do the job.

It is enough to show that $B_\alpha \subseteq \cup_j B(c_{\alpha_j}, 3r_{\alpha_j})$ for every α. Fix an α. If $\alpha = \alpha_j$ for some j then we are done. If $\alpha \notin \{\alpha_j\}$, then let j_0 be the first index with $B_{\alpha_j} \cap B_\alpha \neq \emptyset$ (there must be one, otherwise the process would not have stopped). Then $r_{\alpha_{j_0}} \geq r_\alpha$; otherwise we selected $B_{\alpha_{j_0}}$ incorrectly. But then clearly $B(c_{\alpha_{j_0}}, 3r_{\alpha_{j_0}}) \supseteq B(c_\alpha, r_\alpha)$, as desired. $\qquad\square$

COROLLARY 8.4.3   Let $K \subseteq \partial\Omega$ be compact, and let $\{B_\alpha \cap \partial\Omega\}_{\alpha \in A}, B_\alpha = B(c_\alpha, r_\alpha)$, be an open covering of K by balls with centers in $\partial\Omega$. Then there is a pairwise disjoint subcover $B_{\alpha_1}, B_{\alpha_2}, \ldots, B_{\alpha_m}$ such that $\cup_j \{B(c_\alpha, 3r_\alpha) \cap \partial\Omega\} \supseteq K$.

Proof   The set K is a compact subset of $\mathbb{R}^N$ that is covered by $\{B_\alpha\}$. Apply the preceding Lemma 8.4.2 and restrict to $\partial\Omega$. $\qquad\square$

LEMMA 8.4.4   If $f \in L^1(\partial\Omega)$, then

$$\sigma\{x \in \partial\Omega : M_1 f(x) > \lambda\} \leq C \frac{\|f\|_{L^1}}{\lambda},$$

all $\lambda > 0$.

Proof   Let $S_\lambda = \{x \in \partial\Omega : M_1 f(x) > \lambda\}$. Let K be a compact subset of S_λ. It suffices to estimate $\sigma(K)$. Now for each $x \in K$, there is a ball B_x centered at x such that

$$\frac{1}{\sigma(B_x \cap \partial\Omega)} \int_{B_x \cap \partial\Omega} |f(t)| \, d\sigma(t) > \lambda. \qquad (8.4.4.1)$$

The balls $\{\partial\Omega \cap B_x\}_{x \in K}$ cover K. Choose, by Corollary 8.4.3, disjoint balls $B_{x_1}, B_{x_2}, \ldots, B_{x_m}$ such that $\{\partial\Omega \cap 3B_{x_j}\}$ cover K, where $3B_{x_j}$ represents the threefold dilate of B_{x_j} (with the same center). Then

$$\begin{aligned}
\sigma(K) &\leq \sum_{j=1}^{m} \sigma(3B_{x_j} \cap \partial\Omega) \\
&\leq C(N,\Omega) \sum_{j=1}^{m} \sigma(B_{x_j} \cap \partial\Omega),
\end{aligned}$$

where the constant C will depend on the curvature of $\partial\Omega$. But (8.4.4.1) implies that the last line is majorized by

$$C(N,\partial\Omega) \sum_{j} \frac{\int_{B_{x_j} \cap \partial\Omega} |f(t)| \, d\sigma(t)}{\lambda} \leq C(N,\partial\Omega) \frac{\|f\|_{L^1}}{\lambda}.$$

This completes the proof of the theorem.   □

8.5   Boundary Values of Holomorphic Functions in $\mathbb{C}^n$

Everything in Section 8.4 applies a fortiori to domains $\Omega \subseteq \mathbb{C}^n$. However, on the basis of our experience in the classical case, we expect $H^p(\Omega)$ functions to also have pointwise boundary values for $0 < p \leq 1$. That this is indeed the case is established in this section by two different arguments.

First, if $\Omega \subset\subset \mathbb{C}^n$ is a C^2 domain and $f \in H^p(\Omega)$, we shall prove through an application of Fubini's theorem (adapted from a paper of Lempert [1]) that f has pointwise boundary limits in a rather special sense at σ-almost every $\zeta \in \partial\Omega$.

This argument is self-contained. After that, we derive some more powerful results using a much broader perspective. We shall not further develop this second methodology here, but we present an introduction to it for background purposes. Further details may be found in E. M. Stein [2] and C. Fefferman and E. M. Stein [1]. The logical progression of ideas in this chapter will proceed from the first approach based on Lempert [1].

PROPOSITION 8.5.1   Let $\Omega \subset\subset \mathbb{C}^n$ have C^2 boundary. Let $0 < p < \infty$ and $f \in H^p(\Omega)$. Write $\Omega = \cup_{j=1}^k \Omega_j$ as in (8.3.3) through (8.3.5), and let $\nu_1, \ldots, \nu_k$ be the associated normal vectors. Then, for each $j \in \{1, \ldots, k\}$, it holds that

$$\lim_{\epsilon \to 0^+} f(\zeta - \epsilon \nu_j) \equiv \tilde{f}(\zeta)$$

exists for σ-almost every $\zeta \in \partial\Omega_j \cap \partial\Omega$.

Proof   We may suppose that $p < \infty$. Fix $1 \leq j \leq k$. Assume for convenience that $\nu_j = \nu_P = (1 + i0, 0, \ldots, 0), P \in \partial\Omega_j$, and that $P = 0$. If $z \in \mathbb{C}^n$, write $z = (z_1, \ldots, z_n) = (z_1, z')$. We may assume that $\Omega_j = \cup_{|z'|<1}\{(z_1, z') : z_1 \in \mathcal{D}_{z'}\}$, where $\mathcal{D}_{z'} \subseteq \mathbb{C}$ is a diffeomorph of $D \subseteq \mathbb{C}$ with C^2 boundary. For each $|z'| < 1, k \in \mathbb{N}, 0 < 1/k < \epsilon_0$, let $b_{z'}^k = (\mathcal{D}_{z'} \times \{z'\}) \cap \{z \in \Omega_j : \text{dist}(z, \partial\Omega_j) = 1/k\}$. Define $B^k = \cup_{|z'|<1} b_{z'}^k$. Now a simple variant of Lemma 8.3.2 implies that

$$\sup_k \int_{B^k} |f(\zeta)|^p d\sigma_k \leq C_0 < \infty, \tag{8.5.1.1}$$

where σ_k is surface measure on B^k. Formula (8.5.1.1) may be rewritten as

$$\sup_k \int_{|\zeta'|<1} \int_{b_{\zeta'}^k} |f(\zeta_1, \zeta')|^p \, d\tilde{\sigma}_k(\zeta_1) \, dV_{2n-2}(\zeta') \leq C_0, \tag{8.5.1.2}$$

where $\tilde{\sigma}_k$ is surface (= linear) measure on $b_{\zeta'}^k$. If $M > 0, k \geq k_0 > 1/\epsilon_0$, we define

$$S_k^M = \left\{ \zeta' : |\zeta'| < 1, \int_{b_{\zeta'}^k} |f(\zeta_1, \zeta')|^p d\tilde{\sigma}_k(\zeta_1) > M \right\}. \tag{8.5.1.3}$$

Then (8.5.1.2), (8.5.1.3), and Chebycheff's inequality together yield

$$V_{2n-2}(S_k^M) \leq \frac{C_0}{M}, \qquad \text{all } k.$$

Now let

$$
\begin{aligned}
S^M &= \left\{ \zeta' : |\zeta'| < 1, \int_{b_{\zeta'}^k} |f(\zeta_1, \zeta')|^p d\tilde{\sigma}_k(\zeta_1) \le M \text{ for only finitely many } k \right\} \\
&= \bigcup_{\ell=k_0}^{\infty} \bigcap_{k=\ell}^{\infty} S_k^M.
\end{aligned}
$$

Then $V_{2n-2}(S^M) \le C_0/M$. Since M may be made arbitrarily large, we conclude that for V_{2n-2}-almost every $\zeta' \in D^{n-1}(0,1)$, there exist $k_1 < k_2 < \cdots$ such that

$$
\int_{b_{\zeta'}^{k_m}} |f(\zeta_1, \zeta')|^p d\tilde{\sigma}_{k_m}(\zeta_1) = O(1) \qquad \text{as} \quad m \to \infty.
$$

It follows from Lemma 2.1.17 that the functions $f(\cdot, \zeta') \in H^p(\mathcal{D}_{\zeta'})$ for V_{2n-2} almost every $\zeta' \in D^{n-1}(0,1)$. Now Theorem 8.1.21 yields the desired result. $\square$

At this point we *could* prove that f tends nontangentially to the function $\tilde{f}$ constructed in Proposition 8.5.1. However, we do not do so because a much stronger result is proved in the next section. The remainder of the present section consists of a digression to introduce the reader to the second point of view mentioned in the introduction. This second point of view involves far-reaching ideas arising from the "real-variable" school of complex analysis. This methodology provides a more natural—and much more profound—approach to the study of boundary behavior.

The results that we present are due primarily to E. M. Stein [2]. To present Stein's ideas, we first need an auxiliary result of A. P. Calderón (see Stein [2]) and K. O. Widman [1]. Although it is well within the scope of this book to prove this auxiliary result on a half-space, a complete proof on smoothly bounded domains would entail a number of tedious ancillary ideas (such as the maximum principle for second-order elliptic operators). Hence we only state the needed result and refer the reader to the literature for details.

Let $\Omega \subset\subset \mathbb{R}^N$ be a domain with C^2 boundary and let $u : \Omega \to \mathbb{C}$ be harmonic. If $P \in \partial\Omega$, then we say that f is *nontangentially bounded* at P if there is an $\alpha > 1$ and a $C_\alpha < \infty$ such that

$$
\sup_{x \in \Gamma_\alpha(P)} |f(x)| < C_\alpha.
$$

Now the result is as follows:

THEOREM 8.5.2 Let $\Omega \subset\subset \mathbb{R}^N$ have C^2 boundary. Let $u : \Omega \to \mathbb{C}$ be harmonic. Let $E \subseteq \partial\Omega$ be a set of positive σ-measure. Suppose that u is nontangentially bounded at σ-almost every $P \in E$. Then u has nontangential limits at σ-almost every $P \in E$.

Remarks:

1. Obviously if u has a nontangential limit at $P \in \partial\Omega$, then u is nontangentially bounded at P.

2. Theorem 8.5.2 has no analogue for radial boundedness and radial limits, even on the disc. For let $E \subseteq \partial D$ be a F_σ set of first category and measure 2π. Let $f : D \to \mathbb{C}$ be given by $f(z) = \sin[1/(1 - |z|)]$. Then f is continuous, bounded by 1 on D, and does not possess a radial limit at any point of ∂D. Of course f is *not* holomorphic. But by a theorem of Bagemihl and Seidel [1] (see also Exercise 8 at the end of this chapter), there is a holomorphic (!) function u on D with

$$\lim_{r \to 1^-} |u(re^{i\theta}) - f(re^{i\theta})| = 0$$

for every $e^{i\theta} \in E$. □

THEOREM 8.5.3   Let $\Omega \subset\subset \mathbb{C}^n$ have C^2 boundary. Let $f \in H^p(\Omega), 0 < p \leq \infty$. Then f has nontangential boundary limit at almost every $P \in \partial\Omega$.

Proof   We may assume that $p < \infty$. The function $|f|^{p/2}$ is subharmonic and uniformly square integrable over $\partial\Omega_\epsilon, 0 < \epsilon < \epsilon_0$. So $|f|^{p/2}$ has a harmonic majorant $h \in \mathbf{h}^2(\Omega)$. Since $h \in \mathbf{h}^2$, it follows that h has almost everywhere nontangential boundary limits. Therefore, h is nontangentially bounded at almost every point of $\partial\Omega$. As a result, $|f|^{p/2}$, and therefore $|f|$ itself, is nontangentially bounded at almost every point of $\partial\Omega$. So, by 8.5.2, f has a nontangential boundary limit $\tilde{f}$ defined at almost every point of $\partial\Omega$. □

We conclude this section with a recasting of the ideas in the proof of 8.5.3 to make more explicit the role of the maximal function. We first need two lemmas.

LEMMA 8.5.4   Let $\{X, \mu\}$ be a measure space with $\mu \geq 0$. Let $f \geq 0$ on X be measurable and $0 < p < \infty$. Then

$$\int_X f(x)^p d\mu(x) = \int_0^\infty ps^{p-1}\mu_f(s)\, ds = -\int_0^\infty s^p d\mu_f(s),$$

where $\mu_f(s) \equiv \mu\{x : f(x) \geq s\}, 0 \leq s < \infty$.

Proof   We have

$$
\begin{aligned}
\int_X f(x)^p d\mu(x) &= \int_X \int_0^{f(x)^p} dt\, d\mu(x) \\
&\overset{\text{(Fubini)}}{=} \int_0^\infty \int_{\{x:f(x)^p \geq t\}} d\mu(x)\, dt \\
&\overset{(t=s^p)}{=} \int_0^\infty ps^{p-1}\mu\{x:f(x) \geq s\}\, ds \\
&\overset{\text{(parts)}}{=} -\int_0^\infty s^p\, d\mu_f(x). \qquad\qquad \square
\end{aligned}
$$

LEMMA 8.5.5   The maximal operator M_1 is bounded from $L^2(\partial\Omega)$ to $L^2(\partial\Omega)$.

Proof   We know that M_1 maps L^∞ to L^∞ (trivially) and is weakly bounded on L^1. After normalizing by a constant, we may suppose that

$$\|M_1 f\|_{L^\infty} \leq \frac{1}{3}\|f\|_{L^\infty}, \qquad \text{all } f \in L^\infty(\partial\Omega). \qquad (8.5.5.1)$$

Also we may assume that

$$\sigma\{\zeta \in \partial\Omega : |M_1 f(\zeta)| > \lambda\} \leq \frac{\|f\|_{L^1}}{\lambda}, \qquad \text{all } \lambda > 0, \quad f \in L^1(\partial\Omega). \qquad (8.5.5.2)$$

If $f \in L^2(\partial\Omega), \|f\|_{L^2} = 1, 0 < t < \infty$, we write

$$f(\zeta) = f(\zeta) \cdot \chi_{\{\zeta:|f(\zeta)|<t\}} + f(\zeta) \cdot \chi_{\{\zeta:|f(\zeta)|\geq t\}} \equiv f_1^t(\zeta) + f_2^t(\zeta).$$

Then

$$
\begin{aligned}
\|M_1 f\|_{L^2}^2 &= 2\int_0^\infty \sigma_{M_1 f}(\lambda)\lambda\, d\lambda \\
&\leq \int_0^\infty \sigma_{M_1 f_2^\lambda}(\lambda/2)\lambda\, d\lambda + \int_0^\infty \sigma_{M_1 f_1^\lambda}(\lambda/2)\lambda\, d\lambda \\
&\equiv I + II.
\end{aligned}
$$

Now

$$
\begin{aligned}
I &\leq 2\int_0^\infty (\|f_2^\lambda\|_{L^1}/\lambda)\lambda\, d\lambda = 2\int_0^\infty \int_\lambda^\infty \sigma_f(t)\, dt\, d\lambda \\
&= 2\int_0^\infty \sigma_f(t)\int_0^t d\lambda\, dt = 2\int_0^\infty \sigma_f(t) \cdot t\, dt = \|f\|_{L^2}^2.
\end{aligned}
$$

Furthermore, notice that since $|f_1^\lambda(\zeta)| < \lambda$ for all ζ, it follows that $|M_1 f_1^\lambda(\zeta)| \leq \lambda/3$. Hence $\sigma_{M_1 f_1^\lambda}(\lambda/2) \equiv 0$. Therefore, $II = 0$. That finishes the proof. $\square$

Let us now examine more closely the interplay between harmonic majorants and maximal functions to obtain a quantitative version of the nontangential boundedness. Let notation be as in Theorem 8.5.3. Let h be the least harmonic majorant for $|f|^{p/2}$ (assuming that $p < \infty$). Notice that the dominated convergence theorem (or Fatou's lemma) implies that $\tilde{f} \in L^p(\partial\Omega)$. (*Exercise:* For a rigorous proof, you must either consider the Jacobian of π_ϵ^{-1} or restrict attention to the Ω_j's). Now, for $\alpha > 1$ fixed and $P \in \partial\Omega$, we have

$$f_1^{*,\alpha}(P)^{p/2} \equiv \sup_{x\in\Gamma_\alpha(P)} |f(x)|^{p/2} \leq \sup_{x\in\Gamma_\alpha(P)} |h(x)| \leq C_\alpha M_1 \tilde{h}(P).$$

Therefore,

$$
\begin{aligned}
\|f_1^{*,\alpha}\|_{L^p(\partial\Omega)}^p &\leq C_\alpha \|M_1\tilde{h}\|_{L^2(\partial\Omega)}^2 \\
&\leq C \cdot C_\alpha \|\tilde{h}\|_{L^2(\partial\Omega)}^2 \\
&\leq C_\alpha' \sup_\epsilon \int_{\partial\Omega_\epsilon} |f(\zeta)|^p \, d\sigma(\zeta) \\
&\leq C_\alpha'' \int_{\partial\Omega} |\tilde{f}(\zeta)|^p \, d\sigma(\zeta) \\
&= C_\alpha'' \|\tilde{f}\|_{L^p(\partial\Omega)}^p.
\end{aligned}
\tag{8.5.6}
$$

Inequality (8.5.6) is valid for $0 < p < \infty$. It is central to the so-called real variable theory of H^p spaces. For instance, one has:

THEOREM 8.5.7   Let u be a real harmonic function on $D \subseteq \mathbb{C}$. Let $0 < p < \infty$. Then u is the real part of an $f \in H^p(D)$ if and only if, for some $\alpha > 1$,

$$\|u_1^{*,\alpha}\|_{L^p(\partial D)} < \infty.$$

Under these circumstances, $\|u_1^{*,\alpha}\|_{L^p(\partial\Omega)} \cong \|f\|_{H^p(D)}$.

Theorem 8.5.7 was originally proved by methods of Brownian motion (D. Burkholder, R. Gundy, and M. Silverstein [1]). C. Fefferman and E. M. Stein [1] gave a real-variable proof and extended the result in an appropriate sense to $\mathbb{R}^N$. The situation in several complex variables is rather more complicated (see J. Garnett and R. Latter [1] and S. G. Krantz and D. Ma [1]).

This ends our digression about the real-variable aspects of boundary behavior of harmonic and holomorphic functions. In the next section, we pick up the thread of Proposition 8.5.1 and prove a theorem that is strictly stronger. This requires a new notion of convergence.

8.6 Admissible Convergence

Let $B \subseteq \mathbb{C}^n$ be the unit ball. The Poisson kernel for the ball has the form

$$P(z, \zeta) = c_n \frac{1 - |z|^2}{|z - \zeta|^{2n}},$$

whereas the Poisson-Szegö kernel has the form

$$\mathcal{P}(z, \zeta) = c_n \frac{(1 - |z|^2)^n}{|1 - z \cdot \bar{\zeta}|^{2n}}.$$

As we know, an analysis of the convergence properties of these kernels entails dominating them by appropriate maximal functions. The maximal function involves the use of certain balls, and the shape of the ball should be compatible with the singularity of the kernel. That is why, when we study the *real analysis* of the Poisson kernel, we consider balls of the form

$$\beta_1(\zeta, r) = \{\xi \in \partial B : |\xi - \zeta| < r\}, \quad \zeta \in \partial B, r > 0.$$

In studying the *complex analysis* of the Poisson-Szegö kernel (equivalently, the Szegö kernel), it is appropriate to use the balls

$$\beta_2(\zeta, r) = \{\xi \in \partial B : |1 - \xi \cdot \bar{\zeta}| < r\}, \quad \zeta \in \partial B, \quad r > 0.$$

These new nonisotropic balls are fundamentally different from the classical (or *isotropic*) balls β_1, as we shall now see. Assume without loss of generality that $\zeta = \mathbb{1} = (1, 0, \ldots, 0)$. Write $z' = (z_2, \ldots, z_n)$. Then

$$\beta_2(\mathbb{1}, r) = \{\xi \in \partial B : |1 - \xi_1| < r\}.$$

Notice that, for $\xi \in \partial B$,

$$
\begin{aligned}
|\xi'|^2 &= 1 - |\xi_1|^2 \\
&= (1 - |\xi_1|)(1 + |\xi_1|) \\
&\leq 2|1 - \xi_1|;
\end{aligned}
$$

hence

$$\beta_2(\mathbb{1}, r) \subseteq \{\xi \in \partial B : |1 - \xi_1| < r, |\xi'| < \sqrt{2r}\}.$$

A similar computation shows that

$$
\begin{aligned}
\beta_2(\mathbb{1}, r) &= \{\xi \in \partial B : |1 - \xi_1| < r, |\xi'| = \sqrt{1 - |\xi_1|^2}\} \\
&\supseteq \partial B \cap \{\xi : |\operatorname{Im}\xi_1| < r/2, 1 - r/2 < \operatorname{Re}\xi_1 < 1, |\xi'| < \sqrt{r}\}.
\end{aligned}
$$

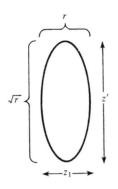

FIGURE 8.6

In short, the balls we are considering have dimension $\sim r$ in the complex space containing ν_1 and dimension $\sim \sqrt{r}$ in the orthogonal complement (see Figure 8.6). The word nonisotropic means that we have different geometric behavior in different directions.

In the classical setup, we considered *cones* modeled on the balls β_1:

$$\Gamma_\alpha(P) = \{z \in B : |z - P| < \alpha(1 - |z|)\}, \quad P \in \partial B, \alpha > 1.$$

In the new situation we consider *admissible regions* modeled on the balls β_2 :

$$\mathcal{A}_\alpha(P) = \{z \in B : |1 - z \cdot \bar{P}| < \alpha(1 - |z|)\}.$$

The reader should calculate that $\mathcal{A}_\alpha(\mathcal{P})$ provides *nontangential* approach to P in complex normal directions but *parabolic* approach in complex tangential directions.

Our new theorem about boundary limits of H^p functions is as follows:

THEOREM 8.6.1   Let $f \in H^p(B), 0 < p \leq \infty$. Let $\alpha > 1$. Then the limit

$$\lim_{\mathcal{A}_\alpha(P) \ni z \to P} f(P) \equiv \tilde{f}(P)$$

exists for σ-almost every $P \in \partial B$.

Since the Poisson-Szegö kernel is known explicitly on the ball, then for $p \geq 1$ the proof is deceptively straightforward: One defines, for $P \in \partial B$ and $g \in L^1(\partial B)$,

$$M_2 g(P) = \sup_{r>0} \frac{1}{\sigma(\beta_2(P,r))} \int_{\beta_2(P,r)} |g(\zeta)| d\sigma(\zeta).$$

Also set $g(z) = \int_{\partial B} \mathcal{P}(z,\zeta)g(\zeta)\,d\sigma(\zeta)$ for $z \in B$. Then, by explicit computation similar to the proof of Proposition 8.1.10,

$$g_2^{*,\alpha}(P) \equiv \sup_{z \in \mathcal{A}_\alpha(P)} |g(z)| \leq C_\alpha M_2 g(P), \qquad \text{all } g \in L^1(\partial B).$$

This crucial fact, together with appropriate estimates on the operator M_2, enables one to complete the proof along classical lines for $p \geq 1$. For $p < 1$, matters are more subtle.

We forgo the details of the preceding argument on B and instead develop the machinery for proving an analogue of Theorem 8.6.1 on an arbitrary C^2 bounded domain in $\mathbb{C}^n$. In this generality, there is no hope of obtaining an explicit formula for the Poisson-Szegő kernel; indeed, there are no known techniques for obtaining estimates for this kernel on arbitrary domains (however, see C. Fefferman [1] and A. Nagel, J. P. Rosay, E. M. Stein, and S. Wainger [1] for estimates on strongly pseudoconvex domains and on domains of finite type in $\mathbb{C}^2$). Therefore, we must develop more geometric methods that do not rely on information about kernels. The results that we present were proved on the ball and on bounded symmetric domains by A. Koranyi [1,2]. Many of these ideas were also developed independently in Gong Sheng [1, 2]. All the principal ideas for arbitrary Ω are due to E. M. Stein [2].

Our tasks, then, are as follows: (1) to define the balls β_2 on the boundary of an arbitrary Ω; (2) to define admissible convergence regions $\mathcal{A}_\alpha$; (3) to obtain appropriate estimates for the corresponding maximal function; and (4) to couple the maximal estimates, together with the fact that "radial" boundary values are already known to exist (see Proposition 8.5.1) to obtain the admissible convergence result.

If z, w are vectors in $\mathbb{C}^n$, we continue to write $z \cdot \bar{w}$ to denote $\sum_j z\bar{w}_j$. (*Warning:* It is also common in the literature to use the notation $z \cdot w = \sum_j z_j \bar{w}_j$.) Also, for $\Omega \subseteq \mathbb{C}^n$ a domain with C^2 boundary, $P \in \partial\Omega$, we let ν_P be the unit outward normal at P. Let $\mathbb{C}\nu_P$ denote the complex line generated by $\nu_P : \mathbb{C}\nu_P = \{\zeta\nu_P : \zeta \in \mathbb{C}\}$.

By dimension considerations, if $T_P(\partial\Omega)$ is the $(2n-1)$-dimensional real tangent space to $\partial\Omega$ at P, then $\ell = \mathbb{C}\nu_P \cap T_P(\partial\Omega)$ is a (one-dimensional) real line. Let

$$\begin{aligned}\mathcal{T}_P(\partial\Omega) &= \{z \in \mathbb{C}^n : z \cdot \bar{\nu}_P = 0\} \\ &= \{z \in \mathbb{C}^n : z \cdot \bar{w} = 0 \ \forall w \in \mathbb{C}\nu_P\}.\end{aligned}$$

A fortiori, $\mathcal{T}_P(\partial\Omega) \subseteq T_P(\partial\Omega)$. If $z \in \mathcal{T}_P(\partial\Omega)$, then $iz \in \mathcal{T}_P(\partial\Omega)$. Therefore, $\mathcal{T}_P(\partial\Omega)$ may be thought of as an $(n-1)$-dimensional complex subspace of $T_P(\partial\Omega)$. Clearly, $\mathcal{T}_P(\partial\Omega)$ is the complex subspace of $T_P(\partial\Omega)$ of maximal dimension. It contains all complex subspaces of $T_P(\partial\Omega)$. (The reader should check that $\mathcal{T}_P(\partial\Omega)$ is the same complex tangent space that was introduced when we first studied the Levi form.) We may think of $\mathcal{T}_P(\partial\Omega)$ as the real orthogonal complement in $T_P(\partial\Omega)$ of ℓ.

Now let us examine the matter from another point of view. The complex structure is nothing other than a linear operator J on $\mathbb{R}^{2n}$ that assigns to $(x_1, x_2, \ldots, x_{2n-1}, x_{2n})$ the vector $(-x_2, x_1, -x_4, x_3, \ldots, -x_{2n}, x_{2n-1})$ (think of multiplication by i). With this in mind, we have that $J : \mathcal{T}_P(\partial\Omega) \to \mathcal{T}_P(\partial\Omega)$ both injectively and surjectively. Notice that $J\nu_P \in T_P(\partial\Omega)$, where $J(J\nu_P) = -\nu_P \notin T_P(\partial\Omega)$. We call $\mathbb{C}\nu_P$ the *complex normal* space to $\partial\Omega$ at P and $\mathcal{T}_P(\partial\Omega)$ the *complex tangent space* to $\partial\Omega$ at P. Let $\mathcal{N}_P = \mathbb{C}\nu_P$. Then we have $\mathcal{N}_P \perp \mathcal{T}_P$ and

$$\mathbb{C}^n = \mathcal{N}_P \oplus_{\mathbb{C}} \mathcal{T}_P$$
$$T_P = \mathbb{R}J\nu_P \oplus_{\mathbb{R}} \mathcal{T}_P.$$

EXAMPLE   Let $\Omega = B \subseteq \mathbb{C}^n$ be the unit ball and $P = 1 \in \partial\Omega$. Then $\mathbb{C}\nu_P = \{(z_1, 0, \ldots, 0) : z_1 \in \mathbb{C}\}$ and $\mathcal{T}_P = \{(0, z') : z' \in \mathbb{C}^{n-1}\}$.   □

Exercises for the Reader

1. Let $\Omega \subseteq \mathbb{C}^n$ be a domain.   Let J be the real linear operator on $\mathbb{R}^{2n}$ that gives the complex structure. Let $P \in \partial\Omega$. Let $z = (z_1, \ldots, z_n) = (x_1 + iy_1, \ldots, x_n + iy_n) = (x_1, y_1, \ldots, x_n, y_n)$ be an element of $\mathbb{C}^n \cong \mathbb{R}^{2n}$. The following are equivalent:
   (i) $w \in \mathcal{T}_P(\Omega)$.
   (ii) $Jw \in \mathcal{T}_P(\Omega)$.
   (iii) $Jw \perp \nu_P$ and $w \perp \nu_P$.

2. With notation as in the previous exercise, let $A = \sum_j a_j(z)\partial/\partial z_j$, $B = \sum_j b_j(z)\partial/\partial z_j$ satisfy $A\rho|_{\partial\Omega} = 0$, $B\rho|_{\partial\Omega} = 0$, where ρ is any defining function for Ω. Then the vector field $[A, B]$ has the same property. (However, note that $[A, \bar{B}]$ does not annihilate ρ on $\partial\Omega$ if Ω is the ball, for instance.) Therefore, the holomorphic part of $\mathcal{T}_P$ is integrable (see G. B. Folland and J. J. Kohn [1]).

3. If $\Omega = B \subseteq \mathbb{C}^2, P = (x_1 + iy_1, x_2 + iy_2) \approx (x_1, y_1, x_2, y_2) \in \partial B$, then $\nu_P = (x_1, y_1, x_2, y_2)$ and $J\nu_P = (-y_1, x_1, -y_2, x_2)$. Also $\mathcal{T}_P$ is spanned over $\mathbb{R}$ by $(y_2, x_2, -y_1, -x_1)$ and $(-x_2, y_2, x_1, -y_1)$.

The next definition is best understood in light of the foregoing discussion and the definition of $\beta_2(P, r)$ in the boundary of the unit ball B. Let $\Omega \subset\subset \mathbb{C}^n$ have C^2 boundary.   For $P \in \partial\Omega$, let $\pi_P : \mathbb{C}^n \to \mathcal{N}_P$ be (real or complex) orthogonal projection.

DEFINITION 8.6.2   If $P \in \partial\Omega$, let

$$\beta_1(P, r) = \{\zeta \in \partial\Omega : |\zeta - P| < r\};$$
$$\beta_2(P, r) = \{\zeta \in \partial\Omega : |\pi_P(\zeta - P)| < r, |\zeta - P| < r^{1/2}\}.$$

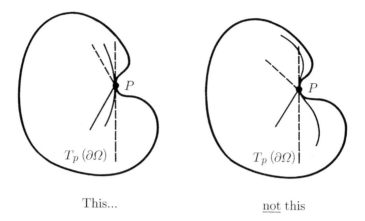

This...   not this

FIGURE 8.7

Exercise for the Reader

The ball $\beta_2(P, r)$ has diameter $\sim \sqrt{r}$ in the $(2n - 2)$ complex tangential directions and diameter $\sim r$ in the one (normal) direction. Therefore, $\sigma(\beta_2(P, r)) \approx (\sqrt{r})^{2n-2} \cdot r \approx C r^n$.

If $z \in \Omega, P \in \partial\Omega$, we let

$$\delta_P(z) = \min\{\text{dist}\,(z, \partial\Omega), \text{dist}\,(z, T_P(\Omega))\}.$$

Notice that if Ω is convex, then $\delta_P(z) = \delta_\Omega(z)$.

DEFINITION 8.6.3   If $P \in \partial\Omega, \alpha > 1$, let

$$\mathcal{A}_\alpha = \{z \in \Omega : |(z - P) \cdot \bar{\nu}_P| < \alpha\delta_P(z), |z - P|^2 < \alpha\delta_P(z)\}.$$

Notice that δ_P is used because near nonconvex boundary points we still want $\mathcal{A}_\alpha$ to have the fundamental geometric shape of (paraboloid $\times$ cone), as shown in Figure 8.7.

DEFINITION 8.6.4   If $f \in L^1(\partial\Omega)$ and $P \in \partial\Omega$, then we define

$$M_j f(P) = \sup_{r>0} \sigma(\beta_j(P, r))^{-1} \int_{\beta_j(P,r)} |f(\zeta)| d\sigma(\zeta), \qquad j = 1, 2.$$

DEFINITION 8.6.5   If $f \in C(\Omega), P \in \partial\Omega$, then we define

$$f_2^{*,\alpha}(P) = \sup_{z \in \mathcal{A}_\alpha(P)} |f(z)|.$$

The first step of our program is to prove an estimate for M_2. This will require a covering lemma (indeed, it is known that weak type estimates for operators like M_j are logically equivalent to covering lemmas—see A. Cordoba and R. Fefferman [1]). We exploit a rather general paradigm due to K. T. Smith [1].

DEFINITION 8.6.6   Let X be a topological space equipped with a positive Borel measure m, and suppose that for each $x \in X$, each $r > 0$, there is a "ball" $B(x, r)$. The "K. T. Smith axioms" for this setting are as follows.

(8.6.6.1) Each $B(x, r)$ is an open set of finite measure that contains x.

(8.6.6.2) If $r_1 \le r_2$, then $B(x, r_1) \subseteq B(x, r_2)$.

(8.6.6.3) There is a constant $c_0 > 0$ such that if $B(x, r) \cap B(y, s) \ne \emptyset$ and $r \ge s$, then $B(x, c_0 r) \supseteq B(y, s)$.

(8.6.6.4) There is a constant K such that $m(B(x_0, c_0 r)) \le K m(B(x_0, r))$ for all r.

Now we have the following.

THEOREM 8.6.7   Let the topological space X, measure m, and balls $B(x, r)$ be as in the definition. Let K be a compact subset of X and $\{B(x_\alpha, r_\alpha)\}_{\alpha \in A}$ a covering of K by balls. Then there is a finite pairwise disjoint subcollection $B(x_{\alpha_1}, r_{\alpha_1}), \ldots, B(x_{\alpha_m}, r_{\alpha_m})$ such that $K \subseteq \cup_{j=1}^{k} B(x_{\alpha_j}, c_0 r_{\alpha_j})$.

It follows that if we define

$$Mf = \sup_{r > 0} (B(x, r))^{-1} \int_{B(x, r)} |f(t)| \, dm(t), \qquad f \in L^1(X, dm),$$

then

$$m\{x : Mf(x) > \lambda\} \le C \frac{\|f\|_{L^1}}{\lambda}.$$

Proof   The proof is left as an exercise for the reader. Imitate the proofs of Lemmas 8.4.2 and 8.4.4. $\square$

Thus we need to see that the $\beta_2(P, r)$ on $X = \partial\Omega$ with $m = \sigma$ satisfy (8.6.6.1)—(8.6.6.4). Now (8.6.6.1) and (8.6.6.2) are trivial. Also, (8.6.6.4) is easy if one uses the fact that $\partial\Omega$ is C^2 and compact (use the exercise for the reader following Definition 8.6.2). Thus it remains to check (8.6.6.3) (in many applications, this is the most difficult property to check).

Suppose that $\beta_2(x, r) \cap \beta_2(y, s) \ne \emptyset$. Thus there is a point $a \in \beta_2(x, r) \cap \beta_2(y, s)$. We may assume that $r = s$ by (8.6.6.2). We thus have $|x - a| \le r^{1/2}, |y - a| \le r^{1/2}$; hence $|x - y| \le 2r^{1/2}$. Let the constant $M \ge 2$ be chosen so that $\|\pi_x - \pi_y\| \le M|x - y|$. (We must use here the fact that the boundary is C^2.) We claim that $\beta_2(x, (3 + 4M)r) \supseteq \beta_2(y, r)$. To see this, let $v \in \beta_2(y, r)$.

The easy half of the estimate is

$$|x - v| \leq |x - y| + |y - v| \leq 2r^{1/2} + r^{1/2} = 3r^{1/2}.$$

Also,

$$\pi_x(x - v) = \pi_x(x - a) + \pi_y(a - v) + \{\pi_x - \pi_y\}(a - v).$$

Therefore,

$$
\begin{aligned}
|\pi_x(x - v)| &\leq r + 2r + \|\pi_x - \pi_y\| |a - v| \\
&\leq 3r + M|x - y| \cdot |a - v| \\
&\leq 3r + M2r^{1/2}(|a - y| + |y - v|) \\
&\leq (3 + 4M)r.
\end{aligned}
$$

This proves (8.6.6.3). Thus we have the following.

COROLLARY 8.6.8   If $f \in L^1(\partial\Omega)$, then

$$\sigma\{\zeta \in \omega : M_2 f(\zeta) > \lambda\} \leq C \frac{\|f\|_{L^1(\partial\Omega)}}{\lambda}, \qquad \text{all} \quad \lambda > 0.$$

Proof   Apply the theorem.   □

COROLLARY 8.6.9   The operator M_2 maps $L^2(\partial\Omega)$ to $L^2(\partial\Omega)$ boundedly.

Proof   The proof is left as an exercise. Use the technique of Lemma 8.5.5.   □

Remark: In fact, there is a general principal at work here. If a linear operator T is bounded from L^∞ to L^∞ and is also weak type $(1, 1)$, then it is bounded on $L^p, 1 < p < \infty$. Of course M_2 is *not* linear. Instead it is *sublinear*: $M_2(f+g)(x) \leq M_2 f(x) + M_2 g(x)$. This is sufficient for the result stated.

The results mentioned in the last paragraph are instances of "interpolation" theorems for operators. For a more thorough discussion of this topic, see E. M. Stein and G. Weiss [1] or J. Bergh and J. Löfström [1].   □

The next lemma is the heart of the matter: It is the technical device that allows us to estimate the behavior of a holomorphic function in the interior (in particular, on an admissible approach region) in terms of a maximal function on the boundary. The argument comes from Stein [2] and Barker [1].

LEMMA 8.6.10   Let $u \in C(\bar{\Omega})$ be nonnegative and plurisubharmonic on Ω. Define $f = u|_{\partial\Omega}$. Then

$$u_2^{*,\alpha}(P) \leq C_\alpha M_2(M_1 f)(P)$$

for all $P \in \partial\Omega$ and any $\alpha > 1$.

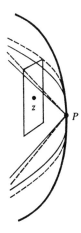

FIGURE 8.8

Proof   After rotating and translating coordinates, we may suppose that $P = 0$ and $\nu_P = (1 + i0, 0, \ldots, 0)$. Let $\alpha' > \alpha$. Then there is a small positive constant k such that if $z = (x_1 + iy_1, z_2, \ldots, z_n) \in \mathcal{A}_\alpha(P)$, then $\mathcal{D}(z) = D(z_1, -kx_1) \times D^{n-1}((z_2, \ldots, z_n), \sqrt{-kx_1}) \subseteq \mathcal{A}_{\alpha'}(P)$ (see Figure 8.8).

We restrict attention to $z \in \Omega$ so close to $P = 0$ that the projection along ν_P given by

$$z = (x_1 + iy_1, \ldots, x_n + iy_n) \to (\tilde{x}_1 + iy_1, x_2 + iy_2, \ldots, x_n + iy_n) \equiv \tilde{z} \in \partial\Omega$$

makes sense. (Observe that points z that are far from $P = 0$ are trivial to control using our estimates on the Poisson kernel.) The projection of $\mathcal{D}(z)$ along ν_P into the boundary lies in a ball of the form $\beta_2(\tilde{z}, Kx_1)$—*this observation is crucial.*

Notice that the subharmonicity of u implies that $u(z) \leq Pf(z)$. Also, there is a $\beta > 1$ such that $z \in \mathcal{A}_{\alpha'}(0) \Rightarrow z \in \Gamma_\beta(\tilde{z})$. Therefore, the standard argument leading up to (8.4.1.1) yields that

$$|u(z)| \leq |Pf(z)| \leq C_\alpha M_1 f(\tilde{z}). \tag{8.6.10.1}$$

Now we bring the complex analysis into play. For we may exploit the plurisubharmonicity of $|u|$ on $\mathcal{D}(z)$ by invoking the subaveraging Property 8 of Theorem 2.1.4 in each dimension in succession. Thus

$$
\begin{aligned}
|u(z)| &\leq \left(\pi |kx_1|^2\right)^{-1} \cdot \left(\pi (\sqrt{-kx_1})^2\right)^{-(n-1)} \int_{\mathcal{D}(z)} |u(\zeta)| \, dV(\zeta) \\
&= Cx_1^{-n-1} \int_{\mathcal{D}(z)} |u(\zeta)| \, dV(\zeta).
\end{aligned}
$$

Notice that if $z \in \mathcal{A}_\alpha(P)$, then each ζ in the last integrand is in $\mathcal{A}_{\alpha'}(P)$. Thus the last line is less than or equal to

$$
\begin{aligned}
C'x_1^{-n-1} \int_{\mathcal{D}(z)} M_1 f(\tilde{\zeta})\, dV(\zeta) &\leq C''x_1^{-n-1} \cdot x_1 \int_{\beta_2(\tilde{z},Kx_1)} M_1 f(t)\, d\sigma(t) \\
&\leq C'''x_1^{-n} \int_{\beta_2(0,K'x_1)} M_1 f(t)\, d\sigma(t) \\
&\leq C''''\left(\sigma\left(\beta_2(0,K'x_1)\right)\right)^{-1} \\
&\quad \times \int_{\beta_2(0,K'x_1)} M_1 f(t)\, d\sigma(t) \\
&\leq C''''M_2(M_1 f)(0). \qquad \square
\end{aligned}
$$

Now we can prove our main result:

THEOREM 8.6.11   Let $0 < p \leq \infty$. Let $\alpha > 1$. If $\Omega \subset\subset \mathbb{C}^n$ has C^2 boundary and $f \in H^p(\Omega)$, then, for σ-almost every $P \in \partial\Omega$,

$$
\lim_{\mathcal{A}_\alpha(P)\ni z \to P} f(z)
$$

exists.

Proof   We already know that the limit exists almost everywhere *in the special sense of* Proposition 8.5.1. Call the limit function $\tilde{f}$. We need consider only the case $p < \infty$. Let $\Omega = \cup_{j=1}^k \Omega_j$ as usual. It suffices to concentrate on Ω_1. Let $\nu = \nu_1$ be the outward normal given by (8.3.5). Then, by (8.5.1), the Lebesgue dominated convergence theorem implies that, for $\partial\tilde\Omega_1 \equiv \partial\Omega \cap \partial\Omega_1$,

$$
\lim_{\epsilon \to 0} \int_{\partial\tilde\Omega_1} |f(\zeta - \epsilon\nu) - \tilde{f}(\zeta)|^p d\sigma(\zeta) = 0. \tag{8.6.11.1}
$$

For each $j,k \in \mathbb{N}$, consider the function $f_{j,k} : \Omega_1 \to \mathbb{C}$ given by

$$
f_{j,k}(z) = |f(z - \nu/j) - f(z - \nu/k)|^{p/2}.
$$

Then $f_{j,k} \in C(\bar\Omega_1)$ and is plurisubharmonic on Ω_1. Therefore, a trivial variant of Lemma 8.6.10 yields

$$
\begin{aligned}
\int_{\partial\tilde\Omega_1} |(f_{j,k})_2^{*,\alpha}(\zeta)|^2 d\sigma(\zeta) &\leq C_\alpha \int_{\partial\tilde\Omega_1} |M_2(M_1 f_{j,k}(\zeta))|^2 d\sigma(\zeta) \\
&\leq C'_\alpha \int_{\partial\tilde\Omega_1} |M_1 f_{j,k}(\zeta)|^2 d\sigma(\zeta) \\
&\leq C''_\alpha \int_{\partial\tilde\Omega_1} |f_{j,k}(\zeta)|^2 d\sigma(\zeta),
\end{aligned}
$$

where we have used Corollary 8.6.9, Lemma 8.5.5, and the *proof* of Lemma 8.5.5. Now let $j \to \infty$ and apply (8.6.11.1) to obtain

$$\int_{\partial \tilde{\Omega}_1} \sup_{z \in \mathcal{A}_\alpha(\zeta)} |f(z) - f(z - \nu/k)|^p d\sigma(\zeta) \leq C_\alpha'' \int_{\partial \tilde{\Omega}_1} |\tilde{f}(\zeta) - f(\zeta - \nu/k)|^p d\sigma(\zeta).$$

$$(8.6.11.2)$$

Let $\epsilon > 0$. Then

$$\sigma\{\zeta \in \partial \tilde{\Omega}_1 : \limsup_{\mathcal{A}_\alpha(\zeta) \ni z \to \zeta} |f(z) - \tilde{f}(\zeta)| > \epsilon\}$$

$$\leq \quad \sigma\{\zeta \in \partial \tilde{\Omega}_1 : \limsup_{\mathcal{A}_\alpha(\zeta) \ni z \to \zeta} |f(z) - f(z - \nu/k)| > \epsilon/3\}$$

$$+ \sigma\{\zeta \in \partial \tilde{\Omega}_1 : \limsup_{\mathcal{A}_\alpha(\zeta) \ni z \to \zeta} |f(z - \nu/k) - f(\zeta - \nu/k)| > \epsilon/3\}$$

$$+ \sigma\{\zeta \in \partial \tilde{\Omega}_1 : \limsup_{\mathcal{A}_\alpha(\zeta) \ni z \to \zeta} |f(\zeta - \nu/k) - \tilde{f}(\zeta)| > \epsilon/3\}$$

$$\leq \quad C \int_{\partial \tilde{\Omega}_1} \sup_{z \in \mathcal{A}_\alpha(\zeta)} |f(z) - f(z - \nu/k)|^p d\sigma(\zeta)/\epsilon^p$$

$$+ 0$$

$$+ C \int_{\partial \tilde{\Omega}_1} |\tilde{f}(\zeta) - f(\zeta - \nu/k)|^p d\sigma(\zeta)/\epsilon^p$$

where we have used (the proof of) Chebycheff's inequality. By 8.6.11.2, the last line does not exceed

$$C' \int_{\partial \tilde{\Omega}_1} |\tilde{f}(\zeta) - f(\zeta - \nu/k)|^p \, d\sigma(\zeta)/\epsilon^p.$$

Now (8.6.11.1) implies that, as $k \to \infty$, this last quantity tends to 0. Since $\epsilon > 0$ was arbitrary, we conclude that

$$\limsup_{\mathcal{A}_\alpha(\zeta) \ni z \to \zeta} |f(z) - \tilde{f}(\zeta)| = 0$$

almost everywhere. $\qquad\square$

The theorem says that f has "admissible limits" at almost every boundary point of Ω. The considerations in the next section (indeed, an inspection of the arguments in the present section) suggest that Theorem 8.6.11 is best possible only for strongly pseudoconvex domains. At the boundary point $(1,0)$ of the domain $\{(z_1, z_2) : |z_1|^2 + |z_2|^{2m} < 1\}$, the natural interior polydiscs to study are of the form

$$\{(1 - \delta + \xi_1, \xi_2) : |\xi_1| < c \cdot \delta, |\xi_2| < c \cdot \delta^{1/2m}\}.$$

This observation, together with an examination of the proof of 8.6.10, suggests that the aperture in complex tangential directions of the approach regions should vary from boundary point to boundary point—and this aperture should depend on the Levi geometry of the point. A theory of boundary behavior for H^p functions taking these observations into account, for a special class of domains in $\mathbb{C}^2$, is enunciated in A. Nagel, E. M. Stein, and S. Wainger [1]. A more general paradigm for theories of boundary behavior of holomorphic functions is developed in S. G. Krantz [17]. Related ideas also appear in S. G. Krantz [16]. The key tool in the last two references is the Kobayashi metric, a geometric device that we study in Chapter 11.

8.7   The Lindelöf Principle

In this section we formulate a version of the Lindelöf principle for domains in $\mathbb{C}^1$ and $\mathbb{C}^n$. It is not optimal but serves to illustrate the main ideas. For further details, see E. Hille [1], J. Cima and S. G. Krantz [1], E. Chirka [1], and O. Lehto and K. I. Virtanen [1].

The classical Lindelöf principle is as follows.

THEOREM 8.7.1   Let $D \subseteq \mathbb{C}$ be the unit disc. Let $f \in H^\infty(D)$. Suppose that $\lim_{r \to 1^-} f(re^{i\theta}) = \ell \in \mathbb{C}$ exists. Then f has nontangential limit ℓ at $e^{i\theta}$.

Proof   Fix $\alpha > 1$. We may assume that $\ell = 0$ and $\theta = 0$. Define, for $j \in \mathbb{N}$,

$$\Omega_j = \{z \in \mathbb{C} : 1 - 2^{-j+1}/(2\alpha) \leq \operatorname{Re} z \leq 1 - 2^{-j-1}/(2\alpha), |\operatorname{Im} z| < 2^{-j+1}\}.$$

For $j_0(\alpha)$ sufficiently large,

$$D \supseteq \bigcup_{j=j_0}^{\infty} \Omega_j \supseteq \Gamma_\alpha(1) \cap \{z \subset D : \operatorname{Re} z \geq 1 - 2^{-j_0+1}/(2\alpha)\}.$$

Also, the map $\phi_j : \Omega_j \to \Omega_{j_0}$ given by $\phi_j(z) = 2^{j-j_0}(z-1) + 1$ is biholomorphic. There is a K compact in Ω_{j_0} such that

$$\bigcup_{j \geq j_0} \phi_j^{-1}(K) \supseteq \Gamma_\alpha(1) \cap \{z \in D : \operatorname{Re} z \geq 1 - 2^{-j_0-1}/(2\alpha)\}.$$

Now the functions $\{f \circ \phi_j^{-1}\}$ form a normal family on Ω_{j_0}. Choose a subsequence $f \circ \phi_{j_k}^{-1}$ that converges uniformly on K. Call the limit function f_0. By hypothesis, $f_0(x + i0) = 0$, all $x + i0 \in K$. Thus $f_0 \equiv 0$. Therefore, $f \circ \phi_j^{-1} \to 0$

uniformly on K. Hence

$$\lim_{\Gamma_\alpha(\mathbb{1}) \ni z \to \mathbb{1}} f(z) = 0. \qquad \qquad \square$$

It is natural to conjecture that, on the ball in $\mathbb{C}^n$, there is an analogous theorem with nontangential convergence replaced by admissible convergence. This is, unfortunately, false.

EXAMPLE   Let $B \subseteq \mathbb{C}^2$ be the ball. Let $f(z_1, z_2) = z_2^2/(1 - z_1)$ for $(z_1, z_2) \in B$. Then

$$
\begin{aligned}
|f(z_1, z_2)| &\leq \frac{|z_2|^2}{|1 - z_1|} \\
&\leq \frac{1 - |z_1|^2}{1 - |z_1|} \\
&\leq 1 + |z_1| \\
&\leq 2.
\end{aligned}
$$

Hence $f \in H^\infty(B)$. Also $\lim_{r \to 1^-} f(r\,\mathbb{1}) = 0$, where as usual $\mathbb{1} = (1, 0) \in \partial B$. But the sequence

$$z_j \equiv \left(1 - \frac{1}{j}, \frac{1}{\sqrt{2j}} \right)$$

approaches $\mathbb{1}$ admissibly (that is, through an admissible approach region) and

$$\lim_{j \to \infty} f(z_j) = \lim_{j \to \infty} \frac{1/(2j)}{1/j} = \frac{1}{2} \neq 0.$$

Thus f has no admissible limit at $\mathbb{1}$.

We conclude that there is no Lindelöf principle for admissible convergence.
$\qquad \qquad \square$

The correct substitute for admissible convergence is *hypoadmissible convergence*.

DEFINITION 8.7.2   Let $B \subseteq \mathbb{C}^n$ be the unit ball and $P \in \partial B$. Assume that $\{z^j\} \subseteq B$ satisfies $\lim_{j \to \infty} z^j = P$. We say that $\{z^j\}$ converges *hypoadmissibly* to P if there is an $\alpha > 1$ such that

$$\frac{|1 - z^j \cdot \bar{P}|}{1 - |z^j|} < \alpha \qquad \text{for all } \ j$$

and

$$\limsup_{j \to \infty} \frac{|z^j - (z^j \cdot \bar{P})P|^2}{|1 - z^j \cdot \bar{P}|} = 0.$$

In other words, $z^j \to P$ hypoadmissibly means that the smallest approach region containing $\{z^j\}$ is nontangential in the complex normal direction and is asymptotically smaller than any admissible approach region in the complex tangential directions. Now we have the following.

THEOREM 8.7.3   Let $B \subseteq \mathbb{C}^n$ be the unit ball. Let $f \in H^\infty(B)$ and fix $P \in \partial B$. If $\lim_{r \to 1^-} f(rP) = \ell \in \mathbb{C}$ exists, then, for any sequence $\{z^j\}_{j=1}^\infty \subseteq B$ that approaches P hypoadmissibly, we have

$$\lim_{j \to \infty} f(z^j) = \ell.$$

Proof   Assume without loss of generality that $P = \mathbf{1}$. For each j, write $z^j = (z_1^j, \tau^j)$, where $\tau^j = (z_2^j, \ldots, z_n^j)$. Let $n^j = (z_1^j, 0, \ldots, 0)$. Fix $\alpha > 1$ as in Definition 8.7.2.

Pick $\epsilon > 0$. Let $\omega = B \cap \{z_2 = \cdots = z_n = 0\}$ and $\hat{f} = f|_\omega$. Then $\hat{f}$ is a bounded holomorphic function on the disc ω. Also, the points $\omega \ni n^j \to 1$ nontangentially. Since $\hat{f}$ has radial limit ℓ at 1, Theorem 8.7.1 implies that there is a J so large that $j \geq J$ implies that $|f(n^j) - \ell| < \epsilon/2$.

Let $\delta^j = 1 - |n^j|$. Then the $(n-1)$-dimensional polydisc

$$\mathbf{d}^j = \left\{ (z_1^j, z_2, \ldots, z_n) : |z_i| < \sqrt{\alpha\delta^j/(2n)}, i = 2, \ldots, n \right\}$$

lies in B. Let $M = \sup_B |f|$. Then, applying the Cauchy estimates to f on $\mathbf{d}^j$, we have (since $|z^j - n^j| < \sqrt{\alpha\delta^j/(8n)}$ for j sufficiently large) that

$$
\begin{aligned}
|f(z^j) - f(n^j)| &\leq |z^j - n^j| \cdot \sup_{0 \leq t \leq 1} \left| \frac{d}{dt} f\left((1-t)z^j + tn^j\right) \right| \\
&\leq |z^j - n^j| \cdot \frac{M}{\sqrt{\alpha\delta^j/(8n)}} \\
&\leq C \cdot \frac{|z^j - (z^j \cdot \bar{P})P|}{\sqrt{|1 - z^j \cdot \bar{P}|}} \to 0
\end{aligned}
$$

as $j \to \infty$. Therefore, there is a K so large that $j \geq K$ implies

$$|f(z^j) - f(n^j)| < \frac{\epsilon}{2}.$$

Thus for $j \geq \max(J, K)$ we have

$$
\begin{aligned}
|f(z^j) - \ell| &\leq |f(z^j) - f(n^j)| + |f(n^j) - \ell| \\
&< \frac{\epsilon}{2} + \frac{\epsilon}{2} = \epsilon.
\end{aligned}
$$

$\square$

Exercise for the Reader
Let $\Omega_k \subseteq \mathbb{C}^2$ be given by $\Omega_k = \{z \in \mathbb{C}^2 : |z_1|^2 + |z_2|^{2k} < 1\}$. If $P = (e^{i\theta}, 0) \in \partial\Omega_k$, then show that there is a Lindelöf principle for P in which the approach region can be taken to be (asymptotically smaller than) size $|1 - z \cdot \bar{P}|^{1/(2k)}$ in the complex tangential directions.

Results of the type in Theorem 8.7.3 were first considered by E. Chirka [1]. Generalizations to arbitrary domains are treated in J. Cima and S. G. Krantz [1]. In the latter work, as with the work described in S. G. Krantz [16, 17], the key is to use the language of the Kobayashi metric.

8.8 Additional Tangential Phenomena: Lipschitz Spaces

In this section we prove a special case of a remarkable result due to E. M. Stein [3] (see also Greiner and Stein [1]) that reflects the same aspects of the complex structure as does the existence of admissible limits. Explicit connections between the results of this section and those in Sections 8.6 and 8.7 are explored in S. G. Krantz [16, 17] using the language of the Kobayashi metric.

Recall that if $U \subseteq \mathbb{R}^N$ is any open set, $0 < \alpha < 1$, then

$$\Lambda_\alpha(U) = \left\{ f \in C(U) : \sup_{x, x+h \in U} |f(x+h) - f(x)||h|^\alpha + \|f\|_{L^\infty(U)} \right.$$

$$\equiv \|f\|_{\Lambda_\alpha(U)} < \infty \left. \right\}.$$

If $k \in \mathbb{N}, 0 < \alpha < 1$, then

$$\Lambda_{\alpha+k}(U) = \left\{ f \in C^k(U) : \sum_{j=1}^N \left\| \frac{\partial f}{\partial x_j} \right\|_{\Lambda_{\alpha+k-1}(U)} + \|f\|_{L^\infty(U)} \right.$$

$$\equiv \|f\|_{\Lambda_{\alpha+k}(U)} < \infty \left. \right\}.$$

The definition of Λ_α when $\alpha \in \mathbb{N}$ is more technical:

$$\Lambda_1(U) = \left\{ f \in C(U) : \sup_{x, x+h, x-h \in U} \frac{|f(x+h) + f(x-h) - 2f(x)|}{|h|} \right.$$

$$+ \|f\|_{L^\infty(U)}$$

$$\equiv \|f\|_{\Lambda_1(U)} < \infty \left. \right\}.$$

This definition is ostensibly different from the more classical

$$
\mathrm{Lip}_1(U) \;=\; \left\{ f \in C(U): \; \sup_{x,x+h\in U} \frac{|f(x+h)-f(x)|}{|h|} + \|f\|_{L^\infty(U)} \right.
$$

$$
\left. \equiv \|f\|_{\mathrm{Lip}_1(U)} < \infty \right\}.
$$

In fact $\mathrm{Lip}_1(U) \subsetneq \Lambda_1(U)$. Although the space Lip_1 is rather natural in geometric applications (see H. Federer [1]), it is the space Λ_1 that is the right space for harmonic analysis and operator theory. A detailed study of Lipschitz spaces appears in S. G. Krantz [10].

We conclude this introductory material by recording the definition of Λ_k, $k \in \mathbb{N}$:

$$
\Lambda_k = \left\{ f \in C^{k-1}(U): \sum_{j=1}^{N} \left\| \frac{\partial f}{\partial x_j} \right\|_{\Lambda_{k-1}(U)} + \|f\|_{L^\infty(U)} \equiv \|f\|_{\Lambda_k} < \infty \right\}.
$$

Now we turn to the principal topic of this section: nonisotropic Lipschitz spaces on domains in $\mathbb{C}^n$. The spirit of the theorem, due to E. M. Stein [3], is that a holomorphic Λ_α function on a C^2 bounded domain in $\mathbb{C}^n$ is, in fact, twice as smooth in complex tangential directions near the boundary. In order to give this result a precise formulation, we need some notation and terminology.

Let $\Omega \subset\subset \mathbb{C}^n$ be a domain with C^2 boundary. Define

$$
C^k(\Omega) = \{ \gamma: [0,1] \to \Omega \text{ such that } \gamma \in C^k \text{ and } |\gamma'(t)| \le 1, \ldots, |\gamma^{(k)}(t)| \le 1 \}.
$$

Let $U \supseteq \partial\Omega$ be a tubular neighborhood of $\partial\Omega$: Each $z \in U$ has a unique nearest point in $\partial\Omega$ (see J. Munkres [1], M. Hirsch [1], or Section 8.2 for a consideration of these matters). Let $\pi: U \cap \Omega \to \partial\Omega$ be normal projection. As in Section 8.6, write $\mathbb{C}^n = \mathcal{N}_P \oplus \mathcal{T}_P$ at each $P \in \partial\Omega$. Extend this decomposition to $P \in U \cap \Omega$ using $\pi^{-1}: \mathcal{N}_P \equiv \mathcal{N}_{\pi(P)}, \mathcal{T}_P \equiv \mathcal{T}_{\pi(P)}$. Now define

$$
C_1^k(U \cap \Omega) = \left\{ \gamma \in C^k : \gamma'(t) \in \mathcal{T}_{\gamma(t)}, 0 \le t \le 1 \right\}.
$$

The elements of C_1^k are called *normalized complex tangential curves.*

DEFINITION 8.8.1   Let $\Omega \subset\subset \mathbb{C}^n$ have C^2 boundary. Fix a neighborhood $U \supseteq \partial\Omega$ as above. Let $0 < \alpha < \beta < \infty$. We say that $f \in \Gamma_{\alpha,\beta}(\Omega)$ if $f \in \Lambda_\alpha(\Omega)$ and

$$
\|f\|_{\Lambda_\alpha(\Omega)} + \sup_{\gamma \in C^{[\beta]+1}(U\cap\Omega)} \|f \circ \gamma\|_{\Lambda_{\beta([0,1])}} \equiv \|f\|_{\Gamma_{\alpha,\beta}(\Omega)} < \infty.
$$

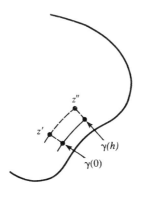

FIGURE 8.9

Remark: Roughly speaking, $f \in \Gamma_{\alpha,\beta}(\Omega)$ means that $f \in \Lambda_\alpha$ in every direction and f is Λ_β in complex tangential directions. Although the definition of $\Gamma_{\alpha,\beta}$ depends on U and on the use of the Euclidean projection (as a method of retraction to the boundary), the choice of a different U or a different method of retraction results in the same space (exercise).   □

The theorem of Stein is as follows.

THEOREM 8.8.2   Let $\Omega \subset\subset \mathbb{C}^n$ have C^2 boundary. Let f be holomorphic on Ω. Then

$$f \in \Lambda_\alpha(\Omega) \Leftrightarrow f \in \Gamma_{\alpha,2\alpha}(\Omega).$$

Proof   We treat the case $0 < \alpha < \frac{1}{2}$. Details for larger α may be found in Greiner and Stein [1] and S. G. Krantz [16]. Only the direction $\Rightarrow$ is nontrivial.

Let U be fixed as in the definition of $\Gamma_{\alpha,2\alpha}$. Let $\delta_0 > 0$ be so small that $\{z \in \Omega : \delta_\Omega(z) < 5\delta_0\} \subseteq U$. Let $z \in \Omega \cap U$, and let $\gamma \in \mathcal{C}_1^1(\Omega \cap U)$ satisfy $\gamma(0) = z$. It suffices to estimate

$$|f(\gamma(h)) - f(\gamma(0))| \qquad (8.8.2.1)$$

when $0 < h < \delta_0$. If $\delta_\Omega(z) \geq 3\delta_0$ and

$$M = \sup_{\delta_\Omega(z) \geq \delta_0} |\nabla f(z)| < \infty, \qquad (8.8.2.2)$$

then (8.8.2.1) does not exceed Mh. Therefore, we may restrict attention to the case $\delta_\Omega(z) < 3\delta_0$. Let $\pi : U \to \partial\Omega$ be normal projection. Define

$$
\begin{aligned}
z' &= \gamma(0) - h^2 \nu_{\pi(z)} \\
z'' &= \gamma(h) - h^2 \nu_{\pi(\gamma(h))}
\end{aligned}
$$

See Figure 8.9.

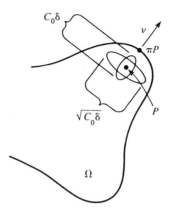

FIGURE 8.10

Then (8.8.2.1) is majorized by

$$|f(\gamma(h)) - f(z'')| + |f(z'') - f(z')| + |f(z') - f(\gamma(0))|$$

$$\equiv I + II + III.$$

In order to analyze I, we let $P \in \Omega, \delta = \delta_\Omega(P) \leq 3\delta_0$. Let $\nu = \nu_{\pi(P)}$ and let τ be a unit element of $\mathcal{T}_{\pi}P$. Elementary geometry shows that there is a $C_0 > 0$, independent of P and τ, such that the bidisc

$$\left\{ P - \lambda_1\nu + \lambda_2\tau : |\lambda_1| \leq C_0\delta, |\lambda_2| \leq \sqrt{C_0\delta} \right\} \subset\subset \Omega.$$

See Figure 8.10.
    Define

$$g(\lambda_1, \lambda_2) = f(P - \lambda_1\nu + \lambda_2\tau).$$

Then g is holomorphic and we may use the Cauchy integral formula to write

$$g(\lambda_1, \lambda_2) = \frac{1}{2\pi} \oint_{\partial D^1(0, C_0\delta)} \frac{g(\zeta, \lambda_2)}{\zeta - \lambda_1} d\zeta.$$

Then

$$\left| \frac{\partial}{\partial \lambda_1} g(\lambda_1, \lambda_2) \right| = \left| \frac{1}{2\pi i} \oint_{\partial D^1(0, C_0\delta)} \frac{g(\zeta, \lambda_2)}{(\zeta - \lambda_1)^2} d\zeta \right|.$$

But the integrand $1/(\zeta - \lambda_1)^2$ integrates to zero. So we may subtract a constant inside the integrand and write

$$\left|\frac{\partial}{\partial \lambda_1}g(\lambda_1,\lambda_2)\right| = \left|\frac{1}{2\pi i}\oint_{\partial D^1(0,C_0\delta)}\frac{g(\zeta,\lambda_2)-g(\lambda_1,\lambda_2)}{(\zeta-\lambda_1)^2}d\zeta\right|.$$

Now the Lipschitz hypothesis on f and elementary estimates show that

$$\left|\frac{\partial}{\partial \lambda_1}g(\lambda_1,\lambda_2)\right| \leq C\delta^{\alpha-1}, \qquad (8.8.2.3)$$

all $|\lambda_1| \leq C_0\delta/2$.

We apply this derivative estimate as follows:

$$\begin{aligned}|I| &= \int_0^{h^2}\frac{d}{dt}f\left(\gamma(h)-t\nu_{\pi(\gamma(h))}\right)dt\\ &\leq \int_0^{h^2}C\cdot t^{\alpha-1}dt\\ &\leq C\cdot h^{2\alpha}.\end{aligned}$$

Thus I and, by symmetry, III are estimated.

We turn our attention to II. First,

$$II = |f(\tilde\gamma(h))-f(\tilde\gamma(0))|,$$

where $\tilde\gamma(t) \equiv \gamma(t)-h^2\nu_{\pi(\gamma(t))}$. This last does not exceed

$$h\cdot\sup_{0\leq t\leq h}\left|\frac{d}{dt}f\circ\tilde\gamma(t)\right|. \qquad (8.8.2.4)$$

We claim that

$$\left|\frac{d}{dt}f\circ\mu(t)\right| \leq C\delta^{\alpha-1/2}, \qquad 0\leq t\leq 1, \qquad (8.8.2.5)$$

where μ is any element of $\mathcal{C}_1^1$ and $\delta \leq 3\delta_0$ is its distance from $\partial\Omega$. This will complete the proof, for then (8.8.2.4) is majorized by $hC(h^2)^{\alpha-1/2} = Ch^{2\alpha}$. So it suffices to prove the claim.

To do so, we apply the Cauchy estimates to the estimate (8.8.2.3) on the disc $D(0,\sqrt{C_0\delta/2})$ in the λ_2 variable to obtain

$$\left|\frac{\partial}{\partial\lambda_2}\frac{\partial}{\partial\lambda_1}g(\lambda_1,\lambda_2)\right| \leq C\delta^{\alpha-1}(\sqrt\delta)^{-1}$$

or

$$\left| \frac{\partial}{\partial \xi_1} \left(\frac{\partial}{\partial \lambda_2} g(\lambda_1, \lambda_2) \right) \right| \leq C\delta^{\alpha - 3/2},$$

all $|\lambda_1| \leq C_0 \delta/8, |\lambda_2| \leq \sqrt{C_0 \delta/8}, \xi_1 = \mathrm{Re}\lambda_1$.
  In particular, setting $\lambda = 0$ yields

$$\left| \frac{\partial}{\partial \nu} \left(\left(\frac{\partial}{\partial \tau} \right) f \right) \Big|_P \right| \leq C\delta^{\alpha - 3/2}. \tag{8.8.2.6}$$

  Now if $Q \in \Omega$ satisfies $\delta_\Omega(Q) \leq 3\delta_0$, we write

$$\left| \left(\frac{\partial}{\partial \tau} \right) f(Q) \right| = \left| \int_{-2\delta_0}^0 \frac{d}{ds} \left\{ \left(\frac{\partial}{\partial \tau} \right) f(Q + s\nu) \right\} ds + \left(\frac{\partial}{\partial \tau} \right) f(Q - 2\delta_0 \nu) \right|.$$

Estimate the first term with (8.8.2.6), letting $P = Q + s\nu$, and estimate the second term with (8.8.2.2). The result is

$$\left| \left(\frac{\partial}{\partial \tau} \right) f(Q) \right| \leq C \int_{-2\delta_0}^0 (\delta_\Omega(Q) + s)^{\alpha - 3/2} ds + M \leq C\delta_\Omega(Q)^{\alpha - 1/2}. \tag{8.8.2.7}$$

This proves our claim. $\square$

  The following result was proved by W. Rudin [4] on the ball and generalized to arbitrary domains in S. G. Krantz [9]:

THEOREM 8.8.3   Let $\Omega \subset\subset \mathbb{C}^n$ be a domain with C^2 boundary. Suppose that $0 < \alpha < \infty$. Let $f : \Omega \to \mathbb{C}$ be holomorphic and suppose that for each $P \in \partial\Omega$, the function

$$t \mapsto f(P - t\nu_P)$$

is in $\Lambda_\alpha(0, \delta_0)$ with uniform Lipschitz bound C. Then $f \in \Gamma_{\alpha,2\alpha}(\Omega)$.

  So, it suffices to check smoothness of f along real normals. In S. G. Krantz [3] these ideas were refined even further:

THEOREM 8.8.4   Let $\Omega \subset\subset \mathbb{C}^n$ have C^k boundary, $k \geq 2$. Let $0 < \alpha < k$. Let $\gamma : [0,1] \to \bar{\Omega}$ be called special complex normal if $\gamma : [0,1] \to \partial\Omega$ satisfies $\gamma'(t) \in \mathcal{N}_{\gamma(t)}$, all t. Let $f \in C(\bar{\Omega})$ be holomorphic on Ω. Suppose that for each special complex normal curve $\gamma : [0,1] \to \partial\Omega$ with $\|\gamma\|_{C^k([0,1])} \leq 1$ we have $\| f|_{\partial\Omega} \circ \gamma \|_{\Lambda_\alpha([0,1])} \leq C$. Then $f \in \Gamma_{\alpha,2\alpha}(\bar{\Omega})$.

Remark: The complex normal curves locally foliate the boundary in a nice way (see Chapter 11, especially the exercises at the end). So the theorem says that it is enough to check $f|_{\partial\Omega}$ in *just one* set of directions. Using Exercise 6 at the end of this chapter, the hypothesis that f be continuous on $\bar{\Omega}$ may be weakened. □

Theorem 8.8.2 is optimal only on strongly pseudoconvex domains. For instance, near a strongly *pseudoconcave* point, all holomorphic functions extend C^∞ (indeed holomorphically) past the boundary. And at weakly pseudoconvex points, the degree of increased smoothing in the tangential direction will always be greater than 2. The paper S. G. Krantz [16] presents a natural way to study theorems like 8.8.2 and to derive the optimal estimates. The key to this work is the Kobayashi metric.

EXERCISES

1. A k-dimensional real manifold $M \subseteq \mathbb{C}^n$ is called *totally real* if whenever $P \in M$ and $w \in T_P(M)$, then $Jw \notin T_P M$ (here J is the complex structure tensor). In other words, M is totally real if no tangent space $T_P(M)$ has a nontrivial complex subspace. Prove that if $U \subseteq \mathbb{C}^n$ is an open set, $M \subseteq U$ is a totally real submanifold of real dimension n, $f : U \to \mathbb{C}$ is holomorphic, and $f|_M = 0$, then $f \equiv 0$ on U. (*Hint:* First consider the special case $M = \{(x_1 + i0, \ldots, x_n + i0)\} \subseteq B(0,1) = U$. Treat the general case by considering the power series expansion of f.)

2. Refer to Exercise 1 for terminology. Prove that there are no totally real manifolds in $\mathbb{C}^n$ of real dimension $k > n$.

3. Let $U \subseteq \mathbb{C}^n$ be open. Let $K_j : U \times U \to \mathbb{C}, j = 1,2$, satisfy $K_j(\cdot, \zeta)$ is holomorphic for each ζ and $K_j(z, \cdot)$ is conjugate holomorphic for each z. If $K_1(z,z) = K_2(z,z)$, all $z \in U$, then prove that $K_1 \equiv K_2$.

4. Prove the following result of Carathéodory: If f is holomorphic on the disc $D \subseteq \mathbb{C}$ and f has nontangential limit 0 on a set $E \subseteq \partial D$ of positive one-dimensional measure, then $f \equiv 0$. (*Hint:* Let $E_0 \subseteq E$ be closed and have positive measure. Let $\Omega = \cup_{\zeta \in E_0} \Gamma_1(\zeta)$. Map Ω conformally to D. This mapping extends to the boundary in such a way that sets of zero measure are mapped to sets of zero measure. Now invoke the standard boundary uniqueness theorem for functions continuous on $\bar{D}$ and holomorphic in D.) See Koosis [1] for details.

5. Extend the method of Lemma 8.5.5 to prove the *Marcinkiewicz interpolation theorem*:
   a. If $(X, \mu), (Y, \nu)$ are measure spaces and T maps
$$L^p(X, \mu) \to \{\text{measurable functions on } Y\},$$
   we say that T is *sublinear* if $|T(f + g)(x)| \leq |Tf(x)| + |Tg(x)|$ for all f, g.

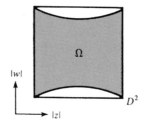

FIGURE 8.11

b. Let T be a sublinear operator as in (a). If $1 \leq p_0 < p_1 \leq \infty$ and T is of weak types (p_0, p_0) and (p_1, p_1), then prove that T maps L^p to L^p, all $p_0 < p < p_1$.

6. Let $B \subseteq \mathbb{C}^2$ be the unit ball and $f : B \to \mathbb{C}$ a bounded holomorphic function. Let $\gamma : [0, 1] \to \partial B$ be a C^2 *complex normal* curve, that is, $\gamma'(t) \notin \mathcal{T}_{\gamma((t)}$ for all $t \in [0, 1]$. The f has radial limits almost everywhere (with respect to one-dimensional measure) along γ. Complete the following outline to prove this result of A. Nagel and W. Rudin [1]:

   a. Restrict γ to a subarc if necessary and reparametrize so that $\gamma : [0, 1] \to \partial B$ and $i\delta\gamma'(t) + \gamma(t) \in B$, all $0 \leq t \leq 1, 0 < \delta$ small.
   b. Define $\phi(x + iy) = f(\gamma(x) + iy\gamma'(x)), 0 \leq x \leq 1, 0 < y < \delta$. Then $\alpha = \partial\phi/\partial\bar{z}$ is bounded.
   c. There is a u on $U = \{x + iy : 0 < x < 1, 0 < y < \delta\}$ such that $\bar{\partial}u = \alpha, u \in \Lambda_\epsilon$, some $\epsilon > 0$.
   d. The function $\phi - u$ is in $H^\infty(U)$. Therefore, $\phi - u$ has boundary values almost everywhere on $\partial U \cap \{y = 0\}$. Thus ϕ has the same property.

7. Complete the following outline to construct an example (N. Sibony [1]) of a pseudoconvex domain $\Omega \subsetneq D^2(0, 1)$ with the property that (i) Ω equals the interior of $\bar{\Omega}$, (ii) $\bar{\Omega} \neq \bar{D}^2(0, 1)$, and (iii) every bounded holomorphic function on Ω analytically continues to $D^2(0, 1)$ (see Figure 8.11). *Outline of construction*

   a. Let $\{a_j\} \subseteq D$ have no accumulation points in D but have every point of ∂D as an accumulation point. Define $u(z) = \sum_j \frac{1}{j^2} \log |(z - a_j)/2|$.
   b. Let $v(z) = \exp u(z)$. Then v is continuous and subharmonic on D and $0 \leq v < 1$ on D.
   c. Let $\Omega = \{(z, w) \in \mathbb{C}^2 : |z| < 1, |w| < \exp(-v(z))\}$. Then Ω satisfies properties (i) and (ii) above and Ω is pseudoconvex. (*Note:* The function $(z, w) \to |w| \exp v(z)$ is plurisubharmonic.)

d. Let f be bounded and holomorphic on Ω; say that $\|f\|_{L^\infty} \leq 1$. Then f has a power series expansion

$$f(z, w) = \sum_{j \geq 0} f_j(z) w^j,$$

where

$$f_j(z) = (j!)^{-1} \frac{\partial^j f}{\partial w^j}(z, 0).$$

Each f_j is holomorphic on D.

e. Write, for $0 < r < 1$,

$$f_j(z) = \frac{1}{2\pi i} \int \frac{f\left(z, r e^{i\theta} \exp(-v(z))\right)}{r^j e^{ij\theta} \exp(-jv(z))} i\, d\theta.$$

Conclude that $|f_j(z)| \leq |\exp jv(z)|$, in particular, $|f_j(a_\ell)| \leq 1$, all j and ℓ.

f. Each $f_j \in H^\infty(D)$. If $\tilde{f}_j$ is the boundary limit function of f_j, then $|\tilde{f}_j(e^{i\theta})| \leq 1$ almost everywhere, hence $|f_j| \leq 1$.

g. The series representation in (d) converges normally for points $(z, w) \in D^2(0, 1)$.

8. Let $E \subseteq \partial D$ be an F_σ of first category and ϕ a continuous function on D. F. Bagemihl and W. Seidel [1] proved that there exists a holomorphic function f on D such that $\lim_{r \to 1^-} |\phi(r\zeta) - f(r\zeta)| = 0$ for all $\zeta \in E$. We omit the rather technical proof.

   Give an example to show that the Bagemihl-Seidel theorem cannot hold if "radial convergence" is replaced by nontangential convergence. Conversely, give an example to show that the result of Exercise 4 cannot hold if nontangential convergence is replaced by radial convergence (feel free to use Bagemihl-Seidel here).

9. Prove the following classical result of Hardy and Littlewood (see Goluzin [1, p. 411]).

   Theorem Let $f : D \to \mathbb{C}$ be holomorphic and bounded. Let $0 < \alpha < \infty$. Then $f \in \Lambda_\alpha(D)$ if and only if there are an $\mathbb{N} \ni k > \alpha$ and a $C_k > 0$ such that

$$|f^{(k)}(z)| \leq C_k(1 - |z|)^{\alpha - k}, \qquad \text{all} \quad z \in D. \tag{$*$}$$

   Prove that $(*)$ holds for one $k > \alpha$ if and only if it holds for all $k > \alpha$.
   The following result is due to W. Rudin [4] and E. M. Stein [3] and can be proved by methods of Section 8.8 (although Rudin takes advantage of the special nature of the ball to obtain a more elegant proof).

Theorem Let $B \subseteq \mathbb{C}^n$ be the unit ball. Define $Rf(z) = \sum_{j=1}^n z_j(\partial/\partial z_j)f(z)$, any $f \in C^1(B)$. Let $f : B \to \mathbb{C}$ be holomorphic and bounded. Then $f \in \Lambda_\alpha(B)$ (hence in $\Gamma_{\alpha,2\alpha}(B)$ by Theorem 8.8.4) if and only if there is an $\mathbb{N} \ni k > \alpha$ and a $C_k > 0$ such that

$$|R^k f(z)| \leq C_k(1 - |z|)^{\alpha-k}. \qquad (**)$$

Inequality $(**)$ holds for some $k > \alpha$ if and only if it holds for all $k > \alpha$.

10. This problem is philosophically related to the preceding one. The classical result, once again, is due to G. H. Hardy and J. E. Littlewood [1, 2].

Theorem Let $f : D \to \mathbb{C}$ be holomorphic. Let $0 < p < 1$. If $f' \in H^p(D)$, then $f \in H^q(D)$ where $q = p/(1 - p)$. If $f' \in H^1$, then $f \in A(D)$; indeed, $f|_{\partial D}$ is absolutely continuous. (Here f' is the derivative of f.)

It is even possible to prove results for $p > 1$: If $f' \in H^p, 1 < p \leq \infty$, then $f \in \Lambda_\alpha(D)$, where $\alpha = 1 - 1/p$. Although the reader should have no difficulty verifying the assertion for $p = \infty$, he or she may encounter trouble with $0 < p < \infty$. Nevertheless, for $1 < p < \infty$, the problem is accessible if we use some Sobolev space ideas:

Identify f with its boundary function $\tilde{f}$. Prove that the hypotheses imply that $(d/d\theta)\tilde{f}(e^{i\theta}) \in L^p([0, 2\pi])$ in the sense of distributions. Prove that, for $p > 1$, this implies that $\tilde{f} \in \Lambda_{1-1/p}(\partial D)$. Hence $f \in \Lambda_{1-1/p}(\bar{D})$. When $p = 1$, matters are even simpler.

The analogous result on the ball in $\mathbb{C}^n$ has a similar flavor but is much more difficult to prove:

Theorem Let $f : B \to \mathbb{C}$ be holomorphic, $B \subseteq \mathbb{C}^n$. Assume that $Rf(z) \in H^p(B), 0 < p < \infty$ (see Exercise 9 for notation). Then $f \in H^q(B)$, where $q = np/(n - p), 1 < p < n$. For $n < p \leq \infty, f \in \Lambda_\alpha(B)$ with $\alpha = 1 - n/p$. Hence $f \in \Gamma_{\alpha,2\alpha}$. If $p = n$, then f is not necessarily in H^∞ but is a function in BMOA (see the next two exercises).

The result for $0 < p < n$ was proved by I. Graham [2] and S. G. Krantz [6]. That for $p \geq n$ was proved by Krantz. The fact that f need not be in H^∞ when $Rf \in H^n$ is due to Graham.

11. Let $U \subseteq \mathbb{R}^N$ be open. A function $f \in L^1_{\text{loc}}(U)$ is said to be of *bounded mean oscillation* on U (denoted $f \in BMO(U)$) if

$$\sup_{Q \subseteq U} \frac{1}{|Q|} \int_Q |f(x) - f_Q|dx \equiv \|f\|_* < \infty.$$

Here the supremum is taken over cubes Q in U with sides parallel to the axes (although this last restriction is superfluous). The notation f_Q denotes the average $(1/|Q|) \int_Q f(t)dt$.

a. The function $\| \ \|_*$ is only a *seminorm* on $BMO(U)$. It is a norm on $BMO(U)/\mathbb{C}$, where $\mathbb{C}$ is identified with the constant functions.

b. Let $f \in L^1_{loc}(U)$. Suppose there is a $C > 0$ such that for each cube $Q \subseteq U$, there is a constant c_Q with

$$\frac{1}{|Q|} \int_Q |f(x) - c_Q|\, dx \leq C.$$

Then $f \in BMO(U)$. Indeed, it follows that

$$\frac{1}{|Q|} \int_Q |f(x) - f_Q|\, dx \leq 2C.$$

c. We have $L^\infty(U) \subseteq BMO(U)$. However the spaces are not equal. Indeed, if $0 \in U$, then

$$\log |x| \in BMO \setminus L^\infty.$$

d. The space $BMO(\mathbb{R}^N)/\mathbb{C}$ is complete. Identifying ∂D with $[0, 2\pi]/(0 \sim 2\pi)$, we have that $BMO(\partial D)/\mathbb{C}$ is complete.

e. The space $BMO(\mathbb{R})$ is not an algebra.

f. Call an interval $I \subseteq \mathbb{R}$ *dyadic* if $I = [j/2^k, (j+1)/2^k]$ for some $k \in \mathbb{Z}^+$, $j \in \mathbb{Z}$. Define $f \in L^1_{loc}$ to be in *dyadic* BMO (written $BMO_d(\mathbb{R})$) if

$$\sup_{I \text{ dyadic}} \frac{1}{|I|} \int_I |f(x) - f_I|\, dx < \infty.$$

Then $BMO_d(\mathbb{R}) \supseteq BMO(\mathbb{R})$ but the spaces are unequal.

g. Suppose that $f \in BMO_d(\mathbb{R})$. Suppose further that if I_1 and I_2 are adjacent dyadic intervals of the same lengths, then $|f_{I_1} - f_{I_2}| \leq C$, with C being independent of I_1, I_2. Then $f \in BMO(\mathbb{R})$.

h. It is a deep theorem of F. John and L. Nirenberg [1] that if $Q \subseteq \mathbb{R}^N$ is fixed, $f \in BMO(Q)$, then there are $C_1, C_2 > 0$ such that $m\{x \in Q : |f(x) - f_Q| > \lambda\} \leq C_1 e^{-C_2 \lambda/\|f\|_*} \cdot |Q|$. Prove that this implies that $f \in L^p(Q)$ for all $1 \leq p < \infty$.

i. The space BMO is a limiting case of the Lipschitz spaces. More precisely, define for $\alpha > 0$ the Campanato-Morrey space

$$\mathcal{L}^{1+\alpha}(\mathbb{R}^N) = \left\{ f \in L^1(\mathbb{R}^N) : \sup_{Q \in \mathbb{R}^N} \inf_{c \in \mathbb{C}} \frac{1}{|Q|^{1+\alpha}} \int_Q |f(x) - c|\, dx \right.$$

$$\equiv \left. \|f\|_{\mathcal{L}^{1+\alpha}(\mathbb{R}^N)} < \infty \right\}.$$

It can be proved that $f \in \mathcal{L}^{1+\alpha}_{loc}(\mathbb{R}^N)$ if and only if $f \in \Lambda^{loc}_{N\alpha}, 0 < \alpha < 1/N$. There are analogues for large α: The constant c must in that case be replaced by a polynomial of sufficiently high order. The reader should be able to prove the "if" part for any $\alpha \notin \mathbb{Z}$. See S. G. Krantz [5], [10] for details on these matters.

j. The John-Nirenberg theorem mentioned in (h) implies that, for $1 \le p < \infty$, those f satisfying

$$\sup_{Q \subseteq \mathbb{R}} \frac{1}{|Q|} \int_Q |f(x) - f_Q|^p dx < \infty$$

comprise precisely $BMO(\mathbb{R}^N)$. (*Hint:* Use Lemma 8.5.4.)

12. (Refer to the remarks at the end of Section 8.7, as well as the preceding exercise.) If $f \in L^1(\partial D)$, then $Pf(re^{i\theta})$ is a harmonic function on D with almost everywhere radial limit f on ∂D. Let $Qf(re^{i\theta})$ be the unique harmonic conjugate to Pf on D that vanishes at 0. Let $F(re^{i\theta}) = Pf(re^{i\theta}) + iQf(re^{i\theta})$. Then F is holomorphic on D. If f is real, $f \ge 0$, then $G(z) \equiv e^{-F(z)} \in H^\infty(D)$. Therefore, G has almost everywhere radial boundary limits on ∂D. Hence $Qf(re^{i\theta})$ has almost everywhere radial boundary limits on ∂D. Call this boundary function $Hf(e^{i\theta})$ (the *Hilbert transform* of f). Extend H to all L^1 by linearity. If $f \in L^1(\partial D)$, then Hf need not be in L^1 (consider $f_j \in L^1(\partial D), \|f_j\|_{L^1} = 1, f_j$ tending weak-$*$ to the Dirac δ mass at 1). Define

$$H^1_{\mathrm{Re}}(\partial D) = \{f \in L^1(\partial D) : Hf \in L^1(\partial D)\}.$$

Norm the space by

$$\|f\|_{H^1_{\mathrm{Re}}} = \|f\|_{L^1(\partial D)} + \|Hf\|_{L^1(\partial D)}.$$

Then the map

$$\Phi : H^1_{\mathrm{Re}}(\partial D) \to H^1(D)/i\mathbb{R}$$

given by

$$f \mapsto Pf(re^{i\theta}) + iQf(re^{i\theta})$$

is a surjective Banach space isomorphism. It is a deep theorem (see C. Fefferman and E. M. Stein [1]) that $(H^1_{\mathrm{Re}})^* = BMO$.

13. Prove that the Lindelöf principle fails for functions in $H^p, p < \infty$. Call a holomorphic function f on the disc *normal* if whenever $\{\phi_j\}$ are conformal self-maps of the disc, then $\{f \circ \phi_j\}$ is a normal family. Give an example of a normal function that is not bounded. Prove that the Lindelöf principle is valid for normal functions.

The correct generalization of this concept to several variables appears in J. Cima and S. G. Krantz [1].

14. If $f \in H^p(D), 0 < p < \infty$, let $\tilde{f}$ denote the boundary function. Define

$$BMOA(D) = \{f \in H^2(D) : \tilde{f} \in BMO(\partial D)\}.$$

Prove that this space is unchanged if we replace the exponent 2 by any $p \geq 1$. Use Fefferman's theorem (see Exercise 12) to show that

$$\left(H^1(D)\right)^* = BMOA(D).$$

On the ball in $\mathbb{C}^n$, $n > 1$, matters are more complicated because there are two types of balls in the boundary. Let

$$BMO_j(\partial B) \;=\; \left\{ f \in L^1(\partial B) : \sup_{\substack{P \in \partial B \\ r > 0}} \frac{1}{\sigma(\beta_j(P,r))} \right.$$
$$\left. \times \int_{\beta_j(P,r)} |f(\zeta) - f_{\beta_j(P,r)}| d\sigma(\zeta) \equiv \|f\|_{BMO_j(\partial B)} < \infty \right\}$$

for $j = 1, 2$. In view of the results in Lipschitz spaces, and because BMO is a limiting case of the Lipschitz condition (see Exercise 11), one might conjecture that $BMOA_1 = BMOA_2$. In fact it turns out that $BMOA_1 \subseteq BMOA_2$, but the spaces are unequal (see S. G. Krantz [8]). It is the space $BMOA_2$ that is the dual of $H^1(B)$. See Krantz and Ma [1] for details.

15. Construct an example of a bounded analytic function f on the disc, f not identically zero, such that for almost every P in the boundary of the disc, there is a tangential sequence $\{z_j\}$ in D approaching P along which f tends to 0. Here a sequence in D is "tangential" if it escapes every nontangential approach region. Consider variants of this result: Can you replace the sequence $\{z_j\}$ by a curve at each point? If you replace "bounded" by H^p, then can you strengthen the example? A good reference for this sort of result is I. Priwalow [1].

16. Let $\Omega \subset\subset \mathbb{C}^n$ be a domain with C^1 boundary. Let ρ be a C^1 defining function for Ω and set $\omega = i\bar{\partial}\rho$. Show that $\omega|_{\partial\Omega}$ is *real* in the following sense: If α is a vector field defined on a relatively open set $U \subseteq \partial\Omega$, then we have $\langle \omega|_U, \alpha \rangle = \langle \bar{\omega}|_U, \alpha \rangle$.

17. Show that the standard rigid imbedding of the 2-torus into $\mathbb{C}^2$ renders (the imbedded) $\mathbb{T}^2$ as a totally real submanifold. That is, the tangent space at any P in the boundary of the imbedded torus never contains a complex line in $\mathbb{C}^2$.

18. Let Ω be a bounded domain in $\mathbb{C}^n$ that has C^2 boundary and $P \in \partial\Omega$. Formulate a notion of hypoadmissible curve at P. Now let $f \in H^\infty(\Omega)$. Suppose that γ is a hypoadmissible curve at $P \in \partial\Omega$ such that $\lim_{t\to 1-} f(\gamma(t)) = \ell \in \mathbb{C}$.

Then prove that for any hypoadmissible sequence $\{w^j\}$ at P it holds that $\lim_{j\to\infty} f(w^j) = \ell$. Part of this problem is to define "hypoadmissible" on an arbitrary Ω. The second part is to prove the theorem.

19. Let $M \subseteq \mathbb{C}^2$ be a smooth, real, k-dimensional regularly imbedded submanifold, $1 \leq k \leq 4$. If $k \geq 3$, then M cannot be totally real. If $k = 2$ then, near a point $P \in M$, we may choose coordinates such that M is locally given by $\{(z, r(z)) : z \in U\}$, some open $U \subseteq \mathbb{C}$. Show then that M is totally real at $(z, r(z))$ if and only if $(\partial r/\partial \bar{z})(z) \neq 0$.

   An abstract version of these ideas is as follows. Let $M \subseteq \mathbb{C}^2$ be a smooth real hypersurface. Let $\mathbb{P}_1$ be the projective space of one-dimensional complex lines in $\mathbb{C}^2$. Let $\psi : M \to \mathbb{C}^2 \times \mathbb{P}_1$ be given by $\psi(z) = (z, \mathcal{T}_z(M))$. Prove that $\psi(M)$ is totally real in $\mathbb{C}^2 \times \mathbb{P}_1$ at the point $\psi(z)$ if and only if z is a point of strong Levi pseudoconvexity in M. The same result is true in $\mathbb{C}^n$ but is a little more difficult to prove (see S. Webster [1]).

20. This problem provides a glimpse of some of the ideas of Lelong.

   a. Examine the proof of the Poisson integral formula (Theorem 1.3.12) to see that it also yields the *Riesz decomposition* for a subharmonic function: If $\Omega \subset\subset \mathbb{R}^N$ has C^2 boundary, $u \in C^2(\bar{\Omega})$, and u is subharmonic on Ω, then

$$u(x) = \int_{\partial\Omega} P_\Omega(x, y)u(y)\, d\sigma(y) - \int_\Omega (\Delta u(y)) G_\Omega(x, y)\, dV(y).$$

   Use a limiting argument to extend this formula to all subharmonic functions in $C(\bar{\Omega})$.

   b. Let $f \in C^2(\bar{D})$, D the disc in $\mathbb{C}$. Let $f|_D$ be holomorphic. Assume that the zeros of f are $\{\alpha_1, \ldots, \alpha_k\} \subseteq D$. Let δ_k be the unit Dirac mass at α_k. Then $\Delta \log |f| = C \cdot \sum_k \delta_k$ in the sense that

$$\int_D \phi(z)\Delta \log |f(z)|\, dV(z) = C \cdot \sum \phi(\alpha_k), \qquad \text{all} \quad \phi \in C_c^\infty(D).$$

   Use Green's formula to prove this result.

   c. Use a limiting argument to extend (b) to elements of the Nevanlinna class.

   d. Lelong has proved that if f is holomorphic on $\Omega \subset\subset \mathbb{C}^n$, then $\omega_0 = \Delta \log |f|$ is a nonnegative measure supported on $\mathcal{N} \equiv \{z \in \Omega : f(z) = 0\}$. If $E \subseteq \Omega$ is Borel, then

$$\mathcal{H}^{2n-2}(E \cap \mathcal{N}) = C \cdot \int_\Omega E\, d\omega_0.$$

   Taking (a) through (d) for granted, we now prove that if $\Omega \subset\subset \mathbb{C}^n$ has C^2 boundary and f is in the Nevanlinna class on Ω, then the zero set of f satisfies an analogue of the Blaschke condition (Lemma 8.1.16) on Ω.

e. Let ρ be a defining function for Ω and, for $0 < \epsilon < \epsilon_0$, let $\Omega_\epsilon, \rho_\epsilon, d\sigma_\epsilon$ be as in Section 8.3. Let P_ϵ be the Poisson kernel for Ω_ϵ and G_ϵ the Green's function for Ω_ϵ. Fix an f in the Nevanlinna class on Ω. Let $\Delta \log |f| = \omega_0$. Let Γ be the fundamental solution for the Laplacian and define

$$V_0(x) = \int \Gamma(x - y) d\omega_0(y).$$

Then V_0 is subharmonic.

f. Let $H(z) = \log |f(z)| - V_0(z)$. Then H is harmonic on Ω. It follows that, for $0 < \epsilon < \epsilon_0, z_0 \in \Omega \setminus \{f(z) = 0\}$, we have

$$H(z_0) = \int_{\partial\Omega_\epsilon} P_\epsilon(z_0, \zeta) H(\zeta) \, d\sigma_\epsilon(\zeta),$$

$$\int_{\Omega_\epsilon} G_\epsilon(z_0, \zeta) \, d\omega_0(\zeta) + V_0(\zeta) = \int_{\partial\Omega_\epsilon} P_\epsilon(z_0, \zeta) V_0(\zeta) \, d\sigma_\epsilon(\zeta).$$

g. Conclude from (f) that

$$\int_{\Omega_\epsilon} G_\epsilon(z_0, \zeta) \, d\omega_0(\zeta) + \log |f(z_0)| = \int_{\partial\Omega_\epsilon} P_\epsilon(z_0, \zeta) \log |f(\zeta)| \, d\sigma_\epsilon(\zeta).$$

h. Let $\epsilon \to 0$ to obtain

$$\int_\Omega G(z_0, \zeta) \, d\omega_0(\zeta) + \log |f(z_0)| \le C_{z_0} \|f\|_N.$$

i. Conclude from (h) that

$$\int_\Omega \delta_\Omega(\zeta) \, d\omega_0(\zeta) < \infty.$$

This is the Blaschke condition.

j. G. M. Henkin [6] and H. Skoda [4] have proved that if Ω is strongly pseudoconvex with smooth boundary (and satisfies a mild topological condition), if $\mathcal{N} \subseteq \Omega$ is the zero set of *some* holomorphic function, and if $\mathcal{N}$ satisfies the Blaschke condition in (i), then $\mathcal{N}$ is the zero set of a function in the Nevanlinna class.

k. Notice that some sort of topological condition on $\mathcal{N}$ is necessary. For any compact set in Ω satisfies the Blaschke condition trivially, but it could never be the zero set of a holomorphic function.

21. Let $\Omega \subset\subset \mathbb{C}^n$ be a domain with C^2 boundary. Let $f \in H^p(\Omega), 0 < p < \infty$. Suppose that f has radial boundary limit 0 on a set $E \subseteq \partial\Omega$ of positive σ measure. Then $f \equiv 0$. (*Hint:* Apply the Riesz decomposition to the

subharmonic function $\log^+ |f|$.) Does the result prevail if we merely assume that $f \in \mathbf{h}^p$? (See S. G. Krantz [8] for more on this matter.)

22. Let Ω_1, Ω_2 be bounded domains in $\mathbb{C}^n$ with C^2 boundary. Let $\phi : \Omega_1 \to \Omega_2$ be a biholomorphic map which extends C^2 to $\bar{\Omega}_1$ and such that ϕ^{-1} extends C^2 to $\bar{\Omega}_2$. Prove that there is a $C > 0$ such that if $P_1 \in \partial \Omega_1$ and $\mathcal{A}_\alpha(P_1)$ is an admissible region in Ω_1, then

$$\mathcal{A}_{\alpha/C}(\phi(P_1)) \subseteq \phi\left(\mathcal{A}_\alpha(P_1)\right) \subseteq \mathcal{A}_{C\alpha}(\phi(P_1)).$$

Which of the hypotheses on ϕ are really needed here?

23. Consider the mapping $\phi : S^3 \to \mathbb{C}^3$ given by $(z, w) \mapsto (z, w, \phi(z, w))$. Prove that the image of this mapping is a totally real manifold if and only if

$$w \frac{\partial \phi}{\partial \bar{z}} - z \frac{\partial \phi}{\partial \bar{w}} \neq 0$$

for all $(z, w) \in S^3$.

9 Constructive Methods

9.1 Story of the Inner Functions Problem

Recall from Section 8.1 that Blaschke products played a central role in the classical theory of the boundary behavior of holomorphic functions on the disc. By the use of this device, the question of boundary limits for H^p functions, $p \leq 1$, was reduced to the easier case $p = 2$. In later sections we realized that the device of Blaschke products was unavailable in higher dimensions, and we learned instead to exploit subharmonicity.

Blaschke products are a special instance of *inner functions*. An inner function on a domain in $\mathbb{C}^n$ (in particular on the disc) is a bounded analytic function with radial boundary limits having modulus 1 almost everywhere. Of course any bounded analytic function has almost everywhere radial boundary limits, so the inner functions form a subclass of the H^∞ functions. Besides finite and infinite Blaschke products, another interesting class of inner functions on the disc consists of the *singular inner functions*. These are functions of the form

$$\phi(\zeta) = \exp\left\{ -\int_0^{2\pi} \frac{e^{i\theta} + \zeta}{e^{i\theta} - \zeta}\, d\mu(\theta) \right\}, \tag{9.1.1}$$

where μ is a positive Borel measure that is singular with respect to Lebesgue measure. We claim that φ is an inner function.

It is not difficult to see, using the techniques of Section 8.1, that the Poisson integral of a finite Borel measure has radial boundary limits agreeing with the *absolutely continuous part* of the measure almost everywhere (see [K. Hoffman [1] and J. Garnett [1]). It follows that the Cauchy integral of a singular measure has radial limit 0 almost everywhere. Moreover, the real part of the Poisson integral of a positive measure will be positive (since it is the integral of the measure against the Poisson kernel). Thus the expression in braces in equation (9.1.1) has negative real part. We conclude that the function ϕ, as defined above, is bounded in absolute value (by 1) and has radial boundary limits of unit modulus almost everywhere. Thus ϕ is inner, justifying the name "singular inner function."

Now the pretty fact, which we shall not prove, is that *every* inner function factors as a singular inner function and a Blaschke product. But there is more.

Define an *outer function* on the disc to be a holomorphic function of the form

$$F(z) = c \cdot \exp\left\{ \frac{1}{2\pi} \int_0^{2\pi} \frac{e^{i\theta} + \zeta}{e^{i\theta} - \zeta} \log \psi(\theta)\, d\theta \right\},$$

where c is a unimodular constant and ψ is a positive measurable function whose logarithm is integrable. The fundamental factorization theorem for H^p functions is that any $f \in H^p(D)$ can be written in the form

$$f(z) = \mathcal{O} \cdot \mathcal{B} \cdot \mathcal{S},$$

where $\mathcal{O}$ is an outer function, $\mathcal{B}$ is a Blaschke product, and $\mathcal{S}$ is a singular inner function. This circle of ideas is treated beautifully in J. Garnett [1] and K. Hoffman [1]. It has no analogue in higher dimensions. This factorization theorem has played a central role in shaping the types of questions that can be asked, and answered, in the function theory of the disc.

In the mid-1960s, Walter Rudin and Anatoli Vitushkin independently raised the question as to whether there are any nonconstant inner functions—that is, bounded analytic functions with unimodular boundary values almost everywhere—on the unit ball in $\mathbb{C}^n$. There quickly amassed considerable evidence that such functions would be so pathological that they could not exist. Here is simple argument to support this contention.

First observe that if f is a nonconstant inner function on the unit disc, then the image of f is dense in the unit disc. To see this, suppose not. Then the image of f omits a small disc $D(a, \epsilon)$. After postcomposing f with a Möbius transformation, we may suppose that $a = 0$. But then $1/f$ is a bounded holomorphic function with unimodular boundary values. Therefore, by our theory in Section 8.1, $|1/f|$ is bounded above by 1. By the maximum principle, it is strictly smaller than 1 inside the disc. But then $|f|$ exceeds 1 in the disc, and that is absurd.

We exploit the observation in the last paragraph as follows: Let ϕ be a nonconstant inner function on the ball in $\mathbb{C}^2$. Let $P \in \partial B$ be any point. After a rotation we may as well suppose that $P = (1, 0) = 1$. Then, by elementary measure theory, the restriction of ϕ to almost every analytic disc

$$\mathbf{d}_\delta \equiv \left\{ (1 - \delta, \zeta), |\zeta| < \sqrt{1 - \delta^2} \right\},$$

$0 < \delta < 1$, will be an inner function of one complex variable. Let ϕ_δ denote the restriction of ϕ to $\mathbf{d}_\delta$. Then, by the result of the last paragraph, almost every

ϕ_δ will have range that is dense in the unit disc. Thus the *cluster set* of ϕ at $P \in \partial B$—that is, the set of all possible limits of sequence $\phi(z_j)$ where $z_j \to P$—is the closed unit disc.

We see that an inner function ϕ on the ball in dimension two is highly discontinuous at *every* boundary point. We know from Chapter 0 that every level set of ϕ must escape to the boundary, but for an inner function matters are vastly more complicated.

In W. Rudin [7], besides a wealth of other material, Rudin formulates a number of interesting problems whose resolution would have resolved the inner functions problem in several complex variables in the negative. It is safe to say that when that book appeared everyone believed that there were certainly no nonconstant inner functions above dimension one, but the problem appeared to be intractable.

The construction of inner functions is the product of work of A. B. Aleksandrov [1], M. Hakim and N. Sibony [2], and Erik Løw [1] in the fall of 1981. There are delicate priority matters involved that we shall not touch upon (but that are discussed somewhat more discursively in W. Rudin [6]). Suffice it to say that all four authors contributed decisively to the final solution of the problem.

What Hakim and Sibony did was to construct a bounded holomorphic function on the unit ball with (almost everywhere) boundary values of modulus between $1/2$ and 1. This provided powerful encouragement that inner functions existed and also created the principal technical tool that Løw used in his construction.

Aleksandrov's construction is interesting in that it is ingenious but completely elementary. In fact, he constructs (by hand) a nonnegative singular measure on ∂B that has Poisson integral which is pluriharmonic (recall the definition of singular inner function on the disc for motivation of what now follows). Call that Poisson integral u. By Proposition 2.2.3, there is a pluriharmonic function v such that $u + iv$ is holomorphic on the ball. Define

$$\phi(z) = \exp\{-u - iv\}.$$

Then clearly u is holomorphic and bounded by 1. Moreover, the function $\phi = e^{-u}$ will have unimodular boundary limits almost everywhere. Thus ϕ is inner.

We should also mention that a third approach to the inner functions problem was developed by A. B. Aleksandrov [2]. This construction uses *Ryll-Wojtaszczyk* polynomials. These homogeneous, holomorphic polynomials provide building blocks (similar in spirit to spherical harmonics, for which see E. M. Stein and G. Weiss [1]) that may be summed to produce inner functions. It would be interesting to know the analogue of Ryll-Wojtaszczyk polynomials on more general domains.

Of the three methodologies, those of Hakim/Løw/Sibony and the second method of Aleksandrov have proved to be most flexible and to have utility in other contexts. In Section 9.2 we shall present the Hakim/Løw/Sibony approach to the subject. We have made this choice for two reasons. First, the Ryll-Wojtaszczyk approach has been covered both thoroughly and elegantly in W. Rudin [6]. Secondly, Løw [2] has carried the program further than other authors by constructing inner functions on an arbitrary strongly pseudoconvex domain. We shall reproduce that result in the next section.

Some experience with function algebras leads one to suspect, in view of the Hakim/Sibony result, that there are nonconstant inner functions. By the same token, one would suspect that the construction would allow one to create holomorphic functions ϕ on the ball and on strongly pseudoconvex domains with virtually arbitrary prescribed radial boundary behavior for $|\phi|$ (after all, what can be special about 1?). This, in fact, turns out to be the case. We shall consider questions of this type in Section 9.3.

It is apparent from examining the proof that the crucial ingredient in the construction of inner functions on a domain Ω is the existence at each boundary point of a peak function that approaches the value 1 at a certain rate. That strongly pseudoconvex domains possess such functions will follow easily from our work in Section 5.2. However there are more general classes of domains (such as the finite-type domains treated in Section 11.5) for which these tools are available. In fact, it follows from deep work of D. Catlin [3] that, on a pseudoconvex domain of finite type in any dimension, the weakly pseudoconvex points form a set of measure zero. As a result, the methods of Section 9.2 can be modified to produce inner functions on any finite type domain in any dimension. Unfortunately, the details of the proof of this last assertion must be considered part of the folklore. The ambitious reader may try it as an exercise.

Finally, we should mention that a number of interesting function-algebraic results are a consequence of the inner functions construction. We shall say nothing about them here but instead refer the reader to W. Rudin [6].

W. Rudin [6] notes that there are not as many inner functions in higher dimensions as there are on the disc. On the disc, the closed linear span of the inner functions (in the $L^\infty(\partial D)$ topology) is all of H^∞ (see D. E. Marshall [1]). Nothing of the sort is true in dimension two or greater. And the inner functions will, therefore, probably not prove to be as useful in this new context. Nevertheless, the methods for constructing inner functions comprise one of the most significant new devices for constructing holomorphic functions of several variables that has been produced in the last 20 years. In several complex variables we do not have Blaschke factors and Weierstrass polynomials to use as building blocks. One might hope that, along with Ryll-Wojtaszczyk polynomials, the inner functions construction will develop into a useful analytic tool. These types of constructions are currently being used to construct pathological proper holomorphic maps (see A. Noell and B. Stensønes [1], for instance). But the full significance of the constructive methods that we discuss in this chapter has yet to be realized.

9.2 The Hakim/Løw/Sibony Construction of Inner Functions

For the sake of simplicity, we shall give up some generality and concentrate just on constructing inner functions. The same construction gives more general results, but those are treated in the next section. The presentation here is taken from E. Løw [2].

The idea of the construction we present is to begin with a small function, of modulus bounded above by $\frac{1}{2}$, and to add peak functions in a fashion that pushes the image out toward the boundary of the unit disc. Of course, this has to be done systematically, and in such a fashion that the $(n+1)^{\text{st}}$ push does not destroy what was achieved by the nth push. Thus a great deal of care must be exercised in the preliminary lemmas. Throughout this section, Ω will be a fixed strongly pseudoconvex domain with C^2 boundary.

PROPOSITION 9.2.1   There exist positive constants C_3, C_4, C_5 and r_0 and a function $\Phi(z, w)$ in $C^1(\bar{\Omega} \times \partial\Omega)$ such that the following properties hold. For every $w \in \partial\Omega$, the function $\Phi(\cdot, w)$ is a holomorphic function in Ω and we also have

1. $\Phi(w, w) = 0$ for every $w \in \partial\Omega$.

2. $\operatorname{Re}\Phi(z, w) \leq C_4|z - w|^2$ for all $z \in \partial\Omega, w \in \partial\Omega$.

3. $\operatorname{Re}\Phi(z, w) \geq C_5|z - w|^2$ for all $z \in \bar{\Omega}, w \in \partial\Omega$.

4. For each $m \in \{1, 2, \ldots\}, 0 < a < 1, w \in \partial\Omega$, and $0 < r \leq r_0$, there is a closed subset $V \subseteq \beta_1(w, r)$ such that $\cos(m \operatorname{Im}\Phi(z, w)) \geq a$ for $z \in V$ and

$$\mathcal{H}^{2n-1}(V) \geq C_3(\cos^{-1} a)r^{2n-1}.$$

Here $\beta_1(w, r)$ denotes the usual Euclidean ball in the boundary (see Section 8.6) and $\mathcal{H}^{2n-1}$ is $(2n-1)$- dimensional Hausdorff measure. The estimate from below on $\mathcal{H}^{2n-1}(V)$ is uniform in m.

Proof   Of course the function Φ is nothing other than the Henkin separating function that we constructed in Section 5.2. Thus properties 1—3 are known to us from that section.

It remains to treat property 4. This is merely a calculus exercise—it has nothing to do with complex analysis. We may introduce local coordinates centered about the point w such that $(\partial/\partial x_1)\Phi(z, w) \neq 0$ in the ball $\beta_1(w, r)$ for small r. Thus we may, by the Implicit Function Theorem, think of $\operatorname{Im}\Phi(\cdot, w)$ as the function icx_1 plus a small error.

But the set of points in $\beta_1(w, r)$ in which $\cos mx_1 \geq a$ is clearly a set of bands in the ball β_1; now it is a simple calculation to see that the required estimate from below on the size of V is true. We leave the details as an exercise.   □

Remark: If Ω is strictly convex with defining function ρ, then the function

$$\Phi(z,w) = \sum_{j=1}^{n} \frac{\partial \rho}{\partial z_j}(w)(z_j - w_j)$$

is easily seen to satisfy the conclusions of the proposition. The case for general strictly pseudoconvex domains is then immediate from the Fornæss imbedding theorem. $\qquad\qquad\square$

In the remainder of the construction we shall use only the proposition and the fact (discussed in Section 8.6) that isotropic balls $\beta_1(w,r)$ have $\mathcal{H}^{2n-1}$ measure comparable to r^{2n-1} : We have $C_1 r^{2n-1} \leq \mathcal{H}^{2n-1}(\beta_1(w,r)) \leq C_2 r^{2n-1}$.

Next we present the Hakim-Sibony approximation scheme. Our function Φ will be used to prove this scheme; after that, we use only the approximation scheme itself. Let μ denote the Hausdorff measure $\mathcal{H}^{2n-1}$ on $\partial\Omega$ normalized to have total mass 1. We continue to use constants C_1, C_2 as above.

LEMMA 9.2.2   There exist positive numbers ϵ_0, C, and A such that the following properties hold: Let $\frac{1}{2} < a < 1, 0 < \epsilon \leq \epsilon_0 < \frac{1}{2}$. Fix a compact subset E of Ω. Let f be a continuous, complex-valued function on $\partial\Omega$ such that $|f| < 1$. Assume that $V \subseteq \partial\Omega$ is a closed, proper subset with $|f| \geq a$ on V. Then there is a function $g \in A(\Omega)$ and a closed set $W \subseteq \partial\Omega$ such that

a. $|f(z) + g(z)| \leq 1 + 2\epsilon$ for all $z \in \partial\Omega$;

b. $\max_E |g| < \epsilon$;

c. $|f(z) + g(z)| \geq a - 3\epsilon$ on $V \cup W$;

d. $V \cap W = \emptyset$ and $\mu(W) \geq C(\cos^{-1} a) \left(\frac{\log 1/a}{\log A/\epsilon} \right)^{(2n-1)/2} (1 - \mu(V))$.

Remark: We see that this lemma is our device for pushing the image of a function outward. At the nth stage of our construction, we will have a function doing approximately what we want on a set V. We will then do some modifications on a new set W. The aim is enlarge the set on which our function is under control from V to W, while not destroying what we have already achieved on V. Part c says, in a quantitative fashion, that $f + g$ is large on $V \cup W$. Part a says that, globally, the function that we are building is still under control. $\qquad\square$

Proof of the Lemma   If $\gamma > 0$, then let

$$W^\gamma = \{z \in \partial\Omega : \text{dist}\,(z,V) > \gamma\}.$$

Then

$$\lim_{\gamma \to 0^+} \mu(W^\gamma) \nearrow \mu(\partial\Omega \setminus V) = 1 - \mu(V).$$

Thus there exists a number $\gamma_1 > 0$ such that if $0 < \gamma < \gamma_1$, then $\mu(W^\gamma) > \frac{1}{2}(1 - \mu(V))$. By the uniform continuity of f, there is a number $\gamma_2 > 0$ such that when $|z - z'| < \gamma_2$, then $|f(z) - f(z')| < \epsilon$. Let $0 < r < \min(\gamma_1, \gamma_2, r_0)$ (the number r_0 comes from the preceding proposition).

We choose a maximal pairwise disjoint collection $\{\beta_1(w_j, r)\}$ of closed balls with $w_j \in W^r$, $j = 1, \ldots, N_r$. By design, these balls are all contained in $\partial\Omega \setminus V$; therefore,

$$1 - \mu(V) \geq \sum_{j=1}^{N_r} \mu\left(\beta_1(w_j, r)\right) \geq N_r C_1 r^{2n-1}. \tag{9.2.2.1}$$

The balls $\beta_1(w_j, 3r)$ must cover W^r (remember the covering lemma from Section 8.4). Therefore,

$$N_r C_2 3^{2n-1} r^{2n-1} \geq \sum_{j=1}^{N_r} \mu\beta_1(w_j, 3r) \geq \mu(W^r) \geq \frac{1}{2}(1 - \mu(V)). \tag{9.2.2.2}$$

Combining (9.2.2.1) and (9.2.2.2) gives

$$\frac{C_6}{r^{2n-1}}(1 - \mu(V)) \leq N_r \leq \frac{C_7}{r^{2n-1}}(1 - \mu(V))$$

for suitable constants C_6, C_7.

Now we will estimate the number of points w_j that lie at a given distance from a point $z \in \partial\Omega$. Let $M_r = [(\operatorname{diam}\Omega)/r]$, where $[\ ]$ denotes the greatest integer function. For z fixed and $k \in \{0, 1, \ldots, M_r\}$, let

$$V_k(z) = \{w_j : kr \leq d(z, w_j) < (k+1)r\}.$$

Set $N^k(z) = \operatorname{card} V_k(z)$. Then $N^0(z) \leq 1$. If $w_j \in V_k(z)$, then $\beta_1(w_j, r) \subseteq \beta_1(z, (k+2)r)$; hence

$$N^k(z)C_1 r^{2n-1} \leq \sum_{w_j \in V_k(z)} \mu[\beta_1(w_j, r)] \leq \mu[\beta_1(z, (k+2)r)] \leq C_2(k+2)^{2n-1} r^{2n-1},$$

which implies that

$$N^k(z) < C_8 k^{2n-1} \qquad \text{for} \quad k \geq 1. \tag{9.2.3}$$

Now set

$$g(z) = \sum_{j=1}^{N_r} \tau_j e^{-m\Phi(z, w_j)},$$

where τ_j is a complex number selected so that

$$|f(w_j) + \tau_j| = |f(w_j)| + |\tau_j| = 1$$

and $m > 0$ will be specified momentarily.

Then $g \in A(\Omega)$ and each $|\tau_j| \le 1$. We now show that m and r can be selected so that g has the desired properties.

By part 3 of Proposition 9.2.1 and by (9.2.2.3), we know for $z \in \partial\Omega$ that

$$
\begin{aligned}
|g(z)| &\le \sum_{j=1}^{N_r} |\tau_j| e^{-m\operatorname{Re}\Phi(z,w_j)} \\
&\le \sum_{j=1}^{N_r} e^{-C_5 m|z-w_j|^2} \\
&= \sum_{k=0}^{M_r} \sum_{w_j \in V_k(z)} e^{-C_5 m|z-w_j|^2} \\
&\le 1 + \sum_{k=1}^{M_r} N^k(z) e^{-C_5 m k^2 r^2} \\
&\le 1 + C_8 \sum_{k=1}^{M_r} k^{2n-1} e^{-(C_5 m r^2)k^2}.
\end{aligned}
$$

If we select m and r so that mr^2 is very large, then we see that

$$
\begin{aligned}
|g(z)| &\le 1 + C_8 \sum_{k=1}^{\infty} e^{-(C_5 m r^2)k} \\
&< 1 + 2C_8 e^{-C_5 m r^2} \\
&\equiv 1 + A \cdot e^{-C_5 m r^2}.
\end{aligned}
$$

Thus, if ϵ has been chosen and if we select m and r to satisfy

$$mr^2 = \frac{1}{C_5} \log \frac{A}{\epsilon}, \tag{9.2.2.4}$$

then we have the estimate

$$|g(z)| < 1 + \epsilon \tag{9.2.2.5}$$

and

$$|g(z)| < \epsilon \quad \text{if} z \in \partial\Omega \setminus \left(\bigcup_j \beta_1(w_j, r) \right) \equiv F. \tag{9.2.2.6}$$

(For (9.2.2.6) notice that if $z \notin \cup_j \beta_1(w_j, r)$, then $V_0 = \emptyset$ so that the estimate (9.2.2.5) for g has no 1 on the right side.) Observe that there is a number $\epsilon_0 > 0$ such that if $\epsilon \leq \epsilon_0$, then (9.2.2.4) implies that mr^2 is large and (9.2.2.5), (9.2.2.6) hold. Condition (9.2.2.4) is satisfied by arbitrarily small r and arbitrarily large m.

Now lines (9.2.2.5) and (9.2.2.6) yield conclusion (a) rather quickly. For V is disjoint from all the balls, so $|f(z) + g(z)| \geq a - \epsilon$ in V. Also, if $z \in F$, then $|f(z) + g(z)| < |f(z)| + \epsilon$. If $z \in \beta_1(w_j, r)$ for some j, then

$$
\begin{aligned}
|f(z) + g(z)| &\leq |f(z) - f(w_j)| + \left| f(w_j) + \tau_j e^{-m\Phi(z, w_j)} \right| \\
&\quad + \left| \sum_{k \neq j} \tau_k e^{-m\Phi(z, w_k)} \right| \\
&\leq \epsilon + 1 + \epsilon \\
&= 1 + 2\epsilon.
\end{aligned}
$$

Thus we have established (a).

Set $W' = \cup_{j=1}^{N_r} \beta_1(w_j, r)$. Then W' lies in $\partial\Omega \backslash V$. We now find a distinguished subset W of W' such that $|f(z) + g(z)| \geq a - 3\epsilon$ in W and we give an estimate on its area. We introduce the notation

$$
\begin{aligned}
\alpha &= |f(w_j)| \\
s &= \left| e^{-m\Phi(z, w_j)} \right| = e^{-m\operatorname{Re}\Phi(z, w_j)} \\
\theta &= \arg\left\{ e^{-m\Phi(z, w_j)} \right\} = -m\operatorname{Im}\Phi(z, w_j).
\end{aligned}
$$

Let $z \in \beta_1(w_j, r)$. Then

$$
\begin{aligned}
|f(z) + g(z)| &\geq \left| f(w_j) + \tau_j e^{-m\Phi(z, w_j)} \right| - |f(z) - f(w_j)| \\
&\quad - \left| \sum_{k \neq j} \tau_k e^{-m\Phi(z, w_j)} \right| \\
&\geq \left| f(w_j) + \tau_j e^{-m\Phi(z, w_j)} \right| - 2\epsilon \\
&= \left| f(w_j)\overline{\operatorname{sgn} f(w_j)} + \tau_j \overline{\operatorname{sgn} f(w_j)} e^{-m\Phi(z, w_j)} \right| - 2\epsilon \\
&= \left| |f(w_j)| + |\tau_j| \left| e^{-m\Phi(z, w_j)} \right| \right| - 2\epsilon \\
&= \left| \alpha + (1 - \alpha)s e^{i\theta} \right| - 2\epsilon.
\end{aligned}
$$

It follows that $|f(z) + g(z)| \geq a - 3\epsilon$ if $|\alpha + (1 - \alpha)se^{i\theta}| \geq a$. This last condition is equivalent to

$$\alpha^2 + 2\alpha(1 - \alpha)s \cos\theta + (1 - \alpha)^2 s^2 \geq a^2. \qquad (9.2.2.7)$$

This new condition is easily seen to hold if

1. $s \geq a$; and

2. $\cos\theta \geq a$.

Indeed, if these two conditions hold, then the left side of (9.2.2.7) is greater than or equal to

$$
\begin{aligned}
\alpha^2 + 2\alpha(1 - \alpha)a^2 + (1 - \alpha)^2 a^2 &= a^2(1 - \alpha^2) + \alpha^2 \\
&= a^2 + \alpha^2(1 - a^2) \\
&\geq a^2.
\end{aligned}
$$

Now, by part 2 of Proposition 9.2.1, we know that

$$s \geq e^{-mC_4 |z - w_j|^2}$$

and therefore condition 1 holds in a ball $\beta_1(w_j, \rho)$ with

$$m\rho^2 = \frac{1}{C_4} \log \frac{1}{a}.$$

This all makes sense provided that $\rho < r$, and can be verified using (9.2.2.4).
 Condition 2 says that $\cos(m \mathrm{Im}\, \Phi(z, w_j)) \geq a$. By part 4 of Proposition 9.2.1, there is a a closed subset W_j of $\beta_1(w_j, \rho)$ such that this inequality holds and such that

$$\mu(W_j) \geq C_3(\cos^{-1} a)\rho^{2n-1}.$$

Letting

$$W = \bigcup_j W_j,$$

we see that $V \cap W = \emptyset$. We now get a lower bound on $\mu(W)$ by using the estimates on the measures of the individual W_j's, the estimates on N_r, (9.2.2.4), and the equation defining ρ. We have

$$
\begin{aligned}
\mu(W) &\geq C_3 \left(\cos^{-1} a\right) \rho^{2n-1} \cdot N_r \\
&\geq C_3 \cdot C_6 (\cos^{-1} a) \left(\frac{\rho}{r}\right)^{2n-1} (1 - \mu(V)) \\
&= C_3 \cdot C_6 \left(\frac{C_5}{C_4}\right)^{(2n-1)/2} \cdot \left(\cos^{-1} a\right) \left(\frac{\log 1/a}{\log A/\epsilon}\right)^{(2n-1)/2} (1 - \mu(V)).
\end{aligned}
$$

This proves parts c and d.

Finally, the set E is at a positive distance from $U = \cup_j \beta_1(w_j, r)$ (remember that E is a compact subset of the *interior* of Ω), so part 3 of Proposition 9.2.1 tells us that for $z \in E$ we have

$$
\mathrm{Re}\,\Phi(z, w_j) \geq C_5 |z - w_j|^2 \geq C_5 [\mathrm{dist}\,(E, U)]^2 \equiv B > 0.
$$

Since we may assume that $mr^2 \geq 1$, we see from the estimate on N_r that

$$
|g(z)| \leq N_r e^{-mB} \leq C_7 \frac{1}{r^{2n-1}} e^{-mB} \leq C_7 m^{(2n-1)/2} e^{-mB}.
$$

Choosing m large enough, we may verify part b.

The proof is now complete.   □

Remark: The lemma comprises the technical part of the construction of inner functions. The reader should understand that we added together peaking functions in a very organized fashion in order to bump out the range of the function f. The next parts of the proof use the statement of the lemma, but not the ingredients that went into it.   □

LEMMA 9.2.3   Let Ω, μ, and E be as in Lemma 9.2.2. Let f be a continuous, complex-valued function on $\partial\Omega$ such that $|f| < 1$. Then for each $\epsilon > 0$ there is a function $h \in A(\Omega)$ and a closed set $V \subseteq \partial\Omega$ such that

a. $|f(z) + h(z)| \leq 1 + \epsilon$ for all $z \in \partial\Omega$;

b. $\max_E |h| \leq \epsilon$;

c. $|f(z) + h(z)| \geq 1 - \epsilon$ for all $z \in V$;

d. $\mu(V) > 1 - \epsilon$.

Proof   Before we begin, we observe that the Lemma 9.2.2 is still true if the special number 1 is replaced by any other constant c. We will apply the preceding lemma iteratively, with new ϵ, f, c, V at each iteration. We shall refer to these items as "data" for the lemma. Fix $\epsilon > 0$.

Set $a = 1 - \epsilon/2$ and choose numbers $\epsilon_i > 0$ such that $6 \sum_{i=1}^{\infty} \epsilon_i < \epsilon$. We begin our process by applying Lemma 9.2.2 to the data $\epsilon_1, f, 1, \emptyset$. This produces a function $h_1 \in A(\Omega)$ and a closed set $V_1 \subseteq U$ (here $V_1 = V \cup W$ in the language of Lemma 9.2.2) such that

a$_1$. $|f(z) + h_1(z)| \leq 1 + 3\epsilon_1$ for all $z \in \partial\Omega$;

b$_1$. $\max_E |h_1| \leq \epsilon_1$;

c$_1$. $|f(z) + h_1(z)| \geq a - 3\epsilon_1 = a(1 - \frac{3}{a}\epsilon_1)$ for all $z \in V_1$;

d$_1$. $\sigma_1 \equiv \mu(V_1) \geq C(\cos^{-1} a) \cdot \left(\frac{\log 1/a}{\log A/\epsilon_1} \right)^{(2n-1)/2}$.

We then proceed inductively as follows: Suppose that functions $h_1, \ldots, h_k \in A^1(\Omega)$ have been found and also pairwise disjoint closed sets $V_1, \ldots, V_k$ in U. Let $W_k = \cup_{i=1}^k V_i$ and suppose that we have the following:

a$_k$. $|f(z) + \sum_{i=1}^k h_i(z)| \leq 1 + 3\sum_{i=1}^k \epsilon_i$ for all $z \in \partial\Omega$;

b$_k$. $\max_E |\sum_{i=1}^k h_i| \leq \sum_{i=1}^k \epsilon_i < \epsilon$;

c$_k$. $|f(z) + \sum_{i=1}^k h_i(z)| \geq a - 3\sum_{i=1}^k \epsilon_i = a(1 - \frac{3}{a}\sum_{i=1}^k \epsilon_i)$ for all $z \in W_k$;

d$_k$. $\sigma_k \equiv \mu(V_k) \geq C(\cos^{-1} a) \cdot \left(\frac{\log 1/a}{\log A/\epsilon_k} \right)^{(2n-1)/2} \cdot \left(1 - \sum_{i=1}^{k-1} \sigma_i \right)$.

By c, we may (for our inductive step) apply Lemma 9.2.2 to the data ϵ_{k+1}, $f + \sum_{i=1}^k h_i$, $1 - \frac{3}{a}\sum_{i=1}^k \epsilon_i$, W_k to produce a new function $h_{k+1} \in A(\Omega)$ and a closed set $V_{k+1} \subseteq U$. Defining $W_{k+1} = \cup_{i=1}^{k+1} V_i$, it is straightforward to verify statements a$_{k+1}$–d$_{k+1}$.

The induction is now complete. If we choose a small number $\tau > 0$ and select

$$\epsilon_k = A\tau^{(k^{2/(2n-1)})},$$

then

$$\sum_{k=1}^{\infty} \frac{1}{[\log(A/\epsilon_k)]^{(2n-1)/2}} = \infty. \tag{9.2.3.1}$$

If it were the case that $\sum_{i=1}^{\infty} \sigma_i < 1$, then statements d$_k$ would show that there is a constant C_9 such that

$$\sigma_k \geq C_9 \frac{1}{[\log(A/\epsilon_k)]^{(2n-1)/2}},$$

and that would contradict (9.2.3.1) (the σ_k must sum to a finite number because they are measures of pairwise disjoint subsets of $\partial\Omega$). We conclude that $\sum_{k=1}^{\infty} \sigma_k \geq 1$; hence it must, in fact, equal 1.

It follows that, for k sufficiently large and $V = W_k$, we have

$$\mu(V) = \sum_{i=1}^{k} \sigma_i > 1 - \frac{1}{2}\epsilon.$$

That proves (d). If we set $h = \sum_{i=1}^{\infty} h_i$, then conclusions a and b of our lemma follow immediately from statements a_k and b_k. By statements c_k,

$$|f(z) + h(z)| \geq a - 3\sum_{i=1}^{k} \epsilon_i > a - \frac{\epsilon}{2} = 1 - \epsilon \qquad \text{for all} \quad z \in V.$$

That establishes conclusion c. The proof of the lemma is now complete. □

As in Chapter 8, if f is a bounded holomorphic function on a bounded domain Ω with C^2 boundary, then we let $\tilde{f}$ denote its boundary limit function. We observe that the last lemma remains true with the constant 1 replaced by any other constant c. Now we have the following.

THEOREM 9.2.4 Let Ω be a C^2 strongly pseudoconvex domain and let f be a continuous complex-valued function on $\partial\Omega$ such that $|f(z)| < 1$ for all z. Let $F \subseteq \Omega$ be a compact set. Fix $\epsilon > 0$. Then there is a function $g \in H^\infty(\Omega)$ such that $|f + \tilde{g}| = 1$ almost everywhere on $\partial\Omega$ and $\max_{z \in F} |g(z)| < \epsilon$.

Proof We will apply the last lemma iteratively with variable data f, c, E, ϵ. Here, as in the proof of the lemma, c is a constant that will change at each step. Since $|f|$ must be bounded away from 1 on $\partial\Omega$, we may choose numbers $\{b_j\}$ such that

$$|f| < b_1 < b_2 < \cdots \nearrow 1.$$

Also select numbers ϵ_k such that $b_k + \epsilon_k < b_{k+1}$ for every k and $\sum_{k=1}^{\infty} \epsilon_k < \epsilon$. As usual, let μ be area on $\partial\Omega$ normalized to have total mass 1.

We begin by applying Lemma 9.2.3 to the data f, b_1, F, ϵ_1 to obtain a function $g_1 \in A(\Omega)$ such that

a_1. $|f(z) + g_1(z)| \leq b_1 + \epsilon_1 < b_2$ for all $z \in \partial\Omega$;

b_1. $\max_{z \in F} |g_1(z)| \leq \epsilon_1$;

c_1. $|f(z) + g_1(z)| \geq b_1 - \epsilon_1$ for all $z \in V_1$;

d_1. $\mu(V_1) > 1 - \epsilon_1$.

By a_1, c_1, and uniform continuity of g_1, there must exist a $\delta_1 > 0$ such that

e_1. $b_1 - 2\epsilon_1 \leq |f(z) + g_1(z - \delta_1 \nu_z)| \leq b_1 + 2\epsilon_1$ for all $z \in V_1$.

Here, as in Chapter 8, ν_z denotes the unit outward normal to the boundary of Ω at $z \in \partial\Omega$. Let us denote the set F by E_1 and choose $E_2 \subseteq \Omega$ a compact set such that

f$_1$. $E_1 \subseteq E_2$ and $z - \delta_1 \nu_z \in E_2$ for all $z \in \partial\Omega$.

Proceeding inductively, let us assume that functions $g_1, \dots, g_k \in A(\Omega)$, closed sets $V_1, \dots, V_k \subseteq \partial\Omega$, positive numbers $\delta_1, \dots, \delta_k$, and closed sets $E_1, \dots, E_{k+1} \subseteq \Omega$ have been found such that, setting $h_k = \sum_{i=1}^{k} g_i$, we have

a$_k$. $|f(z) + h_k(z)| \le b_k + \epsilon_k < b_{k+1}$ for all $z \in \partial\Omega$;

b$_k$. $\max_{z \in E_k} |g_k(z)| \le \epsilon_k$;

c$_k$. $|f(z) + h_k(z)| \ge b_k - \epsilon_k$ for all $z \in V_k$;

d$_k$. $\mu(V_k) > 1 - \epsilon_k$;

e$_k$. $b_k - 2\epsilon_k \le |f(z) + h_k(z - \delta_k \nu_z)| \le b_k + 2\epsilon_k$ for all $z \in V_k$;

f$_k$. $E_k \subseteq E_{k+1}$, $E_{k+1} \subset\subset \Omega$, and $z - \delta_k \nu_z \in E_{k+1}$ for all $z \in \partial\Omega$.

By a$_k$ and f$_k$, we may apply Lemma 9.2.3 to the data $f + h_k$, b_{k+1}, E_{k+1}, ϵ_{k+1} and obtain a new function $g_{k+1} \in A(\Omega)$ and a closed set $V_{k+1} \subseteq \partial\Omega$. It is then trivial to verify a$_{k+1}$, b$_{k+1}$, c$_{k+1}$, and d$_{k+1}$ for g_{k+1} and V_{k+1}. By a$_{k+1}$, c$_{k+1}$, and the uniform continuity of h_{k+1}, there is some $\delta_{k+1} > 0$ such that e$_{k+1}$ holds. We then choose E_{k+2} such that f$_{k+1}$ holds.

That completes the induction. There is no loss of generality to assume at this point that $\delta_k \searrow 0$ and that $\Omega = \cup_{k=1}^{\infty} E_k$. By statements b$_k$, the function $g = \lim_{k \to \infty} h_k = \sum_{i=1}^{\infty} g_i$ is well defined and holomorphic on Ω, since the series converges uniformly on compact subsets. It is also immediate that $\max_F |g| \le \sum \max_F |g_i| \le \sum \epsilon_i < \epsilon$. By statements a$_k$, $g \in H^{\infty}(\Omega)$.

Now let

$$W_j = \cap_{k \ge j} V_k.$$

Then $W_j \subseteq W_{j+1}$, each j. The statements d$_k$ then imply that $\mu(W_j) \ge 1 - \sum_{k \ge j} \epsilon_k$; therefore, $\lim_{j \to \infty} \mu(W_j) = 1$ and $W \equiv \cup_{j=1}^{\infty} W_j$ contains almost all of $\partial\Omega$. If $z \in W$, then there is some j such that $z \in V_k$ for all $k \ge j$. For such a k, the statements b$_k$, e$_k$, and f$_k$ imply that

$$\left| b_k - |f(z) + g(z - \delta_k \nu_z)| \right| \le 2\epsilon_k + \sum_{i > k} \epsilon_i;$$

hence

$$\lim_{k \to \infty} |f(z) + g(z - \delta_k \nu_z)| = 1.$$

Thus, in the full measure subset of W where $\tilde{g}(z)$ exists, we may conclude that

$$|f(z) + \tilde{g}(z)| = 1.$$

The theorem is proved.   □

COROLLARY 9.2.5   There exist nonconstant inner functions on any strongly pseudoconvex domain.

Proof   In the theorem take $f = 0$, F any nonempty compact set, and $\epsilon = \frac{1}{2}$. We find that there exists a bounded holomorphic g with unimodular boundary values that is smaller than $\frac{1}{2}$ on F. Thus g is a nonconstant inner function.   □

Remark: Observe that once a domain has one nonconstant inner function, then it has many. For one may take powers of the inner function, postcompose it with Möbius transformations, and perform other function-algebraic tricks to generate more inner functions. It turns out that the inner functions are dense in H^∞ *in the topology of uniform convergence on compact sets*. This is a much weaker statement than Marshall's theorem in the disc that the closed linear span is dense in the L^∞ topology on the boundary. The analogue of Marshall's theorem in higher dimensions is known to fail.   □

9.3 Further Results Obtained with Constructive Methods

It was Erik Løw who first realized the flexibility inherent in the inner functions construction. In fact, one can use it to specify the boundary behavior of holomorphic functions in a variety of interesting ways. We shall indicate some of these in the present section. As in Section 9.2, our domain Ω is strongly pseudoconvex with C^2 boundary.

We begin by stating Løw's version of Lemma 9.2.2.

LEMMA 9.3.1   There exist positive numbers ϵ_0, C, and A such that the following properties hold: Let $\frac{1}{2} < a < 1$, $0 < \epsilon \leq \epsilon_0$. Fix a compact subset E of Ω. Let f be a continuous, complex-valued function on $\partial\Omega$ and ϕ a continuous function on $\partial\Omega$ such that $\phi \leq 1$. Assume that $V \subseteq \partial\Omega$ is a closed set with $|f| \geq a\phi$ on V. Then there is a function $g \in A(\Omega)$ and a closed set $W \subseteq \partial\Omega$ such that

a. $|f(z) + g(z)| \leq \max\{|f(z)|, \phi(z)\} + 3\epsilon$ for all $z \in \partial\Omega$;

b. $\max_E |g| < \epsilon$;

c. $|f(z) + g(z)| \geq a\phi(z) - 3\epsilon$ on $V \cup W$;

d. $V \cap W = \emptyset$ and $\mu(W) \geq C(\cos^{-1} a)\left(\frac{\log 1/a}{\log A/\epsilon}\right)^{(2n-1)/2}(1 - \mu(V))$.

In fact the proof of this version of the lemma requires only formal changes applied to the one that we presented in Section 9.2. We leave it as an exercise for the reader to provide the details, or the reader may consult Løw [3].

This more general version of Lemma 9.2.2 leads directly to the following more general version of 9.2.3.

LEMMA 9.3.2 Let Ω, μ, and E be as in Lemma 9.3.1. Let f be a continuous, complex-valued function on $\partial\Omega$ and let ϕ be a continuous function on $\partial\Omega$ such that $|f| \leq \phi$. Then for each $\epsilon > 0$, there is a function $h \in A(\Omega)$ and a closed set $V \subseteq \partial\Omega$ such that

a. $|f(z) + h(z)| \leq \phi(z) + \epsilon$ for all $z \in \partial\Omega$;

b. $\max_E |h| \leq \epsilon$;

c. $|f(z) + h(z)| \geq \phi(z) - \epsilon$ for all $z \in V$;

d. $\mu(V) > 1 - \epsilon$.

Again, we shall not provide details of the proof (which are just like the details of the proof of 9.2.3), but leave them to the interested reader. The payoff for the new generality we have is the following powerful version of Theorem 9.2.4.

THEOREM 9.3.3   Let Ω be strongly pseudoconvex with C^2 boundary and let ϕ be a positive, continuous function on $\partial\Omega$. Assume that f is a continuous, complex-valued function on $\partial\Omega$ such that $|f(z)| < \phi(z)$ for all z. Let $F \subseteq \Omega$ be a compact set. Fix $\epsilon > 0$. Then there is a function $g \in H^\infty(\Omega)$ such that $|f + \tilde{g}| = \phi$ almost everywhere on $\partial\Omega$ and $\max_{z \in F} |g(z)| < \epsilon$.

Details of the proof of this theorem are left for the reader. Let us derive some consequences. We provide full details for the proofs that follow.

THEOREM 9.3.4 Let $\phi > 0$ be any continuous function on $\partial\Omega$. There exists a function $g \in H^\infty(\Omega)$ such that $|\tilde{g}| = \phi$ almost everywhere on $\partial\Omega$.

Proof   Take $f = 0$ in the preceding theorem.   □

These theorems represent a striking generalization of the existence of inner functions. The methods of sheaf theory, partial differential equations, and integral representations do not even hint that a result such as this is possible.

Notice that in one complex variable the result of Theorem 9.3.4 is straightforward. Let u be the solution of the Dirichlet problem with boundary data $\log \phi$. If v is a harmonic conjugate to u, then $h(\zeta) = e^{u(\zeta)+iv(\zeta)}$ does the job. And, in fact, $|h|$ extends continuously to the entire closure of the domain; it agrees with ϕ at *every* boundary point.

Now we derive a statement about the closure of the inner functions in the topology of normal convergence. In the proof we deviate from strict rules of logic and use a result that is proved in the next chapter by independent reasoning.

PROPOSITION 9.3.5 Let $\Omega \subset\subset \mathbb{C}^n$ be strongly pseudoconvex with C^4 boundary. Then the inner functions are dense in the unit ball of the space $H^\infty(\Omega)$ in the topology of uniform convergence on compact sets.

Proof   Fix a function $h \in H^\infty(\Omega)$ with $\|h\| \leq 1$. By the proof of Theorem 10.4.2 and by diagonalization there is a sequence $h_j \in A(\Omega)$ such that $\|h_j\| < 1$ and $h_j \to h$ normally on $\bar{\Omega}$ (it is to apply Theorem 10.4.2 that we need C^4 boundary).

Let $E_1 \subseteq E_2 \subseteq \cdots \subseteq \Omega$ be a sequence of compact sets that exhaust Ω. Now by Theorem 9.3.3 we may find $f_j \in H^\infty(\Omega)$ such that $|h_j + \tilde{f}_j| = 1$ almost everywhere on $\partial\Omega$ and such that $\max_{E_j} |f_j| < 1/j$. Therefore, $\{h_j + f_j\}$ is a sequence of inner functions converging uniformly on compacta to h. □

Løw was the first to consider the application of constructive methods to the study of the boundary values of functions in $A(\Omega)$. We begin our presentation of some of these results with a fundamental lemma of Aleksandrov. This lemma is the core of *his* construction of inner functions.

LEMMA 9.3.6   Let Ω be a strongly pseudoconvex domain with C^2 boundary. Fix $z \in \Omega$ and $m \in \mathbb{N}$. Then the following hold:

1. There is a constant $\gamma < 1$ such that for any positive function $\phi \in C(\partial\Omega)$, there is a function $f \in A(\Omega)$ such that $|f| < \phi$ on $\partial\Omega$ and

$$\|\phi - \mathrm{Re} f\|_{L^{1/2}(\partial\Omega)}^{1/2} \leq \gamma \|\phi\|_{L^{1/2}(\partial\Omega)}^{1/2}.$$

2. For each $k \in \mathbb{N}$ there is an $h_k \in A(\Omega)$ such that
   a. $\mathrm{Re} h_k < \phi$ on $\partial\Omega$ and $\|\phi - h_k\|_{\sup} \leq 2^k \|\phi\|_{\sup}$;
   b. $\|h_k\|_{\sup} \leq 2^k \|\phi\|_{\sup}$;
   c. $\|\phi - \mathrm{Re} h_k\|_{L^{1/2}(\partial\Omega)}^{1/2} \leq \gamma^k \|\phi\|_{L^{1/2}(\partial\Omega)}^{1/2}$;
   d. If $\epsilon > 0$ and $E = \{z \in \partial\Omega : \phi(z) - \mathrm{Re} h_k(z) \geq \epsilon\}$, then

$$\mu(E) \leq \|\phi\|_{\sup}^{1/2} \gamma^k / \epsilon^{1/2}.$$

Proof   We shall see that $\gamma = 127/128$ will do the job. Fix a positive $\phi \in C(\partial\Omega)$. First, there is an $\epsilon > 0$ such that

$$\int_U \phi^{1/2} d\mu < \frac{1}{128} \int_{\partial\Omega} \phi^{1/2} \, d\mu \qquad \text{whenever } \mu(U) \leq \epsilon.$$

According to Lemma 9.3.2, we can find a function $F \in A(\Omega)$ and a closed set $V \subseteq \partial\Omega$ such that $|F| < \phi$ on $\partial\Omega$, $|F| > \frac{1}{2}\phi$ on V, and $\mu(V) > 1 - \epsilon$. We claim that one of $F, -F, iF$, or $-iF$ is the function that we seek.

To prove this assertion, we will exploit the elementary inequality

$$\frac{1}{2}(1+x)^{1/2} + \frac{1}{2}(1-x)^{1/2} \leq 1 - \frac{1}{8}x^2 \qquad \text{for all } |x| \leq 1.$$

We have

$$\int_{\partial\Omega} (\phi - \text{Re}F)^{1/2} + (\phi + \text{Re}F)^{1/2} + (\phi - \text{Im}F)^{1/2} + (\phi + \text{Im}F)^{1/2} \, d\mu$$

$$= \int_{\partial\Omega} \phi^{1/2} \left[\left(1 - \text{Re}\frac{F}{\phi} \right)^{1/2} \right.$$

$$+ \left(1 + \text{Re}\frac{F}{\phi} \right)^{1/2}$$

$$\left. + \left(1 - \text{Im}\frac{F}{\phi} \right)^{1/2} + \left(1 + \text{Im}\frac{F}{\phi} \right)^{1/2} \right] d\mu$$

$$\leq \int_{\partial\Omega} \phi^{1/2} \left[4 - \frac{1}{4} \left(\text{Re}\frac{F}{\phi} \right)^2 - \frac{1}{4} \left(\text{Im}\frac{F}{\phi} \right)^2 \right] d\mu$$

$$= \int_{\partial\Omega} \phi^{1/2} \left[4 - \frac{1}{4} \left| \frac{F}{\phi} \right|^2 \right] d\mu$$

$$\leq \int_V \phi^{1/2} \left(4 - \frac{1}{16} \right) d\mu + 4 \int_{\partial\Omega \setminus V} \phi^{1/2} d\mu$$

$$< 1 \cdot \left(4 - \frac{1}{16} \right) \int_{\partial\Omega} \phi^{1/2} \, d\mu + 4 \cdot \left(\frac{1}{128} \right) \int_{\partial\Omega} \phi^{1/2} \, d\mu$$

$$= \frac{127}{32} \int_{\partial\Omega} \phi^{1/2} \, d\mu.$$

It follows that one of the four integrals

$$\int_{\partial\Omega} (\phi - \text{Re}F)^{1/2} \, d\mu, \qquad \int_{\partial\Omega} (\phi + \text{Re}F)^{1/2} \, d\mu \ ,$$

$$\int_{\partial\Omega} (\phi - \text{Im}F)^{1/2} \, d\mu, \qquad \int_{\partial\Omega} (\phi + \text{Im}F)^{1/2} \, d\mu$$

must be smaller than

$$\frac{127}{128} \int_{\partial\Omega} \phi^{1/2} \, d\mu \equiv \frac{127}{128} \|\phi^{1/2}\|_{L^{1/2}(\partial\Omega)}^{1/2}.$$

This verifies conclusion 1 of the lemma.

Now we turn to conclusion 2. For $k = 1$, conclusions a, b, and c reduce to conclusion 1. Then we inductively apply (1) to $\phi - \text{Re}h_k$ to produce f_{k+1} and set $h_{k+1} = h_k + f_{k+1}$. Conclusions a, b, and c for $k + 1$ then follow. The induction is complete and we have proved a, b, and c.

To prove (d), we use (c) and Chebycheff's inequality to write

$$\epsilon^{1/2}\mu(E) \leq \int_{\partial\Omega} (\phi - \mathrm{Re}h_k)^{1/2}\, d\mu \leq \gamma^k \|\phi\|_{L^{1/2}(\partial\Omega)}^{1/2} \leq \gamma^k \|\phi\|_{\mathrm{sup}}^{1/2}.$$

That completes the proof of the lemma.   □

COROLLARY 9.3.7   Let $\epsilon' > 0$. Then there are constants $0 < \gamma < 1$ and $K > 0$ such that for any positive $\phi \in C(\partial\Omega)$ and $j \in \mathbb{N}$, there is an $h_j \in A(\Omega)$ such that

a. $\mathrm{Re}h_j < \phi$ on $\partial\Omega$;

b. $\|h_j\|_{\mathrm{sup}} \leq 2^j K \|\phi\|_{\mathrm{sup}}$;

c. If $\epsilon_j > 0$ and $E = \{z \in \partial\Omega : \phi(z) - \mathrm{Re}h_j(z) \geq \epsilon_j\}$, then $\mu(E) \leq \epsilon' \dfrac{\|\phi\|_{\mathrm{sup}}^{1/2}\gamma^j}{\epsilon_j^{1/2}}$.

Proof   This is little more than a restatement of Aleksandrov's lemma. For let γ be as in the lemma and choose a positive integer N such that $\gamma^N < \epsilon'$. We let $K = 2^N$ and apply part 2 of the lemma with $k = N + j$. The corollary follows. □

Now we have reached the main goal of this section. For emphasis and clarity, we state it in two separate parts.

THEOREM 9.3.8   Let Ω be a strongly pseudoconvex domain with C^2 boundary. Let ϕ be any continuous, real-valued function on $\partial\Omega$. Then there is a nonconstant function $f \in A(\Omega)$ such that $\mathrm{Re}f \leq \phi$ on $\partial\Omega$ and

$$\mu\{z \in \partial\Omega : \mathrm{Re}f(z) \neq \phi(z)\} < \epsilon.$$

COROLLARY 9.3.9   With Ω as in the theorem, $\epsilon > 0$, and ϕ a positive, continuous function on $\partial\Omega$, there exists a nonconstant $g \in A(\Omega)$ such that $|g| \leq \phi$ on $\partial\Omega$ and $\mu\{z \in \partial\Omega : |g(z)| \neq \phi(z)\} < \epsilon$.

The corollary follows from the theorem because we may apply the theorem to the function $\psi = \log \phi$. Thus we obtain a function $f \in A(\Omega)$ whose real part agrees with ψ on the boundary except on a set of measure less than ϵ. But then the function $g = \exp f$ satisfies the conclusion of the corollary.

Proof of the Theorem   By adding a large constant to ϕ and then multiplying by a small one, we may assume that $0 < \phi < 1$. We shall iteratively apply Corollary 9.3.7, using $\epsilon_j = 4^{-j}$ and discarding bad points in the boundary at each step.

For $j = 1$ the corollary gives us a function h_1 such that $\mathrm{Re}h_1 < \phi$ on $\partial\Omega$, $\|h_1\|_{\mathrm{sup}} \leq 2K$, and $\mu\{z \in \partial\Omega : \phi - \mathrm{Re}h_1 \geq 1/4\} \leq 2\epsilon'\gamma$. Let $\phi_1 = \min\{\phi - \mathrm{Re}h_1, \frac{1}{4}\}$. Then $\phi_1 > 0$ and we set $g_1 = \phi - \mathrm{Re}h_1 - \phi_1$. It follows that $g_1 \geq 0$ and we see that

a₁. $\mathrm{Re}h_1 + g_1 < \phi$ on $\partial\Omega$ and $\|\phi - \mathrm{Re}h_1 - g_1\|_{\mathrm{sup}} \leq 1/4$;

b$_1$. $\|h_1\|_{\sup} \le 2K$;

c$_1$. $\mu\{z \in \partial\Omega : g_1(z) > 0\| \le 2\epsilon'\gamma$.

Inductively, suppose that we have found $h_1, \ldots, h_k \in A(\Omega)$ and positive, continuous functions $g_1, \ldots, g_k$ on $\partial\Omega$ such that

a$_k$. $\mathrm{Re}\left(\sum_{\ell=1}^{k} h_\ell\right) + \sum_{\ell=1}^{k} g_\ell < \phi$ on $\partial\Omega$ and $\left\|\phi - \mathrm{Re}\left(\sum_{\ell=1}^{k} h_\ell\right) - \sum_{\ell=1}^{k} g_k\right\|_{\sup} \le 4^{-k}$;

b$_k$. $\|h_k\|_{\sup} \le 4K2^{-k}$;

c$_k$. $\mu\{z \in \partial\Omega : g_K(z) > 0\} < 2\epsilon'\gamma^k$.

Then, using (a$_k$), we may apply the corollary of Aleksandrov's lemma to the function $\phi_k = \phi - \mathrm{Re}\left(\sum_{\ell=1}^{k} h_\ell\right) - \sum_{\ell=1}^{k} g_\ell$ to obtain the function h_{k+1}. By b$_k$ we have

$$\|h_{k+1}\|_{\sup} \le 2^{k+1} K \|\phi_k\|_{\sup} \le 2^{k+1} K 4^{-k} \le 4K2^{-(k+1)}.$$

Therefore, (b$_{k+1}$) holds.

Now by (a$_k$) we have

$$\mathrm{Re}\left(\sum_{\ell=1}^{k+1} h_\ell\right) + \sum_{\ell=1}^{k} g_\ell < \phi \qquad \text{on} \quad \partial\Omega.$$

We define ϕ_{k+1} by

$$\phi_{k+1} = \min\left\{\phi - \mathrm{Re}\left(\sum_{\ell=1}^{k+1} h_\ell\right) - \sum_{\ell=1}^{k} g_\ell, 4^{-(k+1)}\right\}$$

and then we define g_{k+1} by the identity

$$\phi_{k+1} = \phi - \mathrm{Re}\left(\sum_{\ell=1}^{k+1} h_\ell\right) - \sum_{\ell=1}^{k+1} g_\ell.$$

It is immediate from these definitions that

$$\begin{aligned}
\phi_{k+1} &> 0, \\
\|\phi_{k+1}\|_{\sup} &\le 4^{-(k+1)}, \\
g_{k+1} &\ge 0.
\end{aligned}$$

Therefore, (a_{k+1}) holds. Also, $g_{k+1} > 0$ if and only if

$$\phi - \text{Re}\left(\sum_{\ell=1}^{k+1} h_\ell\right) - \sum_{\ell=1}^{k} g_\ell > 4^{-(k+1)}.$$

Therefore, by (c_k),

$$\mu\{z \in \partial\Omega : g_{k+1}(z) > 0\} \leq \epsilon' \cdot \frac{\|\phi_k\|_{\sup}^{1/2}}{\epsilon_{k+1}^{1/2}} \cdot \gamma^{k+1} \leq 2\epsilon' \gamma^{k+1}.$$

Thus (c_{k+1}) holds. The induction process is complete.

Now the statements (b_k) tells us that the function $h \equiv \sum_{k=1}^{\infty} h_k$ is in $A(\Omega)$. By the statements (a_k), we see that $g \equiv \sum_{k=1}^{\infty} g_k$ is in $C(\partial\Omega)$, $g \geq 0$, and $(\text{Re} h) + g = \phi$. Also

$$\{z \in \partial\Omega : \text{Re} h(z) \neq \phi(z)\} \subseteq \bigcup_{j=1}^{\infty} \{z \in \partial\Omega : g_j(z) > 0\}.$$

By the statements c_k, the measure of this last set is less than

$$2\epsilon' \sum_{j=1}^{\infty} \gamma^j = \epsilon' \cdot \frac{2\gamma}{1-\gamma}.$$

Now selecting ϵ' small enough proves the theorem.   □

The corollary says that we may prescribe the absolute value of an $A(\Omega)$ function on the boundary *except on a set of measure ϵ*. It is in the nature of things that this is essentially the best we can do. Exercise 5 at the end of the chapter asks you to provide an example to confirm this assertion.

EXERCISES

1. Give details of the proof of Lemma 9.3.2.

2. Give details of the proof of Lemma 9.3.3.

3. Give details of the proof of Theorem 9.3.4.

4. Provide an example to demonstrate that the function g in the conclusion of Theorem 9.3.4 could not be taken to be continuous on $\bar{\Omega}$.

5. Provide an example to demonstrate that we could not hope to prescribe the boundary values of the modulus of an $A(\Omega)$ function on the *entire* boundary.

6. Provide an example to demonstrate that Marshall's theorem fails in dimensions exceeding one.

7. Modify the inner functions construction to show that if Ω is a weakly pseudoconvex domain and $S \subseteq \partial\Omega$ is the set of strongly pseudoconvex boundary points, then there exists a bounded analytic function on Ω with unimodular boundary values almost everywhere *on the set S*.

8. Fix a positive integer m. Prove that if Ω is strongly pseudoconvex and $z_0 \in \Omega$ is a fixed point, then there exists an inner function on Ω that vanishes at z_0 to order m. (*Hint:* In the construction of the inner function, arrange to subtract off the right Taylor polynomial.)

9. How can we be sure that the functions constructed in Theorem 9.3.8 and Corollary 9.3.9 are nonconstant?

10. Apply Corollary 9.3.9 to the function $\phi \equiv 1$. Why does the resulting function $g \in A(\Omega)$ not contradict our argument in Section 9.1 that functions of several variables having unimodular boundary values must have cluster sets of maximal size?

10 Integral Formulas for Solutions to the $\bar{\partial}$ Problem and Norm Estimates

10.1 The Henkin Integral Formula

Throughout this chapter $\Omega \subset\subset \mathbb{C}^n$ will be strongly pseudoconvex with C^4 boundary. We shall do the following five things:

1. Obtain an integral formula for a solution $H_\Omega f = u$ to the equation $\bar{\partial} u = f$ on Ω, when f is the restriction to Ω of a $\bar{\partial}$-closed $(0,1)$ form with C^1 coefficients on a neighborhood of $\bar{\Omega}$.

2. Obtain estimates of the form

$$\|H_\Omega f\|_{L^\infty(\Omega)} \leq C\|f\|_{L^\infty(\Omega)}.$$

3. Extend the results of (1) and (2) to $\bar{\partial}$-closed forms on $\Omega \subseteq \mathbb{C}^2$ with coefficients in $C^1(\Omega)$ (that is, we remove the hypothesis that f is defined on a *neighborhood* of the closure).

4. Obtain, for $\Omega \subseteq \mathbb{C}^2$, estimates of the form

$$\|H_\Omega f\|_{\Lambda_{1/2}(\Omega)} \leq C\|f\|_{L^\infty(\Omega)}.$$

5. Find applications for some of the preceding estimates.

This section is devoted to item 1. Of course, we shall exploit the Henkin integral formula from Chapter 5 and the Bochner-Martinelli formula. Watch how the fact that the Henkin kernel is holomorphic plays a crucial role.

Now let f be as in (1). Let $\hat{\Omega}$ be a strongly pseudoconvex domain containing $\bar{\Omega}$ such that $f \in C^1_{(0,1)}(\hat{\Omega}), \bar{\partial} f = 0$ on $\hat{\Omega}$. Then, by Corollary 4.6.12 and Exercise 19 at the end of Chapter 4, we know that there is a $u \in C^1(\hat{\Omega})$ satisfying $\bar{\partial} u = f$.

By the Bochner-Martinelli formula (Theorem 1.1.4) we know that if $z \in \Omega$, then

$$u(z) = c_n \left\{ \int_{\partial\Omega} \frac{u(\zeta)}{|\zeta - z|^{2n}} \eta(\bar{\zeta} - \bar{z}) \wedge \omega(\zeta) - \int_{\Omega} \frac{\bar{\partial}u(\zeta)}{|\zeta - z|^{2n}} \eta(\bar{\zeta} - \bar{z}) \wedge \omega(\zeta) \right\}$$

where $c_n = 1/(nW(n)) = (-1)^{(n^2-n)/2} n!/(2\pi i)^n$. But this is equal

$$c_n \left\{ \int_{\partial\Omega} \frac{u(\zeta)}{|\zeta - z|^{2n}} \eta(\bar{\zeta} - \bar{z}) \wedge \omega(\zeta) - \int_{\Omega} \frac{f(\zeta)}{|\zeta - z|^{2n}} \wedge \eta(\bar{\zeta} - \bar{z}) \wedge \omega(\zeta) \right\}.$$

Consider the function

$$\tilde{u}(z) = c_n \int_{\partial\Omega} u(\zeta)\eta\left(\frac{P}{\Phi}\right) \wedge \omega(\zeta),$$

where P, Φ are as in Sections 5.2 and 5.3. By the discussion following Corollary 5.3.5, $\tilde{u}$ is holomorphic on Ω. So the function $v = u - \tilde{u}$ satisfies $\bar{\partial}v = \bar{\partial}u = f$. Now

$$\begin{aligned}
v(z) = {} & c_n \left\{ \int_{\partial\Omega} u(\zeta)\eta\left(\frac{\bar{\zeta} - \bar{z}}{|\zeta - z|^2}\right) \wedge \omega(\zeta) \right. \\
& \left. - \int_{\partial\Omega} u(\zeta)\eta\left(\frac{P(z,\zeta)}{\Phi(z,\zeta)}\right) \wedge \omega(\zeta) \right\} \\
& - c_n \int_{\Omega} \frac{f(\zeta)}{|\zeta - z|^{2n}} \wedge \eta(\bar{\zeta} - \bar{z}) \wedge \omega(\zeta).
\end{aligned} \tag{10.1.1}$$

We wish to rewrite the expression in $\{\ \}$. By invoking a little of the theory of differential forms, we can avoid some very messy computations. Following Henkin, with z fixed, we let

$$M = \left\{ (\zeta_1, \ldots, \zeta_n, \mu_1, \ldots, \mu_n) \in \mathbb{C}^{2n} : \sum_{j=1}^n \mu_j \cdot (\zeta_j - z_j) = 1 \right\}.$$

Then, for $0 \leq \lambda \leq 1$, the points

$$\begin{aligned}
G = {} & \left\{ \left(\zeta_1, \ldots, \zeta_n, \lambda\frac{\bar{\zeta}_1 - \bar{z}_1}{|\zeta - z|^2} + (1 - \lambda)\frac{P_1(z,\zeta)}{\Phi(z,\zeta)}, \right.\right. \\
& \left. \ldots, \lambda\frac{\bar{\zeta}_n - \bar{z}_n}{|\zeta - z|^2} + (1 - \lambda)\frac{P_n(z,\zeta)}{\Phi(z,\zeta)} \right), \\
& \left. \zeta \in \partial\Omega, \ 0 \leq \lambda \leq 1 \right\},
\end{aligned}$$

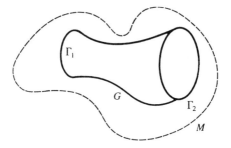

FIGURE 10.1

lie in M. Notice that, for fixed z, G is of real dimension $2n$, whereas M is of real dimension $4n - 2$. The surface G is a sort of cylinder (since $\partial\Omega$ has no boundary). The topological boundary of G is given by the two cycles

$$
\begin{aligned}
\Gamma_1 &= \{(\zeta_1, \ldots, \zeta_n, P_1(z, \zeta), \ldots, P_n(z, \zeta)) : \zeta \in \partial\Omega\} \\
\Gamma_2 &= \left\{\left(\zeta_1, \ldots, \zeta_n, (\bar{\zeta}_1 - \bar{z}_1)/|\zeta - z|^2, \ldots, (\bar{\zeta}_n - \bar{z}_n)/|\zeta - z|^2\right) : \zeta \in \partial\Omega\right\}.
\end{aligned}
$$

See Figure 10.1.

Thus, in this new notation, the expression in $\{\ \ \}$ in (10.1.1) may be rewritten as

$$
\begin{aligned}
\int_{\mu \in \Gamma_2} u(\zeta) \cdot \eta(\mu) \wedge \omega(\zeta) &- \int_{\mu \in \Gamma_1} u(\zeta) \cdot \eta(\mu) \wedge \omega(\zeta) \\
&= \int_{\mu \in \partial G} u(\zeta) \cdot \eta(\mu) \wedge \omega(\zeta) \\
&\overset{\text{(Stokes)}}{=} \int_G d\left[u(\zeta) \wedge \eta(\mu) \wedge \omega(\zeta)\right].
\end{aligned}
$$

Now M can support only $2n - 1$ holomorphic differentials. So a fortiori G can support only $2n - 1$ holomorphic differentials. Since $\eta(\mu) \wedge \omega(\zeta)$ already consists of $2n - 1$ holomorphic differentials, it follows that on G we have

$$
\begin{aligned}
d\left[u(\zeta) \wedge \eta(\mu) \wedge \omega(\zeta)\right] &= du(\zeta) \wedge \eta(\mu) \wedge \omega(\zeta) \\
&= \bar{\partial}u(\zeta) \wedge \eta(\mu) \wedge \omega(\zeta) \\
&= f(\zeta) \wedge \eta(\mu) \wedge \omega(\zeta).
\end{aligned}
$$

Therefore

$$
v(z) = c_n \int_{\partial\Omega \times [0,1]} f(\zeta) \wedge \eta(\mu) \wedge \omega(\zeta) - c_n \int_\Omega \frac{f(\zeta)}{|\zeta - z|^{2n}} \wedge \eta(\bar{\zeta} - \bar{z}) \wedge \omega(\zeta).
$$

We have proved the following theorem.

THEOREM 10.1.1 (Henkin [5])   If $\Omega \subseteq \mathbb{C}^n$ is strongly pseudoconvex with C^4 boundary and f is a $\bar{\partial}$-closed $(0,1)$ form on a neighborhood of $\bar{\Omega}$ with C^1 coefficients, then the function

$$H_\Omega f(z) = v(z) \quad = \quad c_n \int_{\partial\Omega \times [0,1]} f(\zeta) \wedge \eta(\mu) \wedge \omega(\zeta)$$

$$- c_n \int_\Omega \frac{f(\zeta)}{|\zeta - z|^{2n}} \wedge \eta(\bar{\zeta} - \bar{z}) \wedge \omega(\zeta)$$

satisfies $\bar{\partial} H_\Omega f(z) = f(z)$. Here

$$\omega(\zeta) \quad = \quad d\zeta_1 \wedge \cdots \wedge d\zeta_n.$$

$$\eta(w) \quad = \quad \sum_{j=1}^{n} (-1)^{j+1} w_j dw_1 \wedge \cdots \wedge dw_{j-1} \wedge dw_{j+1} \wedge \cdots \wedge dw_n$$

for any $(w_1, \ldots, w_n) \in \mathbb{C}^n$, and

$$\mu = \left(\lambda \frac{\bar{\zeta}_1 - \bar{z}_1}{|\zeta - z|^2} + (1-\lambda) \frac{P_1(z,\zeta)}{\Phi(z,\zeta)}, \ldots, \lambda \frac{\bar{\zeta}_n - \bar{z}_n}{|\zeta - z|^2} + (1-\lambda) \frac{P_n(z,\zeta)}{\Phi(z,\zeta)} \right)$$

Exercise for the Reader

For the case $n = 2$, do the calculations for the rewriting of $\{ \ \}$ in (10.1.1) explicitly.

It is important to understand the genesis of Henkin's integral formula in light of what has gone before. We know from Chapter 4 that a "good" or canonical solution to $\bar{\partial} u = f$ is the one that is orthogonal to holomorphic functions (in an appropriate metric). However, the good solution that we chose in that chapter is of no use to us now because it was selected in a metric that annihilates information at the boundary. What is more appropriate in the present context is to study the solution orthogonal to holomorphic functions in the standard L^2 metric (this method was used by Kohn—see G. B. Folland and J. J. Kohn [1] or Krantz [19]). This solution can be manufactured by beginning with any solution u and subtracting from it the projection Pu onto L^2 holomorphic functions. However we have no accessible formula for Pu. An alternate approach would be to subtract from u its Szegö projection Su (this method was successfully used by D. H. Phong [1]). Again, we have no accessible formula for Su. What we *do* have under control is the Henkin integral of u, which is the function that we called $\tilde{u}$. It is plausible that the Henkin integral of a function on Ω should be very close to the Szegö integral. (As evidence, compare the two on the ball.) Therefore, $u - \tilde{u}$ ought to give a good solution, and that motivates our construction.

It should be noted that the preceding heuristics were made substantive by N. Kerzman and E. M. Stein [2], who showed that on a strongly pseudoconvex domain, the function Su has an asymptotic expansion in which the principal term is the Henkin integral.

Exercise for the Reader
Use the *method* of constructing a homotopy between the Bochner-Martinelli kernel and the Henkin kernel to give a new proof of the Cauchy-Fantappiè formula (Exercise 36 at the end of Chapter 1).

We conclude this section by noting that, independently and simultaneously with Henkin's work, a number of other constructions of integral solutions to the $\bar{\partial}$ problem were discovered. H. Grauert and I. Lieb [1], N. Kerzman [2], and N. Øvrelid [1] should particularly be mentioned. E. Ramirez [1] developed integral reproducing formulas, but did not study the $\bar{\partial}$ problem.

10.2 Estimates for the Henkin Solution on Domains in $\mathbb{C}^2$

To avoid being distracted by numerous technical details, we write out the Henkin solution to $\bar{\partial}u = f$ explicitly on strongly pseudoconvex domains Ω in $\mathbb{C}^2$ only. Then we shall calculate uniform estimates on $H_\Omega f$. The result in higher dimensions is analogous but much more tedious. We continue to use the notation of Section 10.1.

To begin, notice that for $\Omega \subseteq \mathbb{C}^2$, $\partial\Omega$ is three-dimensional. Writing

$$f = f_1 d\bar{\zeta}_1 + f_2 d\bar{\zeta}_2,$$

the first integrand of the Henkin formula becomes

$$(f_1 d\bar{\zeta}_1 + f_2 d\bar{\zeta}_2) \wedge \eta(\mu) \wedge d\zeta_1 \wedge d\zeta_2. \tag{10.2.1}$$

By consideration of dimension, the only differentials in $\eta(\mu)$ that will not force this form to vanish on $\partial\Omega \times [0,1]$ are those involving $d\lambda$ alone. So (10.2.1) on $\partial\Omega \times [0,1]$ equals

$$(f_1 d\bar{\zeta}_1 + f_2 d\bar{\zeta}_2) \wedge \left[\mu_1 \frac{\partial \mu_2}{\partial \lambda} d\lambda - \mu_2 \frac{\partial \mu_1}{\partial \lambda} d\lambda \right] \wedge d\zeta_1 \wedge d\zeta_2. \tag{10.2.2}$$

But the expression in [ ] equals

$$\left(\frac{P_1(z,\zeta)}{\Phi(z,\zeta)} + \lambda\left\{\frac{\bar{\zeta}_1 - \bar{z}}{|\zeta - z|^2} - \frac{P_1(z,\zeta)}{\Phi(z,\zeta)}\right\}\right)\left(\frac{\bar{\zeta}_2 - \bar{z}_2}{|\zeta - z|^2} - \frac{P_2(z,\zeta)}{\Phi(z,\zeta)}\right)d\lambda$$

$$-\left(\frac{P_2(z,\zeta)}{\Phi(z,\zeta)} + \lambda\left\{\frac{\bar{\zeta}_2 - \bar{z}}{|\zeta - z|^1} - \frac{P_2(z,\zeta)}{\Phi(z,\zeta)}\right\}\right)\left(\frac{\bar{\zeta}_1 - \bar{z}_1}{|\zeta - z|^1} - \frac{P_1(z,\zeta)}{\Phi(z,\zeta)}\right)d\lambda$$

$$= \frac{P_1(z,\zeta)(\bar{\zeta}_2 - \bar{z}_2) - P_2(z,\zeta)(\bar{\zeta}_1 - \bar{z}_1)}{\Phi(z,\zeta) \cdot |\zeta - z|^2}\, d\lambda.$$

Substituting this back into (10.2.2), we see that λ integrates out and Henkin's formula becomes

$$
\begin{aligned}
H_\Omega(z) &= \frac{1}{4\pi^2}\Bigg\{\int_\Omega \frac{f_1(\zeta)\cdot(\bar{\zeta}_1 - \bar{z}_1) + f_2(\zeta)\cdot(\bar{\zeta}_2 - \bar{z}_2)}{|\zeta - z|^4} \\
&\qquad\qquad \times d\bar{\zeta}_1 \wedge d\bar{\zeta}_2 \wedge d\zeta_1 \wedge d\zeta_2 \\
&\qquad - \int_{\partial\Omega} \frac{P_1(z,\zeta)(\bar{\zeta}_2 - \bar{z}_2) - P_2(z,\zeta)(\bar{\zeta}_1 - \bar{z}_1)}{\Phi(z,\zeta)|\zeta - z|^2} \\
&\qquad\qquad \times \big(f_1(\zeta)d\bar{\zeta}_1 + f_2(\zeta)d\bar{\zeta}_2\big) \wedge d\zeta_1 \wedge d\zeta_2\Bigg\} \\
&\equiv \int_\Omega f_1(\zeta)K_1(z,\zeta)\,dV(\zeta) + \int_\Omega f_2(\zeta)K_2(z,\zeta)\,dV(\zeta) \\
&\quad + \int_{\partial\Omega} f_1(\zeta)L_1(z,\zeta)\,d\sigma(\zeta) + \int_{\partial\Omega} f_2(\zeta)L_2(z,\zeta)\,d\sigma(\zeta) \qquad (10.2.3)
\end{aligned}
$$

where the identity defines the kernels. We will prove that

$$\int_\Omega |K_j(z,\zeta)|dV(\zeta) \le C_j, \qquad j = 1,2 \qquad (10.2.4)$$

$$\int_{\partial\Omega} |L_j(z,\zeta)|d\sigma(\zeta) \le D_j, \qquad j = 1,2 \qquad (10.2.5)$$

where the bounds are independent of $z \in \Omega$. This will immediately imply that

$$\|H_\Omega f\|_{L^\infty(\Omega)} \le (C_1 + C_2 + D_1 + D_2)\|f\|_{L^\infty(\Omega)}, \qquad (10.2.6)$$

which is the principal result of this section. By symmetry, we need to prove (10.2.4) and (10.2.5) only when $j = 1$.

LEMMA 10.2.7 We have

$$\int_\Omega |K_1(z,\zeta)|\,dV(\zeta) \le C,$$

where C does not depend on z.

Proof Let R be so large that $\Omega \subseteq B(z,R)$, any $z \in \Omega$. Then

$$\int_\Omega |K_1(z,\zeta)|dV(\zeta) \le \int_{B(z,R)} |z-\zeta|^{-3}\,dV(\zeta) = C\int_0^R r^{-3}r^3\,dr = C'. \qquad \square$$

LEMMA 10.2.8 If $\delta_0 > 0$ and $|z-\zeta| \ge \delta_0$ for all $\zeta \in \partial\Omega$, then

$$\int_{\partial\Omega} |L_1(z,\zeta)|\,d\sigma(\zeta) \le C = C(\delta_0).$$

Proof By Lemma 5.2.7 and Proposition 5.2.13, $|\Phi(z,\zeta)| \ge C_0(\delta_0)$, uniformly over such z. Also the functions $P_j(z,\zeta)$ and $|\zeta_j - z_j|$ are bounded over $\bar\Omega \times \partial\Omega$. Finally, $|\zeta - z|^2 \ge \delta_0^2$ for such z. Therefore,

$$\int_{\partial\Omega} |L_1(z,\zeta)|\,d\sigma(\zeta) \le \int_{\partial\Omega} \frac{C}{\delta_0^2 C_0(\delta_0)}\,d\sigma(\zeta) = \frac{C}{\delta_0^2 C_0(\delta_0)}\sigma(\partial\Omega). \qquad \square$$

LEMMA 10.2.9 There is a $\delta_0 > 0$ such that for all $z \in \Omega$ satisfying $\delta_\Omega(z) < \delta_0$ we have

$$\int_{\partial\Omega \cap B(z,\delta_0)} |L_1(z,\zeta)|\,d\sigma(\zeta) \le C.$$

Here C is independent of z.

Proof By Proposition 5.2.13, we may choose a $\delta_0 > 0$ such that the estimate

$$|\Phi(z,\zeta)| \ge C_1 |L_\zeta(z)| \qquad (10.2.9.1)$$

holds for $\zeta \in \partial\Omega, z \in \Omega \cap B(\zeta,\delta_0)$. We may also assume that $B(\zeta,\delta_0)$ lies in a tubular neighborhood of $\partial\Omega$. Notice that, by Lemma 5.2.7,

$$|2\,\mathrm{Re}\,L_\zeta(z)| \ge |\rho(z)| + \gamma|z-\zeta|^2/2. \qquad (10.2.9.2)$$

Thinking of z as fixed, we introduce local coordinates for $\zeta \in \partial\Omega$ as follows.

Call the coordinates t_1, t_2, t_3. Let $\tilde z$ be the normal projection of z to $\partial\Omega$. We specify $t_i(\tilde z) = 0, i = 1,2,3$. Also, set $t_1 = \mathrm{Im}\,L_\zeta(z)$. Since

$$L_\zeta(z) = \sum_{j=1}^n \frac{\partial\rho}{\partial\zeta_j}(\zeta)(z_j - \zeta_j) + \text{(second-order terms)},$$

it follows that t_1 is, up to second order, nothing other than the complex normal component of $(z - \zeta)$ at ζ. Notice that $w \in \mathbb{C}^n$ is *complex tangential* at ζ if and only if $\sum_j \frac{\partial \rho}{\partial \zeta_j}(\zeta)w_j = 0$. It is tangential in the *classical sense* precisely when the real part of this expression is zero.

So we may choose two other coordinates t_1, t_2 to complete the coordinate system. (We are intentionally omitting the details to force the reader to think geometrically. However, details may be found in G. M. Henkin [2].)

It follows from (10.2.9.1) and (10.2.9.2) that

$$|\Phi| \geq C \left(|t_1| + |\rho(z)| + t_2^2 + t_3^2 \right) \tag{10.2.9.3}$$

in the new coordinates. Of course,

$$|\zeta - z|^2 \cong |t_1^2 + t_2^2 + t_3^2| + \rho(z)^2. \tag{10.2.9.4}$$

Finally, introduction of these coordinates in the integral will involve a bounded Jacobian factor. Therefore,

$$\int_{\partial\Omega \cap B(z,\delta_0)} |L_1(z,\zeta)| d\sigma(\zeta)$$

$$\leq \int_{(t_1^2+t_2^2+t_3^2)<R^2} \frac{C\sqrt{t_1^2 + t_2^2 + t_3^2}}{(t_1^2 + t_2^2 + t_3^2)\left[|t_1| + |\rho(z)| + t_2^2 + t_3^2\right]} dt_1\, dt_2\, dt_3.$$

Here $R = R(\delta_0, \Omega)$ is a constant. Now the last line does not exceed

$$C \int_{(t_1^2+t_2^2+t_3^2)<R^2} (t_1^2 + t_2^2 + t_3^2)^{-1/2}(|t_1| + t_2^2 + t_3^2)^{-1}\, dt_1\, dt_2\, dt_3$$

$$= C \int_{\substack{|t|<R \\ |t_1|<t_2^2+t_3^2}} + C \int_{\substack{|t|<R \\ |t_1|\geq t_2^2+t_3^2}} \equiv I + II.$$

We then have

$$I \leq C \int_{\substack{|t|<R \\ |t_1|<t_2^2+t_3^2}} (t_2^2 + t_3^2)^{-1/2}(t_2^2 + t_3^2)^{-1}\, dt_1\, dt_2\, dt_3$$

$$\leq C \int_{t_2^2+t_3^2<R^2} (t_2^2 + t_3^2)^{-1/2}\, dt_2\, dt_3 \leq C.$$

Also,

$$
\begin{aligned}
II \;\le\; & C\int_{\substack{|t|<R \\ |t_1|\ge t_2^2+t_3^2}} (t_2^2+t_3^2)^{-1/2}|t_1|^{-1}\, dt_1\, dt_2\, dt_3 \\
\le\; & C\int_{t_2^2+t_3^2<R} (t_2^2+t_3^2)^{-1/2}\left(|\log R| + |\log(t_2^2+t_3^2)|\right) dt_2\, dt_3 \\
\le\; & C.
\end{aligned}
$$
$\qquad\qquad\qquad\qquad\qquad\qquad\qquad\qquad\qquad\qquad\qquad\qquad\qquad\;\Box$

Lemmas 10.2.8 and 10.2.9, together with symmetry, yield (10.2.5). Lemma 10.2.7 implies (10.2.4). As a result, we have the next theorem.

THEOREM 10.2.10 (Henkin [3])   Let $\Omega \subset\subset \mathbb{C}^2$ be strongly pseudoconvex with C^4 boundary. Let f be a $\bar\partial$-closed $(0,1)$ form with C^1 coefficients on a neighborhood of $\bar\Omega$. Then the Henkin solution $H_\Omega f$ to the $\bar\partial$ equation satisfies $\bar\partial H_\Omega f = f$ on Ω and

$$
\|H_\Omega f\|_{L^\infty(\Omega)} \le C_\Omega \|f\|_{L^\infty(\Omega)}.
$$

We again point out that the result is true in any dimension; we have proved it only in dimension two for the sake of simplicity. Similar results are also due to Grauert and Lieb [1], N. Kerzman [2], and N. Øvrelid [1].

10.3   Smoothness of the Henkin Solution and the Case of Arbitrary $\bar\partial$-Closed Forms

We now face the following technical problem: We have proved sup norm estimates for the $\bar\partial$ problem for a rather restrictive class of forms f. We made this restriction because we needed to guarantee that u is defined and continuous on $\partial\Omega$ so that the integral of u (and of f) over $\partial\Omega$ make sense. A standard technique for passing from these a priori estimates to estimates for arbitrary f is to prove that $H_\Omega f$ enjoys some smoothness property and then to invoke certain compactness theorems (such as the Ascoli-Arzela theorem). That is the approach that we shall take here.

Let the Lipschitz spaces Λ_α be as in Section 8.8. We have the following.

LEMMA 10.3.1   Let $\Omega \subset\subset \mathbb{R}^N$ have C^1 boundary and $g \in L^\infty(\Omega)\cap C^1(\Omega)$. Let $0 < \alpha < 1$. If

$$
|\nabla g(z)| \le C\left(\delta_\Omega(z)\right)^{\alpha-1}, \qquad \text{all } z \in \Omega,
$$

then $g \in \Lambda_\alpha(\Omega)$.

Proof   The proof is left as an exercise for the reader. Use the techniques from the proof of Theorem 8.8.2 or see S. G. Krantz [1, 10]. $\Box$

Remark: The lemma is true when $\alpha = 1$, but then the hypothesis is too strong. In that case the more natural hypothesis is

$$|\nabla^2 g(z)| \leq C\, (\delta_\Omega(z))^{-1}.$$ □

If $\Omega \subset\subset \mathbb{C}^2, \partial\Omega$ is C^5, and f is a $\bar{\partial}$-closed $(0,1)$ form on a neighborhood of $\bar{\Omega}$ with C^1 coefficients, then $H_\Omega f$ is well defined on Ω and is C^1 (just differentiate the formula in Theorem 10.1.2 under the integral sign). We know from Section 10.2 that $H_\Omega f$ is bounded. Refer to (10.2.3). We now prove that

$$\int_\Omega |K_j(z+h, \zeta) - K_j(z, \zeta)|\, dV(\zeta) \;\leq\; C_j |h|^{1/2},$$
$$j = 1, 2, \quad z, z+h \in \Omega \qquad (10.3.2)$$
$$\int_{\partial\Omega} |\nabla_z L_j(z, \zeta)|\, d\sigma(\zeta) \;\leq\; D_j \delta_\Omega(z)^{-1/2}, \qquad (10.3.3)$$

where C_j, D_j are constants independent of z. Now (10.3.2) trivially implies that the terms

$$A_j(z) \equiv \int_\Omega f_j(\zeta) K_j(z, \zeta)\, dV(\zeta)$$

in (10.2.3) are in $\Lambda_{1/2}(\Omega)$. Likewise, with Lemma 10.3.1, estimate (10.3.3) implies that the terms

$$B_j(z) \equiv \int_{\partial\Omega} f_j(\zeta) L_j(z, \zeta) d\sigma(\zeta)$$

in (10.2.3) are in $\Lambda_{1/2}(\Omega)$. By symmetry, we consider only $j = 1$.

LEMMA 10.3.4 For $z, z+h \in \Omega$ we have

$$\int_\Omega |K_1(z+h, \zeta) - K_1(z, \zeta)|\, dV(\zeta) \leq C|h|^{1/2}.$$

Proof Choose $R > 0$ such that $\Omega \subseteq B(z, R)$ for all $z \in \Omega$. Let $0 < |h| < R$. Then

$$\int_\Omega |K_1(z+h, \zeta) - K_1(z, \zeta)|\, dV(\zeta)$$
$$\leq C\int_{B(z,R)} \big||z+h-\zeta|^{-3} - |z-\zeta|^{-3}\big|\, dV(\zeta)$$
$$= C\int_{B(0,R)} \big||-h+\zeta|^{-3} - |\zeta|^{-3}\big|\, dV(\zeta)$$
$$= C\int_{\substack{|\zeta|<2|h| \\ \zeta\in B(0,R)}} + \int_{\substack{|\zeta|\geq 2|h| \\ \zeta\in B(0,R)}} \equiv I + II.$$

Now

$$|I| \leq C \int_0^{3|h|} r^{-3} r^3 \, dr = C|h|.$$

Also

$$
\begin{aligned}
|II| &\leq C \int_{R > |\zeta| \geq 2|h|} \frac{|h||\zeta|^2}{|\zeta - h|^3 |\zeta|^3} \, dV(\zeta) \\
&\leq C \int_{2R > |\zeta| \geq |h|} \frac{|h|}{|\zeta|^4} \, dV(\zeta) \\
&\leq C|h| \int_{|h|}^{R} r^{-4} r^3 \, dr,
\end{aligned}
$$

where $R = \operatorname{diam} \Omega$. But this last line is less than or equal to

$$C|h|\,|\log|h|| \leq C|h|^{1/2}. \qquad \square$$

Now as a preparation for the proof of (10.3.3) in the case $j = 1$, we observe that

$$L_1(z, \zeta) = \alpha(z, \zeta) \left\{ \frac{P_1(z, \zeta)(\bar{\zeta}_2 - \bar{z}_2) - P_2(z, \zeta)(\bar{\zeta}_1 - \bar{z}_1)}{\Phi(z, \zeta)|\zeta - z|^2} \right\}$$

where α is a bounded C^1 factor with bounded derivatives. Thus $\nabla_z L_1$ consists of terms of the form

$$\alpha'(z, \zeta) \cdot \frac{(\bar{\zeta}_i - \bar{z}_i)}{\Phi(z, \zeta)|\zeta - z|^2},$$

$$\alpha(z, \zeta) \cdot \frac{\beta(z, \zeta)}{\Phi(z, \zeta)|\zeta - z|^2},$$

$$\alpha(z, \zeta) \cdot \frac{\gamma(z, \zeta)(\bar{\zeta}_i - \bar{z}_i)}{\Phi^2(z, \zeta)|\zeta - z|^2},$$

$$\alpha(z, \zeta) \cdot \frac{\gamma(z, \zeta)(\bar{\zeta}_i - \bar{z}_i)}{\Phi(z, \zeta)|\zeta - z|^3}.$$

Each of these expressions is majorized in absolute value by one of

$$C \cdot \frac{1}{|\Phi(z, \zeta)||\zeta - z|}, \quad C \cdot \frac{1}{|\Phi(z, \zeta)||\zeta - z|^2}, \quad C \cdot \frac{1}{|\Phi(z, \zeta)|^2|\zeta - z|}.$$

Now the second of these clearly majorizes the first (up to a constant factor). Also, the third majorizes the second (since $|\Phi(z, \zeta)| \leq C|z - \zeta|$) when ζ is near

z and is of no interest otherwise. So to prove (10.3.4) it suffices to prove the following.

LEMMA 10.3.5   We have

$$\int_{\partial\Omega} \frac{1}{|\Phi(z,\zeta)|^2|\zeta - z|} \, dV(\zeta) \leq C\delta_\Omega(z)^{-1/2}, \qquad \text{all} \quad z \in \Omega,$$

where C does not depend on z.

Proof   We use the notation of the proof of Lemma 10.2.9. We can restrict attention to $\delta_\Omega(z) \leq \delta_0/2, |z - \zeta| \leq \delta_0$ (otherwise the result is trivial). Introducing the local coordinates from the proof of Lemma 10.2.9 and using the estimations

$$\begin{aligned}
|\Phi(z,\zeta)| &\geq C\left(|\rho(z)| + |t_1| + t_2^2 + t_3^2\right) \\
|\zeta - z|^2 &\geq C\left(t_1^2 + t_2^2 + t_3^2\right),
\end{aligned}$$

we have

$$\int_{\partial\Omega\cap B(z,\delta_0)} \frac{1}{|\Phi(z,\zeta)|^2|z - \zeta|} \, dV(\zeta)$$

$$\leq \int_{(t_1^2+t_2^2+t_3^2)\leq R^2} \frac{dt_1 \, dt_2 \, dt_3}{\left(|\rho(z)| + |t_1| + t_2^2 + t_3^2\right)^2 \left(t_1^2 + t_2^2 + t_3^2\right)^{1/2}}$$

$$= C \int_{|t_1|<t_2^2+t_3^2} + \int_{t_2^2+t_3^2\leq|t_1|\leq\sqrt{t_2^2+t_3^2}} + \int_{|t_1|>\sqrt{t_2^2+t_3^2}}$$

$$\equiv I + II + III.$$

Now

$$\begin{aligned}
I &\leq C \int_{\substack{|t|\leq R \\ |t_1|<t_2^2+t_3^2}} \frac{dt_1 \, dt_2 \, dt_3}{\left(|\rho(z)| + t_2^2 + t_3^2\right)^2 \left(t_2^2 + t_3^2\right)^{1/2}} \\
&\leq C \int_{(t_2^2+t_3^2)\leq R^2} \frac{dt_2 \, dt_3}{\left(|\rho(z)| + t_2^2 + t_3^2\right)^{3/2}} \\
&\leq C \int_0^R \frac{r \, dr}{\left(|\rho(z)| + r^2\right)^{3/2}} \\
&\leq C|\rho(z)|^{-1/2} \leq C\left(\delta_\Omega(z)\right)^{-1/2}.
\end{aligned}$$

Also

$$II \leq C \int_{\substack{|t| \leq R \\ t_2^2+t_3^2 \leq |t_1| \leq \sqrt{t_2^2+t_3^2}}} \frac{dt_1 \, dt_2 \, dt_3}{(|\rho(z)| + |t_1|)^2 (t_2^2 + t_3^2)^{1/2}}$$

$$\leq C \int_{|t_1| \leq R} \frac{\sqrt{|t_1|} \, dt_1}{(|\rho(z)| + |t_1|)^{3/2}}$$

$$\leq C|\rho(z)|^{-1/2} \leq C(\delta_\Omega(z))^{-1/2}.$$

Finally,

$$III \leq C \int_{\substack{|t| \leq R \\ |t_1| \geq \sqrt{t_2^2+t_3^2}}} \frac{dt_1 \, dt_2 \, dt_3}{(|\rho(z)| + |t_1|)^2 |t_1|}$$

$$\leq C \int_{|t_1| \leq R} \frac{dt_1}{(|\rho(z)| + |t_1|)}$$

$$\leq C\big|\log |\rho(z)|\big|$$

$$\leq C|\rho(z)|^{-1/2} \leq C(\delta_\Omega(z))^{-1/2}. \qquad \square$$

Lemma 10.3.5 proves that $|\nabla B_1(z)| \leq C(\delta_\Omega(z))^{-1/2}$. Likewise, $|\nabla B_2(z)| \leq C(\delta_\Omega(z))^{-1/2}$. It follows that $B_1, B_2 \in \Lambda_{1/2}(\Omega)$ by Lemma 10.3.1. Lemma 10.3.4 proves that $A_1 \in \Lambda_{1/2}(\Omega)$. Likewise, $A_2 \in \Lambda_{1/2}(\Omega)$. Note that our estimates depend on the L^1 norms of certain kernels *and on the L^∞ norm of f.* Thus we have proved:

THEOREM 10.3.6   Let $\Omega \subset\subset \mathbb{C}^2$ be strongly pseudoconvex with C^5 boundary. Let f be a $\bar{\partial}$-closed $(0,1)$ form, with C^1 coefficients, on a neighborhood of Ω. Then $H_\Omega f$ is well defined and

$$\|H_\Omega f\|_{\Lambda_{1/2}(\Omega)} \leq C_\Omega \|f\|_{L^\infty(\Omega)}.$$

Now we remove the restrictions on f. This procedure uses an important stability feature of analysis on strongly pseudoconvex domains. We shall not explicitly prove the stability result but instead invite the reader to review the relevant proofs and verify them (see also S. G. Krantz [1] and R. E. Greene and S. G. Krantz [2]):

STABILITY RESULT Let $\Omega \subset\subset \mathbb{C}^2$ be strongly pseudoconvex with C^5 boundary. Let Ω have defining function ρ defined on a bounded neighborhood U of $\bar{\Omega}$. Then there is a $\delta > 0$ such that if $\tilde{\rho} : \mathbb{C}^2 \to \mathbb{R}$ is a C^5 function with $\|\rho - \tilde{\rho}\|_{C^5(U)} < \delta$ and $\tilde{\Omega} = \{z \in \mathbb{C}^2 : \tilde{\rho}(z) < 0\}$, then Theorem 10.3.6 holds on $\tilde{\Omega}$ with $C_{\tilde{\Omega}} \leq 2C_\Omega$.

Now proceed as follows. Let $\Omega \subseteq \mathbb{C}^2$ be a *fixed* strongly pseudoconvex domain with C^5 boundary. Let $\rho : \mathbb{C}^2 \to \mathbb{R}$ be a C^5 defining function for Ω. For $\epsilon > 0$ small, let $\Omega_\epsilon = \{z \in \mathbb{C}^2 : \rho(z) < -\epsilon\}$. If ϵ is sufficiently small, then Ω_ϵ is still strongly pseudoconvex; because $\rho_\epsilon(z) = \rho(z) + \epsilon$ is a defining function for Ω_ϵ, the stability result will apply. Now let f be a $\bar{\partial}$-closed $(0,1)$-form defined on Ω (not necessarily on a neighborhood of the closure) with bounded, C^1 coefficients. For each sufficiently small $\epsilon > 0$, the form f satisfies the hypotheses of Theorem 10.3.6 on Ω_ϵ. Therefore, $H_{\Omega_\epsilon} f$ is well defined and satisfies $\bar{\partial} H_{\Omega_\epsilon} f = f$ on Ω_ϵ; moreover,

$$\|H_{\Omega_\epsilon} f\|_{\Lambda_{1/2}(\Omega_\epsilon)} \leq C_{\Omega_\epsilon} \|f\|_{L^\infty} \leq 2C_\Omega \|f\|_{L^\infty}.$$

If K is any compact subset of Ω, then we see that the functions $\{H_{\Omega_\epsilon} f\}$ form an equicontinuous family on K when $\epsilon < \mathrm{dist}\,(K, {}^c\Omega)$. Of course, it is also an equi-bounded family. By the Ascoli-Arzela theorem and diagonalization, there is a subsequence $\epsilon_j, j = 1, 2, \ldots,$ such that $H_{\Omega_j} f$ converges uniformly on compacta to some $u \in \Lambda_{1/2}(\Omega)$ with $\bar{\partial} u = f$ on Ω (in the weak sense). Thus we have proved the following.

THEOREM 10.3.7 Let $\Omega \subseteq \mathbb{C}^2$ be strongly pseudoconvex with C^5 boundary. Let f be a $\bar{\partial}$-closed $(0,1)$-form on Ω with bounded, C^1 coefficients. Then there exists a function u on Ω with $\bar{\partial} u = f, u$ bounded, and

$$\|u\|_{\Lambda_{1/2}(\Omega)} \leq 2C_\Omega \|f\|_{L^\infty(\Omega)}.$$

In case f on Ω is a $(0,1)$ form with only L^∞ coefficients satisfying $\bar{\partial} f = 0$ in the weak sense, then an approximation argument involving convolution of the coefficients of f with Friedrichs mollifiers shows that there is a $u \in \Lambda_{1/2}(\Omega)$ such that $\bar{\partial} u = f$ in the weak sense and

$$\|u\|_{\Lambda_{1/2}(\Omega)} \leq C \|f\|_{L^\infty(\Omega)}.$$

Details are left to the reader (or see S. G. Krantz [1]).

Remark: The result of Theorem 10.3.7 is stated for $\bar{\partial}$ construed in the weak sense. But $\bar{\partial}$ is elliptic in the interior. So if f has even Λ_ϵ coefficients, then u will be C^1 on Ω (but not *up to* the boundary). See Kerzman [2], Krantz [19], or Folland and Kohn [1] for details.   □

PROPOSITION 10.3.8 (Kerzman [2])   Let $B \subseteq \mathbb{C}^2$ be the unit ball. There is a $\bar{\partial}$-closed $(0,1)$-form f on B with C^∞, bounded coefficients such that if $\epsilon > 0$ and $\bar{\partial} u = f$ then $u \notin \Lambda_{1/2+\epsilon}(\Omega)$.

Remark: This proposition shows that the estimate in Theorem 10.3.7 is sharp.   □

Proof of the Proposition: Let $v(z) = \bar{z}_2/\log(z_1-1)$. Notice that $\mathrm{Re}(z_1-1) < 0$ on B, so that the principal value of $\log(z_1 - 1)$ is well defined and holomorphic on B. Let

$$f = \bar{\partial}v = \frac{1}{\log(z_1 - 1)}\, d\bar{z}_2.$$

Notice that f has bounded coefficients on B. Suppose that there exists a function $u \in \Lambda_{1/2+\epsilon}(B)$ with $\bar{\partial}u = f$. Then $u - v \equiv h$ is holomorphic. Let $\delta > 0$ be small, and consider

$$F(\delta) \equiv \int_{|z_2|=\sqrt{\delta}} u(1 - 2\delta, z_2) - u(1 - \delta, z_2)\, dz_2.$$

Notice that, for $\delta < 1, |z_2| = \sqrt{\delta}$, it holds that $(1 - 2\delta, z_2), (1 - \delta, z_2)$ lie in B. On the one hand,

$$
\begin{aligned}
|F(\delta)| &\leq \int_{|z_2|=\sqrt{\delta}} |u(1 - 2\delta, z_2) - u(1 - \delta, z_2)|\, d|z_2| \\
&\leq \int_{|z_2|=\sqrt{\delta}} \delta^{1/2+\epsilon}\, d|z_2| \leq \delta^{1+\epsilon}. \qquad (10.3.8.1)
\end{aligned}
$$

On the other hand,

$$
\begin{aligned}
|F(\delta)| &= \left| \int_{|z_2|=\sqrt{\delta}} (v + h)(1 - 2\delta, z_2) - (v + h)(1 - \delta, z_2)\, dz_2 \right| \\
&= \left| \int_{|z_2|=\sqrt{\delta}} v(1 - 2\delta, z_2) - v(1 - \delta, z_2)\, dz_2 \right|
\end{aligned}
$$

by Cauchy's theorem. But this is

$$
\begin{aligned}
\left| \int_{|z_2|=\sqrt{\delta}} \left(\frac{\bar{z}_2}{\log(-2\delta)} - \frac{\bar{z}_2}{\log(-\delta)} \right) dz_2 \right| & \\
= 2\pi\delta \left(\frac{1}{\log(-2\delta)} - \frac{1}{\log(-\delta)} \right) = C \frac{\delta}{\log(-2\delta)\log(-\delta)}. & \qquad (10.3.8.2)
\end{aligned}
$$

Putting (10.3.8.1) and (10.3.8.2) together yields

$$\left| \frac{C\delta}{\log(-2\delta)\log(-\delta)} \right| \leq C\delta^{1+\epsilon}.$$

Letting $\delta \to 0$ yields a conradiction.   □

10.3.1   Further Results

The following are optimal estimates for the $\bar{\partial}$ equation on a strongly pseudoconvex domain with C^4 boundary. Proofs may be found in S. G. Krantz [1, 2, 7] and Greiner and Stein [1].

THEOREM 10.3.9   Let $\Omega \subset\subset \mathbb{C}^n$ be strongly pseudoconvex with C^4 boundary. Let f be a $\bar{\partial}$-closed $(0,1)$-form on Ω with L^p coefficients, $1 \leq p \leq \infty$. Then the Henkin solution $H_\Omega f = u$ of the equation $\bar{\partial} u = f$ satisfies the following estimates:

1. $\|u\|_{L^q} \leq C_q \|f\|_{L^1}$,   any $q < (2n+2)/(2n+1)$.
2. $\|u\|_{L^q} \leq C_p \|f\|_{L^q}$, $1 < p < 2n+2$, $\frac{1}{q} = \frac{1}{p} - \frac{1}{2n+2}$.
3. $\int_\Omega \exp\left[(c|u|/\|f\|_{L^{2n+2}})^{(2n+2)/(2n+1)}\right] dV \leq C$, for $c > 0$ small.
4. $\|u\|_{\Gamma_{\alpha,2\alpha}} \leq C_p \|f\|_{L^p}$, $\alpha = \frac{1}{2} - \frac{n+1}{p}$, $2n+2 < p \leq \infty$.

Notice that the estimates on u form a smooth continuum: through increasingly better L^p spaces (higher L^p is "better" by Sobolev's theorem) to exponential integrability to Lipschitz spaces. All of these estimates may, by methods similar to those in Proposition 10.3.8, be shown to be sharp.

The $\bar{\partial}$-Neumann on a strongly pseudoconvex domain is the first example of a subelliptic system for which a complete theory of estimates has been computed. For more on this matter see Greiner and Stein [1].

10.4   First Cousin Problem with Bounds and Uniform Approximation on Strongly Pseudoconvex Domains in $\mathbb{C}^2$

Let $\Omega \subseteq \mathbb{C}^2$ be strongly pseudoconvex with C^5 boundary. Let $f \in A(\Omega)$. Our goal in this section is to use the results of Section 10.3 to construct a domain $\hat{\Omega} \supseteq \bar{\Omega}$ and a sequence $f_j \in H^\infty(\hat{\Omega})$ such that $f_j \to f$ *uniformly* on $\bar{\Omega}$. The reader should compare this result with Corollary 5.4.3 and be convinced that the present result is essentially stronger. The proof proceeds by way of an auxiliary result that is of independent interest.

THEOREM 10.4.1 (First Cousin Problem with Bounds)   Let $\Omega \subset\subset \mathbb{C}^2$ be strongly pseudoconvex with C^5 boundary. Let $\mathcal{V} = \{V_i\}_{i=1}^M$ be a finite open covering of $\bar{\Omega}$. Set $U_i = V_i \cap \Omega, i = 1, \ldots, M$. Write $\mathcal{U} = \{U_i\}_{i=1}^M$. Then there is a constant $C = C(\Omega, \mathcal{U})$ such that if $\{g_{j,k}\}_{j,k=1}^M$ is a family of holomorphic first Cousin data for $\mathcal{U}$ and if

$$\|g_{j,k}\|_{L^\infty(U_j \cap U_k)} \leq K, \qquad j, k = 1, \ldots, M,$$

then there is a holomorphic solution $\{g_j\}_{j=1}^M$ of the first Cousin problem satisfying

$$\|g_j\|_{L^\infty(U_j)} \le C \cdot K, \qquad j = 1, \ldots, M.$$

Proof   Let $\{\phi_i\}$ be a partition of unity on Ω subordinate to $\{U_i\}$. Define

$$h_j = \sum_i \phi_i g_{i,j} \qquad \text{on } U_j.$$

On $U_j \cap U_k$ we have

$$
\begin{aligned}
\bar\partial(h_k - h_j) &= \bar\partial \sum_i \phi_i(g_{i,k} - g_{i,j}) \\
&= \bar\partial \sum_i \phi_i g_{j,k} \\
&= \left(\sum_i \bar\partial\phi_i\right) g_{j,k} + \left(\sum_i \phi_i\right) \bar\partial g_{j,k} \\
&= \bar\partial g_{j,k} = 0.
\end{aligned}
$$

Therefore, the form

$$f = \bar\partial h_j \qquad \text{on } U_j$$

is well defined, $\bar\partial$-closed, and has coefficients that are C^∞ and bounded by KL, where $L = \sum_j \|\phi_j\|_{C^1(\Omega)}$.

By Theorem 10.3.7, there is a $u \in C^\infty(\Omega), u$ bounded by $2C_\Omega KL$, such that $\bar\partial u = f$. Then the functions $g_j \equiv h_j - u$ will solve the Cousin I problem with bounds. $\square$

Now we have our main result.

THEOREM 10.4.2 (Henkin [3], Kerzman [2], Grauert/Lieb [1])   Let $\Omega \subset\subset \mathbb{C}^2$ be strongly pseudoconvex with C^5 boundary. There is a strongly pseudoconvex domain $\hat\Omega \supseteq \bar\Omega$ such that if $f \in A(\Omega)$, then there is a sequence $\{f_j\}$ of functions holomorphic on $\hat\Omega$ such that $f_j \to f$ uniformly on $\bar\Omega$.

Proof   Let $\Omega = \{\rho(z) < 0\}$. As in (8.3.3)–(8.3.5), cover Ω with subdomains $\{\Omega_k\}_{k=1}^M$ that have associated vectors ν_k. For $0 < \delta \le \delta_0, \delta_0$ small, we let $\Omega_k^\delta = \Omega_k + \delta\nu_k, k = 1, \ldots, M$. We may assume that $\cup_k \Omega_k^\delta \supseteq \bar\Omega$, all $0 < \delta \le \delta_0$. Shrinking δ_0 if necessary, we may find for each $0 < \delta \le \delta_0$ a strongly pseudoconvex Ω^δ with C^4 boundary such that

$$\Omega \subseteq \bar\Omega \subseteq \Omega^\delta \subseteq \overline{\Omega^\delta} \subseteq \bigcup_{k=1}^M \Omega_k^\delta.$$

Note that each Ω^δ may be chosen so that $\Omega^\delta = \{\rho_\delta(z) < 0\}$ and ρ_δ is C^5 close to ρ.

Of course $\cap_{0 < \delta \leq \delta_0}(\cup_{k=1}^M \Omega_k^\delta) = \bar{\Omega}$. Let $U_k^\delta = \Omega_k^\delta \cap \Omega^\delta, k = 1, \ldots, M$. Define

$$f_k^\delta(z) = f(z - \delta\nu_k), \qquad z \in U_k^\delta, k = 1, \ldots, M.$$

Let

$$g_{jk}^\delta(z) = f_k^\delta(z) - f_j^\delta(z), \qquad z \in U_j^\delta \cap U_k^\delta, j, k = 1, \ldots, M.$$

Then

1. $\{g_{jk}^\delta\}$ is a set of holomorphic Cousin I data for the covering $\{U_k^\delta\}$ of Ω^δ.

2. Since f is uniformly continuous on $\bar{\Omega}$, it is possible to make $\|g_{jk}^\delta\|_{L^\infty(U_j^\delta \cap U_k^\delta)} < \epsilon$, for $\epsilon > 0$ any prechosen number, provided that $\delta > 0$ is sufficiently small.

By Theorem 10.4.1, there is a solution $\{g_j^\delta\}$ to the above-posed Cousin I problem with

$$\|g_j^\delta\|_{L^\infty(U_j^\delta)} \leq C\left(\Omega^\delta, \{U_j^\delta\}\right) \cdot \epsilon.$$

By the stability of the Henkin estimates discussed in Section 10.3, the constant C may be taken to be independent of δ. Define

$$\tilde{f}_\delta(z) = f_k^\delta(z) - g_k^\delta(z), \qquad z \in U_k^\delta.$$

Then $\tilde{f}_\delta$ is a well-defined holomorphic function on Ω^δ and

$$\|\tilde{f}_\delta - f\|_{L^\infty(\bar{\Omega})} \leq \max_{1 \leq k \leq M} \|f_k^\delta - f\|_{L^\infty(U_k^\delta \cap \bar{\Omega})} + C \cdot \epsilon.$$

Again, the uniform continuity of f may be invoked to make the last expression less than $2C\epsilon$ simply by making δ small enough.

We have succeeded in approximating f, uniformly on $\bar{\Omega}$, by a function holomorphic on Ω^δ. However, δ will tend to 0 as ϵ tends to 0. The final step is for each $\delta > 0$ to apply Corollary 5.4.3 to select a holomorphic f_δ on $\Omega^{\delta_0} \equiv \hat{\Omega}$ such that $\|\tilde{f}_\delta - f_\delta\|_{L^\infty(\bar{\Omega})} < \epsilon$. The point here is that each f_δ is defined on a fixed $\hat{\Omega} \supseteq \bar{\Omega}$, $\hat{\Omega}$ is strongly pseudoconvex, and $\|f - f_\delta\|_{L^\infty(\bar{\Omega})} \leq (2C + 1)\epsilon$.   □

Exercise for the Reader
Show that Theorem 10.4.2 cannot be proved with the additional conclusion that

$$\|f_\delta\|_{L^\infty(\hat{\Omega})} \leq C\|f\|_{L^\infty(\Omega)}.$$

EXERCISES

1. Let $m = (m_1, \ldots, m_n) \in \mathbb{N}^n$, and define

$$\Omega_m = \{z \in \mathbb{C}^n : |z_1|^{2m_1} + \cdots + |z_n|^{2m_n} < 1\}.$$

Prove that Ω_m is pseudoconvex. It is strongly pseudoconvex at boundary points for which all coordinates are nonzero. Find all other strongly pseudoconvex points. Imitate the construction in Proposition 10.3.8 to prove that any estimate of the form

$$\|u\|_{\Lambda_{\alpha(m)}} \leq C\|f\|_{L^\infty}$$

for solutions of the equation $\bar{\partial}u = f, f$ a $\bar{\partial}$-closed $(0,1)$ form on Ω_m, must satisfy $\alpha(m) \leq \min\{1/(2m_j), j = 1, \ldots, n\}$. The fact that such estimates actually hold was proved by R. M. Range [2] in dimension two and in full generality by K. Diederich, J. E. Fornæss, and J. Wiegerinck [1]. Related estimates appear in A. Cumenge and S. G. Krantz [1]. Refer also to Exercise 8.

2. Let

$$\Omega = \left\{ (z_1, z_2) : |z_1|^2 + 2e^{-1/|z_2|^2} < 1 \right\}.$$

Show that for *no* $\alpha > 0$ can it always hold that

$$\|u\|_{\Lambda_\alpha} \leq C\|f\|_{\Lambda^\infty}$$

for solutions to $\bar{\partial}u = f, f$ a $\bar{\partial}$-closed $(0,1)$ form (see R. M. Range [2]).

3. *Failure of local hypoellipticity for the $\bar{\partial}$ problem*   Let $\Omega \subseteq \mathbb{C}^2$ be a smoothly bounded domain. Let $P = 0 \in \partial\Omega$. Suppose that for some $r > 0$ we have $B(P, r) \cap \partial\Omega = \{z \in B(P, r) : \mathrm{Re}\, z_1 = 0\}$. So the analytic disc $\{P + (0, z_2) : |z_2| < r\}$ is contained in $\partial\Omega$. Let $\phi \in C_c^\infty(B(P, r))$ satisfy $\phi \equiv 1$ on $B(P, r/2)$. Let $f = \bar{\partial}\phi/z_1$. Then f is a $\bar{\partial}$-closed $(0,1)$ form on Ω and $f \in C_{(0,1)}^\infty(\bar{B}(P, r/3) \cap \bar{\Omega})$. Prove that it is impossible to find a u on Ω such that $\bar{\partial}u = f$ and $u \in C^\infty(\bar{B}(P, r/4) \cap \bar{\Omega})$. This example illustrates the phenomenon of "propagation of singularities" along analytic discs in $\partial\Omega$. That is, the function u can have singularities on the disc even at point where f does not. See J. J. Kohn [5] for more on these matters.

4. Use the techniques of Section 10.3 to prove that formula (5.3.5.1) is valid for functions continuous on $\bar{\Omega}$ and holomorphic on Ω.

5. Modify the proof of Theorem 7.2.9 to prove that if $\Omega \subseteq \mathbb{C}^2$ is strongly pseudoconvex with C^4 boundary, $0 \in \Omega$, and $f \in A(\Omega)$ vanishes at 0, then it

is possible to write $f(z) = z_1 f_1(z) + z_2 f_2(z)$ with $f_1, f_2 \in A(\Omega)$. (*Hint:* You will need to consider the sheaf over $\bar{\Omega}$ whose stalk at $z \in \Omega$ is just $\mathcal{O}_z$ but whose stalk over $w \in \partial\Omega$ is the ring of germs of functions continuous up to $\partial\Omega$ and holomorphic on Ω. See N. Kerzman and A. Nagel [1] for help. As an alternate approach, use Exercise 14 at the end of Chapter 5.)

6. Consider the operator on $\mathbb{R}^2$ given by

$$P = \frac{\partial}{\partial x} + ix \frac{\partial}{\partial y}.$$

This is perhaps the simplest example of a partial differential operator with the property that there is an $f \in C^\infty(\mathbb{R}^2)$ such that $Pu = f$ is not locally solvable. More precisely, let $D_j = \bar{D}(j^{-1}, 4^{-j}), j \in \mathbb{N}$. Let $f \in C_c^\infty(\mathbb{R})$ satisfy (i) $f(\cdot, y)$ is even for each y, (ii) supp $f \cap \{x \geq 0\} = \cup_j D_j$, and (iii) $\int_{D_j} f \, dx \, dy > 0$ for all j. Complete the following outline to show that there is no $u \in C^1(\mathbb{R}^2)$ such that $Pu = f$.

a. Suppose that u is a C^1 solution; write $u = u_e + u_o$, where u_e is even in x and u_o is odd in x. The even part of the equation $Pu = f$ is

$$\frac{\partial u_o}{\partial x} + ix \frac{\partial u_o}{\partial y} = f. \tag{$*$}$$

It follows that $u(0, y)$ is constant; hence we may suppose that $u_o(0, y) = 0$.

b. Restrict attention to $\{x \geq 0\}$. Let $s = x^2/2, \partial/\partial s = (1/x)(\partial/\partial x)$. Then $(*)$ becomes

$$\frac{\partial u_o}{\partial s} + i\frac{\partial u_o}{\partial y} = \frac{1}{\sqrt{2s}} f\left(\sqrt{2s}, y\right), \quad s > 0,$$
$$u_o = 0, \quad s = 0.$$

c. The transformed function $u_o(s, y) = u_o(x, y)$ satisfies the Cauchy-Riemann equations in the complex variable $s + iy$ off of $\cup_j D_j$.

d. It follows that $u \equiv 0$ on $\mathbb{R}^2 \setminus \cup_j D_j$.

e. Apply Stokes's theorem to $u_o dy - ixu_o dx$ on D_j to obtain a contradiction.

f. The same proof shows that $Pu = f$ has no C^1 solution in any neighborhood of 0.

The references Nirenberg [1] and Krantz [19] contain more extensive discussions of local solvability.

The fact that f is not real analytic is crucial here. Indeed, the Cauchy-Kowalewski theory guarantees local real analytic solutions (infinitely many of them) in case f is real analytic. Conversely, for some operators of this type, P. Greiner, J. J. Kohn, and E. M. Stein [1] have shown that real analyticity of f is necessary.

The first example of an operator that is not locally solvable is due to H. Lewy [1]. It was discovered in the context of extension phenomena for holomorphic functions on $B \subseteq \mathbb{C}^2$. Indeed, the operator is obtained by restricting the $\bar{\partial}$ operator to ∂B (see G. B. Folland and E. M. Stein [1], J. J. Kohn [4]). Exercise 14 has more on this point of view.
More recently, H. Jacobowitz and F. Treves [1] have shown that partial differential operators that are not locally solvable are, in a precise sense, generic.

7. Let $\Omega = \{z \in \mathbb{C}^3 : \rho(z) = \mathrm{Re} z_3 + |z_1^2 - z_3 z_2|^2 + |z_2^2|^2 < 0\}$. Verify that if $\phi : D \to \mathbb{C}^3$ is univalent, $\phi(0) = 0$, and $\phi'(\zeta) \neq 0$ for all $\zeta \in D$ (so that ϕ is an immersed, nonsingular, analytic disc), then $|\rho(\phi(\zeta))| \geq C|\zeta|^4$ for all $\zeta \in D$. On the other hand, for t small and real, let $\gamma_t(\zeta) = (\zeta, \zeta^2/(it), it)$ for all $\zeta \in D$. Then $\rho(\gamma_t(\zeta)) = O(|\zeta|^8/t^4)$. So the order of tangency of discs *near* 0 is different from the order of tangency of discs *at* 0. This example, due to D'Angelo, is important in the study of subelliptic estimates for the $\bar{\partial}$ problem.

8. Let $\Omega_m \subseteq \mathbb{C}^2$ be given by

$$\Omega_m \equiv \{(z_1, z_2) : |z_1|^2 + |z_2|^{2m} < 1\}.$$

Define $\Phi(z, \zeta) = \sum_{j=1}^n (\partial \rho(\zeta)/\partial \zeta_j)(z_j - \zeta_j)$ for $z \in \Omega_m, \zeta \in \partial \Omega_m$. Use the Cauchy-Fantappiè formula to define a Cauchy integral formula for holomorphic functions that extend to be, say, C^1 on $\bar{\Omega}_m$. Imitate the arguments in Section 10.1 to construct a solution formula for the equation $\bar{\partial} u = f$ when f is a $\bar{\partial}$-closed $(0,1)$ form. It is possible to compute uniform, and even Lipschitz, estimates for this solution. See R. M. Range [2] and Exercise 1.

9. Let $\Omega \subset\subset \mathbb{C}^n$ and $A(\Omega)$ as usual. Call $P \in \partial\Omega$ a *local peak point* if there is a neighborhood U of P and a function $f \in A(\Omega \cap U)$ such that $f(P) = 1$ and $|f(z)| < 1$ for $z \in \bar{\Omega} \cap U \setminus \{P\}$. Suppose that the equation $\bar{\partial} u = f$ has a bounded solution on Ω whenever f is a bounded, $C^1, \bar{\partial}$-closed $(0,1)$ form on Ω. Prove that then every local peak point in $\partial\Omega$ is a (global) peak point for $A(\Omega)$. N. Sibony [2] (see also the newer reference, Sibony [6], which has even more refined counterexamples) has used this idea to construct a smooth pseudoconvex $\Omega \subset\subset \mathbb{C}^3$ for which the equation $\bar{\partial} u = f$ does not have any bounded solution for some smooth $\bar{\partial}$-closed $(0,1)$ form f with bounded coefficients.

10. It is well known and easy to compute that the operator $\bar{\partial}$ is elliptic on relatively compact subsets of any domain $\Omega \subseteq \mathbb{C}^n$ (see Folland and Kohn [1, pp. 10–11] or S. G. Krantz [19] for details). Let us use an idea of N. Kerzman [2] that exploits the interior ellipticity in an elementary manner to give very good estimates for $\bar{\partial}$ in the interior.
Let $\Omega_0 \subset\subset \Omega_1 \subset\subset \Omega \subset\subset \mathbb{C}^n \cong \mathbb{R}^{2n}$. Let $\Gamma(x)$ be the fundamental solution to the Laplacian on $\mathbb{R}^{2n}$ that was constructed in Proposition 1.3.2.

a. Let $u \in C^1(\Omega)$. Let $\phi \in C_c^\infty(\Omega)$ be identically equal to 1 on Ω_1. Then, for $z \in \Omega_0$, we have

$$u(z) = \int_\Omega \Gamma(z - w)\Delta(\phi(w)u(w))\, dV(w).$$

b. Write $\Delta = 4\sum_j(\partial^2/\partial w_j \partial \bar{w}_j)$. Integrate by parts to obtain

$$u(z) = 4\sum_{j=1}^n \int_\Omega \Gamma_{w_j}(z - w)\frac{\partial}{\partial \bar{w}_j}(\phi(w)u(w))\, dV(w).$$

c. Obtain the estimate

$$|u(z)| \le C\int_\Omega |u(w)|\, dV(w) + C\int_\Omega \frac{1}{|z - w|^{2n-1}}|\bar{\partial}u(w)|\, dV(w)$$

for $z \in \Omega_0$. You must use the fact that $\nabla\phi \equiv 0$ on Ω_1.
d. Notice that if $K \subset\subset \mathbb{C}^n$, then

$$\int_K \frac{1}{|\zeta|^{2n-1}}\, dV(\zeta) \le C(K).$$

e. Use Exercise 10 at the end of Chapter 1 to conclude that, for $1 \le p \le \infty$,

$$\|u\|_{L^p(\Omega_0)} \le C(\Omega_0,\Omega_1)\left\{\|u\|_{L^1(\Omega)} + \|\bar{\partial}u\|_{L^p(\Omega)}\right\}.$$

f. The fact that we are estimating only in the interior was used decisively in two places. What are they?
g. When $p = \infty$, refine the estimate to

$$\|u\|_{L^\infty(\Omega_0)} \le C(\Omega_0,\Omega_1,\epsilon)\left\{\|u\|_{L^1(\Omega)} + \|\bar{\partial}u\|_{L^{2n+\epsilon}(\Omega)}\right\}, \qquad \text{any } \epsilon > 0.$$

11. Modify the argument in the last exercise to show that for any $0 \le k \in \mathbb{Z}$, it holds that

$$\|u\|_{C^{k+1}(\Omega_0)} \le C(\Omega_0,\Omega_1)\left\{\|u\|_{L^1(\Omega)} + \|\bar{\partial}u\|_{C^k(\Omega)}\right\}.$$

12. Use the outline that follows to prove this proposition about almost analytic extensions:

PROPOSITION Let $k \in \mathbb{N}$. Let $M \subseteq \mathbb{C}^n$ be a totally real submanifold, that is, $T_P M$ has no nontrivial complex subspace for any $P \in M$ (see also Exercise 1 at the end of Chapter 8). Let $K \subseteq M$ be a compact set. Let $u \in C^k(M)$. Then there is a function $U \in C^1(\mathbb{C}^n)$ such that $U|_K = u$ and $\bar{\partial}U(z) = \mathcal{O}(\text{dist}\,(z,K)^{k-1})$.

In a sense, U is an "almost analytic" extension of U. For the proof, we proceed by induction on k. The case $k = 1$ is trivial. Now let $k = 2$.

a. If each $x \in K$ has a neighborhood W_x such that $u|_{K \cap W_x}$ has the indicated extension, then passing to a finite subcover and using a partition of unity gives the full result. So it suffices to treat the problem locally.

b. Choose $x \in K$. Choose a neighborhood W_x of x and real functions $\rho_1, \ldots, \rho_m$ such that

$$M \cap U = \{z \in U; \rho_1(z) = \cdots = \rho_m(z) = 0\}.$$

Notice that, since M is totally real, $m \geq n$.

c. Shrinking W_x if necessary, there is a sequence $1 \leq j_1, \ldots, j_n \leq m$ such that

$$\left\{\left(\frac{\partial \rho_{j_i}}{\partial \bar{z}_1}(z), \ldots \frac{\partial \rho_{j_i}}{\partial \bar{z}_n}(z)\right)\right\}_{i=1}^{n}$$

is a basis for $\mathbb{C}^n$ over the field $\mathbb{C}$ at each point $z \in W_x$.

d. Shrinking W_x if necessary, we see that $u|_{W_x \cap M}$ has a C^k extension to $\mathbb{C}^n$ (with only bounded increase in norm). Denote this extension also by u.

e. Denote the functions ρ_{j_i} chosen in (c) by $\rho_1, \ldots, \rho_n$. Define functions $h_i(z)$ on W_x by

$$\left(\frac{\partial u}{\partial \bar{z}_1}(z), \ldots, \frac{\partial u}{\partial \bar{z}_n}(z)\right) = \sum_{i=1}^{n} h_i(z)\left(\frac{\partial \rho_i}{\partial \bar{z}_1}(z), \ldots, \frac{\partial \rho_i}{\partial \bar{z}_n}(z)\right).$$

Then $h_i \in C^{k-1}(W_x), i = 1, \ldots, n$. In the standard several complex variables shorthand,

$$\bar{\partial} u = \sum_{i=1}^{n} h_i \bar{\partial} \rho_i.$$

f. Define $u_1 = u - \sum_{i=1}^{n} h_i \rho_i$. Then $u_1 = u$ on $M \cap W_x$ and

$$\bar{\partial} u_1 = -\sum_{i=1}^{n} (\bar{\partial} h_i) \rho_i = \mathcal{O}(\text{dist}(\cdot, M)).$$

This completes the proof when $k = 2$.

g. For $k = 3$, notice that (in the notation of part e) $\{\bar{\partial}\rho_j \wedge \bar{\partial}\rho_i\}_{i<j}$ are linearly independent. Define $h_{ij}(z)$ by the equations

$$\bar{\partial} h_i = \sum_{j=1}^{n} h_{ij}(z) \bar{\partial} \rho_j(z).$$

Then

$$0 = \bar{\partial}^2 u = \sum_{i<j} [h_{ij} - h_{ji}] \bar{\partial}\rho_j \wedge \bar{\partial}\rho_i.$$

h. Conclude that $h_{ij} = h_{ji}$ for $i < j$. Define

$$u_2(z) = u_1(z) + \frac{1}{2!} \sum_{i,j} h_{ij} \rho_i \rho_j.$$

Then $u_2 = u$ on M and

$$\bar{\partial}u_2(z) = \frac{1}{2!} \sum_{i,j} \bar{\partial}h_{ij}(z)\rho_i(z)\rho_j(z) = \mathcal{O}\left((\text{dist}\,(z, M))^2\right).$$

i. Iterate this procedure to construct u_{k-1}, each k.

13. Let $\Omega \subset\subset \mathbb{C}^n$ be a smoothly bounded pseudoconvex domain. J. J. Kohn [3] has proved that for each $k \in \mathbb{N}$ there is an $m(k) \in \mathbb{N}, m(k) \geq k$, and a constant $C(k)$ such that the following property holds: For each $\bar{\partial}$-closed $(0,1)$ form f with $C^{m(k)}$ coefficients on $\bar{\Omega}$ there is a $u_k \in C^k(\bar{\Omega})$ such that $\bar{\partial}u_k = f$. Also

$$\|u_k\|_{C^k(\bar{U})} \leq C(k)\|f\|_{C^{m(k)}(\bar{U})}.$$

It should be understood that $m(k), C(k)$ depend on the geometry on $\partial\Omega$ but *not* on f. It is not well understood how $m(k)$ depends on k. In general, $m(k) >> k$. The proof of Kohn's theorem is too difficult to consider here so we take it for granted.

Use Kohn's theorem and the technique of Exercise 23 at the end of Chapter 1 to prove that, with $\Omega \subset\subset \mathbb{C}^n$ as above and f a $\bar{\partial}$-closed $(0,1)$ form with C^∞ coefficients on $\bar{\Omega}$, there is a $u \in C^\infty(\bar{\Omega})$ such that $\bar{\partial}u = f$. Details of this last assertion appear in J. J. Kohn [4]; see also Exercise 12 at the end of Chapter 4. Here are some hints:

a. It is enough to construct $v_j \in C^j, j = 1, 2, \ldots$, such that $\bar{\partial}v_j = f$ and $\|v_j - v_{j+1}\|_{C^j} < 2^{-j}$.

b. Assume inductively that $v_1, \ldots, v_j$ have been constructed. Let $\tilde{v}_{j+1}$ be a solution in C^{j+1} to $\bar{\partial}u = f$. Imitate the proof of Theorem 10.4.2 to find a holomorphic V on Ω such that $\|(v_j - \tilde{v}_{j+1}) - V\|_{C^j} < 2^{-j}$.

c. Define $v_{j+1} = \tilde{v}_{j+1} + V$.

14. Provide the details for the following argument which provides an elegant way to think about the local solvability question. Let (x, y, t) be coordinates in $\mathbb{R}^3$. Identify (x, y) with $z = x + iy$. Define

$$L = \frac{\partial}{\partial z} + i\bar{z}\frac{\partial}{\partial t}.$$

Refer to Exercise 10 at the end of Chapter 2. Let $\mathcal{U} \subseteq \mathbb{C}^2$ be given by $\mathcal{U} = \{z \in \mathbb{C}^2 : \mathrm{Im}z_1 > |z_2|^2\}$. Map $\mathbb{R}^3$ to $\partial\mathcal{U}$ by

$$(z,t) \mapsto (t + i|z|^2, z).$$

Then L on $\mathbb{R}^3$ is transformed to

$$L' = \frac{\partial}{\partial z_2} + 2i\bar{z}_2 \frac{\partial}{\partial z_1}.$$

Notice that L' is the formal adjoint to

$$A = \frac{\partial}{\partial \bar{z}_2} - 2iz_2 \frac{\partial}{\partial \bar{z}_1}.$$

If $f \in L^2(\partial\mathcal{U})$, let Sf be its Szegö integral on $\mathcal{U}$. We have the following.

Theorem The equation $L'u = f$ has a solution near $P \in \partial\mathcal{U}$ if and only if Sf is real analytic near P.

The "if" part of this theorem exceeds the scope of this exercise. For the "only if" part, proceed as follows. Let $\phi \in C_c^\infty(\partial\mathcal{U})$ satisfy $\phi = 1$ near P. Prove that

$$S(f - A^*(\phi u)) = Sf.$$

But $f - A^*(\phi u)$ equals 0 near P; hence, $S(f - A^*(\phi u))$ extends holomorphically past P.

This argument, together with a beautiful exposition of related matters, comes from J. J. Kohn [4]. Compare this exercise with Exercise 6.

11 Holomorphic Mappings and Invariant Metrics

11.1 Classical Theory of Holomorphic Mappings

In this section we prove a collection of results (mainly due to H. Cartan) that include a characterization of those biholomorphic maps of a circular domain that fix 0. There follows a proof of Poincaré's theorem that the ball and the polydisc are not biholomorphic. Note that, throughout this chapter, the word "map" (or "mapping") usually means *holomorphic* or *biholomorphic* mapping unless explicitly noted otherwise.

Recall that a map $f : \Omega_1 \to \Omega_2$ of domains in $\mathbb{C}^n$ is said to be *biholomorphic* if $f = (f_1, \ldots, f_n)$ is one-to-one, onto, and has a holomorphic inverse. If $\Omega_1 = \Omega_2 = \Omega$, then we say that f is an *automorphism* of Ω and we write $f \in \mathrm{Aut}(\Omega)$.

A mapping $f : X \to Y$ of topological spaces is called *proper* if $f^{-1}(K) \subseteq X$ is compact whenever $K \subseteq Y$ is compact. Trivially, a biholomorphic mapping f is proper since f^{-1} is continuous. For bounded domains $\Omega_1, \Omega_2 \subseteq \mathbb{C}^n$, notice that $g : \Omega_1 \to \Omega_2$ is proper if and only if the following condition is satisfied: Whenever $\{z^j\} \subseteq \Omega_1$ has the property that $z^j \to \partial\Omega_1$, then $g(z^j) \to \partial\Omega_2$.

We begin our study by considering a self-mapping of a bounded domain Ω that possesses a fixed point.

PROPOSITION 11.1.1 Let $\Omega \subseteq \mathbb{C}^n$ be a bounded domain. Let $P \in \Omega$ and suppose that $\phi : \Omega \to \Omega$ satisfies $\phi(P) = P$. If $\mathrm{Jac}_{\mathbb{C}}\phi(P) = \mathrm{id}$, then ϕ is the identity.

Proof We may assume that $P = 0$. Expanding ϕ in a power series about $P = 0$ (and remembering that ϕ is vector-valued hence so is the expansion), we have

$$\phi(z) = z + P_k(z) + O(|z|^{k+1}),$$

where P_k is the first homogeneous polynomial of order exceeding 1 in the Taylor expansion. Defining $\phi^j(z) = \phi \circ \cdots \circ \phi$ (j times), we have

$$\phi^2(z) = z + 2P_k(z) + O(|z|^{k+1})$$
$$\phi^3(z) = z + 3P_k(z) + O(|z|^{k+1})$$
$$\vdots$$
$$\phi^j(z) = z + jP_k(z) + O(|z|^{k+1}).$$

Choose polydiscs $D^n(0, a) \subseteq \Omega \subseteq D^n(0, b)$. Then for $0 \leq j \in \mathbb{Z}$ we know that $D^n(0, a) \subseteq \operatorname{dom} \phi^j \subseteq D^n(0, b)$. Therefore, the Cauchy estimates imply that for any multi-index α with $|\alpha| = k$, we have

$$j|D^\alpha \phi(0)| = |D^\alpha \phi^j(0)| \leq n\frac{b \cdot \alpha!}{a^k}.$$

Letting $j \to \infty$ yields that $D^\alpha \phi(0) = 0$.

We conclude that $P_k = 0$; this contradicts the choice of P_k unless $\phi(z) \equiv z$.
□

Remark: Notice that this proposition is a generalization of the uniqueness part of the classical Schwarz lemma on the disc. In fact a great deal of work has been devoted to generalizations of the Schwarz lemma to more general settings. We refer the reader to H. H. Wu [1], S. T. Yau [1], S. G. Krantz [20, 21], and Burns and Krantz [1] for more on this matter.
□

PROPOSITION 11.1.2 Let $\Omega \subseteq \mathbb{C}^n$ be a bounded circular domain. Assume that $0 \in \Omega$. Let $\phi \in \operatorname{Aut} \Omega$ satisfy $\phi(0) = 0$. Then ϕ is linear.

Remark: This is a special case of a recent and much more powerful result: If Ω_1, Ω_2 are bounded, circular domains that are biholomorphically equivalent, then there is a *linear biholomorphism* between them (see R. Braun, W. Karp, and H. Upmeier [1]).
□

Proof of the Proposition Let $\phi' = \operatorname{Jac}_{\mathbb{C}} \phi(0)$. Let ρ_θ be the map $z \mapsto e^{i\theta} z, 0 \leq \theta < 2\pi$. Define

$$\psi = \rho_{-\theta} \circ \phi^{-1} \circ \rho_\theta \circ \phi.$$

Then

$$\psi' = \rho'_{-\theta} \circ (\phi^{-1})' \circ (\rho_\theta)' \circ \phi'.$$

Now $(\rho_{-\theta})'$ and $(\rho_\theta)'$ are diagonal matrices, hence they are in the center of $GL(n, \mathbb{C})$. Therefore, the last line is just

$$\psi' = (\phi^{-1})' \cdot \phi' = \operatorname{id}. \tag{11.1.2.1}$$

Therefore, ψ satisfies the hypotheses of Proposition 11.1.1. It follows that $\psi(z) \equiv z$. Therefore,

$$\phi \circ \rho_\theta = \rho_\theta \circ \phi$$

for all θ. Since the only monomials that commute with the ρ_θ are the linear polynomials, we conclude that ϕ is linear.   □

PROPOSITION 11.1.3 (Computation of Aut(D^n))   Let $\phi \in \text{Aut } D^n$. Then there exist $a_1, \ldots, a_n \in D$, numbers $\theta_1, \ldots, \theta_n, 0 \le \theta_i < 2\pi$, and a permutation σ on n letters such that

$$\phi(z) = \left(e^{i\theta_1} \frac{z_{\sigma(1)} - a_1}{1 - \bar{a}_1 z_{\sigma(1)}}, \ldots, e^{i\theta_n} \frac{z_{\sigma(n)} - a_n}{1 - \bar{a}_n z_{\sigma(n)}} \right).$$

Proof   Let $\phi(0) = \alpha = (\alpha_1, \ldots, \alpha_n)$ and write

$$\psi(z) = \left(\frac{z_1 - \alpha_1}{1 - \bar{\alpha}_1 z_1}, \ldots, \frac{z_n - \alpha_n}{1 - \bar{\alpha}_n z_n} \right).$$

Then $f = \psi \circ \phi \in \text{Aut}(D^n)$, $f(0) = 0$, and it suffices to show that $f(z) = (e^{i\theta_1} z_{\sigma(1)}, \ldots, e^{i\theta_n} z_{\sigma(n)})$. By Proposition 11.1.2, f is linear. Thus $f(z) = (b_{ij}) \cdot (z), 1 \le i, j \le n$. Let

$$z^{ik} = \left(\left(1 - \frac{1}{k}\right) \overline{\text{sgn } b_{i1}}, \ldots, \left(1 - \frac{1}{k}\right) \overline{\text{sgn } b_{in}} \right).$$

[Here sgn z, the *sign* of the nonzero complex number z, is given by sgn $z = z/|z|$.] Then the ith component of $f(z^{ik})$ is

$$\sum_{j=1}^{n} \left(1 - \frac{1}{k}\right) |b_{ij}|.$$

Since this number must be less than 1, letting $k \to \infty$ yields $\sum_j |b_{ij}| \le 1$. Let $w^{jk} = (0, \ldots, 0, (1 - 1/k), 0, \ldots, 0)$, where only the jth entry is nonzero. Then $w^{jk} \to \partial D^n$ as $k \to \infty$. Since f is proper, the points $f(w^{jk})$ accumulate at ∂D^n. That is,

$$\left(\left(1 - \frac{1}{k}\right) b_{1j}, \ldots, \left(1 - \frac{1}{k}\right) b_{nj} \right)$$

accumulates at ∂D^n. Therefore, $\max_{1 \le i \le n} |b_{ij}| = 1$. So $(|b_{ij}|)$ is a matrix in which each row sums to at most 1 and each column has at least one element of modulus 1. It follows that each column has precisely one element of modulus 1; otherwise some row would have two such elements, contradicting $\sum_j |b_{ij}| \le 1$. For each j, say that $|b_{\eta(j),j}| = 1$. If $\eta(j_1) = \eta(j_2)$ for some $j_1 \ne j_2$, then again $\sum_j |b_{ij}| \le 1$ is contradicted when $i = \eta(j_1) = \eta(j_2)$. So η is a permutation of $\{1, \ldots, n\}$ and $b_{ij} = 0$ if $i \ne \eta(j)$.

Let $\sigma = \eta^{-1}$. Then we may change notation and say that $|b_{\ell,\sigma(\ell)}| = 1, \ell = 1, \ldots, n$. Also, $b_{\ell j} = 0$ if $j \neq \sigma(\ell)$. Write $b_{\ell,\sigma(\ell)} = e^{i\theta_\ell}$. Then $(b_{\ell j}) \cdot (z) = (e^{i\theta_1} z_{\sigma(1)}, \ldots, e^{i\theta_n} z_{\sigma(n)})$ as desired. $\qquad\square$

DEFINITION 11.1.4   Let $\Omega \subseteq \mathbb{C}^n$ be a domain and $P \in \Omega$. Then $I_P \subseteq \mathrm{Aut}\,(\Omega)$ is the subgroup of elements that fix P. We call I_P the *isotropy subgroup* of P.

Exercises for the Reader

1. If $f : \Omega_1 \to \Omega_2$ is biholomorphic, then the map $\omega_f : \sigma \mapsto f \circ \sigma \circ f^{-1}$ is a group isomorphism of $\mathrm{Aut}\,(\Omega_1)$ to $\mathrm{Aut}\,(\Omega_2)$. If the automorphism groups are equipped with the compact-open topology, then ω_f is bicontinuous. If $f(P) = Q$ then ω_f maps I_P isomorphically onto I_Q. Also, ω_f maps the connected component of the identity in I_P isomorphically onto the connected component of the identity in I_Q.

2. The group $I_0 \subseteq \mathrm{Aut}\,(D^n)$ has $n!$ connected components. (*Hint:* What classical group has $n!$ elements?) Let $B \subseteq \mathbb{C}^n$ be the unit ball. It is possible to show (see W. Rudin [7]) that $I_0 \subseteq \mathrm{Aut}\,B, B \subseteq \mathbb{C}^n$, is connected. Conclude that the ball and the polydisc are not biholomorpically equivalent.

3. The connected component of the identity in $I_0 \subseteq \mathrm{Aut}\,(D^n)$ is the group of maps $z \mapsto (e^{i\theta_1} z_1, \ldots, e^{i\theta_n} z_n), 0 \leq \theta_i < 2\pi$. In particular, it is abelian.

4. The connected component of the identity in $I_0 \subseteq \mathrm{Aut}\,B, B \subseteq \mathbb{C}^n$ the unit ball, contains all unitary automorphisms. So it is not abelian. Conclude again that the ball and the polydisc are biholomorphically inequivalent.

Let $\Omega \subseteq \mathbb{C}^n$. We say that $\phi : \Omega \to \mathbb{C}^n$ is an *imbedding* if it is one-to-one and proper *onto its image*. H. Alexander [3] and L. Lempert [4] have proved that the imbedding of the ball into the polydisc that contains the largest $D^n(0,r), r < 1$ is the identity. Likewise, the imbedding of the polydisc into the ball that contains the largest $B(0,r)$ is the map $z \mapsto z/\sqrt{n}$. Both maps are unique extremals (up to composition with unitary mappings). Once again, we may conclude that the ball and the polydisc are not biholomorphic.

It was mentioned earlier that two circular domains that are biholomorphically equivalent are in fact linearly equivalent. Again, it follows that the ball and polydisc cannot be biholomorphically equivalent.

It is known that there is no *proper* holomorphic mapping of the ball to the polydisc or of the polydisc to the ball. The latter assertion is a special case of Theorem 11.1.6. This result will give us a completely self-contained, and geometrically appealing, way to see that the ball and the polydisc cannot be biholomorphically equivalent. The principal ideas in the proof are due to R. Remmert and K. Stein [1]. We also incorporate some ideas of G. M. Henkin [5]. More recently, A. Huckleberry [1] has used similar techniques to prove non-existence of proper maps in rather general circumstances.

DEFINITION 11.1.5   Let $M \subseteq \mathbb{C}^n$ be a real C^k manifold of real codimension r. A *smooth foliation of M by complex manifolds* of real codimension q in M ($q \leq 2n - r, q + r$ even) is a set $\mathcal{F}$ of complex submanifolds of M and a C^k mapping $\mu : M \to \mathbb{R}^q$ with the following properties:

(11.1.5.1) The Frechet derivative μ' of μ has rank q at each point of M.

(11.1.5.2) We have $N \in \mathcal{F}$ if and only if $N = \mu^{-1}(c)$, some c in the image of μ.

(11.1.5.3) Each $N \in \mathcal{F}$ is a regularly imbedded complex submanifold of M of complex dimension $n - (q + r)/2$.

EXAMPLE   Let $\mathbb{C}^2 \supseteq M = \{x_1 = 0\}$. Then $\rho(z_1, z_2) = y_1$ provides a foliation of M by (complex one-dimensional) manifolds of real codimension 1 in M.   □

Exercises for the Reader
1. Let $M \subseteq \mathbb{C}^n$ be a real analytic manifold of real dimension three. Is it necessarily true that M is locally foliated by one-dimensional complex manifolds? (*Hint:* Think of the concept of totally real manifold.)

2. Let
$$\Omega = \{(z_1, z_2, z_3) \in \mathbb{C}^3 : (x_1 - x_2)^2 - (y_1 - y_2)^2 < x_3\}.$$

Then there is a relatively open set $U \subseteq \partial\Omega$ that is foliated by two-dimensional complex manifolds. What is the largest such U?

THEOREM 11.1.6   Let $\Omega \subseteq \mathbb{C}^n$ be an open set. Suppose that there is an analytic disc $\mathbf{d} \subseteq \partial\Omega$. Let $P \in \partial\Omega$ be the center of $\mathbf{d}$. Let $T_P(\mathbf{d})$ be the tangent space to $\mathbf{d}$ at P and $N_{\mathbf{d}}$ be its real orthogonal complement *in* $T_P(\partial\Omega)$. Suppose there is a small neighborhood U of 0 in $N_{\mathbf{d}}$ such that $\mathbf{d} + U \subseteq \partial\Omega$. Let $\Omega' \subset\subset \mathbb{C}^m$ be strongly pseudoconvex with C^4 boundary. Then there is no proper holomorphic mapping $f : \Omega \to \Omega'$.

COROLLARY 11.1.7   There is no proper holomorphic mapping of $D^n \subseteq \mathbb{C}^n$ to $B \subseteq \mathbb{C}^m$, any $m, n \in \mathbb{N}$ with $n > 1$.

Remark: The boundary of a strongly pseudoconvex domain contains no analytic discs (exercise). Some weakly pseudoconvex boundaries contain discs; most do not. In a sense, the more analytic structure $\partial\Omega$ has the further it is from being strongly pseudoconvex.
    By techniques that are beyond the scope of this book, S. Bell [2] has shown that a C^∞ strongly pseudoconvex domain cannot be biholomorphically equivalent to a C^∞ weakly pseudoconvex domain. Indeed, he shows that if there were such a biholomorphism then the map and its inverse must extend smoothly to the boundaries. The Levi form (a differential invariant) then provides a contradiction. See Section 11.4 for more on this matter.   □

Proof of Theorem 11.1.6   There is no loss of generality to suppose that $P = 0 \in \partial\Omega$ and that $T_P(\mathbf{d})$ is spanned by the x_2, y_2 directions. Further assume that the unit outward normal ν_P is given by $(1, 0, \ldots, 0)$. Let $r > 0$ be so small that $B(P, r) \cap \partial\Omega \subseteq \mathbf{d} + U$. Consider the analytic disc

$$\mathbf{d}_0 = \{(\zeta, 0, \ldots, 0) \in \Omega : |\zeta| < r\}.$$

This is a half-disc that is transversal to $\partial\Omega$. We intend to show that if $f : \Omega \to \Omega'$ is proper, then $(\partial/\partial z_2)f$ is zero on $\mathbf{d}_0$. This will lead to a contradiction.

Let $z \in \partial\mathbf{d}_0 \cap \partial\Omega$, z near P. Then $z = P + u$, some $u \in U$. Let $\epsilon_j \to 0$ be any sequence of small positive numbers. Since f is proper, the sequence $z\{f(z - \epsilon_j\nu_P)\}$ has a boundary accumulation point $Q_z \in \partial\Omega'$; say that $f(z - \epsilon_{j_k}\nu_P) \to Q_z$. Let $\psi_z \in A(\Omega')$ be a peaking function for the point Q_z (see Section 5.2). Thus $\psi_z(Q_z) = 1$ and $|\psi_z(w)| < 1$ for all $w \in \bar{\Omega}' \setminus \{Q_z\}$. Let $\phi : D \to \mathbf{d}$ be a holomorphic parametrization of $\mathbf{d}$. Consider the holomorphic functions

$$\begin{aligned} \alpha_k : D &\to \mathbb{C} \\ \xi &\mapsto \psi_z \circ f(\phi(\xi) - \epsilon_{j_k}\nu_P). \end{aligned}$$

The functions α_k form a normal family on D. So there is a convergent subsequence $\alpha_{k_\ell} \to \alpha_z$, with normal convergence on D. But $|\alpha_z(\xi)| \leq 1$ and $\alpha_z(0) = 1$. The maximum modulus theorem then implies that $\alpha_z \equiv 1$. Therefore, $f(\phi(\xi) - \epsilon_{k_\ell}\nu_P) \to Q_z$ normally on D. It follows from the Cauchy estimates that $(\partial/\partial z_2)f(z - \epsilon_{k_\ell}\nu_P) \to 0$ normally on D.

This shows that if we take any $z \in \partial\mathbf{d}_0 \cap \partial\Omega$, z near P, and any sequence $z^j = z - \epsilon_j\nu_P$ approaching z radially, then there is a subsequence $\tilde{z}^\ell \equiv z^{j_{k_\ell}}$ for which $(\partial/\partial z_2)f(\tilde{z}^\ell) \to 0$. So $(\partial/\partial z_2)f$ tends to 0 radially at each point $z \in \partial\mathbf{d}_0 \cap \partial\Omega$, z near P. Thus the holomorphic function $(\partial/\partial z_2)f$ on $\mathbf{d}_0$ has radial boundary limit 0 on a set of one-dimensional positive measure. Thus $(\partial/\partial z_2)\,f|_{\mathbf{d}_0} \equiv 0$.

The same argument works not just at P but at points in $\partial\Omega$ sufficiently near to P. Therefore, $(\partial/\partial z_2)f = 0$ on a small open set $W \cap \Omega$, W a neighborhood of P in $\mathbb{C}^n$. Therefore , $(\partial/\partial z_2)f \equiv 0$. Thus f cannot be proper. □

Exercise for the Reader
Give a proof using ideas from Chapter 1 that there is no proper holomorphic mapping from any domain $\Omega' \subseteq \mathbb{C}^n, n \geq 2$, to any domain $\Omega' \subseteq \mathbb{C}$.

11.2   Invariant Metrics

The Bergman metric is not the only biholomorphically invariant metric on a domain in $\mathbb{C}^n$. The Carathéodory and Kobayashi/Royden metrics, for instance, have invariance properties that are of interest. For many purposes, these last two metrics interact more naturally with function theory than does the Bergman metric. On the other hand, the Bergman metric is a Kähler metric, whereas these last two are only Finsler metrics. In this section, we shall learn about the Carathéodory and Kobayashi/Royden metrics.

Because this is not a differential geometry text, we shall treat these metrics in an ad hoc fashion. For motivation, we first recall the classical setup for metric geometry. Let M be a smooth manifold. A *Riemannian metric* on M is a smoothly varying, positive definite, inner product on $T_x(M), x \in M$. In other words, we may think of a Riemannian metric as a positive definite matrix $(g_{ij})(x)$ with entries that are continuous functions of x. If $\xi \in T_x(M)$, then the length of ξ is given by $\|\xi\|_{M,x} = (\sum_{i,j} g_{ij}(x)\xi_i\xi_j)^{1/2}$. If $\gamma : [0,1] \to M$ is a C^1curve, then the *length* of γ is

$$\|\gamma\| = \int_0^1 \|\gamma'(t)\|_{M,\gamma(t)} \, dt.$$

If $x, y \in M$, we define the distance of x to y to be

$$d(x,y) = \inf_{\Gamma(x,y)} \|\gamma\|,$$

where

$$\Gamma(x,y) = \{\gamma : [0,1] \to M | \gamma \text{ is a piecewise } C^1 \text{ curve}$$

$$\text{with } \gamma(0) = x, \gamma(1) = y\}.$$

For more on the fundamentals of Riemannian geometry, see S. Kobayashi and K. Nomizu [1] and S. Helgason [1]. Our purpose here has been only to set the stage for the Carathéodory and Kobayashi/Royden metrics; these metrics share much, but not all, of the structure just adumbrated.

Now let $U \subseteq \mathbb{C}^n, V \subseteq \mathbb{C}^n$ be open sets. We let $U(V)$ denote the set of all holomorphic mappings from V to U. As usual, $B \subseteq \mathbb{C}^n$ is the unit ball.

DEFINITION 11.2.1 If $\Omega \subseteq \mathbb{C}^n$ is open, then the *infinitesimal Carathéodory metric* is given by $F_C : \Omega \times \mathbb{C}^n \to \mathbb{R}$, where

$$F_C(z,\xi) = \sup_{\substack{f \in B(\Omega) \\ f(z)=0}} |f_*(z)\xi| \equiv \sup_{\substack{f \in B(\Omega) \\ f(z)=0}} \left| \sum_{j=1}^n \frac{\partial f}{\partial z_j}(z) \cdot \xi_j \right|.$$

Remark: In this definition, the $\mathbb{C}^n$ in $\Omega \times \mathbb{C}^n$ should be thought of as the tangent space to Ω at z—in other words, $\Omega \times \mathbb{C}^n$ should be canonically identified with the tangent bundle of Ω at z. We think of $F_C(z,\xi)$ as the length of the tangent vector ξ at the point $z \in \Omega$. In general, $F_C(z,\xi)$ is not given by a quadratic form $(g_{ij}(z))$, hence F_C is not a Riemannian metric. □

It is known (D. Burns, S. Shnider, and R. Wells [1], R. E. Greene and S. G. Krantz [2]) that a generic open set in $\mathbb{C}^n$ that is diffeomorphic to B is not biholomorphic to B. If we recall the proof of the Riemann mapping theorem, we see that the definition of the Carathéodory metric (and of the Kobayashi/Royden metric, to be discussed below) takes the extremal problem from that proof (L. Ahlfors [1], Greene/Krantz [12]) and uses it to measure to what extent the proof fails in general.

Exercise for the Reader
If $f \in B(\Omega), z \in \Omega$, and ϕ is an automorphism of the ball taking $f(z)$ to 0, then $|(\phi \circ f)_*(z)\xi| \geq |f_*(z)\xi|$ (use Exercise 5 at the end of Chapter 1 or Schwarz's lemma). Thus the condition $f(z) = 0$ in Definition 11.2.1 is superfluous.

DEFINITION 11.2.2   Let $\Omega \subseteq \mathbb{C}^n$ be open and $\gamma : [0,1] \to \Omega$ be a C^1 curve. The *Carathéodory length* of γ is defined to be

$$L_C(\gamma) = \int_0^1 F_C(\gamma(t), \gamma'(t)) \, dt.$$

This definition parallels the definition of the length of a curve in a Riemannian metric. It would be natural at this point to define the (integrated) Carathéodory distance between two points to be the infimum of lengths of all curves connecting them. One advantage of defining distance in this fashion is that it is then straightforward to verify the triangle inequality.

However, we shall not take this approach. One of the most important features of the Carathéodory metric is that it is, in a precise sense, the smallest metric under which holomorphic mappings are distance decreasing. The notion of distance suggested in the last paragraph is *not* the smallest. It is fortuitous that the following definition *does* result in the smallest distance-decreasing metric—in particular, it *does* satisfy the triangle inequality.

DEFINITION 11.2.3   Let $\Omega \subseteq \mathbb{C}^n$ be an open set and $z, w \in \Omega$. The *Carathéodory distance* between z and w is defined to be

$$C(z, w) = \sup_{f \in B(\Omega)} \rho(f(z), f(w)),$$

where ρ is the Poincaré-Bergman distance on B.

Notice that, depending on the domain Ω, the Carathéodory distance between two distinct points may be 0.

Exercise for the Reader
Use this outline to calculate the Bergman distance on the ball explicitly: Using our formula for the Poincaré distance in the disc and the fact that the discs $\zeta \mapsto \zeta\xi$ through the origin are totally geodesic in the Bergman metric, one calculates that the Bergman distance of 0 to $r\mathbb{1}$ is $(\sqrt{n+1}/2) \log[(1+r)/(1-r)]$.

Next notice that the map

$$(z_1, \ldots, z_n) \mapsto \left(\frac{z_1 - r}{1 - rz_1}, \frac{\sqrt{1 - r^2}\, z_2}{1 - rz_1}, \ldots, \frac{\sqrt{1 - r^2}\, z_n}{1 - rz_1} \right)$$

is an automorphism of B that takes $(r, 0, \ldots, 0)$ to 0. In particular, one can now construct automorphisms that map any point $r\,\mathbb{1}$ to any other point $s\,\mathbb{1}$. Together with the unitary rotations, we now have enough automorphisms to map any point of the ball to any other.

   Use the automorphisms described in the preceding paragraph, together with the formula for the Bergman distance from 0 to $r\,\mathbb{1}$, to determine that the integrated Bergman distance on the unit ball in $\mathbb{C}^n$ is

$$B(z, w) = \frac{\sqrt{n+1}}{2} \log \left(\frac{|1 - w \cdot \bar{z}| + \sqrt{|w - z|^2 + |z \cdot \bar{w}|^2 - |z|^2 |w|^2}}{|1 - w \cdot \bar{z}| - \sqrt{|w - z|^2 + |z \cdot \bar{w}|^2 - |z|^2 |w|^2}} \right),$$

where $z \cdot \bar{w} \equiv \sum_j z_j \bar{w}_j$.

DEFINITION 11.2.4   Let $\Omega \subseteq \mathbb{C}^n$ be open. Let $e_1 = (1, 0, \ldots, 0) \in \mathbb{C}^n$. The infinitesimal form of the *Kobayashi/Royden metric* is given by $F_K : \Omega \times \mathbb{C}^n \to \mathbb{R}$, where

$$
\begin{aligned}
F_K(z, \xi) \\
&\equiv \inf\{\alpha : \alpha > 0 \text{ and } \exists f \in \Omega(B) \text{ with } f(0) = z, (f'(0))(e_1) = \xi/\alpha\} \\
&= \inf \left\{ \frac{|\xi|}{|(f'(0))(e_1)|} : f \in \Omega(B), (f'(0))(e_1) \text{ is a constant multiple of } \xi \right\} \\
&= \frac{|\xi|}{\sup\{|(f'(0))(e_1)| : f \in \Omega(B), (f'(0))(e_1) \text{ is a constant multiple of } \xi\}}.
\end{aligned}
$$

   We now wish to define an integrated distance based on elements of $\Omega(B)$. The natural analogue for our definition of integrated Carathéodory distance in Definition 11.2.3 does not satisfy a triangle inequality (see Exercise 32 at the end of the chapter). Moreover, we want the Kobayashi/Royden distance to be the *greatest* metric under which holomorphic mappings are distance decreasing. Therefore, we proceed as follows.

DEFINITION 11.2.5   Let $\Omega \subseteq \mathbb{C}^n$ be open and $\gamma : [0, 1] \to \Omega$ a piecewise C^1 curve. The *Kobayashi/Royden length* of γ is defined to be

$$L_K(\gamma) = \int_0^1 F_K(\gamma(t), \gamma'(t))\, dt.$$

DEFINITION 11.2.6   Let $\Omega \subseteq \mathbb{C}^n$ be an open set and $z, w \in \Omega$. The *Kobayashi/Royden distance* between z and w is defined to be

$$K(z, w) = \inf\{L_K(\gamma) : \gamma \text{ is a piecewise } C^1 \text{ curve connecting } z \text{ and } w\}.$$

Recall that we did not implement a definition like this one for the (integrated) Carathéodory distance because we were able to find a *smaller* distance that satisfied the triangle inequality. Now we are at the other end of the spectrum: We want the Kobayashi metric to be as large as possible and to satisfy a triangle inequality as well. Definition 11.2.6 serves that dual purpose.

Exercise for the Reader
Examine the domain $\Omega = D^2(0, 5/4) \setminus \bar{D}^2(0, 3/4)$ and the points $z = (1, 0), w = (-1, 0)$ to see that, in general,

$$C(z, w) \neq \inf_\gamma \int_0^1 F_C(\gamma(t), \gamma'(t))\, dt.$$

Here the infimum is taken over all (piecewise) C^1 curves connecting z to w. The analogous identity *does* hold for $K(z, w)$ by fiat.

We remark in passing that the use of the ball as a model domain when defining the Carathéodory and Kobayashi/Royden metrics is important, but it is not the unique choice. The theory is equally successful if either the disc or the polydisc is used. However, in the current state of the theory, it is essential that the model domain have a transitive group of automorphisms. Further ideas concerning the use of "model domains" appear, for instance, in R. E. Greene and S. G. Krantz [7] and in D. Ma [1].

PROPOSITION 11.2.7 (Distance-Decreasing Properties)   If Ω_1, Ω_2 are domains in $\mathbb{C}^n, z, w \in \Omega_1, \xi \in \mathbb{C}^n$, and if $f : \Omega_1 \to \Omega_2$ is holomorphic, then

$$F_C^{\Omega_1}(z, \xi) \geq F_C^{\Omega_2}(f(z), f_*(z)\xi) \qquad F_K^{\Omega_1}(z, \xi) \geq F_K^{\Omega_2}(f(z), f_*(z)\xi)$$
$$C_{\Omega_1}(z, w) \geq C_{\Omega_2}(f(z), f(w)) \qquad K_{\Omega_1}(z, w) \geq K_{\Omega_2}(f(z), f(w)).$$

Proof   This follows by inspection from the definitions.   □

COROLLARY 11.2.8   If $f : \Omega_1 \to \Omega_2$ is biholomorphic, then f is an isometry in both the Carathéodory and the Kobayashi/Royden metrics.

Proof   Apply the proposition to both f and f^{-1}.   □

COROLLARY 11.2.9   If $\Omega_1 \subseteq \Omega_2 \subseteq \mathbb{C}^n$ then for any $z, w \in \Omega_1$, any $\xi \in \mathbb{C}^n$, we have

$$F_C^{\Omega_1}(z, \xi) \geq F_C^{\Omega_2}(z, \xi) \qquad F_K^{\Omega_1}(z, \xi) \geq F_K^{\Omega_2}(z, \xi)$$
$$C_{\Omega_1}(z, w) \geq C_{\Omega_2}(z, w) \qquad K_{\Omega_1}(z, w) \geq K_{\Omega_2}(z, w).$$

Exercise for the Reader

Prove that, up to a constant multiple, the Bergman, Carathéodory, and Kobayashi/Royden metrics are equal on the ball B.

LEMMA 11.2.10   Let $B \subseteq \mathbb{C}^n$ be the unit ball and $D_1 \subseteq B$ be the disc $\{(\zeta, 0, \ldots, 0) \in B\}$. Fix $\xi = (t, 0, \ldots, 0) \in \mathbb{C}^n$, $z \in D_1$. Then

$$F_C^B(z, \xi) = F_C^{D_1}(z, \xi),$$
$$F_K^B(z, \xi) = F_K^{D_1}(z, \xi).$$

Proof   Since $D_1 \subseteq B$, Corollary 11.2.9 implies the inequality $\leq$ in both cases. For $\geq$, apply the distance-decreasing property to the mapping $\pi_1 : B \to D_1$ given by $(z_1, \ldots, z_n) \mapsto (z_1, 0, \ldots, 0)$. $\square$

DEFINITION 11.2.11   Let $\Omega_1 \subseteq \mathbb{C}^m, \Omega_2 \subseteq \mathbb{C}^n$ be open sets. A family of maps $\mathcal{F} \subseteq \Omega_2(\Omega_1)$ is said to be *normal* if every sequence in $\mathcal{F}$ contains either a subsequence that converges uniformly on compact sets to an element of $\Omega_2(\Omega_1)$ (i.e., a *normally convergent subsequence*) *or* a subsequence $\{f_j\}$ such that for every $K_1 \subset\subset \Omega_1, K_2 \subset\subset \Omega_2$, there is a $J > 0$ that satisfies $f_j(K_1) \cap K_2 = \emptyset$ when $j \geq J$ (i.e., a *compactly divergent subsequence*).

PROPOSITION 11.2.12   Let $\Omega \subseteq \mathbb{C}^n$ be open, $z \in \Omega, \xi \in \mathbb{C}^n$. Then there is an $f \in B(\Omega)$ such that $F_C(z, \xi) = |f_*(z)\xi|$.

Proof   Choose $f^j \in B(\Omega)$ with $|(f^j)_*(z)\xi| \to F_C(z, \xi)$. We may assume, after composing with suitable unitary rotations, that $f_*^j(z)\xi = (\lambda_j \xi_1, 0 \ldots, 0)$, some $\lambda_j \in \mathbb{R}^+$, each $j = 1, 2, \ldots$. If $\pi_1 : B \to D$ is projection on the first coordinate, then $(\pi_1 \circ f^j)$ is a normal family. Since $\pi_1 \circ f^j(z) = 0$, there is a normally convergent subsequence. Let $\tilde{f} \in D(\Omega)$ be the limit function. Then $(\pi_1 \circ f^j)' \to \tilde{f}'$ normally. Finally, let $f(z) = (\tilde{f}(z), 0, \ldots, 0)$. By Lemma 11.2.10, this f does the job. $\square$

Exercise for the Reader

One must be cautious about normal families in complex analysis of several variables (the problem is with holomorphic *mappings* that are not scalar-valued). Show that if $\Omega \subseteq \mathbb{C}^n$ is a bounded open set that is *not* a domain of holomorphy, then the family $\Omega(D)$ is not normal (use (3.3.5.6)). On the other hand, $D(\Omega)$ will be a normal family regardless of what Ω is. N. Kerzman [1] has shown that a C^1 domain of holomorphy Ω has the property that $\Omega(D)$ is normal (see Exercise 10 at the end of the chapter). The proof is similar in spirit to arguments that we saw in Chapter 3. This result was refined in N. Kerzman and J. P. Rosay [1].

It is a beautiful result of H. Alexander [1] that a family of holomorphic functions $\{f_\alpha\}_{\alpha \in A}$ on $B \subseteq \mathbb{C}^n$ is normal if and only if its restriction to each complex line through 0 is normal. There does not seem to be any elementary way to prove this.

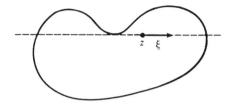

FIGURE 11.1

The foundations for the study of normal families of holomorphic mappings in several variables were laid in the important paper H. H. Wu [1]. This paper has inspired much work in the intervening years. One can even see the germ of the idea for the Kobayashi/Royden metric in it.

Exercise for the Reader
Show that if $\Omega \subseteq \mathbb{C}^n$ is open, $z \in \Omega, \xi \in \mathbb{C}^n$, then it is not necessarily the case that there is an $f \in \Omega(B)$ with $|\xi/f'(0)| = F_K(z,\xi)$. (*Hint:* Use Figure 11.1. By Lemma 11.2.10, it suffices to consider maps in $\Omega(D)$.)

Here is an alternative definition for the integrated Kobayashi/Royden metric that is particularly useful on complex manifolds. For domains in $\mathbb{C}^n$ it is not difficult to see that this new definition is equivalent to the one given above (exercise). The definitions are also equivalent on manifolds (see H. Royden [1]), but this is rather tricky to prove.

DEFINITION 11.2.13   Let $\Omega \subseteq \mathbb{C}^n$ and fix $z, w \in \Omega$. Call a set $\{p_0, \ldots, p_k\}$ *admissible* for $\{z, w\}$ if $p_0 = z, p_k = w$, and there exist $f_j \in \Omega(B)$ and points $u_j^1, u_j^2 \in B, j = 1, \ldots, k$, with $f_j(u_j^1) = p_{j-1}, f_j(u_j^2) = p_j$.

Set

$$K(z,w) = \inf_{\substack{\{p_0,\ldots,p_k\} \\ \text{admissible}}} \sum_{j=1}^{k} B(u_j^1, u_j^2),$$

where B is the Bergman distance on the ball B.

PROPOSITION 11.2.14 Let $\Omega \subseteq \mathbb{C}^n$ be an open set. Let d be a metric on Ω that satisfies $d(f(z), f(w)) \le B(z, w)$ for all $f \in \Omega(B), z, w \in B$ (here $B(z, w)$ is the Bergman distance on the ball). Then $d(z, w) \le K(z, w)$.

Proof   Let $z, w \in \Omega, f_1, \ldots, f_k, p_0, \ldots, p_k, u_1^1, \ldots, u_k^1, u_1^2, \ldots, u_k^2$ be as in the second definition of the integrated Kobayashi/Royden metric (using admissible sequences). Then

$$
\begin{aligned}
d(z,w) &\leq \sum_{j=1}^{k} d(p_{j-1}, p_j) \\
&= \sum_{j=1}^{k} d\big(f_j(u_j^1), f_j(u_j^2)\big) \\
&\leq \sum_{j=1}^{k} B(u_j^1, u_j^2).
\end{aligned}
$$

Passing to the infimum on the right-hand side gives

$$ d(z,w) \leq K(z,w), $$

as desired.   □

PROPOSITION 11.2.15 Let $\Omega \subseteq \mathbb{C}^n$ be an open set. Let d be any metric on Ω that satisfies $d(z,w) \geq B(f(z), f(w))$ for all $f \in B(\Omega)$ and $z, w \in \Omega$. Then $d(z,w) \geq C_\Omega(z,w)$.

Proof   This is a formality.   □

PROPOSITION 11.2.16   Let $\Omega \subseteq \mathbb{C}^n$ be open. Then $K_\Omega(z,w) \geq C_\Omega(z,w)$ for all $z, w \in \Omega$.

Proof   Use one of the preceding two propositions.   □

In the function theory of one complex variable, the Poincaré metric can be transplanted from the disc to any planar domain by way of the uniformization theorem (see S. Fisher [1] or H. Farkas and I. Kra [1]). Thus one does not generally see the Carathéodory or Kobayashi metrics in one variable treatments (however, see S. G. Krantz [20]).

Precious little is known about calculating the Bergman, Carathéodory, or Kobayashi/Royden metrics. For special domains such as the ball, or bounded symmetric domains, the automorphism group is a powerful tool for obtaining explicit formulas. Fefferman obtained information about the asymptotic boundary behavior of the Bergman metric on a strongly pseudoconvex domain in C. Fefferman [1]. No theorem of this sort has ever been proved on a more general class of domains. However, see Blank et al. [1] for some interesting calculations. For the Carathéodory and Kobayashi metrics, one of the most striking results is due to I. Graham [1]. We now describe it.

THEOREM 11.2.17 (I. Graham)   Let $\Omega \subset\subset \mathbb{C}^n$ be a strongly pseudoconvex domain with C^2 boundary. Fix $P \in \partial\Omega$. Let $\xi \in \mathbb{C}^n$, and write $\xi = \xi_T + \xi_N$, the decomposition of ξ into complex tangential and normal components relative to the geometry at the point P.

Let ρ be a defining function for Ω normalized so that $|\nabla\rho(P)| = 1$. Let $\Gamma_\alpha(P)$ be a nontangential approach region at P. If F represents either the Carathéodory or Kobayashi/Royden metric on Ω, then

$$\lim_{z\to P} d_\Omega(z) \cdot F(z,\xi) = \frac{1}{2}|\xi_N|.$$

Here $|\ \ |$ denotes Euclidean length and $d_\Omega(x)$ is the distance of x to the boundary of Ω.

If $\xi = \xi_T$ is complex tangential, then we have

$$\lim_{\Gamma_\alpha(P)\ni z\to P} \sqrt{d_\Omega(z)} \cdot F(z,\xi) = \frac{1}{2}\mathcal{L}(\xi,\xi).$$

Here $\mathcal{L}$ is the Levi form calculated with respect to the defining function ρ.

This result is striking for several reasons. First, it gives a direct connection between the interior geometry of Ω (vis-à-vis the Carathéodory and Kobayashi/Royden metrics) and the boundary geometry (vis-à-vis the Levi form). Second, it provides enough information to see that both of these metrics are *complete* (exercise). Some generalizations and refinements of Graham's theorem appear in G. Aladro [1].

The method of proof of Graham's theorem is an important one: he exploits, in a precise way, the fact that a strongly pseudoconvex point is "very nearly" like a boundary point of the unit ball. We shall not present the details of this argument. However, our proof of Theorem 11.3.4 is very much in the same spirit.

We close this section with a glimpse at one of the most surprising and deepest pieces of work on invariant metrics to appear in the last decade. It is both profound and puzzling, for it uses the classical notion of convexity in a fashion not customary in this subject.

Call a domain $\Omega \subset\subset \mathbb{C}^n$ with C^2 boundary *strongly convex* if all its boundary curvatures are positive (i.e., the real Hessian of some defining function is strictly positive definite). It is convenient, in the statement of this theorem, to assume (as we may) that the Kobayashi metric is defined using maps $f \in \Omega(D)$ rather than in $\Omega(B)$.

THEOREM 11.2.18 (Lempert [2])   Let $\Omega \subset\subset \mathbb{C}^n$ be strongly convex with C^6 boundary. Fix $P \in \Omega$. For each $\xi \in \mathbb{C}^n, |\xi| = 1$, there is a uniquely determined function $\phi_\xi : D \to \Omega$ such that

1. $\phi_\xi'(0)$ is a scalar multiple of ξ;
2. $\phi_\xi(0) = P$;
3. $F_K(P,\xi) = 1/|\phi_\xi'(0)|$.

If $\zeta \in B, \zeta \neq 0$, then we may write ζ uniquely in the form $\zeta = r\xi$, where $0 < r < 1$ and $\xi \in \partial B$. Define a mapping $F : B \to \Omega$ by

$$F(\zeta) = \begin{cases} \phi_\xi(r) & \text{if } \zeta \neq 0 \\ P & \text{if } \zeta = 0. \end{cases}$$

Then F extends to be a C^2 diffeomorphism of $\bar{B}$ to $\bar{\Omega}$.

This is the only known result, for a general class of domains, about the global behavior of extremal discs for the Kobayashi metric. Lempert has used it, together with related ideas, to do the following

- On a convex domain, he proved that the Carathéodory and Kobayashi metrics coincide.

- He has shown that, in a strongly convex domain, there is a holomorphic retraction of the domain onto any Kobayashi extremal disc.

- He has given a new proof of Fefferman's theorem about boundary regularity of biholomorphic mappings (see Section 11.4).

- He has obtained sharp regularity results for the boundary behavior of biholomorphic mappings of strongly pseudoconvex domains.

Sibony [5] has shown that Lempert's results fail categorically on general strongly pseudoconvex domains. (Another approach to some of Lempert's ideas, using a dual extremal problem, appears in H. Royden and P. Wong [1].) It remains a mystery to see what Lempert's results might mean (or even suggest) for a class of domains that is indigenous to the main stream of several complex variables. To this end, see the promising results in M. Y. Pang [1].

For domains that are not strongly pseudoconvex, the most important result about invariant metrics is that of Catlin [4]. He proved that on a domain of finite type in $\mathbb{C}^2$ (see Section 11.5), the Bergman, Carathéodory, and Kobayashi metrics are all comparable. For *strongly pseudoconvex* domains in any dimension, such comparability follows from the results of I. Graham [1] and C. Fefferman [1].

11.3   The Theorem of Bun Wong and Rosay

The section presents a remarkable characterization of the unit ball in terms of its automorphism group. We begin by considering a rather sharp Schwarz lemma that generalizes Proposition 11.1.1. The exercise preceding it records a technical fact that will be needed in the proof.

Exercise for the Reader

Let $U_1 \subseteq \mathbb{C}^m, U_2 \subseteq \mathbb{C}^n$ be open and connected. Let d_1, d_2 be metrics on U_1, U_2, respectively, both of which induce the Euclidean topology. Let $\mathcal{F}$ be the family of functions $f : U_1 \to U_2$ such that

$$d_2(f(z), f(w)) \leq d_1(z, w)$$

for all $z, w \in U_1$. If $p \in U_1, K \subset\subset U_2$, let $\mathcal{F}(p, K) \equiv \{f \in \mathcal{F} : f(p) \in K\}$. Then $\mathcal{F}(p, K)$ is compact in the compact-open topology. (*Hint:* Imitate the proof of the Ascoli/Arzela theorem. Use the fact that a metric ball of radius r is mapped into a metric ball of radius r. Do not fall into the trap of thinking that U_2 must be complete!) See Wu [1] for more on these matters.

THEOREM 11.3.1 (Carathéodory-Cartan-Kaup-Wu)   Let $\Omega \subseteq \mathbb{C}^n$ be open, bounded, and connected. Let $f \in \Omega(\Omega)$ and $P \in \Omega$. Assume that $f(P) = P$. Let $f' = J_{\mathbb{C}} f(P)$ be the differential of f at P. Then:
(11.3.1.1)   The eigenvalues of f' all have modulus not exceeding 1;
(11.3.1.2) $|\det f'| \leq 1$;
(11.3.1.3)   If $|\det f'| = 1$, then $f \in \mathrm{Aut}\,(\Omega)$;
(11.3.1.4)   If $f' = \mathrm{id}$, then $f = \mathrm{id}$.

Remark: This is a sort of Schwarz lemma for holomorphic mappings. For even more general results, see S. T. Yau [1]. Notice that (11.3.1.4) is just Proposition 11.1.1. Alternatively, it follows from (11.3.1.3), since $f' = \mathrm{id}$ implies, by (11.3.1.3), that f is biholomorphic. Thus f is an isometry in the Bergman metric. But an isometry is completely determined by its first-order behavior (its value and differential at a point).   □

Proof of Theorem 11.3.1   We use the Carathéodory metric on Ω. All distances and balls are determined in that metric. Choose $r > 0$ small such that the metric ball $B(P, r)$ is relatively compact in Ω. Let $\mathcal{F}_r$ be the family of distance nonincreasing maps from $B(P, r)$ to $B(P, r)$ that fix P (for small r, such balls are connected). By the exercise, $\mathcal{F}_r$ is compact. If f is as in the hypotheses of the theorem, λ is an eigenvalue of f', and $|\lambda| > 1$, then we obtain a contradiction as follows. The sequence $f_k = (f)^k \equiv f \circ \cdots \circ f \in \mathcal{F}_r$ (composition of f with itself k times) satisfies $(f_k)' = (f')^k$. This mapping has λ^k as an eigenvalue. Since $\mathcal{F}_r$ is compact, the fact that $\lambda^k \to \infty$ is a contradiction. That proves (11.3.1.1). Of course, (11.3.1.2) follows immediately.

For (11.3.1.4), let D be some fixed differential monomial at P. Assume that $Df \neq 0$ and that D is of order at least 2. Then $D\left((f)^k\right) = k \cdot Df$ (the expression simplifies because $f' = \mathrm{id}$). This is a contradiction to the compactness of $\mathcal{F}_r$ as $k \to \infty$. (Compare this argument to the proof of Theorem 11.1.1)

Finally, we address (11.3.1.3). This involves more work. We assume that $|\det f'| = 1$. Then the eigenvalues of f', by (11.3.1.1), each have modulus equal to 1. Put f' in Jordan canonical form. Then we claim that the resulting matrix is diagonal. If it is not, then there is a block of the form

$$
\begin{pmatrix}
\lambda & 1 & & \\
& \lambda & 1 & \\
& & \ddots & 1 \\
& & & \lambda
\end{pmatrix}
$$

where λ is one of the eigenvalues. Then $\left((f)^k\right)'$ has a corresponding block of the form

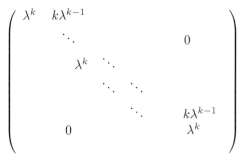

$$
\begin{pmatrix}
\lambda^k & k\lambda^{k-1} & & & & \\
& \ddots & & & 0 & \\
& & \lambda^k & \ddots & & \\
& & & \ddots & \ddots & \\
& & & & \ddots & k\lambda^{k-1} \\
& 0 & & & & \lambda^k
\end{pmatrix}
$$

As $k \to \infty$, the compactness of $\mathcal{F}_r$ is contradicted.

So f', in Jordan canonical form, is diagonal. By the compactness of $\mathcal{F}_r$, there is a sequence f^{k_j} such that $\left(f^{k_j}\right)' \to \mathrm{id}$ (treat each coordinate in succession) and $f^{k_j} \to \tilde{f}$, some $\tilde{f} \in \mathcal{F}_r$. Since the convergence is normal, $\tilde{f}$ is holomorphic on $B(P, r)$, and $\tilde{f}' = \mathrm{id}$. Hence (11.3.1.4) implies that $\tilde{f}\big|_{B(P,r)} = \mathrm{id}$.

We somehow need to spread the last line out to all Ω. Let W be the largest open subset of Ω on which $\{f^k\}_{k=1}^{\infty}$ has a normally convergent subsequence. Changing notation, we may suppose that the f^{k_j} found in the preceding paragraph converge normally to the identity on W. Now let $z \in \overline{W}$. We claim that $z \in W$, so that W is closed. Since W is also open, it follows then that $W = \Omega$. Assuming the claim for the moment, and making the temporary additional assumption that Ω is complete in the Carathéodory metric, let us complete the proof.

We have $\{f^{k_j}\}$ converging normally to id on Ω. Let $\mathcal{G}$ be the family of all distance nonincreasing maps of Ω into Ω that fix P. Then $\mathcal{G}$ is compact. Thus there is a subsequence of $\{f^{k_j-1}\}$ that converges normally to some g on Ω. Call this subsequence f^{k_i-1}. Of course, g is then holomorphic. Finally,

$$
f \circ g = f \circ \left(\lim_{i \to \infty} \left(f^{k_i-1}\right)\right) = \lim_{i \to \infty} f^{k_i} = \mathrm{id}
$$

and

$$
g \circ f = \left(\lim_{i \to \infty} \left(f^{k_i-1}\right)\right) \circ f = \lim_{i \to \infty} f^{k_i} = \mathrm{id}.
$$

It follows that f is invertible and that $f^{-1} = g$.

To prove the claim, notice that, by continuity, f must fix z. Since f is distance nonincreasing, there are connected neighborhoods U, V of z such that $f^{k_j}(U) \subseteq V \subset\subset \Omega$ for all sufficiently large j. Let $\mathcal{F}_{V,U}$ be the set of all distance nonincreasing maps from U into V that fix z. Then $\mathcal{F}_{V,U}$ is compact. Once again, extract a normally convergent subsequence. Since this subsequence must converge to the identity on $U \cap W$, it, in fact, converges to the identity on all of U. Thus $z \in W$ as claimed. We are finished in case Ω is complete in the Carathéodory metric.

In case Ω is not complete, we choose $r > 0$ so small that the metric ball $B(z, r)$ is relatively compact in Ω. Then the family of holomorphic mappings that send $B(z, r)$ to $B(z, r)$ and that fix z is compact. As before, we may argue that a subsequence $\{f^{k_j}\}$ of the maps converges to the identity on $B(z, r)$, and therefore on $W \cup B(z, r)$. This device, of not simply extending the set W, but of extending instead the *functional element* (f, W), enables us to show that $W = \Omega$. As an exercise, the reader should now complete the proof in case Ω is not complete. Complete details may be found in Kobayashi [1, pp. 75–77].   □

We now introduce two new notions that are related to the Carathéodory and Kobayashi metrics. These are holomorphic invariants that correspond, loosely, to volume forms. The exact nature of this correspondence is explored in D. Eisenman [1].

DEFINITION 11.3.2   Let $\Omega \subseteq \mathbb{C}^n$ be open. Define the *Carathéodory volume form* on Ω by

$$M_\Omega^C(z) = \sup\{|\det f'(z)| : f \in B(\Omega), f(z) = 0\}.$$

Define the *Kobayashi-Eisenman volume form* on Ω by

$$M_\Omega^K(z) = \inf\{1/|\det f'(0)| : f \in \Omega(B), f(0) = z\}.$$

Remark: The quantities M^C and M^K transform in a natural manner under automorphisms of Ω. If Φ is such an automorphism, then

$$M_\Omega^C(\Phi(z)) = M_\Omega^C(z)\big|\det J_{\mathbb{C}}\Phi(z)\big|^{-1}$$

and

$$M_\Omega^K(\Phi(z)) = M_\Omega^K(z)\big|\det J_{\mathbb{C}}\Phi(z)\big|^{-1}.$$

Thus M_Ω^K/M_Ω^C is a biholomorphic invariant. This fact has been exploited in Greene and Krantz [7,9,10,11].   □

PROPOSITION 11.3.3 (Bun Wong)   Let $\Omega \subseteq \mathbb{C}^n$ be a bounded, connected open set. Suppose that there is a point $P \in \Omega$ for which

$$M_\Omega^C(P) = M_\Omega^K(P).$$

Then Ω is biholomorphic to the ball in $\mathbb{C}^n$.

Proof   There is a holomorphic map $g \in B(\Omega)$ with $M_\Omega^C(P) = |\det g'(P)|$. We may *not*, in general, choose a corresponding map h for $M_\Omega^K(P)$ (see the discussion in Section 11.2), but we may instead choose a sequence $h_j \in \Omega(B)$ with $|\det h_j'(P)|^{-1} \to M_\Omega^K(P)$. Consider the sequence $G_j = g \circ h_j \in B(B)$. By a previously noted argument, there is a normally convergent subsequence $G_{j_k} \to G \in B(B)$. By hypothesis, it must be that $|\det G'(0)| = 1$. Hence, by (11.3.1.3), $G \in \text{Aut}(B)$. In particular, we may conclude that G is surjective. Therefore g is surjective.

Now consider $H_j = h_j \circ g \in \Omega(\Omega)$. We would like, once again, to extract a normally convergent subsequence. However, we cannot immediately do this. What we *can* do is to restrict attention to a ball $B(P, r)$ in the Carathéodory metric. Applying the exercise at the beginning of this section, we may extract a subsequence H_{j_k} that converges normally to a mapping H on $B(P, r)$. Then $H_{j_k}'(P)$ converges to the identity so that $H' \equiv \text{id}$ on $B(P, r)$. Applying the connectedness argument of the proof of (11.3.1.3), we see that a subsequence $\{H_{j_{k_\ell}}\}$ of $\{H_j\}$ may be extracted that converges normally to some $\tilde{H} \in \Omega(\Omega)$ with $\tilde{H}'(P) = \text{id}$. Thus $\tilde{H} \equiv \text{id}$ by (11.3.1.4). As a result, $h_{j_{k_\ell}} \circ g \to \text{id}$ normally on Ω. It follows that g must be injective.

We have established that $g \in B(\Omega)$ is both surjective and injective. Thus it is a biholomorphism of Ω to B.   □

We now turn to the principal result of this section (see B. Wong [1], J. P. Rosay [1]). It should be compared with the Riemann mapping theorem. We also refer the reader to S. G. Krantz [11], in which the crux of the proof is boiled down for application to one complex variable.

In what follows, when we say that $P \in \partial\Omega$ is a point of strong pseudo-convexity, then it is understood that $\partial\Omega$ is C^2 in a relative neighborhood of P (hence that $\partial\Omega$ is strongly pseudoconvex at points near P).

THEOREM 11.3.4   Let $\Omega \subseteq \mathbb{C}^n$ be any bounded, connected open set with a boundary point $P \in \Omega$ that is strongly pseudoconvex. Suppose further that there exist $K \subset\subset \Omega, z^k \in K$, and $f_k \in \text{Aut}(\Omega)$ with $f_k(z^k) \to P$. Then Ω is biholomorphic to the unit ball in $\mathbb{C}^n$.

COROLLARY 11.3.5   If $\Omega \subseteq \mathbb{C}^n$ has C^2 boundary and has a transitive automorphism group, then Ω is biholomorphic to B.

Proof   Let $P \in \partial\Omega$ be the point in $\partial\Omega$ that is furthest from 0. Then P is a point of strong convexity, hence of strong pseudoconvexity. (To see this, let

$R = \text{dist}\,(0, P)$ and consider the Euclidean ball $B(0, R)$.) Let $K = \{z^0\}$, where z^0 is some fixed interior point of Ω. Let $\Omega \ni w^j \to P$. By hypothesis, there are automorphisms f_j such that $f_j(z^0) = w^j \to P$. The result now follows from the theorem. $\qquad\square$

COROLLARY 11.3.6 Let $\Omega \subseteq \mathbb{C}^n$ be any strongly pseudoconvex set with a noncompact automorphism group. Then Ω is biholomorphic to $B \subseteq \mathbb{C}^n$.

Proof   If $K, L \subset\subset \Omega$, then the set $\mathcal{F}(K, L) = \{f \in \text{Aut}\,(\Omega) : f(K) \subseteq L\}$ is compact (exercise). Therefore, $\text{Aut}\,\Omega$ noncompact implies that there is a $z^0 \in \Omega$ and $f_j \in \text{Aut}\,\Omega$ such that $f_j(z^0) \to P \in \partial\Omega$. Now apply the theorem. $\qquad\square$

The remainder of the section is devoted to a proof of the theorem. The proof proceeds in stages by way of a sequence of preliminary results. Some of these have independent interest.

LEMMA 11.3.7 Let $\Omega \subseteq \mathbb{C}^n$ be open and bounded. Let $P \in \partial\Omega$ be a point of strong pseudoconvexity. Fix $A > 0$. For each $0 < \eta < 1$, there is a $\delta > 0$ such that if $z \in \Omega, |z - P| < \delta$, and $f \in \Omega(B)$ satisfies $f(0) = z$, then $|f(w) - P| < A$ for all w such that $|w| < 1 - \eta$.

Proof   Let L_P be the Levi polynomial at P. Then $\exp(L_P) \equiv g$ satisfies $g(P) = 1$ and $|g(z)| < 1$ for all $|z - P| < \epsilon, z \in \bar{\Omega} \setminus \{P\}$, ϵ sufficiently small.

If the conclusion of the lemma were not true, then there would be an $\eta > 0$ such that for each $j \geq 1/\epsilon$ there is a z_j with $|z_j - P| < 1/j$ and an $f_j \in \Omega(B)$ with $f_j(0) = z_j$ but $|f_j(w_j) - P| \geq A$ for some w_j with $|w_j| < 1 - \eta$. Consider the sequence $G_j = g \circ f_j$. Then $\{G_j\}$ is a normal family. Let G be the normal limit of a subsequence $\{G_{j_k}\}$. Then $G(0) = \lim_{k \to \infty} g \circ f_{j_k}(0) = \lim g(z_{j_k}) = 1$. But $\{f_j\}$ is an equicontinuous family on $\{z : |z| \leq \frac{3}{4}\}$. Thus $|G_j|$, and hence $|G|$, are bounded by 1 on some $\{z : |z| \leq \mu\}$, some $\mu > 0$. Hence $G \equiv 1$ on $\{z : |z| \leq \mu\}$. But then $G \equiv 1$ on all of B. Therefore, $G_{j_k} \to 1$ uniformly on $\{w : |w| \leq 1 - \eta\}$. We may also suppose that $w_{j_k} \to w, |w| \leq 1 - \eta$. But then $|f_{j_k}(w) - P| \geq A/2$ for all large k by equicontinuity. We conclude that $G(w)$ is bounded from 1. That is a contradiction. $\qquad\square$

LEMMA 11.3.8 Let $\Omega \subseteq \mathbb{C}^n$ be open and bounded. Let $P \in \partial\Omega$ be a point of strong pseudoconvexity. If $\{z^j\} \subseteq \Omega$ is any sequence such that $z^j \to P$ and if $A > 0$, then

$$\frac{M_\Omega^K(z^j)}{M_{\Omega \cap B(P,A)}^K(z^j)} \to 1 \qquad \text{as} \quad j \to \infty.$$

Proof   Let $0 < \eta < 1$. Choose $\delta > 0$ according to the last lemma. Thus if $|z^j - P| < \delta$ and $f \in \Omega(B), f(0) = z^j$, then the map $f_\eta : w \mapsto f((1 - \eta)w)$ is in $(B(P, A) \cap \Omega)(B)$ with $f_\eta(0) = z^j$. Thus

$$M_{\Omega \cap B(P,A)}^K(z^j) \leq \frac{1}{|\det f_\eta'(0)|} = \frac{(1 - \eta)^{-n}}{|\det f'(0)|}.$$

Taking the infimum over all such f gives

$$M^K_{\Omega \cap B(P,A)}(z^j) \le (1-\eta)^{-n} M^K_\Omega(z^j).$$

Also

$$M^K_{\Omega \cap B(P,A)}(z^j) \ge M^K_\Omega(z^j)$$

by Corollary 11.2.9. This proves the result. $\square$

LEMMA 11.3.9 Let $\Omega \subseteq \mathbb{C}^n$ be open and bounded. Let $P \in \partial\Omega$ be a point of strong pseudoconvexity. Fix $A > 0, z^0 \in \Omega$, and let $\phi_j \in \mathrm{Aut}\,(\Omega)$ be such that $\phi_j(z^0) \to P$. Then

$$\frac{M^C_\Omega(\phi_j(z^0))}{M^C_{\Omega \cap B(P,A)}(\phi_j(z^0))} \to 1 \qquad \text{as } j \to \infty.$$

Proof Let $\Omega_1 \subset\subset \Omega_2 \subset\subset \cdots \subset \Omega$ satisfy $\cup\Omega_j = \Omega$ and $z^0 \in \Omega_j$ for all j. Choose $f_j \in B(\Omega_j)$ such that $M^C_{\Omega_j} = |\det f'_j(z^0)|$. Now there is an $f \in B(\Omega)$ and a subsequence f_{j_k} such that $f_{j_k} \to f$ normally on Ω. Of course, $f(z_0) = 0$. It follows that $|\det f'_{j_k}(z_0)| \to |\det f'(z^0)|$.

Let $\eta > 0$. By the preceding paragraph, there is a k so large that $M^C_\Omega(z^0) \le M^C_{\Omega_{j_k}}(z^0) \le (1+\eta)M^C_\Omega(z^0)$. By the proof of Lemma 11.3.7, there is an ℓ so large that $\phi_\ell(\Omega_{j_k}) \subseteq B(P,A)$. Therefore,

$$
\begin{aligned}
1 &\le \frac{M^C_{\Omega \cap B(P,A)}(\phi_\ell(z^0))}{M^C_\Omega(\phi_\ell(z^0))} \\
&\le \frac{M^C_{\phi_\ell(\Omega_{j_k})}(\phi_\ell(z^0))}{M^C_\Omega(\phi_\ell(z^0))} \\
&= \frac{M^C_{\Omega_{j_k}}(z^0)}{M^C_\Omega(z^0)} \le 1+\eta.
\end{aligned}
$$

Since $\eta > 0$ was arbitrary, the result follows. $\square$

LEMMA 11.3.10 Let Q be a positive definite quadratic form on $\mathbb{C}^n \times \mathbb{C}^n$. Define $\mathcal{D}_Q = \{w \in \mathbb{C}^n : 2\,\mathrm{Re}\,w_1 < -Q(w,w)\}$. Then $\mathcal{D}_Q$ is biholomorphic to $B \subseteq \mathbb{C}^n$.

Proof Write

$$Q(w,w) = \sum_{i,j} a_{ij} w_i \bar{w}_j.$$

Since Q is positive definite, we can, by a linear change of coordinates, diagonalize $\sum_{i,j=2}^{n} a_{ij} w_i \bar{w}_j$. Therefore, $\mathcal{D}_Q$ is biholomorphic to a region of the form

$$2 \operatorname{Re} w_1 < -\sum_{i=2}^{n} |\alpha_i w_i|^2 - 2 \operatorname{Re} \sum_{i=2}^{n} a_{i1} w_i \bar{w}_1 - a_{11} |w_1|^2.$$

A change of variable of the form

$$
\begin{aligned}
z_1' &= w_1 \\
z_i' &= w_i + \beta_i w_1, \qquad i = 2, \dots, n,
\end{aligned}
$$

eliminates the cross terms, and we arrive at a region of the form

$$|\alpha_0(z_1' - \beta_1)|^2 + \sum_{i=2}^{n} |\alpha_i z_i'|^2 < \gamma.$$

Finally, translation and dilation of z_1', together with dilations of $z_2', \dots, z_n'$, give

$$\sum_{i=1}^{n} |z_i'|^2 < 1.$$

COROLLARY 11.3.11   If $\mathcal{D}_Q$ is as in the preceding lemma and $P \in \mathcal{D}_Q$, then $M_{\mathcal{D}_Q}^C(P) = M_{\mathcal{D}_Q}^K(P)$.

Proof   It suffices to check the assertion at the origin in the ball. This is trivial.   □

Finally, we may give a proof of our main result.

Proof of Theorem 11.3.4   By an easy compactness argument, we may find an $a \in \Omega$ and $f_k \in \operatorname{Aut} \Omega$ such that $f_k(a) \to P \in \partial\Omega$. We may choose a small neighborhood U of P in $\mathbb{C}^n$ such that every point $\zeta \in \partial\Omega \cap U$ is strongly pseudoconvex. If U is small enough, k large, and $f_k(a) \equiv w^k \in U$, then we may select $\zeta^k \in \partial\Omega$ with the following property: When Narasimhan's lemma (Lemma 3.2.2) is applied at the point ζ^k, then, in the new coordinates at ζ^k, the boundary has the defining function

$$\tilde{\rho}_k(\zeta) = 2 \operatorname{Re}\zeta_1 + \mathcal{L}_{\zeta^k}(\zeta) + o(|\zeta|^2)$$

(where $\mathcal{L}_{\zeta^k}$ is the Levi form at ζ^k) and

$$w^k = (a_k, 0, \dots, 0) \quad , \qquad a_k < 0. \tag{11.3.4.1}$$

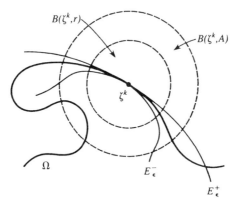

FIGURE 11.2

For this k and $\epsilon > 0$ we then have associated ellipsoids $E_\epsilon^+(\zeta^k)$ and $E_\epsilon^-(\zeta^k)$ given by

$$
\begin{aligned}
E_\epsilon^+(\zeta^k) &= \{2\operatorname{Re}z_1 < -\mathcal{L}_{\zeta^k}(z) + \epsilon|z|^2\}, \\
E_\epsilon^-(\zeta^k) &= \{2\operatorname{Re}z_1 < -\mathcal{L}_{\zeta^k}(z) - \epsilon|z|^2\}.
\end{aligned}
$$

For $\epsilon > 0$ fixed, there exist $A > r > 0$ such that if k is large enough (w^k is near enough to P), then

$$
E_\epsilon^-(\zeta^k) \cap B(\zeta^k, r) \subseteq \Omega \cap B(\zeta^k, A) \subseteq E_\epsilon^+(\zeta^k).
$$

Refer to Figure 11.2.

Notice that A, r depend on ϵ; but by the uniform continuity on U of the second derivatives of any defining function for Ω, they do not depend on ζ^k.

We need to calculate

$$
\frac{M_\Omega^K(a)}{M_\Omega^C(a)} = \frac{M_\Omega^K(w^k)}{M_\Omega^C(w^k)}. \tag{11.3.4.2}
$$

Trivially, the expression is at least 1.

For an inequality in the opposite direction, we apply Lemmas 11.3.8 and 11.3.9 (and the *proof* of 11.3.8) to see that it is enough to estimate

$$
\begin{aligned}
\limsup_{k\to\infty} \frac{M_{\Omega\cap B(\zeta^k,r)}^K(w^k)}{M_{\Omega\cap B(\zeta^k,A)}^C(w^k)} &\leq \limsup_{k\to\infty} \frac{M_{E^-(\zeta^k)\cap B(P,r/2)}^K(w^k)}{M_{E^+(\zeta^k)}^C(w^k)} \\
&= \limsup_{k\to\infty} \frac{M_{E^-(\zeta^k)}^K(w^k)}{M_{E^+(\zeta^k)}^C(w^k)}. \tag{11.3.4.3}
\end{aligned}
$$

Here Corollary 11.2.9 has been applied to obtain the last inequality. We have also used repeatedly the fact that, letting $\tilde{r} = r/2$, it holds that $B(\zeta^k, \tilde{r}) \subseteq B(P, 2\tilde{r}) \subseteq B(\zeta^k, 3\tilde{r})$ for k large.

Now let

$$\Phi_{\epsilon,k}^- : E_\epsilon^-(\zeta^k) \to B$$
$$\Phi_{\epsilon,k}^+ : E_\epsilon^+(\zeta^k) \to B$$

be the biholomorphisms constructed in Lemma 11.3.10. Then $\det J_{\mathbb{C}}\Phi_{\epsilon,k}^-(w^k)$ and $\det J_{\mathbb{C}}\Phi_{\epsilon,k}^+(w^k)$ both depend continuously on ϵ as $\epsilon \to 0^+$. Also their quotient may be made arbitrarily close to 1 for ϵ small, uniformly in k. Finally, notice that since $w^k = (a_k, 0, \ldots, 0)$, it holds that

$$\Phi_{\epsilon,k}^-(w^k) = (a_k(a_{11} + \epsilon) + 1, 0, \ldots, 0),$$
$$\Phi_{\epsilon,k}^+(w^k) = (a_k(a_{11} - \epsilon) + 1, 0, \ldots, 0).$$

It follows that (11.3.4.3) equals

$$\limsup_{k \to \infty} \frac{M_B^K(a_k(a_{11} + \epsilon) + 1, 0, \ldots, 0) \det J_{\mathbb{C}}\Phi_{\epsilon,k}^-(w^k)}{M_B^C(a_k(a_{11} - \epsilon) + 1, 0, \ldots, 0) \det J_{\mathbb{C}}\Phi_{\epsilon,k}^+(w^k)}$$

$$= \limsup_{k \to \infty} \frac{M_B^K(a_k + 1, 0, \ldots, 0) \cdot \mathcal{J}^+ \cdot \det J_{\mathbb{C}}\Phi_{\epsilon,k}^-(w^k)}{M_B^C(a_k + 1, 0, \ldots, 0) \cdot \mathcal{J}^- \cdot \det J_{\mathbb{C}}\Phi_{\epsilon,k}^+(w^k)}.$$

Here $\mathcal{J}^\pm$ are the complex Jacobian determinants of the standard automorphism of B taking $(a_k(a_{11} \pm \epsilon) + 1, 0, \ldots, 0)$ to $(a_k + 1, 0, \ldots, 0)$ (see the exercise for the reader preceding Definition 11.2.4). The last line is as near to 1 as we please if ϵ is small enough.

In conclusion, $M_\Omega^K(a) = M_\Omega^C(a)$. By Proposition 11.3.3, we conclude that Ω is biholomorphic to B.   □

Exercise for the Reader

Compute $\Phi_{\epsilon,k}^+$ and $\Phi_{\epsilon,k}^-$ explicitly and determine their Jacobian determinants exactly. Compute also $\mathcal{J}^+$ and $\mathcal{J}^-$. Use these to render precise the last part of the proof of Theorem 11.3.4.

11.4 Smoothness to the Boundary of Biholomorphic Mappings

Poincaré's theorem that the ball and polydisc are biholomorphically inequivalent shows that there is no Riemann mapping theorem (at least in the traditional sense) in several complex variables. More recent results of D. Burns, S. Shnider, and R. Wells [1] and of R. E. Greene and S. G. Krantz [1, 2] confirm how truly dismal the situation is. First, we need a definition.

DEFINITION 11.4.1   Let ρ_0 be a C^k defining function for a domain $\Omega_0 \subset\subset \mathbb{R}^N, k \geq 2$. We define a neighborhood basis for Ω_0 in the C^k topology as follows: Let $\epsilon > 0$ be so small that if $\|\rho - \rho_0\|_{C^k} < \epsilon$, then ρ has nonvanishing gradient on $\{x : \rho(x) = 0\}$. For any such ρ, let $\Omega_\rho = \{x \in \mathbb{R}^N : \rho(x) < 0\}$. Define

$$\mathcal{U}_\epsilon^k(\Omega_0) = \{\Omega_\rho \subseteq \mathbb{R}^N : \|\rho - \rho_0\|_{C^k} < \epsilon\}.$$

Observe that $\mathcal{U}_\epsilon^k(\Omega_0)$ is a *collection of domains*. Then the sets $\mathcal{U}_\epsilon$ are called neighborhoods of Ω_0 in the C^k topology. Of course, neighborhoods in the C^∞ topology are defined similarly.

THEOREM 11.4.2 (Burns/Shnider/Wells)   Let $k \in \mathbb{N}$. Let $\mathcal{U}_\epsilon^k(B)$ be any neighborhood of the ball $B \subseteq \mathbb{C}^n$ in the C^k topology as defined above. If Ω_1, Ω_2 are domains in $\mathbb{C}^n$, then we say that $\Omega_1 \sim \Omega_2$ if Ω_1 is biholomorphic to Ω_2. If $n \geq 2$, then $\mathcal{U}_\epsilon^k / \sim$ is uncountable, no matter how small $\epsilon > 0$ is or how large k is (even $k = \infty$ or $k = \omega$).

Theorem 11.4.2 is in striking contrast to the situation in $\mathbb{C}^1$, where $\mathcal{U}_\epsilon^k / \sim$ has only one element as soon as $k = 1$ and ϵ is, say, $\frac{1}{5}$. Greene and Krantz [1, 2]) have refined the theorem to show that in fact each of the equivalence classes is closed and nowhere dense.

We now give a brief accounting of some of the differences between $n = 1$ and $n > 1$. A more detailed discussion appears in R. E. Greene and S. G. Krantz [9]. This subject begins with the following breakthrough of C. Fefferman [1].

THEOREM 11.4.3   Let $\Omega_1, \Omega_2 \subseteq \mathbb{C}^n$ be strongly pseudoconvex domains with C^∞ boundary. If $\phi : \Omega_1 \to \Omega_2$ is biholomorphic, then ϕ extends to a C^∞ diffeomorphism of $\bar{\Omega}_1$ onto $\bar{\Omega}_2$.

Fefferman's theorem enables one to see that if Ω_1 and Ω_2 are biholomorphic under ϕ, then there are certain differential invariants of $\partial\Omega_1, \partial\Omega_2$ that must be preserved under ϕ. More precisely, if k is large, then the kth order Taylor expansion of the defining function ρ_1 for Ω_1 (resp. of the defining function ρ_2 for Ω_2) has more coefficients than the k order Taylor expansion for ϕ (the disparity in the number grows rapidly with k). Since ρ_1 is mapped to ρ_2 under composition with ϕ^{-1}, it follows that some of these coefficients, or combinations thereof, must be invariant under biholomorphic mappings. N. Tanaka [1] and S. S. Chern and J. Moser [1] have made these remarks precise and have shown how to calculate these invariants. A more leisurely discussion of these matters appears in R. E. Greene and S. G. Krantz [9].

Now it is easy to see intuitively that two domains Ω and Ω_0 can be close in the C^k topology, any k, and have entirely different Chern/Moser/Tanaka invariants. This notion is made precise, for instance, in D. Burns, S. Shnider, and R. Wells [1], by using a transversality argument. (Note that everything we are saying is vacuous in $\mathbb{C}^1$ because the invariants must live in the complex tangent space to the boundary—which is empty in dimension one.) It is essentially a foregone conclusion that things will go badly in higher dimensions.

If one seeks positive results in the spirit of the Riemann mapping theorem in dimension $n \geq 2$, then one must find statements of a different nature. B. Fridman [3] has constructed a "universal domain" Ω^* which can be used to biomorphically exhaust any other. He has obtained a number of variants of this idea, using elementary but clever arguments. S. Semmes [1] has yet another approach to the Riemann mapping theorem that is more in the spirit of the work of Lempert that was discussed in Section 11.2. We present, mainly for background, a substitute for the Riemann mapping theorem whose statement and proof is more in the spirit of Fefferman's theorem.

THEOREM 11.4.4 (Greene/Krantz [2])   Let $B \subseteq \mathbb{C}^n$ be the unit ball. Let $\rho_0(z) = |z|^2 - 1$ be the usual defining function for B. If $\epsilon > 0$ is sufficiently small, $k = k(n)$ is sufficiently large, and $\Omega \in \mathcal{U}_\epsilon^k(B)$ then either
(11.4.4.1) $\Omega \sim B$ or
(11.4.4.2) Ω is not biholomorphic to the ball *and*

1. Aut Ω is compact;

2. Aut Ω has a fixed point;

3. If $K \subset\subset B, \epsilon > 0$ is sufficiently small (depending on K), and $\Omega \in \mathcal{U}_\epsilon^k(B)$ has the property that its fixed point set lies in K, then there is a biholomorphic mapping $\Phi : \Omega \to \Phi(\Omega) \equiv \Omega' \subseteq \mathbb{C}^n$ such that Aut (Ω') is the *restriction* to Ω' of a subgroup of the group of unitary matrices.

The collection of domains to which (11.4.4.2) applies is both dense and open.

Theorem 11.4.4 shows, in a weak sense, that domains near the ball that have any automorphisms other than the identity are (biholomorphic to) domains with only Euclidean automorphisms. It should be noted that 11.4.4.2(1) is already contained in the theorem of Bun Wong and Rosay and that the denseness of the domains to which (11.4.4.2) applies is contained in the work of Burns-Shnider-Wells. The proof of Theorem 11.4.4 involves a detailed analysis of Fefferman's asymptotic expansion for the Bergman kernel and of the $\bar{\partial}$-Neumann problem and would double the length of this book if we were to treat it in any detail.

The purpose of this lengthy introduction has been to establish the importance of Theorem 11.4.3 and to set the stage for what follows. It may be noted that the proof of the analogous result in $\mathbb{C}^1$, that a biholomorphic mapping of smooth domains extends smoothly to the boundary, was proved in P. Painlevé [1]. The result in one complex dimension has been highly refined, beginning with work of O. Kellogg [2] and more recently of S. Warschawski [1,2,3], Rodin and Warschawski [1], and others. This classical work uses harmonic estimation and the Jordan curve theorem, devices that have no analogue in higher dimensions. A short, self-contained, proof of the one-variable result—using ideas closely related to those presented here—appears in S. Bell and S. G. Krantz [1].

We conclude this section by presenting a short and elegant proof of Fefferman's theorem (Theorem 11.4.3). The techniques are due to S. Bell [2] and S. Bell and E. Ligocka [1]. The proof uses an important and nontrivial fact (known as "Condition R" of Bell and Ligocka) about the $\bar{\partial}$-Neumann problem. We will have to take Condition R on a strongly pseudoconvex domain for granted. (However, Condition R, and more generally the solution of the $\bar{\partial}$-Neumann problem, *is* considered in detail in S. G. Krantz [19].)

Let $\Omega \subset\subset \mathbb{C}^n$ be a domain with C^∞ boundary. We define Condition R as follows.

Condition R (S. Bell [2]) Define an operator on $L^2(\Omega)$ by

$$Pf(z) = \int_\Omega K(z,\zeta)f(\zeta)\,dV(\zeta),$$

where $K(z,\zeta)$ is the Bergman kernel for Ω. Then for each $j > 0$, there is an $m = m(j) > 0$ such that P satisfies the estimates

$$\|Pf\|_{W^j(\Omega)} \le C_j\|f\|_{W^m(\Omega)}$$

for all testing functions f.

The deep fact, which we shall use without proof (but see the aforementioned reference S. G. Krantz [19]) is that *Condition R holds on any strongly pseudoconvex domain.*

We remark in passing that, in general, it does not matter whether $m(j)$ is much larger than j or whether $m(j)$ depends polynomially on j or exponentially on j. It so happens that for a strongly pseudoconvex domain we may take $m = j$. This assertion is proved in S. G. Krantz [19] in detail. On the other hand, Barrett [2] has shown that on the Diederich-Fornaess worm domain, we must take $m(j) > j$.

Now we build a sequence of lemmas leading to Fefferman's theorem. First we record some notation.

If $\Omega \subset\subset \mathbb{C}^n$ is any smoothly bounded domain and if $j \in \mathbb{N}$, we let

$$
WH^j(\Omega) \;=\; W^j(\Omega) \cap \{\text{holomorphic functions on } \Omega\},
$$

$$
WH^\infty(\Omega) \;=\; \bigcap_{j=1}^{\infty} WH^j(\Omega) = C^\infty(\bar\Omega) \cap \{\text{holomorphic functions on } \Omega\}.
$$

Here W^j is the standard Sobolev space on a domain—see the discussion in Chapter 4. Let $W_0^j(\Omega)$ be the W^j closure of $C_c^\infty(\Omega)$. (Exercise: If j is sufficiently large, then the Sobolev imbedding theorem implies trivially that $W_0^j(\Omega)$ is a proper subset of $W^j(\Omega)$.)

Let us say that $u, v \in C^\infty(\bar\Omega)$ agree up to order k on $\partial\Omega$ if

$$
\left(\frac{\partial}{\partial z}\right)^\alpha \left(\frac{\partial}{\partial \bar z}\right)^\beta (u - v)\Bigg|_{\partial\Omega} = 0 \qquad \forall \alpha, \beta \text{ with } |\alpha| + |\beta| \le k.
$$

LEMMA 11.4.5   Let $\Omega \subset\subset \mathbb{C}^n$ be smoothly bounded and strongly pseudoconvex. Let $w \in \Omega$ be fixed. Let K denote the Bergman kernel. There is a constant $C_w > 0$ such that

$$
\|K(w, \cdot)\|_{\sup} \le C_w.
$$

Proof   The function $K(z, \cdot)$ is harmonic. Let $\phi : \Omega \to \mathbb{R}$ be a radial, C_c^∞ function centered at w. Assume that $\phi \ge 0$ and $\int \phi(\zeta)dV(\zeta) = 1$. Then the mean value property implies that

$$
K(z, w) = \int_\Omega K(z, \zeta)\phi(\zeta)\, dV(\zeta).
$$

But the last expression equals $P\phi(z)$. Therefore,

$$
\begin{aligned}
\|K(w, \cdot)\|_{\sup} &= \sup_{z \in \Omega} |K(w, z)| \\
&= \sup_{z \in \Omega} |K(z, w)| \\
&= \sup_{z \in \Omega} |P\phi(z)|.
\end{aligned}
$$

By Sobolev's theorem (see Exercise 7 at the end of Chapter 4), this is

$$
\le C(\Omega) \cdot \|P\phi\|_{WH^{2n+1}}.
$$

By Condition R, this is

$$
\le C(\Omega) \cdot \|\phi\|_{W^{m(2n+1)}} \equiv C_w. \qquad \square
$$

LEMMA 11.4.6 Let $u \in C^\infty(\bar{\Omega})$ be arbitrary. Let $s \in \{0, 1, 2, \ldots\}$. Then there is a $v \in C^\infty(\bar{\Omega})$ such that $Pv = 0$ and the functions u and v agree to order s on $\partial\Omega$.

Proof   After a partition of unity, it suffices to prove the assertion in a small neighborhood U of $z_0 \in \partial\Omega$. After a rotation, we may suppose that $\partial\rho/\partial z_1 \neq 0$ on $U \cap \bar{\Omega}$, where ρ is a defining function for Ω. Define the differential operator

$$\nu = \frac{\mathrm{Re}\left\{\sum_{j=1}^{n} \frac{\partial\rho}{\partial z_j} \frac{\partial}{\partial\bar{z}_j}\right\}}{\sum_{j=1}^{n} \left|\frac{\partial\rho}{\partial z_j}\right|^2}.$$

Notice that $\nu\rho = 1$. Now we define v by induction on s.

For the case $s = 0$, let

$$w_1 = \frac{\rho u}{\partial\rho/\partial\zeta_1}.$$

Define

$$v_1 = \frac{\partial}{\partial\zeta_1} w_1 = u + O(\rho).$$

Then u and v_1 agree to order 0 on $\partial\Omega$. Also,

$$Pv_1(z) = \int K(z,\zeta) \frac{\partial}{\partial\zeta_1} w_1(\zeta)\, dV(\zeta).$$

This equals, by integration by parts,

$$-\int \frac{\partial}{\partial\zeta_1} K(z,\zeta) w_1(\zeta) dV(\zeta).$$

Notice that the integration by parts is valid by Lemma 11.4.5 and because $w_1|_{\partial\Omega} = 0$. Also, the integrand in this last line is zero because $K(z, \cdot)$ is conjugate holomorphic. Thus $Pv_1 \equiv 0$ and the case $s = 0$ is complete.

Suppose inductively that $w_{s-1} = w_{s-2} + \theta_{s-1}\rho^s$ and $v_{s-1} = (\partial/\partial z_1)(w_{s-1})$ have been constructed. We show that there is a w_s of the form

$$w_s = w_{s-1} + \theta_s \cdot \rho^{s+1}$$

such that $v_s = (\partial/\partial z_1)(w_s)$ agrees to order $(s-1)$ with u on $\partial\Omega$. By the inductive hypothesis,

$$
\begin{aligned}
v_s &= \frac{\partial}{\partial z_1} w_s = \frac{\partial w_{s-1}}{\partial z_1} + \frac{\partial}{\partial z_1}\left[\theta_s \cdot \rho^{s+1}\right] \\
&= v_{s-1} + \rho^s \left[(s+1)\theta_s \frac{\partial\rho}{\partial z_1} + \rho \cdot \frac{\partial\theta_s}{\partial z_1}\right]
\end{aligned}
$$

agrees to order $s - 1$ with u on $\partial\Omega$ so long as θ_s is smooth. So we need to examine $D(u - v_s)$, where D is an s-order differential operator. But if D involves a tangential derivative D_0, then write $D = D_0 \cdot D_1$. It follows that $D(u - v_s) = D_0(\alpha)$, where α vanishes on $\partial\Omega$ so that $D_0\alpha = 0$ on $\partial\Omega$. So we need only check $D = \nu^s$.

We have seen that θ_s must be chosen so that

$$\nu^s(u - v_s) = 0 \qquad \text{on } \partial\Omega.$$

Equivalently,

$$\nu^s(u - v_{s-1}) - \nu^s\left(\frac{\partial}{\partial z_1}\right)(\theta_s \rho^{s+1}) = 0 \qquad \text{on } \partial\Omega$$

or

$$\nu^s(u - v_{s-1}) - \theta_s\left(\nu^s \frac{\partial}{\partial z_1} \rho^{s+1}\right) = 0 \qquad \text{on } \partial\Omega$$

or

$$\nu^s(u - v_{s-1}) - \theta_s \cdot (s+1)! \frac{\partial\rho}{\partial z_1} = 0 \qquad \text{on } \partial\Omega.$$

It follows that we must choose

$$\theta_s = \frac{\nu^s(u - v_{s-1})}{(s+1)! \frac{\partial\rho}{\partial z_1}},$$

which is indeed smooth on U. As in the case $s = 0$, it holds that $Pv_s = 0$. This completes the induction and the proof. $\square$

Remark: A retrospection of the proof reveals that we have constructed v by subtracting from u a Taylor-type expansion in powers of ρ. $\square$

LEMMA 11.4.7   For each $s \in \mathbb{N}$, we have $WH^\infty(\Omega) \subseteq P(W_0^s(\Omega))$.

Proof   Let $u \in C^\infty(\bar{\Omega})$. Choose v according to Lemma 11.4.6. Then $u - v \in W_0^s$ and $Pu = P(u - v)$. Therefore,

$$P(W_0^s) \supseteq P(C^\infty(\bar{\Omega})) \supseteq P(WH^\infty(\Omega)) = WH^\infty(\Omega). \qquad \square$$

Henceforth, let Ω_1, Ω_2 be fixed C^∞ strongly pseudoconvex domains in $\mathbb{C}^n$, with K_1, K_2 their Bergman kernels and P_1, P_2 the corresponding Bergman projections. Let $\phi : \Omega_1 \to \Omega_2$ be a biholomorphic mapping, and let $u = \det \mathrm{Jac}_{\mathbb{C}}\phi$. For $j = 1, 2$, let $\delta_j(z) = \delta_{\Omega_j}(z) = \mathrm{dist}\,(z, {}^c\Omega_j)$.

LEMMA 11.4.8   For any $g \in L^2(\Omega_2)$ we have

$$P_1(u \cdot (g \circ \phi)) = u \cdot ((P_2(g)) \circ \phi).$$

Proof   Notice that $u \cdot (g \circ \phi) \in L^2(\Omega_1)$ by change of variables. Therefore,

$$
\begin{aligned}
P_1(u \cdot (g \circ \phi))(z) &= \int_{\Omega_1} K_1(z, \zeta) u(\zeta) g(\phi(\zeta)) \, dV(\zeta) \\
&= \int_{\Omega_1} u(z) K_2(\phi(z), \phi(\zeta)) \overline{u(\zeta)} u(\zeta) g(\phi(\zeta)) \, dV(\zeta)
\end{aligned}
$$

by Proposition 1.4.12. Change of variable now yields

$$
\begin{aligned}
P_1(u \cdot (g \circ \phi))(z) &= u(z) \int_{\Omega_2} K_2(\phi(z), \xi) g(\xi) \, dV(\xi) \\
&= u(z) \cdot [(P_2(g)) \circ \phi](z). \qquad \square
\end{aligned}
$$

LEMMA 11.4.9   Let $\psi : \Omega_1 \to \Omega_2$ be a C^j diffeomorphism that satisfies

$$\left| \frac{\partial^\alpha \psi}{\partial z^\alpha}(z) \right| \le C \cdot (\delta_1(z))^{-|\alpha|}, \tag{11.4.9.1}$$

for all multi-indices α with $|\alpha| \le j \in \mathbb{N}$ and

$$|\nabla \psi^{-1}(w)| \le C(\delta_2(w))^{-1}. \tag{11.4.9.2}$$

Suppose also that

$$\delta_2(\psi(z)) \le C\delta_1(z). \tag{11.4.9.3}$$

Then there is a number $J = J(j)$ such that, whenever $g \in W_0^{j+J}(\Omega_2)$, then $g \circ \psi \in W_0^j(\Omega_1)$.

Proof   The subscript 0 causes no trouble by the definition of W_0^j. Therefore, it suffices to prove an estimate of the form

$$\|g \circ \psi\|_{W_0^j} \le C\|g\|_{W_0^{j+J}}, \quad \text{all } g \in C_c^\infty(\Omega).$$

By the chain rule and Leibniz's rule, if α is a multi-index of modulus not exceeding j, then

$$\left(\frac{\partial}{\partial z} \right)^\alpha (g \circ \psi) = \sum \left[(D^\beta g) \circ \psi \right] \cdot D^{\gamma_1} \psi \cdots D^{\gamma_\ell} \psi,$$

where $|\beta| \le |\alpha|$, $\sum |\gamma_i| \le |\alpha|$, and the number of terms in the sum depends only on α (a classical formula of Faà de Bruno—see S. Roman [1]—actually gives this sum quite explicitly, but we do not require such detail). Note here that $D^{\gamma_i}\psi$ is used to denote a derivative of *some component* of ψ. By hypothesis, it follows that

$$\left|\left(\frac{\partial}{\partial z}\right)^\alpha (g \circ \psi)\right| \leq C \sum |(D^\beta g) \circ \psi| \cdot (\delta_1(z))^{-j}.$$

Therefore,

$$\int_{\Omega_1} \left|\left(\frac{\partial}{\partial z}\right)^\alpha (g \circ \psi)\right|^2 dV \leq C \sum \int_{\Omega_1} |(D^\beta g) \circ \psi|^2 (\delta_1(z))^{-2j} \, dV(z)$$

$$= C \sum \int_{\Omega_2} |D^\beta g(w)|^2 \delta_1 \left(\psi^{-1}(w)\right)^{-2j}$$

$$\times |\det J_{\mathbb{C}} \psi^{-1}|^2 \, dV(w).$$

But (11.4.9.2) and (11.4.9.3) imply that the last line is majorized by

$$C \sum \int_{\Omega_2} |D^\beta g(w)|^2 \delta_2(w)^{-2j} \delta_2(w)^{-2n} \, dV(w). \tag{11.4.9.4}$$

Now if J is large enough, depending on the Sobolev imbedding theorem (see Exercise 7 at the end of Chapter 4), then

$$|D^\beta g(w)| \leq C \|g\|_{W_0^{j+J}} \cdot \delta_2(w)^{2n+2j}.$$

(Remember that g is compactly supported in Ω_2.) Hence (11.4.9.4) is majorized by $C\|g\|_{W_0^{j+J}}$. $\qquad\square$

LEMMA 11.4.10 For each $j \in \mathbb{N}$, there is an integer J so large that if $g \in W_0^{j+J}(\Omega_2)$, then $g \circ \phi \in W_0^j(\Omega_1)$. (Here ϕ is the biholomorphic mapping in 11.4.3.)

Proof   The Cauchy estimates give (since ϕ is bounded) that

$$\left|\frac{\partial^\alpha \phi_\ell}{\partial z^\alpha}(z)\right| \leq C \cdot (\delta_1(z))^{-|\alpha|}, \qquad \ell = 1, \ldots, n \tag{11.4.10.1}$$

and

$$|\nabla \phi^{-1}(w)| \leq C(\delta_2(w))^{-1}, \tag{11.4.10.2}$$

where $\phi = (\phi_1, \ldots, \phi_n)$. We will prove that

$$C \cdot \delta_1(z) \geq \delta_2(\phi(z)). \tag{11.4.10.3}$$

Then Lemma 11.4.9 gives the result.

To prove (11.4.10.3), let ρ be a smooth strictly plurisubharmonic defining function for Ω_1. (See Exercise 20 at the end of Chapter 3.) Then $\rho \circ \phi^{-1}$ is a smooth plurisubharmonic function on Ω_2. Since ρ vanishes on $\partial\Omega_1$ and since ϕ^{-1} is proper, we conclude that $\rho \circ \phi^{-1}$ extends continuously to $\bar\Omega_2$. If $P \in \partial\Omega_2$ and ν_P is the unit outward normal to $\partial\Omega_2$ at P, then Hopf's lemma (Exercise 22 at the end of Chapter 1) implies that the (lower) one-sided derivative $(\partial/\partial\nu_P)(\rho\circ\phi^{-1})$ satisfies

$$\frac{\partial}{\partial\nu_P}(\rho\circ\phi^{-1}(P)) \geq C.$$

So, for $w = P - \epsilon\nu_P, \epsilon$ small, it holds that

$$-\rho\circ\phi^{-1}(w) \geq C \cdot \delta_2(w).$$

These estimates are uniform in $P \in \partial\Omega_2$. Using the comparability of $|\rho|$ and δ_1 yields

$$C\delta_1(\phi^{-1}(w)) \geq \delta_2(w).$$

Setting $z = \phi^{-1}(w)$ now gives

$$C'\delta_1(z) \geq \delta_2(\phi(z)),$$

which is (11.4.10.3). □

Exercise for the Reader
Let $\Omega \subset\subset \mathbb{C}^n$ be a smoothly bounded domain. Let $j \in \mathbb{N}$. There is an $N = N(j)$ so large that $g \in W_0^N$ implies that g vanishes to order j on $\partial\Omega$.

LEMMA 11.4.11 The function u is in $C^\infty(\bar\Omega_1)$.

Proof It suffices to show that $u \in W^j(\Omega_1)$, every j. So fix j. Let $m = m(j)$ as in Condition R. According to (11.4.10.1), $|u(z)| \leq C\delta_1(z)^{-2n}$. Then, by Lemma 11.4.10 and the exercise for the reader following it, there is a J so large that $g \in W_0^{m+J}(\Omega_2)$ implies $u \cdot (g \circ \phi) \in W^m(\Omega_1)$. Choose, by Lemma 11.4.7, a $g \in W_0^{m+J}(\Omega_2)$ such that $P_2 g \equiv 1$. Then Lemma 11.4.8 yields

$$P_1(u \cdot (g \circ \phi)) = u.$$

By Condition R, it follows that $u \in W^j(\Omega_1)$. □

LEMMA 11.4.12 The function u is bounded from 0 on $\bar\Omega_1$.

Proof By symmetry, we may apply Lemma 11.4.11 to ϕ^{-1} and $\det J_{\mathbb{C}}(\phi^{-1}) = 1/u$. We conclude that $1/u \in C^\infty(\bar\Omega_2)$. Thus u is nonvanishing on $\bar\Omega$. □

Proof of Fefferman's Theorem (Theorem 11.4.3)   Use the notation of the proof of Lemma 11.4.11. Choose $g_1, \ldots, g_n \in W_0^{m+J}(\Omega_2)$ such that $P_2 g_i(w) = w_i$ (here w_i is the ith coordinate function). Then Lemma 11.4.8 yields that $u \cdot \phi_i \in W^j(\Omega_1), i = 1, \ldots, n$. By Lemma 11.4.12, $\phi_i \in W^j(\Omega_1), i = 1, \ldots, n$. By symmetry, $\phi^{-1} \in W^j(\Omega_2)$. Since j is arbitrary, the Sobolev imbedding theorem finishes the proof.   □

It is important to understand the central role of Condition R in this proof. With some emendations, the proof we have presented shows that if $\Omega_1, \Omega_2 \subset\subset \mathbb{C}^n$ are smoothly bounded, pseudoconvex, and both satisfy Condition R, then a biholomorphic mapping from Ω_1 to Ω_2 extends smoothly to the boundary (in fact S. Bell [2] has shown that it suffices for just one of the domains to satisfy Condition R). Condition R is known to hold on domains that have real analytic boundaries (see K. Diederich and J. E. Fornæss [4]), and more generally on domains of finite type (see Section 11.5). There are a number of interesting examples of nonpseudoconvex domains on which Condition R fails (see D. Barrett [1] and C. Kiselman [1]). F. M. Christ [5] has recently shown that Condition R fails on the Diederich-Fornæss worm domain.

L. Lempert [3] has derived a sharp boundary regularity result for biholomorphic mappings of strongly pseudoconvex domains with C^k boundary. The correct conclusion turns out to be that there is a loss of smoothness in some directions. So the sharp regularity result is formulated in terms of nonisotropic spaces. It is too technical to describe here.

11.5   Concept of Finite Type

Let us begin with the simplest domain in $\mathbb{C}^n$—the ball. Let $P \in \partial B$. It is elementary to see that no complex line (equivalently no affine analytic disc) can have geometric order of contact with ∂B at P exceeding 2. That is, a complex line may pass through P and also be tangent to ∂B at P, but it can do no better. The boundary of the ball has positive curvature and a complex line is flat. The differential geometric structures disagree at the level of second derivatives. Another way to say this is that if ℓ is a complex line tangent to ∂B at P then, for $z \in \ell$,

$$\text{dist}\,(z, \partial B) = \mathcal{O}\big(|z - P|^2\big) \tag{11.5.1}$$

and the exponent 2 cannot be improved. The number 2 is called the "order of contact" of the complex line with ∂B.

The notion of strongly pseudoconvex point can be viewed as the correct biholomorphically invariant version of the phenomenon described in the first paragraph: No analytic disc can osculate to better than first order tangency to a strongly pseudoconvex boundary point. In fact the positive definiteness of the

Levi form provides the obstruction that makes this statement true. Let us sketch a proof.

Suppose that $P \in \partial\Omega$ is a point of strong pseudoconvexity and that we have fixed a defining function ρ whose complex Hessian at P is positive definite on all of $\mathbb{C}^n$. We may further suppose, by the proof of Narasimhan's lemma, that the only second-order terms in the Taylor expansion of ρ about P are the mixed terms occurring in the complex Hessian.

Let $\phi : D \to \mathbb{C}^n$ be an analytic disc that is tangent to $\partial\Omega$ at P such that $\phi(0) = P$, $\phi'(0) \neq 0$. The tangency means that

$$\mathrm{Re}\left(\sum_{j=1}^{n} \frac{\partial\rho}{\partial z_j}(P)\phi_j'(0) \right) = 0.$$

It follows that if we expand $\rho \circ \phi(\zeta)$ in a Taylor expansion about $\zeta = 0$, then the zero- and first-order terms vanish. As a result, for ζ small,

$$\rho \circ \phi(\zeta) = \left[\sum_{j,k=1}^{n} \frac{\partial^2\rho}{\partial z_j \partial \bar{z}_k}(P)\phi_j'(0)\bar{\phi}_k'(0) \right] |\zeta|^2 + o(|\zeta|^2).$$

But this last is

$$\geq C \cdot |\zeta|^2$$

for ζ small. This gives an explicit lower bound, in terms of the eigenvalues of the Levi form, for the order of contact of the image of ϕ with $\partial\Omega$.

It turns out that the number 2, which we see arises rather naturally from geometric considerations of a strongly pseudoconvex point, has important analytic consequences. For instance, we learned in Chapter 10 that the optimal Λ_α regularity

$$\|u\|_{\Lambda_\alpha(\Omega)} \leq C\|f\|_{L^\infty(\Omega)}$$

for solutions to the $\bar{\partial}$ equation $\bar{\partial}u = f$, f a bounded, $\bar{\partial}$-closed $(0,1)$ form, occurs when $\alpha = \frac{1}{2}$. No such inequality holds for $\alpha > \frac{1}{2}$. We say that the $\bar{\partial}$ problem exhibits "a gain of $1/2$." Thus the best index is the reciprocal of the integer describing the optimal order of contact of varieties with the boundary of the domain in question.

That this is no coincidence is already apparent in the example of Stein from Section 10.3. However it was J. J. Kohn who first appreciated the logical foundations of this geometric analysis. In J. J. Kohn [5], Kohn studied the regularity of the $\bar{\partial}$ equation in a neighborhood of a point at which the maximal order of contact (to be defined precisely below) of one-dimensional complex curves is at most m (this work is in dimension two only). He proved (in the Sobolev topology rather than the Lipschitz topology) that the $\bar{\partial}$ problem near such a point exhibits a gain of $(1/m) - \epsilon$, any $\epsilon > 0$. He conjectured that the correct gain is $1/m$. P. Greiner [1] gave examples that showed that Kohn's conjecture was sharp (see also S. G. Krantz [4] for a different approach and examples in other topologies). Rothschild and Stein [1] proved the sharp estimate.

David Catlin [1, 2, 3] has shown that the $\bar{\partial}$-Neumann problem on a pseudo-convex domain in $\mathbb{C}^n$ exhibits a "gain" in regularity if and only if the boundary admits only finite order of contact of (possibly singular) varieties. This result was made possible by the work of D'Angelo, who laid the algebro-geometric foundations for the theory of order of contact of complex varieties with the boundary of a domain (see J. P. D'Angelo [1–4, 6]).

The purpose of the present section is to acquaint the reader with the circle of ideas that was described in the preceding paragraphs and to indicate the applications of these ideas to the theory of holomorphic mappings.

EXAMPLE   Let m be a positive integer and define

$$\Omega = \Omega_m = \{(z_1, z_2) \in \mathbb{C}^2 : \rho(z_1, z_2) = -1 + |z_1|^2 + |z_2|^{2m} < 0\}.$$

Consider the boundary point $P = (1, 0)$. Let $\phi : D \to \mathbb{C}^2$ be an analytic disc that is tangent to $\partial\Omega$ at P and such that $\phi(0) = P, \phi'(0) \neq 0$. We may in fact assume, after a reparametrization, that $\phi'(0) = (0, 1)$. Then

$$\phi(\zeta) = (1 + 0\zeta + \mathcal{O}(\zeta^2), \zeta + \mathcal{O}(\zeta^2)). \tag{11.5.2}$$

What is the greatest order of contact (measured in the sense of equation (11.5.1), with the exponent 2 replaced by some m) that such a disc ϕ can have with $\partial\Omega$? Obviously the disc $\phi(\zeta) = (1, \zeta)$ has order of contact $2m$ at $P = (1, 0)$, for

$$\rho \circ \phi(\zeta) = |\zeta|^{2m} = \mathcal{O}\big(|(1, \zeta) - (1, 0)|^{2m}\big).$$

The question is whether we can improve upon this estimate with a different curve ϕ. Since all curves ϕ under consideration must have the form (11.5.2), we calculate that

$$\begin{aligned}
\rho \circ \phi(\zeta) &= -1 + \big|1 + \mathcal{O}(\zeta^2)\big|^2 + \big|\zeta + \mathcal{O}(\zeta)^2\big|^{2m} \\
&= -1 + \Big[\big|1 + \mathcal{O}(\zeta^2)\big|^2\Big] + \Big[|\zeta|^{2m}\big|1 + \mathcal{O}(\zeta)\big|^{2m}\Big].
\end{aligned}$$

The second expression in brackets is essentially $|\zeta|^{2m}$, so if we wish to improve on the order of contact, then the first term in brackets must cancel it. But then the first term would have to be $|1 + c\zeta^m + \cdots|^{2m}$. The resulting term of order $2m$ would be positive and, in fact, would *not* cancel the second. We conclude that $2m$ is the optimal order of contact for complex curves with $\partial\Omega$ at P. Let us say that P is of "geometric type $2m$."

Now we examine the domain Ω_m from the analytic viewpoint. Consider the vector fields

$$
\begin{aligned}
L &= \frac{\partial \rho}{\partial z_1} \frac{\partial}{\partial z_2} - \frac{\partial \rho}{\partial z_2} \frac{\partial}{\partial z_1} \\
&= \bar{z}_1 \frac{\partial}{\partial z_2} - m z_2^{m-1} \bar{z}_2^m \frac{\partial}{\partial z_1}
\end{aligned}
$$

and

$$
\begin{aligned}
\bar{L} &= \frac{\partial \rho}{\partial \bar{z}_1} \frac{\partial}{\partial \bar{z}_2} - \frac{\partial \rho}{\partial \bar{z}_2} \frac{\partial}{\partial \bar{z}_1} \\
&= z_1 \frac{\partial}{\partial \bar{z}_2} - m \bar{z}_2^{m-1} z_2^m \frac{\partial}{\partial \bar{z}_1}.
\end{aligned}
$$

One can see from their very definition or can compute directly that both these vector fields are tangent to $\partial\Omega$. That is to say, $L\rho \equiv 0$ and $\bar{L}\rho \equiv 0$. It is elementary to verify that the commutator of two tangential vector fields must still be tangential *in the sense of real geometry*. That is, $[L, \bar{L}]$ must lie in the three-dimensional real tangent space to $\partial\Omega$ at each point of $\partial\Omega$. However, there is no a priori guarantee that this commutator must lie in the *complex* tangent space (as discussed in Sections 3.3 and 8.6); in general, it will not. Take, for example, the case $m = 1$, when our domain is the ball. A calculation reveals that, at the point $P = (1, 0)$,

$$
[L, \bar{L}] \equiv L\bar{L} - \bar{L}L = -i \frac{\partial}{\partial y_1}.
$$

This vector is indeed tangent to the boundary of the ball at P (it annihilates the defining function), but it is equal to the negative of the complex structure tensor J applied to the Euclidean normal $\partial/\partial x_1$; therefore, it is what we call *complex normal*. (There is an excellent opportunity here for confusion. It is common in the literature to say that "the direction $\partial/\partial y_1$ is i times the direction ∂x_1" when what is meant is that when the complex structure tensor is applied to $\partial/\partial x_1$, then one obtains $\partial/\partial y_1$. One must distinguish between the linear operator J and the tensoring of space with $\mathbb{C}$ that enables one to multiply by the scalar i. These matters are laid out in detail in R. O. Wells [2].) Compare with the discussion in Section 8.6.

The reason that it takes only a commutator of order one to escape the complex tangent and have a component in the complex normal direction is that the ball is strongly pseudoconvex—refer to the invariant definition of the Levi form and Cartan's formula in Exercises 4 and 5 at the end of Chapter 5. Calculate for yourself that on our domain Ω_m, at the point $P = (1, 0)$, it requires a commutator

$$[L, [\bar{L}, \dots [L, \bar{L}] \dots]]$$

of length $2m - 1$ (that is, a total of $2m$ L's and $\bar{L}$'s) to have a component in the complex normal direction. We say that P is of "analytic type $2m$."

Thus, in this simple example, a point of geometric type $2m$ is of analytic type $2m$.   □

Next we shall develop in full generality both the geometric and the analytic notions of "type" for domains in complex dimension two. In this low-dimensional context, the whole idea of type is rather clean and simple (misleadingly so). In retrospect we shall see that the reason for this is that the varieties of maximal dimension that can be tangent to the boundary (that is, one-dimensional complex analytic varieties) have no interesting subvarieties (the subvarieties are all zero-dimensional). Put another way, any irreducible one-dimensional complex analytic variety V has a holomorphic parametrization $\phi : D \to V$. Nothing of the kind is true for higher-dimensional varieties.

11.5.1   Finite Type in Dimension Two

We begin with the formal definitions of geometric type and of analytic type for a point in the boundary of a smoothly bounded domain $\Omega \subseteq \mathbb{C}^2$. The main result of this subsection will be that the two notions are equivalent. We will then describe, but not prove, some sharp regularity results for the $\bar{\partial}$ problem on a finite type domain. Good references for this material are J. J. Kohn [5], T. Bloom and I. Graham [1], and S. G. Krantz [4].

DEFINITION 11.5.3   A *first-order commutator* of vector fields is an expression of the form

$$[L, M] \equiv LM - ML.$$

Here the right-hand side is understood according to its action on C^∞ functions:

$$[L, M](\phi) \equiv (LM - ML)(\phi) \equiv L(M(\phi)) - M(L(\phi)).$$

Inductively, an mth order commutator is the commutator of an $(m - 1)$st order commutator and a vector field N. The commutator of two vector fields is again a vector field.

DEFINITION 11.5.4   A holomorphic vector field is any linear combination of the expressions

$$\frac{\partial}{\partial z_1} \quad , \quad \frac{\partial}{\partial z_2}$$

with coefficients in the ring of C^∞ functions.

A conjugate holomorphic vector field is any linear combination of the expressions

$$\frac{\partial}{\partial \bar{z}_1} \quad , \quad \frac{\partial}{\partial \bar{z}_2}$$

with coefficients in the ring of C^∞ functions.

DEFINITION 11.5.5   Let M be a vector field defined on the boundary of $\Omega = \{z \in \mathbb{C}^2 : \rho(z) < 0\}$. We say that M is *tangential* if $M\rho = 0$ at each point of $\partial\Omega$.

Now we define a gradation of vector fields, which will be the basis for our definition of analytic type. Throughout this section $\Omega = \{z \in \mathbb{C}^2 : \rho(z) < 0\}$ and ρ is C^∞. If $P \in \partial\Omega$, then we may make a change of coordinates so that $\partial\rho/\partial z_2(P) \neq 0$. Define the holomorphic vector field

$$L = \frac{\partial \rho}{\partial z_1} \frac{\partial}{\partial z_2} - \frac{\partial \rho}{\partial z_2} \frac{\partial}{\partial z_1}$$

and the conjugate holomorphic vector field

$$\bar{L} = \frac{\partial \rho}{\partial \bar{z}_1} \frac{\partial}{\partial \bar{z}_2} - \frac{\partial \rho}{\partial \bar{z}_2} \frac{\partial}{\partial \bar{z}_1}.$$

Both L and $\bar{L}$ are tangent to the boundary because $L\rho = 0$ and $\bar{L}\rho = 0$. They are both nonvanishing near P by our normalization of coordinates.

The real and imaginary parts of L (equivalently of $\bar{L}$) generate (over the ground field $\mathbb{R}$) the complex tangent space to $\partial\Omega$ at all points near P. The vector field L alone generates the space of all holomorphic tangent vector fields and $\bar{L}$ alone generates the space of all conjugate holomorphic tangent vector fields.

DEFINITION 11.5.6   Let $\mathcal{L}_1$ denote the module, over the ring of C^∞ functions, generated by L and $\bar{L}$. Inductively, $\mathcal{L}_\mu$ denotes the module generated by $\mathcal{L}_{\mu-1}$ and all commutators of the form $[F, G]$ where $F \in \mathcal{L}_1$ and $G \in \mathcal{L}_{\mu-1}$.

Clearly $\mathcal{L}_1 \subseteq \mathcal{L}_2 \subseteq \cdots$. Each $\mathcal{L}_\mu$ is closed under conjugation. *It is not generally the case that $\cup_\mu \mathcal{L}_\mu$ is the entire three-dimensional tangent space at each point of the boundary.* A counterexample is provided by

$$\Omega = \{z \in \mathbb{C}^2 : |z_1|^2 + 2e^{-1/|z_2|^2} < 1\}$$

and the point $P = (1, 0)$. We invite the reader to supply details of this assertion.

DEFINITION 11.5.7   Let $\Omega = \{\rho < 0\}$ be a smoothly bounded domain in $\mathbb{C}^2$ and let $P \in \partial\Omega$. We say that $\partial\Omega$ is of *finite analytic type* m *at* P if $\langle \partial\rho(P), F(P) \rangle = 0$ for all $F \in \mathcal{L}_{m-1}$ while $\langle \partial\rho(P), G(P) \rangle \neq 0$ for some $G \in \mathcal{L}_m$. In this circumstance we call P a point of analytic type m.

Remark: A point is of finite analytic type m if it requires the commutation of m vector fields to obtain a component in the complex normal direction. Such a commutator lies in $\mathcal{L}_m$. This notation is different from that in our source Bloom and Graham [1] but is necessary for consistency with D'Angelo's ideas, which will be presented later.

There is an important epistemological observation that needs to be made at this time. Complex tangential vector fields do not, after being commuted with each other finitely many times, suddenly "pop out" into the complex normal direction. What is really being discussed in this definition is an order of vanishing of coefficients.

For instance, suppose that, at the point P, the complex normal direction is the z_2 direction. A vector field

$$F(z) = a(z)\frac{\partial}{\partial z_1} + b(z)\frac{\partial}{\partial z_2},$$

such that b vanishes to some finite positive order at P and $a(P) \neq 0$, will be tangential at P. But when we commute vector fields *we differentiate their coefficients.* Thus if F is commuted with the appropriate vector fields finitely many times, then b will be differentiated (lowering the order of vanishing by one each time) until the coefficient of $\partial/\partial z_2$ vanishes to order 0. This means that, after finitely many commutations, the coefficient of $\partial/\partial z_2$ does not vanish at P. In other words, after finitely many commutations, the resulting vector field has a component in the normal direction at P.   □

Notice that the condition $\langle \partial\rho(P), F(P) \rangle \neq 0$ is just an elegant way of saying that the vector $G(P)$ has nonzero component in the complex normal direction. As we explained earlier, any point of the boundary of the unit ball is of finite analytic type 2. Any point of the form $(e^{i\theta}, 0)$ in the boundary of $\{(z_1, z_2) : |z_1|^2 + |z_2|^{2m} < 1\}$ is of finite analytic type $2m$. Any point of the form $(e^{i\theta}, 0)$ in the boundary of $\Omega = \{z \in \mathbb{C}^2 : |z_1|^2 + 2e^{-1/|z_2|^2} < 1\}$ is *not* of finite analytic type. We say that such a point is of *infinite analytic type.*

Now we turn to a precise definition of finite geometric type. If P is a point in the boundary of a smoothly bounded domain, then we say that an analytic disc $\phi : D \to \mathbb{C}^2$ is a *nonsingular disc tangent to* $\partial\Omega$ at P if $\phi(0) = P, \phi'(0) \neq 0$, and $(\rho \circ \phi)'(0) = 0$.

DEFINITION 11.5.8 Let $\Omega = \{\rho < 0\}$ be a smoothly bounded domain and $P \in \partial\Omega$. Let m be a nonnegative integer. We say that $\partial\Omega$ is of *finite geometric type* m at P if the following condition holds: There is a nonsingular disc ϕ tangent to $\partial\Omega$ at P such that, for small ζ,

$$|\rho \circ \phi(\zeta)| \leq C|\zeta|^m.$$

But there is no nonsingular disc ψ tangent to $\partial\Omega$ at P such that, for small ζ,

$$|\rho \circ \phi(\zeta)| \leq C|\zeta|^{(m+1)}.$$

In this circumstance we call P a point of finite geometric type m.

We invite the reader to reformulate the definition of geometric finite type in terms of the order of vanishing of ρ restricted to the image of ϕ.

The principal result of this section is the following theorem.

THEOREM 11.5.9 Let $\Omega = \{\rho < 0\} \subseteq \mathbb{C}^2$ be smoothly bounded and $P \in \partial\Omega$. The point P is of finite geometric type $m \geq 2$ if and only if it is of finite analytic type m.

Proof   We may assume that $P = 0$. Write ρ in the form

$$\rho(z) = 2\mathrm{Re}\, z_2 + f(z_1) + \mathcal{O}(|z_1 z_2| + |z_2|^2).$$

We do this, of course, by examining the Taylor expansion of ρ and using the theorem of E. Borel to manufacture f from the terms that depend on z_1 only. Notice that

$$L = \frac{\partial f}{\partial z_1}\frac{\partial}{\partial z_2} - \frac{\partial}{\partial z_1} + (\text{error terms}).$$

Here the error terms arise from differentiating $\mathcal{O}(|z_1 z_2| + |z_2|^2)$.

Now it is a simple matter to notice that the best order of contact of a one-dimensional nonsingular complex variety with $\partial\Omega$ at 0 equals the order of contact of the variety $\zeta \mapsto (\zeta, 0)$ with $\partial\Omega$ at 0, which is just the order of vanishing of f at 0.

On the other hand,

$$
\begin{aligned}
[L, \bar{L}] &= \left[-\frac{\partial^2 \bar{f}}{\partial z_1 \partial \bar{z}_1}\frac{\partial}{\partial \bar{z}_2}\right] - \left[-\frac{\partial^2 f}{\partial \bar{z}_1 \partial z_1}\frac{\partial}{\partial z_2}\right] + (\text{error terms}) \\
&= 2i\,\mathrm{Im}\left[\frac{\partial^2 f}{\partial \bar{z}_1 \partial z_1}\frac{\partial}{\partial z_2}\right] + (\text{error terms}).
\end{aligned}
$$

Inductively, one sees that a commutator of m vector fields chosen from L, $\bar{L}$ will consist of (real or imaginary parts of) m^{th}-order derivatives of f times $\partial/\partial z_2$ plus the usual error terms. And the pairing of such a commutator with $\partial\rho$ at 0 is just the pairing of that commutator with dz_2; in other words it is just the coefficient of $\partial/\partial z_2$. We see that this number is nonvanishing as soon as the corresponding derivative of f is nonvanishing. Thus the analytic type of 0 is just the order of vanishing of f at 0.

Since both notions of type correspond to the order of vanishing of f, we are done. $\square$

From now on, when we say "finite type" (in dimension two), we can mean either the geometric or the analytic definition.

We say that a domain $\Omega \subseteq \mathbb{C}^2$ is of *finite type* if there is a number M such that every boundary point is of finite type not exceeding M.

Analysis on finite-type domains in $\mathbb{C}^2$ has recently become a matter of great interest. It has been proved, in the works D. C. Chang, A. Nagel, and E. M. Stein [1], C. Fefferman and J. J. Kohn [1, 2], F. M. Christ [1, 2, 4], that the $\bar{\partial}$-Neumann problem on a domain $\Omega \subseteq \mathbb{C}^2$ of finite type M exhibits a gain of order $1/M$ in the Lipschitz space topology. In S. G. Krantz [4] it was proved that this last result was sharp. Finally, the paper S. G. Krantz [4] also provided a way to prove the nonexistence of certain biholomorphic equivalences by using sharp estimates for the $\bar{\partial}$ problem.

11.5.2   Finite Type in Higher Dimensions

The most obvious generalization of the notion of geometric finite type from dimension two to dimensions three and higher is to consider orders of contact of $(n-1)$-dimensional complex manifolds with the boundary of a domain Ω at a point P. The definition of analytic finite type generalizes to higher dimensions almost directly (one deals with tangent vector fields $L_1, \ldots, L_{n-1}$ and $\bar{L}_1, \ldots, \bar{L}_{n-1}$ instead of just L and $\bar{L}$). It is a theorem of T. Bloom and I. Graham [1] that, with these definitions, geometric finite type and analytic finite type are the same in all dimensions.

This is an elegant result and is entirely suited to questions of extension of CR functions and reflections of holomorphic mappings. However, it is not the correct indicator of when the $\bar{\partial}$-Neumann problem exhibits a gain. In the late 1970s and early 1980s, John D'Angelo realized that a correct understanding of finite type in all dimensions requires sophisticated ideas from algebraic geometry—particularly the intersection theory of analytic varieties. And he saw that *nonsingular varieties cannot tell the whole story.* An important sequence of papers, beginning with D'Angelo [1], laid down the theory of domains of finite type in all dimensions. The complete story of this work, together with its broader mathematical context, appears in J. P. D'Angelo [6]. David Catlin [3] validated the significance of D'Angelo's work by proving that the $\bar{\partial}$-Neumann problem has a gain in the Sobolev topology near a point $P \in \partial\Omega$ if and only if the point P is of finite type in the sense of D'Angelo. (It is interesting to note that there are partial differential operators that exhibit a gain in the Sobolev topology but not in the Lipschitz topology—see Guan [1].)

The point is that analytic structure in the boundary of a domain is an obstruction to regularity for the $\bar{\partial}$ problem. We saw in Exercise 3 at the end of Chapter 10 that if the boundary contains an analytic disc then it is possible for the equation $\bar{\partial}u = f$ to have data f that is C^∞ but no smooth solution u. What we now learn is that the order of contact of analytic varieties stratifies the insight of this Exercise 3 into degrees, so that one may make precise statements about the "gain" of the $\bar{\partial}$ problem in terms of the order of contact of varieties at the boundary.

In higher dimensions, matters become technical rather quickly. Therefore, we shall content ourselves with a primarily descriptive treatment of this material. One missing piece of the picture is the following: As of this writing, there is no "analytic" description of finite type, using commutators of vector fields, in dimensions three and higher. We know that the notion of finite type that we are about to describe is the right one for the study of the $\bar{\partial}$-Neumann problem because (i) it enjoys certain important semicontinuity properties (to be discussed below) that other notions of finite type do not, and (ii) Catlin's theorem shows that the definition meshes perfectly with the $\bar{\partial}$-Neumann operator.

Let us begin by introducing some notation. Let $U \subseteq \mathbb{C}^n$ be an open set. A subset $V \subseteq U$ is called a *variety* if there are holomorphic functions $f_1, \ldots, f_k$ on U such that $V = \{z \in U : f_1(z) = \cdots = f_k(z) = 0\}$. A variety is called *irreducible* if it cannot be written as the union of proper nontrivial subvarieties. One-dimensional varieties are particularly easy to work with because they can be parametrized (see Gunning [2]).

PROPOSITION 11.5.10 Let $V \subseteq \mathbb{C}^n$ be an irreducible one-dimensional complex analytic variety. Let $P \in V$. There is a neighborhood W of P and a holomorphic mapping $\phi : D \to \mathbb{C}^n$ such that $\phi(0) = P$ and the image of ϕ is $W \cap V$. When this parametrization is in place, then we refer to the variety as a holomorphic curve.

In general, we cannot hope that the parametrization ϕ will satisfy (nor can it be arranged that) $\phi'(0) \neq 0$. As a simple example, consider the variety

$$V = \{z \in \mathbb{C}^2 : z_1^2 - z_2^3 = 0\}.$$

Then the most natural parametrization for V is $\phi(\zeta) = (\zeta^3, \zeta^2)$. Notice that $\phi'(0) = 0$ and there is no way to reparametrize the curve to make the derivative nonvanishing. This is because the variety has a singularity—a cusp—at the point $P = 0$.

DEFINITION 11.5.11   Let f be a scalar-valued holomorphic function of a complex variable and P a point of its domain. The *multiplicity* of f at P is defined to be the least positive integer k such that the kth derivative of f does not vanish at P. If m is that multiplicity, then we write $v_P(f) = v(f) = m$.

If ϕ is instead a vector-valued holomorphic function of a complex variable, then its multiplicity at P is defined to be the minimum of the multiplicities of its entries. If that minimum is m, then we write $v_P(\phi) = v(\phi) = m$.

In this subsection we will exclusively calculate the multiplicities of holomorphic curves $\phi(\zeta)$ at $\zeta = 0$.

For example, the function $\zeta \mapsto \zeta^2$ has multiplicity 2 at 0; the function $\zeta \mapsto \zeta^3$ has multiplicity 3 at 0. Therefore, the curve $\zeta \mapsto (\zeta^2, \zeta^3)$ has multiplicity 2 at 0.

If ρ is the defining function for a domain Ω, then of course the boundary of Ω is given by the equation $\rho = 0$. D'Angelo's idea is to consider the *pullback* of the function ρ under a curve ϕ:

DEFINITION 11.5.12   Let $\phi : D \to \mathbb{C}^n$ be a holomorphic curve and ρ the defining function for a hypersurface M (usually, but not necessarily, the boundary of a domain). Then the *pullback* of ρ under ϕ is the function $\phi^*\rho(\zeta) = \rho \circ \phi(\zeta)$.

DEFINITION 11.5.13   Let M be a real hypersurface and $P \in M$. Let ρ be a defining function for M in a neighborhood of P. We say that P is a *point of finite type* (or finite 1-type) if there is a constant $C > 0$ such that

$$\frac{v(\phi^*\rho)}{v(\phi)} \leq C$$

whenever ϕ is a nonconstant, one-dimensional holomorphic curve through P such that $\phi(0) = P$.

The infimum of all such constants C is called the type (or 1-type) of P. It is denoted by $\Delta(M, P) = \Delta_1(M, P)$.

This definition is algebro-geometric in nature. We now offer a more geometric condition that is equivalent to it.

PROPOSITION 11.5.14   Let P be a point of the hypersurface M. Let $\mathcal{E}_P$ be the collection of one-dimensional complex varieties passing through P. Then we have

$$\Delta(M, P) = \sup_{V \in \mathcal{E}_P} \sup_{a > 0} \left\{ a \in \mathbb{R}^+ : \lim_{V \ni z \to P} \frac{\mathrm{dist}(z, M)}{|z - P|^a} \ \text{exists} \right\}.$$

We leave the proof of the proposition to the exercises. Notice that its statement is attractive in that it gives a characterization of finite type that makes no reference to a defining function. The proposition, together with the material in the first part of this section, motivates the following definition.

DEFINITION 11.5.15   Let P be a point of the hypersurface M. Let $\mathcal{R}_P$ be the collection of nonsingular, one-dimensional complex varieties passing through P (that is, we consider curves $\phi : D \to \mathbb{C}^n$, $\phi(0) = P$, $\phi'(0) \neq 0$). Then we define

$$\Delta^{\mathrm{reg}}(M, P) = \Delta_1^{\mathrm{reg}}(M, P)$$
$$= \sup_{V \in \mathcal{R}_P} \sup_{a > 0} \left\{ a \in \mathbb{R}^+ : \lim_{V \ni z \to P} \frac{\mathrm{dist}(z, M)}{|z - P|^a} \ \text{exists} \right\}.$$

The number $\Delta_1^{\mathrm{reg}}(M, P)$ measures order of contact of nonsingular complex curves (i.e., one-dimensional complex analytic *manifolds*) with M at P. By contrast, $\Delta_1(M, P)$ looks at all curves, both singular and nonsingular. Obviously $\Delta_1^{\mathrm{reg}}(M, P) \leq \Delta_1(M, P)$. The following example of D'Angelo shows that the two concepts are truly different.

EXAMPLE   Consider the hypersurface in $\mathbb{C}^3$ with defining function given by

$$\rho(z) = 2\,\mathrm{Re}\,z_3 + |z_1^2 - z_2^3|^2.$$

Let the point P be the origin. Then we have the following:

- We may calculate that $\Delta_1^{\mathrm{reg}}(M, P) = 6$. We determine this by noticing that the z_3 direction is the normal direction to M at P; hence any tangent curve must have the form

$$\zeta \mapsto (a(\zeta), b(\zeta), \mathcal{O}(\zeta^2)).$$

  Since we are calculating the "regular type," one of the quantities $a'(0), b'(0)$, must be nonzero. We see that if we let $a(\zeta) = \zeta + \ldots$, then the expression $|z_1^2 - z_2^3|^2$ in the definition of ρ provides the obstruction to the order of contact: The curve cannot have order of contact better than 4. Similar considerations show that if $b(\zeta) = \zeta + \ldots$, then the order of contact cannot be better than 6. Putting these ideas together, we see that a regular curve that exhibits maximum order of contact at $P = 0$ is $\phi(\zeta) = (0, \zeta, 0)$. Its order of contact with M at P is 6. Thus $\Delta_1^{\mathrm{reg}}(M, P) = 6$.

- We may see that $\Delta_1(M, P) = \infty$ by considering the (singular) curve $\phi(\zeta) = (\zeta^3, \zeta^2, 0)$. This curve actually *lies in* M.   □

An appealing feature of the notion of analytic finite type that we learned about in dimension two is that it is upper semicontinuous: If, at a point P, the expression $\langle \partial\rho, F \rangle$ is nonvanishing for some $F \in \mathcal{L}_\mu$, then it will certainly be nonvanishing at nearby points. Therefore, if P is a point of type m it follows that sufficiently nearby points will be of type at most m. It is considered reasonable that a viable notion of finite type should be upper semicontinuous. Unfortunately, this is not the case, as the following example of D'Angelo shows.

EXAMPLE   Consider the hypersurface in $\mathbb{C}^3$ defined by the function

$$\rho(z_1, z_2, z_3) = \mathrm{Re}(z_3) + |z_1^2 - z_2 z_3|^2 + |z_2|^4.$$

Take $P = 0$. Then we may argue as in the last example to see that $\Delta_1(M, P) = \Delta_1^{\mathrm{reg}}(M, P) = 4$. The curve $\zeta \mapsto (\zeta, \zeta, 0)$ gives best possible order of contact.
    But for a point of the form $P = (0, 0, ia)$, a a positive real number, let α be a square root of ia. Then the curve $\zeta \mapsto (\alpha\zeta, \zeta^2, ia)$ shows that $\Delta_1(M, P) = \Delta_1^{\mathrm{reg}}(M, P)$ is at least 8 (in fact, it equals 8 (exercise)).

Thus we see that the number Δ_P is not an upper semicontinuous function of the point P. □

D'Angelo [1] proves that the invariant Δ_1 can be compared with another invariant that comes from intersection-theoretic considerations; that is, he compares Δ_1 with the dimension of the quotient of the ring $\mathcal{O}$ of germs of holomorphic functions by an ideal generated by the components of a special decomposition of the defining function. This latter *is* semicontinuous. The result gives essentially sharp bounds on how Δ_1 can change as the point P varies within M.

We give now a brief description of the algebraic invariant that is used in D'Angelo [1]. Take $P \in M$ to be the origin. Let ρ be a defining function for M near 0. The first step is to prove that one can write the defining function in the form

$$\rho(z) = 2\operatorname{Re}h(z) + \sum_j |f_j(z)|^2 - \sum_j |g_j|^2,$$

where h, f_j, g_j are holomorphic functions. In case ρ is a polynomial, then each sum can be taken to be finite—say $j = 1, \ldots, k$. Let us restrict attention to that case. (See D'Angelo [6] for a thorough treatment of this decomposition and Krantz [18] for auxiliary discussion.) Write $f = (f_1, \ldots, f_k)$ and $g = (g_1, \ldots, g_k)$.

Let U be a unitary matrix of dimension k. Define $\mathcal{I}(U, P)$ to be the ideal generated by h and $f - Ug$. We set $D(\mathcal{I}(U, P))$ equal to the dimension of $\mathcal{O}/\mathcal{I}(U, P)$. Finally, declare $B_1(M, P) = 2\sup D(\mathcal{I}(U, P))$, where the supremum is taken over all possible unitary matrices of order k. Then we have the following.

THEOREM 11.5.16   With M, ρ, P as usual, we have

$$\Delta_1(M, P) \leq B_1(M, P) \leq 2(\Delta_1(M, P))^{n-1}.$$

THEOREM 11.5.17   The quantity $B_1(M, P)$ is upper semicontinuous as a function of P.

We learn from the two theorems that Δ_1 *is* locally finite in the sense that if it is finite at P, then it is finite at nearby points. We also learn by how much it can change; namely, for points Q near P we have

$$\Delta_1(M, Q) \leq 2(\Delta_1(M, P))^{n-1}.$$

In case the hypersurface M is pseudoconvex near P, then the estimate can be sharpened. Assume that the Levi form is positive semidefinite near P and has rank q at P. Then we have

$$\Delta_1(M, Q) \leq \frac{(\Delta(M, P))^{n-1-q}}{2^{n-2-q}}.$$

We conclude this section with an informal statement of the theorem of D. Catlin [3].

THEOREM 11.5.18   Let $\Omega \subseteq \mathbb{C}^n$ be a bounded pseudoconvex domain with smooth boundary. Let $P \in \partial\Omega$. Then the problem $\bar{\partial}u = f$, with f a $\bar{\partial}$-closed $(0,1)$ form, enjoys a gain in regularity near P in the Sobolev topology if and only if P is a point of finite type in the sense that $\Delta_1(M, P)$ is finite.

It is not known how to determine the sharp "gain" in regularity of the $\bar{\partial}$-Neumann problem at a point of finite type in dimensions $n \geq 3$. There is considerable evidence (see D'Angelo [6]) that our traditional notion of "gain" as described here will have to be refined in order to formulate a precise result.

It turns out that to study finite type, and concomitantly gains in Sobolev regularity for the problem $\bar{\partial}u = f$ when f is a $\bar{\partial}$-closed $(0,q)$ form, requires the study of order of contact of q-dimensional varieties with the boundary of the domain. One develops an invariant $\Delta_q(M, P)$. The details of this theory have the same flavor as what has been presented here but are considerably more complicated.

In the theory of the $\bar{\partial}$-Neumann problem (see G. B. Folland and J. J. Kohn [1] and S. G. Krantz [19]), one learns that the Bergman projection P satisfies

$$P = I - \bar{\partial}^* N \bar{\partial}.$$

Here $\bar{\partial}^*$ is the L^2 adjoint of $\bar{\partial}$ and N is the *Neumann operator*—a canonical right inverse for the $\bar{\partial}$-Laplacian. Part of proving Theorem 11.5.18 is to see that, for each s, there is an $\epsilon > 0$ such that $N : W^s(\Omega) \to W^{s+\epsilon}(\Omega)$. But then it is plain (since $\bar{\partial}, \bar{\partial}^*$ are first-order partial differential operators) that $P : W^s(\Omega) \to W^{s-2}(\Omega)$. Thus we see that a finite-type domain satisfies Condition R (see Section 10.4). And we learned in Section 10.4 that a biholomorphic map of domains that satisfy Condition R extends smoothly to the boundary. We summarize with a theorem.

THEOREM 11.5.19   Let Ω_1, Ω_2 be domains of finite type in $\mathbb{C}^n$. If $\Phi : \Omega_1 \to \Omega_2$ is a biholomorphic mapping, then Φ extends to a C^∞ diffeomorphism of $\bar{\Omega}_1$ onto $\bar{\Omega}_2$.

Heartening progress has been made in studying the singularities and mapping properties of the Bergman and Szegö kernels on domains of finite type both in dimension two and in higher dimensions. We mention particularly Nagel, Rosay, Stein, and Wainger [1], Christ [1–4], and McNeal [1, 2].

There is still a great deal of work to be done before we have reached a good working understanding of points of finite type.

11.6 Complex Analytic Dynamics

The subject of complex analytic dynamics was pioneered by P. Fatou and G. Julia in the early part of this century. Fatou [1] and Julia [1] studied iterations of holomorphic functions (in one complex variable) of a domain to itself and discovered remarkable rigidity properties of orbits and of limit functions of the iterated maps. Nowadays, the complex analytic dynamics of one variable is a prospering subject that is receiving great attention. However the complex analytic dynamics of several variables is only in its infancy. As of this writing, there are just a few papers in the subject; there is much work to be done.

In this section we provide a glimpse of the complex analytic dynamics of two complex variables by using iterates of the Henon map to construct a Fatou-Bieberbach domain (see Bochner and Martin [1] for a different approach to this matter). It is a pleasure to thank J. E. Fornæss for teaching me these well-known ideas. We begin with some elementary observations and some definitions.

It is easy to see that if U is a proper subset of the complex plane, then U cannot be biholomorphically (conformally) equivalent to the entire plane $\mathbb{C}$. For U must be simply connected even to be a candidate; and then, by the Riemann mapping theorem, it must be biholomorphically equivalent to the disc. Of course the disc is not biholomorphically equivalent to the plane (by Liouville's theorem, for instance).

It is surprising that in dimensions two and higher matters are quite different. In fact, Fatou and Bieberbach discovered (using techniques similar to those presented here) proper subsets $U \subseteq \mathbb{C}^2$ such that U and $\mathbb{C}^2$ are biholomorphic. Even more surprising is that $\mathbb{C}^2 \setminus U$ is quite large. This shows that a Picard theorem for holomorphic mappings will have to be quite different from a Picard theorem for complex functions of one variable (recall that an entire holomorphic function in dimension one that omits two values must be constant). S. Lefschetz [1] pioneered Picard theorems of several complex variables and discovered that they should be formulated in terms of the omission of complex lines in the image. Some of the best modern work on that subject has been done by M. Green [1]. We can say no more about the Picard theorems here.

DEFINITION 11.6.1   A *Fatou-Bieberbach domain* is a proper subset $U \subseteq \mathbb{C}^2$ such that U is biholomorphic to $\mathbb{C}^2$.

DEFINITION 11.6.2   A *complex Henon map* is a mapping of the form

$$F(z_1, z_2) = F_a(z_1, z_2) = (z_1^2 + az_2, az_1),$$

where a is a complex constant satisfying $0 < |a| < 1$.

Observe that F is a biholomorphic self-mapping of $\mathbb{C}^2$ and that $F(0) = 0$. Set

$$A = \text{Jac}_{\mathbb{C}} F(0) = \begin{bmatrix} 0 & a \\ a & 0 \end{bmatrix}.$$

Thus the eigenvalues of A are $\pm a$.

In the rest of this discussion we shall use the notation F^j to denote the j-fold composition $F \circ \cdots \circ F$.

LEMMA 11.6.3 Fix $a \in \mathbb{C}$ such that $0 < |a| < 1$ and fix $1 > b > |a|$. There is an $\epsilon > 0$ and a $C > 0$ such that if $|(z_1, z_2)| < \epsilon$, then for any integer $j > 0$ we have

$$|F^j(z_1, z_2)| \leq C \cdot b^j.$$

Proof   This is just direct estimation.   □

DEFINITION 11.6.4   The *basin of attraction* for F is defined to be

$$U = \{(z_1, z_2) \in \mathbb{C}^2 : F^j(z_1, z_2) \to 0 \text{ as } n \to \infty\}.$$

The lemma shows that the basin of attraction for the Henon map F is nonempty. We shall show that, in fact, the basin of attraction U is the Fatou-Bieberbach domain that we seek. Our first step is to show that U is biholomorphic to $\mathbb{C}^2$.

Fix $\epsilon > 0$, as in the lemma, once and for all. Set $B = \{z \in \mathbb{C}^2 : |z| < \epsilon\}$. We see that

$$\Omega = \bigcup_j F^{-j}(B).$$

LEMMA 11.6.5 For $z \in B$ we have

$$|A^{-1}F(z) - z| \leq C \cdot |z|^2.$$

Proof   This is an elementary calculation that we leave to the reader.   □

LEMMA 11.6.6 The limit

$$\lim_{j \to \infty} A^{-j} F^j$$

exists for $z \in U$, with the convergence being uniform on compact subsets of U.

Proof   With a fixed as usual, we also choose b such that $b^2 < |a| < b < 1$. Fix a compact set $K \subset\subset U$. Let $z \in K$ and choose j_0 so large that $F^{j_0}(z) \in B = B(0, \epsilon)$. We have

$$|A^{-(j+1)}F^{j+1}(z) - A^{-j}F^j(z)|$$
$$= |A^{-j}(A^{-1}F)F^j(z) - A^{-j}F^j(z)|$$
$$= |A^{-j}(A^{-1}F - I)(F^{j-j_0+j_0}(z))|. \tag{11.6.6.1}$$

Now since $F^{j_0}(z) \in B$, we know that

$$|F^{j-j_0}(F^{j_0}(z))| \leq C \cdot b^{j-j_0}.$$

But then, by Lemma 11.6.5, line (11.6.6.1) does not exceed

$$\|A^{-j}\| \cdot C' \cdot b^{2j-2j_0} \quad = \quad C'' \left(\frac{1}{|a|}\right)^j b^{2j-2j_0}$$
$$\leq \quad C''' \left|\frac{b^2}{a}\right|^j.$$

By our choice of b, this last expression tends to 0 as $j \to \infty$. Noting that our choice of j_0 may be taken uniform over elements of K (by compactness), we see that the proof is complete. $\square$

PROPOSITION 11.6.7   Let Ψ denote the limit of the sequence of mappings $A^{-j}F^j$ that is provided by the last lemma. Then Ψ is a biholomorphic mapping of U into $\mathbb{C}^2$.

Proof   First, $\text{Jac}_\mathbb{C}(A^{-1}F)(0) = \text{id}$; hence $\text{Jac}_\mathbb{C}(A^{-j}F^j)(0) = \text{id}$ for every j. It follows that $\text{Jac}_\mathbb{C}\Psi(0) = \text{id}$. By the inverse function theorem, Ψ is biholomorphic in a neighborhood V of the origin.

If now $z, w \in U$ are any elements with the property that $\Psi(z) = \Psi(w)$, then choose j_0 so large that $F^{j_0}(z) \in V$ and $F^{j_0}(w) \in V$ (remember that U is the basin of attraction for F so that j_0 certainly exists). Then we calculate that $\Psi(z) = \Psi(w)$ implies

$$\lim_{j\to\infty} A^{-j}F^j(z) = \lim_{j\to\infty} A^{-j}F^j(w);$$

hence

$$A^{-j_0}\lim_{j\to\infty} A^{-(j-j_0)}F^{j-j_0}\left[F^{j_0}(z)\right] = A^{-j_0}\lim_{j\to\infty} A^{-(j-j_0)}F^{j-j_0}\left[F^{j_0}(w)\right]$$

or

$$A^{-j_0}\Psi(F^{j_0}(z)) = A^{-j_0}\Psi(F^{j_0}(w)).$$

Since A is one-to-one, we conclude that $\Psi(F^{j_0}(z)) = \Psi(F^{j_0}(w))$. But $F^{j_0}(z)$, $F^{j_0}(w) \in V$, and V is the set on which we already know that Ψ is univalent. We conclude that $F^{j_0}(z) = F^{j_0}(w)$; hence $z = w$. Therefore Ψ is univalent.

□

PROPOSITION 11.6.8   The mapping Ψ is onto $\mathbb{C}^2$.

Proof   We know that $A^{(-j+1)}F^{j+1}$ converges normally to Ψ. Rewriting the expression as $A^{-1}[A^{-j}F^j]F$ and letting $j \to \infty$, we see that the expression also converges normally to $A^{-1}\Psi F$. We conclude that

$$A^{-1}\Psi F = \Psi.$$

Remember that F is a biholomorphism of all of $\mathbb{C}^2$. In conclusion, we see that

$$A^{-1}(\text{range of } \Psi) = \text{range of } \Psi. \qquad (11.6.8.1)$$

But A^{-1} is a dilation of space, and we already know that the range of Ψ contains a ball around the origin. By iterating (11.6.8.1), we see that the range of Ψ must be all of space.

□

The map Ψ that we have constructed is a biholomorphism of the basin of attraction U to $\mathbb{C}^2$. But it is possible that U is just $\mathbb{C}^2$ itself. We shall, in fact, show that possibility fails dramatically.

PROPOSITION 11.6.9   Let $W = \{(z_1, z_2) : |z_1| > 100, |z_2| < 1\}$. Then $U \cap W = \emptyset$.

Proof   We will prove something even stronger: If $|z_1| > 100$ and $|z_2| < |z_1|$, then we will show that $F^j(z_1, z_2) \to \infty$. In fact for such a point (z_1, z_2) we have

$$|z_1{}^2 + az_2| > 2|z_1|$$

and

$$|az_1| < |z_1{}^2 + az_2|.$$

Thus we see that when the mapping F is applied to such a point, it produces a point of the same type with larger modulus and whose first entry is at least twice as large. It follows then that $F^j(z_1, z_2) \to \infty$.

□

The scheme of constructing biholomorphic mappings by iteration has proved to be a powerful and versatile tool in recent years. The papers B. Wong [1], J. P. Rosay [1], R. E. Greene and S. G. Krantz [7, 9], P. G. Dixon and J. Esterle [1], J. P. Rosay and W. Rudin [1], S. Frankel [1], A. Kodama [1–4], K. T. Kim [1], and Bedford and Pinchuk [1–3] illustrate some of the applications of these ideas.

EXERCISES

1. Compute the Carathéodory metric for $\mathbb{C}^n$ and for $\mathbb{C}^n$ less finitely many points.

2. Compute the Carathéodory and Kobayashi metrics for the polydisc.

3. Compute the Carathéodory and Kobayashi metrics for $B(0, R) \setminus \bar{B}(0, r)$ whenever $r < R$.

4. Prove that if $\Omega \subset\subset \mathbb{C}^n$ is any path-connected open set and if $z, w \in \Omega$, then there is a map $f \in \Omega(B)$ and $\zeta^1, \zeta^2 \in B$ such that $f(\zeta^1) = z, f(\zeta^2) = w$.

5. The infinitesimal form of the three invariant metrics that we have studied on $B \subseteq \mathbb{C}^n$ is given by

$$F(z, \xi)^2 = C \cdot \left| \frac{\|\xi\|^2}{1 - \|z\|^2} + \frac{z \cdot \bar{\xi}}{(1 - \|z\|^2)^2} \right|,$$

where $\| \ \|$ represents Euclidean length.

6. Let $B \subseteq \mathbb{C}^n$ be the ball. Let $0 \neq a \in B$. Let $P : \mathbb{C}^n \to \mathbb{C}a$ be Hermitian orthogonal projection on the complex line generated by a. This projection is given by

$$Pz = \frac{\langle z, a \rangle}{|a|^2} a.$$

Here $\langle z, w \rangle = z \cdot \bar{w} = \sum_j z_j \bar{w}j$. Let $Qz = z - Pz$. Then the mapping

$$\phi_a(z) = \frac{a - Pz - \sqrt{1 - |a|^2} Qz}{1 - \langle z, a \rangle}$$

is an automorphism of B with $\phi_a \circ \phi_a = $ id. If ℓ is a complex line, then ϕ_a carries $\ell \cap B$ to $\ell' \cap B$ for some other complex line ℓ'.

Using the one-dimensional Schwarz lemma, prove that if $\phi : B \to B$ and $\phi(0) = 0$, then $|\phi(z)| \leq |z|$. If, in addition, $|\det \phi'(0)| = 1$, then prove by an elementary argument that ϕ is a unitary rotation.

Use the ideas presented in this exercise to give a complete description of the elements of Aut (B).

7. *H. Alexander's theorem* [2] Let $P, Q \in \partial B \subseteq \mathbb{C}^n$, $N > 1$. Let $U \subseteq \mathbb{C}^n$ be an open neighborhood of P and $V \subseteq \mathbb{C}^n$ an open neighborhood of Q. Suppose that $\psi : U \cap \bar{B} \to V \cap \bar{B}$ is biholomorphic on $U \cap B$, a homeomorphism of $U \cap \bar{B}$ to $V \cap \bar{B}$, and that $\psi(U \cap \partial B) \subseteq V \cap \partial B$. Then there is a $\Psi \in$ Aut B such that $\Psi|_{U \cap \bar{B}} = \psi$. In particular, ψ is linear fractional.

Complete the following outline to obtain Rudin's elegant proof of this result (actually Rudin proves much more—see W. Rudin [8]).

a. If $z \in B$, let $D_z = \{\lambda z \in B : \lambda \in \mathbb{C}\}$. What is D_z geometrically? Suppose that $\tilde{U}, \tilde{V} \subseteq B$ are connected open neighborhoods of $0 \in \mathbb{C}^n$ and that $F : \tilde{U} \to \tilde{V}$ is a biholomorphism with $F(0) = 0$. Finally, suppose that there is a $\hat{U} \subseteq \tilde{U}$ such that for all $z \in \hat{U}$ we have $D_z \subseteq \tilde{U}, D_{F(z)} \subseteq \tilde{V}$. Then F is the restriction to $\tilde{U}$ of a unitary map. (*Hint:* Apply Schwarz's lemma on each $D_z, D_{F(z)}$ to conclude that $|F(z)|^2 = |z|^2$ on $\hat{U}$. Conclude that F is an automorphism on $B(0,r), r$ small, that fixes 0.)

b. Assume without loss of generality that $P = Q = 1$. Let $a_j = (1 - 1/j)\,\mathbf{1}$, $b_j = \psi(a_j)$. With notation as in Exercise 6, let

$$F_j = \phi_{b_j} \circ \psi \circ \phi_{a_j},$$

$\tilde{U}_j = \phi_{a_j}(U \cap B), \tilde{V}_j = \phi_{b_j}(V \cap B)$. For j sufficiently large, the functions F_j satisfy the hypotheses of part a. Hence F_j is unitary.

8. Let $\Omega \subseteq \mathbb{C}^n$ be strongly pseudoconvex. Let $\Omega' \subseteq \mathbb{C}^n$ have C^2 boundary. Fix $P' \in \partial\Omega'$ and suppose that the Levi form at P' has a negative eigenvalue. Then prove that Ω and Ω' cannot be biholomorphically equivalent (refer to Exercise 14 at the end of Chapter 3).

9. Refer to Exercise 11 at the end of Chapter 5 for the definition of Runge domain. Prove the following sequence of assertions:
a. The bidisc $D^2(0,1) \subseteq \mathbb{C}^2$ is Runge.
b. Let $K = \{(z,w) \in \mathbb{C}^2 : w = \bar{z}, |\mathrm{Re}z| \leq 1, |\mathrm{Im}z| \leq 1\}$. Then K has a neighborhood basis $\{U_j\}$ of open sets such that each U_j is biholomorphic to $D^2(0,1)$.
c. Define a mapping $\Phi(z,w) = (z, P(z,w))$, where

$$P(z,w) = (1+i)w - izw^2 - z^2w^3.$$

Then the holomorphic Jacobian determinant of Φ is nonvanishing on all $\mathbb{C}^2$.
d. The mapping Φ is one-to-one on a neighborhood of K.
e. For j sufficiently large, Φ gives a holomorphic imbedding of U_j onto a neighborhood of $\Phi(K)$.
f. Let γ be the curve $\{(e^{i\theta}, e^{i\theta}) \in \mathbb{C}^2 : 0 \leq \theta \leq 2\pi\}$. Then $\Phi(\gamma) = \{(e^{i\theta}, 0) \in \mathbb{C}^2 : 0 \leq \theta \leq 2\pi\}$. Hence the polynomially convex hull of $\Phi(\gamma)$ contains

$$\{(z,0) \in \mathbb{C}^2 : |z| \leq 1\}.$$

g. Observe that

$$z \cdot P(z,\bar{z}) = |z|^2 \left[(1 - |z|^4) + i(1 - |z|^2)\right]$$

so that $P(z,\bar{z}) \neq 0$ on $\{z \in \mathbb{C} : 0 < |z| < 1\} \cup \{z \in \mathbb{C} : |z| > 1\}$.

h. If $a \in D \setminus \{0\}$ is fixed, then $(a,0) \notin \Phi(K)$ by (g); hence $(a,0) \notin \Phi(U_j)$ for j sufficiently large. Therefore, $\Phi(U_j)$ is not polynomially convex. Parts b through h comprise an example due to Wermer [1] of a Runge domain whose biholomorphic image is not Runge.

10. H. H. Wu [1] introduced the concept of taut domain. If $\Omega \subset\subset \mathbb{C}^n$, we say that Ω is *taut* provided that any family $\{f_\alpha\}$ of holomorphic mappings of the disc D into Ω is normal (this is actually a reformulation of Wu's original definition). Complete the following outline (from (b) on) to prove a partial converse to (a) (see also N. Kerzman [1] and N. Kerzman and J. P. Rosay [1]).

a. Use (3.3.5.6) to prove that a taut domain must be pseudoconvex.
b. Let $\Omega \subset\subset \mathbb{C}^n$ be pseudoconvex with C^1 boundary. Let $\{f_\alpha\}_{\alpha \in A}$ be a family of maps from the disc into Ω, and assume that it is not normal. It will still be the case that $\{f_\alpha\}_{\alpha \in A}$ is normal as a family of mappings from D into $\mathbb{C}^n$, so there is a subsequence $\{f_{\alpha_j}\}$ that converges to a limit disc

$$f_0 : D \to \mathbb{C}^n.$$

c. Since Ω is not taut, we know that $f_0 : D \to \bar{\Omega}$, $f_0(D) \not\subset \Omega$, $f_0(D) \not\subset \partial\Omega$.
d. Assume without loss of generality that $f_0(0) \in \partial\Omega$. Let ν be the unit outward normal to $\partial\Omega$ at $f(0)$. Then $-\log\delta(f_0(\cdot) - \epsilon\nu)$ is subharmonic when $\epsilon > 0$ is small. In particular,

$$-\log\delta(f_0(0) - \epsilon\nu) \le \frac{1}{2\pi} \int_0^{2\pi} -\log\delta(f_0(re^{i\theta}) - \epsilon\nu) \, d\theta, \qquad 0 < r < 1.$$

e. The last line yields a contradiction as $r \to 1^-, \epsilon \to 0^+$.
(A different proof appears in N. Kerzman and J. P. Rosay [1].)

11. Let $\Omega_1 \subseteq \mathbb{C}^m, \Omega_2 \subseteq \mathbb{C}^n$ be domains. Let $K_j(\cdot, \cdot)$ be the (integrated) Kobayashi distance on $\Omega_j, j = 1, 2$. Let $K_{1,2}$ be the Kobayashi distance on $\Omega_1 \times \Omega_2$. If $p_j, q_j \in \Omega_j$, then

$$\begin{aligned} K_1(p_1, q_1) + K_2(p_2, q_2) &\ge K_{1,2}\left((p_1, p_2), (q_1, q_2)\right) \\ &\ge \max\left\{K_1(p_1, q_1), K_2(p_2, q_2)\right\}. \end{aligned}$$

12. Show that the Bergman metric on D^n does *not* coincide (even up to a constant multiple) with the Kobayashi metric unless $n = 1$. Do so by proving that on D^2 the rightmost inequality in Exercise 11 is an equality.

13. Let $\Omega \subset\subset \mathbb{C}^n$ be a domain. Let K be the Kobayashi/Royden distance on Ω. Then K is sequentially complete if and only if for each $p \in \Omega$ and each $0 < r < \infty$, the closed metric ball $\bar{B}(P, r)$ is compact in Ω. Prove an analogous result for the Carathéodory metric. (*Hint:* You may find it useful to prove the following fact: let $B(p, r)$ be a ball of center p and radius r in

either the Kobayashi/Royden or Carathéodory metric. Let $d(x,y)$ be either $K(x,y)$ or $C(x,y)$. Then

$$\{q : d(x,q) < r' \text{ for some } x \in B(p,r)\} = B(p,r+r').)$$

A domain or manifold on which the Kobayashi/Royden metric is nondegenerate (i.e., the distance between two distinct points is always positive) is called *hyperbolic*. If the metric is also complete, then it is called *complete hyperbolic*.

14. Let $\Omega = \{z \in \mathbb{C}^n : 1 < |z| < 2\}, n \geq 2$. Then the Kobayashi metric on Ω is not complete. The domain is also not taut. J. P. Rosay [2] and N. Sibony (unpublished) have constructed a nonsmoothly bounded taut domain for which the Kobayashi metric is not complete. It is an open problem to construct such an example that is *smoothly bounded*.

    The Kobayashi/Royden metric on the unit ball *is* complete.

15. If $\Omega \subset\subset \mathbb{C}^n$ has complete Kobayashi/Royden metric, then Ω is a domain of holomorphy. It is known, but not obvious, that both the Carathéodory and Kobayashi metrics on a strongly pseudoconvex domain are complete (see I. Graham [1]). See subsequent exercises for more on this matter.

16. Let $\Omega_1 \subset\subset \mathbb{C}^m, \Omega_2 \subset\subset \mathbb{C}^n$ be domains. Let $C_j(\cdot,\cdot)$ be the integrated Carathéodory distance on $\Omega_j, j = 1,2$. Let $C_{1,2}$ be the Carathéodory distance on $\Omega_1 \times \Omega_2$. If $p_j, q_j \in \Omega_j$, then

$$\begin{aligned} C_1(p_1,q_1) + C_2(p_2,q_2) &\geq C_{1,2}\left((p_1,p_2),(q_1,q_2)\right) \\ &\geq \max\left\{C_1(p_1,q_1), C_2(p_2,q_2)\right\}. \end{aligned}$$

    Prove that when Ω_1, Ω_2 are polydiscs, then the second inequality is an equality. Conclude that the Carathéodory and Kobayashi/Royden metrics coincide on the polydisc. Refer to Exercise 11.

17. Let $\Omega = \{z \in \mathbb{C}^n : \frac{1}{2} < |z| < 1\}$, B the unit ball, $n > 1$. Then the Carathéodory metric on Ω is that of the ball restricted to Ω. For the Kobayashi metric, this statement is false.

18. Prove that if $\Omega_1, \Omega_2 \subset\subset \mathbb{C}^n$ have complete Carathéodory or Kobayashi/Royden metrics, then so does $\Omega_1 \times \Omega_2$.

19. Assume that $\Omega_1, \Omega_2, \ldots$ have complete Carathéodory or Kobayashi/Royden metric and set $\Omega = \cap_j \Omega_j$. If Ω is an open domain, then prove that Ω has complete Carathéodory or Kobayashi/Royden metric.

20. Prove that an analytic polyhedron has complete Carathéodory metric. (*Hint:* Let the polyhedron be

$$P = \{z \in \mathbb{C}^n : |f_j(z)| < 1, j = 1, \ldots, k\}.$$

Let $p \in P$. Given $a > 0$, choose $b, 0 < b < 1$, such that

$$\{z \in D : \rho(f_j(p), z) \le a, j = 1, \ldots, k\} \subseteq \{z \in D : |z| \le b\}.$$

Here ρ is the Poincaré metric on the disc D. Then

$$\{q \in P : C(p,q) \le a\} \subseteq \{q \in P : |f_j(q)| \le b, \quad j = 1, \ldots, k\}.$$

Conclude that $\{q \in P : C(p,q) \le a\}$ is compact in P.) Can you give an alternate proof using Exercise 19?

21. If $\Omega \subset\subset \mathbb{C}^n$ has complete Carathéodory metric, then Ω is a domain of holomorphy. (*Hint:* The domain Ω must be convex with respect to the family of bounded holomorphic functions on Ω.)

22. Use Exercise 21 to show that $D \setminus \{0\}$ does not have complete Carathéodory metric. If $\Omega \subseteq \mathbb{C}^n$ is a domain of holomorphy, then $\Omega \times (D \setminus \{0\})$ is also a domain of holomorphy but does *not* have complete Carathéodory metric. The question as to whether smoothly bounded domains of holomorphy have complete Carathéodory metric is open.

23. The Bergman metric on a strongly pseudoconvex domain is complete.

24. If $\Omega \subset\subset \mathbb{C}^n$ has complete Carathéodory metric and

$$f : D \setminus \{0\} \to \Omega$$

is holomorphic, then f extends to a holomorphic mapping of D to Ω. (*Hint:* The function f is distance decreasing.)

25. If $\Omega \subseteq \mathbb{C}^n$ has complete Carathéodory metric, then it has complete Kobayashi/Royden metric. What can you then conclude about the metric of Kobayashi/Royden on an analytic polyhedron (see Exercise 20)?

26. Complete the following outline to prove that if $\Omega \subset\subset \mathbb{C}^n$ has complete Kobayashi/Royden metric, then $\Omega' \equiv \{z \in \Omega : f(z) \ne 0\}$ has complete Kobayashi/Royden metric for any bounded holomorphic f on Ω.
   a. Without loss of generality, $f : \Omega \to D$. Let $D^* \equiv D \setminus \{0\}$. Fix $p \in \Omega', a > 0$. It is necessary to show that the closure of $\{q \in \Omega' : K_{\Omega'}(p,q) < a\}$ is compact in Ω'.
   b. There is a $b > 0$ so small that

$$\{\mu \in D : |\mu| \ge b\} \supseteq \{\mu \in D^* : K_{D^*}(f(p), \mu) \le a\}.$$

   c. Let

$$\begin{aligned} A &= \{q \in \Omega : K_\Omega(p,q) \le a\}, & A' &= \{q \in \Omega' : K_{\Omega'}(p,q) \le a\}, \\ B &= \{q \in \Omega : |f(q)| \ge b\}, & B' &= \{q \in \Omega' : |f(q)| \ge b\}. \end{aligned}$$

   Conclude that $A \supseteq A', B = B'$.

d. Prove that $A' \subseteq B' = B$.

e. Use the fact that Ω is complete to prove that $A \cap B$ is compact in Ω.

f. It holds that $A \cap B \subseteq \Omega'$, hence $A \cap B$ is compact in Ω'.

g. It holds that $A' \cap (A \cap B)$ is compact in Ω'.

h. We see that $A' \cap (A \cap B) = A'$, thus concluding the proof.

27. Show that Ω' in the preceding exercise *never* has complete Carathéodory metric unless $\Omega = \Omega'$.

28. Prove that if Ω has complete Kobayashi metric, then Ω is taut.

29. It is fairly obvious that if two C^2 domains, Ω_1, Ω_2, in $\mathbb{R}^N$ are convex and disjoint and if $U \subseteq \partial\Omega_1 \cap \partial\Omega_2$ is a relatively open set, then U must be *flat* (that is, U is the isometric image of an open subset of $\mathbb{R}^{N-1}$). This is immediate from the differential characterization of convexity. What can you say if $\mathbb{R}^N$ is replaced by $\mathbb{C}^n$ and the term *convex* is replaced by *pseudoconvex*?

30. Let $\Omega = \{(z_1, z_2) \in \mathbb{C}^2 : \mathrm{Re}\, z_1 > 0\}, f(z_1, z_2) = (z_1, z_2 + \sqrt{z_1})$. Then f maps Ω to Ω biholomorphically. However, f does not extend C^1 to the boundary of Ω. Of course, Ω is unbounded. There is no known bounded analogue for this example (see also Exercise 48).

31. Let $\Omega \subset\subset \mathbb{C}^2$ be given by

$$\Omega = \{(z_1, z_2) \in \mathbb{C}^2 : |z_2| > |z_1|^\pi\}.$$

Let ϕ be a conformal map of the disc D to $\{\zeta \in \mathbb{C} : 0 < \arg\zeta < 7\pi/4, \frac{3}{4} < |\zeta| < \frac{5}{4}\}$. For any $k \in \mathbb{N}$, consider

$$\mathcal{F}_k = \{(\phi(\zeta)^k, \phi(\zeta)^{\pi k})\} \subseteq \partial\Omega.$$

Then each $\mathcal{F}_k$ is a complex submanifold of $\partial\Omega$ and $\mathcal{F}_1 \subseteq \mathcal{F}_2 \cdots$. Use Zorn's lemma to see that there is a "largest" connected complex manifold in $\partial\Omega$ that contains every $\mathcal{F}_k$. This manifold is dense in $\partial\Omega$. It also foliates $\partial\Omega$.

32. For $\epsilon > 0$, let

$$D_\epsilon = \{z \in \mathbb{C}^2 : |z_1| < 2, |z_2| < 2, |z_1 z_2| < \epsilon\}.$$

For $z, w \in D_\epsilon$, define

$$\delta_{D_\epsilon}(z, w) = \inf\{\rho(\zeta, \xi) : f(\zeta) = z, f(\xi) = w, f \in D_\epsilon(B)\}.$$

This is the "one-disc" Kobayashi/Royden distance on D_ϵ. For ϵ small, show that $\delta_D(z, w) \neq K_{D_\epsilon}(z, w)$. Indeed, show that for ϵ small, δ_{D_ϵ} does not satisfy the triangle inequality. (*Hint:* Let $z = (1, 0), w = (0, 1), 0 = (0, 0)$. Then $\delta_{D_\epsilon}(z, 0), \delta_{D_\epsilon}(0, w)$ remain bounded as $\epsilon \to 0^+$ while $\delta_{D_\epsilon}(z, w) \to \infty$.)

33. Let $\Omega = B \subseteq \mathbb{C}^2$ be the unit ball. Define $\tilde{\Omega} = B \cap \{z_2 = 0\}$. Prove that $\tilde{\Omega}$ is a totally geodesic submanifold of Ω in both the Kobayashi/Royden and the Carathéodory metrics. That is, if $z, w \in \tilde{\Omega}$, then $K_\Omega(z, w) = K_{\tilde{\Omega}}(z, w)$ and $C_\Omega(z, w) = C_{\tilde{\Omega}}(z, w)$. It follows that the geodesic in K_Ω or C_Ω that connects z to w actually lies in $\tilde{\Omega}$.

34. It has already been noted that the Carathéodory and Kobayashi/Royden metrics may be defined using the disc or polydisc instead of the ball as a model. Prove that the very same metric results if the disc is used (this is so because the map $z \mapsto (z, 0, \ldots, 0)$ maps the disc into the ball so that its image is a totally geodesic submanifold). Does exactly the same metric result when the polydisc is used?

35. Let $\Omega_0 = \{z \in \mathbb{C}^2 : 1 < |z| < 2\}$. For $0 < \delta < 1$, let $P_\delta = (-1 - \delta, 0) \in \partial\Omega$. Then there exist constants C, C' such that

$$C'\delta^{-3/4} \leq F_K^{\Omega_0}\big(P_\delta, (1, 0)\big) \leq C\delta^{-3/4}.$$

See Krantz [23] for details. Conclude that if Ω is *any* bounded domain with C^2 boundary, z is near $\partial\Omega$, and ν_z is the unit outward normal, then

$$F_K^\Omega(z, \nu_z) \geq C'' \cdot \delta(z)^{-3/4}.$$

Here the constant C'' depends on the curvatures of $\partial\Omega$ at πz, the projection of z to the boundary.

36. Recall the rank theorem: Let $N, M \in \mathbb{N}$ and $0 \leq r \leq N$. Let $U \subseteq \mathbb{R}^N$ be an open set and $F : U \to \mathbb{R}^M$ a C^1 mapping. Assume that $J_\mathbb{R}F(x)$ has rank $r \leq M$ at each $x \in U$. Let $x_0 \in U$ be fixed, and let A denote the matrix $J_\mathbb{R}F(x_0)$. Define $\mathcal{I} = \text{image } A$ and $\mathcal{K} = \mathcal{I}^\perp$. Then there exist open sets $\tilde{V}, \tilde{U} \subseteq \mathbb{R}^N$ with $x_0 \in \tilde{U} \subseteq U$ and C^1 mappings $G : \tilde{V} \to \tilde{U}$ and $\phi : A(\tilde{V}) \to \mathcal{K}$ such that G is surjective and $F(G(x)) = Ax + \phi(Ax)$, all $x \in \tilde{V}$. (See W. Rudin [1] for a proof.)

   Prove that the theorem says that U is foliated, by way of F, by $(N - r)$-dimensional manifolds. Prove that the manifolds may be taken to be the level sets of F. In case F is C^∞, show that G, ϕ, and the foliation may be taken to be C^∞. In case F is real analytic, then so are G, ϕ, and the foliation (this is more difficult).

37. Let $\Omega \subset\subset \mathbb{C}^n$ be a domain. Let $f : \Omega \to f(\Omega) \subseteq \mathbb{C}^n$ be a univalent holomorphic mapping. Complete the following outline to prove that f has a holomorphic inverse. (J. E. Fornæss taught me this argument; some of the ideas in it can also be found in S. Bochner and W. Martin [1] and R. Narasimhan [1]). This shows that the Jacobian condition in the definition of "biholomorphic" in Section 1.4 is superfluous.

a. Induct on dimension. The result is true for $n = 1$. Assume now that it is proved on domains in $\mathbb{C}^{n-1}$.

b. Now let $\Omega \subseteq \mathbb{C}^n$. It is enough to prove that $u(z) \equiv \det J_{\mathbb{C}} f(z) \neq 0$ for all $z \in \Omega$.

c. If (b) is not true, let $S = \{u(z) = 0\} \neq \emptyset$. Then S is closed.

d. Let $p \in S$ be a point where $J_{\mathbb{C}} f$ has maximal rank r over $\mathbb{C}$. By the rank theorem (Exercise 36), $r = n - 1$.

e. There is a relatively open neighborhood $\tilde{U}$ of p in S such that rank $J_{\mathbb{C}} f \equiv n - 1$ on $\tilde{U}$.

f. Restrict f to a small neighborhood U of p in Ω so that $U \cap S = \tilde{U}$. By (e), we may change coordinates so that (shrinking U if necessary) $U \cap S = \{z \in U : z_1 = 0\}$. Assume that $p = 0$.

g. Abbreviate $z' = (z_2, \ldots, z_n)$. The inductive hypothesis implies that the map $z' \mapsto (f_2(0, z'), \ldots, f_n(0, z'))$ on $S \cap U$ has nonvanishing Jacobian determinant and hence is biholomorphic.

h. It follows from (g) that the map $z \mapsto (z_1, f_2(z_1, z'), \ldots, f_n(z_1, z'))$ has nonvanishing Jacobian on U and hence is a biholomorphic mapping.

i. Part (h) implies that the map

$$(z_1, \ldots, z_n) \mapsto (z_1, f_2(z_1, z'), \ldots, f_n(z_1, z'))$$

is a biholomorphic change of coordinates when z_1 is small. It follows that the map $(z_1, \ldots, z_n) \mapsto (f_1(z_1, 0, \ldots, 0), z_2, \ldots, z_n)$ is one-to-one.

j. It follows from (i) that $(\partial f_1/\partial z_1)(z_1, 0, \ldots, 0)|_{z_1 = 0} \neq 0$. Otherwise, $z_1 \mapsto f_1(z_1, 0, \ldots, 0)$ is not one-to-one. Thus the map in (i) would not be one-to-one.

k. From (i) and (j) we see that $\det J_{\mathbb{C}} f(0) \neq 0$. This is a contradiction.

38. Let $\Omega \subset\subset \mathbb{C}^n$ be a connected open set. Let $E_\Omega(z, \xi) = \sqrt{\sum_j |\xi_j|^2}$ be the Euclidean metric on Ω. If $K \subset\subset \Omega$, show that there is a constant $0 < M$ such that

$$\frac{1}{M} \leq \frac{E_\Omega(z, \xi)}{F_K(z, \xi)} \leq M,$$

$$\frac{1}{M} \leq \frac{E_\Omega(z, \xi)}{F_C(z, \xi)} \leq M,$$

all $z \in A$, all $\xi \neq 0$.

39. Let $\Omega_1, \Omega_2 \subset\subset \mathbb{C}$ be simply connected domains. Let $\phi : \Omega_1 \to \Omega_2$ be an isometry in the Kobayashi/Royden (resp. Carathéodory or Bergman) metric. Prove that ϕ is then either biholomorphic or conjugate biholomorphic. (*Hint:* You may assume that Ω_1, Ω_2 are the disc. Also assume that ϕ takes 0 to 0. Then ϕ preserves circles centered at 0. Either ϕ preserves orientation or it does not. Proceed from there.)

40. *Proper holomorphic maps of B_n into B_{n+1}.*

   a. *The case $n \geq 3$.* For this case, S. Webster [2] has proved that the only proper holomorphic maps of B_n into B_{n+1} that continue to be C^3 to the boundary are linear fractional. More precisely, if f is such a proper mapping then there is a biholomorphism $\phi : B_n \to B_n$ and a biholomorphism $\psi : B_{n+1} \to B_{n+1}$ such that $f = \psi \circ i \circ \phi$. Here $i : B_n \to B_{n+1}$ is the inclusion mapping $i(z_1, \ldots, z_n) = (z_1, \ldots, z_n, 0)$.

   b. *The case $n = 2$.* Webster's proof breaks down for $n = 2$. Amazingly, so does the theorem. Faran [1] has shown that if $F : B_2 \to B_3$ is a proper holomorphic mapping that extends C^3 to the boundary, then there is a biholomorphism $\phi : B_2 \to B_2$ and a biholomorphism $B_3 \to B_3$ such that $f = \psi \circ j \circ \phi$. Here j is *one of the following four maps:*

   i. $j(z_1, z_2) = (z_1, z_2, 0)$,
   ii. $j(z_1, z_2) = (z_1^2, \sqrt{2}z_1z_2, z_2^2)$,
   iii. $j(z_1, z_2) = (z_1, z_1z_2, z_2^2)$,
   iv. $j(z_1, z_2) = (z_1^3, \sqrt{3}z_1z_2, z_2^3)$.

   Verify that these *are* proper holomorphic maps as claimed.

   c. *The case $n = 1$.* Of course, $B_1 = D$, the disc. Let $a_1 : \partial D \to \mathbb{R}, a_2 : \partial D \to \mathbb{R}$ be C^4 functions such that $\exp(2a_1) + \exp(2a_2) \equiv 1$. Then a_1, a_2 have harmonic extensions A_1, A_2 to $\bar{D}$. Verify (by integration by parts) that $A_1, A_2 \in C^3(\bar{D})$. Let $\tilde{A}_1, \tilde{A}_2$ be harmonic conjugates of A_1, A_2. It is an old result of Privalov that $\tilde{A}_1, \tilde{A}_2$ extend C^3 to $\bar{D}$. Define $f_j = \exp(A_j + i\tilde{A}_j), j = 1, 2$. Then $f = (f_1, f_2) : D \to B_2$ is a C^3 proper holomorphic mapping of $\bar{D}$ into $\bar{B}_2$. In short, there is no classification of proper mappings of the disc into B_2.

   The work of Webster and Faran has inspired many others to look at proper maps among domains of different dimensions. In J. Cima, S. G. Krantz, and T. Suffridge [1], a generalization of some of these results was found for strongly pseudoconvex domains. The proof in that paper is not complete; however it was corrected in Forstneric [1]. The latter paper also has some results about existence and nonexistence of proper mappings. J. Cima and T. Suffridge [1] reduced the smoothness hypothesis from 3 to 2 and also introduced more elementary techniques. D'Angelo [5] gives a scheme for classifying polynomial proper mappings among balls in different dimensions.

41. Let $\Omega \subset\subset \mathbb{C}^n$ be a domain. Suppose that every $P \in \partial\Omega$ is a peak point for $A(\Omega)$. Prove that Ω is complete in the Carathéodory metric and hence in the Kobayashi/Royden metric. E. Bedford and J. E. Fornæss [1] have proved that this condition holds on domains in $\mathbb{C}^2$ with real analytic boundary. The paper K. Diederich and J. E. Fornæss [5] generalizes the result, using different techniques, to higher dimensions.

42. Even for domains that are topologically trivial, the map that realizes the Kobayashi metric for a given point z and direction ξ may not be unique. For instance, consider the bidisc in $\mathbb{C}^2, z = 0, \xi = (1, 0)$. Show that there are

(at least) two distinct maps $f \in D^2(B)$ such that $f'(0) = \lambda\xi, \lambda > 0$, and λ maximal. However, uniqueness does obtain when the domain is $B \subseteq \mathbb{C}^2$.

43. Let $\Omega = \{z \in \mathbb{C} : |z + 1| < 1\}, \Phi(z) = z/\log z, W = \Phi(\Omega)$. Then $\partial\Omega$ is $C^1, \partial W$ is C^1, and Φ is biholomorphic from Ω to W. However Φ does not extend C^1 to $\partial\Omega$. This example was noticed by S. Webster.

There has been considerable interest recently in boundary regularity of conformal mappings (of one complex variable) of domains with rougher than C^1 boundary. Some references are F. D. Lesley [1,2], Näkki and Palka [1], Rodin and Warschawski [1], and Smith and Stegenga [1].

B. Fridman [2] has given a clever example of a domain in $\mathbb{C}^2$ that is biholomorphic to the bidisc but has a rather rough boundary such that no biholomorphic mapping of the domain to itself extends even continuously to the boundary. He has produced more examples, with slightly smoother boundaries, in B. Fridman [4].

44. Prove Royden's lemma (H. Royden [1], I. Graham [1]): Let $\Omega \subset\subset \mathbb{C}^n$ be a domain with C^2 boundary. Let $P \in \partial\Omega$, and let U and V be open neighborhoods of P in $\mathbb{C}^n$ with $V \subset\subset U$. Then there is a constant $C = C(U,V,\Omega) > 0$ such that

$$F_K^\Omega(z,\xi) \geq C \cdot F_K^{\Omega \cap U}(z,\xi),$$

all $z \in V \cap \Omega$, all $\xi \in \mathbb{C}^n$. Notice that an inequality in the opposite direction is trivial.

45. Complete the following outline to obtain the proof due to L. Nirenberg, S. Webster, and P. Yang [1] that a biholomorphic map $F : \Omega_1 \to \Omega_2$ of C^2 strongly pseudoconvex domains extends $\Lambda_{1/2}$ to the closure of Ω_1. (This assertion, and more general statements, were first proved by G. M. Henkin [5].)

a. Let ρ_1, ρ_2 be strictly plurisubharmonic defining functions for Ω_1, Ω_2 (see Exercise 19 at the end of Chapter 3). Use them, together with the Hopf lemma, to prove that there is a $C > 0$ such that

$$\frac{1}{C} \leq \frac{\delta_{\Omega_1}(z)}{\delta_{\Omega_2}(F(z))} \leq C,$$

all $z \in \Omega_1$.

b. The problem is local. If $P' \in \partial\Omega_2$ then, after a suitable change of coordinates, we may assume that there is an $r > 0$ such that $B(P',r) \cap \partial\Omega_2$ is strictly convex. There are positive numbers $0 < s < r < t$ such that for $z' \in B(P',s)$ and $\pi z' \in \partial\Omega_2 \cap B(P',r)$ the orthogonal projection, it holds that $B'_z \equiv B(\pi z' - t\nu_{\pi z'}, t) \supseteq \Omega_2 \cap B(P',r)$.

c. There is a $u > 0$ such that if $z \in \Omega_1$ is sufficiently close to $\partial\Omega_1$, then $z \in B(\pi z - u\nu_{\pi z}, u) \equiv B_z \subseteq \Omega_1$.

d. Conclude that if z is as in (c), $F(z) = z'$ is as in (b), $\xi \in \mathbb{C}^n, \xi' = F_*(z)\xi$, then

$$F_K^{B_z}(z, \xi) \geq C \cdot F_K^{\Omega_2 \cap B'_{z'}}(z', \xi') \geq C \cdot F_K^{B'_{z'}}(z', \xi').$$

You will need Royden's lemma (see Exercise 44) for the first of the inequalities.

e. The estimates in (d) are uniform in z near $\partial\Omega_1$.

f. Use Exercise 5 to conclude that

$$C\left\{\frac{\|\xi_T\|}{\sqrt{\delta_{\Omega_1}(z)}} + \frac{\|\xi_N\|}{\delta_{\Omega_1}(z)}\right\} \geq \left\{\frac{\|\xi'_T\|}{\sqrt{\delta_{\Omega_2}(z')}} + \frac{\|\xi'_N\|}{\delta_{\Omega_2}(z')}\right\},$$

where $\xi = \xi_T \oplus \xi_N$ (resp. $\xi' = \xi'_T \oplus \xi'_N$) is the decomposition of ξ (resp. ξ') into tangential and normal components at z (resp. z'). Here $\| \ \|$ denotes the Euclidean length.

g. Conclude that

$$C \cdot \frac{\|\xi\|}{\delta_{\Omega_1}(z)} \geq \frac{\|\xi'\|}{\sqrt{\delta_{\Omega_2}(z')}}.$$

h. Conclude that

$$|\nabla F(z)| \leq C \cdot \delta_{\Omega_1}(z)^{-1/2}.$$

i. Apply Lemma 10.3.1.

j. Show that the hypotheses on Ω_1 may be drastically weakened. Does F have to be biholomorphic?

46. Complete the following outline to prove (see K. T. Hahn [1]) that the Bergman metric always dominates the Carathéodory metric. (Related results appear in J. Burbea [1] and Lu Qi-Keng [2].) In this problem, for convenience, we use the Carathéodory metric modeled on elements of $D(\Omega)$ rather than $B(\Omega)$.

a. Fix $\Omega \subset\subset \mathbb{C}^n$. Let A_1^2 be the unit ball in $A^2(\Omega)$. For $z \in \Omega$, let f_z be that element of A_1^2 that has greatest (real) value at z. Then $f_z(\cdot) = \overline{K(z, \cdot)}$, where K is the Bergman kernel for Ω.

b. Let $z \in \Omega$ and let $\xi \in \mathbb{C}^n$ be thought of as a holomorphic tangent vector at z. Let $h_{z,\xi}$ be that element of A_1^2 such that $h_{z,\xi}(z) = 0$ and $|\partial_\xi h_{z,\xi}(z)|^2$ is maximal. Here ∂_ξ represents directional differentiation.

c. We have

$$\sum_{i,j} g_{ij}(z)\xi_i\bar{\xi}_j = \frac{|\partial_\xi h_{z,\xi}(z)|^2}{|f_z(z)|^2}.$$

d. Let ϕ be holomorphic on $\Omega, \phi(z) = 0, \|\phi\|_{\sup} \leq 1$. Then ϕf_z is a competitor for $h_{z,\xi}$. Therefore,

$$|\partial_\xi h_{z,\xi}(z)|^2 \geq |\partial_\xi \phi(z)|^2 |f_z(z)|^2.$$

e. Conclude that

$$\sum_{i,j} g_{ij}(z)\xi_i \bar{\xi}_j \geq F_C(z,\xi).$$

K. Diederich and J. E. Fornæss [3] have shown that, in general, the Bergman and Kobayashi metrics are not comparable. On domains of finite type in $\mathbb{C}^2$ (see D. Catlin [4]), it is known that the Bergman, Carathéodory, and Kobayashi/Royden metrics are all comparable. On convex domains, the last two metrics are precisely *equal* (see Lempert [5]).

47. Use an invariant metric to prove the last part of Exercise 28 at the end of Chapter 3.

48. Every few years the following incorrect counterexample to the mapping conjecture surfaces. Find the error. Let

$$\Omega = \left\{ (z_1, z_2) : \mathrm{Re} z_1 > \exp\left(-1/(|z_2|^2 + (\mathrm{Im} z_1)^2)\right) \right\}.$$

Define $\Phi(z_1, z_2) = (z_1, z_2 + \sqrt{z_1})$. Set $\Omega' = \Phi(\Omega)$. Notice that Ω, Ω' are both pseudoconvex. Then Φ is a biholomorphic map of Ω to Ω' that does not continue smoothly to the boundary. These domains may both be truncated to give bounded domains, thus providing a smooth counterexample to the mapping conjecture.

49. Prove that the notion of finite type of a point is invariant under a biholomorphic change of coordinates in a neighborhood of that point.

50. True or false: If ρ is the defining function of a smoothly bounded domain Ω and if ρ is a polynomial, then Ω is of finite type.

51. Prove that if a domain in $\mathbb{C}^2$ has real analytic boundary, then it is of finite type. (*Hint:* You may have to use the theorem of Lojaciewicz. See S. Lojaciewicz [1, 2], or S. G. Krantz and H. R. Parks [2].)

52. Prove that if a point P in the boundary of a smoothly bounded domain in $\mathbb{C}^2$ is of finite Bloom/Graham/Kohn type m, then all nearby boundary points must be of finite type not exceeding m.

53. Prove that P is a point of pseudoconvexity and is of finite type 2, then P is strongly pseudoconvex.

54. Prove that if U is a boundary neighborhood of a smoothly bounded domain $\Omega \subseteq \mathbb{C}^2$, then it is impossible for every point of U to have finite type greater than 2. (*Hint:* Use the result on foliations in Exercise 9 at the end of Chapter

7.) This shows that, on a pseudoconvex domain, strongly pseudoconvex points are generic. It follows from work of Catlin [2, 3] that, in the boundary of a finite type domain in any dimension, the strongly pseudoconvex points form a dense open set of full measure.

55. Prove that the definition of finite type given in Section 5 is independent of the choice of defining function.

56. Give an example of a domain with a nonpseudoconvex point that is of type 2.

57. Let the dimension be at least three. Prove that if a point P in the boundary of a domain in $\mathbb{C}^n$ is of finite type in the sense of Section 5, then it is of finite analytic type in the sense that there are finitely many complex tangential vector fields whose commutator has a complex normal component. The converse is not, in general, true.

58. Is there a "real variable" analogue of the concept of finite type? What about in the context of convex sets?

59. Let U be the Fatou-Bieberbach domain constructed in Section 11.6. Prove that $U \cap \{(z_1, 0)\}$ is a bounded, relatively open, set. Prove that each of its connected components is simply connected.

60. Let $\Omega \subseteq \mathbb{C}^n$ be a smoothly bounded domain. Assume that $f : \Omega \to \mathbb{C}^m$ and $g : \Omega \to \mathbb{C}^m$ are mappings satisfying $|f(z)| = |g(z)|$ for every z. How are f and g related?

Appendix I   Manifolds

Let X be a paracompact Hausdorff space. Let $0 \le N \in \mathbb{Z}$. Suppose that for each $x \in X$, there is a neighborhood $U = U_x$ of x in X, an open set $W \subseteq \mathbb{R}^N$, and a homeomorphism $\phi : U \to \phi(U) = W$. Then X is said to be a *manifold* of dimension N. The pairs (ϕ, U) are called *coordinate charts* or *coordinate pairs*. The set $\{(\phi, U)\}$ is called an *atlas*.

Let $0 \le k \in \mathbb{Z}$. Suppose that for any two coordinate pairs $(\phi, U), (\phi', U')$, it holds that

$$
\begin{aligned}
\phi' \circ \phi^{-1} : \phi(U \cap U') &\to \phi'(U \cap U') \\
\phi \circ \phi'^{-1} : \phi'(U \cap U') &\to \phi(U \cap U')
\end{aligned}
\qquad (*)
$$

are C^k. Then X is said to be a C^k *manifold*. If the two maps in $(*)$ are C^∞, for all pairs of coordinate charts, then X is a C^∞ (or *smooth*) manifold. If the maps are real analytic, then X is said to be *real analytic*. In case $N = 2n$, $\mathbb{R}^N$ is identified with $\mathbb{C}^{2n}$ in the usual way, and the maps in $(*)$ are holomorphic, then X is a *complex analytic manifold* (of complex dimension n).

Let X be a C^k manifold. Then $f : X \to \mathbb{C}$ is said to be C^k if $f \circ \phi^{-1} : \phi(U) \to \mathbb{C}$ is C^k for every coordinate chart (ϕ, U). Likewise, C^∞, real analytic, and holomorphic functions are defined on C^∞, real analytic, and complex analytic manifolds, respectively.

Let X and Y be C^k manifolds and let $F : X \to Y$ be a mapping. We say that F is C^k if for every coordinate chart (ϕ, U) on X and every coordinate chart (ψ, V) on Y, we have that $\psi \circ F \circ \phi^{-1}$ is a C^k mapping (whenever this composition makes sense).

Let $X, \tilde{X}$ be N-dimensional C^k manifolds and $F : X \to \tilde{X}$ a homeomorphism. We call F a C^k *diffeomorphism* provided that F is invertible and, for every choice of coordinate charts (ϕ, U) on X and $(\tilde{\phi}, \tilde{U})$ on $\tilde{X}$, it holds that $\tilde{\phi} \circ F \circ \phi^{-1} : \phi(U) \to \tilde{\phi}(U)$ and $\phi \circ F^{-1} \circ \tilde{\phi}^{-1} : \tilde{\phi}(U) \to \phi(U)$ are C^k maps. Of course C^∞, real analytic, and complex analytic diffeomorphisms (for manifolds of the corresponding type) are defined similarly.

Let $\Omega \subseteq \mathbb{R}^N$ have C^k boundary. Then $M = \partial\Omega$ is the zero set of a C^k defining function ρ. Therefore the implicit function theorem makes it clear that M is a C^k manifold of dimension $N-1$. The Cartesian product of two C^k (resp. C^∞, real analytic, complex analytic) manifolds is C^k (resp. C^∞, real analytic, complex analytic). However, the union of two such manifolds need not be (in fact it is *generically not*) such a manifold.

Let X be a C^k manifold of dimension N and $Y \subseteq X$ a C^k manifold of dimension $M < N$. We say that Y is a *regularly imbedded submanifold* of X if each $y \in Y$ has a neighborhood $W \subseteq X$ with a homeomorphism $\psi : W \to \psi(W) \subseteq \mathbb{R}^N$ satisfying the conditions: (1) (ψ, W) is a coordinate chart for X; (2) $\psi(W \cap Y) = \psi(W) \cap \{x_{M+1} = \cdots = x_N = 0\}$; and (3) $W \cap Y$ is a coordinate chart for Y with coordinate pair $(\psi|_{W \cap Y}, W \cap Y)$. If $Y \subseteq \mathbb{R}^N$, then Y is a regularly imebedded C^k submanifold of dimension $N-1$ if and only if there is an open set $U \subseteq \mathbb{R}^N$ and a C^k function $\rho : U \to \mathbb{R}$ such that $\nabla\rho \neq 0$ on Y and $Y = \{x \in U : \rho(x) = 0\}$. More generally, $Y \subseteq \mathbb{R}^N$ is a regularly imbedded C^k submanifold of dimension $M < N$ in $\mathbb{R}^N$ if and only if there exist an open set $U \subseteq \mathbb{R}^N$ and $N - M$ functions $\rho_1, \ldots, \rho_{N-M} : U \to \mathbb{R}$ such that $Y = \{x \in U : \rho_j(x) = 0, j = 1, \ldots, N - M\}$ and such that

$$\begin{pmatrix} \nabla\rho_1 \\ \cdot \\ \cdot \\ \cdot \\ \nabla\rho_{N-M} \end{pmatrix}$$

has rank $N - M$ at each point of Y.

If X is a C^∞, real analytic, or complex analytic manifold, then a regularly imbedded C^∞, real analytic, or complex analytic submanifold $Y \subseteq X$ is defined in an obvious fashion.

Appendix II Area Measures

If $\Omega \subset\subset \mathbb{R}^N$ has C^1 boundary, then we use the symbol $d\sigma$ to denote $(N-1)$-dimensional area measure on $\partial\Omega$. This concept is fundamental; we discuss, but do not prove, the equivalence of several definitions for $d\sigma$. A thorough consideration of geometric measures on lower-dimensional sets may be found in the two masterpieces H. Federer [1] and H. Whitney [1].

First we consider Carathéodory's construction. Let $S \subseteq \mathbb{R}^N$ and $\delta > 0$. Let $\mathcal{U} = \{U_\alpha\}_{\alpha \in A}$ be an open covering of S. Call $\mathcal{U}$ a δ-admissible covering if each U_α is an open Euclidean N-ball of radius $0 < r_\alpha < \delta$. If $0 \leq k \in \mathbb{Z}$, let M_k be the usual k-dimensional Lebesgue measure of the unit ball in $\mathbb{R}^k$ (e.g., $M_1 = 2, M_2 = \pi, M_3 = 4\pi/3$, etc.). Define

$$\mathcal{H}_\delta^k(S) = \inf \left\{ \sum_{\alpha \in A} M_k r_\alpha^k : \mathcal{U} = \{U_\alpha\}_{\alpha \in A} \text{ is } \delta\text{-admissible} \right\}.$$

Clearly, $\mathcal{H}_\delta^k(S) \leq \mathcal{H}_{\delta'}^k(S)$ if $0 < \delta' < \delta$. Therefore, $\lim_{\delta \to 0} \mathcal{H}_\delta^k(S)$ exists in the extended real number system. The limit is called the *k-dimensional Hausdorff measure* of S and is denoted by $\mathcal{H}^k(S)$. The function $\mathcal{H}^k$ is an outer measure.

Exercises for the Reader

1. If $I \subseteq \mathbb{R}^N$ is a line segment, then $\mathcal{H}^1(I)$ is the usual Euclidean length of I. Also, $\mathcal{H}^0(I) = \infty$ and $\mathcal{H}^k(I) = 0$ for all $k > 1$.

2. If $S \subseteq \mathbb{R}^N$ is Borel, then $\mathcal{H}^N(S) = \mathcal{L}^N(S)$, where $\mathcal{L}^N$ is Lebesgue N-dimensional measure.

3. If $S \subseteq \mathbb{R}^N$ is a discrete set, then $\mathcal{H}^0(S)$ is the number of elements of S.

4. Define $M_\alpha = \Gamma(1/2)^\alpha / \Gamma(1+\alpha/2)$ for $\alpha > 0$ (note that this is consistent with the preceding definition of M_k). Then define H^α for any $\alpha > 0$ by using Carathéodory's construction. Let $S \subseteq \mathbb{R}$ be the Cantor set. Compute $\alpha_0 = \sup\{\alpha > 0 : \mathcal{H}^\alpha(S) = \infty\}$. Also compute $\alpha_1 = \inf\{\alpha > 0 : \mathcal{H}^\alpha(S) = 0\}$. Then $\alpha_0 = \alpha_1$. What is its specific value? This number is called the *Hausdorff dimension* of S. What is the Hausdorff

dimension of a regularly imbedded, k-dimensional, C^1 manifold in $\mathbb{R}^N$? It is, in fact, a deep theorem (see H. Federer [1]) that any rectifiable set $S \subseteq \mathbb{R}^N$ has the property that $\alpha_0 = \alpha_1$.

The measure $\mathcal{H}^{N-1}$ gives one reasonable definition of $d\sigma$ on $\partial\Omega$ when $\Omega \subseteq \mathbb{R}^N$ has C^1 boundary. Now let us give another. If $S \subseteq \mathbb{R}^N$ is closed and $x \in \mathbb{R}^N$, let $\mathrm{dist}\,(x, S) = \inf\{|x - s| : s \in S\}$. Then $\mathrm{dist}\,(x, S)$ is finite, and there is a (not necessarily unique) $s_0 \in S$ with $|s_0 - x| = \mathrm{dist}\,(x, S)$. (*Exercise:* Prove these assertions.) Suppose that $M \subseteq \mathbb{R}^N$ is a regularly imbedded C^1 manifold of dimension $k < N$. Let $E \subseteq M$ be compact and define, for $\epsilon > 0$, $E_\epsilon = \{x \in \mathbb{R}^N : \mathrm{dist}\,(x, E) < \epsilon\}$. Define

$$\sigma_k(E) = \limsup_{\epsilon \to 0^+} \frac{\mathcal{L}^N(E_\epsilon)}{M_{N-k}\epsilon^{N-k}},$$

where $\mathcal{L}^N$ is Lebesgue volume measure on $\mathbb{R}^N$. It can, in fact, be shown that *lim sup* may be replaced by *lim*. The resulting set function σ_k is an outer measure. When $E \subseteq M$ is compact, then it can be proved that $\mathcal{H}^k(E) = \sigma_k(E)$. The measure σ_k may be extended to more general subsets of M by the usual exhaustion procedures.

Our third definition of area measure is as follows. Let $M \subseteq \mathbb{R}^N$ be a regularly imbedded C^1 submanifold of dimension $k < N$. Let $p \in M$, and let (ϕ, U) be a coordinate chart for $M \subseteq \mathbb{R}^N$, as in the definition of "regularly imbedded submanifold." If $E \subseteq U \cap M$ is compact, define

$$m_k(E) = \int_{\phi(E)} \mathrm{vol}\langle J_{\mathbb{R}}\phi^{-1}(x)e_1, \ldots, J_{\mathbb{R}}\phi^{-1}(x)e_k\rangle d\mathcal{L}^k(x).$$

Here e_j is the jth unit coordinate vector, and the integrand is simply the k-dimensional volume of the k-parallelepiped determined by the vectors $J_{\mathbb{R}}\phi^{-1}(x)e_j, j = 1, \ldots, k$. We know from calculus (Spivak [1]) that this gives a definition of surface area on compact sets $E \subseteq U \cap M$ that coincides with the preceding definitions. The new definition may be extended to all of M with a partition of unity and to more general sets E by inner regularity.

Finally, we mention that a k-dimensional, C^1 submanifold M of $\mathbb{R}^N$ may be given (locally) in *parametrized form*. That is, for $P \in M$ there is a neighborhood $P \in U_P \subseteq \mathbb{R}^N$ and we are given functions $\phi_1, \ldots, \phi_N$ defined on an open set $W_P \subseteq \mathbb{R}^k$ such that the mapping

$$\Phi = (\phi_1, \ldots, \phi_N) : W_P \to U_P \cap M$$

is C^1, one-to-one, and onto, and the real Jacobian of this mapping has rank k at each point of W_P. In this circumstance, we define

$$\tau_k(U_P \cap M) = \int_{x \in W_P} |N_x| \, d\mathcal{L}^k(x),$$

where N_x is defined to be the standard k-dimensional volume of the image of the unit cube in $\mathbb{R}^k$ under the linear mapping $J_{\mathbb{R}} \Psi(x)$. (The object N_x can be defined rather naturally using the language of differential forms—see Appendix IV. The definition we have given has some intuitive appeal.) For imbedded manifolds the construction of $\mathcal{T}_k$ is virtually the same as that of m_k.

On a k-dimensional regularly imbedded submanifold of $\mathbb{R}^N$, we have that

$$\mathcal{H}^k = \sigma_k = m_k = \tau_k.$$

Appendix III   Exterior Algebra

Let $V_1, \ldots, V_p$ be vector spaces over $\mathbb{R}$. Let F consist of all *formal* finite linear combinations over $\mathbb{R}$ of elements of $V_1 \times \cdots \times V_p$. So $f \in F$ means that

$$f = \sum_{j=1}^{k} a_j \left(v_1^j, \ldots, v_p^j \right)$$

for some $k \in \mathbb{N}$, some $v_i^j \in V_i, 1 \leq j \leq k, 1 \leq i \leq p$. If we declare $(0, \ldots, 0)$ to be the additive identity and also $1 \cdot (v_1, \ldots, v_p) = (v_1, \ldots, v_p)$ and $0 \cdot (v_1, \ldots, v_p) = (0, \ldots, 0)$, then F is a real vector space in an obvious fashion. Let G be the vector subspace generated by elements of the form

$$(v_1, \ldots, v_{i-1}, x, v_{i+1}, \ldots, v_p) + (v_1, \ldots, v_{i-1}, y, v_{i+1}, \ldots, v_p)$$
$$-(v_1, \ldots, v_{i-1}, x+y, v_{i+1}, \ldots, v_p)$$

and

$$c(v_1, \ldots, v_{i-1}, v_i, v_{i+1}, \ldots, v_p) - (v_1, \ldots, v_{i-1}, cv_i, v_{i+1}, \ldots, v_p).$$

Then we define

$$V_1 \otimes \cdots \otimes V_p = F/G.$$

Let us denote the residue of an element $(v_1, \cdots, v_p) \in V_1 \times \cdots \times V_p$ in F/G by $v_1 \otimes \cdots \otimes v_p$. These elements are called *simple p-tensors*. Then any element of $V_1 \otimes \cdots \otimes V_p$ can be written as a finite sum of expressions of the form $v_1 \otimes \cdots \otimes v_p$. The algebraic operations in $V = V_1 \otimes \cdots \otimes V_p$ are

$$(v_1 \otimes \cdots \otimes v_{i-1} \otimes x \otimes v_{i+1} \otimes \cdots \otimes v_p) + (v_1 \otimes \cdots \otimes v_{i-1} \otimes y \otimes v_{i+1} \otimes \cdots \otimes v_p)$$

$$= v_1 \otimes \cdots \otimes v_{i-1} \otimes (x+y) \otimes v_{i+1} \otimes \cdots \otimes v_p$$

and

$$
\begin{aligned}
c(v_1 \otimes \cdots \otimes v_p) &= cv_1 \otimes v_2 \otimes \cdots \otimes v_p \\
&= v_1 \otimes cv_2 \otimes \cdots \otimes v_p \\
&= \cdots \\
&= v_1 \otimes \cdots \otimes cv_p.
\end{aligned}
$$

Observe that these algebraic operations are nothing other than a restatement of the meaning of the quotient space F/G.

If V is a vector space over $\mathbb{R}, 1 \le p \in \mathbb{Z}$, we let

$$
\otimes_p V = \underbrace{V \otimes \cdots \otimes V}_{p \text{ times}}.
$$

Define $\otimes_0 V = \mathbb{R}$. If $\alpha \in \otimes_p V, \beta \in \otimes_q V$, then $\alpha \otimes \beta$ is an element of $\otimes_{p+q} V$ in a natural way. With this notion of multiplication (extended by linearity), the space

$$
\otimes_* V \equiv \oplus_{p=0}^{\infty} \otimes_p V
$$

becomes a (graded) algebra.

Let $\mathcal{J} \subseteq \otimes_* V$ be the two-sided ideal generated by all elements of the form $v \otimes v \in \otimes_2 V$. Then we define

$$
\begin{aligned}
\wedge_p V &= \otimes_p V / (\otimes_p V \cap \mathcal{J}) \\
\wedge_* V &= \otimes_* V / \mathcal{J}.
\end{aligned}
$$

Clearly

$$
\bigwedge{}_* V = \sum_{p=0}^{\infty} \bigwedge{}_p V.
$$

Now $\bigwedge_p V$ is clearly generated, under addition, by elements that are images under the quotient of the elements $v_1 \otimes \cdots \otimes v_p$. We denote these generators by $v_1 \wedge \cdots \wedge v_p$.

Since $\mathcal{J}$ contains elements of the form

$$
(x + y) \otimes (x + y) - (x \otimes x) - (y \otimes y) = x \otimes y + y \otimes x,
$$

it follows that $v_1 \wedge v_2 = -v_2 \wedge v_1$ in $\bigwedge_2 V$. Thus, in particular, $v \wedge v = 0$ in $\bigwedge_2 V$. More generally, if $k \in \mathbb{N}, v_1, \ldots, v_k \in V$, and σ is a permutation on k letters, then

$$
v_1 \wedge \cdots \wedge v_k = \epsilon(\sigma) \cdot v_{\sigma(1)} \wedge \cdots \wedge v_{\sigma(k)}.
$$

Here $\epsilon(\sigma)$ is the signature of the permutation σ. Of course, $v_1 \wedge \cdots v_k = 0$ if $v_i = v_j$ for some $i \neq j$. If V is finite dimensional and $\{e_1, \ldots, e_N\}$ is a basis for V, then $\{e_{i_1} \wedge \cdots \wedge e_{i_p} : i_1 < \cdots < i_p\}$ is a basis for $\bigwedge_p V$ provided that $p \leq N$. If $p > N$, then $\bigwedge_p V = \{0\}$.

Let V^* be the vector space dual of V. Then we may define $\bigwedge_* V^*$. Let $E^1, \ldots, E^N$ be the canonical dual basis for V^*, that is, $E^i(e_j) = \delta_{ij}$. Then $\{E^{i_1} \wedge \cdots \wedge E^{i_p} : i_1 < \cdots < i_p\}$ is a basis for $\bigwedge_p V^*$, each $1 \leq p \leq N$. We may extend the pairing between V^* and V to one between $\bigwedge_* V^*$ and $\bigwedge_* V$ as follows. Set

$$(E^{i_1} \wedge \cdots \wedge E^{i_p})(e_{i_1} \wedge \cdots \wedge e_{i_p}) = 1$$

and all other pairings to be zero. Extend to $\bigwedge_* V^*$ and $\bigwedge_* V$ by multilinearity. If $\alpha \in \bigwedge_* V^*, \beta \in \bigwedge_* V$, it is sometimes convenient, because of symmetry considerations, to write

$$\alpha(\beta) = \langle \alpha, \beta \rangle.$$

Now we define an operation that is dual to the wedge product—namely, if $0 \leq p \leq q, \beta \in \bigwedge_p V$, and $\alpha \in \bigwedge_q V^*$, we define $\beta \rfloor \alpha$ to be the unique element of $\bigwedge_{q-p} V^*$ such that

$$\langle \gamma, \beta \rfloor \alpha \rangle = \langle \gamma \wedge \beta, \alpha \rangle, \quad \text{all} \quad \gamma \in \bigwedge_{q-p} V.$$

Likewise, if $\mu \in \bigwedge_q V$ $\zeta \in \bigwedge_p V^*$, then $\mu \lfloor \zeta$ is defined to be the unique element of $\bigwedge_{q-p} V$ such that

$$\langle \mu \lfloor \zeta, \xi \rangle = \langle \mu, \zeta \wedge \xi \rangle, \quad \text{all} \quad \xi \in \bigwedge_{q-p} V^*.$$

Both $\beta \rfloor \alpha$ and $\mu \lfloor \zeta$ are understood to be 0 when $p > q$. It is easy to check the following formulas: If $p \leq q$ and $\beta = e_{i_1} \wedge \cdots \wedge e_{i_p}, \alpha = E^{j_1} \wedge \cdots \wedge E^{j_q}$, $j_1 = i_1, \ldots, j_p = i_p$, then $\beta \rfloor \alpha = E^{j_{p+1}} \wedge \cdots \wedge E^{j_q}$. If

$$\{i_1, \ldots, i_p\} \not\subseteq \{j_1, \ldots, j_q\}$$

then

$$\beta \rfloor \alpha = 0.$$

It is often convenient to denote $\bigwedge_p V^*$ by $\bigwedge^p V$ and $\bigwedge_* V^*$ by $\bigwedge^* V$.

Appendix IV Vectors, Covectors, and Differential Forms

Let $U \subseteq \mathbb{R}^N$ be an open set and let $P \in U$. Denote by $\mathcal{C}_P$ the collection of C^1 curves $\gamma : (-1, 1) \to \mathcal{U}$ with $\gamma(0) = P$. Call two elements $\gamma_1, \gamma_2 \in \mathcal{C}_P$ *equivalent* if

$$\left. \frac{d}{dt} f(\gamma_1(t)) \right|_{t=0} = \left. \frac{d}{dt} f(\gamma_2(t)) \right|_{t=0}$$

for every $f \in C^1(U)$. The equivalence classes are called *tangent vectors* at $P \in U$. Denote the set of tangent vectors by $T_P(U)$. Clearly the definition of $T_P(U)$ does not depend on U. If $[\gamma], [\eta] \in T_P(U)$ and $f \in C^1(U)$, we define

$$([\gamma] + [\eta])(f) = \left. \frac{d}{dt} f(\gamma(t)) \right|_{t=0} + \left. \frac{d}{dt} f(\eta(t)) \right|_{t=0} .$$

Also

$$(c[\gamma])(f) = c \left. \frac{d}{dt} f(\gamma(t)) \right|_{t=0}$$

for any $c \in \mathbb{R}$. With these operations, $T_P(U)$ becomes a real vector space.

We claim that $T_P(U)$ has dimension N. To see this, notice that if $[\gamma] \in T_P(U), f \in C^1(U)$, then

$$\left. \frac{d}{dt} f(\gamma(t)) \right|_{t=0} = \sum_{j=1}^{N} \frac{\partial \gamma_j}{\partial t}(0) \cdot \left. \frac{\partial}{\partial x_j} f \right|_P .$$

Hence $\{\partial/\partial x_1, \ldots, \partial/\partial x_N\}$ is a basis for T_P, where we have identified $\partial/\partial x_j$ with the curve $t \mapsto P + t e_j$ and e_j is the jth coordinate vector.

Let $T_P^*(U)$ denote the vector space dual of $T_P(U)$. Let the canonical dual basis be denoted by $dx_1, \ldots, dx_N$. Thus $dx_i(\partial/\partial_j) = \delta_{ij}$. We call $T_P^*(U)$ the N-dimensional vector space of *cotangent vectors* (or *covectors*, for short).

The elements of $\bigwedge_* T_P(U)$ are called *multivectors*, and the elements of $\bigwedge_* T_P^*(U) \equiv \bigwedge^* T_P(U)$ are called *multicovectors*.

A *vector field* on U is the continuous assignment of an element of $T_P(U)$ to each $P \in U$. Formally, each $T_P(U)$ is disjoint from any other $T_{P'}(U), P \neq P'$, so the use of the word *continuous* here is not correct. However matters are clarified either by topologizing the tangent bundle or, as we do here, using the fact that the same elements $\{\partial/\partial x_j\}$ serve as a basis for every $T_P(U), P \in U$. Thus a vector field is a function of the form

$$\alpha(x) = \sum_j a_j(x) \frac{\partial}{\partial x_j},$$

where each $a_j \in C(U)$. The vector field is called C^k if each of the coefficient functions a_j lies in $C^k(U)$. A C^k *multivector field* on U is a function of the form

$$\alpha(x) = \sum_I a_I \left(\frac{\partial}{\partial x} \right)_I,$$

where

$$\left(\frac{\partial}{\partial x} \right)_I \equiv \left(\frac{\partial}{\partial x_{i_1}} \right) \wedge \cdots \wedge \left(\frac{\partial}{\partial x_{i_p}} \right), \qquad I = (i_1, \ldots, i_p).$$

Likewise, a C^k differential form (or *multi-covector field*) on U is a function

$$\beta(x) = \sum_I a_I(x) dx^I,$$

where

$$dx^I \equiv dx_{i_1} \wedge \cdots \wedge dx_{i_p}, \qquad I = (i_1, \ldots, i_p).$$

Let $\bigwedge_* T(U) = \cup_{P \in U} \bigwedge_* T_P(U)$ and $\bigwedge^* T(U) = \cup_{P \in U} \bigwedge^* T_P(U)$, and give these sets obvious vector bundle structures over U (for details on the basics of vector bundles, see D. Husemoller [1], E. Spanier [1], and N. Steenrod [1]). (These bundles are, in a natural way, graded unions of sub-bundles $\bigwedge_p T(U)$ and $\bigwedge^p T(U)$.) Then a C^k multivector field is a C^k section of $\bigwedge_* T(U)$ and a C^k differential form is a C^k section of $\bigwedge^* T(U)$. So multivector fields and differential forms are intrinsically defined objects, *independent of coordinates*. Let $\bigwedge_p(U)$ denote the continuous sections of $\bigwedge_p T(U)$; these are called *p-vector* fields. Let $\bigwedge^p(U)$ denote the continuous sections of $\bigwedge^p T(U)$; these are called p-forms. Naturally, we may define

$$\bigwedge_*(U) \equiv \bigcup_p \bigwedge_p(U), \qquad \bigwedge^*(U) \equiv \bigcup_p \bigwedge^p(U).$$

We shall not explore all the natural and obvious relationships among the bundles introduced, but instead refer the reader to the aforementioned works from the bibliography.

The pairing between multivector fields and differential forms (multicovector fields) on $U \subseteq \mathbb{R}^N$ is *not* intrinsically defined. Let dV be Lebesgue measure on U. For $\alpha \in \bigwedge_*(U), \beta \in \bigwedge^*(U)$, we set

$$\langle \alpha, \beta \rangle \equiv \int_U \langle \alpha(x), \beta(x) \rangle \, dV(x).$$

(Clearly a different choice of measure would result in a different inner product.) Now we wish to consider all the preceding notions on a C^1 manifold X. If (ϕ, U) is a coordinate chart on X and $P \in U$, then the objects $C^1(U), \mathcal{C}_p, T_P(U)$, and $T_P^*(U)$ are defined as in the Euclidean case. The algebraic structures $\bigwedge_* T_P(U)$ and $\bigwedge^* T_P(U)$ are then immediately defined. So the bundles $\bigwedge_* T(U)$ and $\bigwedge^* T(U)$ are defined. Now a multivector field is a continuous section of $\bigwedge_* T(U)$ and a differential form is a continuous section of $\bigwedge^* T(U)$. All these ideas extend to the full manifold X through a partition of unity (or simply by the way functions are defined on manifolds).

All our constructs transform in a natural way under smooth mappings. Let $X, \tilde{X}$ be C^1 manifolds and $F : X \to \tilde{X}$ a C^1 diffeomorphism. Then there is an induced map

$$\begin{aligned} \hat{F} : C^1(\tilde{X}) &\to C^1(X) \\ f &\mapsto f \circ F. \end{aligned}$$

If $P \in X$, then there is a map

$$\begin{aligned} F_* : T_P(X) &\to T_{F(P)}(\tilde{X}) \\ \alpha &\mapsto F_*(\alpha), \end{aligned}$$

where $(F_*(\alpha))(\psi) = \alpha(\hat{F}\psi)$ for all $\psi \in C^1(\tilde{X})$. Now there is a map

$$\begin{aligned} F^* : T_{F(P)}^*(\tilde{X}) &\to T_P^*(X) \\ \omega &\mapsto F^*(\omega), \end{aligned}$$

where $(F^*(\omega))(\alpha) = \omega(F_*(\alpha))$ for all $\alpha \in T_P(X)$. The functions F_*, F^* extend naturally to multivectors and to forms.

The preceding paragraph suggests how we might integrate forms and vector fields on a manifold. Let X be a C^1 manifold and let (ϕ, U) be a coordinate patch on X. Let $\alpha \in \bigwedge_*(U)$ and $\beta \in \bigwedge^*(U)$. Define

$$\langle \alpha, \beta \rangle \equiv \langle (\phi_* \alpha, (\phi^{-1})^* \beta \rangle.$$

If $\tilde{\phi}$ is another coordinate map on U, then one easily checks that

$$\langle \alpha, \beta \rangle \equiv \langle (\tilde{\phi}_*) \alpha, (\tilde{\phi}^{-1})^* \beta \rangle.$$

So the definition is independent of the choice of coordinates. Now the pairing may be unambiguously extended to forms defined on all of X through a partition of unity.

As a special case of the preceding ideas, let $U, V \subseteq \mathbb{R}^N$ and $F : U \to V$ a C^1 map. Fix $\omega(t) = w(t) \, dt_1 \wedge \cdots \wedge dt_N \in \bigwedge^N T(V)$. Then

$$(F^* \omega)(x) = (w \circ F)(x) \det \left(\frac{\partial F}{\partial x} \right) dx_1 \wedge \cdots \wedge dx_N.$$

More generally, if $\eta(t) = n(t) \, dt_{i_1} \wedge \cdots \wedge dt_{i_p}$, then

$$(F^* \eta)(x) = \sum_{1 \le j_1 < \cdots < j_p \le N} w \circ F(x) \det \left(\frac{\partial(f_{i_1}, \ldots, f_{i_p})}{\partial(x_{j_1}, \ldots, x_{j_p})} \right) dx_{j_1} \wedge \cdots \wedge dx_{j_p}.$$

Now we define integration of a differential form on a manifold. Begin with the simple manifold $U \subseteq \mathbb{R}^N$ (an open set). Let $\omega = w(x) dx_1 \wedge \cdots \wedge dx_N$ be an N-form on U. We define

$$\int_U \omega = \int_U w(x) dV(x). \tag{$*$}$$

This immediately defines $\int_U \tilde{\omega}$ for any simple N-form $\tilde{\omega}$. For write $\tilde{\omega} = \tilde{w} dx_{i_1} \wedge \cdots \wedge dx_{i_N}$. Let σ be the permutation taking $(1, \ldots, N)$ to $(i_1, \ldots, i_N)$ and let $\epsilon(\sigma)$ be its signature. Then

$$\int_U \tilde{\omega} = \epsilon(\sigma) \int_U \tilde{w}(X) dV(x).$$

Define $\int_U \beta = 0$ for any form $\beta \notin \bigwedge^N(U)$. Extend these definitions to all elements of $\bigwedge^*(U)$ by multilinearity. Notice that line $(*)$, which completely determines what follows, amounts to choosing an orientation on U (for more on this, see de Rham [1]).

Now if X is a C^1 manifold of dimension N, (ϕ, U) is a coordinate chart, and ω is an N-form on U, then define

$$\int_U \omega = \int_{\phi(U)} (\phi^{-1})^* \omega.$$

One checks that this definition is independent of the choice of ϕ. The definition is extended to all of X by means of a partition of unity.

Now we define the exterior derivative. First consider $U \subseteq \mathbb{R}^N$ open. If $\omega = w(x) dx^I$ is a C^1 form on U, then define

$$d\omega = \sum_{j=1}^{N} \frac{\partial w}{\partial x_j} dx_j \wedge dx^I.$$

Extend d by linearity to all forms on U. It is immediate that if ω is a p-form, then $d\omega$ is a $(p+1)$-form. It is also easily calculated that $d(d\omega) = 0$.

If $\omega(x) = w(x) dx^I$ and $\tilde\omega(x) = \tilde w(x) dx^J$ are forms on U with C_c^∞ coefficients and both I and J having indices in increasing order, then we let

$$\langle \omega, \tilde\omega \rangle \equiv \begin{cases} 0 & \text{if } I \neq J \\ \int_U w(x) \tilde w(x) dV(x) & \text{if } I = J. \end{cases}$$

Then δ is defined to be the formal adjoint operator to d with respect to this inner product in the following sense: $\langle d\omega, \tilde\omega \rangle = \langle \omega, \delta\tilde\omega \rangle$ for all $\omega, \tilde\omega \in \bigwedge^*(U)$, $\tilde\omega$ of degree at least 1. The reader may check that if $\omega = w(x) dx_{i_1} \wedge \cdots \wedge dx_{i_p}$ then

$$\delta\omega(x) = \sum_{j=1}^{k} (-1)^{j-1} \frac{\partial w(x)}{\partial x_{i_j}} dx_{i_1} \wedge \cdots \wedge dx_{i_{j-1}} \wedge dx_{i_{j+1}} \wedge \cdots \wedge dx_{i_p}.$$

If X is a C^1 manifold, (ϕ, U) a coordinate patch, and ω a C^1 compactly supported form on U, define $d\omega = \phi^*(d((\phi^{-1})^*\omega))$; if ω is of degree at least 1, set $\delta\omega = \phi^*(\delta((\phi^{-1})^*\omega))$. Extend these notions to arbitrary forms with a partition of unity.

We say that ω is d-closed if $d\omega \equiv 0$. It is d-exact if there is a form γ such that $d\gamma = \omega$.

If α and β are 1-vector field and θ is a 1-covector field, then one may verify that

$$\langle \alpha \wedge \beta, d\theta \rangle = \alpha \langle \beta, \theta \rangle - \beta \langle \alpha, \theta \rangle - \langle [\alpha, \beta], \theta \rangle.$$

This formula, due to Cartan, gives a coordinate-free way to *define* exterior derivative. On the other hand, from our point of view the simplest way to verify the formula is to introduce local coordinates and to verify the formula for vector and covector fields chosen from a convenient basis.

List of Notation

We have made every effort to use standard notation in this book, so much of what follows should be familiar. We provide it for ease of reference. In the last column, an "E" following a number means that the term is introduced in the Exercise section of the chapter denoted by the number. An "A" followed by a roman number indicates that the term is introduced in the corresponding Appendix.

Notation	Meaning	Where Defined
$\mathbb{R}, \mathbb{R}^N$	Real Euclidean space	0.1
$\mathbb{C}, \mathbb{C}^n$	Complex Euclidean space	0.1
$\frac{\partial}{\partial z_j}, \frac{\partial}{\partial \bar{z}_j}$	Complex partial derivatives	0.1
$dz_j, d\bar{z}_j$	Complex differentials	0.1
dV	Volume form	0.1
$B(z^0, r)$	Euclidean ball	0.1
$D^n(z^0, r)$	Polydisc	0.1
D	Unit disc	0.1
$\mathbb{Z}^+$	Nonnegative integers	0.1
$\mathbb{N}$	Positive integers	0.1
$z^\alpha, \bar{z}^\alpha$	Multi-index exponent	0.1
$\left(\frac{\partial}{\partial z}\right)^\alpha, \left(\frac{\partial}{\partial \bar{z}}\right)^\alpha$	Multi-index differentiation	0.1
$\alpha!$	Multi-index factorial	0.1
$|\alpha|$	Magnitude of a multi-index	0.1
$\alpha < \beta$	Multi-index inequality	0.1
$dz^\alpha, d\bar{z}^\beta$	Multi-index differentials	0.1
Ω	Domain (connected open set)	0.1
ρ	Defining function	0.1
$\partial\Omega$	Boundary of Ω	0.1
L^1_{loc}	Locally integrable functions	0.2
$\mathcal{N}f$	Zero set of f	0.3
$A(\Omega)$	The "Ω-algebra"	0.3
C^∞_c	C^∞ functions with compact support	0.3

Notation	Meaning	Where Defined
$\partial \omega$	Exterior holomorphic derivative	1.1
$\bar{\partial} \omega$	Exterior antiholomorphic derivative	1.1
(p, q)	Grading of differential forms	1.1
supp f	Support of f	1.1
$\omega(z)$	Holomorphic volume form	1.1
$\eta(z)$	Leray form	1.1
$f * g$	Convolution	1.1
$\subset\subset$	Relatively compact in	1.1
Δ	Laplacian	1.3
ν	Unit outward normal	1.3
$d\sigma$	Area measure	1.3
ω_{N-1}	Area measure of unit sphere	1.3
Γ_N	Fundamental solution of the Laplacian	1.3
$G(x, y)$	Green's function	1.3
$P(x, y)$	Poisson kernel	1.3
$Pf(x)$	Poisson integral of f	1.3
$A^2(\Omega)$	Bergman space	1.4
$K(z, \zeta)$	Bergman kernel	1.4
$J_{\mathbb{C}} f$	Holomorphic Jacobian matrix	1.4
$J_{\mathbb{R}} f$	Real Jacobian matrix	1.4
$\mathbb{1}$	$(1, 0, \ldots, 0)$	1.4
$z \cdot \bar{\zeta}$	$\sum_j z_j \bar{\zeta}_j$	1.4
$H^2(\partial\Omega)$	Hardy space	1.5
$S(z, \zeta)$	Szegö kernel	1.5
$\mathcal{P}(z, \zeta)$	Poisson-Szegö kernel	1.5
Δ^B	Bergman Laplace-Beltrami operator	1E
$\mathcal{U}$	Siegel upper half-space	1E
$\mathbf{H}_n$	Heisenberg group	1E
u.s.c.	Upper semicontinuous	2.1
l.s.c.	Lower semicontinuous	2.1
psh	Plurisubharmonic	2.2
$s(x)$	Silhouette of x	2.3
$\mathcal{C}$	Set of convergence for a power series	2.3
$\mathcal{B}$	Abel domain for a power series	2.3
$\log \|S\|$	Log modulus of a set	2.3
$T_P(\partial\Omega)$	Tangent space to $\partial\Omega$ at P	3.1
$\hat{K}_{\mathcal{F}}, \hat{K}, \hat{K}_\Omega$	Hull of K with respect to the family $\mathcal{F}$ on Ω	3.1
$\mathbf{d}$	Analytic disc	3.2
$\mathcal{P}(\mathcal{A})$	Peak points for $\mathcal{A}$	3.2
μ	Distance function	3.3

Notation	Meaning	Where Defined
μ_Ω	Distance to boundary	3.3
δ_Ω	Distance to boundary	3.3
T_ω	Tube domain over ω	3.5
$\operatorname{sgn} z$	Sign of z	3.5
$\mathcal{D}_T$	Domain of T	4.2
$\mathcal{G}_T$	Graph of T	4.2
T^*	Adjoint of T	4.2
Σ'	Sum over increasing multi-indices	4.3
D	C_c^∞	4.3
$L^2_{(p,q)}(\Omega)$	(p,q) forms with L^2 coefficients	4.3
ϵ^β_α	Signature of the permutation $\alpha \leftrightarrow \beta$	4.3
$\square$	The $\bar{\partial}$ Laplacian	4.3
$L^2(\Omega,\phi)$	Weighted L^2 space	4.3
$D(p,q)$	(p,q) forms with C_c^∞ coefficients	4.3
$\|f\|_\phi$	Weighted L^2 norm	4.3
$\langle\ ,\ \rangle_\phi$	Weighted inner product	4.4
δ_j	Formal adjoint of $\partial/\partial\bar{z}_j$	4.4
$W^s(\Omega)$	Sobolev space	4.6
$W^s_{\mathrm{loc}}(\Omega)$	Local Sobolev space	4.6
$W^s_{(p,q)}(\Omega)$	(p,q) forms with Sobolev space coefficients	4.6
D'	Distributions	4E
$\hat{f}$	Fourier transform	4E
S	Schwartz space	4E
S'	Schwartz distributions	4E
$p(x,\xi)$	Symbol	4E
T_p	Pseudodifferential operator	4E
$\mathcal{E}_{\omega,\Omega}$	Extension operator	5.1
Ω_δ	Neighborhood of $\bar{\Omega}$	5.2
U_δ	Neighborhood of $\partial\Omega$	5.2
$L_P(z)$	Levi polynomial	5.2
$\Phi(z,P)$	Henkin separating function	5.2
e_z	The point evaluation functional	5.2
$P_i(z,\zeta)$	Coordinate factors of $\Phi(z,\zeta)$	5.3
$P(\Omega)$	Plurisubharmonic functions on Ω	5.4
$\mathcal{L}(L,M)$	Levi form	5E
$g_{j,k}$	Cousin data	6.1
$(\mathcal{F},X,\pi)$	A sheaf	6.2
$\Gamma(U,\mathcal{F})$	Sections of a sheaf	6.2
$\mathcal{F}_x$	Stalk of a sheaf	6.2
C^∞_m	Ring of germs of C^∞ functions	6.2
$\gamma_m f, [f]_m$	Germ	6.2

Notation	Meaning	Where Defined
$\mathcal{U}$	Open covering	6.2
$C^r(\mathcal{U}, \mathcal{F})$	r-cochains	6.2
δ	Coboundary operator	6.2
$Z^r(\mathcal{U}, \mathcal{F})$	r-cocycles	6.2
$B^r(\mathcal{U}, \mathcal{F})$	r-coboundaries	6.2
$H^r(\mathcal{U}, \mathcal{F})$	r-cohomology with respect to the cover $\mathcal{U}$ with coefficients in $\mathcal{F}$	6.2
$H^r(X, \mathcal{F})$	r-cohomology of X with coefficients in $\mathcal{F}$	6.2
$C_0^p(\mathcal{U}, \mathcal{F})$	Acyclic cochains	6.2
$Z_0^p(\mathcal{U}, \mathcal{F})$	Acyclic cocycles	6.2
$B_0^p(\mathcal{U}, \mathcal{F})$	Acyclic coboundaries	6.2
$H_0^p(\mathcal{U}, \mathcal{F})$	Acyclic cohomology	6.2
δ^*	Induced coboundary operator	6.2
$\mathcal{O}$	Sheaf of germs of holomorphic functions	6.3
$H_D^p(\Omega)$	Dolbeault cohomology	6.3
$H_{DR}^p(\Omega)$	De Rham cohomology	6.3
$H_{DR\mathcal{H}}^p(\Omega)$	Holomorphic de Rham cohomology	6.3
$\mathcal{M}, \mathcal{M}^*$	Sheaf of germs of meromorphic functions	6.5
$\mathcal{O}^*$	Invertible elements of $\mathcal{O}$	6.5
$\mathbb{CP}^n$	Complex projective space	6E
$R(f_1, \ldots, f_k)$	Sheaf of relations	7.1
$\mathcal{I}(M)$	Ideal sheaf of M	7.2
$H^p(D), H^\infty(D)$	Hardy space	8.1
$\mathbf{h}^p(D), \mathbf{h}^\infty(D)$	p-class of harmonic functions	8.1
Mf	Hardy-Littlewood maximal function	8.1
$\Gamma_\alpha(e^{i\theta})$	Stolz angle	8.1
$B_a(z)$	Blaschke factor	8.1
$N(D)$	Nevanlinna class	8.1
$B(c_y, r), \tilde{B}(\tilde{c}_y, \tilde{r})$	Osculating balls	8.2
$\mathbf{h}^p(\Omega), \mathbf{h}^\infty(\Omega)$	p-class of harmonic functions	8.3
$H^p(\Omega), H^\infty(\Omega)$	Hardy class	8.3
$\Gamma_\alpha(P)$	Nontangential approach region	8.4
$M_1 f$	Hardy-Littlewood maximal function	8.4, 8.6
$f_1^{*,\alpha}(P)$	Nontangential maximal function	8.5
$\beta_1(\zeta, r), \beta_2(\zeta, r)$	Balls in ∂B	8.6
$\mathcal{A}_\alpha(P)$	Admissible approach region	8.6
$M_2 f$	Nonisotropic maximal function	8.6

Notation	Meaning	Where Defined
$f_2^{*,\alpha}(P)$	Admissible maximal function	8.6
$\mathcal{T}_P(\partial\Omega)$	Holomorphic tangent space	8.6
$\mathcal{N}_P(\partial\Omega)$	Complex normal space	8.6
J	Complex structure map	8.6
$\Lambda_\alpha, \mathrm{Lip}_1$	Lipschitz spaces	8.8
$\mathcal{C}_1^k$	Complex tangential curves	8.8
$\Gamma_{\alpha,\beta}$	Nonisotropic Lipschitz space	8.8
BMO	Functions of bounded mean oscillation	8E
$\mathcal{O}\cdot\mathcal{B}\cdot\mathcal{S}$	Canonical factorization	9.1
$H_\Omega(f)$	Henkin solution to the $\bar\partial$ problem	10.1
$\mathrm{Aut}(\Omega)$	Biholomorphic self-maps of Ω	11.1
I_P	Isotropy subgroup at P	11.1
$g_{ij}(x)$	Riemannian metric	11.2
$U(V)$	Holomorphic maps from V to U	11.2
$F_C(z,\xi)$	Infinitesimal form of the Carathéodory metric	11.2
$C(z,w)$	Carathéodory distance	11.2
$L_C(\gamma)$	Carathéodory length	11.2
$F_K(z,\xi)$	Infinitesimal form of the Kobayashi/Royden metric	11.2
$K(z,w)$	Kobayashi/Royden distance	11.2
$L_K(\gamma)$	Kobayashi/Royden length	11.2
M_Ω^C	Carathéodory "volume element"	11.3
M_Ω^K	Kobayashi "volume element"	11.3
W_0^j, WH^j	Sobolev spaces	11.4
$L, \bar{L}$	Complex tangent vectors	11.5
$[L, M]$	Commutator of vector fields	11.5
$\mathcal{L}_j$	Stratification of tangent space	11.5
$\Delta(M, P)$	Type of a point	11.5
Δ_1^{reg}	Regular type of a point	11.5
$B_1(M, P)$	Algebraic type of a point	11.5
$F_a(z)$	Complex Henon map	11.6
(ϕ, U)	Coordinate chart	AI
$\mathcal{H}^k$	Hausdorff measure	AII
$V_1 \otimes V_2$	Tensor product	AIII
$\otimes_p V, \otimes_* V$	Tensor product	AIII
$V_1 \wedge V_2$	Wedge product	AIII
$\bigwedge_p V, \bigwedge_* V$	Wedge product	AIII
$\lfloor,\rfloor$	Contraction	AIII
$\bigwedge^p V, \bigwedge^* V$	Dual exterior algebra	AIII
$\frac{\partial}{\partial x_i}, dx_i$	Differentials	AIV

Notation	Meaning	Where Defined
$\bigwedge_* T(U), \bigwedge^* T(U)$	Multivector and multi-covector bundles	AIV
$\bigwedge_p(U), \bigwedge^p(U)$	Multivector fields and multicovector fields	AIV
$\bigwedge_*(U), \bigwedge^*(U)$	Exterior algebras over U	AIV
F^*, F_*	Pull back, push forward	AIV
δ, d	Exterior derivatives	AIV

Bibliography

R. Adams

   1. *Sobolev Spaces*, Academic Press, 1975.

L. Ahlfors

   1. *Complex Analysis*, 3rd ed., McGraw-Hill, New York, 1979.

G. Aladro

   1. The comparability of the Kobayashi approach region and the admissible approach region, *Illinois Jour. Math.* 33 (1989), 42–63.

A. B. Aleksandrov

   1. The existence of inner functions in the ball, *Math. USSR Sbornik* 46 (1983), 143–159.

   2. Inner functions on compact spaces, *Functional Analysis and its Applications* 18 (1984), 87–98.

H. Alexander

   1. Volume images of varieties in projective space and in Grassmannians, *Trans. Am. Math. Soc.* 189 (1974), 237–249.

   2. Holomorphic mappings from the ball and polydisc, *Math. Ann.* 209 (1974), 249–256.

   3. Extremal holomorphic imbeddings between the ball and polydisc, *Proc. Am. Math. Soc.* 68 (1978), 200–202.

A. Andreotti and H. Grauert

1. Theorèmes de finitude pour la cohomologie des espaces complexes, *Bull. Soc. Math. France* 90 (1962), 193–259.

A. Andreotti and E. Vesentini

1. Carleman estimates for the Laplace-Beltrami equations on complex manifolds, *Pub. Math., Inst. Hautes Etudes Scientifiques* 25 (1965), 81–130.

N. Aronszajn

1. Theory of reproducing kernels, *Trans. Am. Math. Soc.* 68 (1950), 337–404.

J. M. Ash

1. Multiple trigonometric series, in *Studies in Harmonic Analysis*, J. M. Ash, ed., The Mathematical Association of America, Washington, D. C., 1976.

M. Atiyah and I. MacDonald

1. *Introduction to Commutative Algebra*, Addison-Wesley, Reading, 1969.

F. Bagemihl and W. Seidel

1. Some boundary properties of analytic functions, *Math. Zeitschr.* 61 (1954), 186–199.

M. S. Baouendi and L. P. Rothschild

1. CR mappings and their holomorphic extension, *Journées "Équations aux derivées partielles"* (Saint Jean de Monts, 1987), *Exp. No.* XXIII, 6 pp. *École Polytechn.*, Palaiseau, 1987.

2. Normal forms for generic manifolds and holomorphic extension of CR functions, *J. Differential Geom.* 25 (1987), 431–467.

S. R. Barker

1. Two theorems on boundary values of analytic functions, *Proc. A. M. S.* 68 (1978), 48–54.

D. Barrett

1. Irregularity of the Bergman projection on a smooth bounded domain in $\mathbb{C}^2$, *Annals of Math.* 119 (1984), 431–436.

2. The behavior of the Bergman projection on the Diederich-Fornaess worm, *Acta Math.*, in press.

R. Basener

1. Peak points, barriers, and pseudoconvex boundary points, *Proc. Am. Math. Soc.* 65 (1977), 89–92.

F. Beatrous

1. L^p estimates for extensions of holomorphic functions, *Mich. Jour. Math.* 32 (1985), 361–380.

E. Bedford

1. The Dirichlet problem for some overdetermined systems on the unit ball in $\mathbb{C}^n$, *Pac. Jour. Math.* 51 (1974), 19–25.

2. A short proof of the classical edge of the wedge theorem, *Proc. Am. Math. Soc.* 43 (1974), 485–486.

3. Holomorphic continuation at a totally real edge, *Math. Ann.* 230 (1977), 213–225.

4. Proper holomorphic maps, *Bull. A.M.S.* 10 (1984), 157–175.

E. Bedford and P. Federbush

1. Pluriharmonic boundary values, *Tohoku Math. Jour.* 26 (1974), 505–511.

E. Bedford and J. E. Fornæss

1. A construction of peak functions on weakly pseudoconvex domains, *Ann. Math.* 107 (1978), 555–568.

2. Biholomorphic maps of weakly pseudoconvex domains, *Duke Math. Jour.* 45 (1978), 711–719.

3. Approximation on pseudoconvex domains, *Complex Approximation (Proc. Conf. Quebec, 1978)*, pp. 18–31, *Progr. Math.* 4, Birkhäuser, Boston, 1980.

E. Bedford and S. Pinchuk

1. Domains in $\mathbb{C}^2$ with noncompact group of automorphisms, *Math. Sb. Nov. Ser.* 135 (1988), No. 2, 147–157.

2. Domains in $\mathbb{C}^{n+1}$, with noncompact automorphism group, *Journal of Geometric Analysis*, in press.

3. Convex domains with noncompact group of automorphisms, *Mat. Sb.* 185 (1994), 3–26 (Russian).

E. Bedford and B. A. Taylor

1. The Dirichlet problem for a complex Monge-Ampère equation, *Invent. Math.* 37 (1976), 1–44.

2. Variational properties of the complex Monge-Ampère equation. I. Dirichlet principle, *Duke Math. J.* 45 (1978), 375–403.

3. Variational properties of the complex Monge-Ampère equation. II. Intrinsic norms, *Amer. J. Math.* 101 (1979), 1131–1166.

4. A new capacity for plurisubharmonic functions, *Acta Math.* 149 (1982), 1–40.

S. Bell

1. Analytic hypoellipticity of the $\bar{\partial}$-Neumann problem and extendability of holomorphic mappings, *Acta Math.* 147 (1981), 109–116.
2. Biholomorphic mappings and the $\bar{\partial}$ problem, *Ann. Math.*, 114 (1981), 103–113.

S. Bell and H. Boas

1. Regularity of the Bergman projection in weakly pseudoconvex domains, *Math. Annalen* 257 (1981), 23–30.

S. Bell and S. G. Krantz

1. Smoothness to the boundary of conformal maps, *Rocky Mt. Jour. Math.* 17 (1987), 23–40.

S. Bell and E. Ligocka

1. A simplification and extension of Fefferman's theorem on biholomorphic mappings, *Invent. Math.* 57 (1980), 283–289.

J. Bergh and J. Löfström

1. *Interpolation Spaces: An Introduction*, Springer, Berlin, 1976.

S. Bergman

1. *The Kernel Function and Conformal Mapping*, Am. Math. Soc., Providence, R.I., 1970.

S. Bergman and M. Schiffer

1. *Kernel Functions and Elliptic Differential Equations in Mathematical Physics*, Academic Press, New York, 1953.

C. Bernstein, D. C. Chang, and S. G. Krantz

1. *Analysis on the Heisenberg Group*, unpublished manuscript.

L. Bers

1. *Introduction to Several Complex Variables*, New York Univ. Press, New York, 1964.

L. Bers, F. John, and M. Schechter

1. *Partial Differential Equations*, Interscience, New York, 1964.

O. Besov, V. Il'in, and S. Nikol'skii

1. *Integral Representation of Functions and Imbedding Theorems*, Vols. I, II, Wiley, New York, 1978.

L. Bieberbach

1. Beispiel zweier ganzer Funktionen zweier komplexer Variablen, welche eine schlicht volumetreue Abbildung des $\mathbb{R}_4$ auf einen Teil seiner selbst vermitteln, *Preuss. Akad. Wiss. Sitzungsber*, (1933), 476–479.

E. Bishop

1. A general Rudin-Carleson theorem, *Proc. Am. Math. Soc.* 13 (1962), 140–143.

2. Mappings of partially analytic spaces, *Am. J. Math.* 83 (1961), 209–242.

B. Blank, Fan Dashan, David Klein, S. G. Krantz, Daowei Ma, and M. Y. Pang

1. The Kobayashi metric of a complex ellipsoid in $\mathbb{C}^2$, *Experimental Math.* 1 (1992), 47–55.

T. Bloom

1. C^∞ peak functions for pseudoconvex domains of strict type, *Duke Math. J.* 45 (1978), 133–147.

T. Bloom and I. Graham

1. A geometric characterization of points of type m on real submanifolds of $\mathbb{C}^n$, *J. Diff. Geom.* 12 (1977), 171–182.

H. Boas

1. A geometric characterization of the ball and the Bochner-Martinelli kernel, *Math. Ann.* 248 (1980), 275–278.

2. Counterexample to the Lu Qi-Keng conjecture, *Proc. Am. Math. Soc.* 97 (1986), 374–375.

S. Bochner

1. Orthogonal systems of analytic functions, *Math. Z.* 14 (1922), 180–207.

S. Bochner and W. Martin

1. *Several Complex Variables*, Princeton Univ. Press, Princeton, N.J., 1948.

A. Boggess

1. *CR Functions and the Tangential Cauchy-Riemann Complex*, CRC Press, Boca Raton, Fla., 1991.

N. N. Bogoliubov and D. V. Shirkov

1. *Introduction to the Theory of Quantized Fields*, GITTL, Moscow 1957; English translation, Interscience, New York, 1959.

N. N. Bogoliubov and V. Vladimirov

1. On some mathematical problems of quantum field theory, *Proc. Internat. Congress Math.* (Edinburgh, 1958), Cambridge University Press, New York, 1960, 19–32.

P. Bonneau

1. La reconstruction des fonctions holomorphes dans des domaines de $\mathbb{C}^n$, *Indag. Math.* 5 (1994) 381–395.

P. Bonneau and A. Cumenge

1. Approximation d'operateurs d'extension minimale et de division, *Math. Zeit.* 207 (1991), 37–63.

P. Bonneau and K. Diederich

1. Integral solution operators for the Cauchy-Riemann equations on pseudoconvex domains, *Math. Annalen* 286 (1990), 77–100.

T. Bonneson and W. Fenchel

1. *Theorie der konvexen Körper*, Springer Verlag, Berlin, 1934.

L. Boutet de Monvel

1. On the index of Toeplitz operators of several complex variables, *Invent. Math.* 50 (1978/9), 249–272.

L. Boutet de Monvel and J. Sjöstrand

1. Sur la singularité des noyaux de Bergman et Szegö, *Soc. Mat. de France Asterisque* 34-35 (1976), 123–164.

R. Braun, W. Karp, and H. Upmeier

1. On the automorphisms of circular and Reinhardt domains in complex Banach spaces, *Manuscripta Math.* 25 (1978), 97–133.

H. Bremerman

1. Über die Äquivalenz der pseudoconvex Gebiete und der Holomorphie-gebiete in Raum von n Komplexen Veränderlichen, *Math. Ann.* 128 (1954), 63–91.

2. On the conjecture of the equivalence of the plurisubharmonic functions and the Hartogs functions, *Math. Ann.* 131 (1956), 76–86.

3. On a generalized Dirichlet problem for plurisubharmonic functions and pseudoconvex domains. Characterization of Šilov boundaries, *Trans. Am. Math. Soc.* 91 (1959), 246–276.

L. Bungart

1. Holomorphic functions with values in locally convex spaces and applications to integral formulas, *Trans. Am. Math. Soc.* 111 (1964), 317–344.

J. Burbea

1. The Carathéodory metric and its majorant metrics, *Can. Jour. Math.* 29 (1977), 771–780.

D. Burkholder, R. Gundy, and M. Silverstein

1. A maximal function characterization of the class H^p, *Trans. Am. Math. Soc.* 157 (1971), 137–153.

D. Burns and S. Krantz

1. Rigidity of holomorphic mappings and a new Schwarz lemma at the boundary, *J. Amer. Math. Soc.* 7 (1994), 661–676.

D. Burns, S. Shnider, and R. Wells

1. On deformations of strictly pseudoconvex domains, *Invent. Math.* 46 (1978), 237–253.

P. Butzer and H. Berens

1. *Semi-groups of Operators and Approximation Theory*, Springer, Berlin, 1967.

A. P. Calderón

1. On the behavior of harmonic functions near the boundary, *Trans. Am. Math. Soc.* 68 (1950), 47–54.

S. Campanato

1. Proprieta di una famiglia di spazi funzionali, *Ann. Scuola Norm. Sup. Pisa* 18 (1964), 137–160.

L. Carleson

1. Interpolation by bounded analytic functions and the corona problem, *Ann. Math.* 76 (1962), 542-559.

2. On the support of harmonic measure for sets of Cantor type, *Ann. Acad. Sci. Fenn.*, 10 (1985), 113–123.

G. Carrier, M. Crook, and C. Pearson

1. *Functions of a Complex Variable*, McGraw-Hill, New York, 1966.

D. Catlin

1. Global regularity for the $\bar{\partial}$-Neumann problem, *Proc. Symp. Pure Math.* v. 41 (Y. T. Siu ed.), Am. Math. Soc., Providence, 1984.
2. Necessary conditions for subellipticity of the $\bar{\partial}$-Neumann problem, *Ann. Math.* 117 (1983), 147–172.
3. Subelliptic estimates for the $\bar{\partial}$-Neumann problem, *Ann. Math.* 126 (1987), 131–192.
4. Estimates of invariant metrics on pseudoconvex domains of dimension two, *Math. Z.* 200 (1989), 429–466.

U. Cegrell

1. *Capacities in Complex Analysis*, Vieweg, Braunschweig, 1988.

D. C. Chang and S. G. Krantz

1. Holomorphic Lipschitz functions and applications to the $\bar{\partial}$-problem, *Colloquium Mathematicum*, 62 (1991), 227–256.

D. C. Chang, A. Nagel, and E. M. Stein

1. Estimates for the $\bar{\partial}$-Neumann problem in pseudoconvex domains of finite type in $\mathbb{C}^2$, *Acta Math.*, 169 (1992), 153–228.

S. Y. Cheng and S. T. Yau

1. On the existence of a complete Kähler metric on noncompact complex manifolds and the regularity of Fefferman's equation, *Comm. Pure and Appl. Math.* 33 (1980), 507–544.

S. S. Chern and J. Moser

1. Real hypersurfaces in complex manifolds, *Acta Math.* 133 (1974), 219–271.

C. Chevalley

1. *Theory of Lie Groups*, Princeton University Press, Princeton, N.J., 1946.

E. Chirka

1. The theorems of Lindelöf and Fatou in $\mathbb{C}^n$, *Mat. Sb.* 92 (134) (1973), 622–644; *Math. U.S.S.R. Sb.* 21 (1973), 619–639.

F. M. Christ

1. Regularity properties of the $\bar{\partial}_b$-equation on weakly pseudoconvex CR manifolds of dimension 3, *Jour. A. M. S.* 1(1988)
2. On the $\bar{\partial}_b$ equaton and Szegö projection on CR manifolds, *Harmonic Analysis and PDE's* (El Escorial, 1987), Springer Lecture Notes vol. 1384, Springer Verlag, Berlin, 1989.

3. Embedding compact three-dimensional CR manifolds of finite type in $\mathbb{C}^n$, *Ann. of Math.* 129 (1989), 195–213.

4. Estimates for fundamental solutions of second-order subelliptic differential operators, *Proc. A.M.S.* 105 (1989), 166–172.

5. Global C^∞ irregularity of the $\bar{\partial}$-Neumann problem for worm domains, *Jour. of A.M.S.* 9 (1996), 1171–1185.

J. Cima and S. G. Krantz

1. A Lindelöf principle and normal functions in several complex variables, *Duke Math. Jour.* 50 (1983), 303–328.

J. Cima, S. G. Krantz, and T. Suffridge

1. A reflection principle for proper holomorphic mappings of strongly pseudoconvex domains and applications, *Math. Z* 186 (1984), 1–8.

J. Cima and T. Suffridge

1. A reflection principle with applications to proper holomorphic mappings, *Math. Ann.* 265 (1983), 489–500.

A. Cordoba and R. Fefferman

1. On differentiation of integrals, *Proc. Nat. Acad. Sci. U.S.A.* 74 (1977), 2211–2213.

R. Courant and D. Hilbert

1. *Methods of Mathematical Physics*, 2nd ed., Interscience, New York, 1966.

J. Dadok and Paul Yang

1. Automorphisms of tube domains and spherical hypersurfaces, *Am. Jour. Math.* 107 (1985), 999–1013.

H. G. Dales

1. Boundaries and peak points for Banach function algebras, *Proc. London Math. Soc.* 22 (1971), 121–136.

J. P. D'Angelo

1. Real hypersurfaces, orders of contact, and applications, *Annals of Math.* 115 (1982), 615–637.

2. Intersection theory and the $\bar{\partial}$-Neumann problem, *Proc. Symp. Pure Math.* 41 (1984), 51–58.

3. Finite type conditions for real hypersurfaces in $\mathbb{C}^n$, in *Complex Analysis Seminar*, Springer Lecture Notes vol. 1268, Springer Verlag, 1987, 83–102.

4. Iterated commutators and derivatives of the Levi form, in *Complex Analysis Seminar*, Springer Lecture Notes vol. 1268, Springer Verlag, 1987, 103–110.

5. Proper polynomial mappings between balls, *Duke Math. Jour.* 57 (1988), 211–219.

6. *Several Complex Variables and Geometry*, CRC Press, Boca Raton, Fla., 1992.

A. M. Davie and N. Jewell

1. Toeplitz operators in several complex variables, *J. Funct. Anal.* 26 (1977), 356–368.

K. M. Davis and Yang-Chun Chang

1. *Lectures on Bochner-Riesz Means*, Cambridge University Press, Cambridge, England, 1987.

J. P. Demailly

1. Un exemple de fibré holomorphe non de Stein à fibre $\mathbb{C}^2$ ayant pour base le disque ou le plan, *Inventiones Math.* 48 (1978), 293–302.

G. de Rham

1. *Varieties Differentiables*, 3rd ed., Hermann, Paris, 1973.

K. Diederich and J. E. Fornæss

1. Pseudoconvex domains: Bounded strictly plurisubharmonic exhaustion functions, *Invent. Math.* 39 (1977), 129–141.

2. Pseudoconvex domains: An example with nontrivial Nebenhülle, *Math. Ann.* 225 (1977), 275–292.

3. Comparison of the Bergman and Kobayashi metric, *Math. Ann.* 254 (1980), 257–262.

4. Pseudoconvex domains with real-analytic boundary, *Annals of Math.* 107 (1978), 371–384.

5. Proper holomorphic maps onto pseudoconvex domains with real-analytic boundary, *Annals of Math.* 110 (1979), 575–592.

K. Diederich, J. E. Fornæss, and J. Wiegerinck

1. Sharp Hölder estimates for $\bar{\partial}$ on ellipsoids, *Manuscripta Math.* 56 (1986), 399–417.

P. G. Dixon and J. Esterle

1. Michael's problem and the Poincaré-Fatou-Bieberbach phenomenon, Bull. A. M. S. 15 (1986), 127–187.

R. Douglas

1. *Banach Algebra Techniques in Operator Theory*, Academic Press, New York, 1972.

P. Duren

1. *The Theory of H^p Spaces*, Academic Press, New York, 1970.

L. Ehrenpreis

1. A new proof and an extension of Hartogs's theorem, *Bull. Am. Math. Soc.* 67 (1961), 507–509.

2. *Fourier Analysis in Several Complex Variables*, Interscience, New York, 1970.

D. Eisenman

1. *Intrinsic Measures on Complex Manifolds and Holomorphic Mappings*, a Memoir of the American Mathematical Society, Providence, R.I., 1970.

B. Epstein

1. *Orthogonal Families of Functions*, Macmillan, New York, 1965.

J. J. Faran

1. Maps from the two-ball to the three-ball, *Invent. Math.* 68 (1982), 441–475.

2. The linearity of proper holomorphic maps in the low codimension case, *J. Diff. Geom.* 24 (1986), 15–17.

H. Farkas and I. Kra

1. *Riemann Surfaces*, Springer, Berlin, 1979.

P. Fatou

1. Sur les équations functionelles, *Bull. Soc. Math. France* 47 (1919), 161–271.

2. Sur le fonctions meromorphes de deux variables, *C. R. Acad. Sciences Paris* 175 (1922), 862–865.

3. Sur certaines fonctions uniformes de deux variables, *C. R. Acad. Sciences Paris* 175 (1922), 1030–1033.

H. Federer

1. *Geometric Measure Theory*, Springer, Berlin, 1969.
2. Curvature measures, *Trans. Am. Math. Soc.* 93 (1959), 418–491.

C. Fefferman

1. The Bergman kernel and biholomorphic mappings of pseudoconvex domains, *Invent. Math.* 26 (1974), 1–65.
2. Parabolic invariant theory in complex analysis, *Adv. Math.* 31 (1979), 131–262.

C. Fefferman and J. J. Kohn

1. Hölder estimates on domains of complex dimension two and on three dimensional CR-manifolds, *Adv. in Math.* 69 (1988), 223–303.
2. Estimates of kernels on three dimensional CR manifolds, *Rev. Mat. Iboamericana*, 4 (1988), 355–405.

C. Fefferman and E. M. Stein

1. H^p spaces of several variables, *Acta Math.* 129 (1972), 137–193.

W. Fenchel

1. Convexity through the ages, in *Convexity and its Applications*, Birkhäuser, Basel, Switzerland, 1983, 120–130.

P. C. Fenton

1. Sufficient conditions for harmonicity, *Trans. Am. Math. Soc.* 253 (1979), 139–147.

M. Field

1. *Several Complex Variables and Complex Manifolds*, Cambridge University Press, Cambridge, England, 1982.

S. Fisher

1. *Function Theory on Planar Domains: A Second Course in Complex Analysis*, Wiley, New York, 1983.

G. B. Folland

1. Spherical harmonic expansion of the Poisson-Szegö kernel for the ball, *Proc. Am. Math. Soc.* 47 (1975), 401–408.

G. B. Folland and J. J. Kohn

1. *The Neumann Problem for the Cauchy-Riemann Complex*, Princeton University Press, Princeton, N.J., 1972.

G. B. Folland and E. M. Stein

1. Estimates for the $\bar{\partial}_b$ complex and analysis on the Heisenberg group, *Comm. Pure Appl. Math.* 27 (1974), 429–522.

F. Forelli

1. Pluriharmonicity in terms of harmonic slices, *Math. Scand.* 41 (1977), 358–364.

J. E. Fornæss

1. Strictly pseudoconvex domains in convex domains, *Am. J. Math.* 98 (1976), 529–569.

2. Peak points on weakly pseudoconvex domains, *Math. Ann.* 227 (1977), 173–175.

3. Biholomorphic mappings between weakly pseudoconvex domains, *Pac. J. Math.* 74 (1978), 63–66.

4. Sup Norm Estimates for $\bar{\partial}$ in $\mathbb{C}^2$, *Annals of Math.* 123 (1986), 335–346.

J. E. Fornæss and S. G. Krantz

1. Continuously varying peaking functions, *Pac. J. Math.* 83 (1979), 341–347.

J. E. Fornæss and J. McNeal

1. Personal communication.

J. E. Fornæss and A. Nagel

1. The Mergelyan property for weakly pseudoconvex domains, *Manuscr. Math.* 22 (1977), 199–208.

J. E. Fornæss and N. Øvrelid

1. Finitely generated ideals in $A(\Omega)$, *Ann. Inst. Fourier (Grenoble)* 33 (1983), 77–85.

J. E. Fornæss and N. Sibony

1. L^p estimates for $\bar{\partial}$, *Proc. Symp. Pure Math.*, vol. 52 (E. Bedford, J. D'Angelo, R. Greene, and S. Krantz eds.), American Mathematical Society, Providence, R.I., 1991.

2. Construction of plurisubharmonic functions on weakly pseudoconvex domains, *Duke Math. J.* 58 (1989), 633–655.

F. Forstneric

1. On the boundary regularity of proper mappings, *Ann. Scuola Norm. Sup. Pisa Cl Sci.* 13 (1986), 109–128.

S. Frankel

1. Complex geometry of convex domains that cover varieties, *Acta Math.* 163 (1989), 109–149.

M. Freeman

1. Local complex foliation of real submanifolds, *Math. Ann.* 209 (1974), 1–30.

M. Freeman and R. Harvey

1. A compact set which is locally holomorphically convex but not holomorphically convex, *Pac. J. Math.* 48 (1973), 77–81.

B. Fridman

1. Biholomorphic invariants of a hyperbolic manifold and some applications, *Trans. Am. Math. Soc.* 276 (1983), 685–698.

2. One example of the boundary behavior of biholomorphic transformations, *Proc. Am. Math. Soc.* 89 (1983), 226–228.

3. A universal exhausting domain, *Proc. Am. Math. Soc.* 98 (1986), 267–270.

4. Biholomorphic transformations that do not extend biholomorphically to the boundary, *Michigan Math. J.* 38 (1992).

B. A. Fuks

1. *Introduction to the Theory of Analytic Functions of Several Complex Variables* Translations of Mathematical Monographs, American Mathematical Society, Providence, R.I., 1963.

T. W. Gamelin

1. *Uniform Algebras*, Prentice Hall, Englewood Cliffs, N.J., 1969.

2. *Uniform Algebras and Jensen measures*, Cambridge University Press, Cambridge, England, 1978.

P. Garabedian

1. *Partial Differential Equations*, Wiley, New York, 1964.

J. Garnett

1. *Bounded Analytic Functions*, Academic Press, New York, 1981.

J. Garnett and Peter W. Jones

   1. The Corona theorem for Denjoy domains, *Acta. Math.* 155 (1985), 27–40.

J. Garnett and R. Latter

   1. The atomic decomposition for Hardy spaces in several complex variables, *Duke Math. J.* 45 (1978), 815–845.

E. Gavosto

   1. Thesis, Washington Univ., 1990.

B. Gelbaum and J. Olmsted

   1. *Counterexamples in Analysis*, Holden-Day, San Francisco, 1964.

S. Gilbarg and N. Trudinger

   1. *Elliptic Partial Differential Equations of Second Order*, Springer, Berlin, 1977.

A. Gleason

   1. Finitely generated ideals in Banach algebras, *J. Math. Mech.* 13 (1964), 125–132.

   2. The abstract theorem of Cauchy-Weil, *Pac. J. Math.* 12 (1962), 511–525.

G. M. Goluzin

   1. *Geometric Theory of Functions of a Complex Variable*, American Mathematical Society, Providence, R.I., 1969.

Gong Sheng

   1. *Singular Integrals in Several Complex Variables*, Oxford University Press, Oxford, England, 1991.

   2. *Harmonic Analysis on Classical Groups*, Springer Verlag, 1991.

C. R. Graham

   1. The Dirichlet problem for the Bergman Laplacian I, *Comm. Partial Diff. Eqs.* 8 (1983), 433–476; II, ibid. 8 (1983), 563–641.

I. Graham

   1. Boundary behavior of the Carathéodory and Kobayashi metrics on strongly pseudoconvex domains in $\mathbb{C}^n$ with smooth boundary, *Trans. Am. Math. Soc.* 207 (1975), 219–240.

   2. The radial derivative, fractional integrals, and comparative growth of means of holomorphic functions on the unit ball in $\mathbb{C}^n$, *Annals of Math. Studies*, vol. 100, J. Fornæss, ed., Princeton University Press, Princeton, N.J., 1981.

H. Grauert

1. Approximationssätze für holomorphe Funktionen mit Werten in komplexen Räumen, *Math. Ann.* 133 (1957), 139–159.

2. On Levi's problem and the imbedding of real-analytic manifolds, *Ann. Math.* 68 (2) (1958), 460–472.

H. Grauert and K. Fritzsche

1. *Several Complex Variables*, Springer, Berlin, 1976.

H. Grauert and I. Lieb

1. Das Ramirezsche Integral und die Gleichung $\bar{\partial}u = \alpha$ im Bereich der beschränkten Formen, *Rice University Studies* 56 (1970), 29–50.

M. Green

1. Holomorphic maps into complex projective space omitting hyperplanes, *Trans. A. M. S.* 169 (1972), 89–103.

R. E. Greene and S. G. Krantz

1. Stability properties of the Bergman kernel and curvature properties of bounded domains, *Recent Progress in Several Complex Variables*, Princeton University Press, Princeton, N.J., 1982.

2. Deformation of complex structures, estimates for the $\bar{\partial}$ equation, and stability of the Bergman kernel, *Adv. Math.* 43 (1982), 1–86.

3. The automorphism groups of strongly pseudoconvex domains, *Math. Annalen* 261 (1982), 425–446.

4. The stability of the Bergman kernel and the geometry of the Bergman metric, *Bull. Am. Math. Soc.* 4 (1981), 111–115.

5. Stability of the Carathéodory and Kobayashi metrics and applications to biholomorphic mappings, *Proc. Symp. in Pure Math.*, 41 (1984), 77–93.

6. Normal families and the semicontinuity of isometry and automorphism groups, *Math. Zeitschrift* 190 (1985), 455–467.

7. Characterizations of certain weakly pseudo-convex domains with noncompact automorphism groups, in *Complex Analysis Seminar* , Springer Lecture Notes 1268 (1987), 121–157.

8. Characterization of complex manifolds by the isotropy subgroups of their automorphism groups, *Indiana Univ. Math. J.* 34 (1985), 865–879.

9. Biholomorphic self-maps of domains, *Complex Analysis II* (C. Berenstein, ed.), Springer Lecture Notes, vol. 1276, 1987, 136–207.

10. Techniques for Studying the Automorphism Groups of Weakly Pseudoconvex Domains, Proceedings of the Special Year at the Mittag-Leffler Institute (J. E. Fornæss and C. O. Kiselman, eds.) *Math. Notes* 38, Princeton Univ. Press, Princeton, N.J., 1993, 389–410.

11. Invariants of Bergman geometry and results concerning the automorphism groups of domains in $\mathbb{C}^n$, Proceedings of the 1989 Conference in Cetraro (D. Struppa, ed.), Mediterranean Press, 1989, 107–136

12. *Function Theory of One Complex Variable*, Wiley, New York, 1997.

P. Greiner

1. Subelliptic estimates for the $\bar{\partial}$-Neumann problem in $\mathbb{C}^2$, *J. Diff. Geom.* 9 (1974), 239–250.

P. Greiner, J. J. Kohn, and E. M. Stein

1. Necessary and sufficient conditions for solvability of the Lewy equation, *Proc. Nat. Acad. Sci.* 72 (1975), 3287–3289.

P. Greiner and E. M. Stein

1. *Estimates for the $\bar{\partial}$-Neumann Problem*, Princeton University Press, Princeton, N.J., 1977.

P. Griffiths and J. Harris

1. *Principles of Algebraic Geometry*, Wiley, New York, 1978.

L. Gruman and P. Lelong

1. *Entire Functions of Several Complex Variables*, Springer Verlag, Berlin, 1986.

P. Guan

1. Hölder regularity of subelliptic pseudodifferential operators, *Duke Math. Jour.* 60 (1990), 563–598.

V. Guilleman, M. Kashiwara, and T. Kawai

1. *Seminar on Microlocal Analysis*, Princeton University Press, Princeton, N.J., 1979.

R. C. Gunning

1. *Riemann Surfaces*, Princeton Lecture Notes, Princeton, N.J., 1966.
2. *Lectures on Complex Analytic Varieties: The Local Parametrization Theorem*, Princeton University Press, Princeton, N.J., 1970.
3. *Introduction to Holomorphic Functions of Several Variables*, 3 vols., Wadsworth & Brooks/Cole, Pacific Grove, Calif., 1990.

R. C. Gunning and H. Rossi

1. *Analytic Functions of Several Complex Variables*, Prentice Hall, Englewood Cliffs, N.J., 1965.

K. T. Hahn

1. Inequality between the Bergman metric and Carathéodory differential metrics, *Proc. Am. Math. Soc.* 68 (1978), 193–194.

M. Hakim and N. Sibony

1. Quelques conditions pour l'existence de fonctions pics dans les domains pseudoconvexes, *Duke Math. J.* 44 (1977), 399–406.
2. Fonctions holomorphes bornees sur la boule unite de $\mathbb{C}^n$, *Invent. Math.* 67 (1982), 213–222.
3. Spectre de $A(\bar{\Omega})$ pour des domaines bornés faiblement pseudoconvexes réguliers, *J. Funct. Anal.* 37 (1980), 127–135.

G. H. Hardy and J. E. Littlewood

1. Theorems concerning mean values of analytic or harmonic functions, *Q. J. Math. Oxford Ser.* 12 (1941), 221–256.
2. Some properties of functional integrals II, *Math. Zeit.* 4 (1932), 403–439.

F. Hartogs

1. Zur Theorie der analytischen Functionen mehrerer unabhangiger Veränderlichen insbesondere über die Darstellung derselben durch Reihen, welche nach Potenzen einer Veränderlichen fortschreiten, *Math. Ann.* 62 (1906), 1–88.

R. Harvey

1. Holomorphic chains and their boundaries, *Proc. Symp. Pure Math.*, Part 1 (1977), 309–382.

R. Harvey and J. Polking

1. Fundamental solutions in complex analysis I and II, *Duke Math. J.* 46 (1979), 253–300 and 46 (1979), 301–340.

T. Hatziafratis

1. On certain integrals associated to CR-functions, *Trans. A.M.S.* 314 (1989), 781–802.

W. Hayman

1. *Subharmonic Functions*, vol. 2, Academic Press, London, 1989.

W. Hayman and P. B. Kennedy

1. *Subharmonic Functions*, Academic Press, New York, 1976.

E. R. Hedrick

1. Functions that are nearly analytic, *Bull. A.M.S.* 23 (1917), 213.

M. Heins

1. *Complex Function Theory*, Academic Press, New York, 1968.

S. Helgason

1. *Differential Geometry and Symmetric Spaces*, Academic Press, New York, 1962.

G. M. Henkin

1. A uniform estimate for the solution of the $\bar{\partial}$-problem in a Weil region, *Uspekhi Math. Nauk.* 26 (1971), 211–212.

2. Integral representations of functions holomorphic in strictly pseudoconvex domains and some applications, *Mat. Sb.* 78 (120) (1969), 611–632; *Math. U.S.S.R. Sb.* 7 (1969), 597–616.

3. Integral representations of functions holomorphic in strictly pseudoconvex domains and applications to the $\bar{\partial}$ problem, *Mat. Sb.* 82 (124), 300–308 (1970); *Math. U.S.S.R. Sb.* 11 (1970), 273–281.

4. The Lewy equation and analysis on pseudoconvex manifolds, *Uspekhi Math. Nauk* 32:3 (1977), 57–118; *Russ. Math. Surv.* 32:3 (1977), 59–130.

5. An analytic polyhedron is not holomorphically equivalent to a strictly pseudoconvex domain (Russian), *Dokl. Acad. Nauk, SSSR* 210 (1973), 1026–1029.

6. Solutions with estimates of the H. Lewy and Poincaré-Lelong equations, the construction of functions of a Nevanlinna class with given zeros in a strictly pseudoconvex domain, *Dok. Akad. Nauk. SSSR* 224 (1975), 771–774; *Sov. Math. Dokl.* 16 (1975), 1310–1314.

7. The approximation of functions in strongly pseudoconvex domains and a theorem of Z. L. Leibenzon, *Bull. Acad. Polon. Sci. Sér. Sci. Math. Astronom. Phys.* 19 (1971), 37–42.

G. M. Henkin and A. Romanov

1. Exact Hölder estimates of solutions of the $\bar{\partial}$ equations, *Izvestija Akad. SSSR, Ser. Mat.* 35 (1971), 1171–1183, *Math. U.S.S.R. Sb.* 5 (1971), 1180–1192.

I. Herstein

1. *Topics in Algebra*, Xerox, Lexington, 1975.

M. Hervé

1. *Analytic and Plurisubharmonic Functions in Finite and Infinite Dimensional Spaces*, Springer Lecture Notes, vol. 198, Springer Verlag, Berlin, 1971.

E. Hille

1. *Analytic Function Theory*, Ginn, Boston, 1959.

H. Hironaka

1. Resolution of singularities of an algebraic variety over a field of characteristic zero, I, II, *Ann. Math.* 79 (1964), 109–203; *ibid.* 79 (1964), 205–326.

M. Hirsch

1. *Differentiable Topology*, Springer, Berlin, 1976.

K. Hoffman

1. *Banach Spaces of Holomorphic Functions*, Prentice Hall, Englewood Cliffs, N.J., 1962.

L. Hörmander

1. L^2 estimates and existence theorems for the $\bar{\partial}$ operator, *Acta Math.* 113 (1965), 89–152.

2. *Linear Partial Differential Operators*, Springer, Berlin, 1969.

3. *Introduction to Complex Analysis in Several Variables*, North Holland, Amsterdam, 1973.

4. Hypoelliptic second-order differential equations, *Acta Math.* 119 (1967), 147–171.

L. Hörmander and J. Wermer

1. Uniform approximation on compact sets in $\mathbb{C}^n$, *Math. Scand.* 23 (1968), 5–21.

L. Hua

1. *Harmonic Analysis of Functions of Several Complex Variables in the Classical Domains*, American Mathematical Society, Providence, R.I., 1963.

A. Huckleberry

1. Holomorphic fibrations of bounded domains, *Math. Ann.* 277 (1977), 61–66.

W. Hurewicz and H. Wallman

1. *Dimension Theory*, Princeton University Press, Princeton, N.J., 1948.

D. Husemoller

1. *Fibre Bundles*, Springer Verlag, Berlin, 1975.

H. Jacobowitz and F. Treves

1. Nowhere solvable homogeneous partial differential equations, *Bull. A.M.S.* 8 (1983), 467–469.

N. Jewell and S. G. Krantz

1. Toeplitz operators and related function algebras on certain pseudoconvex domains, *Trans. Am. Math. Soc.* 252 (1979), 297–312.

F. John and L. Nirenberg

1. On functions of bounded mean oscillation, *Comm. Pure Appl. Math.* XIV (1961), 415–426.

P. W. Jones

1. Quasiconformal mappings and extendability of functions in Sobolev spaces, *Acta Math.* 147 (1981), 71–88.

P. Jones and T. Wolff

1. Hausdorff dimension of harmonic measures in the plane, *Acta Math.* 161 (1988), 131–144.

B. Josefson

1. On the equivalence between locally polar and globally polar sets for plurisubharmonic functions on $\mathbb{C}^n$, *Ark. Mat.* 16 (1978), 109–115.

G. Julia

1. Memoire sur l'Itération des Fonctions Rationelles, *Journal de Mathematiques Pure et Appliques* 4 (1918), 47–245.

S. Kaneyuki

1. *Homogeneous Domains and Siegel Domains*, Springer Lecture Notes #241, Berlin, 1971.

Y. Katznelson

1. *An Introduction to Harmonic Analysis*, Wiley, New York, 1968.

Keldych Lavrentieff

1. Sur une evaluation de la fonction de Green, *Doklady Acad. USSR* 24 (1939), 102.

O. Kellogg

1. *Foundations of Potential Theory*, Dover, New York, 1953.

2. Harmonic functions and Green's integral, *Trans. Am. Math. Soc.* 13 (1912), 109–132.

N. Kerzman

1. Taut manifolds and domains of holomorphy in $\mathbb{C}^n$, *Notices Am. Math. Soc.* 16 (1969), 675.

2. Hölder and L^p estimates for solutions of $\bar{\partial}u = f$ on strongly pseudoconvex domains, *Comm. Pure Appl. Math.* XXIV (1971), 301–380.

3. The Bergman kernel function. Differentiability at the boundary, *Math. Ann.* 195 (1972), 149–158.

4. *Topics in Complex Analysis*, unpublished notes.

N. Kerzman and A. Nagel

1. Finitely generated ideals in certain function algebras, *J. Funct. Anal.* 7 (1971), 212–215.

N. Kerzman and J. P. Rosay

1. Fonctions plurisubharmonique d'exhaustion bornées et domaines taut, *Math. Ann.* 257 (1982), 171–184.

N. Kerzman and E. M. Stein

1. The Cauchy kernel, the Szegö kernel, and the Riemann mapping function, *Math. Ann.* 236 (1978), 85–93.

2. The Szegö kernel in terms of Cauchy-Fantappiè kernels, *Duke Math. J.* 45 (1978), 197–224.

K. T. Kim

1. Complete localization of domains with noncompact automorphism group, *Trans. A. M. S.* 319 (1990), 139–153.

C. Kiselman

1. A study of the Bergman projection in certain Hartogs domains, *Proc. Symposia Pure Math.*, vol. 52 (E. Bedford, J. D'Angelo, R. Greene, and S. Krantz eds.), American Mathematical Society, Providence, R.I., 1991.

P. Klembeck

1. Kähler metrics of negative curvature, the Bergman metric near the boundary and the Kobayashi metric on smooth bounded strictly pseudoconvex sets, *Indiana Univ. Math. J.* 27 (1978), 275–282.

M. Klimek

1. *Pluripotential Theory*, Oxford University Press, Oxford, England, 1991.

S. Kobayashi

1. The geometry of bounded domains, *Trans. A.M.S.* 92 (1959), 267–290.

2. Intrinsic metrics on complex manifolds, *Bull. A.M.S.* 73 (1967), 347–349.

3. *Hyperbolic Manifolds and Holomorphic Mappings*, Dekker, New York, 1970.

4. Intrinsic distances, measures and geometric function theory, *Bull. A.M.S.* 82 (1976), 357–416.

S. Kobayashi and K. Nomizu

1. *Foundations of Differential Geometry*, Vols. I and II, Interscience, New York, 1963, 1969.

K. Kodaira

1. *Complex Manifolds and Deformation of Complex Structures*, Springer Verlag, New York, 1986.

A. Kodama

1. On the structure of a bounded domain with a special boundary point, *Osaka Jour. Math.* 23 (1986), 271–298.

2. On the structure of a bounded domain with a special boundary point II, *Osaka Jour. Math.* 24 (1987), 499–519.

3. Characterizations of certain weakly pseudoconvex domains $E(k,\alpha)$ in $\mathbb{C}^n$, *Tôhoku Math. J.* 40 (1988), 343–365.

4. A characterization of certain domains with good boundary points in the sense of Greene-Krantz, *Kodai Math. J.* 12 (1989), 257–269.

J. J. Kohn

1. Harmonic integrals on strongly pseudoconvex manifolds I, *Ann. Math.* 78 (1963), 112–148; II, ibid. 79 (1964), 450–472.

2. Sufficient conditions for subellipticity on weakly pseudoconvex domains, *Proc. Nat. Acad. Sci. (USA)* 74 (1977), 2214–2216.

3. Global regularity for $\bar{\partial}$ on weakly pseudoconvex manifolds, *Trans. A.M.S.* 181 (1973), 273–292.

4. Methods of partial differential equations in complex analysis, *Proc. Symp. Pure Math.* 30, Part 1 (1977), 215–237.

5. Boundary behavior of $\bar{\partial}$ on weakly pseudoconvex manifolds of dimension two, *J. Diff. Geom.* 6 (1972), 523–542.

6. Subellipticity of the $\bar{\partial}$-Neumann problem on pseudoconvex domains: Sufficient conditions, *Acta Math.* 142 (1979), 79–122.

7. A Survey of the $\bar{\partial}$-Neumann problem, *Proceedings of Symposia in Pure Mathematics*, vol. 41, American Mathematical Society, Providence, R.I., 1984, 137–145.

J. J. Kohn and L. Nirenberg

1. A pseudo-convex domain not admitting a holomorphic support function, *Math. Ann.* 201 (1973), 265–268.

2. On the algebra of pseudo-differential operators, *Comm. Pure Appl. Math.* 18 (1965), 269–305.

P. Koosis

1. *Lectures on H^p Spaces*, Cambridge University Press, Cambridge, England, 1980.

W. Koppelman

1. The Cauchy integral formula for functions of several complex variables, *Bull A.M.S.* 73 (1967), 373–377.

A. Koranyi

1. Harmonic functions on Hermitian hyperbolic space, *Trans. A.M.S.* 135 (1969), 507–516.

2. Boundary behavior of Poisson integrals on symmetric spaces, *Trans. A.M.S.* 140 (1969), 393–409.

A. Koranyi and S. Vági

1. Singular integrals on homogeneous spaces and some problems of classical analysis, *Ann. Scuola Norm. Sup. Pisa* 25 (1971), 575–648.

2. Cauchy-Szegö integrals for systems of harmonic functions, *Ann. Scuola Norm. Sup. Pisa* 26 (1972), 181–196.

J. Korevaar and J. Wiegerinck

1. A lemma on mixed derivatives with applications to edge-of-the-wedge, Radon transformation and a theorem of Forelli, International symposium on complex analysis and applications (Arandjelovac, 1984). *Mat. Vesnik* 37 (1985), 145–157.

S. G. Krantz

1. Optimal Lipschitz and L^p estimates for the equation $\bar\partial u = f$ on strongly pseudoconvex domains, *Math. Ann.* 219 (1976), 233–260.

2. Structure and interpolation theorems for certain Lipschitz classes and applications to the $\bar\partial$ equation, *Duke Math. Jour.* 43 (1976), 417–439.

3. Boundary values and estimates for holomorphic functions of several complex variables, *Duke Math. J.* 47 (1980), 81–98.

4. Characterization of various domains of holomorphy via $\bar\partial$ estimates and applications to a problem of Kohn, *Ill. J. Math.* 23 (1979), 267–286.

5. Geometric Lipschitz spaces and applications to complex function theory and nilpotent groups, *J. Funct. Anal.* 34 (1979), 456–471.

6. Analysis on the Heisenberg group and estimates for functions in Hardy classes of several complex variables, *Math. Ann.* 244 (1979), 243–262.

7. Estimates for integral kernels of mixed type, fractional integration operators, and optimal estimates for the $\bar\partial$ operator, *Manuscripta Math.* 30 (1979), 21–52.

8. Holomorphic functions of bounded mean oscillation and mapping properties of the Szegö projection, *Duke Math. J.* 47 (1980), 743–761.

9. Intrinsic Lipschitz classes on manifolds with applications to complex function theory and estimates for the $\bar\partial$ and $\bar\partial_b$ equations, *Manuscripta Math.* 24 (1978), 351–378.

10. Lipschitz spaces, smoothness of functions, and approximation theory, *Expositiones Math.* 3 (1983), 193–260.

11. Characterizations of smooth domains in $\mathbb{C}$ by their biholomorphic self-maps, *Am. Math. Monthly 90* (1983), 555–557.

12. What is several complex variables? *Am. Math. Monthly,* 94 (1987), 236–256.

13. Functions of Several Complex Variables and Analytic Spaces, *The Encyclopedia of Physical Science and Technology,* Academic Press, vol. 5 (1986), 698–722.

14. *Real Analysis and Foundations*, CRC Press, Boston, 1991.

15. Integral formulas in complex analysis, a chapter in *The Beijing Lectures in Harmonic Analysis,* Annals of Mathematics Studies, Princeton University Press, Princeton, vol. 112 (1986), 185–240.

16. On a theorem of Stein I, *Trans. A.M.S.* 320 (1990), 625–642.

17. Invariant metrics and the boundary behavior of holomorphic functions on domains in $\mathbb{C}^n$, *Jour. Geometric. Anal.* 1 (1991).

18. Convexity in complex analysis, *Proc. Symp. in Pure Math.*, vol. 52, (E. Bedford, J. D'Angelo, R. Greene, and S. Krantz eds.), American Mathematical Society, Providence, R.I.,1991.

19. *Partial Differential Equations and Complex Analysis*, CRC Press, Boca Raton, Fla., 1992.

20. *Complex Analysis: The Geometric Viewpoint*, A CARUS Monograph of the Mathematics Association of America, Washington, D.C., 1990.

21. A compactness principle in complex analysis, *Division de Matematicas, Univ. Autonoma de Madrid Seminarios*, vol. 3, 1987, 171–194.

22. Recent progress and future directions in several complex variables, in *Complex Analysis Seminar*, S. Krantz ed., Springer Lecture Notes vol. 1268, Springer Verlag, 1987, 1–23.

23. On the boundary behavior of the Kobayashi metric, *Rocky Mt. Jour. Math.* 22 (1992), in press.

S. G. Krantz and H. R. Parks

1. Distance to C^k manifolds, *Jour. Diff. Equations* 40 (1981), 116–120.

2. *A Primer of Real Analytic Functions*, Birkhäuser Verlag, Basel, 1992.

C. Kuratowski

1. *Topologie*, Vol. I, Z. Subwencji Funduszu Kultury Narodowej, Warsawa-Lwów, 1933.

S. Lang

1. *Algebra*, Addison-Wesley, Reading, Mass., 1965.

S. R. Lay

1. *Convex Sets and their Applications*, Wiley, New York, 1982.

S. Lefschetz

1. *L'analysis situs et la géometrie algébrique*, Gauthier-Villars, Paris, 1924.

2. *Introduction to Topology*, Princeton University Press, Princeton, N.J., 1949.

3. *Topology*, American Mathematical Society, Providence, R.I., 1930.

O. Lehto and K. I. Virtanen

1. Boundary behavior and normal meromorphic functions, *Acta Math.* 97 (1957), 47–65.

Z. L. Leibenzon

1. Personal communication with H. Rossi.

L. Lempert

1. Boundary behavior of meromorphic functions of several complex variables, *Acta Math.* 144 (1980), 1–26.

2. La metrique Kobayashi et las representation des domains sur la boule, *Bull. Soc. Math. France* 109 (1981), 427–474.

3. A precise result on the boundary regularity of biholomorphic mappings, *Math. Z.* 193 (1986), 559–579.

4. A note on mapping polydiscs into balls and vice versa, *Acta Mathematica Scientiarum Hungaricae* 34 (1979), 117–119.

5. Holomorphic retracts and intrinsic metrics in convex domains, *Analysis Mathematica* 8 (1982), 257–261.

6. Intrinsic distances and holomorphic retracts, *Complex Analysis and Applications*, Sofia, 1984.

F. D. Lesley

1. On interior and conformal mappings of the disk, *J. London Math. Society* 20 (1979), 67–78.

2. Hölder continuity of conformal mappings at the boundary via the strip method, *Indiana Univ. Math. Jour.* 31 (1982), 341–354.

H. Lewy

1. An example of a smooth linear partial differential equation without solution, *Ann. Math.* 66 (1957), 155–158.

2. On the boundary behavior of holomorphic mappings, *Acad. Naz. Lincei* 35 (1977), 1–8.

E. Ligocka

1. The Sobolev spaces of harmonic functions, *Studia Math.* 54 (1986). 79–87.

2. The Hölder duality for harmonic functions, *Studia Math.* 84 (1986), 269–277.

3. On orthogonal projections onto spaces of pluriharmonic functions and duality, *Studia Math.* 84 (1986), 279–295.

S. Lojaciewicz

1. Sur la problème de la division, *Studia Math.*, 18 (1959), 87–136.
2. *Complex Analytic Geometry*, Birkhäuser, Boston, 1991.

L. Loomis and S. Sternberg

1. *Advanced Calculus*, Addison-Wesley, Reading, Mass., 1968.

E. Løw

1. A construction of inner functions on the unit ball in $\mathbb{C}^p$, *Inventiones Math.* 67 (1982), 223–229.
2. Inner functions and boundary values in $H^\infty(\Omega)$ and $A(\Omega)$ in smoothly bounded pseudoconvex domains, thesis, Princeton Univ., 1983.
3. Inner functions and boundary values in $H^\infty(\Omega)$ and $A(\Omega)$ in smoothly bounded pseudoconvex domains, *Math. Zeitschrift* 185 (1984), 194–210.

Lu Qi-Keng (= K. H. Look)

1. On Kähler manifolds with constant curvature, *Acta Math. Sinica* 16 (1966), 269–281. (Chinese);(=*Chinese Math.* 9 (1966)), 283–298.
2. The estimation of the intrinsic derivatives of the analytic mappings of bounded domains, *Sci. Sinica*, Special Issue II on Math., 1–17.

Daowei Ma

1. Washington Univ., thesis, 1990.

N. Makarov

1. Distortion of boundary sets under conformal mappings, *Proc. London Math. Soc.* 51 (1985), 369–384.

D. E. Marshall

1. *Approximation and Interpolation by Inner Functions*, University of California, thesis, Los Angeles, 1976.

J. McNeal

1. Boundary behavior of the Bergman kernel function in $\mathbb{C}^2$, *Duke Math. J.* 58 (1989), 499–512.
2. Local geometry of decoupled pseudoconvex domains, *Proceedings of a Conference for Hans Grauert*, Vieweg, 1990.
3. Holomorphic sectional curvature of some pseudoconvex domains, *Proc. Am. Math. Soc.* 107 (1989), 113–117.

B. Mityagin and E. M. Semenov

1. The space C^k is not an interpolation space between C and $C^n, 0 < k < n$, *Sov. Math. Dokl.* 17 (1976), 778–782.

C. B. Morrey

1. *Multiple Integrals in the Calculus of Variations*, Springer, Berlin, 1966.

J. Munkres

1. *Elementary Differential Topology*, Princeton University Press, Princeton, N.J., 1963.

A. Nagel, J. P. Rosay, E. M. Stein, and S. Wainger

1. Estimates for the Bergman and Szegö kernels in $\mathbb{C}^2$, *Ann. Math.* 129 (1989), 113–149.

A. Nagel and W. Rudin

1. Local boundary behavior of bounded holomorphic functions, *Can. Jour. Math.* 30 (1978), 583–592.

A. Nagel, E. M. Stein, and S. Wainger

1. Boundary behavior of functions holomorphic in domains of finite type, *Proc. Nat. Acad. Sci. USA* 78 (1981), 6596–6599.

R. Näkki and B. Palka

1. Quasiconformal circles and Lipschitz classes, *Comment. Math. Helv.* 55 (1980), 485–498.

R. Narasimhan

1. *Several Complex Variables*, University of Chicago Press, Chicago, 1971.
2. Holomorphically complete convex spaces, *Am. J. Math.* 82 (1961), 917–934.

L. Nirenberg

1. *Lectures on Linear Partial Differential Equations*, CBMS, American Mathematical Society, Providence, R.I., 1970.

L. Nirenberg, S. Webster, and P. Yang

1. Local boundary regularity of holomorphic mappings, *Comm. Pure Appl. Math.* 33 (1980), 305–338.

L. Nirenberg and R. O. Wells

1. Holomorphic approximation on real submanifolds of a complex manifold, *Bull. A.M.S.* 73 (1967), 378–381.

A. Noell

1. Local vs. global convexity of pseudoconvex domains, *Proc. Symp. Pure Math.*, vol. 52 (E. Bedford, J. D'Angelo, R. Greene, and S. Krantz eds.), American Mathematical Society, Providence, R.I., 1991.

A. Noell and B. Stensønes

1. Proper holomorphic maps from weakly pseudoconvex domains, *Duke Math. J.* 60 (1990), 363–388.

F. Norguet

1. Sur les domaines d'holomorphie des fonctions uniformes de plusieurs variables complexes, *Bull. Soc. Math. France* 82 (1954), 137–159.

K. Oka

1. Sur les fonctions de plusieurs variables. II. Domaines d'holomorphie, *J. Sc. Hiroshima Univ.* 7 (1937), 115–130.

2. Sur les fonctions de plusieurs variables. III. Deuxieme problème de Cousin, *J. Sc. Hiroshima Univ.* 9 (1939), 7–19.

3. Sur les fonctions de plusieurs variables. IX. Domaines finis sans points critique interieur, *Jap. J. Math.* 23 (1953), 97–155.

4. *Collected Papers*, Springer, Berlin, 1984.

B. O'Neill

1. *Elementary Differential Geometry,* Academic Press, New York, 1966.

W. F. Osgood

1. *Lehrbuch der Funktionentheorie*, vols. 1 and 2, B. G. Teubner, Leipzig, 1912.

N. Øvrelid

1. Integral representations formulas and L^p estimates for the $\bar{\partial}$ equation, *Math. Scand.* 29 (1971), 137–160.

P. Painlevé

1. Sur les lignes singulières des functions analytiques, *Thèse*, Gauthier-Villars, Paris, 1887.

M. Y. Pang

1. On the infinitesimal behavior of the Kobayashi distance, *Pacific Jour. Math.* 162 (1994), 121–141.

P. Pflug

1. Über polynomiale Funktionen auf Holomorphie gebieten, *Math. Z.* 139 (1974), 133–139.

D. H. Phong

1. On Hölder and L^p estimates for the $\bar{\partial}$ equation on strongly pseudoconvex domains, Princeton Univ., thesis, 1977.

D. H. Phong and E. M. Stein

1. Estimates for the Bergman and Szegö projections on strongly pseudoconvex domains, *Duke Math. J.* 44 (1977), 695–704.

S. Pinchuk

1. On the analytic continuation of holomorphic mappings, *Mat. Sb.* 98 (140) (1975), 375–392; *Mat. U.S.S.R. Sb.* 27 (1975), 416–435.

I. Priwalow

1. *Randeigenschaften Analytischer Funktionen*, Deutsch Verlag der Wissenschaften, Berlin, 1956.

I. P. Ramadanov

1. Sur une propriétie de las fonction de Bergman, *C. R. Acad. Bulgare des Sci.* 20 (1967), 759–762.

E. Ramirez

1. Divisions problem in der komplexen analysis mit einer Anwendung auf Rand integral darstellung, *Math. Ann.* 184 (1970), 172–187.

R. M. Range

1. Hölder estimates for $\bar{\partial}$ on convex domains in $\mathbb{C}^2$ with real analytic boundary, *Proc. Symposia in Pure Math.* 30 (1977), 31–33.

2. On Hölder estimates for $\bar{\partial}u = f$ on weakly pseudoconvex domains, *Several Complex Variables* (Cortona, 1976/7), pp. 247–267, *Scuola Norm. Sup. Pisa* 1978.

3. *Holomorphic Functions and Integral Representations in Several Complex Variables*, Springer Verlag, Berlin, 1986.

4. A remark on bounded strictly plurisubharmonic exhaustion functions, *Proc. Am. Math. Soc.* 81 (1981), 220–222.

5. An elementary integral solution operator for the Cauchy-Riemann equations on pseudoconvex domains in $\mathbb{C}^n$, *Trans. Am. Math. Soc.* 274 (1982), 809–816.

R. M. Range and Y. T. Siu

1. Uniform estimates for the $\bar{\partial}$ equation on domains with piecewise smooth strictly pseudoconvex boundaries, *Math. Annalen* 206 (1973), 325–354.

R. Remmert and K. Stein

1. Eigentliche holomorphie Abbildungen, *Math. Zeit.* 73 (1960), 159–189.

B. Rodin and S. Warschawski

1. Estimates of the Riemann mapping function near a boundary point, in *Romanian-Finnish Seminar on Complex Analysis*, Springer Lecture Notes, vol. 743, 1979, 349–366.

S. Roman

1. The formula of Faà di Bruno, *Am. Math. Monthly* 87 (1980), 805–809.

J. P. Rosay

1. Sur une characterization de la boule parmi les domains de $\mathbb{C}^n$ par son groupe d'automorphismes, *Ann. Inst. Four. Grenoble* XXIX (1979), 91–97.

2. Un example d'ouvert borné de $\mathbb{C}^3$ "taut" mais non hyperbolique complet, *Pac. Jour Math.* 98 (1982), 153–156.

J. P. Rosay and W. Rudin

1. Holomorphic maps from $\mathbb{C}^n$ to $\mathbb{C}^n$, *Trans. Am. Math. Soc.* 310 (1988), 47–86.

A. Rosenblatt

1. Sur la fonction de Green ordinaire de l'espace a trois dimensions, *C. Rendus* 201 (1936).

H. Rossi

1. Holomorphically convex sets in several complex variables, *Ann. Math.* 74 (1961), 470–493.

L. P. Rothschild and E. M. Stein

1. Hypoelliptic differential operators and nilpotent groups, *Acta Math.* 137 (1976), 247–320.

H. Royden

1. Remarks on the Kobayashi Metric, *Several Complex Variables II*, Maryland 1970, Springer, Berlin, 1971, 125–137.

H. Royden and P. Wong

1. Carathéodory and Kobayashi metrics on convex domains, unpublished preprint.

L. A. Rubel and A. Shields

1. The failure of interior-exterior factorization in the polydisc and the ball, *Tôhoku Math. J.* 24 (1972), 409–413.

W. Rudin

1. *Principles of Mathematical Analysis*, 3rd ed., McGraw-Hill, New York, 1976.

2. *Lectures on the Edge-of-the-Wedge Theorem*, CBMS # 6, American Mathematical Society, Providence, R.I., 1970.

3. Zeros of holomorphic functions in balls, *Indag. Math.* 38 (1976), 57–65.

4. Holomorphic Lipschitz functions in balls, *Comment. Math. Helvet.* 53 (1978), 143–147.

5. Peak-interpolation sets of class C^1, *Pac. J. Math.* 75 (1978), 267–277.

6. New Constructions of Functions Holomorphic in the Unit Ball of $\mathbb{C}^n$, CBMS Series, American Mathematical Society, Providence, R.I., 1986.

7. *Function Theory in the Unit Ball of* $\mathbb{C}^n$, Grundlehren der Mathematischen Wissenschaften in Einzeldarstellungen, Springer, Berlin, 1980.

8. Holomorphic maps that extend to automorphisms of a ball, *Proc. A.M.S.* 81 (1981), 429–432.

9. *Real and Complex Analysis*, McGraw-Hill, New York, 1966.

10. *Function Theory in Polydiscs*, Benjamin, New York, 1969.

11. *Functional Analysis*, McGraw-Hill, New York, 1973.

A. Sadullaev

1. Inner functions in $\mathbb{C}^n$, *Mat. Zametki* 19 (1976), 63–66 (Russ.); *Math. Notes* 19 (1976), 37–38.

I. J. Schoenberg

1. The elementary cases of Landau's problem of inequalities between derivatives, *Am. Math. Monthly* 80 (1973), 121–158.

S. Semmes

1. A generalization of Riemann mappings and geometric structures on a space of domains in $\mathbb{C}^n$, *Memoirs of the American Mathematical Society*, 1991.

N. Sibony

1. Prolongement des fonctions holomorphes bornées et Métrique de Cara-théodory, *Inventiones Math.* 29 (1975), 205–230.

2. Un exemple de domain pseudoconvexe regulier ou l'equation $\bar\partial u = f$ n'admet pas de solution bornee pour f bornee, *Invent. Math.* 62 (1980), 235–242.

3. Problème de la couronne pour des domaines pseudoconvex à bord lisse, *Annals of Math.* 126 (1987), 675–682.

4. Sur le plongement des domaines faiblement pseudoconvexes dans des domaines convexes, *Math. Ann.* 273 (1986), 209–214.

5. Unpublished notes.

6. Personal communication.

N. Sibony and J. Wermer

1. Generators for $A(\Omega)$, *Trans. A.M.S.* 194 (1974), 103–114.

Y. T. Siu

1. The Levi problem, *Proc. Symposia on Pure Math.*, vol. XXX, Americal Mathematical Society, Providence, R.I., 1977.

H. Skoda

1. Solution à croissance du second problème de Cousin dans $\mathbb{C}^n$, *Ann. Inst. Fourier (Grenoble)* 21 (1971), 11–23.

2. Application des techniques L^2 à la théorie des idéaux d'une algèbre de fonctions holomorphes avec poids, *Ann. Sci. École Norm. Sup.* 5 (1972), 545–579.

3. Fibrés holomorphes à base et à fibre de Stein, *Inventiones Math.* 43 (1977), 97–107.

4. Valeurs au bord pour les solutions de l'operateur d'' et caracterization des zeros des fonctions de la classe Nevanlinna, *Bull. Soc. Math. de France* 104 (1976), 225–229.

M. Skwarczynski

1. The distance in the theory of pseudo-conformal transformations and the Lu Qi-King conjecture, *Proc. A.M.S.* 22 (1969), 305–310.

K. T. Smith

1. A generalization of an inequality of Hardy and Littlewood, *Can. J. Math.* 8 (1956), 157–170.

W. Smith and D. Stegenga

1. A geometric characterization of Hölder domains, *J. London Math. Soc.* 35 (1987), 471–480.

R. Solovay

1. A model of set theory in which every set is Lebesgue measurable, *Ann. Math.* 92 (1970), 1–56.

E. Spanier

1. *Algebraic Topology*, McGraw-Hill, New York, 1966.

D. C. Spencer

1. Overdetermined systems of linear partial differential equations, *Bull. Am. Math. Soc.* 75 (1969), 179–239.

M. Spivak

1. *Calculus on Manifolds*, Benjamin, New York, 1965.

N. Steenrod

1. *The Topology of Fibre Bundles*, Princeton University Press, Princeton, N.J., 1951.

E. M. Stein

1. *Singular Integrals and Differentiability Properties of Functions*, Princeton University Press, Princeton, N.J., 1970.

2. *Boundary Behavior of Holomorphic Functions of Several Complex Variables*, Princeton University Press, Princeton, N.J., 1972.

3. Singular integrals and estimates for the Cauchy-Riemann equations, *Bull. A.M.S.* 79 (1973), 440–445.

E. M. Stein and G. Weiss

1. *Introduction to Fourier analysis on Euclidean Space*, Princeton University Press, Princeton, N.J., 1971.

G. Stolzenberg

1. A hull with no analytic structure, *J. Math. Mech.* 12 (1963), 103–111.

E. L. Stout

1. H^p functions on strictly pseudoconvex domains, *Am. J. Math.* 98 (1976), 821–852.

2. Two theorems concerning functions holomorphic on multiply connected domains, *Bull. Am. Math. Society* 69 (1963), 527–530.

N. Suita and A. Yamada

1. On the Lu Qi-Keng conjecture, *Proc. A.M.S.* 59 (1976), 222–224.

T. Sunada

1. Holomorphic equivalence problem for bounded Reinhardt domains, *Math. Ann.* 235 (1978), 111–128.

G. Szegö

1. Über orthogonalsysteme von Polynomen, *Math. Z.* 4 (1919), 139–151.

N. Tanaka

1. On generalized graded Lie algebras and geometric structures, I, *J. Math. Soc. Japan* 19 (1967), 215–254.

M. E. Taylor

1. *Pseudodifferential Operators*, Princeton University Press, Princeton, N.J., 1981.

F. Treves

1. *Introduction to Pseudodifferential and Fourier Integral Operators*, 2 vols., Plenum, New York, 1980.

M. Tsuji

1. *Potential Theory in Modern Function Theory*, Maruzen, Tokyo, 1959.

F. A. Valentine

1. *Convex Sets*, McGraw-Hill, New York, 1964.

N. Th. Varopoulos

1. BMO functions and the $\bar{\partial}$-equation, *Pac. Jour. Math.* 71 (1977), 221–273.

2. BMO functions in complex analysis, *Proceedings of Symposia in Pure Mathematics*, XXXV: *Harmonic Analysis in Euclidean Space*, American Mathematical Society, Providence, R.I., 1979, 43–62.

3. Zeros of H^p functions in several complex variables, *Pac. J. Math.* 87 (1980), 189–246.

U. Venugopalkrishna

1. Fredholm operators associated with strongly pseudoconvex domains in $\mathbb{C}^n$, *J. Funct. Anal.* 9 (1972), 349–373.

V. Vladimirov

1. *Methods of the Theory of Functions of Several Complex Variables*, MIT Press, Cambridge, Mass., 1966.

B. L. van der Waerden

1. *Algebra*, Vols. I and II, F. Ungar, New York, 1970.

S. Warschawski

1. On the boundary behavior of conformal maps, *Nagoya Math. Jour.* 30 (1967), 83–101.

2. On boundary derivatives in conformal mapping, *Ann. Acad. Sci. Fenn.* Ser. A I no. 420 (1968), 22 pp.

3. Hölder continuity at the boundary in conformal maps, *Jour. Math. Mech.* 18 (1968/9), 423–427.

S. Webster

1. On the reflection principle in several complex variables, *Proc. A.M.S.* 71 (1978), 26–28.

2. On mapping an n-ball into an $(n + 1)$-ball in complex space, *Pac. Jour. Math.* 81 (1979), 267–272.

R. O. Wells

1. Function theory on differentiable submanifolds, *Contributions to Analysis*, Academic Press, New York, 1974.

2. *Differential Analysis on Complex Manifolds*, 2nd ed., Springer Verlag, 1979.

J. Wermer

1. On a domain equivalent to the bidisc, *Math. Ann.* 248 (1980), 193–194.

2. An example concerning polynomial convexity, *Math. Ann.* 139 (1959), 147–150; Addendum: *Math. Ann.* 140 (1959), 322–323.

H. Whitney

1. *Geometric Integration Theory*, Princeton University Press, Princeton, N.J., 1957.

E. Whittaker and G. Watson

1. *A Course of Modern Analysis*, 4th ed., Cambridge Univ. Press, London, 1935.

K. O. Widman

1. On the boundary behavior of solutions to a class of elliptic partial differential equations, *Ark. Mat.* 6 (1966), 485–533.

J. Wiegerinck

1. Domains with finite dimensional Bergman space, *Math. Z.* 187 (1984), 559–562.

2. Separately subharmonic functions need not be subharmonic, *Proc. A.M.S.* 104 (1988), 770–771.

3. Entire functions of Paley-Wiener type in $\mathbb{C}^n$, Radon transforms, and problems of holomorphic extension, *Academisch Proefschrift*, Centrum voor Wiskunde en Informatica, Amsterdam, 1985.

B. Wong

1. Characterizations of the ball in $\mathbb{C}^n$ by its automorphism group, *Invent. Math.* 41 (1977), 253–257.

J. D. M. Wright

1. All operators on a Hilbert space are bounded, *Bull. A.M.S.* 79 (1973), 1247–1250.

H. H. Wu

1. Normal families of holomorphic mappings, *Acta Math.* 119 (1967), 193–233.

P. Yang

1. Automorphisms of tube domains, *Am. Jour. Math.* 104 (1982), 1005–1024.

S. T. Yau

1. A generalized Schwarz lemma for Kähler manifolds, *Am. Jour. Math.* 100 (1978), 197–204.

J. Yu

1. Thesis, Washington University, 1993.

A. Zygmund

1. *Trigonometric Series*, Cambridge University Press, Cambridge, England, 1968.

Index